Heat Exchangers
Volume I

Heat Exchangers Volume I: Classification, Selection, and Thermal Design discusses heat exchangers and their various applications, such as refrigeration, air conditioning, automobiles, gas turbines, process industries, refineries, and thermal power plants.

With a focus on thermal design methods, including rating and sizing, the book covers thermohydraulic fundamentals and thermal effectiveness charts for various flow configurations and shell and tube heat exchangers. It provides construction details, geometrical features and correlations, and thermo-hydraulic details for tube-fin, plate fin, air-cooled, shell and tube, microchannel, and plate heat exchangers and thermal design methods like rating and sizing. The book explores additive manufacturing of heat exchangers, printed circuit heat exchangers, and heat transfer augmentation methods. The book also describes recuperators and regenerators of gas turbine cycles, waste heat recovery devices, and phase change phenomena including boiling, condensation and steam generation.

The book serves as a useful reference for researchers, graduate students, and engineers in the field of heat exchanger design, including heat exchanger manufacturers.

Heat Exchangers
Volume I

Classification, Selection, and Thermal Design

Third Edition

Kuppan Thulukkanam

CRC Press is an imprint of the
Taylor & Francis Group, an **informa** business

Designed cover image: Shutterstock

Third edition published 2024
by CRC Press
2385 Executive Center Drive, Suite 320, Boca Raton, FL 33431

and by CRC Press
4 Park Square, Milton Park, Abingdon, Oxon, OX14 4RN

CRC Press is an imprint of Taylor & Francis Group, LLC

First edition published by Marcel-Dekker 2000
Second edition published by CRC Press 2013

Library of Congress Cataloging-in-Publication Data
Names: Kuppan, T., 1957– author.
Title: Heat exchangers. Classification, selection, and thermal design / Kuppan Thulukkanam.
Other titles: Heat exchanger design handbook. Selections (Classification)
Description: Third edition. | Boca Raton, FL : CRC Press, 2024. |
Includes revised material previously published in the author's Heat exchanger
design handbook. | Includes bibliographical references and index.
Identifiers: LCCN 2023034442 (print) | LCCN 2023034443 (ebook) |
ISBN 9781032399324 (hbk) | ISBN 9781032399331 (pbk) | ISBN 9781003352044 (ebk)
Subjects: LCSH: Heat exchangers–Design and construction–Handbooks, manuals, etc.
Classification: LCC TJ263 .K872 2024b (print) | LCC TJ263 (ebook) |
DDC 621.402/5–dc23/eng/20230731
LC record available at https://lccn.loc.gov/2023034442
LC ebook record available at https://lccn.loc.gov/2023034443

ISBN: 978-1-032-39932-4 (hbk)
ISBN: 978-1-032-39933-1 (pbk)
ISBN: 978-1-003-35204-4 (ebk)

DOI: 10.1201/9781003352044

Typeset in Times
by Newgen Publishing UK

Dedicated to

---

my parents, S. Thulukkanam and T. Senthamarai,
my wife, Tamizselvi Kuppan,
and my mentor, Dr. Ramesh K. Shah

Contents

Preface

CHAPTER 1

Heat Exchangers: Introduction, Classification, and Selection

A heat exchanger is a heat transfer device that is used for the transfer of heat between two or more fluids available at different temperatures. This chapter discusses basics of heat exchangers, heat transfer methods, heat exchanger construction details, design standards and codes, classification of heat exchangers based on construction, heat transfer processes, surface compactness, flow and pass arrangements, phase of the process fluids, heat transfer mechanisms, etc.; details and basics of tubular heat exchangers—double pipe, shell and tube, and coiled tube type, plate heat exchangers such as gasketed, brazed, welded, spiral, panel coil, lamella, etc., extended surface heat exchangers such as tube-fin and plate-fin, brazed aluminum heat exchangers, regenerators—fixed matrix, rotary matrix, shell and tube heat exchangers and their design standards, air cooled heat exchangers, aluminum microchannel and internally grooved copper microfin tube heat exchangers are discussed; flow arrangements like parallelflow, counterflow, and crossflow, and direct contact heat exchangers are discussed; condensers and evaporators, other types of heat exchangers like microscale and printed circuit heat exchangers, perforated plate heat exchangers, scraped surface heat exchangers, heat sinks, materials-based heat exchangers like graphite, ceramic, metal foam, and plastic heat exchanger are also discussed. heat exchanger market analysis, heat exchanger selection criteria, heat exchanger specifications, choice of unit type for intended applications, location and siting considerations, storage, installation, commissioning, operation and maintenance, etc. are also covered.

CHAPTER 2

Heat Exchanger Thermohydraulic Fundamentals and Thermal Design of Heat Exchangers

This chapter discusses thermohydraulic fundamentals and thermal design of heat exchangers. Relationships between the heat transfer rate, surface area, fluid terminal temperatures, and flow rates of a heat exchanger and heat transfer rate equations are presented; heat exchanger performance parameters such as thermal effectiveness and allowable pressure drop for the fluids involved in heat exchange are discussed. The four thermal design methods to calculate the thermal effectiveness of heat exchangers, the ε-NTU, P-NTU$_t$, LMTD, and ψ-P methods, are discussed also. To elucidate thermohydraulic fundamentals, definitions of parameters such as symmetry and flow reversibility, temperature approach, temperature meet, and temperature cross are discussed. The thermal relations formulas and thermal effectiveness for various flow arrangements and pass arrangements and for all TEMA shell types, including multiple $E_{1\text{-}2}$ shells in series, are discussed and graphical thermal effectiveness charts are presented.

This chapter also discusses the basics of thermal design of heat exchangers including rating and sizing, pressure drop analysis for the fluids involved, temperature-dependent fluid properties, thermal performance failures, flow maldistribution, and uncertainties in thermal design of heat exchangers. The basics of the computer design method for thermal design of compact and shell and tube heat exchangers and salient features of this computer program are also outlined.

CHAPTER 3

Compact Heat Exchangers

This chapter discusses the characteristics and classification of compact heat exchangers such as tube-fin and plate-fin, their construction details, surface selection criteria, primary and secondary surfaces, factors influencing performance, contact resistance, thermohydraulic fundamentals, heat transfer and friction factor correlations, the *j* and *f* factors, plate-fin heat exchangers, application, advantages and limitations, construction details and flow arrangements, fin geometry, brazed aluminum plate-fin heat exchanger, ALPEMA standard, mechanical design, fin efficiency, basic heat transfer relations, pressure losses, operation and maintenance, asset integrity management, OEM services and support, aluminum heat exchangers of automobiles, mechanically assembled and brazed aluminum heat exchangers, rating and sizing of a compact exchanger, etc. It also discusses air-cooled heat exchanger (ACHE) design, forced draft and induced draft, air versus water cooling, construction, operation and maintenance, optimization of performance, operation and control, problems with heat exchangers in low-temperature environments, extreme temperature controls, design for viscous liquids, wind milling effect, anti-rotation devices, hot air recirculation, fan noise, thermal design of ache, fabrication, performance control, and air-side performance evaluation. It also discusses microchannel heat exchangers, chillers, CuproBraze heat exchangers, microgroove copper tube heat exchangers, printed circuit heat exchangers, and 3D printing of heat exchangers.

CHAPTER 4

Shell and Tube Heat Exchanger Design

The most commonly used heat exchanger is the shell and tube heat exchanger (STHE). This is the "workhorse" of industrial processes. It has many applications in the power generation, petroleum refinery, chemical, and process industries. This chapter discusses STHE types, their selection and applications, design standards such as TEMA and construction codes such as the ASME Code. TEMA designations are discussed, as well as shellside construction details, TEMA shell types, tube, low-finned tubes, tube count, U-tube, tube pitch, tube layout, baffles and their classifications, tubesheets, tubesheet connection with the shell and channel, double tubesheets, clad tubesheets, tube bundles, spacers, tie-rods, and sealing devices, tube-to-tubesheet joints, shellside and tubeside passes, end channel and channel covers, expansion joints, nozzles and impingement protection, fluid properties and allocation, TEMA shell types (*E*, *F*, *G*, *H*, *J*, *K*, and *X* shells), front and rear head designs, bolted flange joints (BFJs), heat exchanger component material selection, shellside clearances, bypass lanes, and thermal design procedures such as sizing. Outlines are also provided for TEMA specification sheets, design by Bell–Delaware method, shellside and tubeside heat transfer and pressure drop calculations, and software for thermal design. Nonsegmental baffles heat exchangers such as Phillips RODbaffle, Embaffle®, Helixchanger® and Twisted Tube® Heat Exchanger, Breech-Lock™ and Taper-Lok end closures are described, and calculation of crossflow velocity as per the Tinker method, thermal design of disk and doughnut heat exchangers and radial flow gas heat exchangers, details of closed feedwater heaters, and steam surface condensers are discussed also.

CHAPTER 5

Boiling, Condensation, and Steam Generation

Boiling, condensation, and evaporation heat transfer occur in many engineering applications, such as in power plant condensers, boilers, and steam generators, which are all important components in conventional and nuclear power stations. This chapter discusses boiling heat transfer, modes of boiling such as nucleate boiling, bulk boiling, film boiling, departure from nucleate boiling, critical heat flux, etc.; boiling in a vertical tube, flow boiling inside horizontal plain tubes, nucleate boiling correlations, boiling heat transfer enhancement methods using microfin tubes, twisted tape inserts, corrugated tubes, porous surface-coated tubes, heat transfer enhancement methods for pool boiling, etc. It also discusses falling film evaporator, its application areas, falling film evaporation on a single horizontal tube and tube bundle and thermal design considerations, evaporators of refrigeration systems, types of reboiler, kettle reboilers and their thermal performance; condensation phenomena such as dropwise, bulk and surface condensation, heat transfer due to condensation on a plane vertical surface, modes of condensation, laminar film condensation on vertical plates, horizontal tubes, horizontal tube bundles, condensation on low-finned tubes and tube bundles, and flow regimes for condensation in horizontal tubes. It also outlines the parameters of two-phase fluid flow such as void fraction, pressure drop, head loss due to fluid friction, flow instability, instabilities in steam-generating systems, pipe whips, water hammers, pressure spikes, and steam hammers; and also the steam generation principle, steam generators like steam drums and once-through steam generators (OTSGs), surface condensers, etc.

CHAPTER 6

Regenerators and Waste Heat Recovery Devices

Recovery of waste heat from flue gas by means of heat recovery devices, regenerators, and heat exchangers is known as regeneration, which can improve plant thermal efficiency and conserve fossil fuels. This chapter discusses types and salient features of fixed-bed and rotary regenerators, their construction, surface geometries, and size, rotary regenerators as air preheaters, regenerator design features, heating elements, seals, gas leakage, operational issues such as corrosion and gas side fouling, considerations in establishing a heat recovery system, economic benefits and maintainability, and strength and stability at the operating temperature. It also discusses thermal-hydraulic fundamentals of regenerators, thermal design theory, regenerator solution techniques, thermal effectiveness, heat transfer, ε-ntu$_0$ method and reduced length–reduced period (λ–π) method, fluids bypass and carryover, rating and sizing of rotary regenerators, mechanical design, seal design, the drive for the rotor, thermal distortion, and seal leakages. Waste heat recovery systems, devices, their limitations, commercially available heat recovery devices such as air preheaters, fluid-bed regenerative heat exchangers, vortex-flow direct contact heat exchangers, bayonet tube heat exchangers, regenerative burners, radiation recuperators, heat pipe exchangers, economizers, waste heat recovery boilers, direct contact heat exchangers, thermocompressors, combined heat and power cycles, thermoelectric generators, plate heat exchanger energy banks, rotary heat exchangers for space heating, and key barriers to waste heat recovery and materials selection for regenerators are discussed.

CHAPTER 7

Plate Heat Exchangers and Spiral Plate Heat Exchangers

A plate heat exchanger (PHE) is usually comprised of a stack of corrugated or embossed metal plates in mutual contact, and the heat transfer between the two fluids takes place through the plates. This chapter discusses PHE construction details, types, applications, flow patterns and pass arrangements, application areas, benefits and their limitations, with a comparison between a PHE and a shell and tube heat exchanger, plate corrugation including *chevron* or *herringbone* patterns, plate materials, gasket selection, gasket material, and PHE installation. PHE types such as brazed plate heat exchangers, and other types of plate heat exchangers such as all-welded, wide-gap, free-flow, semiwelded or twin-plate heat exchangers, double-wall, Compabloc fully welded heat exchangers, Diabon graphite plate heat exchangers, sanitary heat exchangers, silicon carbide plate heat exchangers are also discussed. Thermal design, including thermo-hydraulic fundamentals, high- and low-theta plates, thermal mixing of plates, thermal design methods such as LMTD and ε-NTU, heat transfer and pressure-drop calculations are discussed. The chapter also outlines the specification sheet for PHE, mechanical design, maintenance practices, gasket failures, performance deterioration, corrosion and fouling, conditions causing water hammer, cleaning of PHE such as manual and chemical cleaning, cleaning-in-place, leak testing, plate heat exchanger services by OEMs, spiral plate heat exchangers and their construction details (types 1, 2, 3, and 4), flow arrangements and applications, construction material, and thermal and mechanical design.

CHAPTER 8

Heat Transfer Augmentation

The study of improved heat transfer performance is referred to as heat transfer enhancement or augmentation. Augmentation of convective heat transfer will minimize the size of the heat exchanger. This chapter discusses the principles of single-phase convective heat transfer in laminar flow and turbulent flow, principles of single-phase and phase change heat transfer enhancement techniques both for internal and external flow, and also for turbulent and laminar flow. Enhancement methods such as the passive or active method, heat transfer enhancement devices and techniques such as extended surfaces, swirl flow devices, laminar flow displacement devices, twisted tape insert and wire mesh, corrugated tubes, internal helical fins, internally ribbed or finned tubes, internally ribbed or finned tubes, internal surface roughness, boundary layer displacement devises, swirl flow devices, turbulators, Calgavin's hitran® thermal systems for heat transfer enhancement are also discussed. The chapter also discusses pertinent issues such as friction, performance testing methods, fouling, performance evaluation criteria, market factors, mechanical design and construction considerations, cost savings, thermal design and optimization considerations, and major areas of applications including refrigeration and air-conditioning systems, lubricating oil coolers, power plant heat exchangers, heat exchangers for automobiles, etc.

The preparation of this book was facilitated by the great volume of existing literature contributed by many scholars in this field and sources in websites. I have tried to acknowledge all the sources and have sought the necessary permissions wherever reproduced. If omissions have been made, I offer my sincere apologies. Fabricators and heat exchanger manufacturers are acknowledged for their permission to use figures from their websites.

This edition is abundantly illustrated with over 500 drawings, diagrams, photos, and tables. *Heat Exchanger Design, Volume 1* is an excellent resource for mechanical, chemical, and petrochemical engineers; heat exchanger, process equipment and pressure vessel designers and manufacturers, consultants, industry professionals and upper-level undergraduate and graduate students in these disciplines.

DISCLAIMER

The text of this book is based upon open literature resources such as Standards, Codes, authentic books on heat exchangers and pressure vessels, technical literature from leading heat exchanger and pressure vessel manufacturers, technical articles and technical information from many websites, etc. No Indian Railways-related technical information is adopted in this book.

Acknowledgments

A large number of my colleagues from Indian Railways, well-wishers, and family members have contributed immensely toward the preparation of this book. I mention a few of them here, as follows: Jothimani Gunasekaran, V. R. Ventakaraman, Amitab Chakraborty (ADG), O. P. Agarwal (ED), and M. Vijayakumar (Director), RDSO Lucknow; Member (Mechanical), and senior officials of the Railway Board, New Delhi, and T. Adikesavan of Southern Railway; K. Narayanan for CAD drawings, Satheeh kumar S., Sundar Raj A., V. Baskaran, Dr. K. Kalaiarasan, Dr. Kumudhini Kumaran, and Er. K. Praveen for their assistance. I have immensely benefited from the contributions of scholars such as Dr. Ramesh K. Shah. Dr. K. P. Singh, and Dr. J. P. Gupta; I also acknowledge the computer facilities of the Engine Development Directorate of RDSO, Lucknow, Ministry of Railways (1992–93), and the library facilities of IIT-M, IIT-K, IIT-D, and RDSO, Lucknow. A large number of heat exchanger manufacturers and research organizations have provided photos and figures, and their names are acknowledged in the respective figure captions.

About the Author

Kuppan Thulukkanam, Indian Railway Service of Mechanical Engineers (IRSME), Ministry of Railways, retired as Principal Executive Director, CAMTECH, Gwalior, RDSO. He has authored an article in the *ASME Journal of Pressure Vessel Technology*. His various roles have included being an experienced administrator, staff recruitment board chairman for a zonal railway, and joint director, Engine Development Directorate of RDSO, Lucknow (Ministry of Railways). He also has been involved in the design and performance evaluation of various types of heat exchangers used in diesel electric locomotives and has served as chief workshop engineer for the production of rolling stocks including coaches, diesel and electric multiple units, wagons, electric locomotive, etc. and as Director, Public Grievances (DPG) to the Minister of State for Railways, Railway Board, Government of India. Kuppan received his BE (Hons) in 1980 from the PSG College of Technology, Coimbatore, Madras University, and his MTech in production engineering in 1982 from the Indian Institute of Technology, Madras, India.

1 Heat Exchangers
Introduction, Classification, and Selection

1.1 INTRODUCTION

A heat exchanger is a heat transfer device that is used for the transfer of internal thermal energy between two or more fluids available at different temperatures. The fluids can be liquids, gases, two fluids, etc. In most heat exchangers, the fluids are separated by a heat transfer surface, and ideally they do not mix. Heat exchangers are used in the process, power, petroleum, transportation, air-conditioning, refrigeration, cryogenic, heat recovery, alternate fuels, and other industries. Common examples of heat exchangers familiar to us in day-to-day use are automobile radiators, condensers, evaporators, oil coolers and air preheaters, economizers, and surface condensers of boiler and steam power plants. Heat exchangers can be classified in many different ways.

1.2 HEAT TRANSFER METHODS

Heat can be transferred by three methods.

i. Conduction—Energy is transferred between solids in contact
ii. Convection—Convection is the mechanism of heat transfer through a fluid in the presence of bulk fluid motion. Convection is classified as natural (or free) and forced convection depending on how the fluid motion is initiated. In natural convection, any fluid motion is caused by natural means such as the buoyancy effect, whereas in forced convection, the fluid is forced to flow over a surface or through a tube by external means such as a pump or fan.
iii. Radiation—Energy is transferred by electromagnetic radiation. An example is the heating of the Earth by the sun.

1.3 BASIC HEAT TRANSFER THEORY

The natural laws of physics always allow the driving energy in a system to flow until equilibrium is reached. Heat leaves the warmer body or the hottest fluid, as long as there is a temperature difference, and will be transferred to the cold medium. A heat exchanger follows this principle in its endeavor to reach equalization. In a heat exchanger, the heat conducts through the surface, which separates the hot medium from the cold one. It is therefore possible to heat or cool fluids or gases

DOI: 10.1201/9781003352044-1

which have minimal energy levels. The theory of heat transfer from one medium to another, or from one fluid to another, is determined by several basic rules:

i. Heat is always transferred from a hot medium to a cold medium.
ii. There must always be a temperature difference between the media, i.e., thermal gradient.
iii. The heat lost from the hot medium is equal to the amount of heat gained by the cold medium.

1.3.1 Heat Transfer Rate

The heat transfer rate, also known as the heat capacity or heat load, is a measure of the heat energy transferred in the heat exchanger per unit time. This is the most fundamental specification for describing heat exchanger performance, and must be known by the user before selecting a heat exchanger or sending a selection form to a manufacturer. The general heat transfer equation can be used to calculate the heat load given the fluid temperature change (of either fluid channel), the fluid flow rate, and the specific heat:

$$q = C_{\mathrm{h}}\left(t_{\mathrm{h,i}} - t_{\mathrm{h,0}}\right) = C_{\mathrm{c}}\left(t_{\mathrm{c,o}} - t_{\mathrm{c,i}}\right) \tag{1.1}$$

where

C_{c} is the capacity rate of the cold fluid, $(Mc_{\mathrm{p}})_{\mathrm{c}}$
C_{h} is the capacity rate of the hot fluid, $(Mc_{\mathrm{p}})_{\mathrm{h}}$
$t_{\mathrm{c,i}}$ and $t_{\mathrm{c,o}}$ are cold fluid terminal temperatures (inlet and outlet)
$t_{\mathrm{h,i}}$ and $t_{\mathrm{h,o}}$ are hot fluid terminal temperatures (inlet and outlet)

1.4 CONSTRUCTION OF HEAT EXCHANGERS

A heat exchanger consists of heat-exchanging elements such as a core or matrix containing the heat transfer surface, and fluid distribution elements such as headers or tanks, inlet and outlet nozzles or pipes, etc. Usually, there are no moving parts in the heat exchanger; however, there are exceptions, such as a rotary regenerator in which the matrix is driven to rotate at some design speed and a scraped surface heat exchanger in which a rotary element with scraper blades continuously rotates inside the heat transfer tube. The heat transfer surface is in direct contact with fluids through which heat is transferred by conduction. The portion of the surface that separates the fluids is referred to as the primary or direct contact surface. To increase the heat transfer area, secondary surfaces known as fins may be attached to the primary surface. Figure 1.1 shows a collection of a few types of heat exchangers and the concept of fluids separation by a primary heat transfer surface for a few types of heat exchangers is shown in Figure 1.2.

1.4.1 Construction Standards and Codes

1.4.1.1 Codes

Standards and codes were established primarily to ensure safety against failure. A code is a system of regulations or a systematic book of law often given statutory force by state or legislative bodies. Codes are intended to set forth engineering requirements deemed necessary for safe design and construction of pressure vessels, piping, etc., e.g., ASME Code for Boilers and Pressure Vessels [1,2] for the construction of boilers and pressure vessels including heat exchangers is the most widely used and is the generally referred to code around the world today.

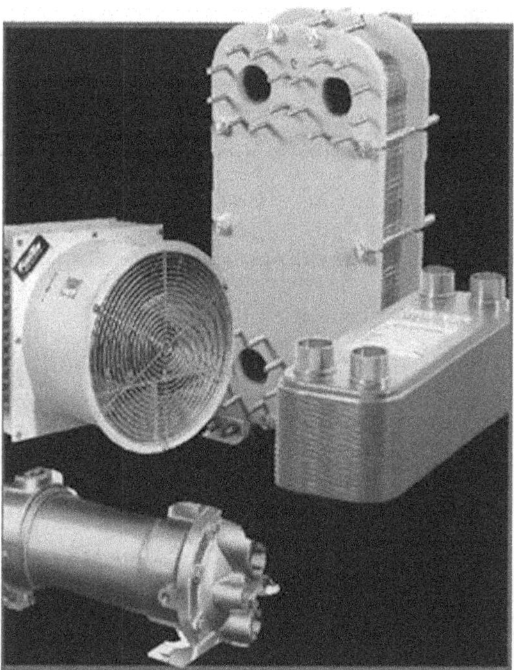

FIGURE 1.1 Collection of few types of heat exchangers. (Courtesy of ITT STANDARD, Cheektowaga, NY.)

1.4.1.2 Standards

A standard can be defined as a set of technical definitions and guidelines, or how-to instructions for designers and manufacturers. A standard is developed by the consensus process, and involves technical experts from the producers and users. Standards are mostly voluntary in nature. They serve as guidelines but do not themselves have the force of law. Standards help to reduce the cost of products and processes. At the design level, rationalization of design procedure, drawings, and specifications takes place. This avoids the repetition of detailed design analysis for either identical or similar jobs. Standards help in complete interchangeability and uniformity of fundamental design, tools, gauges, tool accessories, etc. The standards can be of the following major four types:

1. Company standards
2. Trade or manufacturer's association standards
3. National standards
4. International standards

1.5 CLASSIFICATION OF HEAT EXCHANGERS

In general, industrial heat exchangers have been classified according to (1) construction; (2) heat transfer processes; (3) degrees of surface compactness; (4) flow arrangements; (5) pass arrangements; (6) phase of the process fluids; and (7) heat transfer mechanisms. These classifications are briefly discussed here. For more details on heat exchanger classification and construction, refer to Shah [3,4], Gupta [5], and Graham Walker [6]. For classification and systematic procedure for selection of heat exchangers, refer to Larowski et al. [7a,7b]. For trends and innovation in heat exchanger design and market research refer to Ref. [8]

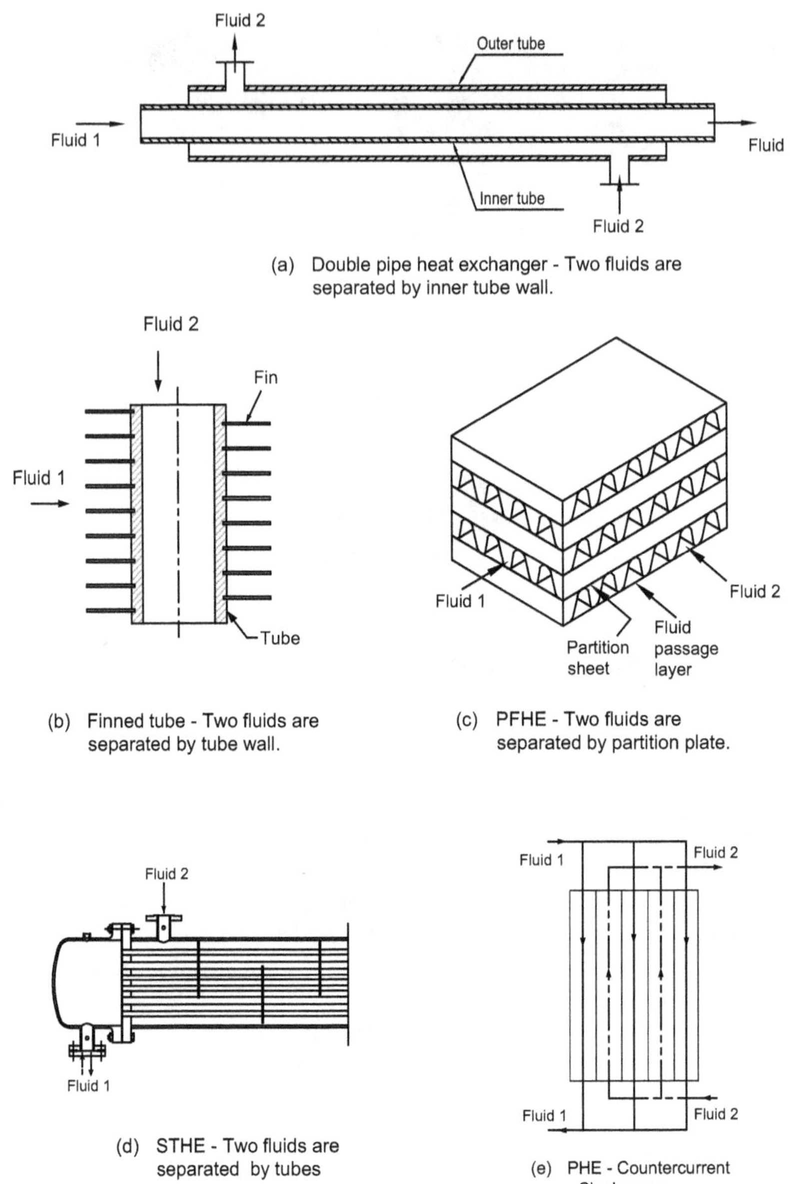

FIGURE 1.2 Concept of fluids separation by a primary heat transfer surface for few types of heat exchangers.

1.5.1 Classification According to Construction

According to constructional details, heat exchangers are classified as [3] follows:

i. Tubular heat exchangers—double pipe, shell and tube, coiled tube
ii. Plate heat exchangers (PHEs)—gasketed, brazed, welded, spiral, panel coil, lamella, etc.
iii. Extended surface heat exchangers—tube-fin, plate-fin
iv. Regenerators—fixed matrix, rotary matrix

Heat exchanger classification according to construction is shown in Figure 1.3.

TABLE 1.1
Major Types of Heat Exchangers, Their Construction Details, and Performance Parameters

Type of heat exchanger	Constructional features/Performance features and Application areas	advantages
Double pipe (Hair pin) heat exchanger	A double pipe heat exchanger has two concentric pipes, usually in the form of a U-bend design. U-bend design is known as Hair Pin heat exchangers. Hairpin Exchangers are available in single tube (Double Pipe) or multiple tubes within a hairpin shell (Multitube), bare tubes, finned tubes, U-tubes, straight tubes (with rod-thru capability), fixed tubesheets and removable bundle. The flow arrangement is pure countercurrent. The surface area ranges from 300 to 6,000 square feet (Finned tubes). Pressure capabilities are full vacuum to over 14,000 psi (limited by size, material, and design condition) and temperature from -100 to 600°C (-150 to 1100°F). Design Standard: TEMA Standards and ASME Code Sec. VIII Div. 1	Applicable services The process results in a temperature crossHigh pressure stream on tubeside A low allowable pressure drop is required on one side When an augmentation device to enhance the heat transfer coefficient is desired When the exchanger is subject to thermal shocks When flow induced vibration may be a problem When solid particultes or slurries are present in the process stream. Advantages Easy to obtain counterflow Can handle high pressure stream Modular construction Easy to maintain and repair
Shell and tube heat exchanger 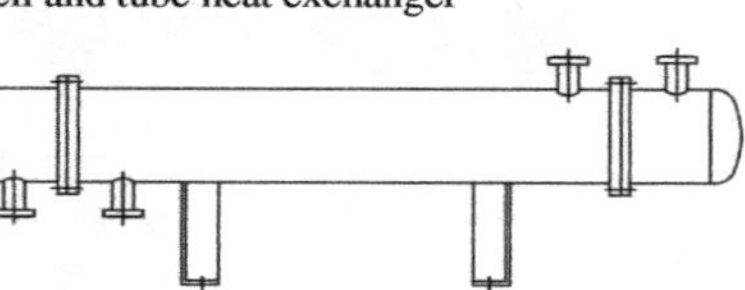	The most commonly used heat exchanger. It is the "workhorse" of industrial process heat transfer. Can be used both for sensible heating/cooling and phase change applications. It has many applications in the power generation, petroleum refinery, chemical industries, and process industries. They are used as oil coolers, surface condensers, feed water heaters, etc. The major components of a shell and tube exchanger are tubes, baffles, shell, front head, rear head, and nozzles. Expansion joint is an important component in the case of fixed tubesheet exchanger for certain design conditions	Advantages Extretmely fltexible and robust design Easy to maintain and repair Can be designed to be dismantled for cleaning many suppliers worldwide Disadvantages Requires large site (footprint) area for installation and often needs extra space to remove the bundle

(continued)

TABLE 1.1 (Continued)
Major Types of Heat Exchangers, Their Construction Details, and Performance Parameters

Type of heat exchanger	Constructional features/Performance features and Application areas	advantages
	Design standard: TEMA Standard, ASME Code Sec. VIII, Did.1, ANSI/API Standard 660 Other types of shell and tube heat exchangers (STHE) based on non-segmental baffle, include Rod baffle-, EM™ baffle-, Koch Twisted tube™-, Helixchanger™ heat exchanger, etc.	Construction is heavy PHE may be cheaper for pressure below 16 bar (230 psi) and temperature below 200°C (392° F)
Coiled tube heat exchanger(CTHE)	Construction of these heat exchangers involves winding a large number of small-bore ductile tubes in helix fashion around a central core tube, with each exchanger containing many layers of tubes along both the principal and radial axes. Different fluids may be passed in counterflow to the single shellside fluid. Advantages, especially when dealing with low-temperature applications where simultaneous heat transfer between more than two streams is desired.	Because of small bore tubes on both sides, CTHEs do not permit mechanical cleaning and therefore are used to handle clean, solid-free fluids or fluids whose fouling deposits can be cleaned by chemicals. Materials are usually aluminum alloys for cryogenics, and stainless steels for high-temperature applications.
Finned-tube heat exchanger 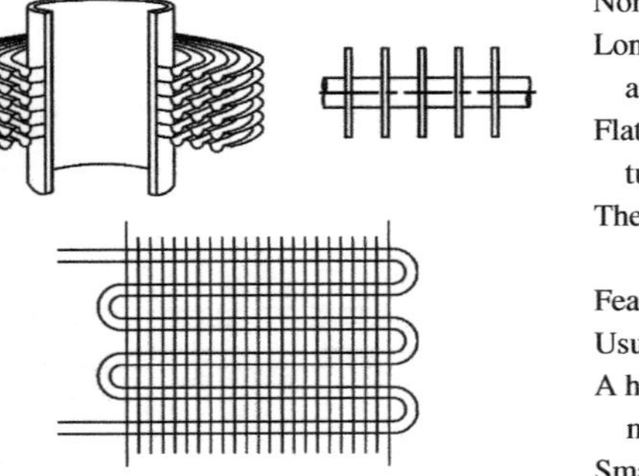	Construction Normal fins on individual tubes referred to as individually finned tubes. Longitudinal fins on individual tubes, which are generally used in condensing applications and for viscous fluids in double-pipe heat exchangers. Flat or continuous (plain, wavy, or interrupted) external fins on an array of tubes (either circular or flat tube). The tube layout pattern is mostly staggered. Features Usually with extended surfaces. A high heat transfer surface area per unit volume of the core, i.e., about 700 m^2/m^3 on gas/air side. Small hydraulic diameter. Fluids must be clean and relatively nonfouling because of small hydraulic diameter (D_h) flow passages and difficulty in cleaning. The fluid pumping power (i.e., pressure drop) consideration is as important as the heat transfer rate.	Merits Small inventory Low weight Easier transport Less foundation Better temperature control Applications Condensers and evaporators of air conditioners, radiators for internal combustion engines, charge air coolers, and intercoolers for cooling supercharged engine intake air of diesel engines, etc.

Plate fin heat exchanger (PFHE) 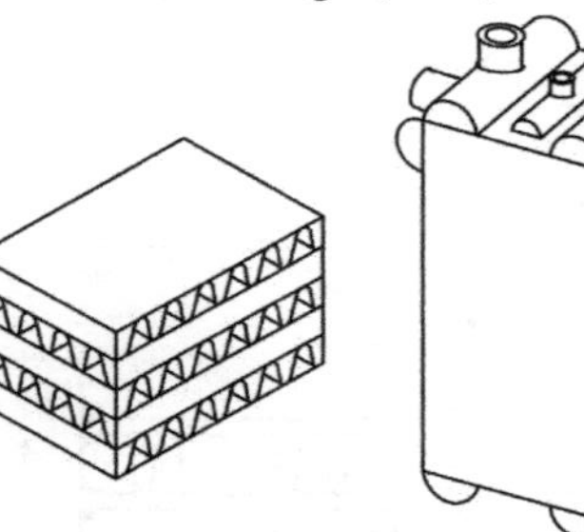	Plate fin heat exchangers (PFHEs) are a form of compact heat exchanger consisting of a stack of alternate flat plates called "parting sheets" and fin corrugations, brazed together as a block. Different fins (such as the plain triangular, louver, perforated, or wavy fin) can be used between plates for different applications. Plate-fin surfaces are commonly used in gas-to-gas exchanger applications. They offer high area densities (up to about 6000 m^2/m^3 or 1800 ft^2/ft^3). Designed for low-pressure applications, with operating pressures limited to about 1000 kPa g (150 psig) and operating temperature from cryogenic to 150°C (all-aluminum PFHE) and about 700°C–800°C (1300°F–1500°F) (made of heat-resistant alloys).	1. PFHE offers superior in thermal performance compared to extended surface heat exchangers. 2. PFHE can achieve temperatured approaches as low as 1°C between single-phase streams and 3°C between multiphase streams. 3. With their high surface compactness, ability to handle multiple streams, and with aluminum's highly desirable low-temperature properties, brazed aluminum plate fins are an obvious choice for cryogenic applications. 4. Very high thermal effectiveness can be achieved; for cryogenic applications, effectiveness of the order of 95% and above is common. Limitations: Narrow passages in plate-fin exchangers make them susceptible to fouling and they cannot be cleaned by mechanical means. This limits their use to clean applications like handling air, light hydrocarbons, and refrigerants
Air-cooled heat exchanger (ACHE) 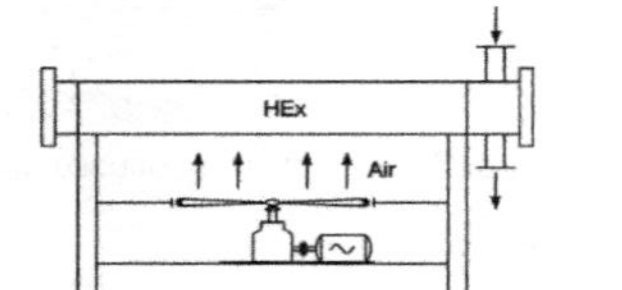	Construction 1. Individually finned tube bundle. 2. An air pumping device (such as an axial flow fan or blower) across the tube bundle which may be either forced draft or induced draft. 3. A support structure high enough to allow air to enter beneath the ACHE. Merits Design of ACHEs simpler compared to STHE, since the airside pressure and temperature pertain to ambient conditions. Tubeside design is same as STHE Usage: Sensible cooling and condensing services Design standard: ASME Code Sec. VIII, Did.1, ANSI/API Standard 660 Embedded fins can withstand temperatures up to 310°C (590°F), whereas extruded fins can withstand up to 330°C (650°F)	Advantages of ACHEs Air is always available free of cost Maintenance costs is normally less than for water-cooled systems The mechanical design is simpler since the air side is closer to atmospheric pressure The fouling on the air side can be cleaned easily. Disadvantages of ACHEs ACHEs require large heat transfer surfaces because of the low heat transfer coefficient on the air side and the low specific heat of air. Water coolers require much less heat transfer surface.

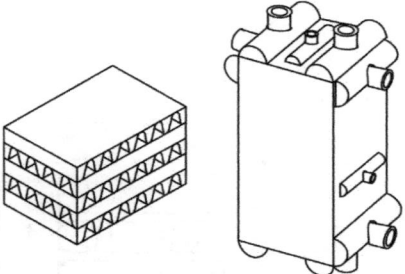

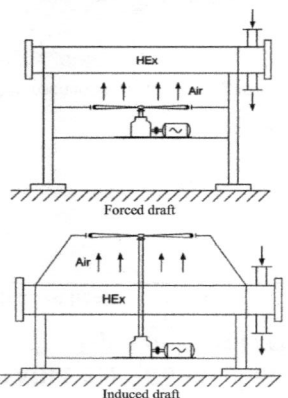

(*continued*)

TABLE 1.1 (Continued)
Major Types of Heat Exchangers, Their Construction Details, and Performance Parameters

Type of heat exchanger	Constructional features/Performance features and Application areas	advantages
		ACHEs are affected by hailstorms and may be affected by cyclonic winds. Noise is a factor with ACHEs. May need special controls for cold weather protection. Cannot cool the process fluid to the same low temperature as cooling water
Microchannel heat exchanger 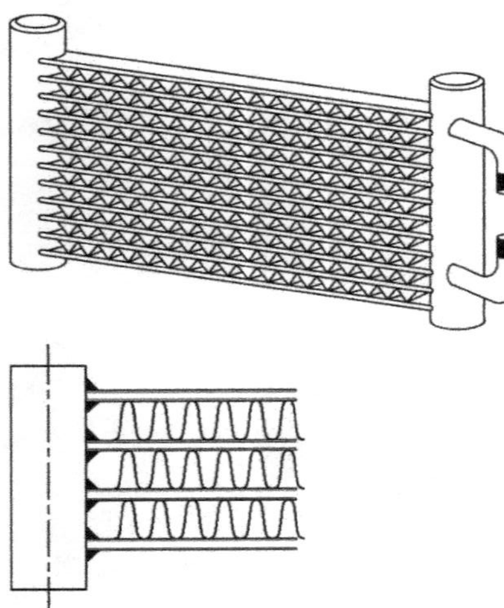	Brazed microchannel coils are all aluminum coils. Used in the automotive and HVAC industry	• Smaller/compact coil, less refrigerant volume, same cooling capacity. • Enhanced fins transfer heat more efficiently. • Small tube/channel size reduces refrigerant volume. • Compact all-aluminum microchannel coils help to reduce the unit weight. • All-aluminum construction improves recyclability and minimizes galvanic corrosion. • In applications where the microchannel coil may be susceptible to a corrosive environment coated coil option is available. • Lower raw material cost (aluminum vs. copper) • Easy recyclability

Internally grooved microfin tubeheat exchanger 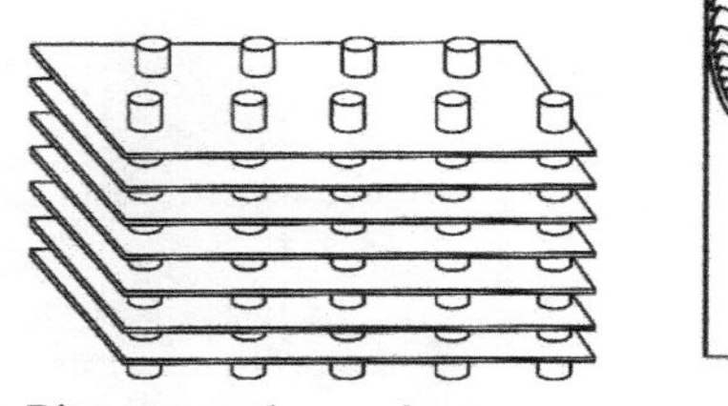	A small diameter copper coil technology for modern air conditioning and refrigeration systems. Grooved coils facilitate more efficient heat transfer than smooth coils. Small diameter coils have better rates of heat transfer Can be made with copper or aluminum. Copper fins are better due to the better corrosion resistance of copper and its antimicrobial benefits	Merits Cost-effective fabrication and assembly Smaller size, less weight and lower material costs Higher heat transfer coefficients Well suited for new refrigerants Uses less refrigerant Overall reduction in system cost
Direct contact heat exchanger 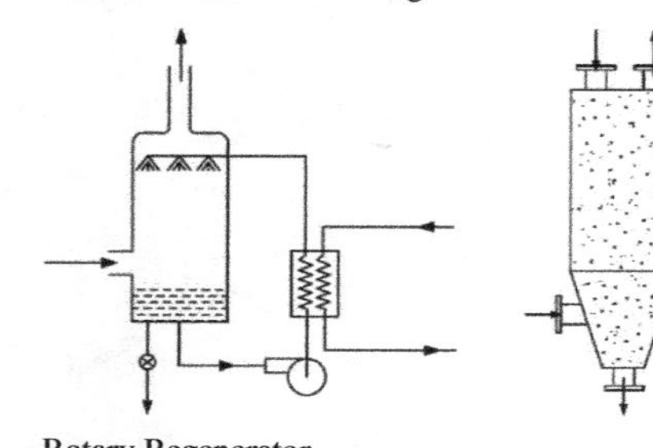		
Rotary Regenerator 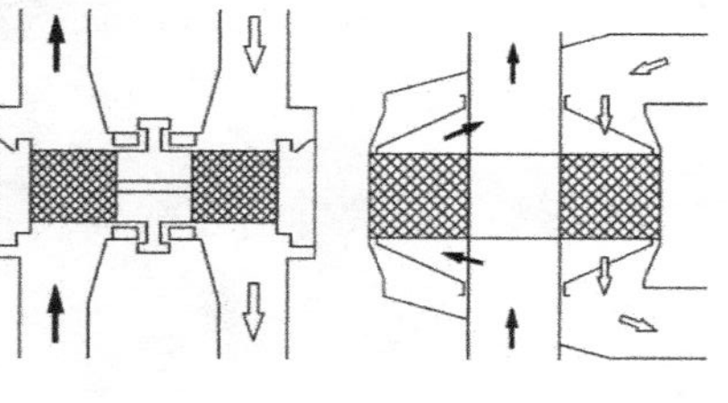	The heat transfer matrix is alternatively heated by hot fluid and cooled by the cold fluid. The regenerators are classified as (1) fixed matrix or fixed bed or (2) rotary regenerators Rotary regenerator A more compact size ($\beta = 8800$ m^2/m^3 for rotating type and 1600 m^2/m^3 for fixed matrix type). Application to both high temperatures (800–1100°C) for metal matrix, and 2000°C for ceramic regenerators for services like gas turbine applications, melting furnaces or steam power plant heat recovery and cryogenic applications (–20°) Operating pressure of 5–7 bar for gas turbine applications and low pressure of 1–1.5 bar for air dehumidifier and waste heat recovery applications.	Usage Air preheater—preheating the combustion air (High temperature heat recovery system) Space heating—Rotary heat exchanger (wheel) is mainly used in building ventilation or in the air supply/discharge system of conditioning equipment.

(continued)

TABLE 1.1 (Continued)
Major Types of Heat Exchangers, Their Construction Details, and Performance Parameters

Type of heat exchanger	Constructional features/Performance features and Application areas	advantages
	Regenerators have self-cleaning characteristics and hence compact regenerators have minimal fouling. If heavy fouling is anticipated, regenerators are not used. Normally a laminar flow condition prevails, due to the small hydraulic diameter. The absence of a separate flow path like tubes or plate walls but the presence of seals to separate the gas stream in order to avoid mixing due to pressure differential.	
Heat pipe heat exchanger 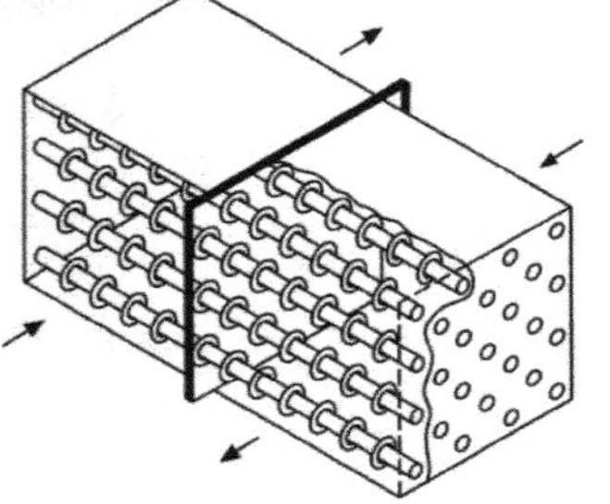	The heat pipe exchanger is a lightweight compact heat recovery system (fin-tube heat exchanger). It virtually does not need mechanical maintenance, as there are no moving parts. It does not need input power for its operation. The heat pipe heat recovery systems are capable of operating at a temperature of 300–315°C with 60 to 80% heat recovery capability	Application The heat pipes are used for (i) heat recovery from process fluid to preheating of air for space heating, (ii) HVAC application—waste heat recovery from the exhaust air to heat the incoming process air
Plate heat exchanger (PHE)	A plate heat exchange is usually comprised of a stack of corrugated or embossed metal plates in mutual contact, each plate having four apertures serving as inlet and outlet ports, and seals designed so as to direct the fluids in alternate flow passages. Standard Performance Limits Maximum operating pressure 25 bar (360 psi) With special construction 30–40 bar (435–600psi) Maximum temperature 160°C (320°F) With special gaskets 200°C (390°F) Maximum flow rate 3600 m^3/h (950,000 USG/min) Heat transfer coefficient 3500–8700 $W/m^2 \cdot °C$ (600–1500) BTU/$ft^2 \cdot hr \cdot °F$)	Merits True counterflow, high turbulence, and high heat transfer performance. Close approach temperature Reduced fouling. Cross-contamination eliminated Multiple duties with a single unit Expandable Easy to inspect and clean, and less maintenance Low liquid volume and quick process control

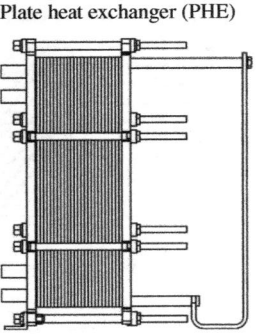

	Heat transfer area 0.1–2200 m^2 (1–24,000 ft^2) Surface area/plate 0.04–3.0 m^2 Maximum connection size 450 mm (18 in) Temperature approach as low as 1°C Thermal effectiveness as high as 93–95% Applications Plate heat exchanger is universally used in many fields: heating and ventilating, breweries, dairy, food processing, pharmaceuticals and fine chemicals, petroleum and chemical industries, cooling water applications, etc. Other types of plate heat exchangers Brazed plate heat exchanger (BHE) Shell & plate Heat Exchanger Welded plate heat exchanger Wide-Gap Plate Heat Exchanger Free-flow plate heat exchanger Semi-welded or Twin-Plate Heat exchanger Double-Wall Plate Heat Exchanger Diabon F Graphite Plate Heat Exchanger Compablock plate heat exchanger	Lower cost Disadvantages The maximum operating temperature and pressure are limited by gasket materials. The gaskets cannot handle corrosive or aggressive media Gaskets always increase the leakage risk. Other varieties of PHEs overcome limitations in respect of pressure, temperature, leakage, contamination, gasket material, maximum allowable particle size that can be handled, etc.
Brazed plate heat exchanger (BPHE) 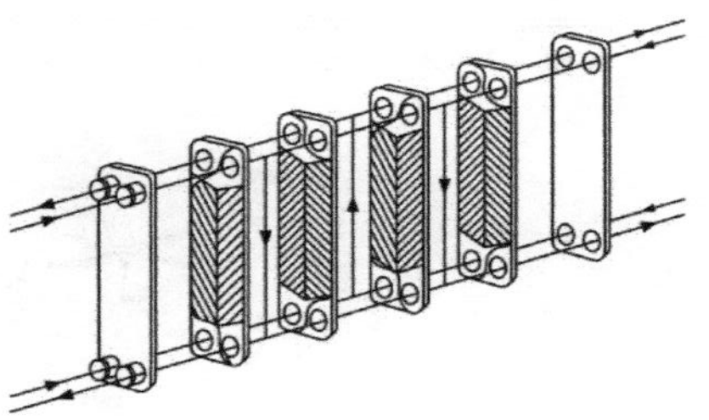	Brazed plate heat exchanger (BPHE) It is constructed of a series of corrugated metal plates brazed together in a vacuum oven to form a complete pressure – resistant unit. Absence of gaskets, tightening bolts, frame, or carrying and guide bars.	A sealed, brazed compact system, high temperature and pressure capability, gasket-free construction, high thermal efficiency and ideal for refrigeration and process applications. The units are not expandable

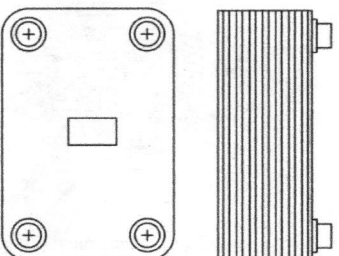

(*continued*)

TABLE 1.1 (Continued)
Major Types of Heat Exchangers, Their Construction Details, and Performance Parameters

Type of heat exchanger	Constructional features/Performance features and Application areas	advantages
Spiral plate heat exchanger (SPHE)	SPHE is fabricated by rolling a pair of relatively long strips of plate to form a pair of spiral passages. Channel spacing is maintained uniformly along the length of the spiral passages by means of spacer studs welded to the plate strips prior to rolling. Heat transfer surface in one single spiral body is 0.5 to 350 m^2. The maximum design pressure is normally limited to 150 psi. Max. Design temp. 450°C (840°F)	To handle slurries and liquids with suspended fibers, and mineral ore treatment where the solid content is up to 50%. The SPHE is the first choice for extremely high viscosities, say up to 500,000 cp, especially in cooling duties Applications: Applications in reboiling, condensing, heating or cooling of viscous fluids, slurries, and sludge.
Printed circuit heat exchangers (PCHEs) 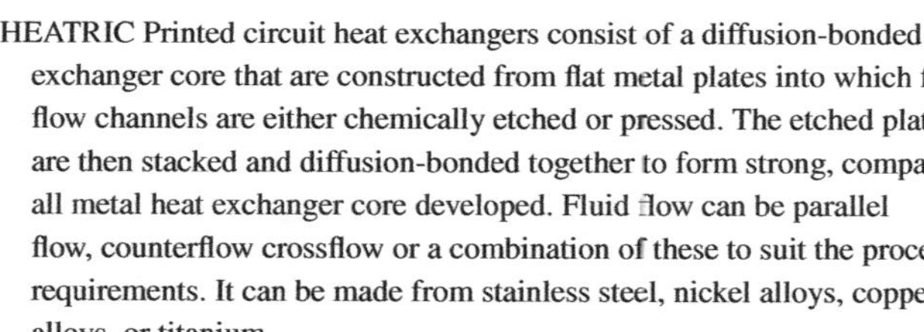 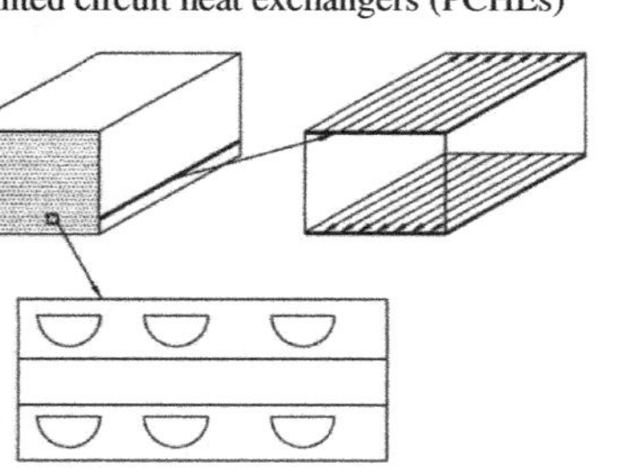	HEATRIC Printed circuit heat exchangers consist of a diffusion-bonded heat exchanger core that are constructed from flat metal plates into which fluid flow channels are either chemically etched or pressed. The etched plates are then stacked and diffusion-bonded together to form strong, compact, all metal heat exchanger core developed. Fluid flow can be parallel flow, counterflow crossflow or a combination of these to suit the process requirements. It can be made from stainless steel, nickel alloys, copper alloys, or titanium.	Merits They can withstand pressure of 600 bar (9000 psi) with extreme temperatures, ranging from cryogenic to 700°C (1650°F). Thermal effectiveness is of the order of 98% in a single unit. They can incorporate more than two process streams into a single unit.
Plate coil heat exchanger 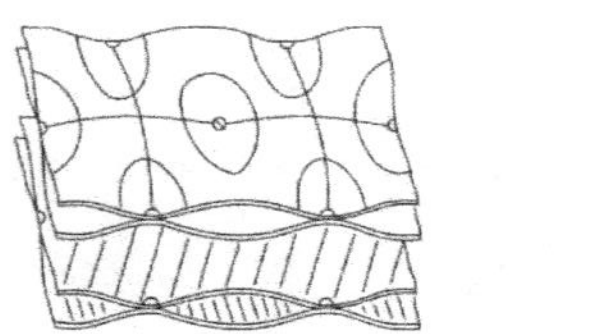 	Fabricated from two metal sheets, one or both of which are embossed. When welded together, the embossings form a series of well-defined passages through which the heat transfer media flows.	A variety of standard PLATECOIL® fabrications, such as pipe coil, half pipe, jacketed tanks and vessels, clamp-on upgrades, immersion heaters and coolers, heat recovery banks, storage tank heaters, drum warmers, etc. are also available. Easy access to panels and robust cleaning surfaces reduce maintenance burdens.

Term	Description	Application
Scraped surface heat exchanger 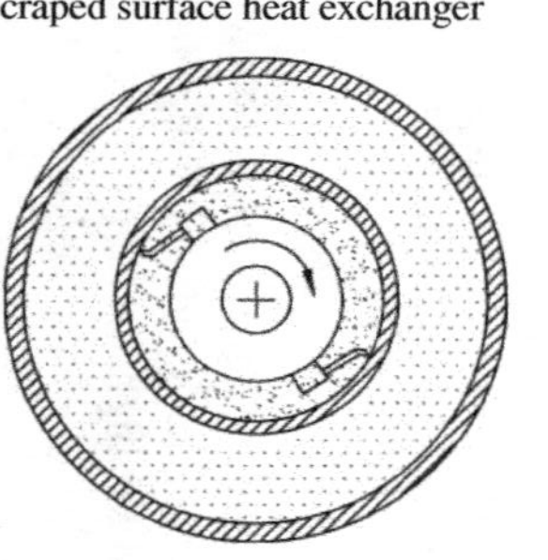	Scraped surface heat exchangers are essentially double pipe construction with the process fluid in the inner pipe and the cooling (water) or rotating element is contained within the tube and is equipped with spring loaded blades. In operation the rotating shaft scraper blades continuously scraped product film from the heat transfer tube wall, thereby enhancing heat transfer and agitating the product to produce a homogeneous mixture.	Scrapped surface heat exchangers are used for processes likely to result in the substantial deposition of suspended solids on the heat transfer surface. Suspended solids on the heat transfer surface. Scraped surface heat exchangers can be employed in the continuous, closed processing of virtually any pumpable fluid or slurry involving cooking, slush freezing, cooling, crystallizing, mixing, plasticizing, gelling, polymerizing, heating, aseptic processing, etc. Use of a scraped surface exchanger prevents the accumulation of significant buildup of solid deposits
Heat sink exchanger 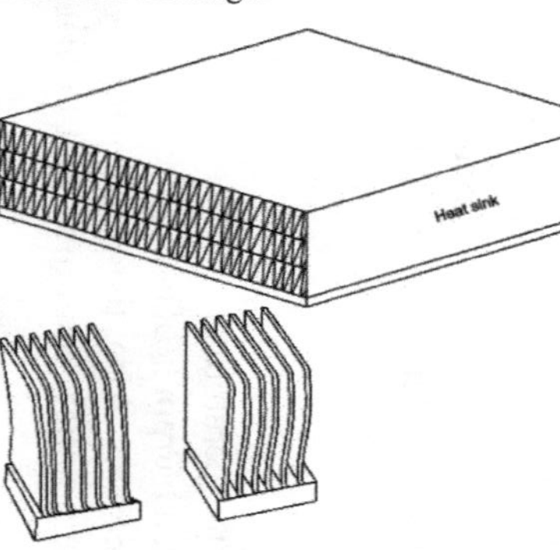	A **heat sink** is a passive heat exchanger that transfers the heat generated by an electronic or a mechanical device to a fluid medium, often air or a liquid coolant, where it is dissipated away from the device, thereby allowing regulation of the device's temperature. It is designed to maximize its surface area in contact with the cooling medium surrounding it. The fluid medium is frequently air.	In computers, heat sinks are used to cool CPUs, GPUs, and some chipsets and RAM modules. Heat sinks are used with high-power semiconductor devices such as power transistors and optoelectronics such as lasers and light-emitting diodes (LEDs).

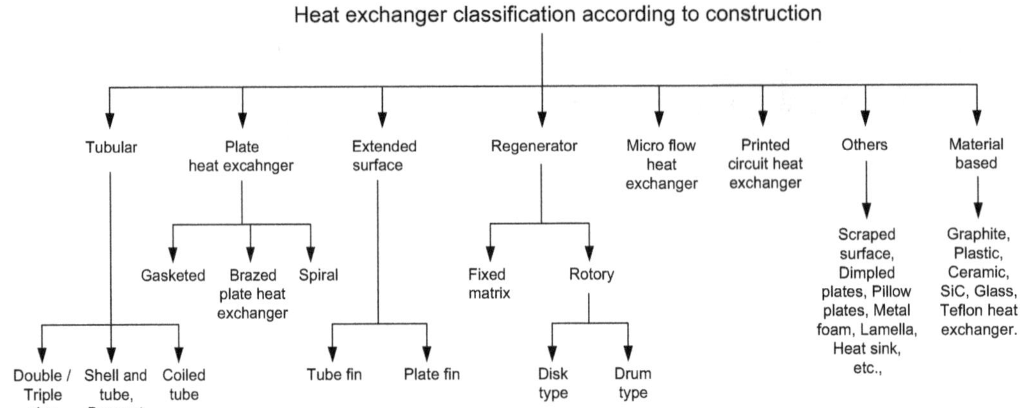

FIGURE 1.3 Heat exchanger classifications according to construction and materials of construction.

1.5.2 Tubular Heat Exchanger

1.5.2.1 Double-Pipe Exchangers

A double-pipe heat exchanger has two concentric pipes, usually in the form of a U-bend design. Double-pipe heat changers with U-bend design are known as hairpin heat exchangers. One fluid flows through the inner pipe (tubeside), and the other flows through the annular space (annulus). The inner pipe is connected by U-shaped return bends enclosed in a return-bend housing. The major use of these heat exchangers is the sensible heating or cooling process of fluids where small heat transfer areas are required. The outer surface of the inner tube can be finned longitudinally, and in another type, there are multiple tubes, finned or bare, inside a larger pipe. The fins increase the heat transfer surface per unit length and reduce the size and, therefore, the number of hairpins required for a given heat duty. Integrally, resistance welded longitudinal fins have proven to be the most efficient for double-pipe heat exchangers. Fins are most efficient when the film coefficient is low. The performance is also improved by using heat transfer enhancement devices like turbulators, inserts, modifying the geometry of channels, etc.

Double-pipe heat exchangers are used in many industrial processes, cooling technology, refrigeration device, sustainable energy applications, and other fields. The fluid flow arrangements can be parallel or counterflow, but are mostly counterflow. The major use of the double-pipe heat exchanger is the sensible heating or cooling process of fluids where small heat transfer areas (up to 50 m^2) are required. This configuration is also very suitable for one or both of the fluids at high pressure because of the smaller diameter of the pipes. The major disadvantage is that they are bulky and expensive per unit of heat transfer surface area. Figures 1.4 through 1.6 show double-pipe heat exchangers.

Design pressures and temperatures are broadly similar to shell and tube heat exchangers (STHEs). The design is straightforward and is carried out using the method of Kern [9] or proprietary programs. The Koch Heat Transfer Company LP, USA, is the pioneer in the design of hairpin heat exchangers.

Application

When the process calls for a temperature cross (when the hot fluid outlet temperature is below the cold fluid outlet temperature), a hairpin heat exchanger is the most efficient design and will result in fewer sections and less surface area. Also, they are commonly used for high-fouling services such as slurries and for smaller heat duties. Multitube heat exchangers are used for larger heat duties (Table 1.1).

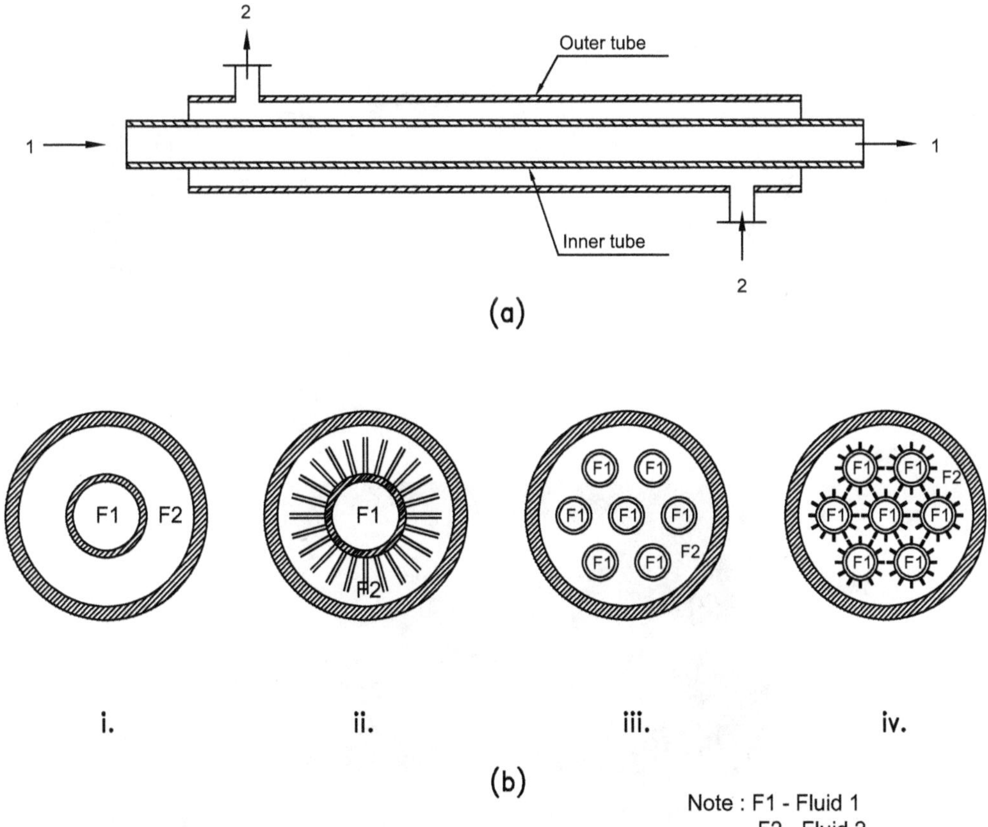

FIGURE 1.4 Double pipe/twin pipe hairpin heat exchanger. (a) Schematic of the unit, (b): (i) double pipe with bare internal tube, (ii) double pipe with finned internal tube, (iii) double pipe with multibare internal tubes, and (iv) double pipe with multifinned internal tubes. (Courtesy of Peerless Mfg. Co., Dallas, TX, Makers of Alco and Bos-Hatten brands of heat exchangers.)

1.5.2.2 Triple Concentric Tube Heat Exchanger (TCTHE)

In the TCTHE, there are three sections: central tube, inner annular space, and outer annular space. Heat transfer medias are passed through the central tube and outer annular space and a thermal fluid is passed through the inner annular space. This type of heat exchanger can be used for higher viscosity fluids.

The triple-tube heat exchanger contributes to higher heat exchanger effectiveness and more energy saving compared with the double-tube heat exchanger per unit length. A schematic of triple concentric tube heat exchangers is shown in Figure 1.7.

1.5.3 Shell and Tube Heat Exchanger

In process industries, shell and tube heat exchangers are used in great numbers, far more than any other type of exchanger. More than 90% of heat exchangers used in industry are of the shell and tube type [10]. They are the most common type of heat exchanger in oil refineries and other large chemical processes, and are suited for higher pressure applications. STHEs are the "workhorses" of industrial process heat transfer [11]. They are the first choice because of well-established procedures for design and manufacture from a wide variety of materials, many years of satisfactory service, and

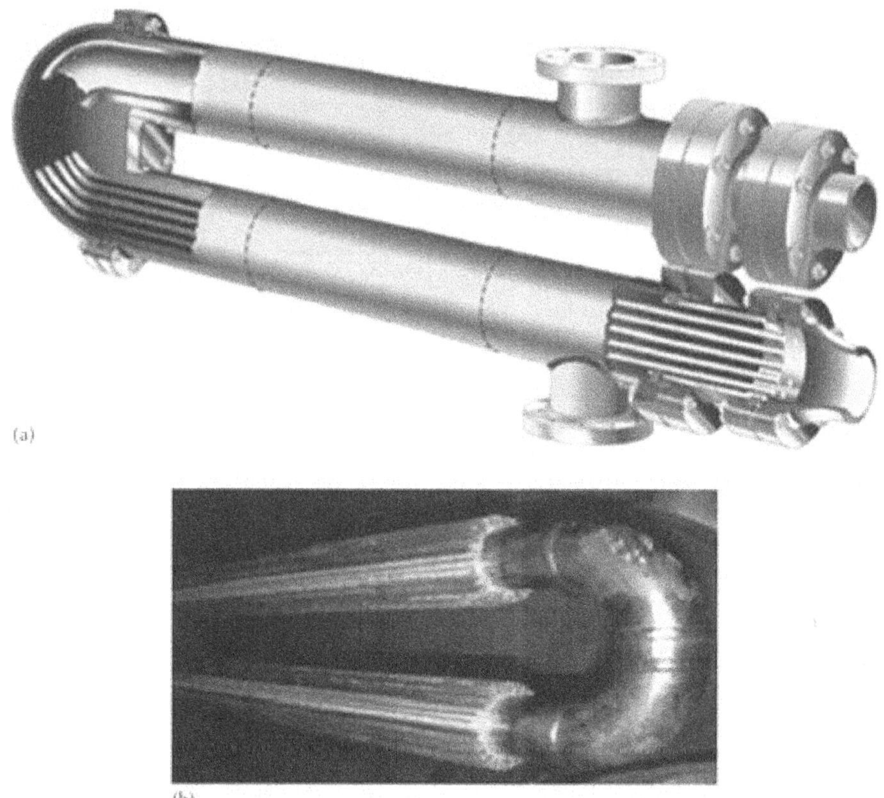

(a)

(b)

FIGURE 1.5 Double pipe/hairpin heat exchanger. (a) 3-D view and (b) tube bundle with longitudinal fins. (Courtesy of Peerless Mfg. Co., Dallas, TX, Makers of Alco and Bos-Hatten brands of heat exchangers.)

the availability of codes and standards for design and fabrication. Based on their functions, STHEs can be classified as condensers, evaporators, reboilers, feedwater heaters, etc. They are produced in the widest variety of sizes and styles. Figure 1.8 shows a cross-sectional view of an STHE and Figure 1.9 shows a disassembled STHE.

1.5.3.1 STHE types

1. Fixed tubesheet heat exchanger. This is a heat exchanger with two stationary tubesheets, each attached to the shell and channel. The heat exchanger contains a bundle of straight tubes connecting both tubesheets.
2. U-tube heat exchanger. This is a heat exchanger with one stationary tubesheet attached to the shell and channel. The heat exchanger contains a bundle of U-tubes attached to the tubesheet.
3. Floating tubesheet heat exchanger. This is a heat exchanger with one stationary tubesheet attached to the shell and channel, and one floating tubesheet that can move axially. The heat exchanger contains a bundle of straight tubes connecting both tubesheets.

1.5.3.2 Construction

This type of heat exchanger consists of a shell with an internal tube bundle that is typically supported by tubesheet(s) at the ends and intermittent tube support plates known as baffles. Two fluids, of different inlet temperatures, flow through the heat exchanger. One flows through the tubes (the tubeside) and the other flows outside the tubes but inside the shell (the shellside). The tube bundle

FIGURE 1.6 Hairpin heat exchanger. (a) Separated head closure using separate bolting on shellside and tubeside and (b) Hairpin exchangers for high-pressure and high-temperature applications and (c) multitubes (bare) bundle. (Photo courtesy of Heat Exchanger Design, Inc., Indianapolis, IN.)

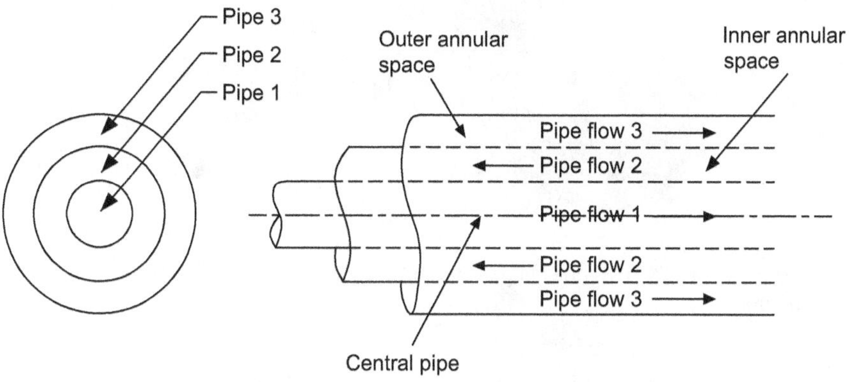

FIGURE 1.7 Triple concentric tubes heat exchanger.

may be composed of several types of tubes: plain, longitudinally finned, low finned, twisted tube, etc. The tube pitch (center-to-center distance of adjoining tubes) is typically a minimum of 1.25 times the tube outer diameter. Typically, the ends of each tube are terminated through holes in a tubesheet. The tubes are mechanically rolled or welded into tube sheet face. Tubes may be straight or bent in the shape of a U, called U-tubes. The baffle configuration shall be segmental or nonsegmental baffles. Based on the baffles configuration, the types of STHEs are classified as shown in Figure 1.10. Refer to Yokkel [12] for construction details and guidelines on the thermal design of STHEs.

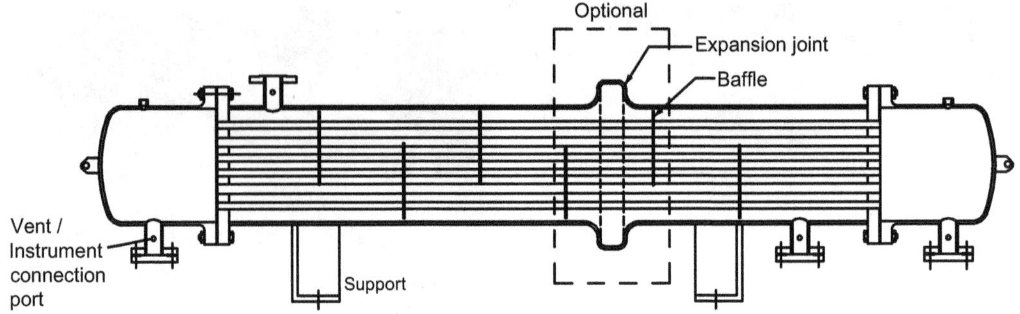

FIGURE 1.8 Cross sectional view of STHE.

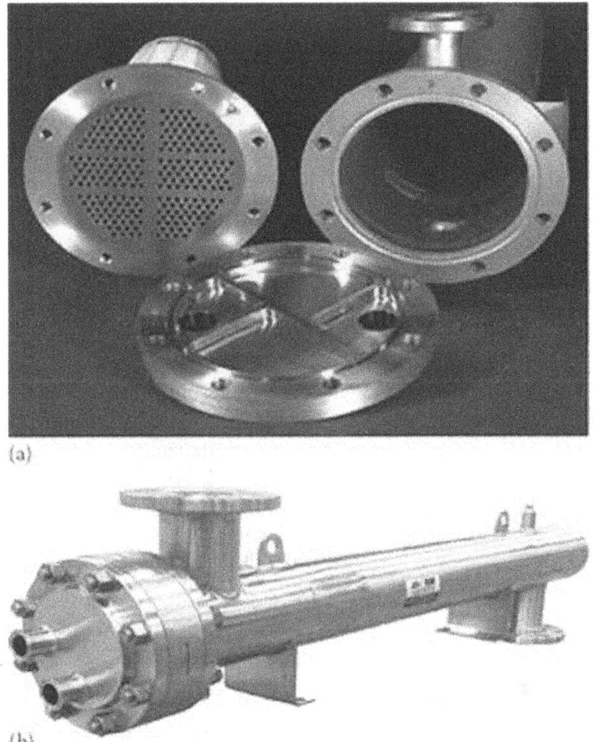

FIGURE 1.9 Disassembled STHE- (a) Components and (b) heat exchanger. (Courtesy of Allegheny Bradford Corporation, Bradford, PA.)

Refer to Chapter 4 for the classification, construction details, and nomenclature for STHEs and thermal design.

1.5.3.3 Design Standards used for Thermal and Mechanical Design of STHEs

Some design standards used for the mechanical design of heat exchangers include the following: TEMA [13], HEI [14], and API [15]. TEMA Standards founded in 1939, the Tubular Exchanger Manufacturers Association, Inc., is a group of leading manufacturers of shell and tube heat exchangers who have pioneered the research and development of heat exchangers for over 80 years. TEMA Standards are followed in most countries around the world for the design of shell and tube heat exchangers.

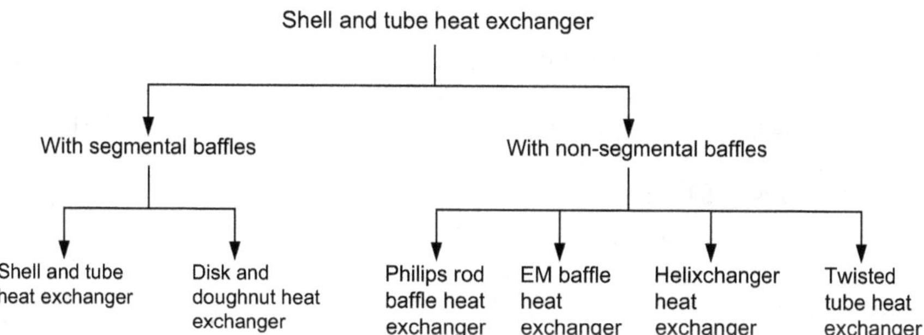

FIGURE 1.10 Types of STHEs based on baffles configuration. (*Note*: There is no baffle in twisted tube heat exchanger.)

1.5.3.4 Standards for the Design of STHE

1. TEMA Standard

   The TEMA Standards are applicable to STHEs which do not exceed any of the following criteria:
   (1) inside diameters of 100 in. (2540 mm)
   (2) product of nominal diameter, in. (mm) and design pressure, psi (kPa) of 100,000(17.5 κ 10^6)
   (3) a design pressure of 3,000 psi (20,684 kPa)

   The intent of these parameters is to limit the maximum shell wall thickness to approximately 3 in. (76 mm), and the maximum stud diameter to approximately 4 in. (102 mm). Criteria contained in these standards may be applied to units which exceed the above parameters.
2. Heat Exchange Institute Standards

   The HEI, Cleveland, Ohio, is an association of manufacturers of heat transfer equipment used in power generation. The association promotes improved designs by developing equipment design standards. It publishes standards for tubular heat exchangers used in power generation. Such exchangers include surface condensers, feedwater heaters, and other power plant heat exchangers [14]. Refer to Chapter 4 for details.
3. API 660-2015, *Shell-and-Tube Heat Exchangers*

   This standard specifies requirements and gives recommendations for the mechanical design, material selection, fabrication, inspection, testing, and preparation for shipment of shell-and-tube heat exchangers for the petroleum, petrochemical, and natural gas industries. This standard is applicable to the following types of shell-and-tube heat exchangers: heaters, condensers, coolers, and reboilers [15].
4. ISO 16812-2019 Petroleum, petrochemical and Natural Gas Industries–Shell and Tube Heat Exchangers
5. PIP VESSM001-2017

   Specification for Small Pressure Vessels and Heat Exchangers with Limited Design Conditions
6. API Std 661-2013(R2018), Petroleum, Petrochemical, and Natural Gas Industries—Air Cooled Heat Exchangers

   Requirements for the design materials fabrication inspection, testing, and shipment preparation for shipment of air-cooled heat exchangers for use in the petroleum, petrochemical, and natural gas industries.
7. API Std 663-2014, Hairpin-type Heat Exchangers

   This standard specifies requirements and gives recommendations for the mechanical design, materials selection, fabrication, inspection, testing, and preparation for shipment of hairpin heat exchangers for use in the petroleum, petrochemical, and natural gas industries. Hairpin heat exchangers include double-pipe and multitube type heat exchangers.

8. API Std 664-2014, Spiral Plate Heat Exchangers
   This standard specifies requirements and gives recommendations for the mechanical design, materials selection, fabrication, inspection, testing, and preparation for shipment of spiral plate heat exchangers for the petroleum, petrochemical, and natural gas industries.

1.5.3.5 Applications of STHE

Some examples of STHEs are given below.

a. Surface condenser

A surface condenser or steam condenser is a water-cooled shell and tube heat exchanger used to condensate the exhaust steam from the last-stage low-pressure steam turbine in thermal power stations. In steam surface condensers there is no mixing of exhaust steam and cooling water.

b. Closed feedwater heater

A feedwater heater is an unfired shell and tube heat exchanger designed to preheat feedwater by means of condensing steam extracted or bleed from a steam turbine. The heater is classified as closed, since the tubeside fluid remains in a closed circuit and does not mix with the shellside condensate, as is the case with open feedwater heaters. They are unfired since the heat transfer within the vessel does not occur by means of combustion, but by convection and condensation. The majority use U-tubes, which are relatively tolerant to the thermal expansion during operation. They are designed as per the HEI standard for closed feed water heaters. Based on operating pressure it can be as classified as:

i. Low-pressure heater
ii. High-pressure heater.

1.5.3.6 Bayonet Tube Heat Exchanger

A bayonet (tube-in-tube) tube heat exchanger (BTHE) is a tubular form consisting of two concentric tubes, with a cap attached to the end of the outer tube, whereas the inner tube opens at both ends and is positioned inside the outer tube which is open only at one end. In either case, during the operation, constructional elements can move independently of one another, enabling the elimination of thermal stresses. This is the reason why BTHEs are well suited to extremely large temperature differences between the two fluids, such as in the case of high-temperature recuperators. The fluid can either flow by entering the inner tube and exiting the annulus termed as flow, or flow through the annulus and exit the inner tube, where the fluid flow is driven by the pressure difference between the inlet and outlet of bayonet tube, and is suitable when the fluid to be heated or cooled is accessible from one side only and is free from bending and axial compressive stresses. A horizontal shell and tube bayonet heat exchanger is shown in Figure 1.11 [16,17].

The bayonet (tube-in-tube) heat exchangers are generally used in general-purpose evaporative processes, and are available for vertical or horizontal installations. The heat exchangers are installed in evaporators, which may be constructed from glass or glass-lined steel. In a vertical unit, the steam enters along the inner tube and then flows into the outer tube where it condenses, supplying heat for vaporization of the moisture in the shell. This means that hot steam is present along the whole length of the tube, preventing freezing of the condensate. Bayonet tubes are made from stainless steels, carbon steel, high-nickel alloys, etc. Common applications of these vaporizers include ammonia, carbon dioxide, chlorine, di/trichlorosilane, methanol, ethanol, silicon tetrachloride, sulfur-dioxide, etc. [18]. A vertical bayonet evaporator is shown in Figure 1.12[19].

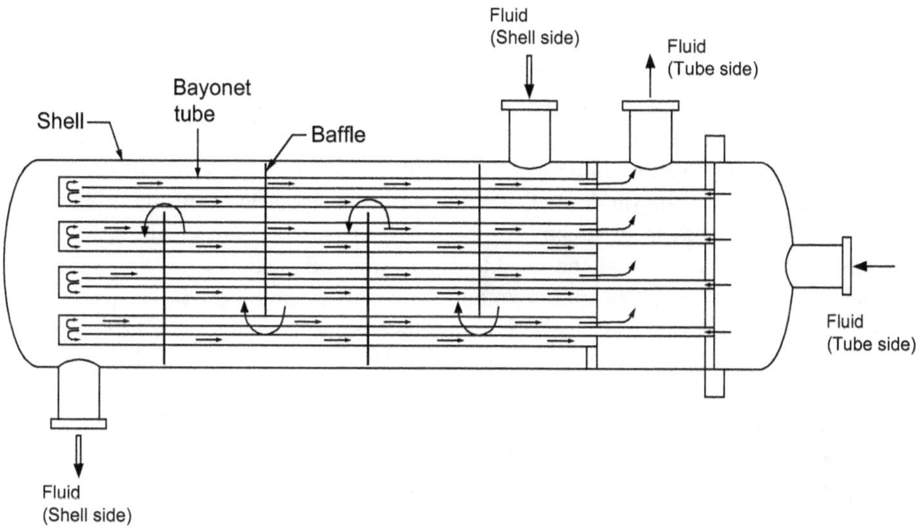

FIGURE 1.11 Horizontal shell and tube bayonet heat exchanger.

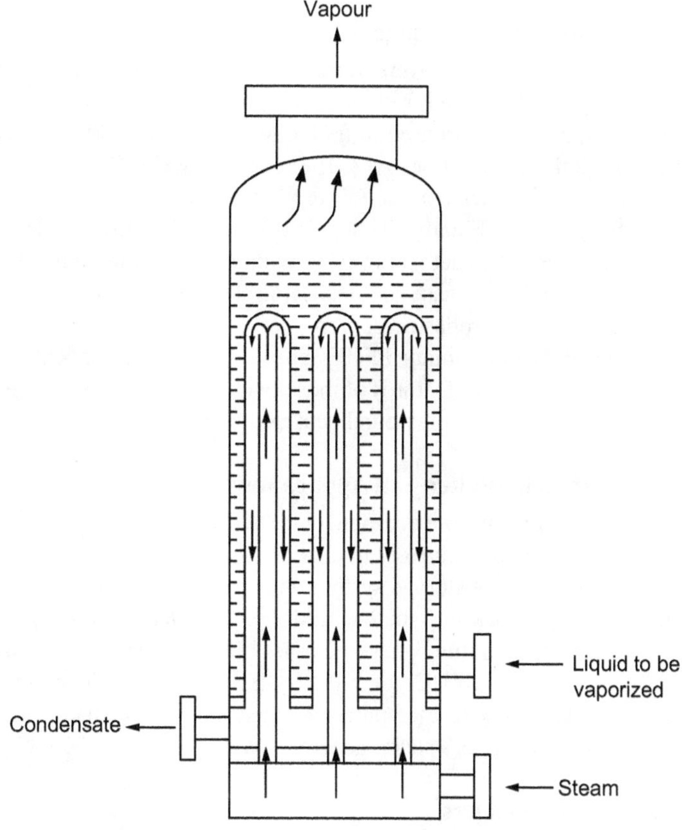

FIGURE 1.12 Vertical evaporative bayonet tube heat exchanger.

1.5.4 Coiled Tube Heat Exchanger

Construction of a coiled tube heat exchanger (CTHE) involves winding a large number of small-bore ductile tubes in helix fashion around a central core tube, with each exchanger containing many layers of tubes along both the principal and radial axes. The tubes in individual layers or groups of layers may be brought together into one or more tube plates through which different fluids may be passed in counterflow to the single shellside fluid. The construction details have been explained in Refs. [7,20]. The high-pressure stream flows through the small-diameter tubes, while the low-pressure return stream flows across the outside of the small-diameter tubes in the annular space between the inner central core tube and the outer shell. The materials used are usually aluminum alloys for cryogenics and stainless steel for high-temperature applications. CTHE offers unique advantages, especially when dealing with low-temperature applications for the following cases [20]:

i. Simultaneous heat transfer between more than two streams is desired. One of the three classical heat exchangers used today for large-scale liquefaction systems is CTHE.
ii. A large number of heat transfer units are required.
iii. High-operating pressures are involved.

The construction cost of CTHE is not cheap because of the material costs, high labor input in winding the tubes, and the central mandrel, which is not useful for heat transfer but increases the shell diameter [7].

1.5.4.1 Linde Coil-Wound Heat Exchangers

Generally, a CWHE comprises multiple layers of tubes spirally wound around a central pipe called a mandrel. The tubes are typically wound from tube coils which allows for a design with comparatively long tube lengths. Spacer bars are used to fix the tube position and adjust layer spacing. The winding direction is changed for each tube layer. The tubes are welded to one or more tube sheets at the end of the exchanger and after completion the bundle is wrapped into a shroud to reduce bypass flows at the exchanger shell. Finally, the bundle is inserted into the prefabricated shell. Linde coil-wound heat exchangers are compact and reliable with a broad temperature and pressure range and suitable for both single- and two-phase streams. Multiple streams can be accommodated in one exchanger. They are known for their robustness, particularly during start-up and shut-down or plant-trip conditions. Both the brazed aluminum PFHEs and CWHEs find application in liquefication processes. A comparison of the salient features of these two types of heat exchangers is provided in Chapter 3. Figure 1.13 shows a Linde coil wound heat exchanger.

1.5.4.2 CWHEs in Solar Field Molten Salt Applications

The molten salt heat exchanger is the core component in thermal energy storage (TES) plants. Here, the CWHE is typically operated consecutively in charge and discharge mode, with the result that it transfers excessive heat energy to the storage system on the one hand, and, on the other, discharges energy from the storage system back into the energy loop. The main characteristics of molten salt exchangers are high working temperatures, in some cases high pressure, wide load range variations during the day, and quick startups. CWHEs are ideal for heat storage applications due to their high mechanical flexibility, allowing very fast temperature changes in the media (typical of solar power plants). Figure 1.14 shows a CWHE as a steam generator in a solar power thermal power plant [21].

1.5.4.3 Glass Coil Heat Exchangers

Two basic types of glass coil heat exchangers are (i) coil type and (ii) STHE with glass or MS shells in combination with a glass tube as standard material for tube. Glass coil exchangers have a coil fused to the shell to make a one-piece unit. This prohibits leakage between the coil and shellside

FIGURE 1.13 Coiled tube heat exchanger. (a) End section of a tube bundle, (b) tube bundle under fabrication, and (c) construction details. (From Linde AG, Engineering Division. With permission.)

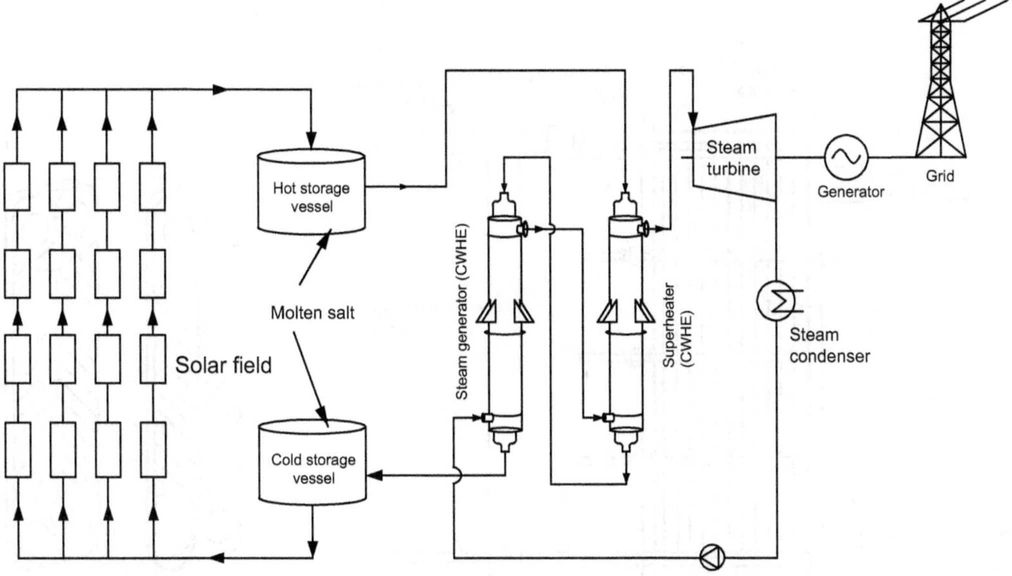

FIGURE 1.14 Coil wound heat exchanger (CWHE) as steam generator in a solar power thermal power plant. (Adapted from Linde AG Engineering Division, Pullach, Germany [21].)

fluids [22]. The reduced heat transfer coefficient of borosilicate glass equipment compares favorably with many alternate tube materials. This is due to the smooth surface of the glass that improves the film coefficient and reduces the tendency for fouling.

1.6 PLATE HEAT EXCHANGERS

PHEs are less widely used than tubular heat exchangers but offer certain important advantages over STHEs. PHEs can be classified into four principal groups:

1. Plate and frame or gasketed PHEs used as an alternative to tube and shell exchangers for low- and medium-pressure liquid–liquid heat transfer applications.
2. Brazed plate heat exchangers (BPHEs).
3. Spiral heat exchanger used as an alternative to shell and tube exchangers where low maintenance is required, particularly with fluids tending to sludge or containing slurries or solids in suspension.
4. Panel heat exchangers made from embossed plates to form a conduit or coil for liquids coupled with fins.

1.6.1 Plate and Frame or Gasketed Plate Heat Exchangers

A PHE essentially consists of a number of corrugated metal plates in mutual contact, each plate having four apertures serving as inlet and outlet ports, and seals designed to direct the fluids in alternate flow passages. The plates are clamped together in a frame that includes connections for the fluids. Since each plate is generally provided with peripheral gaskets to provide sealing arrangements, PHEs are called gasketed PHEs. PHEs are shown in Figure 1.15 and are covered in detail in Chapter 7.

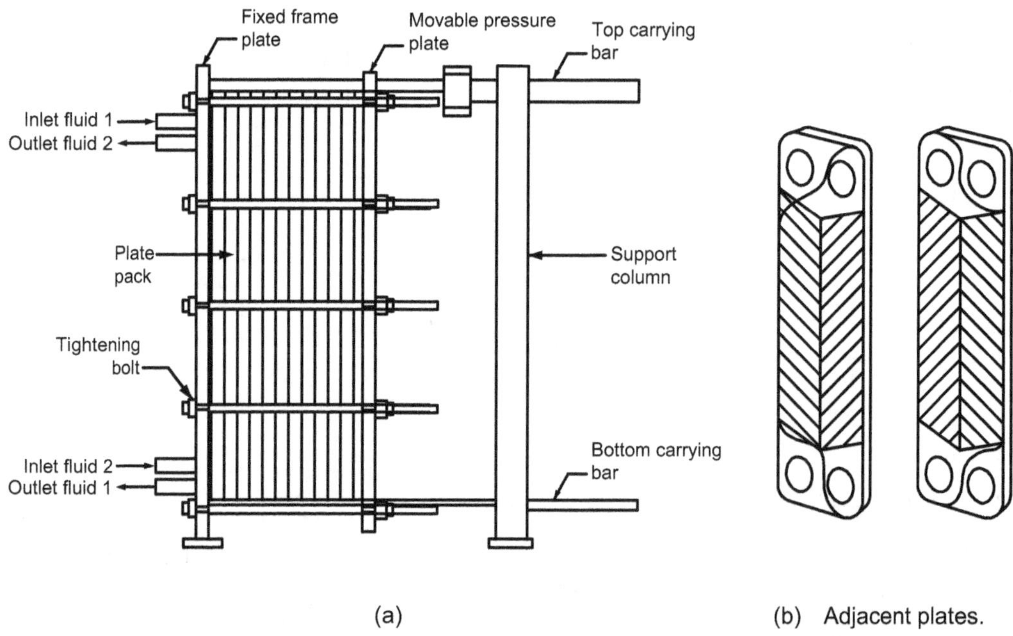

FIGURE 1.15 Construction of a PHE.

1.6.1.1 Sanitary Heat Exchangers

Sanitary heat exchangers are often required in the food and pharmaceutical industries. Some industries have high sanitary standards and require equipment that maintains these standards. This is common in the pharmaceutical, food, beverage, and dairy industries. For these sensitive fluids, sanitary heat exchangers are required. This premium range of plate heat exchangers optimizes process hygiene and energy efficiency. Specifically developed for hygienic applications, they raise the bar on hygiene overall while reducing the carbon footprints for food, beverage, pharmaceutical, and other manufacturers who demand outstanding performance and total efficiency [23].

1.6.2 Spiral Plate Heat Exchanger

SPHEs have been used since the 1930s, when they were originally developed in Sweden for heat recovery in pulp mills. They are classified as a type of welded PHE. An SPHE is fabricated by rolling a pair of relatively long strips of plate around a split mandrel to form a pair of spiral passages. Channel spacing is maintained uniformly along the length of the spiral passages by means of spacer studs welded to the plate strips prior to rolling. Figure 1.16 shows a SPHE. For most applications, both flow channels are closed by alternate channels welded at both sides of the spiral plate. In some services, one of the channels is left open, whereas the other is closed at both sides of the plate. These two types of construction prevent the fluids from mixing. SPHEs are finding applications in reboiling, condensing, heating, or cooling of viscous fluids, slurries, and sludge [24]. More details on SPHEs are furnished in Chapter 7.

1.6.3 Plate or Panel Coil Heat Exchanger

These exchangers are called panel coils, plate coils, or embossed panel or jacketing. The panel coil serves as a heat sink or heat source, depending upon whether the fluid within the coil is being cooled or heated. Panel coil heat exchangers are relatively inexpensive and can be made into any desired shape and thickness for heat sinks and heat sources under varied operating conditions. Hence, they

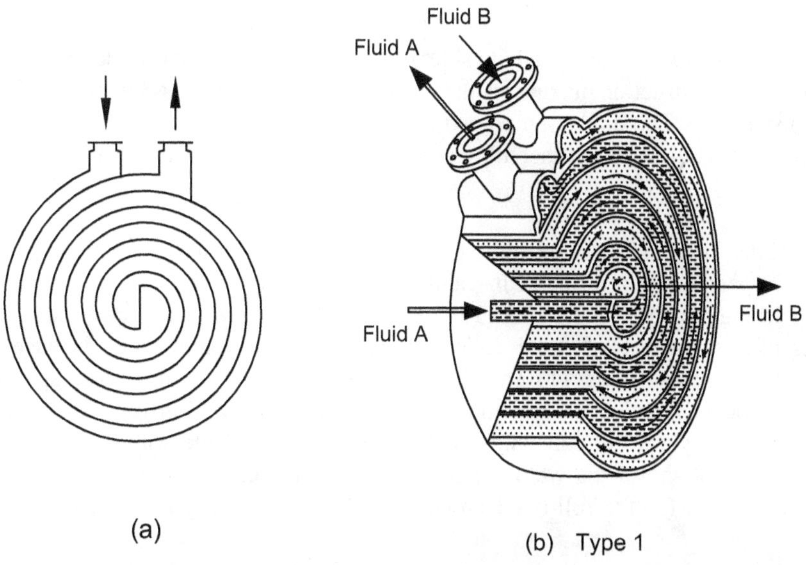

FIGURE 1.16 Spiral plate heat exchanger.

FIGURE 1.17 Temp-Plate® heat transfer surface. (Courtesy of Mueller, Heat Transfer Products, Springfield, MO.)

have been used in many industrial applications such as cryogenics, chemicals, fibers, food, paints, pharmaceuticals, and solar absorbers.

Construction details of a panel coil: A few types of panel coil designs are shown in Figure 1.17. The panel coil is used in such industries as plating, metal finishing, chemical, textile, brewery, pharmaceutical, dairy, pulp and paper, food, nuclear, beverage, waste treatment, and many other industries. The construction details of panel coils are discussed next. M/s Paul Muller Company, Springfield, MO, and Tranter, Inc., TX, are among the leading manufacturers of panel coil/plate coil heat exchangers.

Single-embossed surface: The single-embossed heat transfer surface is an economical type to utilize for interior tank walls, conveyor beds, and when a flat side is required. The single embossed design uses two sheets of material of different thickness and is available in stainless steel, other alloys, carbon steel, and in many material gages and working pressures.

Double-embossed surface: Inflated on both sides using two sheets of material and the same thickness, double-embossed construction maximizes the heating and cooling process by utilizing both sides of the heat transfer plate. The double-embossed design is commonly used in immersion applications and is available in stainless steel, other alloys, carbon steel, and in many material gages and working pressures.

Dimpled surface: This surface is machine punched and swaged, prior to welding, to increase the flow area in the passages. It is available in stainless steel, other alloys, carbon steel, in many material gages and working pressures, and in both MIG plug-welded and resistance spotwelded forms.

Methods of manufacture of panel coils: Basically, three different methods have been used to manufacture the panel coils. (1) They are usually welded by the resistance spot-welding or seamwelding process. An alternate method now available offers the ability to resistance spot-weld the dimpled jacket-style panel coil with a perimeter weldment made with GMAW or resistance welding. Figure 1.18 shows the weld details of a vessel jacket. Other methods are (2) the die-stamping process and (3) the roll-bond process. In the die-stamping process, flow channels are

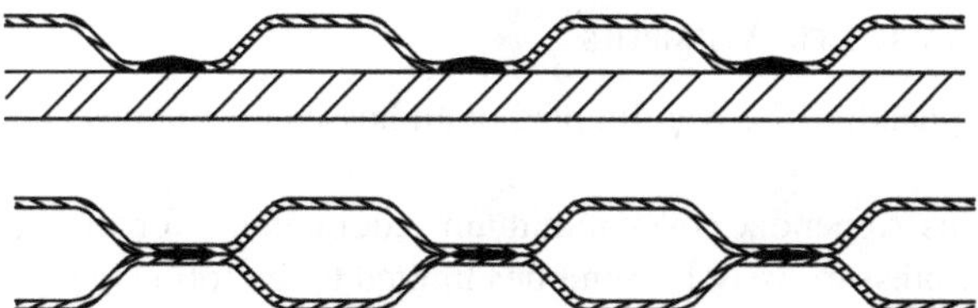

FIGURE 1.18 Welded dimpled jacket template. (a) Gas metal arc welded and (b) resistance welded.

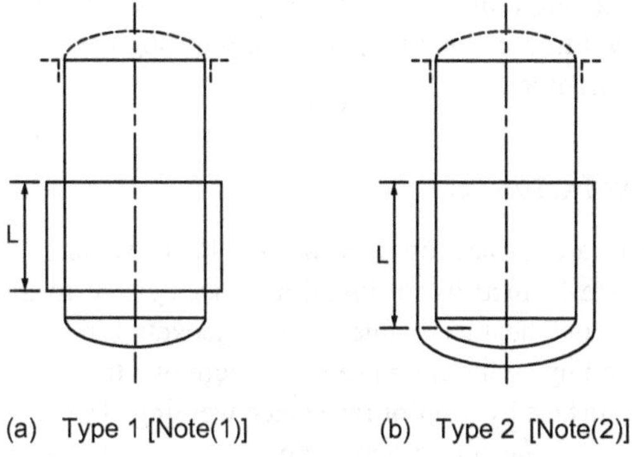
L
L
(a) Type 1 [Note(1)]
(b) Type 2 [Note(2)]
Notes :
(1) Jacket of any length confined entirely to cylindrical shell.
(2) Jacket covering a portion of cylindrical shell and one head.

FIGURE 1.19 Jacketed vessel.

die-stamped on either one or two metal sheets. When one sheet is embossed and joined to a flat (unembossed) sheet, it forms a single-sided embossed panel coil. When both sheets are stamped, it forms a double-sided embossed panel coil.

The advantages of panel coils include that they provide the optimum method of heating and cooling process vessels in terms of control, efficiency, and product quality [25].

1.6.4 Jacketed Heat Exchangers

Jacketed heat exchangers use an enclosure to surround the vessel that contains the heated product. Jacketed heat exchangers are practical for batch processes and for product types that tend to foul or clog tube bundles or coils. Two types of jacketed vessel are shown in Figure 1.19

Types of jackets: Jacketing of process vessels is usually accomplished using one of the three main available types: conventional jackets, dimple jackets, and half-pipe coil jackets [25].

1.6.4.1 ASME Code Sec VIII Div 1 rules for Jacketed Vessels

The rules in Mandatory Appendix 9 cover minimum requirements for the design, fabrication, and inspection of the jacketed portion of a pressure vessel. The jacketed portion of the vessel is defined as the inner and outer walls, the closure devices, and all other penetrations or parts within the jacket which are subjected to pressure stresses. Parts such as nozzle closure members and stiffening or stay rings are included.

1.6.5 Dimpled or Embossed Assemblies

Rules in Mandatory Appendix 17 cover dimpled or embossed assemblies [1].

(a) The rules in this Appendix cover minimum requirements for the design, fabrication, and examination of pressure vessel assemblies limited to the following types:
  (1) dimpled or embossed prior to welding;
  (2) dimpled or embossed form achieved by using hydraulic or pneumatic pressure after welding.

(b) Welding processes covered under the rules of this Appendix include "weld-through" processes in which welding is done by penetrating through one or more members into, but not through, another member.

1.6.6 Pillow-Plate Heat Exchanger

Pillow-plate heat exchangers are a class of fully welded heat exchanger design, which exhibit a wavy, "pillow-shaped" surface formed by an inflation process. Due to their geometric flexibility, they are used also as "plate-type" heat exchangers and as jackets for cooling or heating of vessels. Pillow plates are manufactured by an inflation process, where two thin metal sheets are spot-welded to each other over the entire surface by laser or resistance welding. The sides of the plates are sealed by seam welding, other than the connecting ports. Finally, the gap between the thin metal sheets is pressurized by a hydraulic fluid causing a plastic forming of the plates, known as the hydroforming process, which eventually leads to their characteristic wavy surface. A pillow-plate is shown in Figure 1.20.

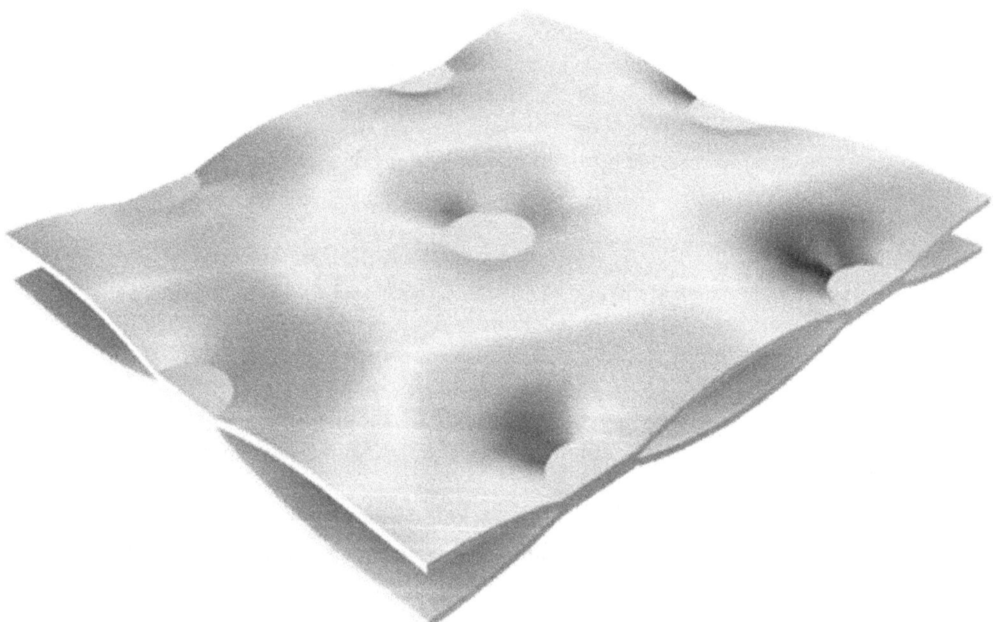

FIGURE 1.20 Pillow-plate for heat exchanger constrction.

(Source. https://commons.wikimedia.org/wiki/File:Pp_double_embossed.png)

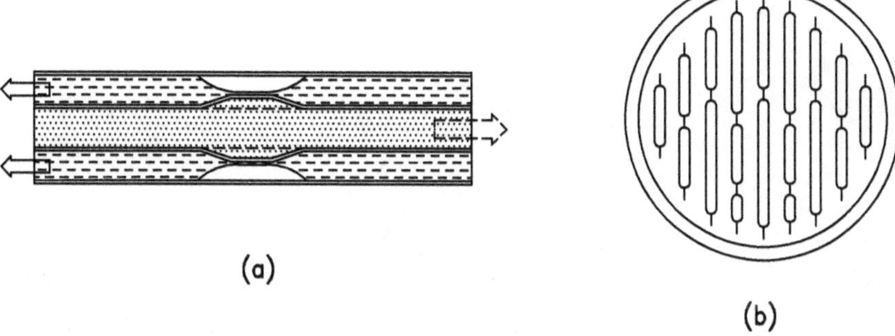

FIGURE 1.21 Lamella heat exchanger. (a) Counterflow concept and (b) lamella tube bundle.

1.6.7 Lamella Heat Exchanger

The lamella heat exchanger is an efficient, and compact, heat exchanger. The principle was originally developed around 1930 by Ramens Patenter. Later Ramens Patenter was acquired by Rosenblads Patenter and the lamella heat exchanger was marketed under the Rosenblad name. In 1988, Berglunds acquired the product and continued to develop it. A lamella heat exchanger normally consists of a cylindrical shell surrounding a number of heat-transferring lamellas. The design can be compared to a tube heat exchanger but with the circular tubes replaced by thin and wide channels, lamellas which are edge welded to provide long narrow channels, and banks of these elements of varying width are packed together to form a circular bundle and fitted within a shell. A cross section of a lamella heat exchanger is shown schematically in Figure 1.21. With this design, the flow area on the shellside is a minimum and similar in magnitude to that of the inside of the bank of elements; due to this, the velocities of the two liquid media are comparable [26]. The flow is essentially longitudinal countercurrent "tubeside" flow of both tube and shell fluids [6]. Lamella heat exchangers can be fabricated from carbon steel, stainless steel, titanium, Incolly, and Hastelloy. They can handle most fluids, with large volume ratios between fluids. The floating nature of the bundle usually limits the working pressure to 300 psi. Lamella heat exchangers are generally used only in special cases. The design is usually done by the vendors.

Sondex Tapiro Oy Ab Pikkupurontie 11, FIN-00810 Helsinki, Finland, markets lamella heat exchangers worldwide.

1.7 EXTENDED-SURFACE HEAT EXCHANGERS

In a heat exchanger with gases or some liquids, if the heat transfer coefficient is quite low, a large heat transfer surface area is required to increase the heat transfer rate. This requirement is served by fins attached to the primary surface. Tube-fin heat exchangers, which are air- or gas-cooled heat exchangers, and plate-fin heat exchangers are the most common examples of extended-surface heat exchangers.

1.7.1 Air-Cooled Heat Exchangers

Air-cooled heat exchangers (ACHEs), or "fin-fans," contain a tube bundle of straight, single pass/multi-pass finned tubes that terminate in header boxes. The tube-fin geometries, namely, attachment of fins to the tubes, include extruded, embedded, and welded—single-footed or double-footed. The fin most commonly used is aluminum, tension wound, and footed. Core tubes may be carbon steel, stainless steel, or various alloys, and are usually of 1 inch in diameter. Header boxes are normally rectangular in cross-section and may have either threaded plugs opposite each tube, or removable

FIGURE 1.22 Air-cooled heat exchanger/condenser. (Courtesy of GEA Iberica S.A., Vizcaya, Spain.)

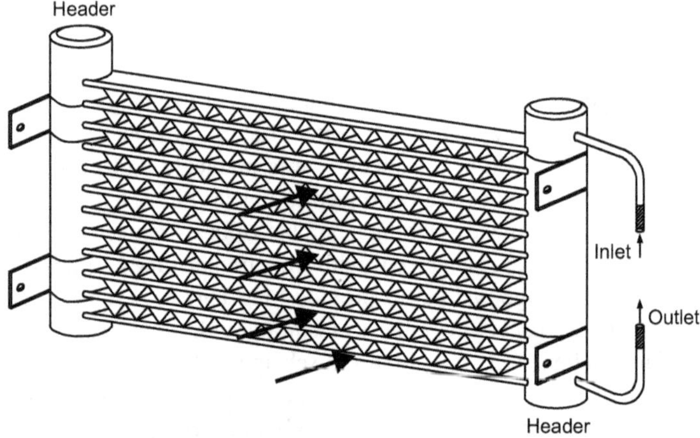

FIGURE 1.23 Aluminum microchannel heat exchanger.

cover plates for out-of-service cleaning and inspection access. In an ACHE, hot process fluid enters one end of the ACHE and flows through tubes, while ambient air flows over and between the tubes, which typically have externally finned surfaces. The air is moved over the tubes in a single cross-flow pass by axial flow fans by means of either a forced or induced draft, as shown in Figure 1.22. ACHEs can be used in all climates. They are increasingly found in a wide spectrum of applications including chemical, process, petroleum refining, and other industries. The ACHE design standard is API 660[27]. ACHE design is further discussed in Chapter 3.

1.7.2 Microchannel Heat Exchangers (MCHEs)

MCHEs are fin and tube heat exchangers with one fluid, usually refrigerant or water, flowing through tubes or enclosed channels either plain or enhanced, while air flows crosscurrent through the connected fins. The hydraulic diameter of the channels is less than 1 mm. Aluminum microchannel heat exchangers have advantages such as high thermal performance and significant weight reduction compared to conventional heat transfer coils. The development of extruded microchannel tubes for refrigerant passage has significantly improved the thermal performance of heat exchangers. A typical aluminum microchannel heat exchanger has three major components: header, multiport microchannel tube, and louvered fin. These components are joined together by brazing. Using a single material for construction produces a consistent rate of heat transfer. A MCHE is illustrated in Figure 1.23. Since MCHEs are so much more efficient, they can be smaller (up to 30%) and

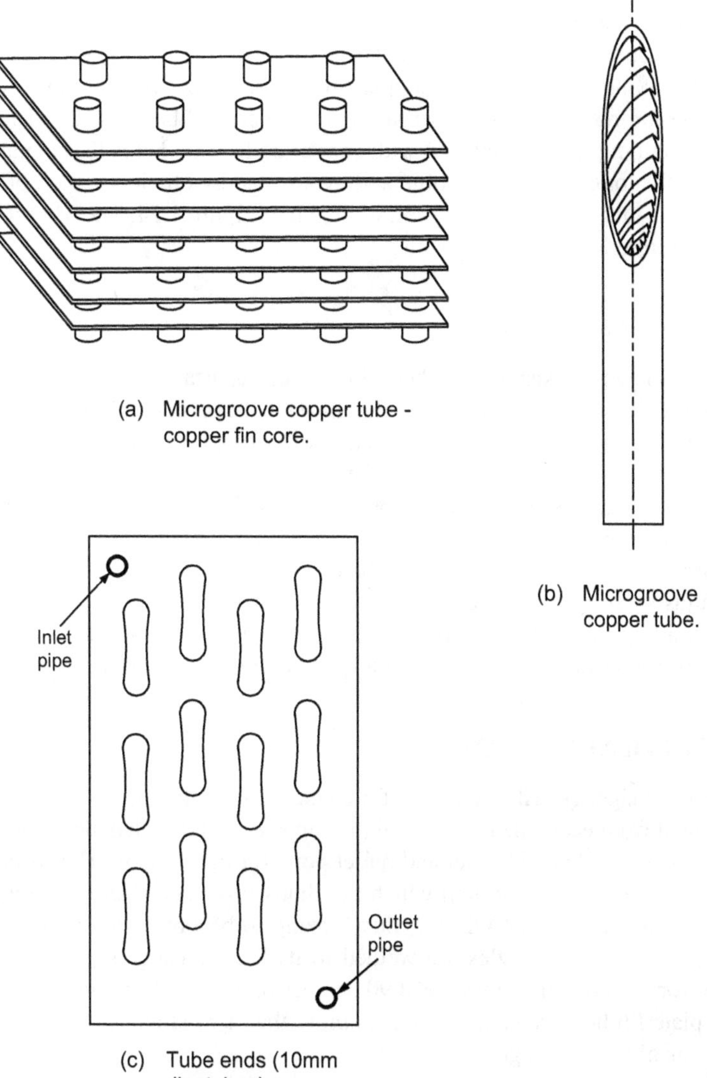

FIGURE 1.24 MicroGroove tube heat exchanger coil.

weigh less (60% less) than comparable heat exchangers [28]. MCHE design is further discussed in Chapter 3.

1.7.3 Internally Grooved Copper Microfin Tube Heat Exchanger

Internally grooved copper tubes, also known as "microfin tubes," are a small-diameter coil technology for modern air conditioning and refrigeration systems. Grooved coils facilitate more efficient heat transfer than smooth coils. Small-diameter coils have better rates of heat transfer than conventional-sized condenser and evaporator coils with round copper tubes and aluminum or copper fins that have been the standard in the HVAC industry for many years. Small-diameter coils can withstand higher pressures required by the new generation of environmentally friendlier refrigerants. They have lower material costs because they require less refrigerant, fin, and coil materials. And they enable the design of smaller and lighter high-efficiency air conditioners and refrigerators because the evaporators and condensers coils are smaller and lighter. Tubes with MicroGroove technology can be made with copper or aluminum. Copper fins are an attractive alternative to aluminum due to the better corrosion resistance of copper and its antimicrobial benefits. Copper is an antimicrobial material. Bio buildup can be reduced with copper coils. This helps to maintain high levels of energy efficiency for longer periods of time and avoids energy efficiency drop off over time [29–31]. A MicroGroove tube heat exchanger coil is shown in Figure 1.24. The MicroGroove tube heat exchanger coil is further discussed in Chapter 3.

1.7.4 Flat-Tube Heat Exchanger

Flat-tube heat exchangers consist of flat tubes with extended surface channels coupled with external fins. The tubes provide additional heat transfer through large internal surface area known as a microchannel tube that is in contact with the fluid. Flow through external fins is optimized to maximize heat transfer. Header manifolds designed for flat-tube heat exchangers result in a low pressure drop that enables smaller, less expensive pumps to be used. This flat-tube and extended fin structure results in an efficient, rugged, and lightweight solution ideal for challenging applications. A flat-tube oil cooler features aluminum flat-tube heat exchangers engineered for high performance with poor heat transfer fluids such as oils and ethylene glycol/water mixtures. By vacuum brazing flat-tube heat exchangers, a high-quality and reliable heat exchanger is obtained. Microchannel tubes [32] and a flat-tube and round-tube construction details and comparison are shown in Figure 1.25 [33]

1.8 PLATE FIN HEAT EXCHANGERS

A plate fin heat exchanger (PFHE) consists of a block of alternating layers of corrugated fins. The layers are separated from each other by special parting sheets and sealed along the edges by means of side bars, and are provided with inlet and outlet ports for the streams. The space in between the parting sheet forms the passage through which the fluids flow. The stacked assembly is brazed in a fluxless vacuum furnace or NOCOLOK® flux brazing to become a rigid core. To complete the heat exchanger, headers with nozzles are welded to it. The exchangers can be set up to provide pure counterflow or such that the flows are at 90 degrees to one another (a cross-flow arrangement). A multistream plate-fin heat exchanger contains more than two streams flowing through different layers and sections of the exchanger.

Various forms of fin surface are available. In duties involving phase change the use of perforated plates is common. For single-phase duties where the flow is expected to be highly turbulent, plain surfaces are often used. There are two basic types of plain surface: rectangular and triangular. For other single-phase duties there are louvered plates, off-set strip fin units, and wavy surfaces. A plate-fin heat exchanger is shown in Figure 1.26.

1.8.1 Brazed Aluminum Heat Exchangers

Brazed aluminum heat exchangers (BAHXs) are custom-designed compact heat-exchange devices manufactured as a composite brazed pressure vessel with welded tanks and nozzles. They are chiefly applied in a variety of cryogenic gas to gas and gas to liquid heat transfer processes, including LNG, industrial gas production, nitrogen rejection, NGL, ethylene production, and hydrogen recovery. They offer advantages such as a high heat transfer surface area per unit volume and the capability of combining multiple process streams into a single unit. The manufacturers' association standard for PFHE design is ALPEMA [34]. PFHE design is further discussed in Chapter 3.

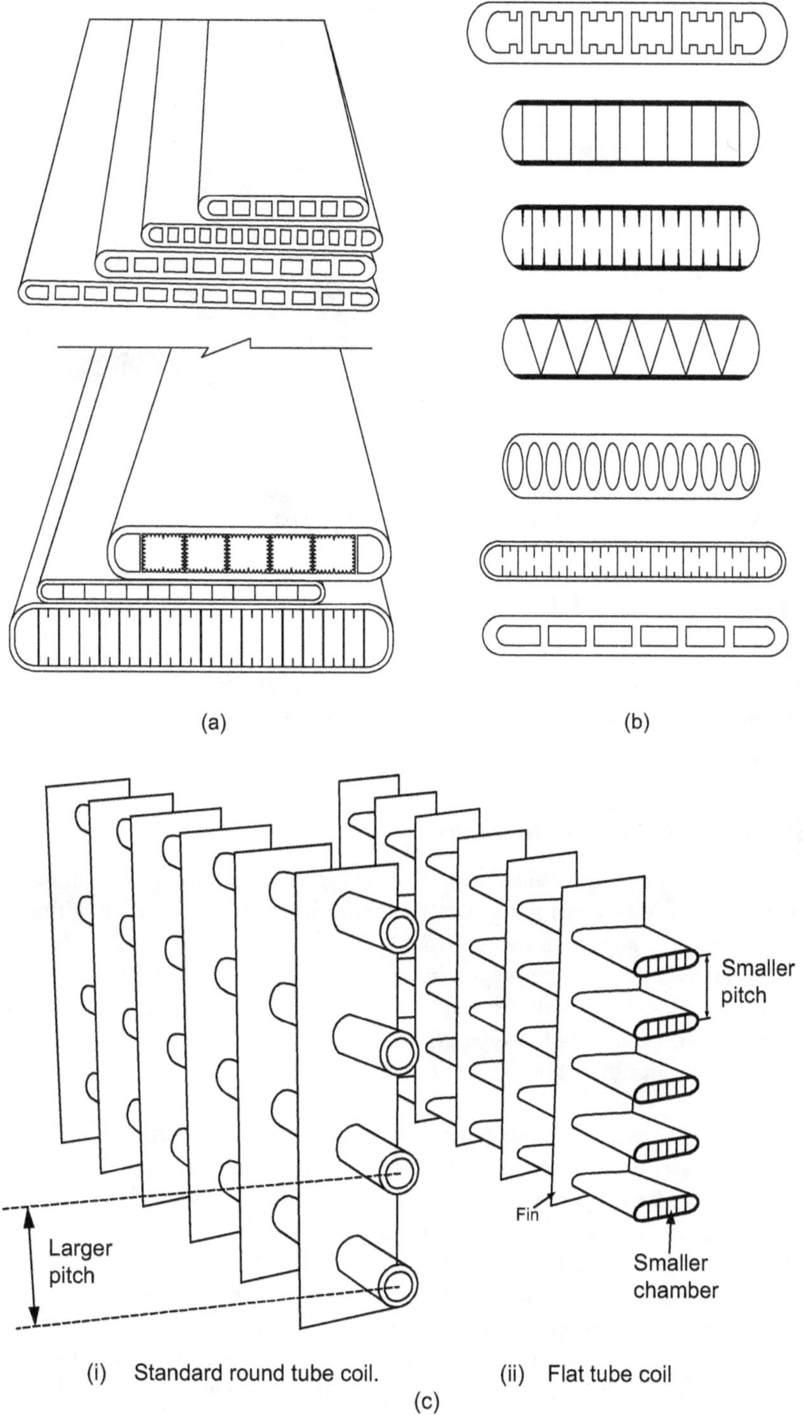

FIGURE 1.25 Flat heat exchanger tubes (a, b) Microchannel tubes, adapted from www.hydro.com/enproducts-and-services/precision-tubes/micro-channel-tubes/ and (c) Flat tube and round tube construction details. Adapted from www.boydcorp.com/thermal/liquid-cooling/liquid-heat-exchangers/flat-tube-heat-exchangers.html

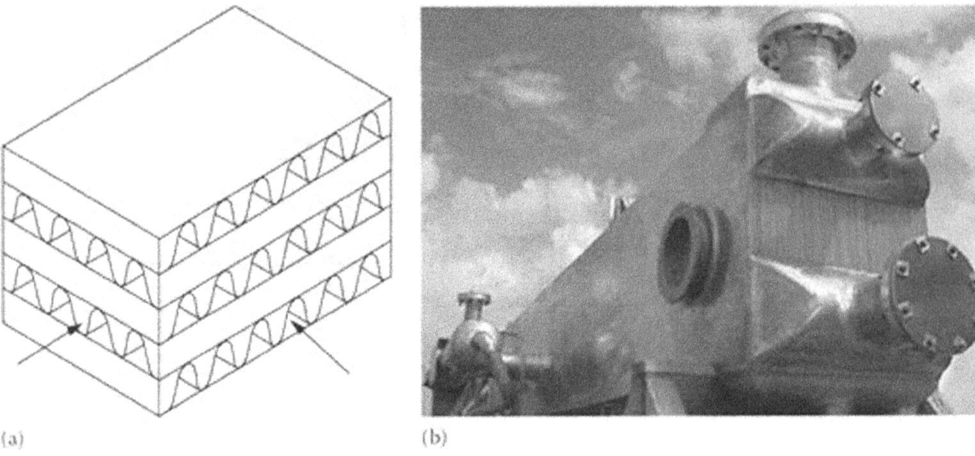

FIGURE 1.26 Plate-fin heat exchanger. (a) Schematic of exchanger and (b) brazed aluminum plate-fin heat exchanger. (From Linde AG, Engineering Division. With permission.)

1.9 REGENERATIVE HEAT EXCHANGERS

Recovery of waste heat from the exhaust gas by means of heat exchangers known as regenerators can improve the overall plant efficiency. For example, the addition of regeneration as a thermodynamic principle by preheating incoming combustion air improves the overall performance of gas turbine power plants and steam power plants.

1.9.1 Gas Turbine Cycle with Regeneration

In a simple gas turbine plant consisting only of the compressor, combustion chamber, and turbine, the single improvement that gives the greatest increase in thermal efficiency is the addition of a regenerator for transferring thermal energy from the hot turbine exhaust gas to the air leaving the compressor, especially when it is employed in conjunction with intercooling during compression. The addition of a regenerator results in a flat fuel economy versus load characteristic, which is highly desirable for the transportation-type prime movers like gas turbine locomotives, marine gas turbine plants, and the aircraft turboprops [35].

Types of regenerators: Regenerators are generally classified as fixed-matrix and rotary regenerators. Further classifications of fixed and rotary regenerators are shown in Figure 1.27. In the former, regeneration is achieved with periodic and alternate blowing of a hot and cold stream through a fixed matrix. During the hot flow period, the matrix receives thermal energy from the hot gas and transfers it to the cold stream during the cold stream flow. In the latter, the matrix revolves slowly with respect to two fluid streams. The rotary regeneration principle is achieved by two means: (1) the flow through the matrix is periodically reversed by rotating the matrix, and (2) the matrix is held stationary whereas the headers are rotated continuously. Both approaches are rotary because, for either design approach, heat transfer performance, pressure drop, and leakage considerations are the same, and rotary components must be designed for either system

The rotary regenerator is commonly employed in gas turbine power plants where the waste heat in the hot exhaust gases is utilized for raising the temperature of compressed air before it is supplied to the combustion chamber. A rotary regenerator working principle is shown in Figure 1.28, and Figure 1.29 shows the Rothemuhle regenerative air preheater of Babcock and Wilcox Company. Rotary regenerators fall in the category of compact heat exchangers since the heat transfer surface area to regenerator volume ratio is very high. Regenerators are discussed in further detail in Chapter 6.

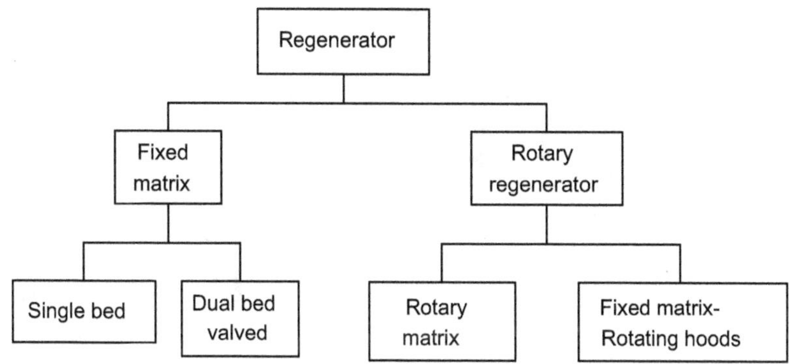

FIGURE 1.27 Classification of regenerators.

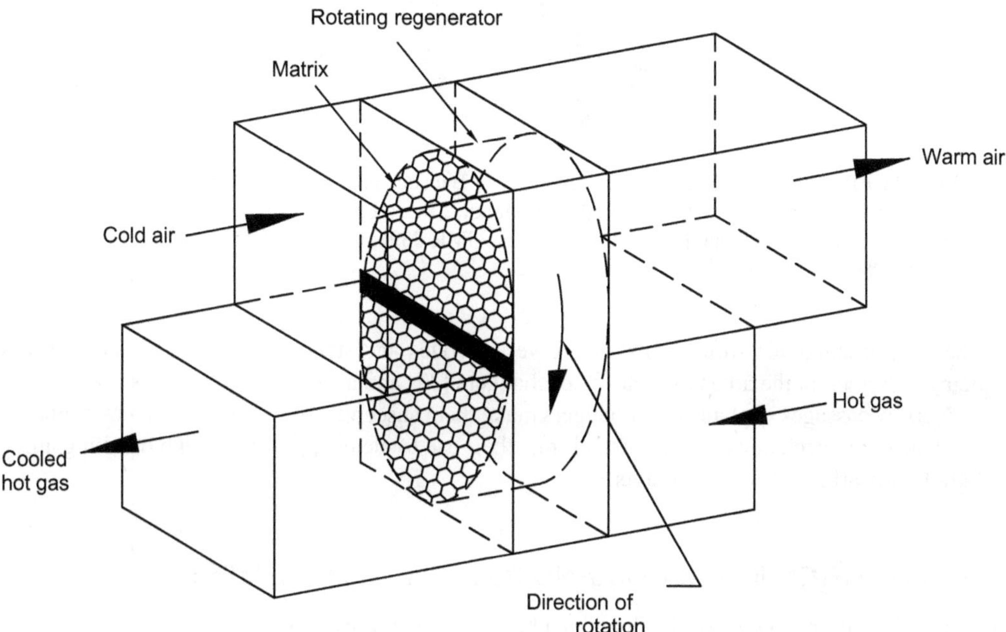

FIGURE 1.28 Rotary regenerator: working principle.

1.10 CLASSIFICATION ACCORDING TO SURFACE COMPACTNESS

Compact heat exchangers are important when there are restrictions on the size and weight of exchangers. A compact heat exchanger incorporates a heat transfer surface having a high area density, β, somewhat arbitrarily 700 m^2/m^3 (200 ft^2/ft^3) and higher [3]. The area density, β, is the ratio of heat transfer area A to its volume V. A compact heat exchanger employs a compact surface on one or more sides of a two-fluid or multifluid heat exchanger. They can often achieve higher thermal effectiveness than shell and tube exchangers (95% vs. the 60%–80% typical for STHEs), which makes them particularly useful in energy-intensive industries [36]. For the least capital cost, the size of the unit should be minimal. There are some additional advantages to small volume including small inventory, making them good for handling expensive or hazardous materials, low weight, easier transport, less foundation, and better temperature control. Some barriers to the use of compact heat exchangers include the following [36]:

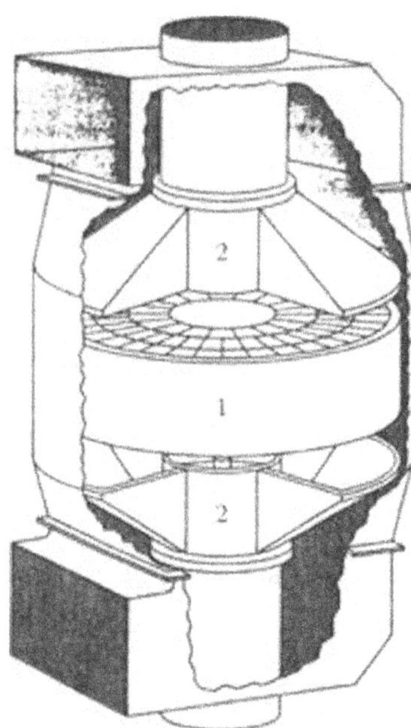

FIGURE 1.29 Rothemuhle regenerative air preheater of Babcock and Wilcox Company—stationary matrix (part 1) and revolving hoods (part 2). (Adapted from Mondt, J.R., Regenerative heat exchangers: The elements of their design, selection and use, Research Publication GMR-3396, General Motors Research Laboratories, Warren, MI, 1980.)

The lack of standards similar to pressure vessel codes and standards, although this is now being redressed in the areas of plate-fin exchangers [37] and air-cooled exchangers [38].

Narrow passages in plate-fin exchangers make them susceptible for fouling and they cannot be cleaned by mechanical means. This limits their use to clean applications like handling air, light hydrocarbons, and refrigerants.

1.11 CLASSIFICATION ACCORDING TO FLOW ARRANGEMENT

The basic flow arrangements of the fluids in a heat exchanger are as follows:

i. Parallelflow
ii. Counterflow
iii. Crossflow

The choice of a particular flow arrangement is dependent upon the required exchanger effectiveness, fluid flow paths, packaging envelope, allowable thermal stresses, temperature levels, and other design criteria. These basic flow arrangements are discussed next.

1.11.1 Parallelflow Exchanger

In this type, both fluid streams enter at the same end, flow parallel to each other in the same direction, and leave at the other end (Figure 1.30a). (For fluid temperature variations, idealized as one-dimensional, refer to Figure 2.2 in Chapter 2.) This arrangement has the lowest exchanger effectiveness among the single-pass exchangers for the same flow rates, capacity rate (mass ×

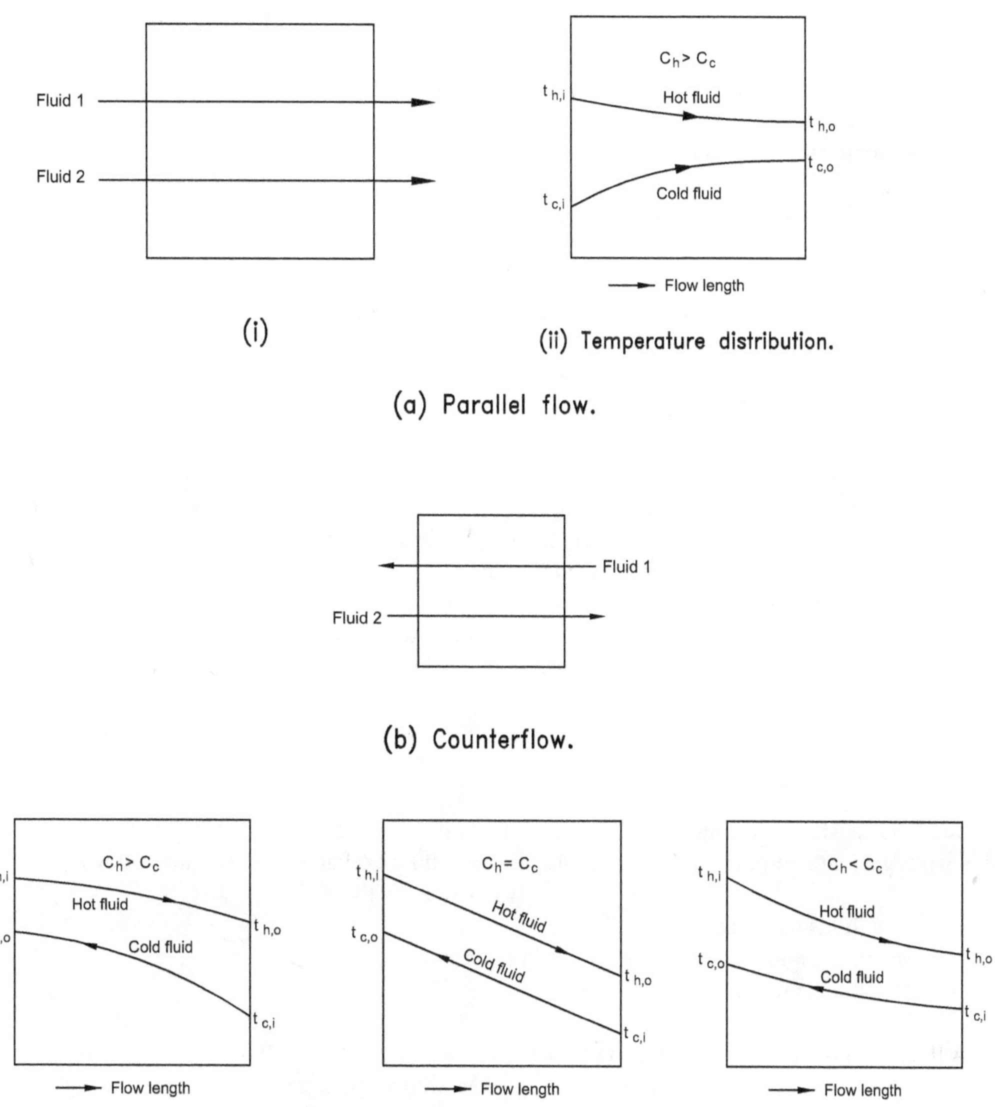

FIGURE 1.30 Fluids flow arrangements – (a) Parallelflow, (b) counterflow arrangement and (c) temperature distribution. (*Note*: C_h and C_c are the heat capacity rate of hot fluid and cold fluid respectively, i refers to inlet, o refers to outlet conditions and t refers to fluid temperature.)

specific heat) ratio, and surface area. Moreover, the existence of large temperature differences at the inlet end may induce high thermal stresses in the exchanger wall at inlet. Parallelflows are advantageous for the following reasons. (a) In heating very viscous fluids, parallelflow provides for rapid heating. The quick change in viscosity results in reduced pumping power requirements through the heat exchanger; (b) where the more moderate mean metal temperatures of the tube walls are required; and (c) where the improvements in heat transfer rates compensate for the lower LMTD. Although this flow arrangement is not used widely, it is preferred for the following reasons [4]:

1. When there is a possibility that the temperature of the warmer fluid may reach its freezing point.
2. It provides early initiation of nucleate boiling for boiling applications.
3. For a balanced exchanger (i.e., heat capacity rate ratio $C^* = 1$), the desired exchanger effectiveness is low and is to be maintained approximately constant over a range of NTU values.
4. The application allows piping only suited to parallelflow.
5. Temperature-sensitive fluids such as food products, pharmaceuticals, and biological products are less likely to be "thermally damaged" in a parallelflow heat exchanger.
6. Certain types of fouling such as chemical reaction fouling, scaling, corrosion fouling, and freezing fouling are sensitive to temperature. Where control of temperature-sensitive fouling is a major concern, it is advantageous to use parallelflow.

1.11.2 Counterflow Exchanger

In a counterflow heat exchanger, as shown in Figure 1.30b, the two fluids flow parallel to each other but in opposite directions, and its temperature distribution may be idealized as shown in Figure 1.30c. Ideally, this is the most efficient of all flow arrangements for single-pass arrangements under the same parameters. Since the temperature difference across the exchanger wall at a given cross-section is the lowest, it produces minimum thermal stresses in the wall for equivalent performance compared to other flow arrangements. In certain types of heat exchangers, the counterflow arrangement cannot be achieved easily, due to manufacturing difficulties associated with the separation of the fluids at each end, and the design of the inlet and outlet header is complex and difficult [4].

1.11.3 Crossflow Exchanger

In this type, as shown in Figure 1.31, the two fluids flow normal to each other. Important types of flow arrangement combinations for a single-pass crossflow exchanger include the following:

i. Both fluids unmixed
ii. One fluid unmixed and the other fluid mixed
iii. Both fluids mixed.

A fluid stream is considered "unmixed" when it passes through an individual flow passage without any fluid mixing between adjacent flow passages. Mixing implies that a thermal averaging process takes place at each cross section across the full width of the flow passage. A tube-fin exchanger with flat (continuous) fins and a plate-fin exchanger wherein the two fluids flow in separate passages (e.g., wavy fin, plain continuous rectangular or triangular flow passages) represent the unmixed–unmixed case. A crossflow tubular exchanger with bare tubes on the outside would be treated as the unmixed–mixed case, that is, unmixed on the tubeside and mixed on the outside. The thermal effectiveness for the crossflow exchanger falls between those of the parallelflow and counterflow arrangements.

1.11.3.1 Unmixed–Unmixed Case

This is the most common flow arrangement used for extended surface heat exchangers because it greatly simplifies the header design. Examples include continuous finned tube-fin heat exchanger like radiator and plate finned heat exchanger (PFHE). For the unmixed–unmixed case, fluid temperature variations are idealized as two-dimensional only for the inlet and outlet sections; this is shown in Figure 1.32. If the desired heat exchanger effectiveness is generally more than 80%, the size penalty for crossflow may become excessive. In such a case, a counterflow unit is preferred [4]. In shell and tube exchangers, the crossflow arrangement is used in the TEMA X shell having a single tube pass.

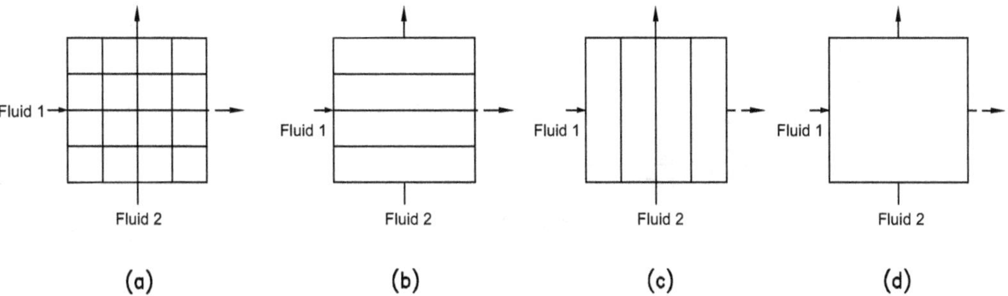

FIGURE 1.31 Crossflow arrangement: (a) unmixed–unmixed, (b) fluid 1 is unmixed–fluid 2 is mixed, (c) fluid 1 is mixed–fluid 2 is mixed and (d) mixed–mixed.

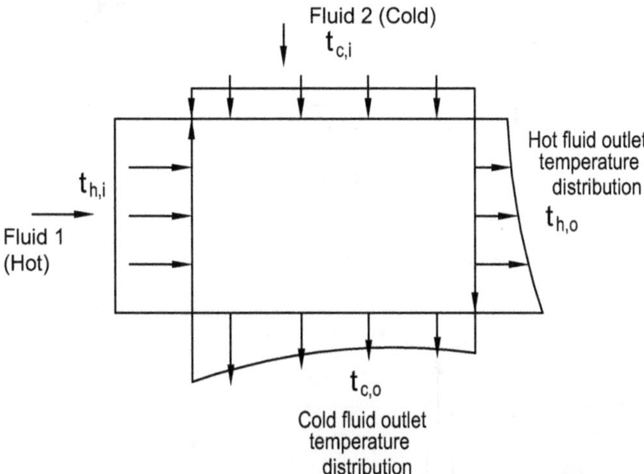

FIGURE 1.32 Temperature distribution for unmixed–unmixed crossflow arrangement.

1.11.3.2 Unmixed–Mixed Case

This flow arrangement refers to a bare tube heat exchanger in which the tubeside fluid is unmixed throughout and tube-outside fluid is mixed throughout.

1.11.3.3 Mixed–Mixed Case

An example for the both fluids mixed case is a jet condenser (direct contact condenser) of a steam-based thermal power plant and represents a limiting case of some multipass shell and tube exchangers (e.g., TEMA E and J shell). In a jet condenser, the process steam from the steam turbine exhaust is condensed and condensate is pumped to an external air-cooled heat exchanger. A mixed–mixed case of a direct contact condenser is shown in Figure 1.33 and a direct contact flue gas washing chamber is shown in Figure 1.34.

1.11.3.4 Direct Contact Heat Exchanger

Direct contact heat exchangers involve heat transfer between hot and cold streams of two phases in the absence of a separating wall. Thus such heat exchangers can be classified as:

i. Gas–liquid
ii. Immiscible liquid–liquid
iii. Solid–liquid or solid–gas.

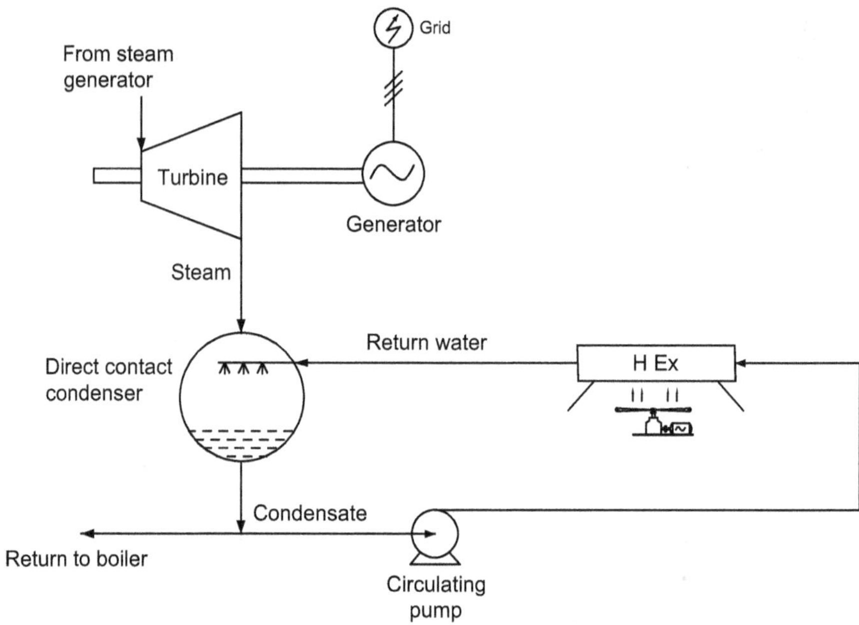

FIGURE 1.33 A mixed-mixed case of direct contact condenser.

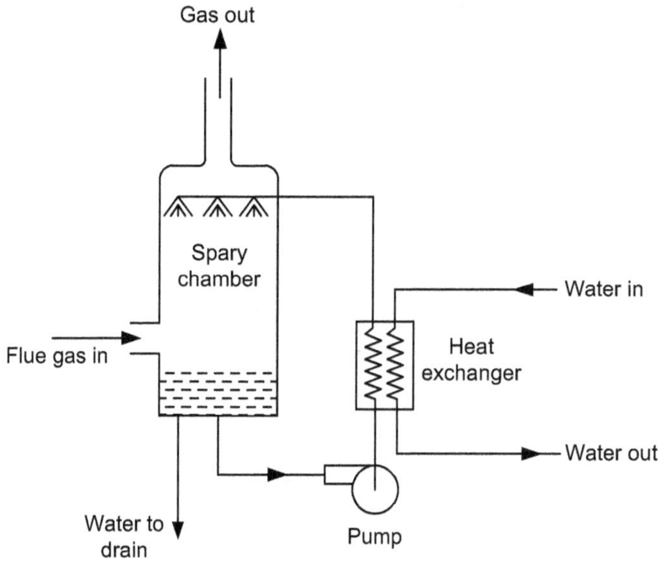

FIGURE 1.34 A direct contact flue gas washing chamber.

Gas–solid heat exchangers. Gas-solid heat exchange is facilitated by bed-type heat exchangers, most commonly by solid granules forming a bed. Moving beds, fluidized beds, and conveyor belts are examples.

Spray column. A spray tower (or spray column or spray chamber) is a gas–liquid contactor used to achieve mass and heat transfer between a continuous gas phase (that can contain dispersed solid particles) and a dispersed liquid phase.

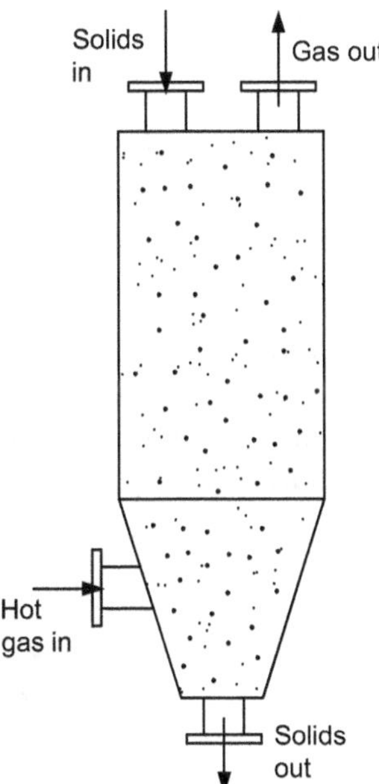

FIGURE 1.35 Direct contact heat exchanger.

Heating or cooling bulk solids. Heating or cooling bulk solids may use a direct-contact, bulk solids heat exchanger. These devices are designed as hoppers, bins, or silos that have been modified to allow for the injection of air or another gas. Heat transfer takes place between the bulk solids and gas when the two streams are fed at different temperatures. The cooling or heating gas can be injected from the bottom of the bulk-solids heat exchanger, such that it passes upward through the moving bed of solids (countercurrent), or it can be injected through the walls of the vessel so that it flows perpendicular to the solids flow (cross-current). A direct-contact heat exchanger is shown in Figure 1.35 [39].

1.12 CLASSIFICATION ACCORDING TO PASS ARRANGEMENTS

These are either single pass or multipass. A fluid is considered to have made one pass if it flows through a section of the heat exchanger through its full length once. In a multipass arrangement, a fluid is reversed and flows through the flow length two or more times. Refer to Chapter 2 for various cases of single-pass and multipass arrangements.

1.12.1 Multipass Exchangers

When the design of a heat exchanger results in either extreme length, significantly low velocities, or low effectiveness, or due to other design criteria, a multipass heat exchanger or several single-pass exchangers in series or a combination of both is employed. Specifically, multipassing is resorted to so as to increase the exchanger thermal effectiveness over the individual pass effectiveness. As the number of passes increases, the overall direction of the two fluids approaches that of a pure counterflow exchanger. The multipass arrangements are possible with compact, shell and tube, and plate exchangers.

1.13 CLASSIFICATION ACCORDING TO PHASE OF FLUIDS

1.13.1 Gas–Liquid

Gas–liquid heat exchangers are mostly tube-fin-type compact heat exchangers with the liquid on the tubeside. The radiator is by far the major type of liquid–gas heat exchanger, typically cooling the engine jacket water by air. Similar units are necessary for all the other water-cooled engines used in trucks, locomotives, diesel-powered equipment, and stationery diesel power plants. Other examples are air coolers, oil coolers for aircraft, intercoolers and aftercoolers in compressors, and condensers and evaporators of room air-conditioners. Normally, the liquid is pumped through the tubes, which have a very high convective heat transfer coefficient. The air flows in crossflow over the tubes. The heat transfer coefficient on the air side will be lower than that on the liquid side. Fins will be generally used on the outside of the tubes to enhance the heat transfer rate.

1.13.2 Liquid–Liquid

Most liquid–liquid heat exchangers are shell and tube type, and PHEs to a lesser extent. Both fluids are pumped through the exchanger, so the principal mode of heat transfer is forced convection. The relatively high density of liquids results in a very high heat transfer rate, so normally fins or other devices are not used to enhance the heat transfer [6]. In certain applications, low-finned tubes, microfin tubes, and heat transfer augmentation devices are used to enhance the heat transfer.

1.13.3 Gas–Gas

This type of exchanger is found in exhaust gas–air preheating recuperators, rotary regenerators, intercoolers, and/or aftercoolers to cool the supercharged engine intake air of some land-based diesel power packs and diesel locomotives, and cryogenic gas liquefaction systems. In many cases, one gas is compressed so that the density is high, while the other is at low pressure and low density. Compared to liquid–liquid exchangers, the size of the gas–gas exchanger will be much larger, because the convective heat transfer coefficient on the gas side is low compared to the liquid side. Therefore, secondary surfaces are mostly employed to enhance the heat transfer rate.

1.14 CLASSIFICATION ACCORDING TO HEAT TRANSFER MECHANISMS

The basic heat transfer mechanisms employed for heat transfer from one fluid to the other are (1) single-phase convection, forced or free; (2) two-phase convection (condensation or evaporation) by forced or free convection; and (3) combined convection and radiation. Any of these mechanisms individually or in combination could be active on each side of the exchanger. Based on the phase change mechanisms, the heat exchangers are classified as (1) condensers and (2) evaporators.

1.14.1 Condensers

Condensers may be liquid (water) or gas (air) cooled. The heat from condensing streams may be used for heating fluid. Normally, the condensing fluid is routed outside the tubes with a water-cooled steam condenser or inside the tubes with gas cooling, that is, air-cooled condensers of refrigerators and air-conditioners. Fins are normally provided to enhance heat transfer on the gas side. A surface condenser is discussed in Chapter 4.

1.14.2 Evaporators

This important group of tubular heat exchangers can be subdivided into two classes: fired systems and unfired systems.

Fired systems: These involve the products of combustion of fossil fuels at very high temperatures but at ambient pressure (and hence low density) and generate steam under pressure. Fired systems are called boilers. A system may be a fire tube boiler (for small low-pressure applications) or a water tube boiler.

Unfired systems: These embrace a great variety of steam generators extending over a broad temperature range from high-temperature nuclear steam generators to very-low-temperature cryogenic gasifiers for liquid natural gas evaporation. Many chemical and food processing applications involve the use of steam to evaporate solvents, concentrate solutions, distill liquors, or dehydrate compounds.

1.15 OTHER TYPES OF HEAT EXCHANGERS

1.15.1 Microscale Heat Exchanger

Microscale heat exchangers are heat exchangers in which at least one fluid flows in lateral confinements with typical dimensions below 1 mm and are fabricated via silicon micromachining, deep X-ray lithography, or nonlithographic micromachining [40]. The plates are stacked forming "sandwich" structures, as in the "large" plate exchangers. All flow configurations (cocurrent, countercurrent, and crossflow) are possible.

Typically, the fluid flows through a cavity called a microchannel. Microheat exchangers have been demonstrated with high convective heat transfer coefficient. Investigation of microscale thermal devices is motivated by the single-phase internal flow correlation for convective heat transfer:

$$h = \frac{Nu\,k}{D_h} \tag{1.2}$$

where

h is the heat transfer coefficient
Nu is the Nusselt number
k is the thermal conductivity of the fluid
D_h is the hydraulic diameter of the channel or duct

In internal laminar flows, the Nusselt number becomes a constant. As the Reynolds number is proportional to the hydraulic diameter, fluid flow in channels of small hydraulic diameter will predominantly be laminar. This correlation therefore indicates that the heat transfer coefficient increases as the channel diameter decreases.

1.15.1.1 Advantages over Macroscale Heat Exchangers

i. Substantially better performance
ii. Enhanced heat transfer coefficient with a large number of smaller channels
iii. Smaller size that allows for an increase in mobility and uses
iv. Light weight reduces the structural and support requirements
v. Lower cost due to less material being used in fabrication.

1.15.1.2 Applications of Microscale Heat Exchangers

Microscale heat exchangers are being used in the development of fuel cells. They are currently used in automotive industries, HVAC applications, aircraft, manufacturing industries, and electronics cooling.

1.15.1.3 Demerits of Microscale Heat Exchangers

One of the main disadvantages of microchannel heat exchangers is the high pressure loss that is associated with a small hydraulic diameter.

1.15.2 Printed Circuit Heat Exchanger

A printed circuit heat exchanger (PCHE), developed by Heatric Division of Meggitt (UK) Ltd., or a diffusion bonded microchannel heat exchanger (DCHE), is a heat exchanger that can handle several fluids at a time and is used, depending on its material and the size of its flow passage, up to the design pressure of 100 MPa and design temperature of 900ºC. It is extremely compact (the most common design feature to achieve compactness has been small channel size) and has high efficiency, of the order of 98%. It can handle a wide variety of clean fluids. The flow configuration can be either crossflow or counterflow. It maintains parent material strength and can be made from stainless steel, nickel alloys, copper alloys, and titanium. Fluid flow channels are etched chemically on metal plates. It has a typical plate thickness of 1.6 mm, width 600 mm, and length 1200 mm. The channels are semicircular with 1–2 mm diameter. Etched plates are stacked and diffusion bonded together to fabricate as a block. The plate material depends on the application. The complete heat exchanger core is shown in Figure 1.36a. The required configuration of the channels on the plates for each fluid is governed by the operating temperature and pressure-drop constraints for the heat exchange duty and the channels can be of unlimited variety and complexity. Fluid flow can be parallelflow, counterflow, crossflow, or a combination of these to suit the process requirements. Figures 1.36b–f show a HEATRIC PCHE. The mechanical design is normally of ASME VIII Division 1. Other design codes can be employed as required.

1.15.2.1 Construction Materials

The majority of diffusion-bonded heat exchangers are constructed from 300 series austenitic stainless steel. Various other metals that are compatible with the diffusion-bonded process and have been qualified for use include 22 chromeduplex, copper–nickel, nickel alloys, and titanium.

1.15.2.2 Features of PCHE

Diffusion-bonded heat exchangers are highly compact and robust and are well established in the upstream hydrocarbon processing, petrochemical, and refining industries. Various salient constructional and performance features are described next:

1. Compactness: Diffusion-bonded heat exchangers are 1/4–1/6th the size of conventional STHEs of the equivalent heat duty. This design feature has space and weight advantages, reducing exchanger size together with piping and valve requirements. The diffusion-bonded heat exchanger in the foreground of Figure 1.36e undertakes the same thermal duty, at the same pressure drop, as the stack of three shell and tube exchangers behind. PCHE might be judged as a promising compact heat exchanger for the high-efficiency recuperator [41].
2. Process capability: They can withstand pressures of 600 bar (9000 psi) or greater and can cope with extreme temperatures, ranging from cryogenic to 900°C (1650°F).
3. Thermal effectiveness: Diffusion-bonded exchangers can achieve high thermal effectiveness of the order of 98% in a single unit.
4. They can incorporate more than two process streams into a single unit.

PCHE is discussed further in Chapter 3 (Section 3.15 Printed Circuit Heat Exchanger).

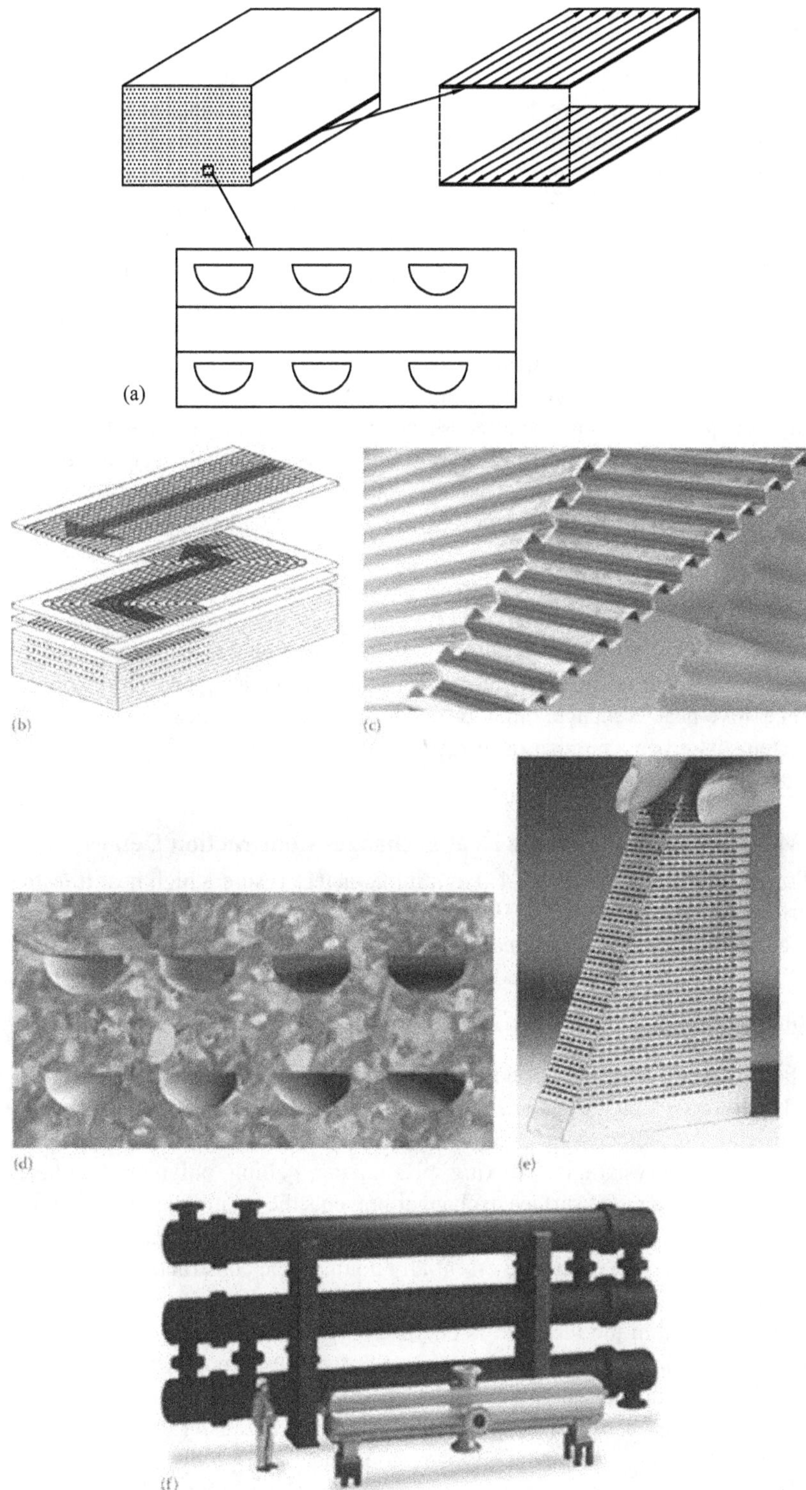

FIGURE 1.36 Printed circuit heat exchanger. (a) Heat exchanger block with flow channel. (b) Flow channel, (c) and (d) section through flow channel, (e) diffusion bonded core, (f) comparison of size of PCHE shell and tube heat exchanger (smaller size) with a conventional exchanger (bigger size) for similar duty. (Courtesy of Heatric UK, Dorset, U.K.)

1.15.3 Perforated Plate Heat Exchanger as Cryocoolers

High-efficiency compact heat exchangers are needed in cryocoolers to achieve very low temperatures. One approach to meet the requirements for compact and efficient cooling systems is perforated plate heat exchangers. Perforated plate matrix heat exchangers are widely used in cryocoolers, helium liquefiers, and in aerospace applications. Perforated plate heat exchangers are made up of a large number of parallel, perforated plates of high-thermal conductivity metal (copper or aluminum) in a stacked array, with gaps between plates being provided by spacers. The spacers, being of low thermal conductivity material, also help in reducing axial conduction and consequent deterioration of performance, alternating with low thermal conductivity spacers (plastics, stainless steel). The packet of alternate high and low thermal conductivity materials is bonded together to form leak-free passages for the streams exchanging heat between one another. The gaps between the plates ensure uniform flow distribution and create turbulence which enhances heat transfer [42–44]. Gas flows longitudinally through the plates in one direction and the other stream flows in the opposite direction through separated portions of the plates. Heat transfer takes place laterally across the plates from one stream to the other. The operating principles of this type of heat exchanger have been described by Fleming [45]. The device employs plates of 0.81 mm thickness with holes of 1.14 mm diameter and a resulting length-to-diameter ratio in the range of 0.5–1.0. The device is designed to operate from room temperature to 80 K. In order to improve operation of a compact cryocooler, much smaller holes, in the low-micron-diameter range, and thinner plates with high length-to-diameter ratio are needed. As per US Patent 5101894 [46], uniform, tubular perforations having diameters down to the low-micron-size range can be obtained. Various types of heat exchange devices including recuperative and regenerative heat exchangers may be constructed in accordance with the invention for use in cooling systems based on a number of refrigeration cycles such as the Linde–Hampson, Brayton, and Stirling cycles.

1.15.3.1 Matrix or Perforated Plate Heat Exchanger Construction Details

The small flow passages (typically 0.3–1.0 mm in diameter) ensure a high heat transfer coefficient and high surface area density (up to 6000 m^2/m^3) [42–46]. A matrix heat exchanger or perforated plate heat exchanger is shown schematically in Figure 1.37

1.15.4 Scraped Surface Heat Exchanger

Scraped surface heat exchangers are used for processes likely to result in the substantial deposition of suspended solids on the heat transfer surface. Scraped surface heat exchangers can be employed in the continuous, closed processing of virtually any pumpable fluid or slurry involving cooking, slush freezing, cooling, crystallizing, mixing, plasticizing, gelling, polymerizing, heating, aseptic processing, etc. Use of a scraped surface exchanger prevents the accumulation of significant buildup of solid deposits. The construction details of scraped surface heat exchangers are explained in Ref. [6]. Scraped surface heat exchangers are essentially double-pipe construction with the process fluid in the inner pipe and the cooling (water) or heating medium (steam) in the annulus. A rotating element is contained within the tube and is equipped with spring-loaded blades. In operation, the rotating shaft scraper blades continuously scrape product film from the heat transfer tube wall, thereby enhancing heat transfer and agitating the product to produce a homogeneous mixture. For most applications, the shaft is mounted in the center of the heat transfer tube. An off-centered shaft mount or *eccentric* design is recommended for viscous and sticky products. This shaft arrangement increases product mixing and reduces the mechanical heat load. Oval tubes are used to process extremely viscous products. All pressure elements are designed in accordance with the latest ASME code requirements. The principle of working of scraped surface heat exchangers is shown in Figure 1.38. For scraped surface exchangers, operating costs are high and applications are highly

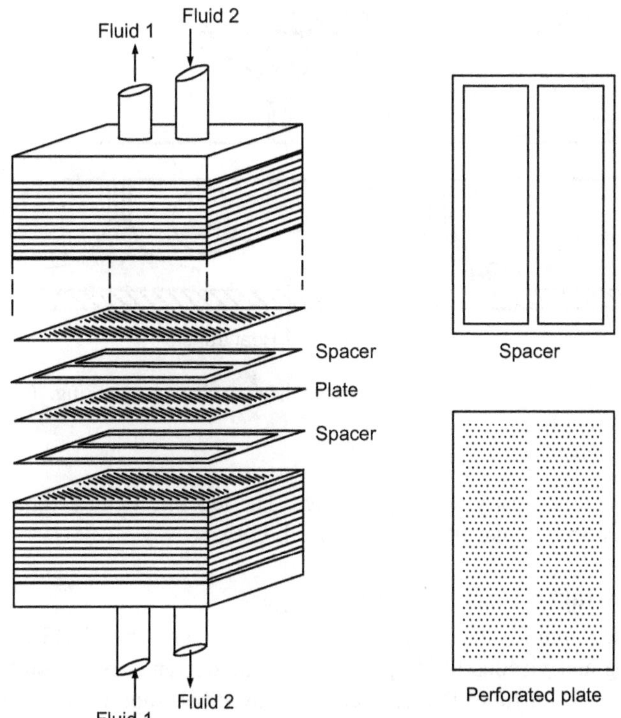

FIGURE 1.37 Matrix (perforated plate) Heat Exchanger construction details. (Adapted from Krishnakumar, K., Venkatarathnam, G. (2007) Heat transfer and flow friction characteristics of perforated plate matrix heat exchangers. *International Journal of Heat Exchangers* 8(1): 45–60.)

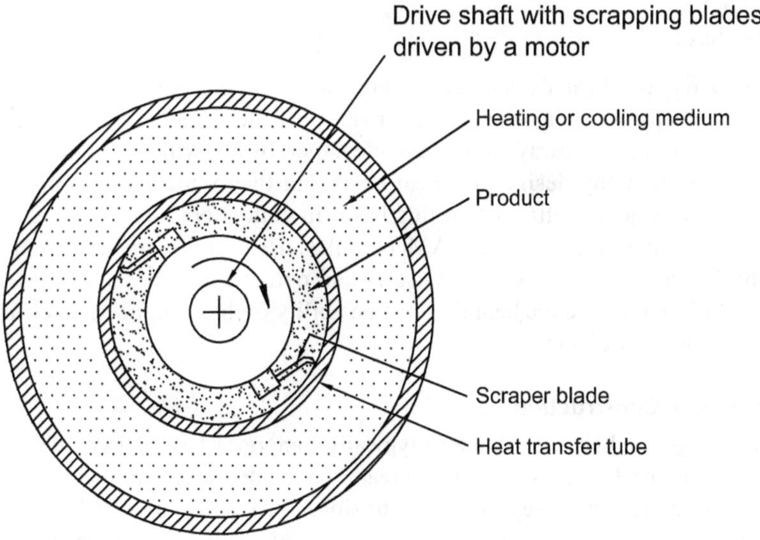

FIGURE 1.38 Scraped surface heat exchanger: principle.

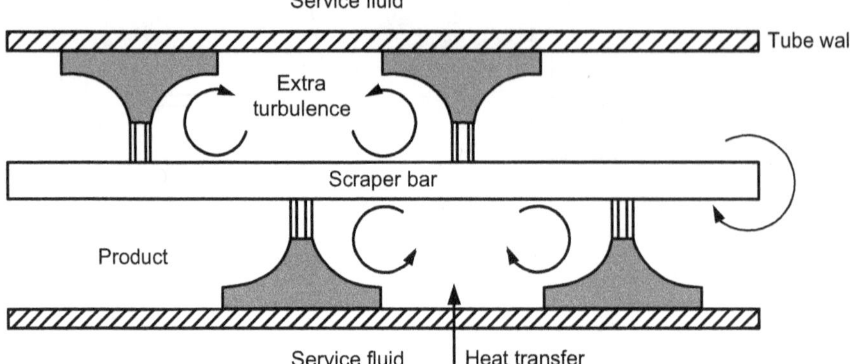

FIGURE 1.39 Principle of Unicus scraped surface heat exchanger working. (Courtesy of HRS Heat Exchangers Ltd, Herts, U.K.)

specific [7]. The design is mostly done by vendors. The leading manufacturers include HRS Heat Exchangers, Ltd., UK and Waukesha Cherry-Burrell, USA.

1.15.4.1 Unicus Scraped Surface Heat Exchanger

Unicus™ is the trade name for the scraped surface heat exchanger of HRS Heat Exchangers Ltd., UK, for high-fouling and viscous fluid applications. The design is based on STHE with scraping elements inside each interior tube. The scrapers are moved back and forth by hydraulic action. The scraping action has two very important advantages: any fouling on the tube wall is removed and the scraping movement introduces turbulence in the fluid, increasing heat transfer. The scraping system consists of a stainless steel rod to which the scraping elements are fitted, as shown in Figure 1.39, and Figure 1.40 shows Unicus scraped surface heat exchangers. The figures show the various types of scrapers that can be applied. For each application, the optimal scraper is selected and fitted.

1.15.5 Heat Sink

A heat sink is a passive heat exchanger, without any moving parts, that transfers the heat generated by an electronic or a mechanical device to a fluid medium, often air or a liquid coolant, where it is dissipated away from the device, thereby allowing regulation of the device's temperature. There are many designs for heat sinks, but they typically comprise a base and a number of protrusions or fins attached to this base. In computers, heat sinks are used to cool CPUs, GPUs, and some chipsets and RAM modules. Heat sinks are used with high-power semiconductor devices such as power transistors and optoelectronics such as lasers and light-emitting diodes (LEDs), where the heat dissipation ability of the component itself is insufficient to moderate its temperature [47].

1.15.5.1 Heat Sink Construction

There are many designs for heat sinks, but they typically comprise a base and a number of protrusions called fins attached to this base. The base is the feature that interfaces with the device to be cooled. Heat is conducted through the base into the protrusions. Heat sinks are usually constructed from copper or aluminum. Copper has a very high thermal conductivity, which means the rate of heat transfer through copper heat sinks is also very high. Whilst lower than that of copper, aluminum's thermal conductivity is still high and it has the added benefits of lower cost and lower density, making it useful for applications where weight is a major concern [48,49]. A heat sink is shown in Figure 1.41

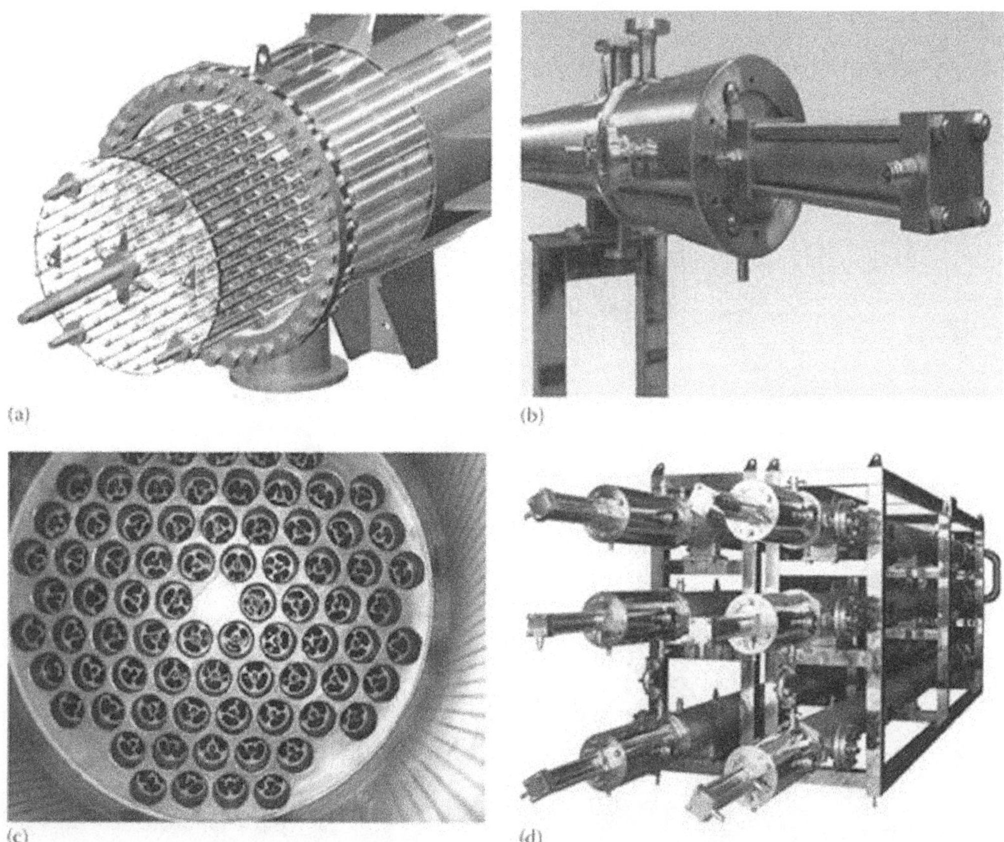

FIGURE 1.40 Unicus dynamic scraped surface heat exchanger. (a) 3-D model, (b) hydraulic cylinder head, (c) a unit of Unicus, and (d) multiple Unicus units. (Courtesy of HRS Heat Exchangers Ltd, Herts, U.K.)

1.15.5.2 Thermal Performance

The heat from a device is transferred to the heat sink by conduction. The primary mechanism of heat transfer from the heat sink is convection, although radiation also has a minor influence. There are two distinct types of convection:

i. Natural convection—where the movement of the fluid particles is caused by the local changes in density due to transfer of heat from the surface of a solid to the fluid particles in close proximity.
ii. Forced convection—where the movement of the fluid particles is caused by an additional device such as a fan or blower.

There are a few general rules that should be followed when using heat sinks [50]:

i. Always use some form of heat sink grease or thermally conductive pad between the heat sink and the device. This will increase the thermal transfer between the two parts, however, excessive quantities of heat sink grease will decrease performance.
ii. Mount fins in the vertical plane for optimum natural convective cooling.
iii. If thermal demands are particularly high, consider using forced convection (i.e., a small fan directed at the heat sink).

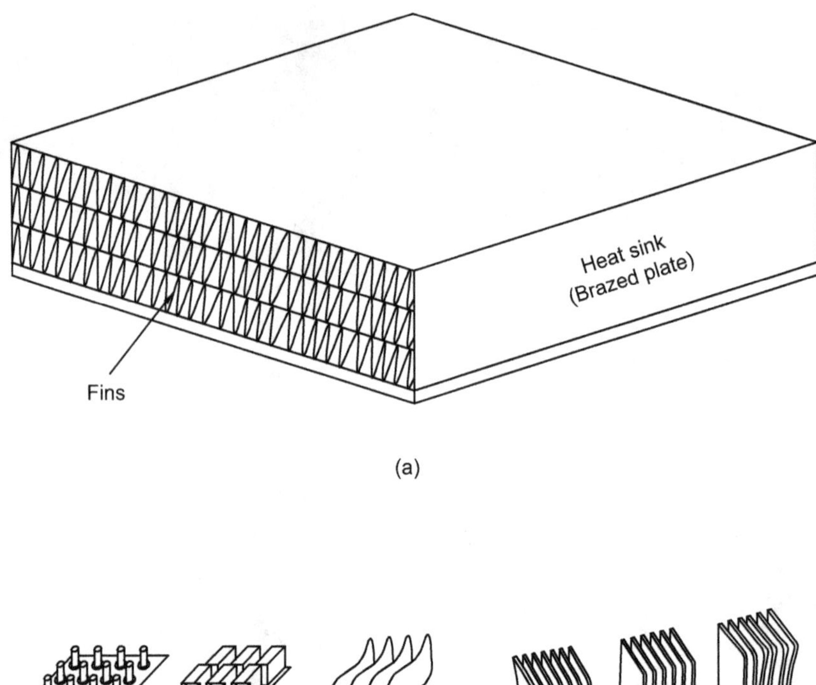

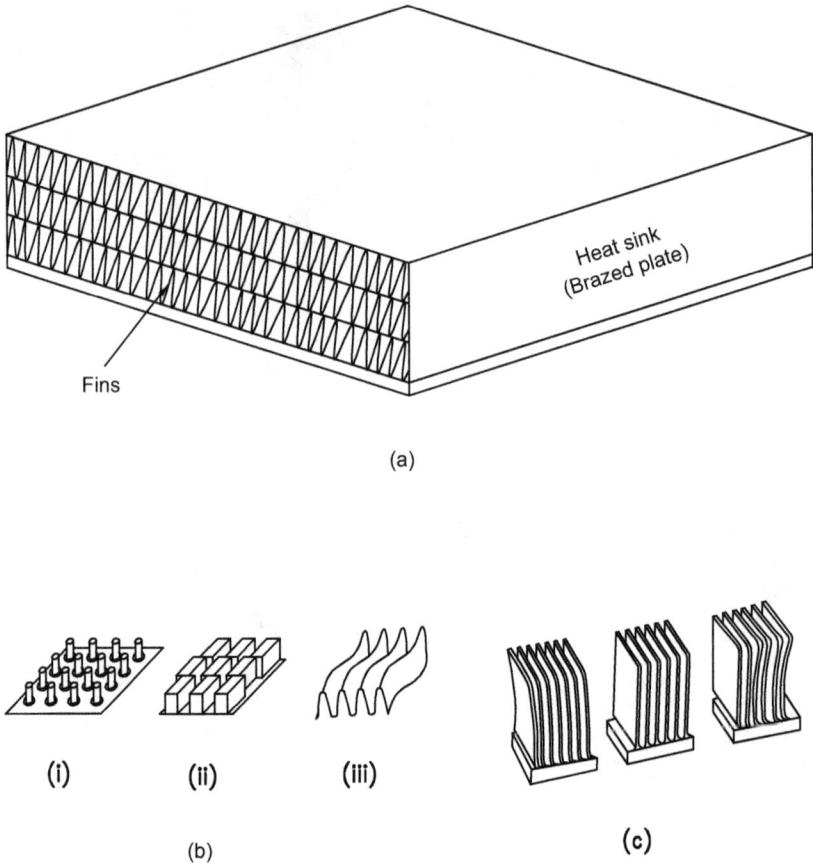

FIGURE 1.41 Heat sink – brazed plate fin type.

1.16 MATERIALS-BASED HEAT EXCHANGER CLASSIFICATION

1.16.1 Graphite Heat Exchanger

Impervious graphite as a heat exchanger material is used for the construction of various types of heat exchangers such as STHE, cubic block heat exchanger, and plate and frame or gasketed heat exchangers (PHEs). Graphite tubes are used in STHE (refer to Chapter 2, Volume 2) and plates are used in PHEs (refer to Chapter 7) for special-purpose applications. It resists a wide variety of inorganic and organic chemicals. Graphite heat exchangers are employed as boilers and condensers in the distillation by evaporation of hydrochloric acid and in the concentration of weak sulfuric acid and of rare earth chloride solutions. Since a cubic heat exchanger cannot be treated in the categorization of an extended surface heat exchanger, the same is covered next. Figure 1.42 shows a graphite heat exchanger block.

Cubic heat exchanger: This is similar to the compact crossflow heat exchanger, consisting of drilled holes in two perpendicular planes. They are suitable when both process streams are corrosive. With a cubic exchanger, a multipass arrangement is possible. It is manufactured by assembling of accurately machined and drilled graphite plates bonded together by synthetic resins, oven cured and sintered. Gasketed headers with nozzles are assembled on both sides of the block to form a block heat exchanger and are clamped together, as shown in Figure 1.43.

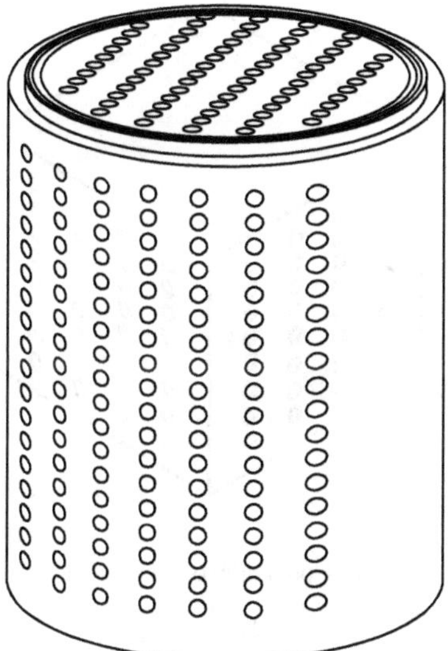

FIGURE 1.42 Graphite heat exchanger – fluid flow holes drilled cylindrical block.

Modular-block cylindrical exchanger: In this arrangement, solid impervious graphite blocks have holes drilled in them. These blocks can be multistacked in a cylindrical steel shell that has gland fittings. The process holes are axial and the service holes are transverse. The units are designed as evaporators and reboilers. Graphite as a heat exchanger material selection is discussed in detail in Chapter 2, Volume 2.

1.16.2 Ceramic Heat Exchanger

A ceramic is a material that is neither metallic nor organic. It may be crystalline, glassy, or both crystalline and glassy. Ceramics are typically hard and chemically non-reactive and can be formed or densified with heat. Depending on their method of formation, ceramics can be dense or lightweight. Typically, they will demonstrate excellent strength and hardness properties; however, they are often brittle in nature. For high-temperature heat exchangers, material temperature limits are a major constraining factor. For metallic materials in use above 649°C (1200°F), the choice is essentially limited to SSs, nickel- and cobalt-base superalloys, and heat-resistant cast alloys. Structural ceramics are used to provide mechanical strength at elevated temperatures, usually in the range of 600°C–1600°C (1110°F–2910°F). Ceramic materials such as silicon carbide and silicon nitride exhibit excellent high-temperature mechanical strength and are used for high-temperature heat exchanger applications. Because of their high temperature capability and oxidation resistance, ceramics are obvious materials for high-temperature heat exchangers of tubular construction and plate fin construction [51].

The main advantages of ceramic materials over traditional metallic materials in HE construction are their extremely high temperature stability, low material cost, and excellent corrosion resistance. However, the major obstacles in the improvement of ceramic CHEs mainly is embodied in their intrinsic brittleness in tension and catastrophic failure without much warning, difficulties in shaping and sealing, and thus high manufacturing costs. By fracturing without reaching plastic deformation, ceramics have low toughness. Additionally, the porosity of ceramic materials further degrades their properties, due to pores acting as stress concentrators within the material [52].

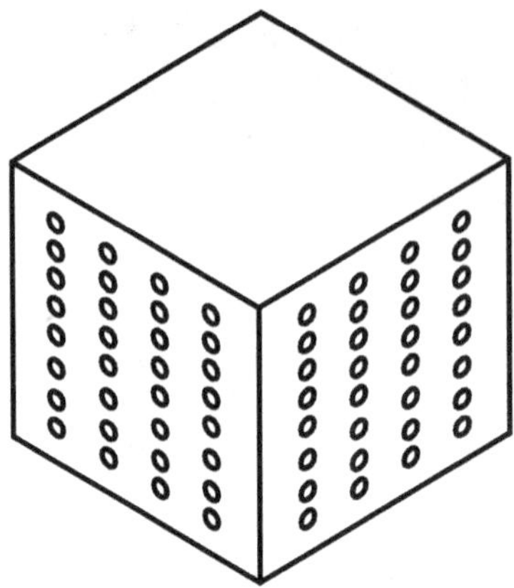

(a) Holes drilled cubic block.

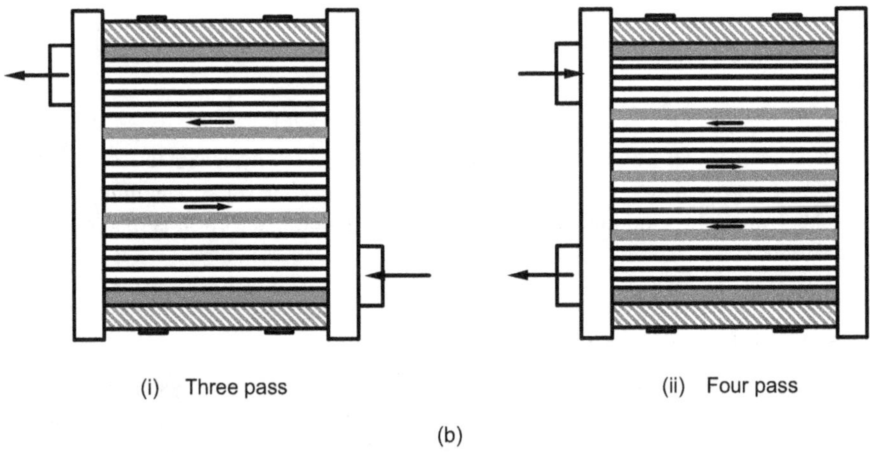

(i) Three pass (ii) Four pass

(b)

FIGURE 1.43 NK series multipass cubic graphite heat exchanger. (a) Fluid flow holes drilled cubic block, (b) multipass flow pattern. (Courtesy of MERSEN, Paris La Défense, France.)

1.16.2.1 Sintered Silicon Carbide (SiC) Ceramic Heat Exchanger

Sintered silicon carbide (SiC) ceramic has become the material of choice for heat exchangers in the harshest process applications because of its extremely high resistance to corrosion and erosion, and superior thermal conductivity. Furthermore, the materials used to construct them are highly resistant to fouling. A ceramic heat exchanger combines a shell and tube design with sintered silicon carbide ceramic tubes using a highly engineered individual tube sealing system coupled with PTFE tubesheets. This results in a durable and innovative product that can perform across a broad range of environments. Should a tube require replacement, this can be quickly and easily accomplished without disturbing any other tubes.

1.16.2.2 Hexoloy® Silicon Carbide Heat Exchanger Tube

An alternative to metals, glass, and other tube materials for enhanced heat transfer, uptime, and reliability for high corrosive environment is Hexoloy silicon carbide (SiC) heat exchanger tubes supplied by Saint-Gobain. A Hexoloy silicon carbide heat exchanger is shown in Figure 1.44a. Hexoloy SiC's thermal conductivity is almost equal to that of commonly used graphite tubes. Its thermal conductivity is two times that of tantalum, five times that of SS, 10 times that of Hastelloy,

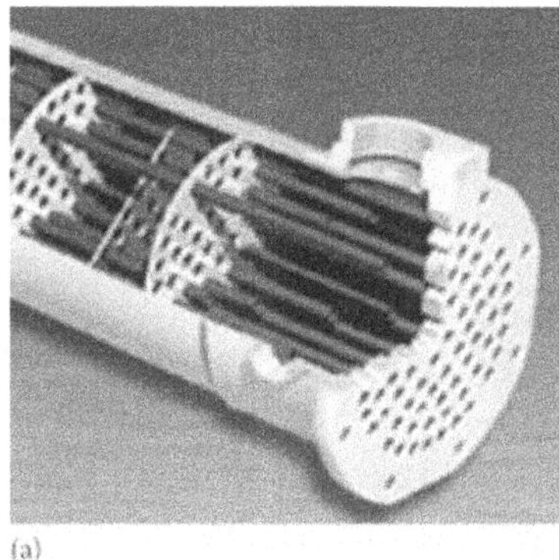

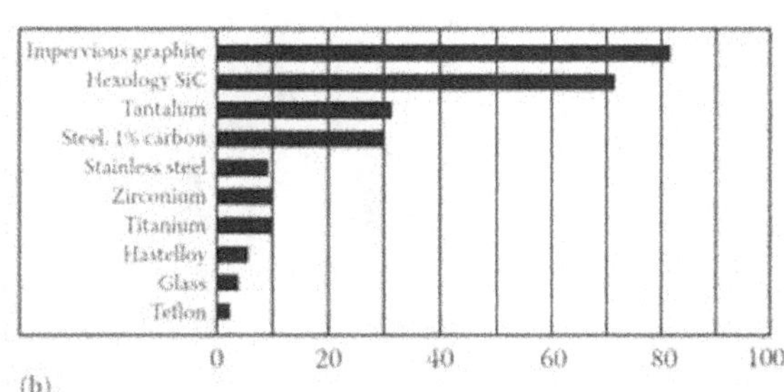

(a) (b)

FIGURE 1.44 Hexoloy heat exchanger tube- (a) Hexoloy silicon carbide heat exchanger unit and (b) comparison of thermal conductivity of Hexoloy SiC with few heat exchanger tube materials. (Courtesy of Saint-Gobain Advanced Ceramics, Niagara Falls, NY.)

and 15 times that of glass. The result is higher efficiency while requiring less heat transfer area. Figure 1.44b shows a comparison of the thermal conductivity of Hexoloy SiC with a few heat exchanger tube materials.

1.16.3 Fluoropolymer Heat Exchanger

Fluoropolymer heat exchangers have been used in process applications for more than 30 years. They are more durable and less brittle than glass and graphite and more corrosion-resistant than most metals, making them a suitable choice for use in a number of process applications [53]. In addition to chemical resistance, fluoropolymer exchangers are versatile. The cost of a fluoropolymer exchanger is similar to those made of glass and graphite. Generally, a fluoropolymer unit is designed to operate at pressures up to 40 psig and temperatures up to 300°F (149°C); however, it can operate above those limits if it is manufactured using proprietary materials [53].

Fluoropolymer shell and tube units have been used successfully as dilution coolers for sulfuric acid, bleach coolers, and cooler absorbers for recovery of hydrochloric acid. Fluoropolymer heat exchangers can be designed with nonmetallic shells to eliminate any metal presence in proximity to the fluid0handling processes. Fluoropolymer immersion coils are used in metal finishing and chemical processing, where the resins' inherent characteristics are especially attractive. Those characteristics include corrosion resistance, long service life, minimum maintenance, high thermal efficiency, and durability under heat and pressure [54].

1.16.4 Metal Foam Heat Exchanger

A metal foam is a cellular structure consisting of a solid metal (frequently aluminum) with gas-filled pores comprising a large portion of the volume. The pores can be sealed (closed-cell foam) or interconnected (open-cell foam). The defining characteristic of metal foams is high porosity: typically only 5–25% of the volume is the base metal. The strength of the material is due to the square–cube law.

Foam-based heat exchangers (typically aluminum or copper foams) operate on the same fundamentals as all heat exchangers, but with increased surface area and turbulence to maximize heat transfer, resulting in better performance in a smaller package. Metal foams with a smaller pore size (40 ppi) have a larger heat transfer coefficient compared to foams with a larger pore size (5 ppi). However, foams with larger pores result in relatively smaller pressure gradients. The foam's high

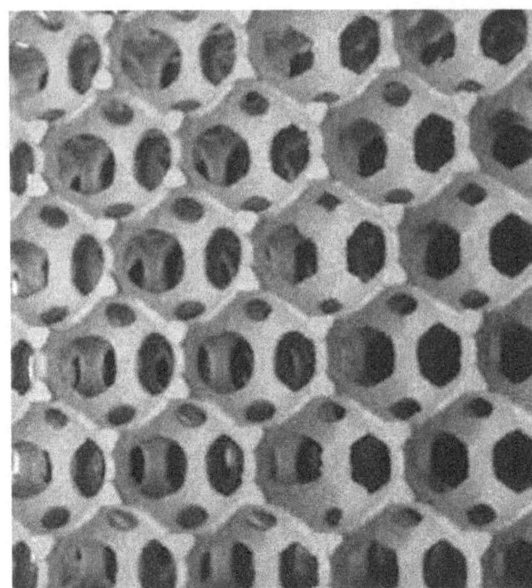

FIGURE 1.45 Metal foam structure[55].

surface area to volume ratio and ligament structure interrupts boundary layers, creating turbulent flow and enhanced mixing, resulting in higher heat transfer compared to conventional technologies. Metal foam can be applied to all types of heat exchangers as an alternative to traditional fins or microchannels. As new products and technologies push the limits of electronics, heat exchangers are required to remove more heat in smaller volumes. Metal foam structure is shown in Figure 1.45 [55] and a metal foam heat exchanger unit is shown in Figure 1.46 [56].

1.16.5 Plastic Heat Exchanger

Plastic heat exchangers have been used in various fields in various industries. They are made from heat-resisting polyethylene (HTPE), polyvinyl di-fluoride (PVDF), and polypropylene (PP). In highly corrosive media, the heat transfer between two fluids is extremely critical. Its success depends not only upon the mechanical properties of materials, but also their resistance to the combined elements of chemical attack, and high temperatures and pressures. There are many exotic materials that can be used to handle corrosive chemicals, such as special glass, titanium, zirconium, tantalum, and chrome/nickel-containing alloys. However, a short service life may be expected [57,58].

Despite their significant advantages, plastic heat exchangers are not suitable for all applications. They cannot be used with refrigerants or high-pressure or high-temperature systems. Plastic is not a suitable heat transfer media for systems with operating temperatures higher than 220°F (105°C) or pressures higher than 150 psi. Plastic additionally cannot be used with any gaseous systems because it does not serve as a sufficient vapor barrier. A block plastic heat exchanger is shown in Figure 1.47

1.17 SELECTION OF HEAT EXCHANGERS

1.17.1 Introduction

In order to achieve optimum process operations, it is essential to use the right type of process equipment in any given process. Heat exchangers are used to transfer energy from one fluid to another, and are no exception. Selecting the wrong type can lead to sub-optimum plant

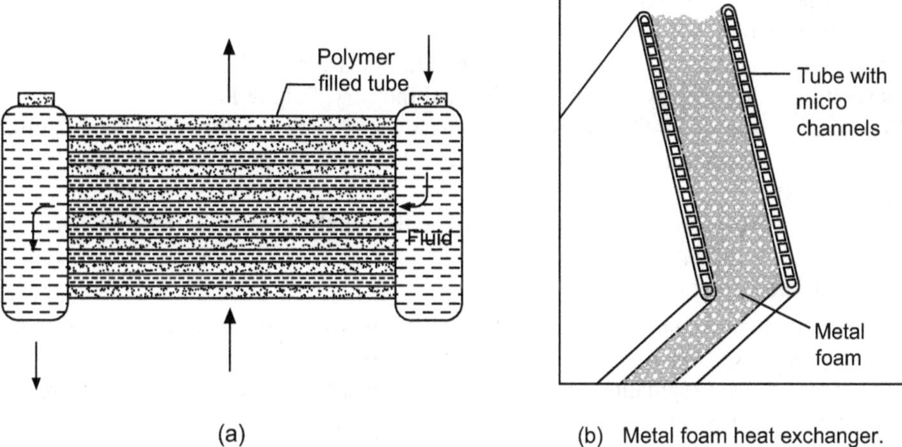

FIGURE 1.46 Metal foam heat exchanger unit[56].

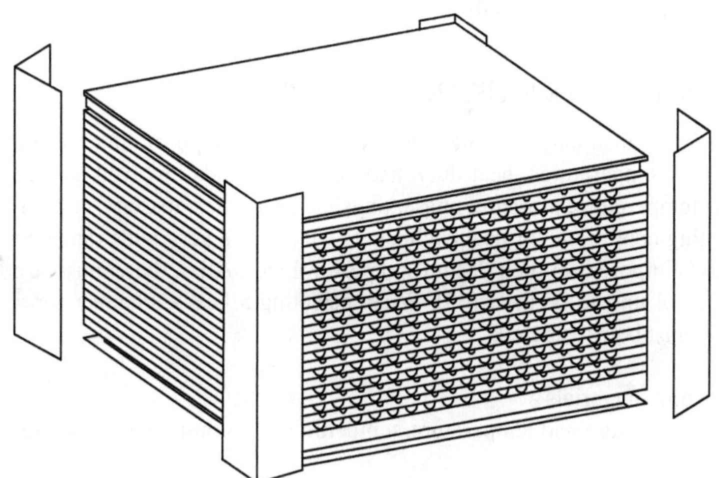

FIGURE 1.47 A block plastic heat exchanger.

performance, operability issues, and equipment failure. Selection is the process in which the designer selects a particular type of heat exchanger for a given application from a variety of heat exchangers. There are a number of alternatives for selecting heat transfer equipment, but only one among them is the best for a given set of conditions. The heat exchanger selection criteria are discussed next.

1.17.2 Market Analysis: Global Liquid Heat Exchanger System Market

The global liquid heat exchanger system market is expected to register a substantial CAGR in the forecast period of 2019–2026 [59–62]. The report contains data from the base year of 2018 and the historic year of 2017 [62]. This rise in market value can be attributed to the massive industrialization activities in the developing regions of the world, along with the advancements of technology in heat exchanging systems. Industry Trends and Forecast to 2026–Global liquid heat exchanger system market by [59–62]:

a. Type—Shell & Tube Heat Exchangers, Plate & Frame Heat Exchangers, Air Cooled Heat Exchangers, Cooling Towers, Others.
b. Application—Chemical, Petrochemicals and Oil & Gas, HVACR, Food & Beverages, Power Generation & Metallurgy, Marine, Mechanical Industry, Pulp & Paper, Others.
c. Construction Material—Carbon Steel, Stainless Steel, Nickel; Others, Graphite (Cubic Heat Exchangers, Graphite Block Heat Exchangers, Polytube Graphite Shell & Tube Block Heat Exchangers).
d. Geography—North America, Europe, Asia-Pacific, South America, Middle East, and Africa.

1.17.2.1 Market Drivers

i. Rapid growth in demand for the reduction of the environmental impact of various industries; this factor is expected to boost the growth of the market.
ii. Presence of strict regulations and compliances regarding the emissions of different industries; this factor is expected to propel the growth of the market.
iii. Increasing focus of different facilities to adopt centralized system for heating and cooling drives the growth of this market.
iv. Significant increase in the price of energy and the need for reducing energy consumption can also enhance the market growth.

1.18 HEAT EXCHANGER SELECTION CRITERIA

The selection criteria are many, but the primary criteria are type of fluids to be handled, operating pressures and temperatures, heat duty, and cost. The fluids involved in heat transfer can be characterized by temperature, pressure, phase, physical properties, toxicity, corrosivity, and fouling tendency. Operating conditions for heat exchangers vary over a very wide range, and a broad spectrum of demands is imposed for their design and performance. All of these must be considered when assessing the type of unit to be used [63]. When selecting a heat exchanger for a given duty, the following points must be considered:

i. Construction materials
ii. Operating pressure and temperature, temperature program, and temperature driving force
iii. Flow rates
iv. Flow arrangements
v. Performance parameters—thermal effectiveness and pressure drops
vi. Fouling tendencies
vii. Accessibility for cleaning and maintenance
viii. Footprint—plot plan and layout constraints
ix. Considerations for future expansions
x. Types and phases of fluids
xi. Maintenance, inspection, cleaning, extension, and repair possibilities
xii. Overall economy
xiii. Fabrication techniques
xiv. Mounting arrangements: horizontal or vertical

Heat exchanger selection criteria are shown in Figure 1.48.

1.18.1 Heat Exchanger Specifications

Heat exchanger specifications provide detailed information of heat exchangers such as construction details, construction codes and standards, process fluid parameters, operating conditions,

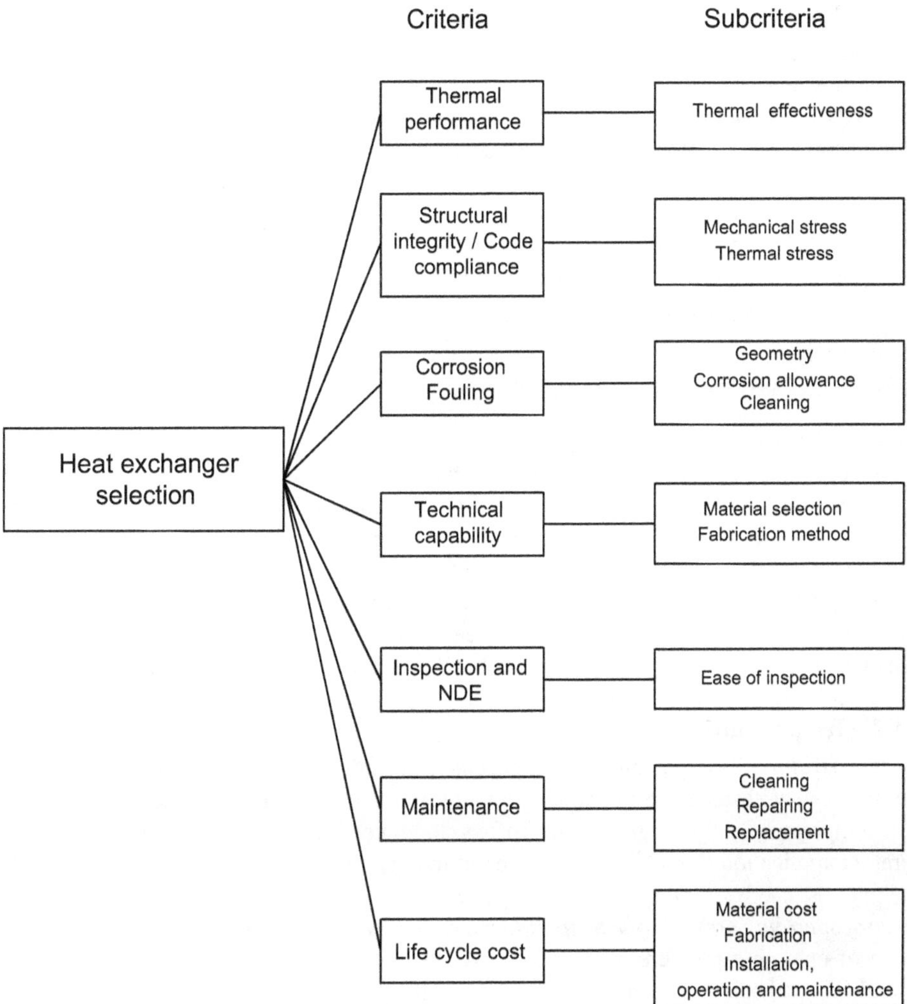

FIGURE 1.48 Heat exchanger selection criteria.

performance requirements, and guidelines for fabrication, testing, and dispatch. Typical specification sheets for ACHEs, STHEs, and PHEs are shown in the respective chapters.

1.18.2 Construction Materials

For reliable and continuous use, the construction materials for pressure vessels and heat exchangers should have a well-defined corrosion rate in the service environment. Furthermore, the material should exhibit sufficient strength to withstand the operating temperature and pressure. STHEs can be manufactured in virtually any material that may be required for corrosion resistance, for example, from nonmetals like glass, Teflon, and graphite to exotic metals like titanium, zirconium, tantalum, etc. Compact heat exchangers with extended surfaces are mostly manufactured from a metal that has drawability, formability, and malleability. Heat exchanger types like PHEs normally require a corrosion-resistant material that can be pressed, welded, or brazed.

1.18.2.1 ASME Code Material Requirements

All materials used for pressure-retaining parts shall be as per ASME codes.

Section II Materials

All materials used for pressure-retaining parts must meet the ASME Code (ASME Boiler and Pressure Vessel Code, Section II—Material Specifications and Section VIII Div 1) [64].

Part A covers Ferrous Material
Part B covers Nonferrous Material
Part C covers Welding Rods, Electrodes, and Filler Metals
Part D covers Material Properties in both Customary and Metric units of measure.

1.18.3 Operating Pressure and Temperature

1.18.3.1 Pressure

The design pressure is important to determine the thickness of the pressure-retaining components. The higher the pressure, the greater will be the required thickness of the pressure-retaining membranes and the more advantage there is to placing the high-pressure fluid on the tubeside. The pressure level of the fluids has a significant effect on the type of unit selected [63]. Operating pressures of the gasketed PHEs and SPHEs are limited because of the difficulty in pressing the required plate thickness, and by the gasket materials in the case of PHEs. The floating nature of floating-head shell and tube heat exchangers and lamella heat exchangers limits the operating pressure.

1.18.3.2 Temperature

Design temperature: This parameter is important as it indicates whether a material at the design temperature can withstand the operating pressure and various loads imposed on the component. For low-temperature and cryogenic applications, toughness is a prime requirement, and for high-temperature applications the material has to exhibit creep resistance.

Influence of operating pressure and temperature on the selection of some types of heat exchangers: The influence of operating pressure and temperature on selection of STHE, compact heat exchanger, gasketed PHE, and spiral exchanger is discussed next.

Shell and tube heat exchanger: STHE units can be designed for almost any combination of pressure and temperature. In extreme cases, high pressure may impose limitations through fabrication problems associated with material thickness, and through the weight of the finished unit. Differential thermal expansion under steady conditions can induce severe thermal stresses, either in the tube bundle or in the shell. Damage due to flow-induced vibration on the shellside is well known. In heat-exchanger applications where high heat transfer effectiveness (close approach temperature) is required, the standard shell and tube design may require a very large amount of heat transfer surface [65]. Depending on the fluids and operating conditions, other types of heat-exchanger design should be investigated.

Compact heat exchanger: Compact heat exchangers are constructed from thinner materials; they are manufactured by mechanical bonding, soldering, brazing, welding, etc. Therefore, they are limited in operating pressures and temperatures.

Gasketed plate heat exchangers and spiral exchangers: Gasketed PHEs and spiral exchangers are limited by pressure and temperature, wherein the limitations are imposed by the capability of the gaskets.

1.18.4 Flow Rate

The flow rate determines the flow area: the higher the flow rate, the higher will be the crossflow area. A higher flow area is required to limit the flow velocity through the conduits and flow passages, and the higher velocity is limited by pressure drop, impingement, erosion, and, in the case of a shell and tube exchanger, by shellside flow-induced vibration. Sometimes, a minimum flow velocity is necessary to improve heat transfer to eliminate stagnant areas and to minimize fouling.

1.18.5 Performance Parameters: Thermal Effectiveness and Pressure Drops

Thermal effectiveness: For high-performance service requiring high thermal effectiveness, brazed plate-fin exchangers (e.g., cryogenic service) and regenerators (e.g., gas turbine applications) are used, tube-fin exchangers are used for slightly less thermal effectiveness in applications, and shell and tube units are used for low-thermal effectiveness service.

Pressure drop: Pressure drop is an important parameter in heat exchanger design. Limitations may be imposed either by pumping cost, process limitations, or both. The heat exchanger should be designed in such a way that unproductive pressure drop is avoided to the maximum extent in areas like inlet and outlet bends, nozzles, and manifolds. At the same time, any pressure-drop limitation that is imposed must be utilized as nearly as possible for an economic design.

1.18.6 Fouling Tendencies

Fouling is defined as the formation on heat exchanger surfaces of undesirable deposits that impede the heat transfer and increase the resistance to fluid flow, resulting in a higher pressure drop. The growth of these deposits causes the thermohydraulic performance of heat exchangers to decline with time. Fouling affects the energy consumption of industrial processes, and it also decides the amount of extra material required to provide extra heat transfer surface to compensate for the effects of fouling. Compact heat exchangers are generally preferred for nonfouling applications. In a shell and tube unit, the fluid with more fouling tendencies should be put on the tubeside for ease of cleaning. On the shellside with cross baffles, it is sometimes difficult to achieve a good flow distribution if the baffle cut is either too high or too low. Stagnation in any regions of low velocity behind the baffles is difficult to avoid if the baffles are cut more than about 20%–25%. PHEs and spiral plate exchangers are better chosen for fouling services. The flow pattern in PHE induces turbulence even at comparable low velocities; in the spiral units, the scrubbing action of the fluids on the curved surfaces minimizes fouling. Also consider the Philips RODbaffle heat exchanger, TWISTED TUBE® heat exchanger, Helixchanger® heat exchanger, or EMbaffle® heat exchanger to improve flow velocity on the shellside, enhance heat transfer performance, and reduce fouling tendencies on shellside.

1.18.7 Maintenance, Inspection, Cleaning, Repair, and Extension Aspects

Consider the suitability of various heat exchangers as regards maintenance, inspection, cleaning, repair, and extension. For example, the pharmaceutical, dairy, and food industries require quick access to internal components for frequent cleaning. Spiral plate exchangers can be made with both sides open at one edge, or with one side open and one closed. They can be made with channels between 5 and 25 mm wide, with or without studs. STHE can be made with fixed tubesheets or with a removable tube bundle, with small- or large-diameter tubes, or small or wide pitch. A lamella heat exchanger bundle is removable and thus fairly easy to clean on the shellside. Inside, the lamella, however, cannot be drilled to remove the hard fouling deposits. Gasketed PHEs are easy to open, especially when all nozzles are located on the stationary end-plate side. The plate arrangement can be changed for other duties within the frame and nozzle capacity.

Repair of some of the shell and tube exchanger components is possible, but repair of the expansion joint is very difficult. Tubes can be renewed or plugged. Repair of compact heat exchangers of fin-tube type is very difficult except by plugging of the tube. Repair of the plate-fin exchanger is generally very difficult. For these two types of heat exchangers, extension of units for higher thermal duties is generally not possible. All these drawbacks are easily overcome in a PHE. It can be easily repaired, and plates and other parts can be easily replaced. Due to modular construction, PHEs possess the flexibility of enhancing or reducing the heat transfer surface area, modifying the pass arrangement, and the addition of more than one duty according to the heat transfer requirements at a future date.

1.18.8 Overall Economy

There are two major costs to consider in designing a heat exchanger: the manufacturing cost and the operating costs, including maintenance costs. In general, the less the heat transfer surface area and less the complexity of the design, the lower is the manufacturing cost. The operating cost is the pumping cost due to pumping devices such as fans, blowers, and pumps. The maintenance costs include costs of spares that require frequent renewal due to corrosion, and costs due to corrosion/fouling prevention and control. Therefore, the heat exchanger design requires a proper balance between thermal sizing and pressure drop.

1.18.9 Fabrication Techniques

Fabrication techniques are likely to be the determining factor in the selection of a heat transfer surface matrix or core. They are the major factors in the initial cost and to a large extent influence the integrity, service life, and ease of maintenance of the finished heat exchanger [66]. For example, shell and tube units are mostly fabricated by welding, plate-fin heat exchangers, automobile aluminum radiators, and microchannel heat exchangers by brazing, copper–brass radiators by soldering, most of the circular tube-fin exchangers by mechanical assembling, etc.

1.18.10 Choice of Unit Type for Intended Applications

According to the intended applications, the selection of heat exchangers will follow the guidelines provided in Table 1.2.

1.18.11 Functional Requirements of Heat Exchangers

The heat exchanger must meet normal process requirements specified through problem specification and service conditions for combinations of the clean and fouled conditions, and uncorroded and corroded conditions. Heat exchangers have to fulfill the following requirements:

i. High thermal effectiveness
ii. Pressure drop as low as possible
iii. Reliability and life expectancy
iv. High-quality product and safe operation
v. Material compatibility with process fluids
vi. Convenient size, easy for installation, reliable in use
vii. Easy maintenance and servicing
viii. Light in weight but strong in construction to withstand the operational pressures and vibrations especially heat exchangers for military applications

TABLE 1.2
Choice of Heat Exchanger Type for Intended Applications

Application	Remarks
Low-viscosity fluids	For high temperatures/pressures, use STHE or double-pipe heat exchanger. Use PHE or LHE for low temperature/pressure applications
Low-viscosity liquid to steam	Use STHE in carbon steel
Medium-viscosity fluids	Use PHE or with high solids content, use SPHE
High-viscosity fluids	PHE offers the advantages of good flow distribution. For extreme viscosities, SPHE is preferred
Fouling liquids	Use STHE with removable tube bundle. SPHE or PHE is preferred due to good flow distribution. Use PHE if easy access is of importance. Also consider Philips RODbaffle heat exchanger, TWISTED TUBE® heat exchanger, Helixchanger® heat exchanger, and EMbaffle® heat exchanger to improve flow velocity on the shellside, enhance heat transfer performance, and reduce fouling tendencies on shellside
Slurries, suspensions, and pulps	SPHE offers the best characteristics. Also consider free flow PHE or wide gap PHE, or scraped surface heat exchanger
Heat-sensitive liquids	PHE fulfills the requirements best. Also consider SPHE
Cooling with air	Extended surface types like tube-fin heat exchanger or PFHE
Gas or air under pressure	Use STHE with extended surface on the gas side or brazed plate-fin exchanger made of stainless steel or nickel alloys
Cryogenic applications	Brazed aluminum plate-fin exchanger, BPHE, coiled tube heat exchangers, or PCHE
Vapor condensation	Surface condensers of STHE in carbon steel are preferred. Also consider SPHE or brazed plate heat exchanger
Vapor/gas partial condensation	Choose SPHE
Refrigeration and air conditioning applications	Finned tube heat exchangers, special types of PHEs, brazed PHE up to 200°C
Air–air or gas–gas applications	Regener ators and plate-fin heat exchangers. Also consider STHE
Viscous products, aseptic products, jam, food, and meat processing, heat-sensitive products, and particulate-laden products	Scraped surface heat exchanger

Note: STHE, shell and tube heat exchanger; PHE, gasketed plate heat exchanger; BPHE, brazed plate heat exchanger, SPHE, spiral plate heat exchanger; LHE, lamella heat exchanger; PCHE, printed circuit heat exchangers; CTHE, coiled tube heat exchanger; PFHE, plate-fin heat exchanger.

ix. Simplicity of manufacture
x. Low cost
xi. Possibility of effecting repair to maintenance problems.

1.19 LOCATION AND SITING CONSIDERATIONS

Consider the following factors when deciding the location where the heat exchanger will be installed:

i. Type of environment, which has a bearing on corrosion of heat exchangers
ii. Wind directions
iii. Vibration from near-by heavy equipment
iv. Space for maintenance activities, space in terms of length for pulling of tube bundle, etc.

1.20 INSTALLATION, OPERATION, AND MAINTENANCE

1.20.1 Storage

If the heat exchanger unit arrives on site several months prior to the actual installation and start-up, it results in the potential for premature corrosion to occur unless the equipment is stored in a proper manner. The unit should be stored so that it is not exposed to dust, industrial contaminants, coastal contaminants, or high levels of humidity and moisture. Improper storage can lead to premature corrosion prior to start-up and can reduce the overall life of the equipment. Extra care should be taken to ensure that equipment located on the ground level remains free from debris prior to start-up.

1.20.2 Installation

1. When transporting the heat exchanger, take precautions to prevent it from tumbling or falling. Install the heat exchanger on a level floor.
2. Ensure that there is sufficient space around the heat exchanger for maintenance. Foundations must be adequate so that exchangers will not settle and impose excessive strains on the exchanger.
3. *Foundation bolts*: Foundation bolts should be loosened at one end of the unit to allow free expansion of shells. Slotted holes in supports are provided for this purpose when saddles are provided.
4. *Safety relief devices*: The ASME code defines the requirements for safety relief devices. When specified by the purchaser, the manufacturer will provide the necessary connections for the safety relief devices. The purchaser will provide and install the required relief devices.

1.20.3 Operation

While starting, ensure that all operating conditions conform to the heat exchanger's specifications. Before placing any exchanger in operation, reference should be made to the exchanger drawings, specification sheet(s), and name plate(s) details for any special instructions. Local safety and health regulations must be considered. Improper startup or shutdown sequences may cause very high thermal stress, catastrophic failures, and may cause a hazard. The unit must be shut down in a manner to minimize differential thermal expansion between various components. Monitor the performance by installing the necessary gages and instruments.

1.20.4 Scheduled Maintenance

Carry out daily check and periodical maintenance as per OEM recommendations. Follow the instructions of the heat exchanger manufacturer during installation, operation, and maintenance. Ensure that the corrosion and fouling control is in place. Overhaul the heat exchanger and clean parts as per schedule. At the same time, check for visible cracks, if any, of heat exchanger components, insulation, etc. Other than the instructions on maintenance, a log of all plant operational events, start-ups, shutdowns, and malfunctions that affect the heat exchanger performance should be kept.

REFERENCES

1. American Society of Mechanical Engineers. (2021) ASME Boiler and *Pressure Vessel Code, Section VIII, Division 1—Pressure Vessels*. American Society of Mechanical Engineers, New York.
2. American Society of Mechanical Engineers. (2021) *ASME Boiler and Pressure Vessel Code, Section VIII, Division 2, Pressure Vessels—Alternative Rules*. American Society of Mechanical Engineers, New York.

3. Shah, R.K. (1981) Classification of heat exchangers. In: *Heat Exchangers: Thermal-Hydraulic Fundamentals and Design*, S. Kakac, A.E. Bergles, F. Mayinger (Eds.), pp. 9–46. Hemisphere, Washington, DC.
4. Shah, R.K., Sekulic, D.P. (2003) *Fundamentals of Heat Exchanger Design*. John & Wiley, New York.
5. Gupta, J.P. (1986) *Fundamentals of Heat Exchanger and Pressure Vessel Technology*. Hemisphere, Washington, DC.
6. Walker, G. (1982) *Industrial Heat Exchangers—A Basic Guide*. Hemisphere/McGraw-Hill, New York.
7a. Larowski, A., Taylor, M.A. (1982) *Systematic Procedures for Selection of Heat Exchangers*, C58/82, pp. 32–56. Institution of Mechanical Engineers, London, UK.
7b. Larowski, A., Taylor, M.A. (1983) Systematic procedure for selection of heat exchangers. *Proc. Inst. Mech. Eng*. 197A: 51–69.
8. https://heat-integration-platform.ispt.eu/heat-exchangers#midway-1.1%20Brazed%20plate%20heat%20exchanger%20(BPHE)-anchor
9. Kern, D.Q. (1950) *Process Heat Transfer.* McGraw-Hill, New York.
10. Chisholm, D. (ed.) (1980) *Developments in Heat Exchanger Technology—1*. Applied Science Publishers, London, England.
11. Minton, P. (1990) Process heat transfer. *Proceedings of the 9th International Heat Transfer Conference*, Paper No. KN–2, 1: 355–362. Heat Transfer 1990–Jerusalem, Israel.
12. Yokell, S. (1990) *A Working Guide to Shell and Tube Heat Exchangers*. McGraw-Hill, New York.
13. Tubular Exchanger Manufacturers Association (2019) *Standards of the Tubular Exchanger Manufacturers Association (TEMA)*, 10th edn. Tubular Exchanger Manufacturers Association, Inc., Tarrytown, NY.
14a. HEI (2017) *Standards for Steam Surface Condensers*, 12th Edition. HEI, Cleveland, Ohio.
14b. HEI (2015) *Standards for Closed Feedwater Heaters*, 9th Edition. HEI, Cleveland, Ohio.
14c. HEI (2013) *Standards for Shell and Tube Exchangers*, 5th edn. Heat Exchange Institute, Cleveland, OH.
15. American Petroleum Institute (2015) *API 660, Shell and Tube Heat Exchangers*, 9th Edition. American Petroleum Institute, Washington, DC.
16. O'Doherty, T., Jolly, A.J., Bates, C.J. (2001) Analysis of a bayonet tube heat exchanger. *Applied Thermal Engineering* 21(1): 1–18.
17. Ivošević, M., Petrović, A., Jaćimović, B ., Genić, S. (2019) Thermal performances and their impact on design of bayonet-tube heat exchangers –single phase plug flow. *Heat and Mass Transfer* 55: 2391–2404.
18. www.armstrong-chemtec.com/blog/the-benefits-of-vertical-bayonet-vaporizers.html
19. www.calgavin.com/articles/heat-exchanger-problems-part-1
20. Timmerhaus, K.D., Flynn, T.M. (1989) *Cryogenic Progress Engineering*. Plenum Press, New York.
21. Linde AG Engineering Division. (2018) *Coil-Wound Heat Exchangers*, pp. 1–15. Linde AG Engineering Division, Pullach, Germany.
22. Muoio, J.M. (1985) Glass as a material of construction for heat transfer equipment. *Industrial Heat Exchangers Conference Proceedings*, A.J. Hayes, W.W. Liang, S.L. Richlen, E.S. Tabb (Eds.), pp. 385–390. American Society for Metals, Metals Park, OH.
23. www.alfalaval.com/products/heat-transfer/plate-heat-exchangers/gasketed-plate-and-frame-heat-exchangers/hygienic-line/?gclid
24. Yilmaz, S., Samuelson, B. (1983) Vertical thermosyphon boiling in spiral plate heat exchangers. In *Heat Transfer—1983, Seattle, Chem. Eng. Prog. Symp. Ser*. 79(225): 47–53.
25. Markovitz, R.E. (1971) Picking the best vessel jacket. *Chem. Eng*. 78(26): 156–162.
26. Usher, J.D., Cattell, G.S. (1980) Compact heat exchangers. In: *Developments in Heat Exchanger Technology—1*, D. Chisholm (Ed.), pp. 127–152. Applied Science Publishers Ltd., London, UK.
27. API (2013) *API STD 661, Petroleum, Petrochemical, and Natural Gas Industries—Air-cooled Heat Exchangers*, 7th Edition.
28. www.therma.com/microchannel-heat-exchangers/
29. www.wikiwand.com/en/Internally_grooved_copper_tube
30. https://microgroove.net/
31. https://microgroove.net/sites/default/files/webinar_iv_slide_show_final.pdf
32. www.hydro.com/en/products-and-services/precision-tubes/micro-channel-tubes/

33(a). http://uni-klima.com/wp-content/uploads/2016/01/R2-sezonal.pdf
33(b). https://www.boydcorp.com/thermal/liquid-cooling/liquid-heat-exchangers/flat-tube-heat-exchangers.html
34. Heat Exchanger Manufacturer's Association (2010) *ALPEMA Standard, The Brazed Aluminium Plate-Fin Heat Exchanger Manufacturer's Association*, 3rd edn. Didcot, Oxon, UK.
35. Reay, D.A. (1980) Heat exchangers for waste heat recovery, International Research and Development Co., Ltd., Newcastle upon Tyne, UK. In: *Developments in Heat Exchanger Technology— 1*, D. Chisholm (Ed.), pp. 233–256. Applied Science Publishers Ltd., London, UK.
36. Butterworth, D., Mascone, C.F. (1991) Heat transfer heads into the 21 century. *Chem. Eng. Prog.* September: 30–37.
37. Taylor, M.A. (ed) (1980) *Plate-Fin Heat Exchangers, Guide to Their Specification and Use.* HTFS (Harwell Laboratory), Oxon, UK.
38. American Petroleum Institute (2002) *API Standard 660, Shell and Tube Exchangers for General Refinery Services*, 7th edn. American Petroleum Institute, Washington, DC.
39. Boehm, R.F. (2014) *Direct Contact Heat Transfer, Chapter 19, Bulk Solids: Operating Direct-Contact Heat Exchangers.* Greg Mehos, Jenike & Johanson, Inc.
40a. Steinke, M.E., Kandlikar, G. (2004) Single-phase heat transfer enhancement techniques in microchannel and minichannel flows. In: *Microchannels and Minichannelts—2004.* ASME, Rochester, NY.
40b. Tuckerman, D.B., Pease, R.F.W. (1981) High-performance heat sinking for VLSI. *IEEE Electron Device Letters* 2(5): 126–129.
41. Nikitin, K., Kato, Y., Ngo, L. (2005) *Thermal-Hydraulic Performance of Printed Circuit Heat Exchanger in Supercritical CO_2 Cycle*. Research Laboratory for Nuclear Reactors, Tokyo Institute of Technology, Okayama, Tokyo, Japan.
42. Krishnakumar, K., Venkatarathnam, G. (2007) Heat transfer and flow friction characteristics of perforated plate matrix heat exchangers. *International Journal of Heat Exchangers* 8(1): 45–60.
43. Krishnakumar, K., Venkatarathnam, G. (2003) Transient testing of perforated plate matrix heat exchangers. *Cryogenics* 43(2): 101–109.
43.1 www.sciencedirect.com/science/article/abs/pii/S0011227503000262
44. Aditya Bhanumurthy, K., Srinivasa Murthy, S., Venkatarathnam, G. (2010) 3rd *International Conference on Thermal Issues in Emerging Technologies Theory and Applications*. 19–22 December 2010.
45. Fleming, R.B. (1969) A compact perforated plate heat exchanger. *Advances in Cryogenic Engineering* 14: 196–204.
46. Hendricks, J.B. (1992) *Perforated plate heat exchanger and method of fabrication.* Patent number: 5101894, Apr 7, 1992.
47. www.arrow.com/en/research-and-events/articles/understanding-heat-sinks-functions-types-and-more#:~:text=A%20heat%20sink%20is%20a,across%20its%20enlarged%20surface%20area. Arrow.com
48. www.radianheatsinks.com/heatsink/
49. https://en.wikipedia.org/wiki/Heat_sink
50. Fiore, J.M. *8.6 Heat Sinks—Engineering LibreTexts.* https://eng.libretexts.org/Bookshelves/Electrical_Engineering/Electronics/Book%3A_Semiconductor_Devices_-_Theory_and_Application_(Fiore)/08%3A_BJT_Class_A_Power_Amplifiers/8.6%3A_Heat_Sinks#:~:text=A%20heat%20sink%20is%20a,feature%20an%20array%20of%20fins.
51. https://depts.washington.edu/matseed/mse_resources/Webpage/Ceramics/ceramics.htm
52. https://cgthermal.com/heat-exchangers/sic-ceramic/
53. *A Practical Guide to Fluoropolymer Heat Exchangers*, March 1, 2001; www.junkosha.com/en/products/ACH-02.
54. www.process-heating.com/articles/84231-a-practical-guide-to-fluoropolymer-heat-exchangers
55. https://commons.wikimedia.org/wiki/File:Alv%C3%A9oles_4.jpg
56. www.electronics-cooling.com/2018/06/metal-foam-heat-exchangers/
57. www.aetnaplastics.com/site_media/media/attachments/aetna_product_aetnaproduct/112/GF%20Heat%20Exchangers%20Brochure.pdf
58. https://www.uk-exchangers.com/heat-exchangers/air-to-air-heat-exchangers/
59. www.databridgemarketresearch.com/reports/global-liquid-heat-exchanger-system-market

60. https://markets.businessinsider.com/news/stocks/global-16-billion-heat-exchangers-market-to-2025-market-trends-key-players-business-and-product-trends-1028563601
61. The "*Heat Exchangers—A Global Market Overview*" report has been added to Research And Markets.com's offering; https://industryexpertsresearch.wordpress.com/tag/global-heat-exchangers-market/.
62. *Global $16 Billion Heat Exchangers Market to 2025: Market Trends, Key Players, Business and Product Trends.* Press release PR Newswire, Dublin, Sep. 30, 2019; www.prnewswire.com/news-releases/global-16-billion-heat-exchangers-market-to-2025-market-trends--key-players-business-and-product-trends-300927555.html.
63. Gollin, M. (1984) Heat exchanger design and rating. In: *Handbook of Applied Thermal Design*, E.C. Guyer (Ed.), Chapter 2, Part 7, pp. 7-24–7-36. McGraw-Hill, New York.
64. ASME (2021) *ASME Boiler and Pressure Vessel Code, Section II Parts A, B, C and D for Materials specification* . ASME.
65. Caciula, L., Rudy, T.M. (1983) Prediction of plate heat exchanger performance. *D Symp. Ser., Heat Transfer—1983*, pp. 76–89. Seattle, WA.
66. Fraas, A.P., Ozisik, M.N. (1965) *Heat Exchanger Design.* John Wiley & Sons, New York.

Suggested Readings

Bell, K.J., Mueller, A.C. (1984) *Wolverine Heat Transfer Data Book II.* Wolverine Division of UOP Inc., Decatur, AWL.

Scaccia, C., Theoclitus, G. (1986) Types, performance and applications. In: *The Chemical Engineering Guide to Heat Transfer, Volume I: Plant Principles*, K.J. McNaughton and the Staff of Chemical Engineering (Eds.), pp. 3–14. Hemisphere/McGraw-Hill, New York.

Shah, R.K. (1983) Classification of heat exchangers. In: *Low Reynolds Number Flow Heat Exchangers*, S. Kakac, R.K. Shah, A.E. Bergles (Eds.), pp. 9–19. Hemisphere, Washington, DC.

Shah, R.K. (1984) What's new in heat exchanger design. *ASME, Mech. Eng.* 106: 50–59.

Sukhatme, S.P., Devotta, S. (1988) Classification of heat transfer equipment. In: *Heat Transfer Equipment Design*, R.K. Shah, E.C. Subbarao, R.A. Mashelikar (Eds.), pp. 7–16. Hemisphere, Washington, DC.

Thome, R.T. (2004) *Wolverine Heat Transfer Engineering Data Book III.* Wolverine Division of UOP Inc., Decatur, AL.

Bibliography

Afgan, N.H., Schlunder, E.U. (Eds.) (1974) *Heat Exchangers: Design and Theory Sourcebook.* McGraw-Hill, New York.

Apblett, W.R. (Ed.) (1982) *Shell and Tube Heat Exchangers.* American Society for Metals, Metals Park, OH.

Bhatia, M.V., Cheremisinoff, N.P. (1980) *Heat Transfer Equipment.* Process equipment series, Vol. 2. Technomic, Westport, CT.

Bryers, R.W. (Ed.) (1983) *Fouling of Heat Exchanger Surfaces.* Engineering Foundation, New York.

Chen, S.S. (1987) *Flow Induced Vibration of Circular Cylindrical Structures.* Hemisphere, Washington, DC.

Cheremisinoff, N. (Ed.) (1986) *Handbook of Heat and Mass Transfer, Vol. 1—Heat Transfer Operation & Vol. 2—Mass Transfer and Reactor Design*, pp. 767–805. Gulf Publishing Company, Houston, TX.

Cheremisinoff, N.P. (1984) *Heat Transfer Pocket Handbook.* Gulf Publishing Company, Books Division, Houston, TX.

Chisholm, D. (Ed.) (1988) *Heat Exchanger Technology.* Elsevier Applied Science, New York.

ESDU International (1987) *Effectiveness Ntu Relationships for Design and Performance Evaluation of Multi-pass Crossflow Heat Exchangers*, Engineering Sciences Data Unit Item 87020. ESDU International, McLean, VA.

ESDU International (1986) *Effectiveness Ntu Relationships for Design and Performance Evaluation of Two-Stream Heat Exchangers*, Engineering Sciences Data Unit Item 86018. ESDU International, McLean, VA.

Foster, B.D., Patton, J.B. (Eds.) (1985) *Ceramic Heat Exchangers.* American Ceramic Society, Columbus, OH.

Fraas, A.P. (1989) *Heat Exchanger Design*, 2nd edn. John Wiley & Sons, New York.

Ganapathy, V. (1982) *Applied Heat Transfer.* PennWell Publishing Co., Tulsa, OK.

Garrett-Price, B.A., Smith, S.A., Watts, R.L., Knudsen, J.C., Marner, W.J., Suitor, J.W. (1985) *Fouling of Heat Exchangers*. Noyes, Park Ridge, NJ.

www.globalspec.com/learnmore/processing_equipment/heat_transfer_equipment/heat_exchangers

Hayes, A.J., Liang, W.W., Richlen, S.L., Tabb, E.S. (Eds.) (1985) *Industrial Heat Exchangers Conference Proceedings*. American Society for Metals, Metals Park, OH.

Hewitt, G.F. (coordinating editor) (1990) *Hemisphere Handbook of Heat Exchanger Design*. Hemisphere, New York.

Hewitt, G.F., Shires, G.L., Bott, T.R. (1994) *Process Heat Transfer*. CRC Press, Boca Raton, FL.

Hewitt, G.F., Whalley, P.B. (1989) *Handbook of Heat Exchanger Calculations*. Hemisphere, Washington, DC.

Hryniszak, W. (1958) *Heat Exchangers: Application to Gas Turbines*. Butterworth Scientific Publications, London, UK.

Hussain, H. (1982) *Heat Transfer in Counterflow, Parallel Flow and Cross Flow*. McGraw-Hill, New York.

Jakob, M. (1957) *Heat Transfer*. John Wiley & Sons, New York.

Kakac, S., Bergles, A.E., Fernandes, E.O. (Eds.) (1988) *Two-Phase Flow Heat Exchangers: Thermal Hydraulic Fundamentals and Design*. Kluwer Academic Publishers, Dordrecht, the Netherlands.

Kakac, S., Bergles, A.E., Mayinger, F. (Eds.) (1981) *Heat Exchangers: Thermal-Hydraulic Fundamentals and Design*. Hemisphere/McGraw-Hill, Washington, DC.

Kakac, S., Shah, R.K., Aung, W. (Eds.) (1987) *Handbook of Single Phase Convective Heat Transfer*. John Wiley & Sons, New York.

Kakac, S., Shah, R.K., Bergles, A.E. (Eds.) (1983) *Low Reynolds Number Flow Heat Exchangers*. Hemisphere, Washington, DC.

Kays, W.M., London, A.L. (1984) *Compact Heat Exchangers*, 3rd edn. McGraw-Hill, New York.

Kern, D.Q., Kraus, A.D. (1972) *Extended Surface Heat Transfer*, pp. 439–641. McGraw-Hill, New York.

Kraus, A.D. (1982) *Analysis and Evaluation of Extended Surface Thermal Systems*. Hemisphere, Washington, DC.

Levenspiel, O. (1984) *Engineering Flow and Heat Exchange*. Plenum Press, New York.

Mahajan, K.K. (1985) *Design of Process Equipments*. Pressure Vessel Handbook Publishing Company, Tulsa, OK.

Manzoor, M. (1984) *Heat Flow through Extended Surface Heat Exchangers*. Springer-Verlag, Berlin, Germany.

Martin, H. (1992) *Heat Exchangers*. Hemisphere, Washington, DC.

McNaughton, K.J. and the Staff of Chemical Engineering (Eds.) (1986) *The Chemical Engineering Guide to Heat Transfer: Vol. I: Plant Principles, Vol. II: Equipment*. Hemisphere/McGraw-Hill, New York.

Melo, L.F., Bott, T.R., Bernardo, C.A. (Eds.) (1988) *Advances in Fouling Science and Technology*. Kluwer Academic Publishers, Dordrecht, the Netherlands.

Mondt, J.R. (1980) *Regenerative Heat Exchangers: The Elements of Their Design, Selection and Use*, Research Publication GMR-3396, General Motors Research Laboratories, Warren, MI.

Mori, Y., Sheindlin, A.E., Afgan, N.H. (Eds.) (1986) *High Temperature Heat Exchangers*, Hemisphere, Washington, DC.

Palen, J.W. (Ed.) (1987) *Heat Exchanger Sourcebook*. Hemisphere, Washington, DC.

Perry, R.H., Chilton, C.H. (Eds.) (1973) *Chemical Engineers' Handbook*, 5th edn. McGraw-Hill, New York.

Reay, D.A. (1979) *Heat Recovery Systems*. E. and F. N. Spon Ltd., London, UK.

Rohsenow, W.M., Hartnett, J.P., Ganic, E.N. (Eds.) (1985) *Handbook of Heat Transfer Applications*. McGraw-Hill, New York.

Saunders, E.A.D. (1989) *Heat Exchangers: Selection, Design and Construction*. Addison Wesley Longman, Reading, MA.

Schlunder, E.U. (editor-in-chief) (1983) *Heat Exchanger Design Handbook*, in five volumes (Vol. 1—*Heal Exchanger Theory*, Vol. 2—*Fluid Mechanics and Heat Transfer*, Vol. 3—*Thermal and Hydraulic Design of Heat Exchangers*, Vol. 4—*Mechanical Design of Heat Exchangers*, and Vol. 5—*Physical Properties*). Hemisphere, Washington, DC.

Schmidt, F.W., Willmott, A.J. (1981) *Thermal Energy Storage and Regeneration*. Hemisphere/McGraw-Hill, Washington, DC.

Shah, R.K., Kraus, A.D., Metzger, D. (eds.) (1990) *Compact Heat Exchangers—A Festschrift for A. L. London*, pp. 31–90. Hemisphere, Washington, DC.

Shah, R.K., London, A.L. (1978) *Laminar Flow Forced Convection in Ducts*. Supplement 1 to advances in heat transfer series. Academic Press, New York.

Shah, R.K., Mueller, A.C. (1989) Heat exchanger, in *Ullmann's Encyclopedia of Industrial Chemistry*, Unit Operations II, Vol. B3, Chapter 2. VCH Publishers, Weinheim, Germany.

Shah, R.K., Subbarao, E.C., Mashelkar, R.A. (Eds.) (1988) *Heat Transfer Equipment Design*. Hemisphere, Washington, DC.

Sheindlin, A.E. (Ed.) (1986) *High Temperature Equipment*. Hemisphere, Washington, DC.

Singh, K.P., Soler, A.I. (1984) *Mechanical Design of Heat Exchangers and Pressure Vessel Components*. Arcturus Publishers, Cherry Hill, NJ.

Somerscales, E.F.C., Knudsen, J.G. (Eds.) (1981) *Fouling of Heat Transfer Equipment*. Hemisphere/McGraw-Hill, Washington, DC.

Stasiulevicius, J., Skrinska, A. (1987) *Heat Transfer of Finned Tube Bundles in Crossflow*. Hemisphere, Washington, DC.

Taborek, J., Hewitt, G.F., Afgan, N. (Eds.) (1983) *Heat Exchangers: Theory and Practice*. Hemisphere/McGraw-Hill, Washington, DC.

Walker, G. (1983) *Cryocoolers, Part 2: Applications*. Plenum Press, New York.

Zukauskas, A.A. (1989) *High Performance Single-Phase Heat Exchangers*, J. Karni (English edition editor). Hemisphere, Washington, DC.

Zukauskas, A., Ulinskas, R. (1988) *Heat Transfer in Tube Banks in Crossflow*. Hemisphere, Washington, DC.

2 Heat Exchanger Thermohydraulic Fundamentals and Thermal Design of Heat Exchangers

2.1 HEAT EXCHANGER THERMAL CIRCUIT AND OVERALL CONDUCTANCE EQUATION

In order to develop relationships between the heat transfer rate q, surface area A, fluid terminal temperatures, and flow rates in a heat exchanger, the basic equations used for analysis are the energy conservation and heat transfer rate equations [1]. The energy conservation equation for an exchanger having an arbitrary flow arrangement is

$$q = C_h\left(t_{h,i} - t_{h,o}\right) = C_c\left(t_{c,o} - t_{c,i}\right) \tag{2.1}$$

and the heat transfer rate equation is

$$q = UA\Delta t_m = \frac{\Delta t_m}{R_o} \tag{2.2}$$

where

- Δt_m is the true mean temperature difference (MTD), which depends upon the exchanger flow arrangement and the degree of fluid mixing within each fluid stream
- C_c is the capacity rate of the cold fluid, $(Mc_p)_c$
- C_h is the capacity rate of the hot fluid, $(Mc_p)_h$
- $t_{c,i}$ and $t_{c,o}$ are cold fluid terminal temperatures (inlet and outlet)
- $t_{h,i}$ and $t_{h,o}$ are hot fluid terminal temperatures (inlet and outlet)

The heat exchanger thermal circuit variables and overall conduction described here are based on Refs. [1,2].

The inverse of the overall thermal conductance UA is referred to as the overall thermal resistance R_o, and is made up of component resistances in series, as shown in Figure 2.1:

$$R_o = R_h + R_1 + R_w + R_2 + R_c \tag{2.3}$$

DOI: 10.1201/9781003352044-2

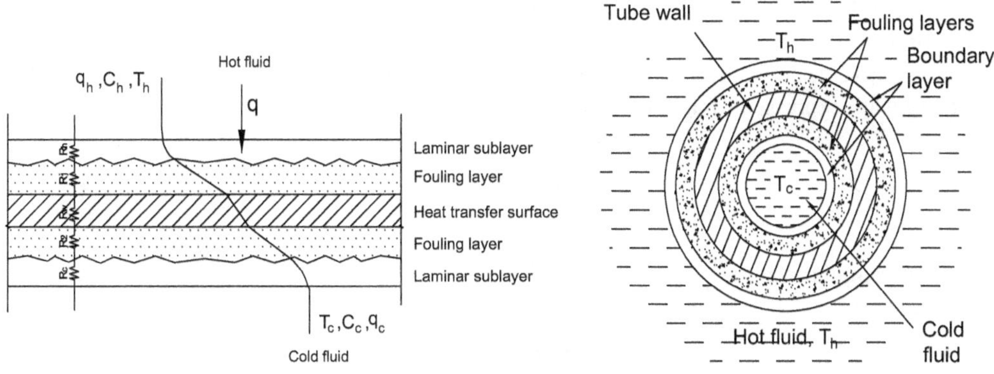

(a) Fouled heat transfer surface.

(b) Fouled tube surface.

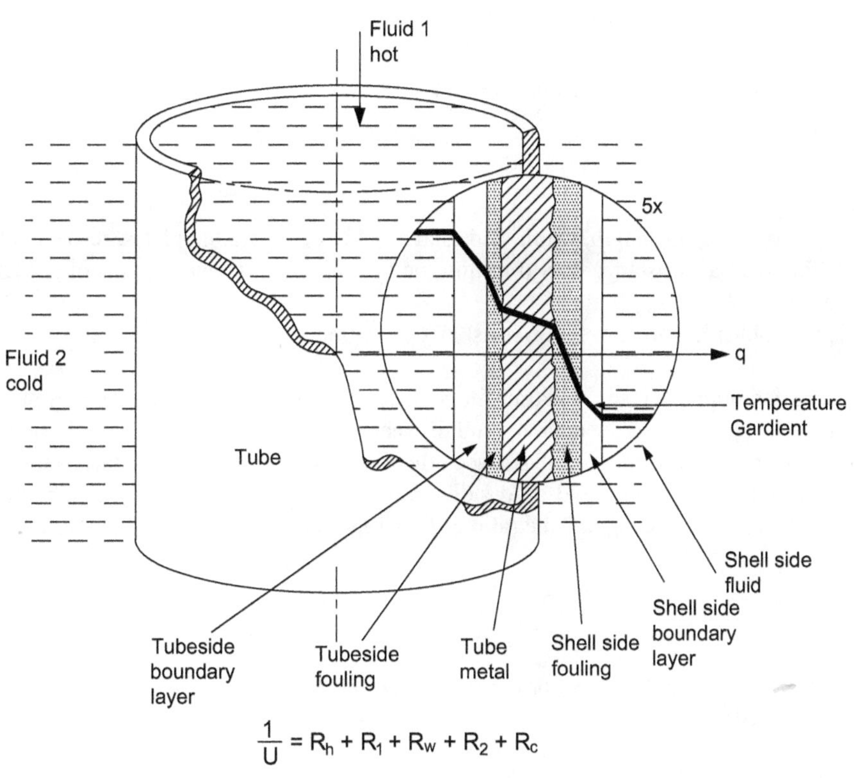

$$\frac{1}{U} = R_h + R_1 + R_w + R_2 + R_c$$

(c) Thermal resistances of fouled tube surface.

FIGURE 2.1 Elements of thermal resistance of a heat exchanger.

where the parameters of the right-hand side of Equation 2.3 are R_h, hot side film convection resistance, $1/(\eta_o hA)_h$; R_1, thermal resistance due to fouling on the hot side given in terms of fouling resistance $R_{f,h}$ (i.e., values tabulated in standards or textbooks), $R_{f,h}/(\eta_o A)_h$; R_w, thermal resistance of the separating wall, expressed for a flat wall by

$$R_w = \frac{\delta}{A_w k_w} \tag{2.4a}$$

and for a circular wall by

$$R_w = \frac{\ln(d/d_i)}{2\pi k_w L N_t} \tag{2.4b}$$

where

- δ is the wall thickness
- A_w is the total wall area for heat conduction
- k_w is the thermal conductivity of the wall material
- d is the tube outside diameter
- d_i is the tube inside diameter
- L is the tube length
- N_t is the number of tubes, and total wall area for heat conduction is given by

$$A_w = L_1 L_2 L_p \tag{2.5}$$

where

- L_1, L_2, and N_p are the length, width, and total number of separating plates, respectively
- R_2 is the thermal resistance due to fouling on the cold side, given in terms of cold side fouling resistance $R_{f,c}$ by $R_{f,c}/(\eta_o A)_c$
- R_c is the cold side film convection resistance, $1/(\eta_o hA)_c$

In these definitions, h is the heat transfer coefficient on the side under consideration, A represents the total of the primary surface area, A_p, and the secondary (finned) surface area, A_f, on the fluid side under consideration, η_o is the overall surface effectiveness of an extended surface, and the subscripts h and c refer to the hot and cold fluid sides, respectively. The overall surface effectiveness η_o is related to the fin efficiency η_f and the ratio of fin surface area A_f to total surface area A as follows:

$$\eta_o = 1 - \frac{A_f}{A}(1 - \eta_f) \tag{2.6}$$

Note that η_o is the unity for an all prime surface exchanger without fins. Equation 2.3 can be alternately expressed as

$$\frac{1}{UA} = \frac{1}{(\eta_o hA)_h} + \frac{R_{f,h}}{(\eta_o A)_h} + R_w + \frac{1}{(\eta_o hA)_c} + \frac{R_{f,c}}{(\eta_o A)_c} \tag{2.7}$$

Since $UA = U_h A_h = U_c A_c$, the overall heat transfer coefficient U as per Equation 2.7 may be defined optionally in terms of either hot fluid surface area or cold fluid surface area. Thus, the option of A_h or A_c must be specified in evaluating U from the product, UA. For plain tubular exchangers, U_o based on tube outside surface is given by

$$\frac{1}{U_o} = \frac{1}{h_o} + R_{f,o} + \frac{d\ln(d/d_i)}{2k_w} + \frac{R_{f,i}d}{d_i} + \frac{d}{h_i d_i} \tag{2.8}$$

The knowledge of wall temperature in a heat exchanger is essential to determine the localized hot spots, freeze points, thermal stresses, local fouling characteristics, or boiling and condensing coefficients. Based on the thermal circuit of Figure 2.1, when R_w is negligible, $T_{w,h} = T_{w,c} = T_w$ is computed from Refs. [1,2] as

$$T_w = \frac{T_h + T_c\left[(R_h + R_1)/(R_c + R_2)\right]}{1 + \left[(R_h + R_1)/(R_c + R_2)\right]} \tag{2.9}$$

When $R_1 = R_2 = 0$, Equation 2.9 further simplifies to

$$T_w = \frac{T_h/R_h + T_c/R_c}{1/R_h + 1/R_c} = \frac{(\eta_o hA)_h T_h + (\eta_o hA)_c T_c}{(\eta_o hA)_h + (\eta_o hA)_c} \tag{2.10}$$

2.2 HEAT EXCHANGER HEAT TRANSFER ANALYSIS METHODS

2.2.1 Energy Balance Equation

The first law of thermodynamics must be satisfied in any heat exchanger design procedure at both the macro- and microlevels. The overall energy balance for any two-fluid heat exchanger is given by

$$m_h c_{p,h}\left(t_{h,i} - t_{h,o}\right) = m_c C_{p,c}\left(t_{c,o} - t_{c,i}\right) \tag{2.11}$$

Equation 2.11 satisfies the "macro" energy balance under the usual idealizations made for the basic design theory of heat exchangers [3].

2.2.2 Heat Transfer

For any flow arrangement, heat transfer for two fluid streams is given by

$$q = C_h\left(t_{h,i} - t_{h,o}\right) = C_c\left(t_{c,o} - t_{c,i}\right) \tag{2.12}$$

and the expression for maximum possible heat transfer rate q_{max} is

$$q_{max} = C_{min}\left(t_{h,i} - t_{c,i}\right) \tag{2.13}$$

The maximum possible heat transfer rate would be obtained in a counterflow heat exchanger with very large surface area and zero longitudinal wall heat conduction, and the actual operating conditions are the same as the theoretical conditions.

2.3 BASIC METHODS TO CALCULATE THERMAL EFFECTIVENESS

There are four design methods to calculate the thermal effectiveness of heat exchangers:

1. ε-NTU method
2. P-NTU$_t$ method

3. LMTD method
4. ψ-*P* method

The basics of these methods are discussed next. For more details on these methods, refer to Refs. [1,2].

2.3.1 ε-NTU Method

The formal introduction of the ε-NTU method for the heat exchanger analysis was in 1942 by London and Seban [4]. In this method, the total heat transfer rate from the hot fluid to the cold fluid in the exchanger is expressed as

$$q = \varepsilon C_{\min}\left(t_{\mathrm{h,i}} - t_{\mathrm{c,i}}\right) \tag{2.14}$$

where ε is the heat exchanger effectiveness. It is nondimensional and for a direct transfer type heat exchanger, in general, it is dependent on NTU, C^*, and the flow arrangement:

$$\varepsilon = \phi\left(\mathrm{NTU}, C^*, \text{flow arrangement}\right) \tag{2.15}$$

These three nondimensional parameters, C^*, NTU, and ε, are defined next.

Heat capacity rate ratio, C^*: This is simply the ratio of the smaller to larger heat capacity rate for the two fluid streams so that $C^* \le 1$.

$$C^* = \frac{C_{\min}}{C_{\max}} = \frac{\left(mc_{\mathrm{p}}\right)_{\min}}{\left(mc_{\mathrm{p}}\right)_{\max}} \tag{2.16}$$

where

C refers to the product of mass and specific heat of the fluid

the subscripts min and max refer to the $C_{\min}$ and $C_{\max}$ sides, respectively

In a two-fluid heat exchanger, one of the streams will usually undergo a greater temperature change than the other. The first stream is said to be the “weak” stream, having a lower thermal capacity rate ($C_{\min}$), and the other with higher thermal capacity rate ($C_{\max}$) is the “strong” stream.

Number of transfer units, NTU: NTU designates the nondimensional “heat transfer size” or “thermal size” of the exchanger. It is defined as a ratio of the overall conductance to the smaller heat capacity rate:

$$\mathrm{NTU} = \frac{\mathrm{UA}}{\mathrm{C}_{\min}} = \frac{1}{\mathrm{C}_{\min}}\int_A U dA \tag{2.17}$$

If *U* is not a constant, the definition of the second equality applies. For constant *U*, substitution of the expression for *UA* results in [1,2]

$$\mathrm{NTU} = \frac{1}{C_{\min}}\left[\frac{1}{1/\left(\eta_{\mathrm{o}} hA\right)_{\mathrm{h}} + R_1 + R_{\mathrm{w}} + R_2 + 1/\left(\eta_{\mathrm{o}} hA\right)_{\mathrm{c}}}\right] \tag{2.18}$$

where R_1 and R_2 are the thermal resistances due to fouling on the hot side and cold side, respectively, as defined in Equation 2.7. In the absence of the fouling resistances, NTU can be given by the expression

$$\frac{1}{\mathrm{NTU}} = \frac{1}{\mathrm{NTU_h}\left(C_h / C_{min}\right)} + R_w C_{min} + \frac{1}{\mathrm{NTU_c}\left(C_c / C_{min}\right)} \tag{2.19}$$

and the number of heat transfer units on the hot and cold sides of the exchanger may be defined as follows:

$$\mathrm{NUT_h} = \frac{(\eta_o hA)_h}{C_h} \quad \mathrm{NTU_c} = \frac{(\eta_o hA)_c}{C_c} \tag{2.20}$$

Heat exchanger effectiveness, ε: Heat exchanger effectiveness, ε, is defined as the ratio of the actual heat transfer rate, q, to the thermodynamically possible maximum heat transfer rate (q_{max}) by the second law of thermodynamics:

$$\varepsilon = \frac{q}{q_{max}} \tag{2.21}$$

The value of ε ranges between 0 and 1. Using the value of actual heat transfer rate q from Equation 2.12 and q_{max} from Equation 2.13, the exchanger effectiveness ε of Equation 2.21 is given by

$$\varepsilon = \frac{C_h\left(t_{h,i} - t_{h,o}\right)}{C_{min}\left(t_{h,i} - t_{c,i}\right)} = \frac{C_c\left(t_{c,o} - t_{c,i}\right)}{C_{min}\left(t_{h,i} - t_{c,i}\right)}$$
$$\text{For} \quad C^* = 1, \quad \varepsilon_h = \varepsilon_c \tag{2.22}$$

Dependence of ε on NTU: At low NTU, the exchanger effectiveness is generally low. With increasing values of NTU, the exchanger effectiveness generally increases, and in the limit it approaches the maximum asymptotic value. However, there are exceptions such that after reaching a maximum value, the effectiveness decreases with increasing NTU.

2.3.2 *P*-NTU$_t$ Method

This method represents a variant of the ε-NTU method. The origin of this method is related to shell and tube exchangers. In the ε-NTU method, one has to keep track of the C_{min} fluid. In order to avoid possible errors, an alternative is to present the temperature effectiveness, P, of the fluid side under consideration as a function of NTU and heat capacity rate of that side to that of the other side, R. Somewhat arbitrarily, the side chosen is the tubeside regardless of whether it is the hot side or the cold side.

General P-NUT$_t$ functional relationship: Similar to the exchanger effectiveness ε, the thermal effectiveness P is a function of NTU$_t$, R, and flow arrangement:

$$P = \phi\left(\mathrm{NTU_t}, R, \text{ flow arrangement}\right) \tag{2.23}$$

where P, NTU$_t$, and R are defined consistently based on the tubeside fluid variables. In this method, the total heat transfer rate from the hot fluid to the cold fluid is expressed by

$$q = PC_t\left(T_1 - t_1\right) \tag{2.24}$$

Thermal effectiveness, P: For a shell and tube heat exchanger, the temperature effectiveness of the tubeside fluid, P, is referred to as the "thermal effectiveness." It is defined as the ratio of the temperature rise (drop) of the tubeside fluid (regardless of whether it is hot or cold fluid) to the difference of inlet temperature of the two fluids. According to this definition, P is given by

$$P = \frac{t_2 - t_1}{T_1 - t_1} \quad \left(P \text{ is referred to tubeside}\right) \tag{2.25}$$

where

t_1 and t_2 refer to tubeside inlet and outlet temperatures, respectively
T_1 and T_2 refer to shellside inlet and outlet temperatures, respectively

Comparing Equations 2.25 and 2.22, it is found that the thermal effectiveness P and the exchanger effectiveness ε are related as

$$P = \frac{C_{min}}{C_t}\varepsilon = \varepsilon \text{ for } C_t = C_{min}$$

$$= \varepsilon C^* \text{ for } C_t = C_{max} \tag{2.26}$$

Note that P is always less than or equal to ε. The thermal effectiveness of the shellside fluid can be determined from the tubeside values by the relationship given by

$$P_s = P\frac{C_t}{C_s} = PR$$
$$\text{For} \quad R^* = 1, \quad P_s = P \; (\text{tubeside}) \tag{2.27}$$

(For TEMA shell types, the thermal effectiveness charts given in this chapter, depicts thermal effectiveness referred to tubeside only.)

Heat capacity ratio, R: For a shell and tube exchanger, R is the ratio of the capacity rate of the tube fluid to the shell fluid. This definition gives rise to the following relation in terms of temperature drop (rise) of the shell fluid to the temperature rise (drop) of the tube fluid:

$$R = \frac{C_t}{C_s} = \frac{T_1 - T_2}{t_2 - t_1} \tag{2.28}$$

where the right-hand-side expressions come from an energy balance and indicate the temperature drop/rise ratios. The value of R ranges from zero to infinity, zero being for pure vapor condensation and infinity being for pure liquid evaporation. Comparing Equations 2.28 and 2.16, R and C^* are related by

$$R = \frac{C_t}{C_s} = C^* \text{ for } C_t = C_{min}$$

$$= \frac{1}{C^*} \text{ for } C_t = C_{max} \tag{2.29}$$

Thus R is always greater than or equal to C^*.

Number of transfer units, NTU_t: For a shell and tube exchanger, the number of transfer units NTU_t is defined as a ratio of the overall conductance to the tubeside fluid heat capacity rate:

$$NTU_t = \frac{UA}{C_t} \tag{2.30}$$

Thus, NTU_t is related to NTU based on C_{min} by

$$NTU_t = NTU\frac{C_{min}}{C_t} = NTU \text{ for } C_t = C_{min}$$

$$= NTUC^* \text{ for } C_t = C_{max} \tag{2.31}$$

Thus NTU_t is always less than or equal to NTU.

2.3.3 Log Mean Temperature Difference Correction Factor Method

The maximum driving force for heat transfer is always the log mean temperature difference (LMTD) when two fluid streams are in countercurrent flow. However, the overriding importance of other design factors causes most heat exchangers to be designed in flow patterns different from true countercurrent flow. The true MTD of such flow arrangements will differ from the logarithmic MTD by a certain factor dependent on the flow pattern and the terminal temperatures. This factor is usually designated as the log MTD correction factor, *F*. The factor *F* may be defined as the ratio of the true MTD to the logarithmic MTD. The heat transfer rate equation incorporating *F* is given by

$$q = UA\Delta t_m = UAF\Delta t_{lm} \tag{2.32}$$

where

Δt_m is the true MTD

Δt_{lm} is the LMTD

The expression for LMTD for a counterflow exchanger is given by

$$\text{LMTD} = \Delta t_{lm} = \frac{\Delta t_1 - \Delta t_2}{\ln(\Delta t_1/\Delta t_2)} \tag{2.33a}$$

where $\Delta t_1 = t_{h,i} - t_{c,o} = T_1 - t_2$ and $\Delta t_2 = t_{h,o} - t_{c,i} = T_2 - t_1$ for all flow arrangements except for parallelflow; for parallelflow $\Delta t_1 = t_{h,i} - t_{c,i}(=T_1 - t_1)$ and $\Delta t_2 = t_{h,o} - t_{c,o}(=T_2 - t_2)$. Therefore, LMTD can be represented in terms of the terminal temperatures, that is, greater terminal temperature difference (GTTD or GTD) and smaller terminal temperature difference (STTD or STD) for both pure parallel- and counterflow arrangements. Accordingly, LMTD is given by

$$\text{LMTD} = \Delta t_{lm} = \frac{\text{GTTD} - \text{STTD}}{\ln(\text{GTTD/STTD})} \tag{2.33b}$$

Note—Some literatures use GTD instead of GTTD and LTD instead of STTD—therefore they are equivalent.

The terminal temperature distribution to calculate LMTD is shown in Figure 2.2a–c.

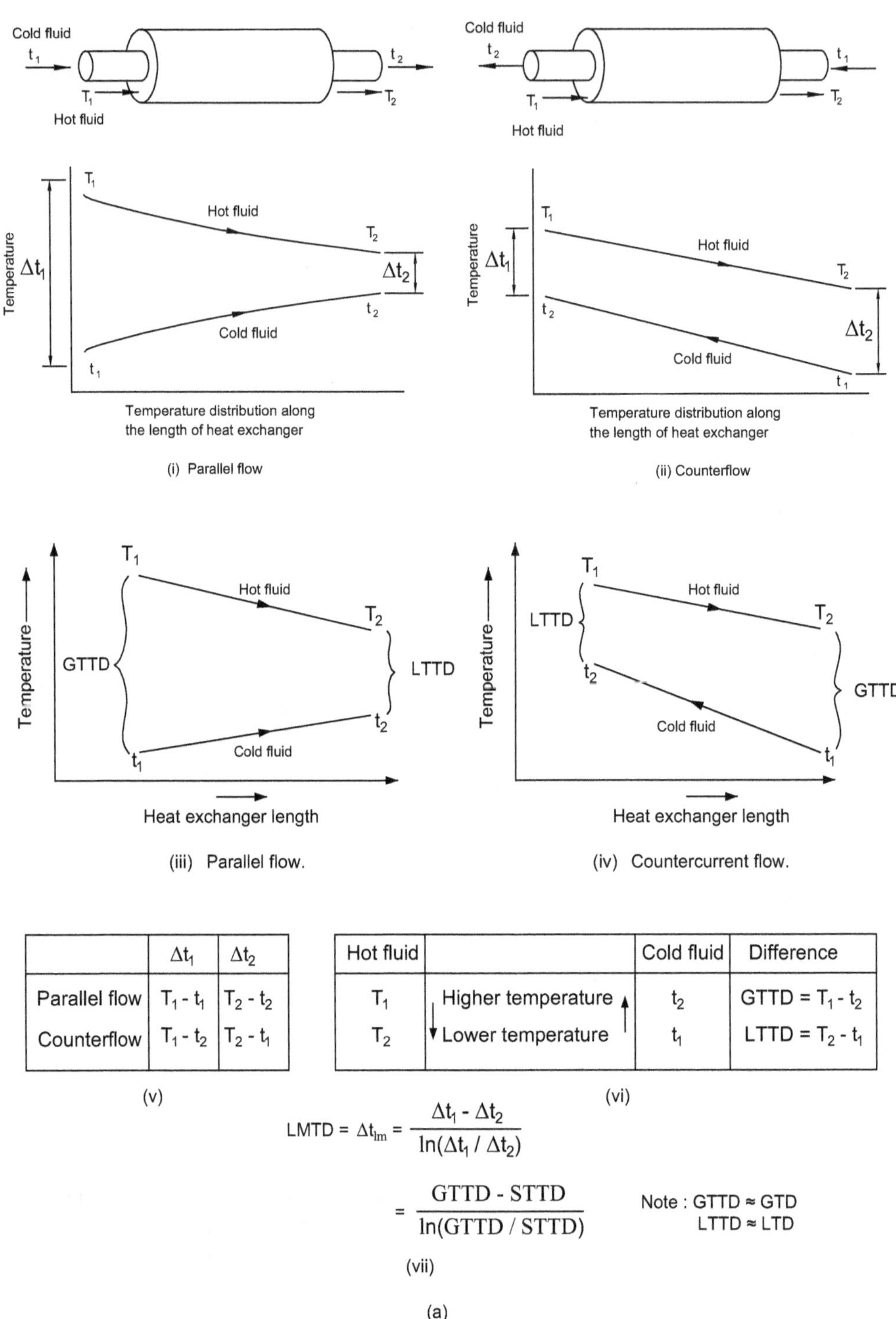

	Δt_1	Δt_2
Parallel flow	$T_1 - t_1$	$T_2 - t_2$
Counterflow	$T_1 - t_2$	$T_2 - t_1$

(v)

Hot fluid		Cold fluid	Difference
T_1	Higher temperature	t_2	GTTD = $T_1 - t_2$
T_2	Lower temperature	t_1	LTTD = $T_2 - t_1$

(vi)

$$\text{LMTD} = \Delta t_{lm} = \frac{\Delta t_1 - \Delta t_2}{\ln(\Delta t_1 / \Delta t_2)} = \frac{\text{GTTD} - \text{STTD}}{\ln(\text{GTTD}/\text{STTD})}$$

(vii)

FIGURE 2.2 LMTD- (a) Terminal temperatures to calculate LMTD; (b) nomogram to find LMTD. (Courtesy of Paul-Muller Company, Springfield, MO.) (*Note*: While referring the nomogram of Figure 2.2b, assume GTTD in place of GTD and STTD in place of STD.)

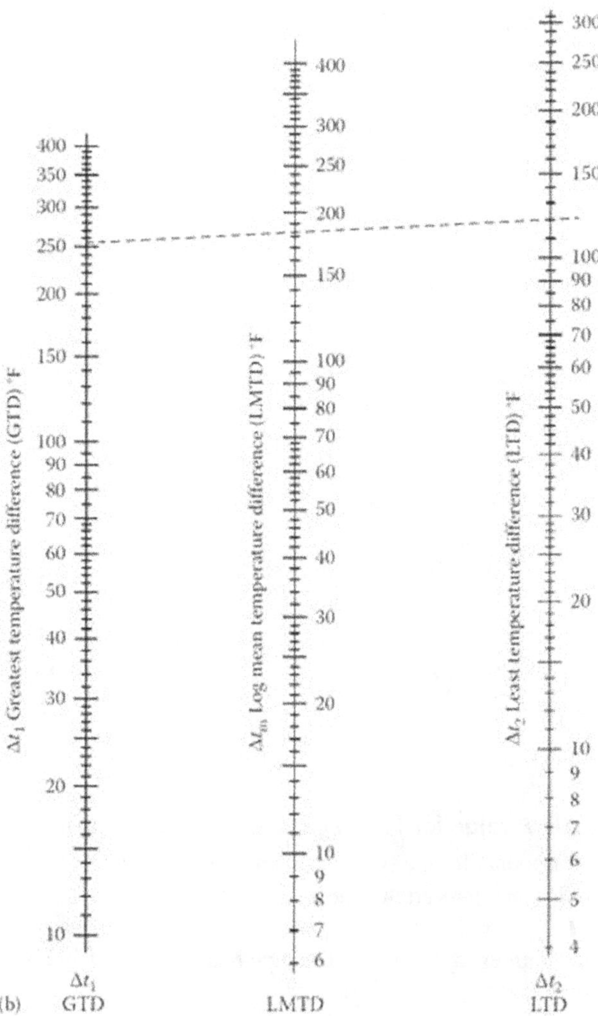

FIGURE 2.2 (Continued)

2.3.3.1 LMTD Correction Factor, *F*

Charts to determine LMTD from the terminal temperature differences are shown in Figure 2.2b. (Note: While referring to the nomogram of Figure 2.2b, assume GTTD in place of GTD and STTD in place of STD.)

From its definition, F is expressed by

$$F = \frac{\Delta t_{\mathrm{m}}}{\Delta t_{\mathrm{lm}}} \tag{2.34}$$

In situations where the heat release curves are nonlinear, the approach just described is not applicable and a "weighted" temperature difference must be determined.

It can be shown that, in general, F is dependent upon the thermal effectiveness P, the heat capacity rate ratio R, and the flow arrangement. Therefore, F is represented by

$$F = \phi\left(P, R, \mathrm{NTU_t}, \text{ flow arrangement}\right) \tag{2.35}$$

and the expression for *F* in terms of *P, R*, and NTU is given by

$$F = \frac{1}{(R-1)NTU}\ln\left[\frac{1-P}{1-PR}\right] \text{ for } R \neq 1$$
$$= \frac{P}{(1-P)NTU} \text{ for } R = 1 \quad (2.36a)$$

$$F = \frac{1}{(1-C^*)NTU}\ln\left[\frac{1-\varepsilon C^*}{1-\varepsilon}\right] \text{ for } C^* \neq 1$$

$$= \frac{\varepsilon}{(1-\varepsilon)NTU} \text{ for } C^* = 1 \quad (2.36b)$$

The factor *F* is dimensionless.

The value of *F* is unity for a true counterflow exchanger, and thus independent of *P* and *R*. For other arrangements, *F* is generally less than unity, and can be explicitly presented as a function of *P*, *R*, and NTU_t by Equation 2.36. The value of *F* close to unity does not mean a highly efficient heat exchanger, but it does mean a close approach to the counterflow behavior for the comparable operating conditions of flow rates and inlet fluid temperatures. Because of the large capital cost involved with a shell and tube exchanger, generally it is designed in the steep region of the P-NTU_t curve (ε-NTU relation for the compact heat exchanger) (ε or *P* < 60%), and as a rule of thumb, the *F* value selected is 0.80 and higher. However, a better guideline for F_{min} is provided in the next section. For more details on heat exchanger thermal design methods, refer to Shah and Sekulic [5] and Ref. [6].

2.3.3.2 Approximate *F* Value for Heat Exchanger Sizing Purpose

This correction factor accounts for the two streams not in counterflow. At the estimation stage, we do not know the detailed flow and pass arrangement so we can assume the following for preliminary sizing:

- *F* = 1.0 for true counterflow, e.g., double-pipe heat exchanger in counterflow arrangement, *F* shell type of shell and tube heat exchanger.
- *F* = 0.7 for crossflow heat exchanger.
- *F* = 0.7 for TEMA *E* shell with single pass on both shellside and tubeside.
- *F* = 0.80 for E_{1-2} shell and tube heat exchanger (refer to Figure 2.28).
- *F* = 0.95 for G_{1-2} (refer Figure 2.34), H_{1-2} shell and tube heat exchanger (refer to Figure 2.36).
- *F* = 0.79 for J_{1-2} shell and tube heat exchanger (refer to Figure 2.38).
- *F* = 0.9 for multi-pass compact heat exchanger and multiple passes on both shellside and tubeside of TEMA *E* shell.
- *F* = 1.0 if one stream is isothermal, $C^* = 0$, $R = 0$ or ∞ (typically boiling or condensation).

Applicability of ε-NTU and LMTD methods: Generally, the ε-NTU method is used for the design of compact heat exchangers. The LMTD method is used for the design of shell and tube heat exchangers. It should be emphasized that either method will yield identical results within the convergence tolerances specified.

2.3.4 ψ-*P* Method

The ψ-*P* method was originally proposed by Smith [7] and modified by Mueller [8]. In this method, a new term ψ is introduced, which is expressed as the ratio of the true MTD to the inlet temperature difference of the two fluids:

TABLE 2.1
General Functional Relationship between Dimensionless Groups of the ε-NTU, *P*-NTU$_t$, and LMTD Methods

Heat transfer parameters	ε-NTU method	*P*-NTU$_t$ method	LMTD method
Heat capacity rate ratio	$C^* = \frac{C_{min}}{C_{max}} = \frac{(mc_p)_{min}}{(mc_p)_{max}}$	$R = \frac{C_t}{C_s} = \frac{T_1 - T_2}{t_2 - t_1}$	$\text{LMTD} = \frac{\Delta t_1 - \Delta t_2}{\ln\left[\frac{\Delta t_1}{\Delta t_2}\right]}$
NTU	$\text{NTU} = \frac{UA}{C_{min}} = \frac{1}{C_{min}}\int_A \text{UdA}$	$\text{NTU}_t = \frac{UA}{C_t}$	LMTD = Δt_{lm} $F = \phi$ (*P*, *R*, NTU$_t$, flow arrangement)
Thermal effectiveness	$\varepsilon = \phi$(NTU, C^*, flow arrangement)	$P = \phi$(NTU$_t$, *R*, flow arrangement)	$F = \frac{\Delta t_m}{\Delta t_{lm}}$
	$\varepsilon = \frac{C_h(t_{h,i} - t_{h,o})}{C_{min}(t_{h,i} - t_{c,i})} = \frac{C_c(t_{c,o} - t_{c,i})}{C_{min}(t_{h,i} - t_{c,i})}$	$P = \frac{t_2 - t_1}{T_1 - t_1}$ $P_s = P\frac{C_t}{C_s} = PR$	$F = \frac{1}{(R-1)\text{NTU}}\ln\left[\frac{1-P}{1-PR}\right]$ for $R \neq 1$ $= \frac{P}{(1-P)\text{NTU}}$ for $R = 1$ $q = UA\Delta t_m = UAF\Delta t_{lm}$
Heat transfer	$q = C_h(t_{h,i} - t_{h,o}) = C_c(t_{c,o} - t_{c,i})$	$q = PC_t(T_1 - t_1)$	

$$\psi = \frac{\Delta t_m}{t_{h,i} - t_{c,i}} = \frac{\Delta t_m}{T_1 - t_1} \tag{2.37}$$

and ψ is related to ε and NTU and *P* and NTU$_t$ as

$$\psi = \frac{\varepsilon}{\text{NTU}} = \frac{P}{\text{NTU}_t} \tag{2.38}$$

and the heat transfer rate is given by

$$q = UA\psi(t_{h,i} - t_{c,i}) \tag{2.39a}$$

$$= UA\psi(T_1 - t_1) \tag{2.39b}$$

Since ψ represents the nondimensional Δt_m, there is no need to compute Δt_{lm} in this method.

Functional relationship between the various thermal design methods: The general functional relationship for the ε-NTU, *P*-NTU$_t$, LMTD, and ψ-*P* methods is shown in Table 2.1, which has been adapted and modified from Ref. [1], and the relationship between the dimensionless groups of these methods is given in Table 2.2.

Thermal design methods for the design of shell and tube heat exchangers: Any of the four methods (ε-NTU, *P*-NTU$_t$, LMTD, and ψ-*P*) can be used for shell and tube exchangers.

TABLE 2.2
Relationship between Dimensionless Groups of the ε-NTU, P-NTU$_t$, and LMTD Methods

$$R = \frac{C_t}{C_s} = C^* \quad \text{for } C_t = C_{\min}$$
$$= \frac{1}{C^*} \quad \text{for } C_t = C_{\max}$$

$$\text{NTU}_t = \text{NTU}\frac{C_{\min}}{C_t} = \text{NTU} \quad \text{for } C_t = C_{\min}$$
$$= \text{NTU}C^* \quad \text{for } C_t = C_{\max}$$

$$F = \frac{1}{(R-1)\text{NTU}}\ln\left[\frac{1-P}{1-PR}\right] \quad \text{for } R \neq 1$$
$$= \frac{P}{(1-P)\text{NTU}} \quad \text{for } R = 1$$

$$\Psi = \frac{\varepsilon}{\text{NTU}} = \frac{P}{\text{NTU}_t}$$

2.4 SOME FUNDAMENTAL RELATIONSHIPS TO CHARACTERIZE THE EXCHANGER FOR "SUBDESIGN" CONDITION

The partial derivatives of the temperature efficiency P with respect to NTU and R enable complete characterization of the exchanger performance around an operating point. Thus, the exchanger performance can be readily predicted for the "subdesign" conditions [9]. Singh [9] developed derivatives of P, F, and NTU. Derivatives for P and F are discussed next.

Dependence of thermal effectiveness: Thermal performance P and thermal effectiveness ε can be represented through R by [9]

$$\begin{aligned} \varepsilon &= \text{P} \qquad \text{R} \leq 1 \\ &= \text{PR} \qquad \text{R} > 1 \end{aligned} \tag{2.40}$$

$$P = f(\text{NTU}, R) \tag{2.41}$$

$$\varepsilon = \phi(NTU, R^*) \tag{2.42}$$

Thus,

$$d\varepsilon = d\phi = \frac{\partial \phi}{\partial \text{NTU}}d\text{NTU} + \frac{\partial \phi}{\partial R}dR \tag{2.43}$$

$$d\varepsilon = dP = \frac{\partial P}{\partial \text{NTU}}d\text{NTU} + \frac{\partial P}{\partial R}dR \quad \text{for } R \leq 1 \tag{2.44}$$

$$d\varepsilon = PdR + RdP \tag{2.45}$$

$$= PdR + R\left(\frac{\partial P}{\partial \text{NTU}}d\text{NTU} + \frac{\partial P}{\partial R}dR\right) \tag{2.46}$$

or

$$d\varepsilon = \left(P + R\frac{\partial P}{\partial R}\right)dR + R\frac{\partial P}{\partial \text{NTU}}d\text{NTU} \quad \text{for } R \geq 1 \tag{2.47}$$

Dependence of LMTD correction factor, F: The derivatives of F with respect to ε, P, and R are given by [9]

$$\frac{\partial F}{\partial \text{NTU}} = -\frac{1}{\text{NTU}^2 (R-1)}\ln\left[\frac{1-P}{1-PR}\right] = \frac{-F}{\text{NTU}} \tag{2.48}$$

$$\frac{\partial F}{\partial P} = \frac{1}{\text{NTU}(1-P)(1-PR)} \tag{2.49}$$

$$\frac{\partial F}{\partial R} = \frac{-F}{(R-1)} + \frac{1}{\text{NTU}(R-1)(1-PR)} \tag{2.50}$$

and

$$dF = \frac{\partial F}{\partial P}dP + \frac{\partial F}{\partial \text{NTU}}d\text{NTU} + \frac{\partial F}{\partial R}dR \tag{2.51}$$

2.5 THERMAL EFFECTIVENESS CHARTS

Broadly speaking, there are two types of heat exchanger problems: rating and sizing. To solve either type of problem from first principles is laborious and time-consuming. However, sizing and rating of heat exchangers are solved easily with the use of performance charts. The graphical charts were introduced many years ago and have gained wide acceptance throughout the industry. Five types of heat exchanger design charts are found in the literature, and the salient features of these charts have been discussed by Turton et al. [10]. These charts are shown schematically in Figures 2.3a–f. The dimensionless variables used in these charts (ε, P, R, C^*, F, NTU, NTU_t) have been defined in Section 2.2.

Figure 2.3a is the most widely used of these charts and was introduced by Bowman et al. [11] in 1940. In this chart, the LMTD correction factor, F, is presented as a function of the effectiveness, P, and the heat capacity rate ratio, R. Using this chart, the design problem where terminal temperatures and flow rates are usually specified but overall U and/or A are unknown can be solved; however, the rating problem can be solved by a trial-and-error solution. Since F compares the true MTD of a given flow arrangement with that of the counterflow arrangement, these charts provide a well-suited means of finding out the best of several possible flow arrangements. The one with the higher F will require the lower NTU, that is, the lower area if U remains constant, operating with the same R and P. Underwood [12] first derived the expression for true MTD for $E_{1\text{-}2}$, $E_{1\text{-}4}$, and $E_{2\text{-}4}$ shell and tube exchangers in 1934. Bowman et al. [11] published a summary of correction factors for exchangers of different flow arrangements. Ten Broeck [13] further constructed charts using dimensionless groups, $UA/(mc_p)_t$, $P = (t_2 - t_1)/(T_1 - t_1)$, and $R = (T_1 - T_2)/(t_2 - t_1)$ for direct calculation of terminal temperatures with known surface area of a heat exchanger. At present, F charts are available for all TEMA shells.

Figures 2.3b and c are due to Kays and London [14] and TEMA [15], respectively. Figure 2.3c is plotted on a semilog paper; generally the applicable NTU range for compact heat exchangers and shell and tube exchangers is about 0.2–3.0. A careful look at the linear graphical presentation of the ε-NTU results of Figure 2.3b indicates that the NTU scale in this range is short and hence one

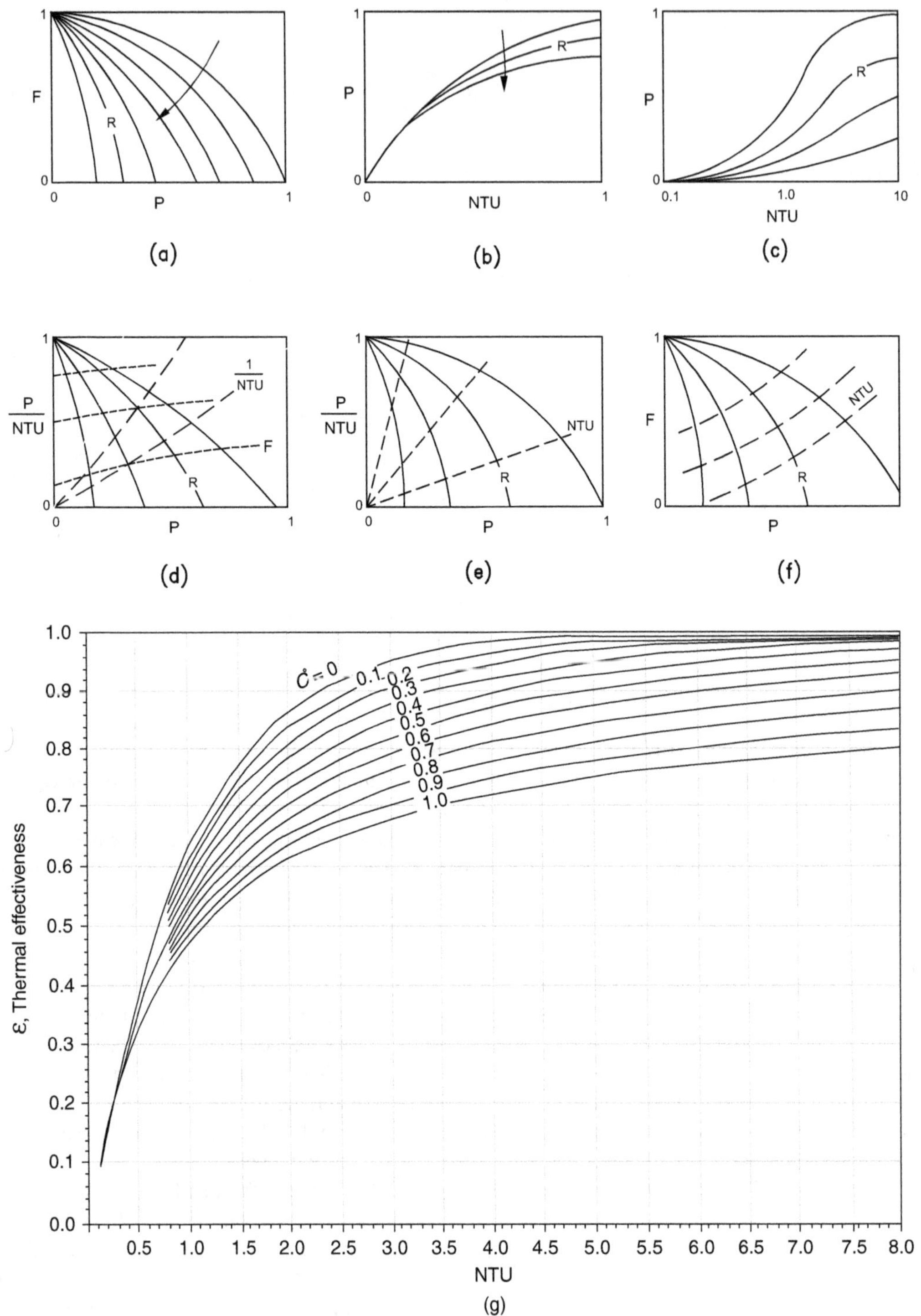

FIGURE 2.3 Thermal effectiveness charts. (a) Bowman chart; (b) Kays and London chart; (c) TEMA chart; (d) *F–P–R*-chart; (e) ψ chart; (f) *F-P-R*-NTU chart (From Turton, R. et al., *Trans. ASME J. Heat Transfer*, 106, 893, 1984); (g) ε-NTU chart for unmixed–unmixed crossflow, as per Eq. T4 of Table 2.4.

cannot obtain the accurate values of ε or NTU from graphs. For better appreciation, this is illustrated through the thermal relation chart (ε-NTU) for a crossflow heat exchanger in Figure 2.3. An alternative is to stretch the NTU scale in the range 0.2–3.0 by using a logarithmic scale. Thus, the P-NTU_t results are generally presented on a semilog paper, as shown for example in Figure 2.3c, in order to obtain more accurate graphical values of P or NTU_t. Using these charts, both the sizing and rating problem can be solved. However, the LMTD correction factor F is not shown in these charts. Hence, it is to be calculated additionally.

Muller [8] proposed the chart in Figure 2.3d with its triple family of curves. This chart can be used to solve both the sizing and rating problems and in addition gives the F values. However, Figure 2.3d is somewhat cramped and difficult to read accurately and introduces yet another parameter, P/NTU_t. The Muller charts have been redrawn recently by Taborek and included in HEDH [16]. The present form of this chart is shown in Figure 2.3e. The main difference between Figures 2.3d and e is that the F parameter curves have been omitted in the latter, and thus the problem of having to separately calculate the F values has been retained.

In a system with four variables, F, P, R, and NTU or NTU_t, any chart displays just one family of curves, such as Figures 2.3a–c, and does not give all the interrelationships directly. On the other hand, a chart with three families of curves, as in Figure 2.3d, has one set that is redundant. To show all the interrelationships between these four variables requires a chart with two families of curves. This is satisfied by Figure 2.3e.

In the graphical presentation, ψ is plotted against P and R as a parameter as shown in Figure 2.3e. The lines of constant R originate at $\psi = 1$ and terminate at $\psi = 0$ so that the asymptotic values of P for NTU tend to infinity. Thus the curves of constant R are similar to those for the F-P charts. In order to tie in with the P-NTU_t and LMTD methods, the lines of constant NTU_t and constant F are also superimposed on this chart. Figure 2.3e also has one limitation: It does not show directly the four parameters of interest.

Constraints due to the charts in Figures 2.3a–e are overcome by a chart, shown in Figure 2.3f, proposed by Turton et al. [10]. The chart in Figure 2.3f extends the easy-to-read Bowman charts of Figure 2.3a to include a second family of curves representing the variable NTU. Both the sizing and rating problems can be solved using this form of chart, and F values can be found directly for both types of problems. Thus to find the exchanger surface area, use P and R to evaluate F and NTU. To find the terminal temperature, use NTU and R to evaluate P and F. Most of the charts included in this book are of the type in Figure 2.3f. Figure 2.2(g) shows thermal effectiveness chart for a crossflow arrangement.

2.6 SYMMETRY PROPERTY AND FLOW REVERSIBILITY AND RELATION BETWEEN THE THERMAL EFFECTIVENESS OF OVERALL PARALLEL AND COUNTERFLOW HEAT EXCHANGER GEOMETRIES

2.6.1 Symmetry Property

The symmetry property relates the thermal behavior of a heat exchange process to that of the reverse process, in which the directions of flow of both fluids are reversed [17]. Figure 2.4 shows four different flow arrangements for the TEMA E_{1-2} shell and tube heat exchanger that are equivalent if complete transverse mixing of the shell fluid is satisfied.

2.6.2 Flow Reversibility

Flow reversibility establishes a relation between the thermal effectiveness of two heat exchanger configurations that differ from each other in the inversion of either one of the two fluids [18]. Although the inversion of both fluids often does not alter the configuration, the inversion of only one of them usually leads from one configuration to an entirely different one, as is the case in going from a pure parallelflow to a pure counterflow arrangement or vice versa. Using this relation, if the expression for the effectiveness, P, of a configuration as a function of the heat capacity rate ratio, R (or C^*), and the number of heat transfer units NTU is known, the corresponding expression for the "inverse" configuration is immediately obtained from the simple relation [18]:

$$P_i(R,\text{NTU}) = \frac{P(-R,\text{NTU})}{1+RP(-R,\text{NTU})} \tag{2.52}$$

where P denotes the effectiveness of a given arrangement, and P_i, that of the same one with fluid direction reversed. The relation is valid under the assumptions of temperature independence of the heat transfer coefficient and heat capacity rates, when one of the fluids proceeds through the exchanger in a single, mixed stream. In some cases with special symmetry, the inversion of both fluids does not alter the geometry, and therefore this property is trivially satisfied. Pignotti [18] illustrates the property of flow reversibility with several examples from the available literature. An example to clarify the meaning of Equation 2.52 is given next. Consider the well-known expression for the effectiveness of a parallelflow configuration:

$$P(R,\text{NTU}) = \frac{\left[1-\exp\left[-\text{NTU}(1+R)\right]\right]}{(1+R)} \tag{2.53}$$

Let us derive from it the expression for the effectiveness of a pure counterflow configuration, which we denote as $P_c(R, \text{NTU})$. Equation 2.52 is applicable, because the counterflow geometry is obtained from parallelflow by inverting the direction of flow of one of the fluids, and the condition that at least one of the fluids should be mixed throughout the exchanger is satisfied. After replacing R by $-R$ in Equation 2.53 and performing the elementary algebraic operations indicated in Equation 2.52, we obtain the expression for the effectiveness of the counterflow configuration:

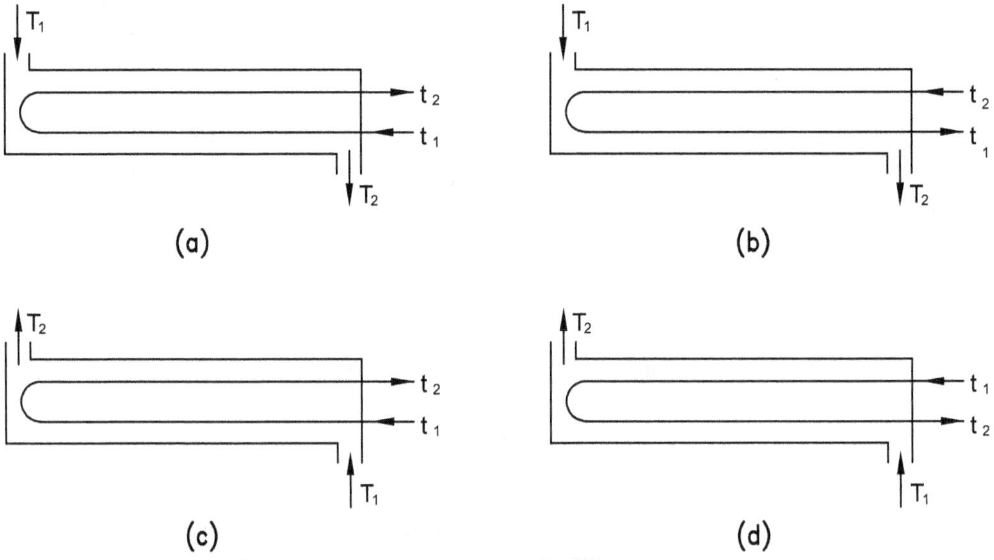

FIGURE 2.4 Flow reversibility principle. (a) Basic $E_{1\text{-}2}$ case, (b) basic case with tube fluid reversed, (c) basic case with shell fluid reversed, and (d) basic case with both shell and tube fluids reversed. (Symmetry operations performed on the TEMA $E_{1\text{-}2}$ shell.) (From Pignotti, A., *Trans. ASME, J. Heat Transfer*, 106, 361, 1984.)

$$P_c(R,\mathrm{NTU}) = \frac{\{1-\exp[-\mathrm{NTU}(1-R)]\}}{\{1-R\exp[-\mathrm{NTU}(1-R)]\}} \tag{2.54}$$

Observe also that the inversion of one fluid leads from a parallelflow connection to a counterflow one, and likewise, from the latter to the former; therefore, Equation 2.52 can be used to go from parallelflow to counterflow and vice versa.

The transformation property of Equation 2.52 can also be expressed in terms of the variables referred to the mixed fluid. For example, if the thermal relation on the shellside or tubeside is known in terms of P_x, R_x, and NTU_x, the thermal relation for the other side P_y, R_y, and NTU_y may be obtained from the relation

$$P_y = R_x P_x, \quad R_y = \frac{1}{R_x}, \quad \mathrm{NTU}_y = R_x \mathrm{NTU}_x \tag{2.55}$$

For example, let the tubeside values of an H_{1-2} exchanger be $P = 0.752$, $R = 0.7$, and NTU = 2.5. Then the shellside values will be $P = 0.7 \times 0.752$, $R = 1/0.7$, and NTU $= 0.7 \times 2.5$. For $R = 1.0$, both the tubeside and shellside values are the same.

When the thermal effectiveness is the same for the original case and the inverted case, it is referred to as stream symmetric. Typical examples for stream symmetric are parallelflow, counterflow, and crossflow unmixed–unmixed and mixed–mixed cases.

2.7 TEMPERATURE APPROACH, TEMPERATURE MEET, AND TEMPERATURE CROSS

The meanings of temperature approach, temperature meet, and temperature cross are as follows. Temperature approach is the difference between the hotside and coldside fluid temperatures at any point of a given exchanger. In a counterflow exchanger or a multipass exchanger, (1) if the cold fluid outlet temperature $t_{c.o}$ is less than the hot fluid outlet temperature $t_{h.o}$, then this condition is referred to as temperature approach; (2) if $t_{c.o} = t_{h,o}$, this condition is referred to as temperature meet; and (3) if $t_{c.o}$ is greater than $t_{h.o}$, the difference $(t_{c.o} - t_{h.o})$ is referred to as the temperature cross or temperature pinch. In this case, the temperature approach $(t_{h.o} - t_{c.o})$ is negative and loses its meaning. Temperature cross indicates a negative driving force for heat transfer between the fluids. It requires either a large area for heat transfer or the fluid velocity to increase the overall heat transfer coefficient. The underlying meanings of these three cases are brought out in Table 2.3 and the same are shown in Figure 2.5a.

The temperature cross is undesirable, particularly for shell and tube exchangers, because the tube surface area is not utilized effectively and hence there is wastage of capital cost. If outlet temperatures form a temperature cross in a multiple tube pass heat exchanger, a lower than desirable LMTD correction factor will occur. A simple way to avoid this is to use more exchanger shells in

TABLE 2.3
Temperature Approach, Temperature Meet, and Temperature Cross

Temperature approach	Temperature meet	Temperature cross
$t_{h,i} \rightarrow t_{h,o}$	$t_{h,i} \rightarrow t_{h,o}$	$t_{h,i} \rightarrow t_{h,o}$
$t_{c,o} \leftarrow t_{c,i}$	$t_{c,o} \leftarrow t_{c,i}$	$t_{c,o} \leftarrow t_{c,i}$
$t_{c,o} < t_{h,o}$	$t_{c,o} = t_{h,o}$	$t_{c,o} > t_{h,o}$

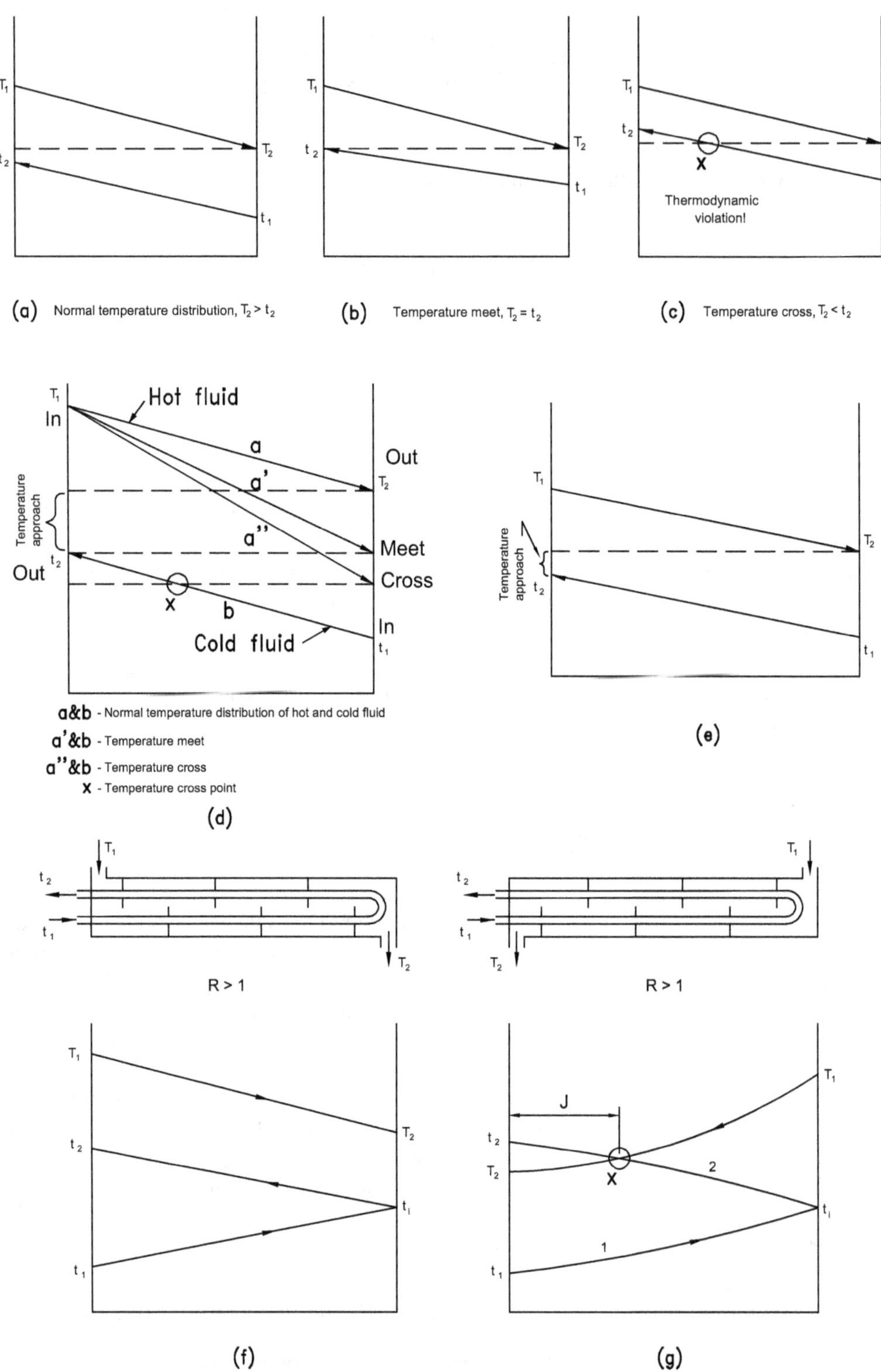

FIGURE 2.5 Principle of temperature approach, temperature meet, and temperature cross. (a) Normal temperature distribution; (b) temperature meet; (c) temperature cross; (d and e) temperature approach, meet, and cross superimposed; (f) temperature distribution in an E_{1-2} exchanger without temperature cross; and (g) temperature distribution in an E_{1-2} exchanger with temperature cross.

series. Other engineers suggest that a small temperature cross may be acceptable and may provide a less expensive design than the more complex alternatives. For a E_{1-2} heat exchanger, the temperature cross occurs around a relatively narrow range of F value about 0.78–0.82. Lower values of F may be taken as an indication that temperature cross will occur.

The concept of the temperature cross or meet at the exchanger outlet can be utilized to determine the number of shells in series required to meet the heat duty without having a temperature cross in any individual shell. Temperature cross is undesirable for a shell and tube heat exchanger because the tube surface area is not utilized cost-effectively. An optimum design would mean that the temperature cross or meet point lies just at the end of the second tube pass. This phenomenon is explained in detail by Shah [2] and is briefly dealt with here with reference to an E_{1-2} exchanger.

For E_{1-2} two possible shell fluid directions with respect to the tube fluid direction are shown in Figure 2.5b. The temperature distributions of Figure 2.5b reveal that there is a temperature cross. In region X, the second tube pass transfers heat to the shell fluid. This is contrary to the design objective, in which ideally the heat transfer should have taken place only in one direction (from the shell fluid to the tube fluid, as shown in Figure 2.5a) throughout the two passes. The reason for this temperature cross is as follows: Although an addition of surface area (a high value of NTU_t, or a low value of LMTD correction factor F) is effective in raising the temperature of the tube fluid and rises in the second pass up to point X, beyond this point the temperature of the shell fluid is lower than that of the tube fluid, since we have considered the shell fluid mixed at a cross section and it is cooled rapidly by the first pass. Thus, the addition of the surface area in the second tube pass left of point X is useless from the thermal design point of view. A "good" design avoids the temperature cross in a shell and tube exchanger. Theoretically, the optimum design would have the temperature cross point just at the end of the second tube pass, which will satisfy the following condition:

$$t_{t,o} = t_{s,o} \quad \text{or} \quad t_{t,o} - t_{s,o} = 0 \tag{2.56}$$

This condition leads to the following formula:

$$P = \frac{1}{1+R} \tag{2.57}$$

Thus for a given R, Equation 2.57 provides the limiting (maximum) value of P. Corresponding to P and R, the limiting (maximum) value of NTU_t beyond which there will be a temperature cross can be determined from its thermal relation formula. Therefore, from P, R, and NTU, F can be calculated. This F value is known as the F_{min} value beyond which there will be a temperature cross. This is illustrated for an E_{1-2} exchanger here. For a known value of R, determine the limiting value of P from Equation 2.57 and NTU from the following equation:

$$NTU_{E_{1-2}} = \frac{1}{\left(1+R^2\right)^{0.5}} \ln\left[\frac{2-P\left[R+1-\left(1+R^2\right)^{0.5}\right]}{2-P\left[R+1+\left(1+R^2\right)^{0.5}\right]}\right] \tag{2.58}$$

For known values of P, R, and NTU, determine F from Equation 2.36.

2.7.1 Temperature Cross for Other TEMA Shells

Temperature cross for other TEMA shells such as G_{1-2}, H_{1-2}, and J_{1-2} can be evaluated from Equation 2.57 [19]. The F_{min} curves for G_{1-2}, H_{1-2}, and J_{1-2} cases are given in the next section.

2.8 THERMAL RELATION FORMULAS FOR VARIOUS FLOW ARRANGEMENTS AND PASS ARRANGEMENTS

The heat exchanger effectiveness is defined as the ratio of the overall temperature drop of the weaker stream to the maximum possible temperature difference between the fluid inlet temperatures. The following assumptions are commonly made in deriving thermal effectiveness:

1. The overall heat transfer coefficient is constant throughout the exchanger.
2. Each pass has the same heat transfer area; that is, unsymmetrical pass arrangements are not considered.
3. There is no phase change.
4. The specific heat of each fluid is constant and independent of temperature.
5. The flow rates of both streams are steady.
6. The flow of both fluids is evenly distributed over both the local and the total transfer areas.
7. Heat losses from the system are negligible.

In this section, thermal relation formulas for (1) various flow arrangements—parallelflow, counterflow, and crossflow—(2) various types of heat exchangers—compact and shell and tube—and (3) multipass arrangements or multiple units of both compact and shell and tube heat exchangers are presented. Most of the formulas are tabulated and the thermal effectiveness charts are given. Mostly counterflow arrangements are considered. For shell and tube exchangers, formulas are given for both parallelflow and counterflow, but thermal effectiveness charts are given only for counterflow arrangements referred to tubeside (similar to TEMA standards [15]). For stream symmetric cases, thermal effectiveness relations referred to the shellside can be derived from the "flow reversibility" principle. From counterflow thermal effectiveness relations, thermal effectiveness relations for parallelflow arrangements can be easily derived (for stream symmetric cases only) from the "flow reversibility" principle. Customarily, the ε-NTU method is employed for compact heat exchangers. In this method, the capacity ratio C^* is always ≤ 1. Hence, thermal effectiveness charts are given in terms of ε–C^*–NTU, and wherever possible, the thermal effectiveness charts are also given in terms of P–R–F–NTU, instead of ε–C^*–NTU.

2.8.1 Parallelflow

For a given set of values of C^* or R, and NTU, (1) the thermal effectiveness is much lower for parallelflow than for counterflow arrangement, except in the limiting case $C^* = R = 0$, where it is the same for both cases and approaches unity as NTU increases to infinity, and (2) at a given value of NTU, the effectiveness increases with decreasing capacity ratio, C^* or R. The formula for thermal effectiveness is given by Equation T1 in Table 2.4, and the thermal effectiveness chart is given in Figure 2.6.

2.8.2 Counterflow

Among the various flow arrangements, counterflow has the highest thermal effectiveness. For counterflow exchangers, at a given value of NTU, the effectiveness increases with decreasing capacity ratio, C^* or R. The formula for thermal effectiveness is given by Equation T2, and the thermal effectiveness chart is given in Figure 2.7.

2.8.3 Crossflow Arrangement

2.8.3.1 Unmixed–Unmixed Crossflow

This is an industrially important arrangement representing the case of a large number of unmixed channels in both sides. The original solution was due to Nusselt [20] and was later reformulated into

TABLE 2.4
Thermal Effectiveness Relations for Basic Cases

Flow arrangement	Equation no./ reference/ Figure no.	General formula	Value for R = 1 and special cases
Parallelflow; stream symmetric	T1 Fig. 2.6	$P=\dfrac{1-e^{-\mathrm{NTU}(1+R)}}{1+R}$	$P=\dfrac{1}{2}\left[1-e^{(-2\mathrm{NTU})}\right]$ for $R=1$ $=1-e^{-\mathrm{NTU}}$ for $R=0$ $P_{max}=50\%$ for $R=1$
Counterflow; stream symmetric	T2 Fig. 2.7	$P=\dfrac{1-e^{-\mathrm{NTU}(1-R)}}{1-Re^{-\mathrm{NTU}(1-R)}}$	$P=\dfrac{\mathrm{NTU}}{1+\mathrm{NTU}}$ for $R=1$ $=1-e^{-NTU}$ for $R=0$
	T3 [21] Fig. 2.8 Fig. 2.9	$P=\dfrac{1}{R\mathrm{NTU}}\sum_{k=0}^{\infty}\left\{\left[1-e^{-\mathrm{NTU}}\sum_{m=0}^{k}\dfrac{\mathrm{NTU}^m}{m!}\right]\left[1-e^{-R\mathrm{NTU}}\sum_{m=0}^{k}\dfrac{(R\mathrm{NTU})^m}{m!}\right]\right\}$ For R = 1, this equation holds.	
Crossflow; both the fluids unmixed; stream symmetric	T4 [22] Fig. 2.4(g)	$\varepsilon=1-e^{[-(1+C^*)\mathrm{NTU}]}\left[I_0\left(2\mathrm{NTU}\sqrt{C^*}\right)+\sqrt{C^*}I_1\left(2\mathrm{NTU}\sqrt{C^*}\right)-\dfrac{1-C^*}{C^*}\sum_{n=2}^{\infty}C^{*n/2}I_n\left(2\mathrm{NTU}\sqrt{C^*}\right)\right]$ For C^* = 1. $\varepsilon=1-e^{-2\mathrm{NTU}}\left[I_0(2\mathrm{NTU})+I_1(2\mathrm{NTU})\right]$	
	T5 [23] Fig. 2.10	$\varepsilon=1-\exp\left\{\dfrac{\mathrm{NTU}^{0.22}}{C^*}\left[\exp\left(-C^*\mathrm{NTU}^{0.78}\right)-1\right]\right\}$	
Crossflow; one fluid mixed and the other fluid unmixed (1) weaker (C_{min}) fluid mixed; (2) stronger (C_{max}) fluid mixed	T6	Weaker (C_{min}) fluid mixed $P_1=[1-\exp(-K/R)]$ $K=1-\exp(-R\mathrm{NTU})$	$P_1=1-e^{-\left(1-e^{-\mathrm{NTU}}\right)}$
	T7 Fig. 2.11	Stronger (C_{max}) fluid mixed $P_1=\left[1-\exp(-KR)\right]/R$ $K=1-\exp(-R\mathrm{NTU})$	$P_1=1-e^{-\left(1-e^{-\mathrm{NTU}}\right)}$
Crossflow; mixed–mixed flow; stream symmetric same as $J_{1-\infty}$	T8	$P=\dfrac{1}{\left(\dfrac{1}{K_1}+\dfrac{R}{K_2}-\dfrac{1}{\mathrm{NTU}}\right)}$ $K_1=1-e^{(-\mathrm{NTU})}$ $K_2=1-e^{(-R\mathrm{NTU})}$	$P=\dfrac{1}{\dfrac{2}{K_1}-\dfrac{1}{\mathrm{NTU}}}$

Note: P_2 can be found using Equation 2.55 or $P_2 = P_1R_1$, $R_2 = 1/R_1$, $\mathrm{NTU}_2 = R_1\mathrm{NTU}_1$
I_0, I_1 and I_n are modified Bessel functions of the first kind.

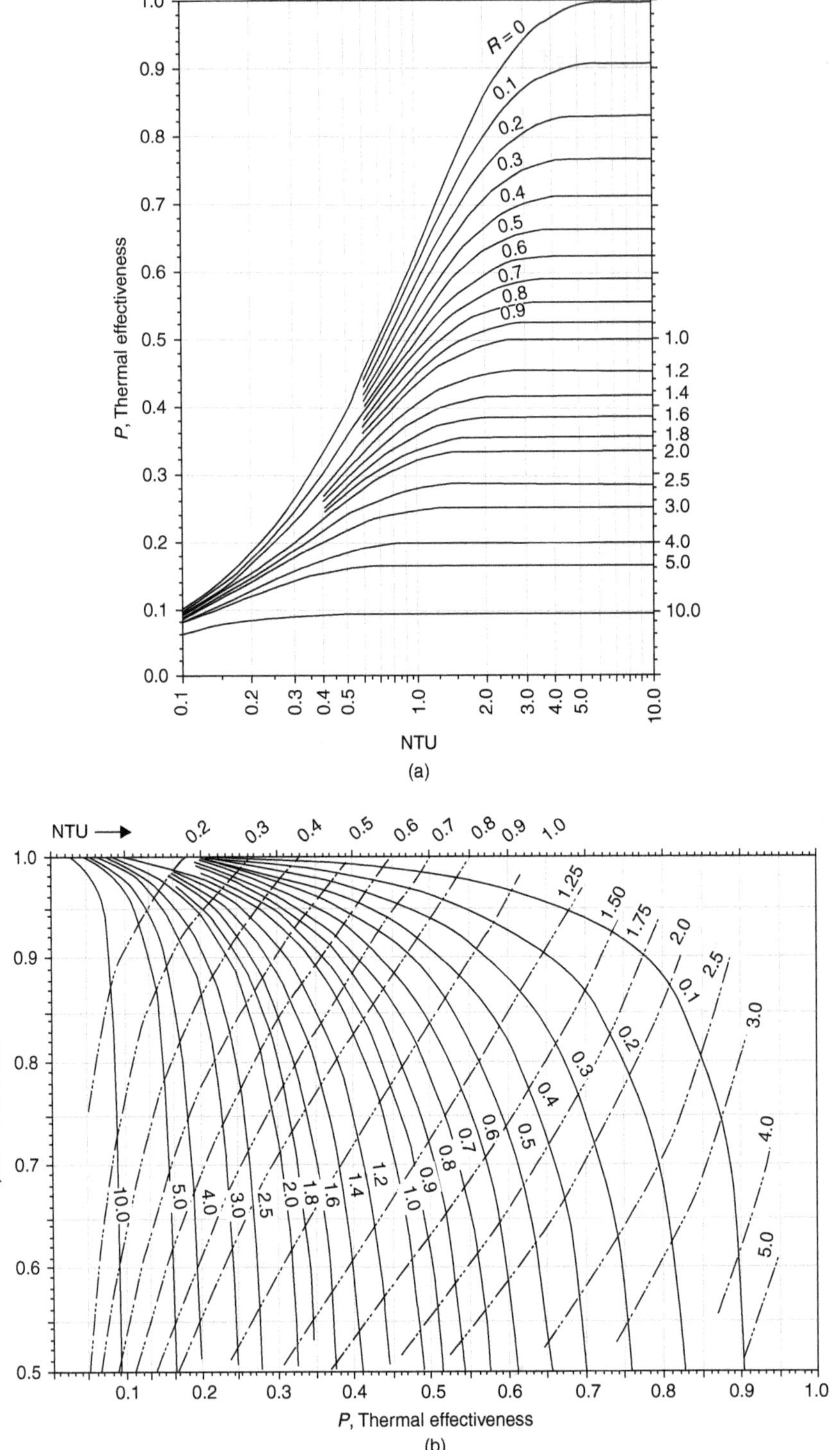

FIGURE 2.6 (a) Thermal effectiveness chart—parallelflow; stream symmetric, R–P–NTU chart (as per Equation T1, Table 2.4); (b) parallelflow; stream symmetric, F–R–P–NTU chart; F as a function of P for constant R (solid lines) and constant NTU (dashed lines) (Equation T1, Table 2.4).

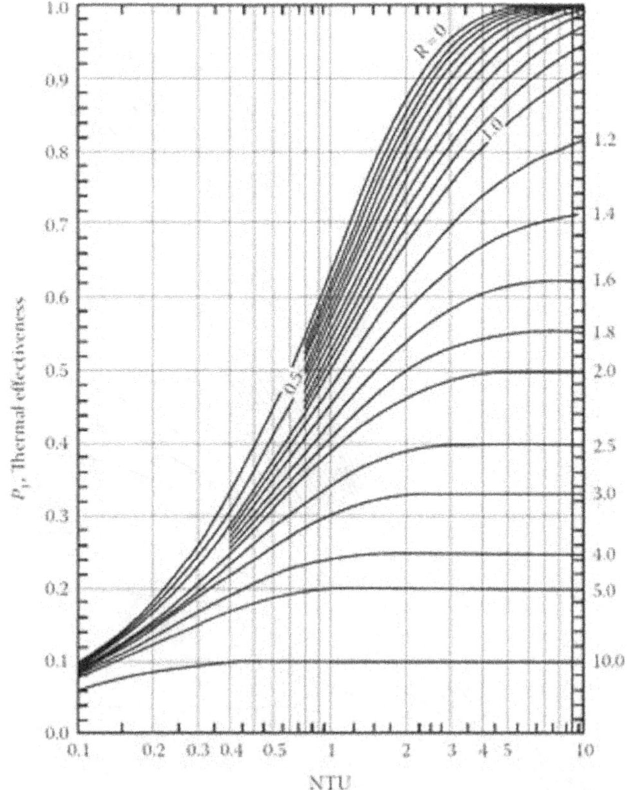

FIGURE 2.7 Thermal effectiveness chart—counterflow; stream symmetric, R–P–NTU chart (as per Equation T2, Table 2.4).

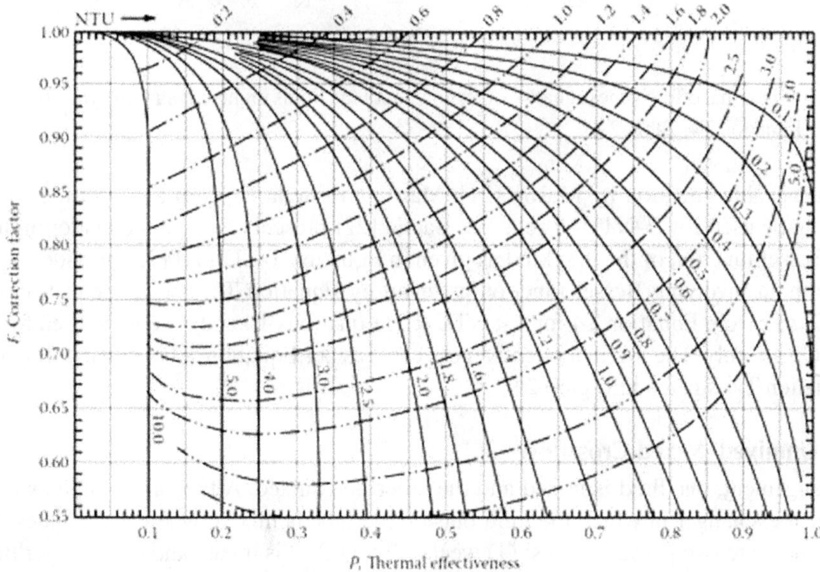

FIGURE 2.8 Thermal effectiveness chart—crossflow; both the fluids unmixed; stream symmetric; F–R–P–NTU chart; F as a function of P for constant R (solid lines) and constant NTU (dashed lines) (as per Equation T3, Table 2.4).

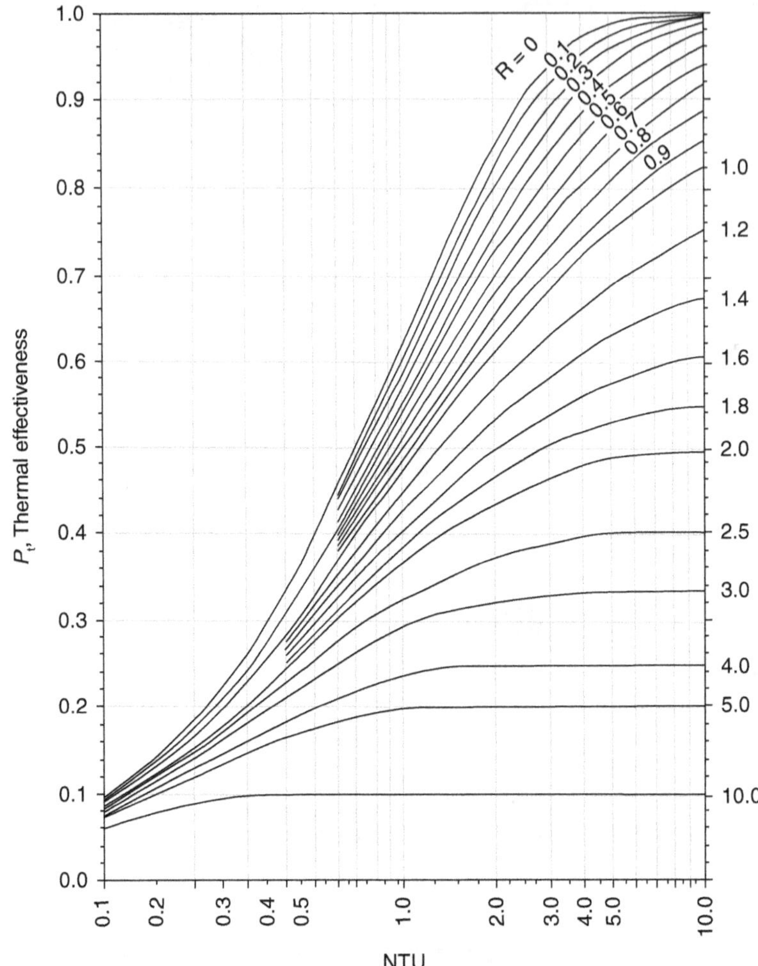

FIGURE 2.9 Thermal effectiveness chart—crossflow; both the fluids unmixed; stream symmetric; *R–P–*NTU chart (as per Equation T4, Table 2.4).

a more manageable equation by Mason [21]. Mason's formula is given by Equation T3, and this equation can be used for *P*–NTU–*R* relation. Baclic [22] presents Nusselt's equation in terms of a modified Bessel function of the first kind as given in Equation T4; Eckert [23] provides a simplified formula without involving Bessel function as given by Equation T5, and this equation predicts ε within ±1% of ε from Equation T4 for $1 < \text{NTU} < 7$; Equations T4 and T5 can be used for formulas involving $C^* \leq 1$ only. The thermal effectiveness chart as per Equation T3 is given in Figure 2.8 and as per Equation T4 is given in Figure 2.9.

2.8.3.2 Unmixed–Mixed Crossflow

In this arrangement, one fluid is mixed and the other is unmixed. A typical example is a bare tube compact heat exchanger in which the fluid outside the tube is mixed, whereas the tubeside fluid is unmixed. There are two possible cases: (1) weaker fluid (C_{min}) is mixed and (2) stronger fluid (C_{max}) is mixed. Formulas for thermal effectiveness for the weaker fluid mixed are given by Equation T6 and for the stronger fluid mixed by Equation T7. The thermal effectiveness charts are given in Figure 2.10 for the weaker fluid mixed and Figure 2.11 for the stronger fluid mixed. For $R = 1$ or $C^* = 1$, the thermal effectiveness is the same for both cases.

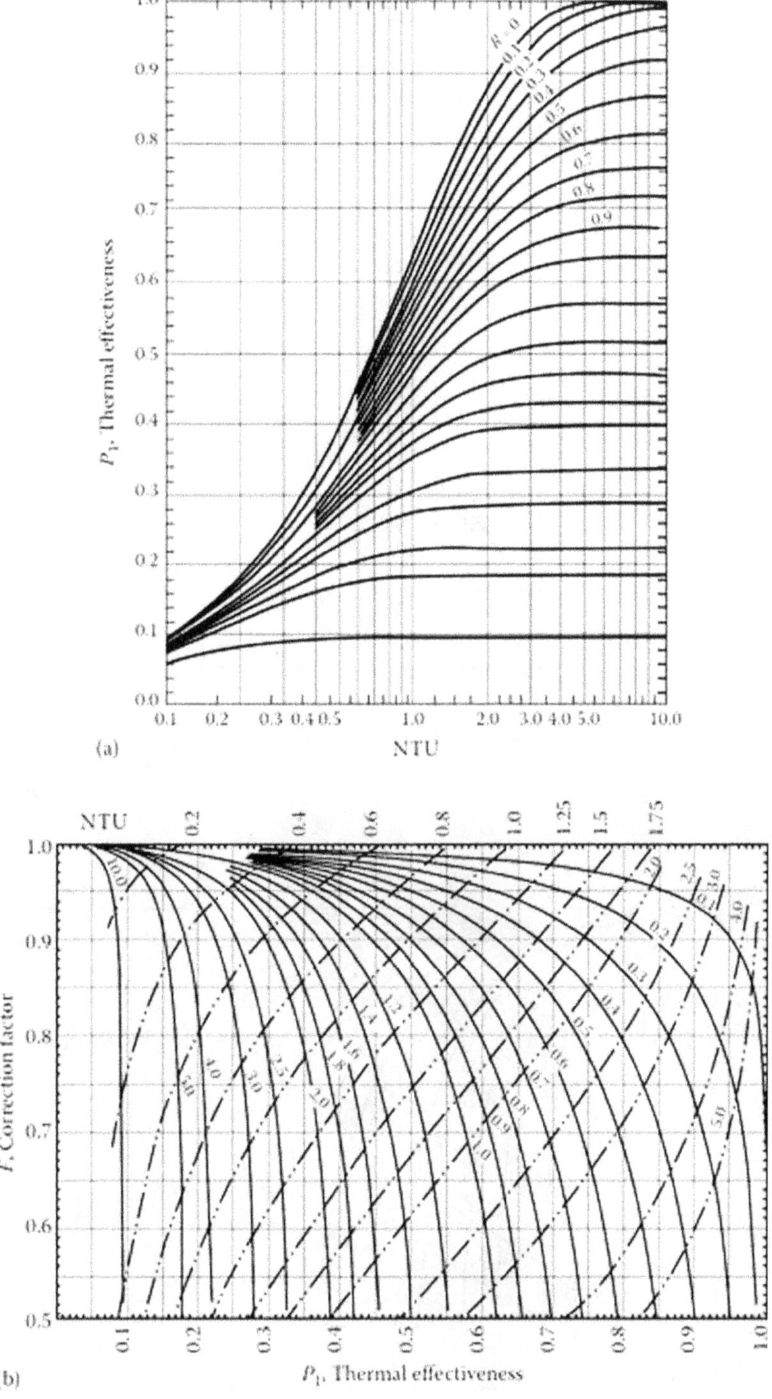

FIGURE 2.10 (a) Thermal effectiveness chart—crossflow: unmixed–mixed—the weaker (C_{min}) fluid mixed, R–P–NTU chart (as per Equation T6, Table 2.4); (b) F–R–P–NTU chart; F as a function of P for constant R (solid lines) and constant NTU (dashed lines) (as per Equation T6, Table 2.4).

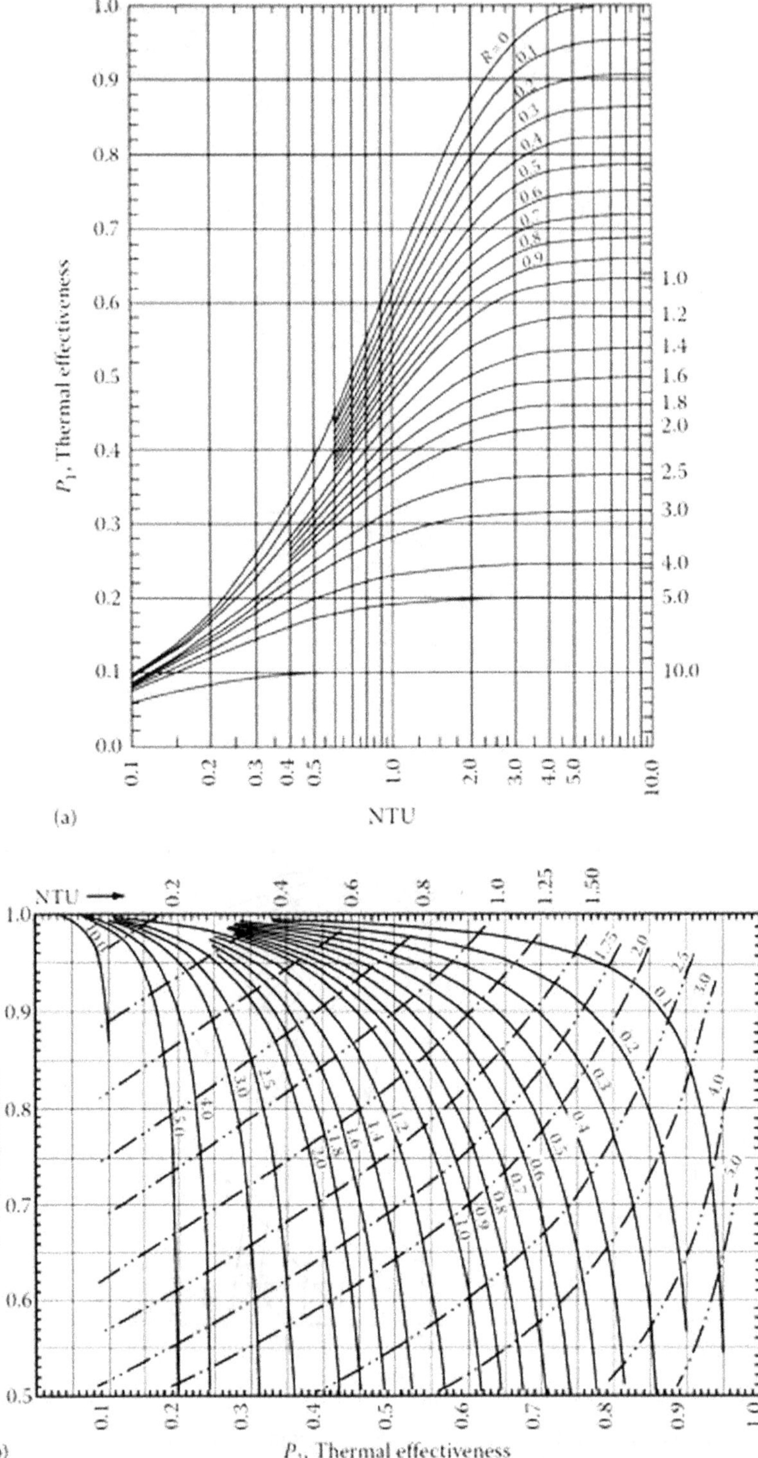

FIGURE 2.11 (a) Thermal effectiveness chart—crossflow: unmixed–mixed—the stronger (C_{max}) fluid mixed, ε-C^*-NTU chart (as per Equation T7, Table 2.4); (b) Thermal effectiveness chart: F–R–P–NTU chart; F as a function of P for constant R (solid lines) and constant NTU (dashed lines) (as per Equation T7, Table 2.4).

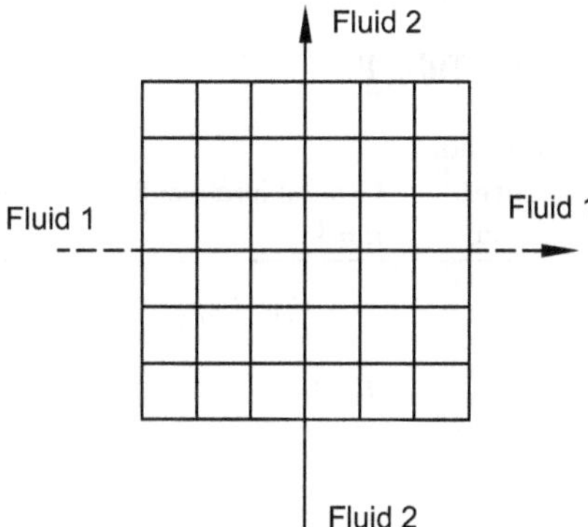

FIGURE 2.12 Unmixed–unmixed crossflow arrangement.

Note: Fluids 1 and 2 flow in two different adjacent layers.

2.8.3.3 Mixed–Mixed Crossflow

This case has no industrial application and is shown here only as an extreme of the crossflow arrangement. The solution is identical to the TEMA *J* shell with infinite tubeside passes. The formula for thermal effectiveness is given by Equation T8.

2.8.3.4 Single or Multiple Rows in Crossflow

Many process heat exchangers provide a crossflow arrangement between the hot (or cold) process fluid that flows through the tubes and the external coolant (or hot air such as supercharged engine intake air), usually air. Because this flow arrangement is not strictly countercurrent, the MTD must be corrected by applying a correction factor, *F*. The factor *F* depends on the terminal temperatures, the number of tube rows per pass, and the number of passes. The basic unmixed–unmixed case shown in Figure 2.12 assumes a large number of flow channels in both streams. For a single tubeside pass with one or more tube rows, the thermal effectiveness formula is different from that of the basic unmixed–unmixed case. Thermal relations for single-pass tube rows arrangements are discussed next.

I_0, I_1, and I_n are modified Bessel functions of the first kind.

2.8.3.5 Single Tubeside Pass, N Rows per Pass, Both Fluids Unmixed

A common header at one end of the tubes distributes the tubeside fluid into a single pass having *N* rows in parallel. A similar header at the other end collects tubeside fluid. For given terminal temperatures, *F* increases with the number of rows per pass and the number of passes being increased and is more sensitive to the latter. Taborek [24], Pignotti and Cordero [25], and Pignotti [26] present values of *F* for a variety of crossflow configurations, applicable to air-cooled heat exchangers.

Schedwill's formula for the thermal effectiveness of *N* rows is given by [27]

$$P = \frac{1}{R}\left\{1 - \left[\frac{Ne^{\mathrm{NKR}}}{1 + \sum_{i=1}^{N-1}\sum_{j=0}^{i}\binom{i}{j}K^{j}e^{-(i-j)\mathrm{NTU}/N}\sum_{k=0}^{j}\frac{(\mathrm{NKR})^{k}}{k!}}\right]\right\}^{-1} \tag{2.59}$$

where

$$\binom{i}{j} = \frac{i!}{(i-j)!\,j!} \tag{2.60}$$

TABLE 2.5
Thermal Effectiveness Relations for Tube Rows with Single Pass Arrangement

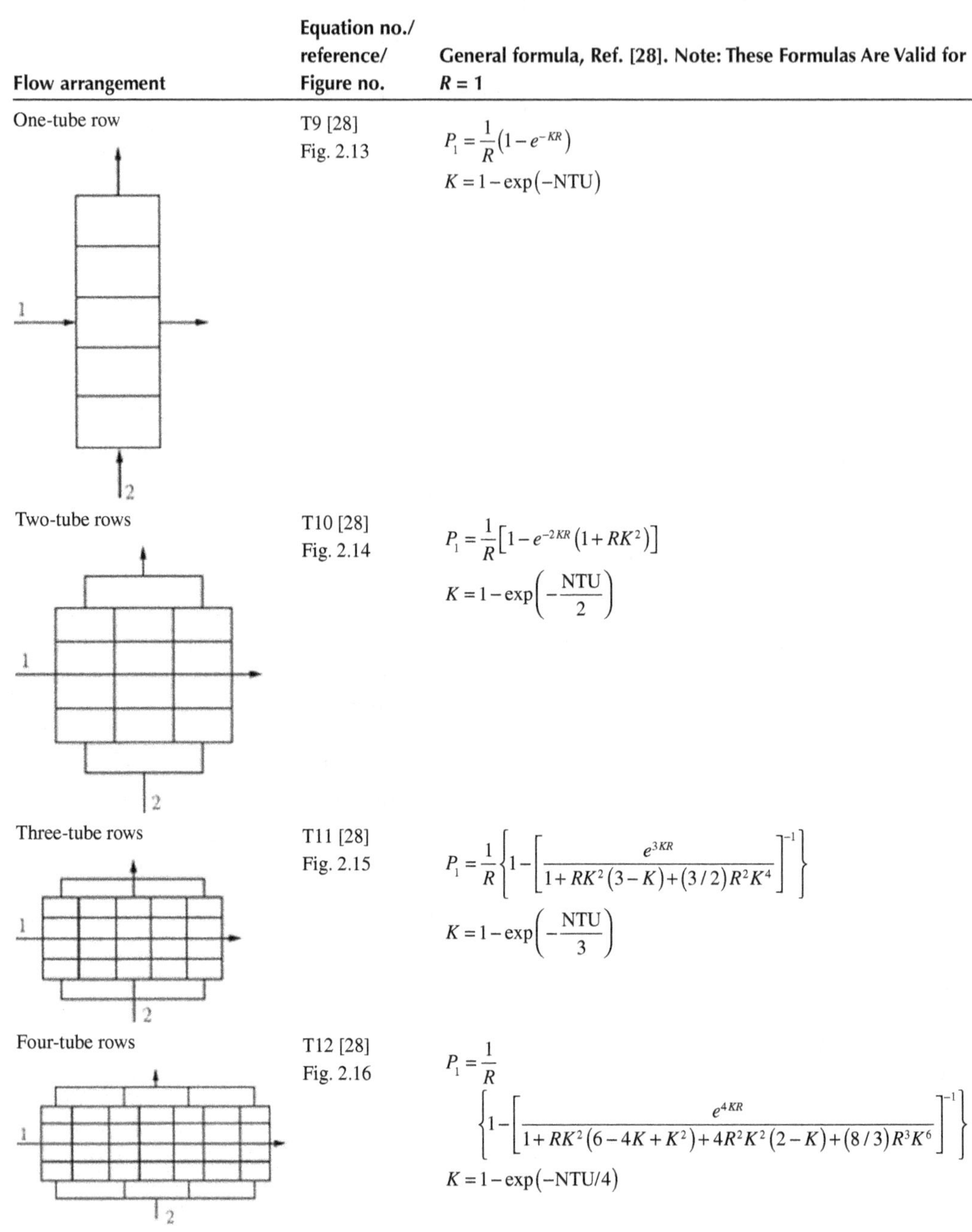

Flow arrangement	Equation no./ reference/ Figure no.	General formula, Ref. [28]. Note: These Formulas Are Valid for $R = 1$
One-tube row	T9 [28] Fig. 2.13	$P_1 = \frac{1}{R}\left(1 - e^{-KR}\right)$ $K = 1 - \exp(-\text{NTU})$
Two-tube rows	T10 [28] Fig. 2.14	$P_1 = \frac{1}{R}\left[1 - e^{-2KR}\left(1 + RK^2\right)\right]$ $K = 1 - \exp\left(-\frac{\text{NTU}}{2}\right)$
Three-tube rows	T11 [28] Fig. 2.15	$P_1 = \frac{1}{R}\left\{1 - \left[\frac{e^{3KR}}{1 + RK^2(3 - K) + (3/2)R^2K^4}\right]^{-1}\right\}$ $K = 1 - \exp\left(-\frac{\text{NTU}}{3}\right)$
Four-tube rows	T12 [28] Fig. 2.16	$P_1 = \frac{1}{R}$ $\left\{1 - \left[\frac{e^{4KR}}{1 + RK^2(6 - 4K + K^2) + 4R^2K^2(2 - K) + (8/3)R^3K^6}\right]^{-1}\right\}$ $K = 1 - \exp(-\text{NTU}/4)$

Note: P_2 can be found using Equation 2.55 or $P_2 = P_1R_1$, $R_2 = 1/R_1$, $\text{NTU}_2 = R_1\text{NTU}_1$.

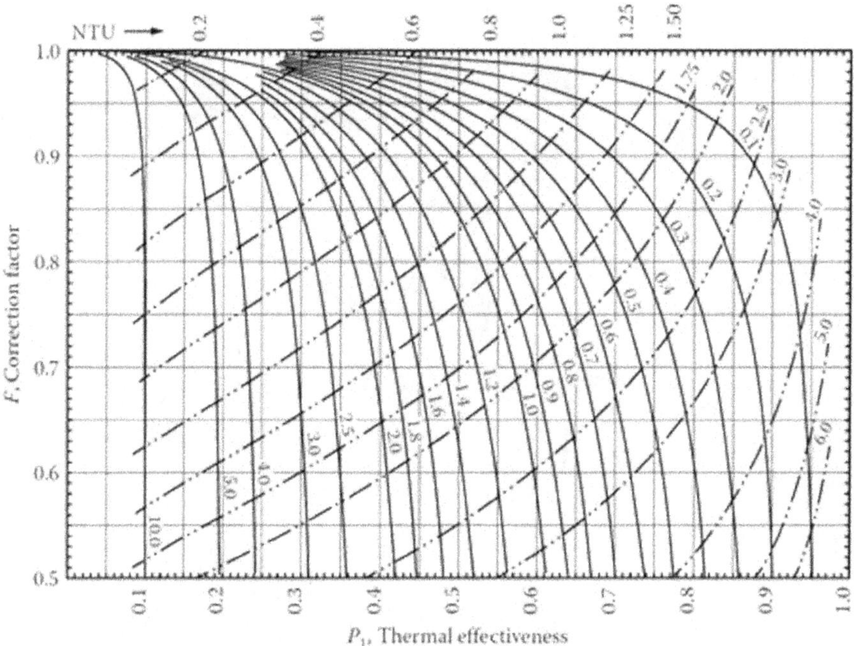

FIGURE 2.13 Thermal effectiveness chart—one-tube row; F as a function of P for constant R (solid lines) and constant NTU (dashed lines) (as per Equation T9, Table 2.5).

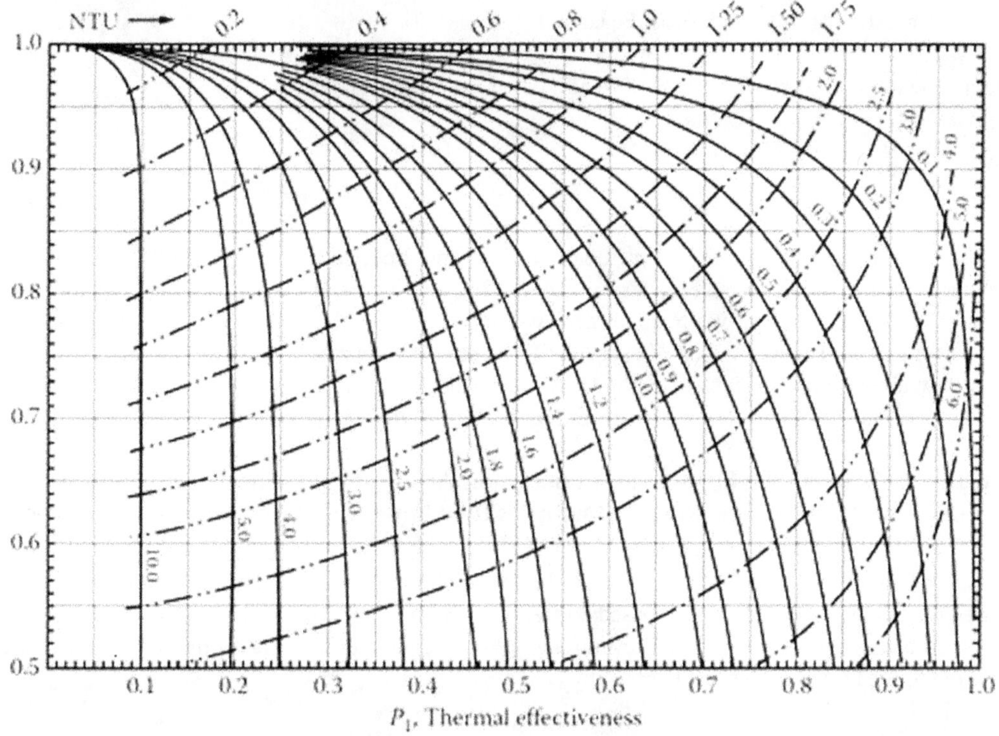

FIGURE 2.14 Thermal effectiveness chart—two-tube rows; F as a function of P for constant R (solid lines) and constant NTU (dashed lines) (as per Equation T10, Table 2.5).

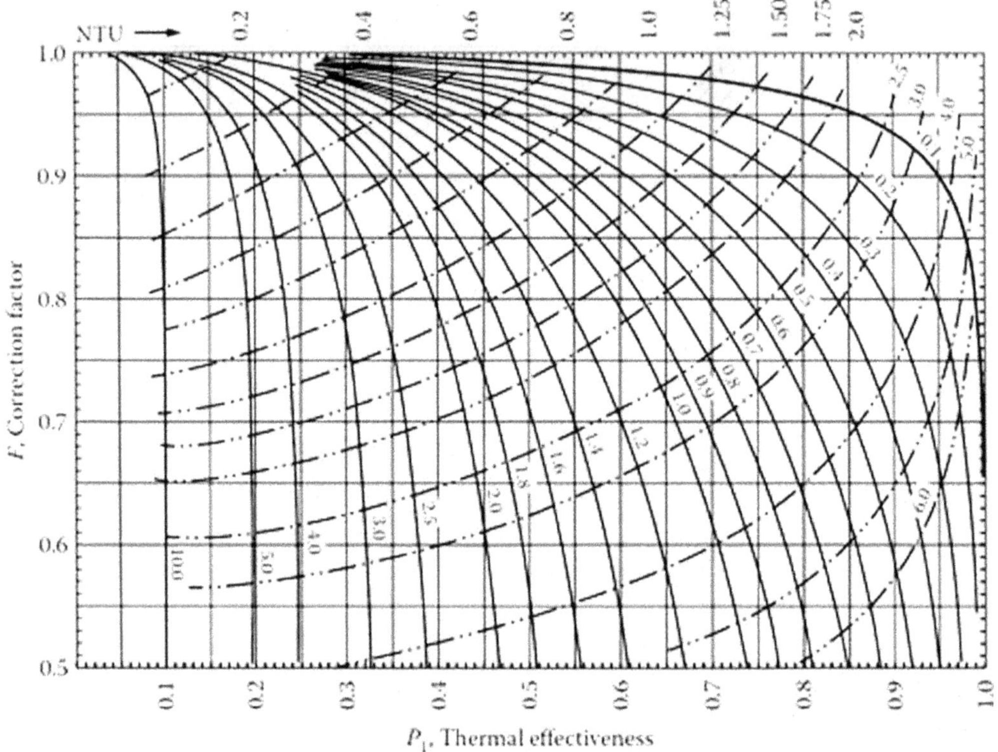

FIGURE 2.15 Thermal effectiveness chart—three-tube rows; F as a function of P for constant R (solid lines) and constant NTU (dashed lines) (as per Equation T11, Table 2.5).

that is, the number of combinations of i and j taken j at a time, and

$$K = 1 - \exp\left(-\frac{\text{NTU}}{N}\right) \tag{2.61}$$

By substituting $N = 1, 2, 3, \ldots$ in Equations 2.59 and 2.61, equations for thermal relations are obtained for the specific arrangements by Nicole [28], and this is given in Table 2.5 (Equations T9 through T12) for one row, two rows, three rows, and four rows. For a larger number of tube rows (for all practical purposes, when N exceeds 5), the solution approaches that of unmixed–unmixed crossflow arrangement. Values of F for $N = 1, 2, 3$, and 4 are shown in Figures 2.13–2.16 and are always less than the basic case of unmixed–unmixed crossflow (Figure 2.8).

2.8.3.6 Multipass Tube Rows Cross-Counterflow Arrangements, Both Fluids Unmixed, and Multiple Tube Rows in Multipass Tube Rows, Cross-Counterflow Arrangements

This would apply to a manifold-type air cooler in which the tubes in one row are connected to the next by U-bends. The solutions are based on Ref. [28]. Solutions for the two rows-two pass and three rows-three pass cases are based on Stevens et al. [29]. The general formula for thermal effectiveness referred to the air side (fin side) is given by [28]

$$P_1 = \frac{1}{R}\left(1 - \frac{1}{\zeta}\right) \tag{2.62}$$

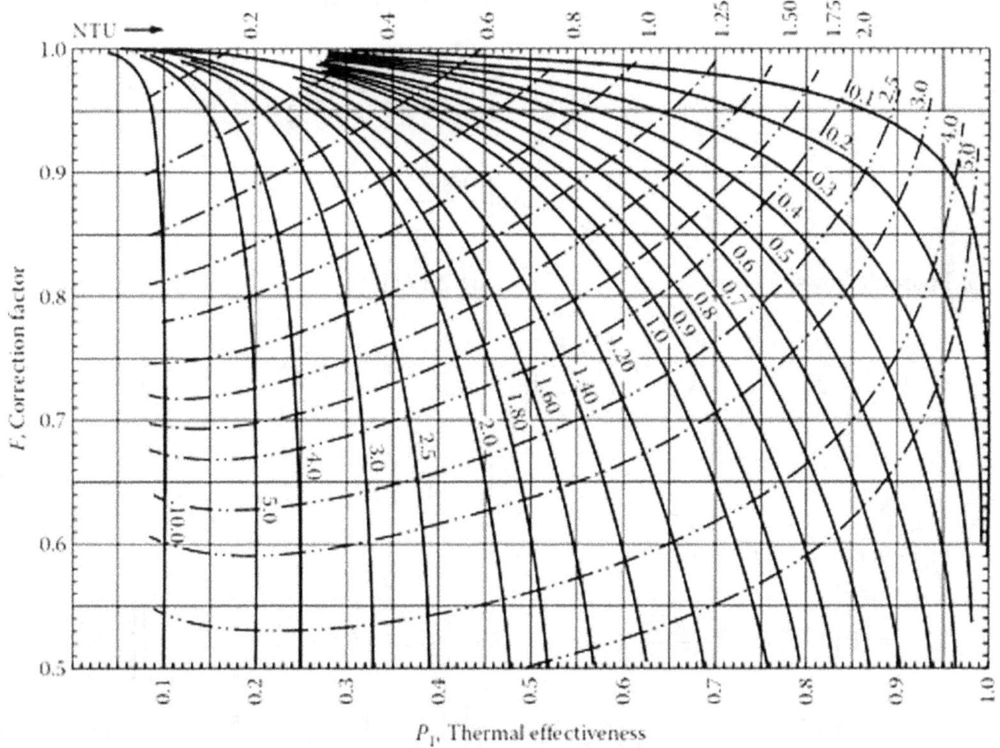

FIGURE 2.16 Thermal effectiveness chart—four-tube rows; F as a function of P for constant R (solid lines) and constant NTU (dashed lines) (as per Equation T12, Table 2.5).

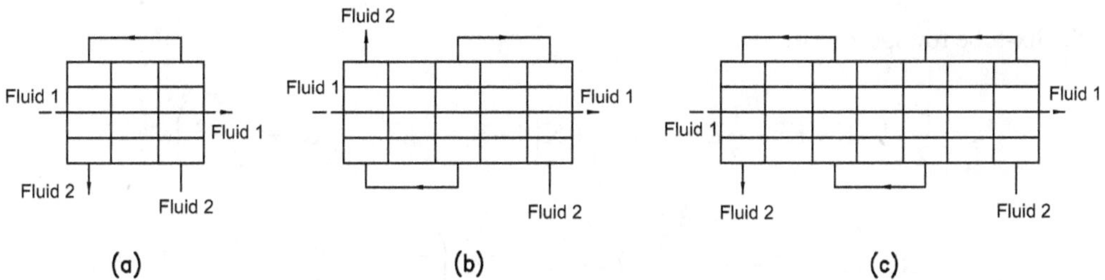

FIGURE 2.17 Multipass tube rows. (a) Two rows-two pass; (b) three rows-three pass; (c) four rows-four pass.

The expressions for ζ for various cases are as follows:

1. Two-tube rows, two passes, as shown in Figure 2.17a [29]

$$\zeta = \frac{K}{2} + \left(1 - \frac{K}{2}\right) e^{2KR} \tag{2.63a}$$

$$K = 1 - \exp\left(-\frac{\text{NTU}}{2}\right) \tag{2.63b}$$

2. Three-tube rows, three passes, as shown in Figure 2.17b [29]

$$\zeta = K\left[1-\frac{K}{4}-RK\left(1-\frac{K}{2}\right)\right]e^{KR}+e^{3KR}\left(1-\frac{K}{2}\right)^2 \quad (2.64a)$$

$$K = 1-\exp\left(-\frac{\text{NTU}}{3}\right) \quad (2.64b)$$

3. Four-tube rows, four passes, as shown in Figure 2.17c [28]

$$\zeta = \frac{K}{2}\left(1-\frac{K}{2}+\frac{K^2}{4}\right)+K\left(1-\frac{K}{2}\right)\left[1-\frac{R}{8}K\left(1-\frac{K}{2}\right)e^{2KR}\right]+e^{4KR}\left(1-\frac{K}{2}\right)^3 \quad (2.65a)$$

$$K = 1-\exp\left(-\frac{\text{NTU}}{4}\right) \quad (2.65b)$$

4. Five-tube rows, five passes [28]

$$\begin{aligned}\zeta = &\left\{K\left(1-\frac{3}{4}K+\frac{K^2}{2}-\frac{K^3}{8}\right)-RK^2\left[1-K+\frac{3}{4}K^2-\frac{1}{4}K^3-\frac{R}{2}K^2\left(1-\frac{K}{2}\right)^2\right]\right\}e^{KR}\\ &+\left[K\left(1-\frac{3}{4}K+\frac{1}{16}K^3\right)-3RK^2\left(1-\frac{K}{2}\right)^3\right]e^{3KR}+\left(1-\frac{K}{2}\right)^4 e^{5KR}\end{aligned} \quad (2.66a)$$

$$K = 1-\exp\left(-\frac{\text{NTU}}{5}\right) \quad (2.66b)$$

5. Six-tube rows, six passes [28]

$$\begin{aligned}\zeta = &\frac{K}{2}\left(1-K+K^2-\frac{1}{2}K^3+\frac{1}{8}K^4\right)+K\left(1-K+\frac{3}{4}K^2-\frac{5}{16}K^3+\frac{1}{32}K^4\right)e^{2KR}\\ &-RK^2\left[2-3K+3K^2-\frac{7}{4}K^3+\frac{3}{8}K^4-RK^2\left(2-3K+\frac{3}{2}K^2-\frac{1}{4}K^3\right)\right]e^{2KR}\\ &+\left[\frac{K}{2}\left(2-2K+\frac{1}{2}K^3-\frac{1}{8}K^4\right)-4KR^2\left(1-\frac{K}{2}\right)^4\right]e^{4KR}+\left(1-\frac{K}{2}\right)^5 e^{6KR}\end{aligned} \quad (2.67a)$$

$$K = 1-\exp\left(-\frac{\text{NTU}}{6}\right) \quad (2.67b)$$

Values of F referred to the fluid outside the tubeside (air side) for N = 2, 3, 4, 5, and 6 are shown in Figures 2.18–2.22, respectively. The F values are always higher than the basic case (Figure 2.8). When N becomes greater than 6, F approaches 1, that is, pure counter current flow.

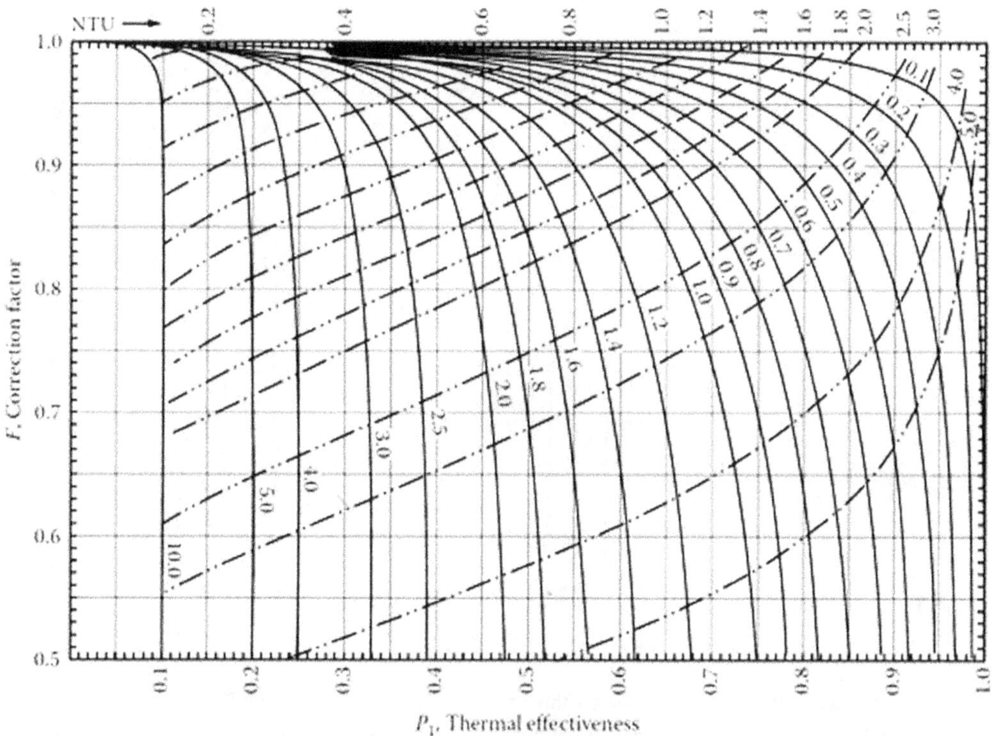

FIGURE 2.18 Thermal effectiveness chart—two-tube rows, two passes. Both fluids unmixed throughout; fluid 1 inverted coupling between passes (for the flow arrangement shown in Figure 2.17a). F as a function of P for constant R (solid lines) and constant NTU (dashed lines) (as per Equations 2.62 and 2.63).

6. Multiple tube rows in multipass cross-counterflow arrangements; one fluid unmixed throughout; other fluid (tube fluid) unmixed in each pass but mixed between passes:
   a. Four rows, two passes, with two rows per pass. Tubeside fluid mixed at the header and the other fluid is unmixed throughout, coupling in inverted order (Figure 2.23a) [28]:

$$P_1 = \frac{1}{R}\left(1-\frac{1}{\zeta}\right) \quad \text{from Equation 2.62}$$

$$\zeta = \left\{\frac{R}{2}K^3\left[4-K+2RK^2\right]+e^{4KR}+K\left[1-\frac{K}{2}+\frac{K^2}{8}\right]\left[1-e^{4KR}\right]\right\}\frac{1}{\left(1+RK^2\right)^2} \tag{2.68a}$$

$$K = 1-\exp\left(\frac{\text{NTU}}{4}\right) \tag{2.68b}$$

   For the above case, the thermal effectiveness chart is shown in Figure 2.24 .

   b. Six rows, two passes, with three rows per pass. Tubeside fluid is mixed between passes, that is, at the header, and the other fluid is unmixed throughout, coupling in inverted order (Figure 2.23b) [26]:

$$P_1 = \frac{2A-RA^2-\delta}{1-R\delta} \tag{2.69a}$$

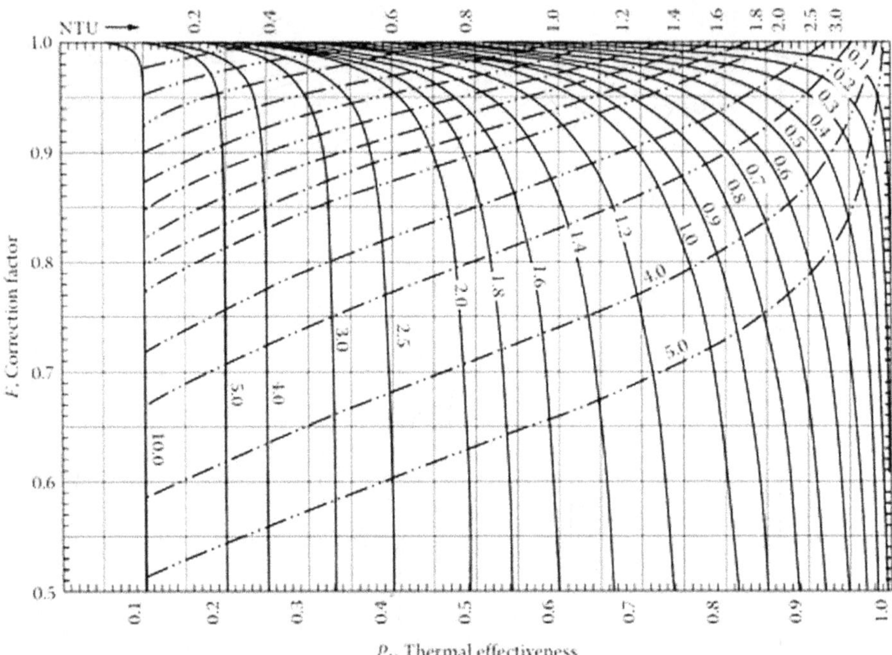

FIGURE 2.19 Thermal effectiveness chart—three-tube rows, three passes. Both fluids unmixed throughout; fluid 1 inverted coupling between passes (for the flow arrangement shown in Figure 2.17b). F as a function of P for constant R (solid lines) and constant NTU (dashed lines) (as per Equations 2.62 and 2.64).

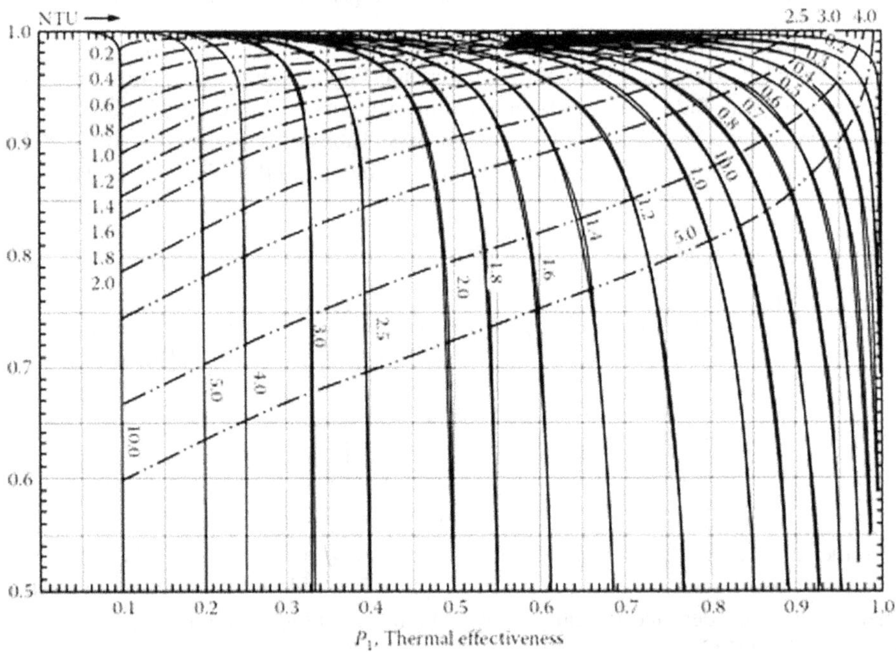

FIGURE 2.20 Thermal effectiveness chart—four-tube rows, four passes. Both fluids unmixed throughout; fluid 1 inverted between passes (for the flow arrangement shown in Figure 2.17c). F as a function of P for constant R (solid lines) and constant NTU (dashed lines) (as per Equations 2.62 and 2.65).

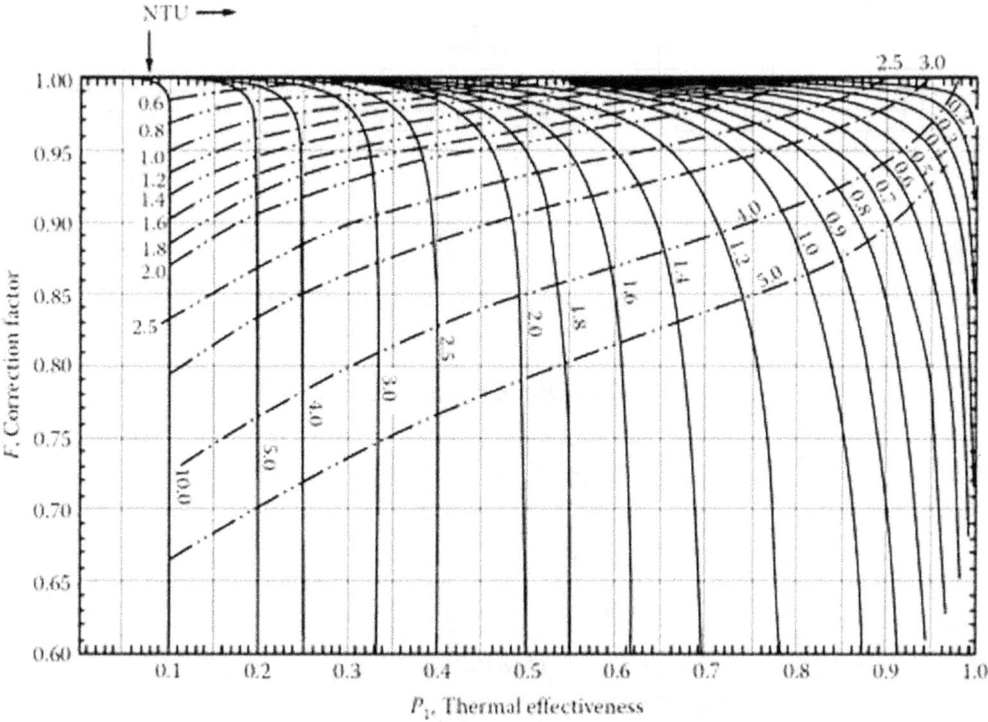

FIGURE 2.21 Thermal effectiveness chart—five-tube rows, five passes. Both fluids unmixed throughout; fluid 1 inverted coupling between passes. F as a function of P for constant R (solid lines) and constant NTU (dashed lines) (as per Equations 2.62 and 2.66).

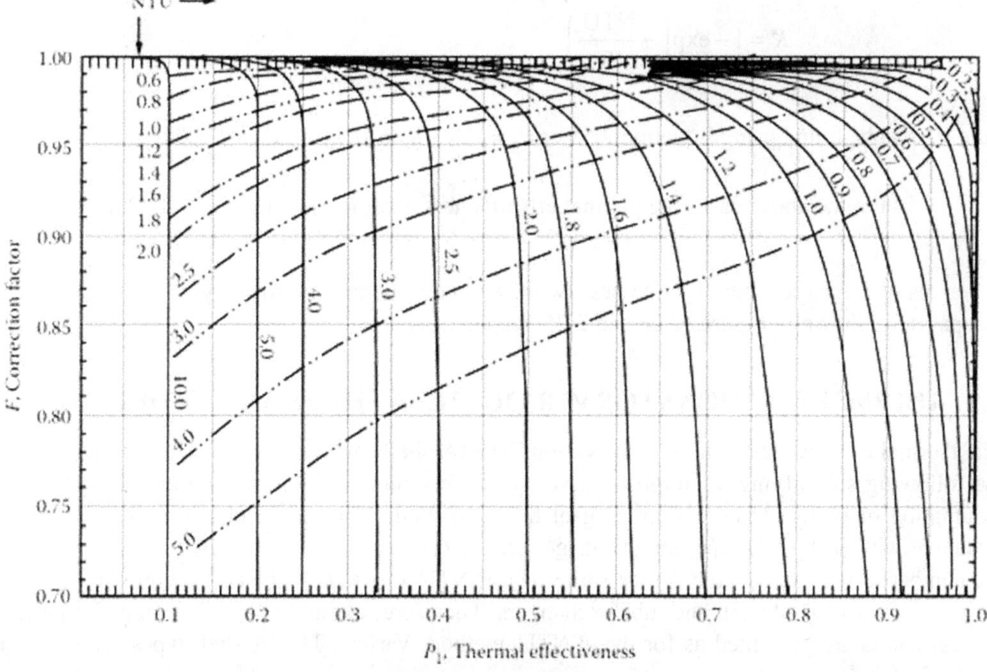

FIGURE 2.22 Thermal effectiveness chart—six-tube rows, six passes. Both fluids unmixed throughout; fluid 1 inverted coupling between passes. F as a function of P for constant R (solid lines) and constant NTU (dashed lines) (as per Equations 2.62 and 2.67).

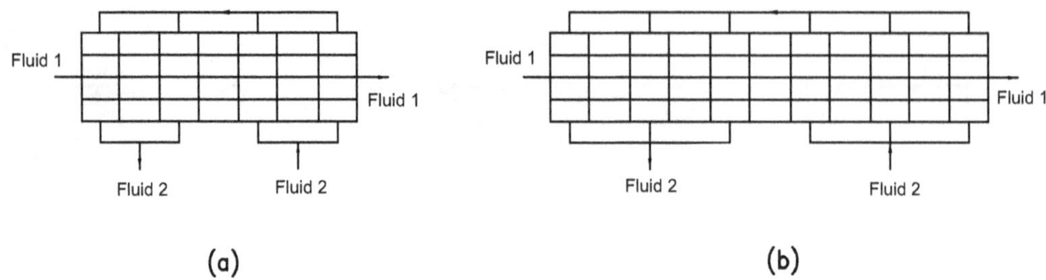

FIGURE 2.23 Tube rows multiples arrangement: (a) four rows-two pass; (b) six rows-two pass.

$$\begin{aligned} A &= a_0C_0(3KR)+a_1C_1(3KR)+a_2C_2(3KR) \\ a_0 &= 1-(1-K)^3 \\ a_1 &= 3RK^3(3-2K) \\ a_2 &= \frac{9R^2K^5}{2} \end{aligned}$$

$$\delta = a_0^2C_0(6KR)+2a_0a_1C_1(6KR)+(a_1^2+2a_0a_2)C_2(6KR) + 2a_1a_2C_3(6KR)+a_2^2C_4(6KR)$$

$$C_n(z)=\frac{n!}{z^{n+1}}\left[1-e^{-z}\left(1+z+\frac{z}{2!}+\cdots+\frac{Z^n}{n!}\right)\right] \tag{2.69b}$$

$$K=1-\exp\left(-\frac{\text{NTU}}{6}\right)$$

Note: P_2 can be found using Equation 2.55 or $P_2 = P_1R_1$, $R_2 = 1/R_1$, $\text{NTU}_2 = R_1\text{NTU}_1$.

For the above case, the thermal effectiveness chart is shown in Figure 2.25.

Values of F for four rows, two passes (two rows in a pass) and six rows, two passes (three rows in a pass) are shown in Figures 2.24 and 2.25, respectively.

2.9 THERMAL RELATIONS FOR VARIOUS TEMA SHELLS AND OTHERS

The thermal effectiveness relations for various TEMA shells and others are presented next, using the following simplifying assumptions given at the beginning of this section and the additional assumption of perfect transverse mixing of the shell fluid. Since the shellside flow arrangement is unique with each shell type, the exchanger effectiveness is different for each shell even though the number of tube passes may be the same. The P-NTU_t or LMTD method is commonly used for the thermal analysis of shell and tube exchangers. Therefore, thermal relation formulas and effectiveness charts are presented as for the P-NTU_t method. Various TEMA shell types are shown in Figure 2.26

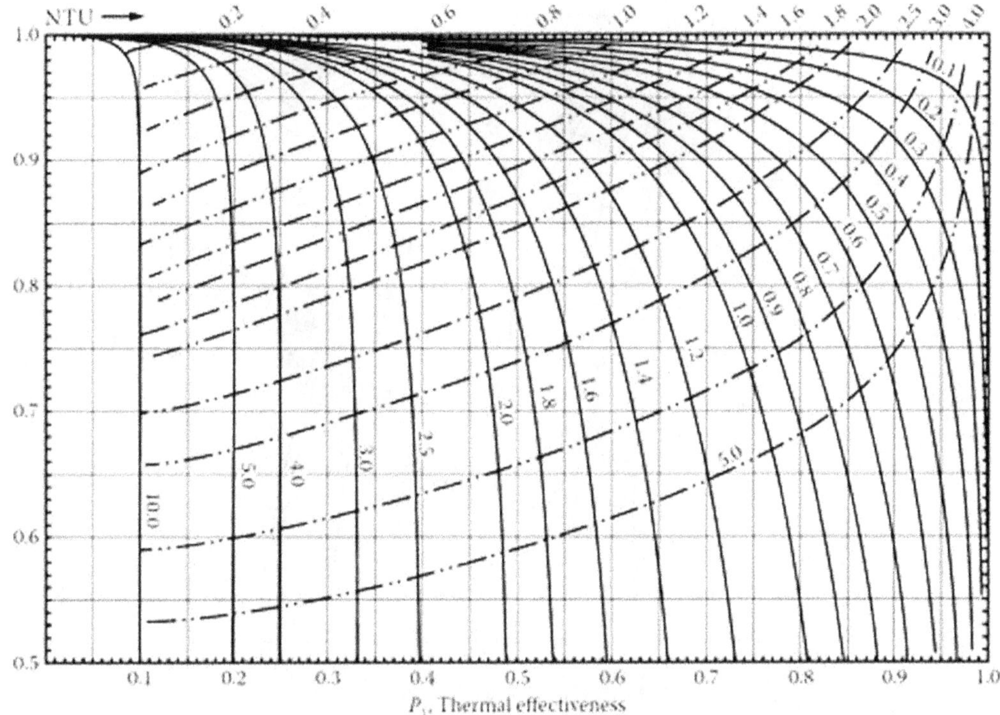

FIGURE 2.24 Thermal effectiveness chart—four rows-two passes with two rows per pass. One fluid is unmixed throughout; tubeside fluid mixed between passes and in each pass unmixed; fluid 1 inverted coupling between passes (for the flow arrangement shown in Figure 2.23a). F as a function of P for constant R (solid lines) and constant NTU (dashed lines) (as per Equations 2.62 and 2.68).

2.9.1 *E* Shell

The basic case of the E shell, one shell pass and one tube pass with parallelflow and counterflow arrangement ($E_{1\text{-}1}$), is shown in Figure 2.27a. For the counterflow case with more than five baffles, the F value can be taken as 1. The thermal effectiveness charts shown in Figure 2.6 can be used for parallelflow arrangement and Figure 2.7 for counterflow arrangement.

2.9.1.1 Multipassing on the Tubeside of E Shell

On the tubeside, any number of odd or even passes is possible. Increasing the even number of tube passes from two to four, six, etc. decreases the exchanger effectiveness slightly, and in the limit when the number of tube passes approaches infinity with one shell pass, the exchanger effectiveness approaches that for a single-pass crossflow exchanger with both fluids mixed. Common tubeside multipass arrangements for TEMA $E_{1\text{-}2}$, $E_{1\text{-}3}$, and $E_{1\text{-}4}$ shells are shown in Figures 2.27b,c and d.

2.9.1.2 Even Number of Tube Passes

One shell pass and two tube passes as shown in Figure 2.4 using a U-tube bundle is one of the most common flow arrangements used in the single-pass TEMA E shell. The heat exchanger with this arrangement is also simply referred to as a conventional 1–2 heat exchanger. If the shell fluid is idealized as well mixed, its temperature is constant at any cross section. In this case, reversing the tube fluid flow direction will not change the idealized temperature distribution and the exchanger thermal effectiveness. Possible flow patterns of the $E_{1\text{-}2}$ shell were already shown in Figure 2.4. This is referred to as "stream symmetric."

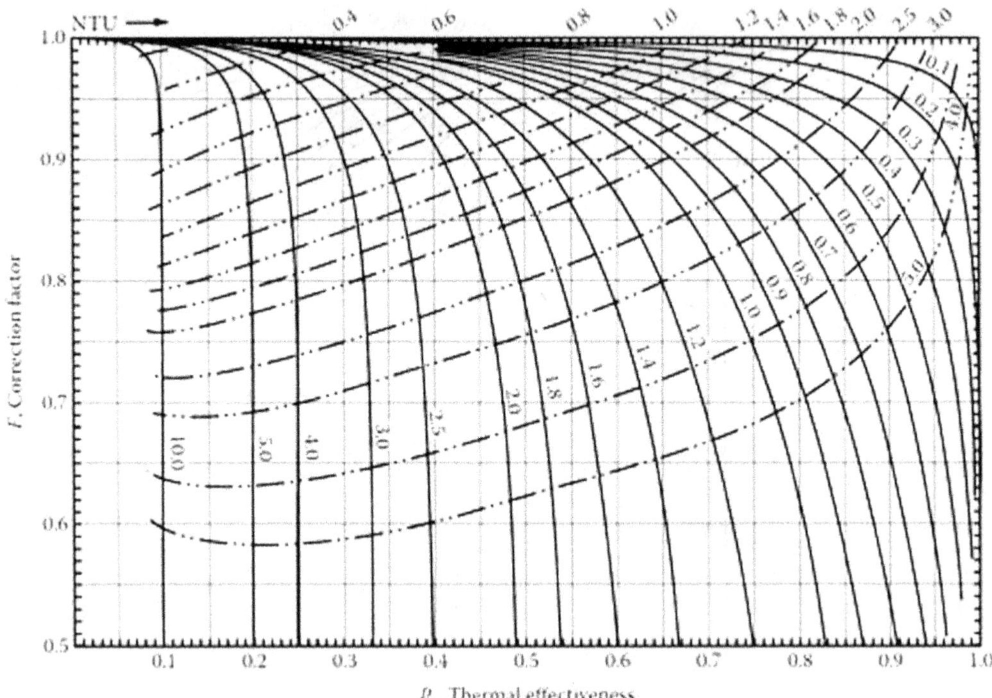

FIGURE 2.25 Thermal effectiveness chart—six rows-two passes with three rows per pass. One fluid is unmixed throughout; tubeside fluid mixed between passes and in each pass unmixed; fluid 1 inverted coupling between passes (for the flow arrangement shown in Figure 2.23b). F as a function of P for constant R (solid lines) and constant NTU (dashed lines) (as per Equations 2.62 and 2.69).

The 1–2 and 1–4 cases were solved long ago by Bowman [11], Underwood [12], and Nagle [30]. The 1–N geometry for an even number of passes was solved by Baclic [31]. The thermal effectiveness formula for the E_{1-2} case is given by Equation T13 in Table 2.6 and the thermal effectiveness chart is given in Figure 2.28 along with the F_{min} curve. The thermal effectiveness formulas for 4 and N tubeside passes are given by Equations T14 and T15, respectively. The thermal effectiveness chart for the E_{1-4} case is shown in Figure 2.29, and this figure may be used for even $N \geq 6$ cases also without loss of much accuracy. Alternate relation (approximate) to find out the F factor for an even number of tubeside passes of E shell (E_{1-2}, E_{1-4}, E_{1-6}, E_{1-2n}) is given as follows (Equation 2.70):

$$F = \frac{\sqrt{R^2+1}\ln\left[\dfrac{1-P}{1-PR}\right]}{(R-1)\ln\left[\dfrac{2-P\left[R+1-\left(1+R^2\right)^{0.5}\right]}{2-P\left[R+1+\left(1+R^2\right)^{0.5}\right]}\right]} \tag{2.70}$$

2.9.1.3 Odd Number of Tube Passes

With the TEMA E shell, an odd number of tube passes can have both parallelflow and counterflow arrangements. The thermal effectiveness relations for overall parallelflow and counterflow arrangements are discussed next.

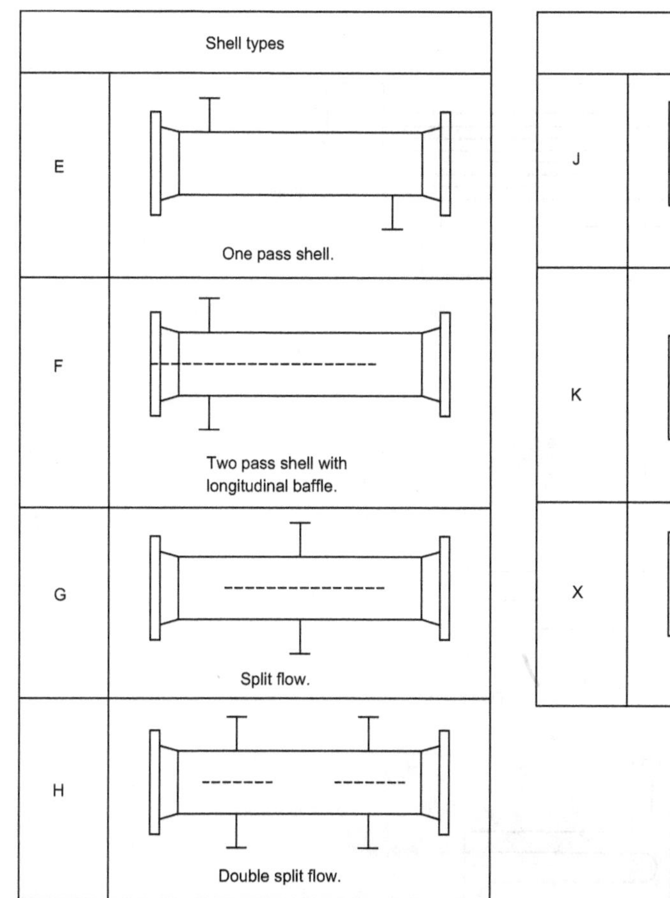

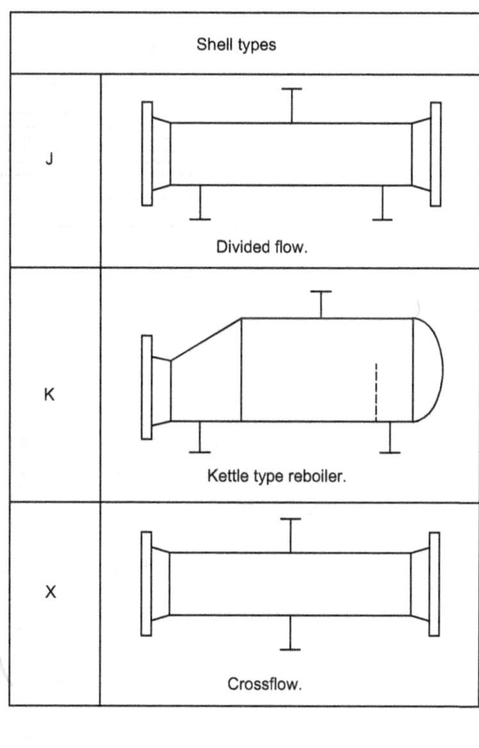

FIGURE 2.26 Various TEMA shell types.

In an E_{1-3} overall counterflow arrangement, two tube passes are in counterflow and one tube pass is in parallelflow, whereas in an E_{1-3} overall parallelflow arrangement, two tube passes are in parallelflow and one tube pass is in counterflow. Similarly, in an E_{1-5} overall counterflow arrangement, three tube passes are in counterflow and two tube passes are in parallelflow. The flow arrangement for E_{1-3} is already shown in Figure 2.27 and for E_{1-5} is shown in Figure 2.30. Pignotti and Tamborenea [32] obtained an analytical solution for the 1–N (odd N) case and obtained effectiveness for the 1–3 and 1–5. Explicit thermal relations are given by Pignotti [18] for E_{1-3} parallelflow and counterflow cases.

For $N = 3$. Thermal effectiveness formulas for E_{1-3} parallelflow arrangement are given by Equation T16 and for E_{1-3} counterflow arrangement by Equation T17 in Table 2.7. The thermal effectiveness chart for E_{1-3} counterflow arrangement is given in Figure 2.31.

For $N = 5$. For the overall counterflow arrangement or opposite end case, the thermal effectiveness is given by Equation T18 in Table 2.8 and the thermal effectiveness chart is given in Figure 2.32.

For the overall parallelflow arrangement or same end case, the configuration can be obtained from that of the opposite end case (T18) by just the inversion of the direction of flow of the tube fluid, which leads to the following expression [32]:

$$P_1 = \left(1 - X_5^*\right) \tag{2.71a}$$

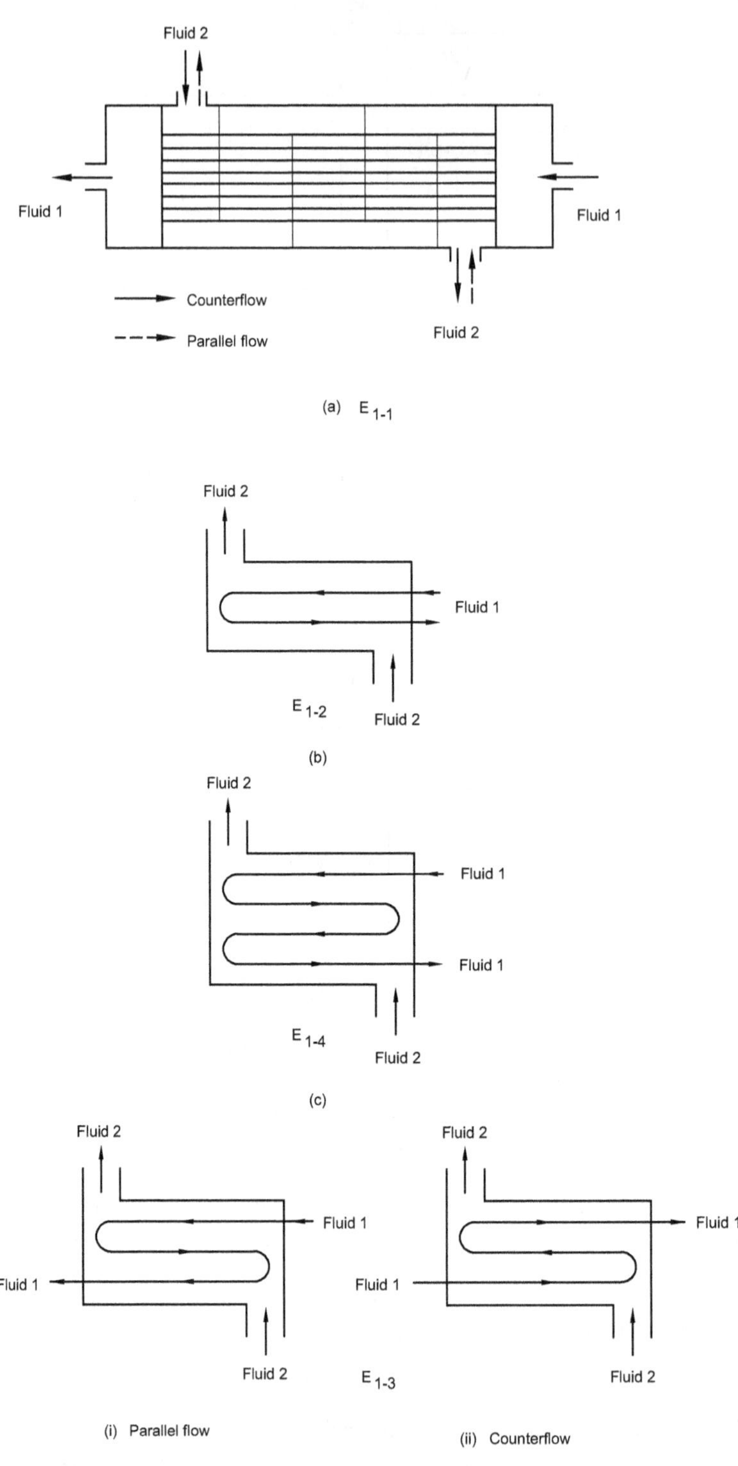

FIGURE 2.27 Common tube pass arrangement for $E_{1\text{-}1}$, $E_{1\text{-}2}$, $E_{1\text{-}4}$, and $E_{1\text{-}3}$ shells.

TABLE 2.6
Thermal Effectiveness Relations for E_{1-3} shell (N even) (Referred to Tubeside)

Flow arrangement	Equation no./ reference/ Fig. no.	General formula	Value for $R = 1$ and special cases
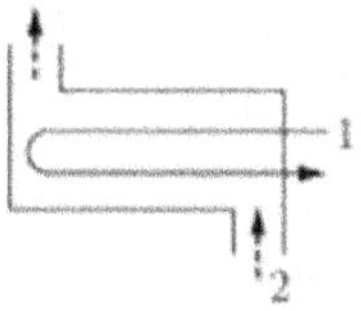 TEMA E_{1-2} shell: shell fluid mixed, tube fluid mixed between passes; stream symmetric	T13 [30] Fig. 2.28	$P = \frac{2}{1 + R + A\coth(A\text{NTU}/2)}$ $A = \sqrt{1 + R^2}$	$P = \frac{1}{1 + \coth(\text{NTU}/\sqrt{2})/\sqrt{2}}$
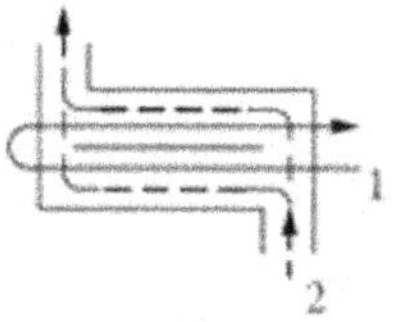 E_{1-2} shell: shell fluid divided into two streams, individually mixed; tube fluid mixed between passes	Same as J_{1-1} shell		Same as J_{1-1} shell
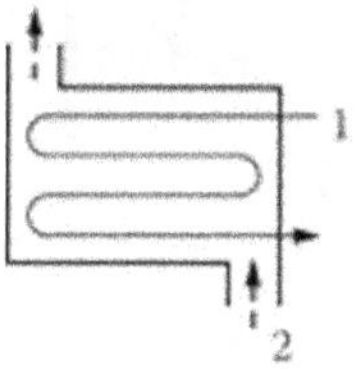 TEMA E_{1-4} shell: shell fluid mixed, tube fluid mixed between passes	T14 [30] Fig. 2.29	$P_1 = \frac{4}{\left[2(1+R) + \gamma\coth\left(\frac{\gamma\text{NTU}}{4}\right) + \tanh\left(\frac{\text{NTU}}{4}\right)\right]}$ $\gamma = \sqrt{(4R^2 + 1)}$	$P = \frac{4}{\left[4 + \sqrt{5}A + B\right]}$ $A = \coth\left(\sqrt{5}\text{NTU}/4\right)$ $B = \tanh(\text{NTU}/4)$
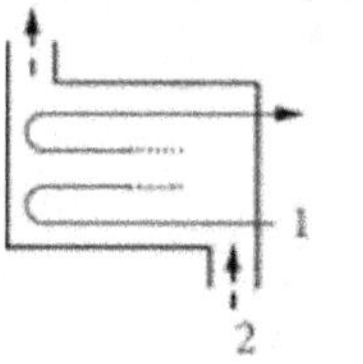 TEMA E_{1-N} (even N) shell: shell fluid mixed and tube fluid mixed between passes	T15 [31] Fig. 2.9	$P_1 = \frac{2}{A + B + C}$ $A = 1 + R + \coth(\text{NTU}/2)$ $B = \frac{-1}{N_1}\coth\left(\frac{\text{NTU}}{2N_1}\right)$ $C = \frac{1}{N_1}\sqrt{1 + N_1^2R^2}\,\coth\left(\frac{\text{NTU}}{2N_1}\sqrt{1 + N_1^2R^2}\right)$ $N_1 = \frac{N}{2}$ $N \to \infty$ results in well-known single-pass crossflow exchanger with both fluids unmixed	Same as Equation T15 with $R = 1$

Note: P_2 (P_s) can be found using Equation 2.55 or P_2 (P_s) = P_1R_1, $R_2 = 1/R_1$, $\text{NTU}_2 = R_1\text{NTU}_1$.

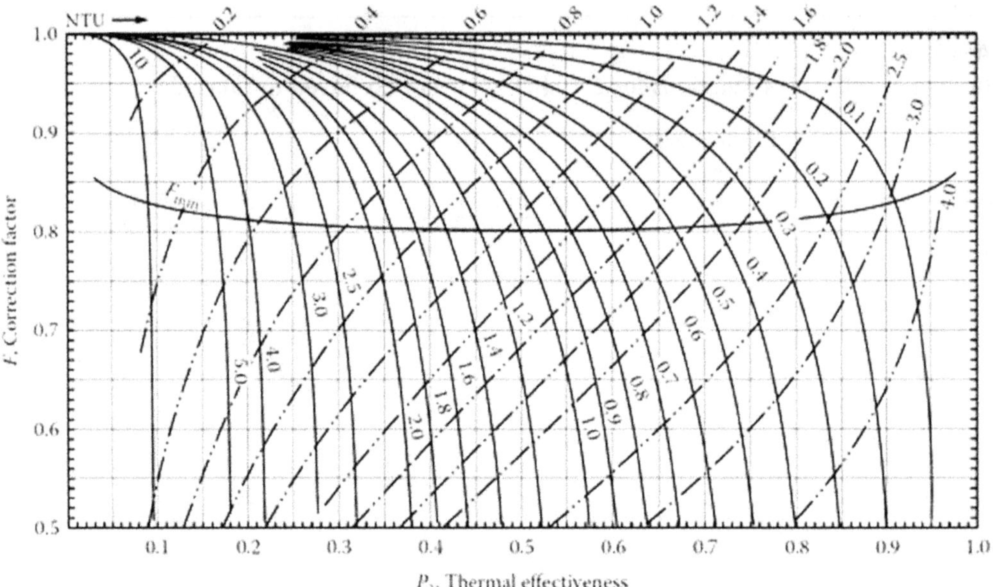

FIGURE 2.28 Thermal effectiveness chart—TEMA E_{1-2} shell; shell fluid mixed, tube fluid mixed between passes. Stream symmetric. F as a function of P for constant R (solid lines) and constant NTU (dashed lines) F_{min} line is also shown (as per Equation T13 an2.6).

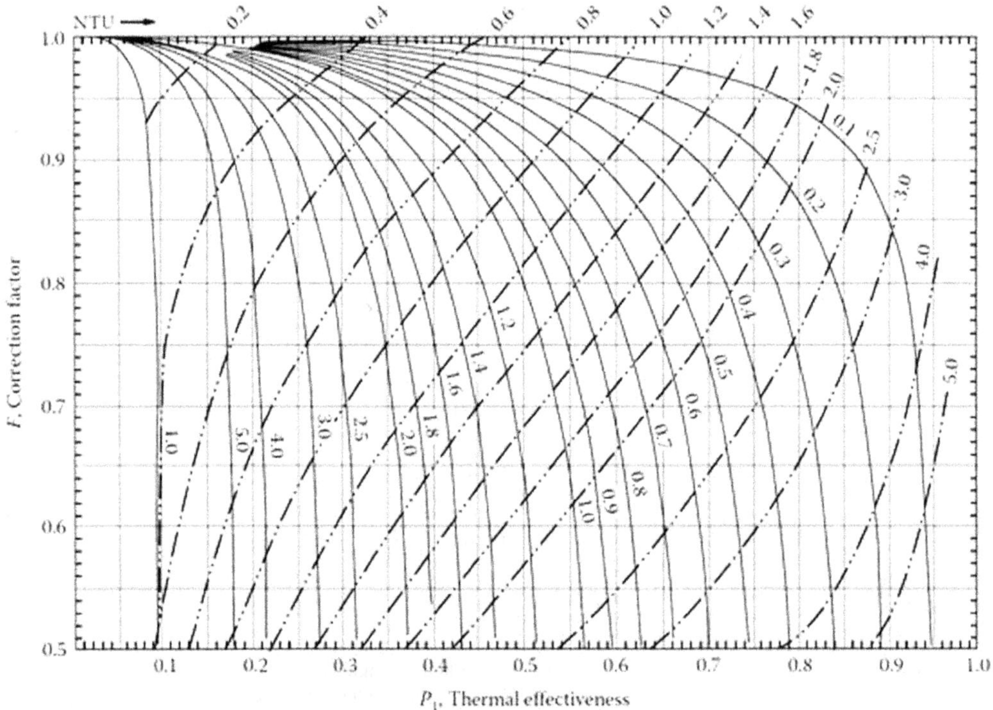

FIGURE 2.29 Thermal effectiveness chart—TEMA E_{1-4} shell. Shell fluid mixed, tube fluid mixed between passes. F as a function of P for constant R (solid lines) and constant NTU (dashed lines) (as per Equation T14, Table 2.6).

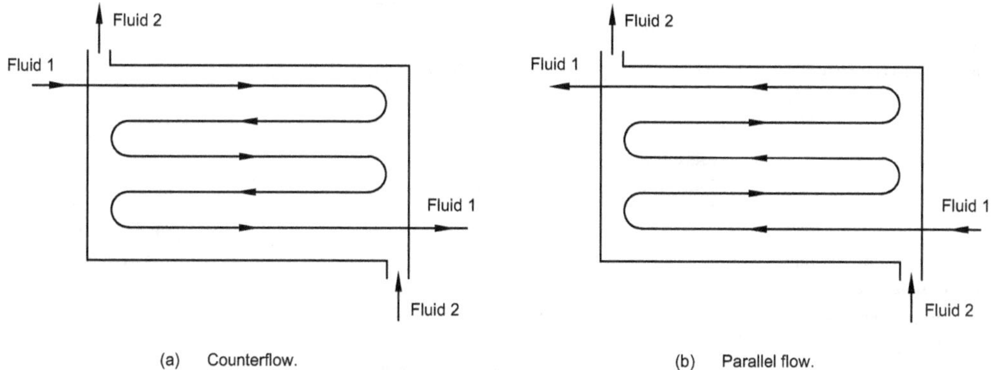

FIGURE 2.30 Flow arrangement of E shell with one shell pass and five tube passes (E_{1-5}). (a) Counterflow; (b) parallelflow.

with

$$X_5^*(R,\text{NTU}) = \frac{1}{X_5(-R,\text{NTU})} \tag{2.71b}$$

2.9.2 TEMA *F* Shell

For the TEMA F shell, as is usual with two tube passes, the arrangement results in pure counterflow with $F = 1$. For four tube passes, the configuration would be treated as two E_{1-2} exchangers in series. As the longitudinal baffle is exposed on the one side to the hot stream and on the other side to the cold stream, there is a conduction heat exchange that reduces the thermal effectiveness and hence the F factor has to be corrected. The effect of thermal leakage by conduction across the baffle has been analyzed by Whistler [33] and in nondimensional form by Rozenman and Taborek [34], with proposed baffle correction methods. The baffle correction factor based on Ref. [34] is presented in Ref. [35].

2.9.3 TEMA *G* Shell or Split-Flow Exchanger

Possible flow arrangements for the TEMA G shell are G_{1-1}, G_{1-2}, and G_{1-4}. The solution of the G_{1-2} counterflow case is due to Schindler and Bates [36] and that of the parallelflow case is due to Pignotti [18]. The thermal effectiveness P of this shell with two tube passes is higher than that of the conventional E_{1-2} exchanger for a given NTU_t and R. It is also possible to have only a single pass on the tubeside, which is stream symmetric [37]. The thermal effectiveness formula for G_{1-1} is given by Equation T19 in Table 2.9, and the thermal effectiveness chart is given in Figure 2.33. For the G_{1-2} arrangement, the thermal relations for the parallelflow and counterflow arrangements are given by Equations T20 and T21, respectively, and the thermal effectiveness chart for the counterflow arrangement is given in Figure 2.34. G_{1-4} arrangement was analyzed by Singh et al. [38], and the thermal effectiveness chart is given at the end of this chapter.

2.9.4 TEMA *H* Shell

The arrangement of the TEMA H shell exchanger resembles a configuration in which two TEMA G shell exchangers are connected side by side. The solution for the TEMA H shell with 1–2 flow

TABLE 2.7
Thermal Effectiveness Relations for E_{1-3} Shell (Referred to Tubeside)

Flow arrangement	Equation no./ reference/ Figure no.	General formula	Value for R = 1
1–3 TMA E shell: two parallelflow and one counterflow passes; shell fluid mixed, tube fluid mixed, tube fluid mixed between passes 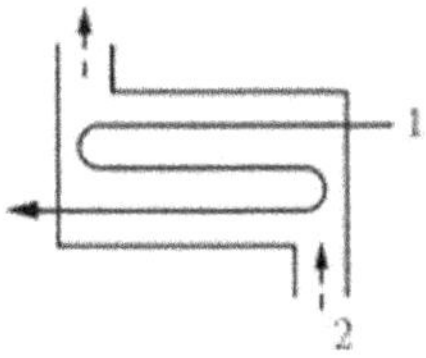	T16 [18]	$P_1 = \left[1 - \frac{AC + B^2}{C}\right]$ $A = \chi_1(1 - R\lambda_1)(1 + R\lambda_2)/2R^2\lambda_1 - E$ $- \chi_2(1 - R\lambda_2)(1 + R\lambda_1)/2R^2\lambda_2 + R/(1+R)$ $B = -\chi_1(1 + R\lambda_2)/R + \chi_2(1 + R\lambda_1)/R + E$ $C = \chi_1(3 - R\lambda_2)/R - \chi_2(3 - R\lambda_1)/R + E$ $E = 0.5e^{(-\text{NTU}/3)}$ $\lambda_{1,2} = (-3 \pm \delta)/2$ $\delta = \frac{\left[9R^2 + 4(1+R)\right]^{0.5}}{R}$ $\chi_{1,2} = \frac{e^{(\lambda_{1,2}R\text{NTU}/3)}}{2\delta}$	Same as Equation T16 with R = 1
1–3 TEMA E shell: one parallelflow and two counterflow passes; shell fluid mixed between passes 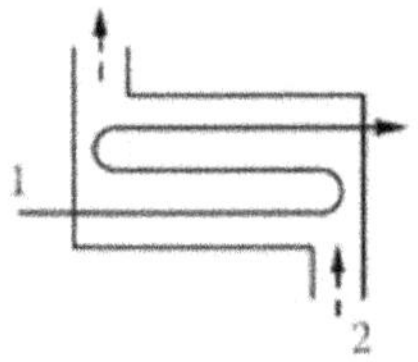	T17 [18] Fig, 2.31	$P_1 = \left[1 - \frac{C}{(AC + B^2)}\right]$ $A = \chi_1(1 + R\lambda_1)(1 - R\lambda_2)/2R^2\lambda_1 - E$ $- \chi_2(1 + R\lambda_2)(1 - R\lambda_1)/2R^2\lambda_2 + R/(R-1)$ $B = \chi_1(1 - R\lambda_2)/R - \chi_2(1 - R\lambda_1)/R + E$ $C = -\chi_1(3 + R\lambda_2)/R + \chi_2(3 + R\lambda_1)/R + E$ $E = 0.5e^{(\text{NTU}/3)}$ $\lambda_{1,2} = (-3 \pm \delta)/2$ $\delta = \frac{\left[9R^2 + 4(1-R)\right]^{0.5}}{R}$ $\chi_{1,2} = \frac{e^{(\lambda_{1,2}R\text{NTU}/3)}}{2\delta}$	Same as Equation T17 with R = 1 and $A = -\frac{e^{-\text{NTU}}}{18} - \frac{e^{\text{NTU}/3}}{2} + \frac{(5+\text{NTU})}{9}$

Note: P_2 (P_s) can be found using Equation 2.55 or P_2 (P_s) = P_1R_1, $R_2 = 1/R_1$, $\text{NTU}_2 = R_1\text{NTU}_1$.

arrangement was analytically derived by Kohei Ishihara and Palen [39]. It is also possible to have only a single pass on the tubeside [40]. The thermal effectiveness relation referred to the shellside for H_{1-1} is given by Equation T22 in Table 2.10, and the thermal effectiveness chart referred to the shellside is given in Figure 2.35. For the H_{1-2} arrangement, the thermal relations for the parallelflow

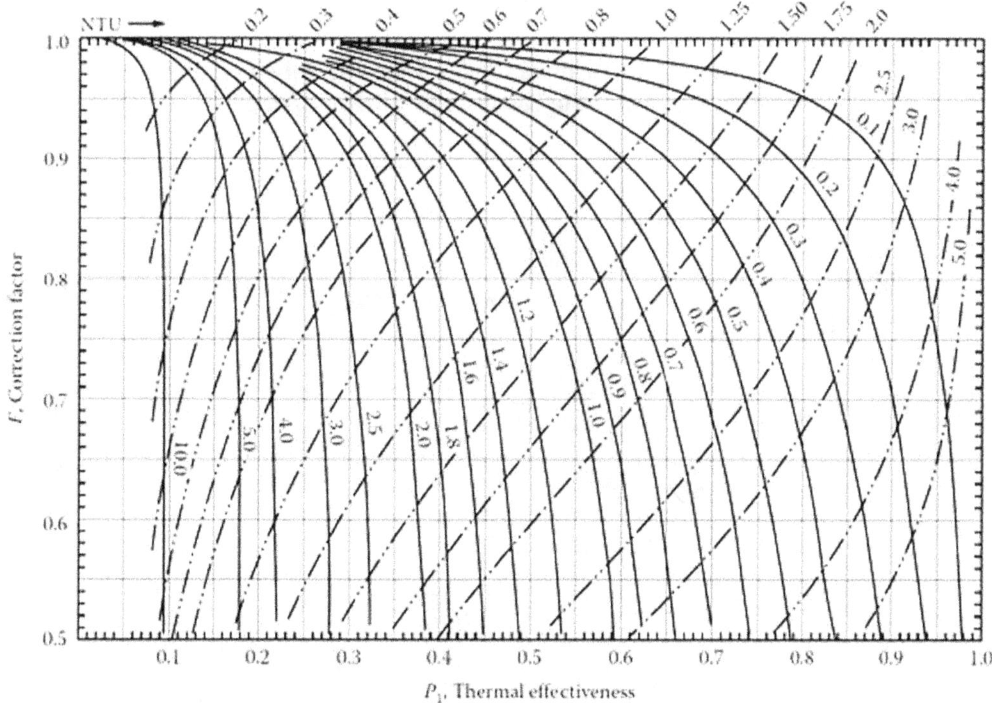

FIGURE 2.31 Thermal effectiveness chart—TEMA E_{1-3} shell. One parallelflow and two counterflow passes (overall counterflow arrangement); shell fluid mixed, tube fluid mixed between passes. F as a function of P for constant R (solid lines) and constant NTU (dashed lines) (as per Equation T17, Table 2.7).

and counterflow arrangements are given by Equations T23 and T24, respectively, and the thermal effectiveness chart for the counterflow arrangement is given in Figure 2.36.

2.9.5 TEMA *J* Shell or Divided-Flow Shell

The use of divided-flow exchangers with one shell pass and one or more tube passes is very common. If the shellside heat transfer resistance is not a limiting factor, and entrance and exit losses are neglected, the shellside pressure loss is approximately one-eighth of that same heat exchanger arranged as the conventional E_{1-2} or E_{1-4} exchanger. The possible flow arrangements are:

1. Divided-flow shell with one tube pass. This arrangement is equivalent to the E_{1-2} exchanger with unmixed shell fluid.
2. Divided-flow shell with two tube passes.
3. Divided-flow shell with four tube passes.
4. Divided-flow shell with infinite number of tube passes.

Jaw [41] analyzed the cases with two and four tube passes. Extending the number of tube passes to infinity, the case becomes identical to that of mixed–mixed crossflow, as derived by Gardner [42a]. Divided flow shell with one tube pass arrangement is equivalent to E_{1-2} exchanger with unmixed shell fluid as presented by Gardner [42b]. The difference in P is negligible for four or more passes compared to the two passes arrangement in the region of interest, i.e., $F > 0.5$. The thermal effectiveness relations for J_{1-1}, J_{1-2}, J_{1-4}, and $J_{1-\infty}$ arrangements are given by Equations T25–T28, respectively,

TABLE 2.8
Thermal Effectiveness Relations for Formula for E_{1-5} Shell and Tube Exchanger with Three Tube Flows in Counterflow and Two Tube Passes in Parallelflow Arrangement; Shell Fluid Mixed (Referred to Tubeside)

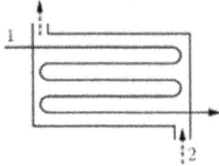

$$P_1 = \frac{(1 - X_N)}{R} \quad \text{for } N = 5 \ \textit{where}$$

$$X_N = X_5 = \frac{(\alpha\gamma - \beta^2)}{\left[(\alpha\gamma - \beta^2)H_{13} + \alpha\xi^2 + \gamma\eta^2 - 2\beta\xi\eta\right]}$$

$$\alpha = H_{24} - H_{11} - 2H_{12}$$

$$\beta = H_{22} - H_{13} - 2H_{12}$$

$$\gamma = H_{24} - H_{13} - 2H_{12}$$

$$\xi = H_{11} + H_{12}$$

$$\eta = H_{13} + H_{12}$$

T18
Fig, 2.32

$$H_{11} = \frac{(1+R\lambda_1)(1-R\lambda_2)}{R^2(N+1)\lambda_1\delta}X_1 - \frac{(1-R\lambda_1)(1+R\lambda_2)}{R^2(N+1)\lambda_2\delta}X_2 + \frac{N-1}{N+1}X_3 + \frac{R}{R-1} \text{ for } R \neq 1$$

$$= \frac{2N - 1 + \text{NTU}}{N^2} + \frac{1-N}{(N+1)N^2}X_2 + \frac{N-1}{N+1}X_3 \text{ for } R = 1$$

$$H_{12} = \frac{(1+R\lambda_1)(1+R\lambda_2)}{R^2(N-1)\lambda_1\delta}X_1 - \frac{(1+R\lambda_1)(1+R\lambda_2)}{R^2(N-1)\lambda_2\delta}X_2 - \frac{R}{R-1} \text{ for R} \neq 1$$

$$= \frac{1 - \text{NTU}}{N^2} - \frac{X_2}{N^2} \text{ for } R = 1$$

$$H_{13} = \frac{(1+R\lambda_1)(1-R\lambda_2)}{R^2(N+1)\lambda_1\delta}X_1 - \frac{(1-R\lambda_1)(1+R\lambda_2)}{R^2(N+1)\lambda_2\delta}X_2 - \frac{2}{N+1}X_3 + \frac{R}{R-1} \text{ for } R \neq 1$$

$$= \frac{2N - 1 + \text{NTU}}{N^2} + \frac{1-N}{(N+1)N^2}X_2 - \frac{2}{N+1}X_3 \text{ for } R = 1$$

$$H_{22} = \frac{(1-R\lambda_1)(1+R\lambda_2)}{R^2(N-1)\lambda_1\delta}X_1 - \frac{(1+R\lambda_1)(1-R\lambda_2)}{R^2(N-1)\lambda_2\delta}X_2 + \frac{N-3}{N-1}X_4 - \frac{R}{R-1} \text{ for } R \neq 1$$

$$= \frac{2N + 1 - \text{NTU}}{N^2} + \frac{1+N}{(N-1)N^2}X_2 - \frac{N-3}{N-1}X_4 \text{ for } R = 1$$

$$H_{24} = \frac{(1-R\lambda_1)(1+R\lambda_2)}{R^2(N-1)\lambda_1\delta}X_1 - \frac{(1+R\lambda_1)(1-R\lambda_2)}{R^2(N-1)\lambda_2\delta}X_2 - \frac{2X_4}{N-1} - \frac{R}{R-1} \text{ for } R \neq 1$$

$$= \frac{2N + 1 - \text{NTU}}{N^2} + \frac{1-N}{(N-1)N^2}X_2 - \frac{2X_4}{N-1} \text{ for } R = 1$$

Sources: Pignotti, A. and Tamborenea, P.I., *Trans. ASME, J. Heat Transfer*, 111, 54, 1988.

Note 1: Expressions for δ, λ_1, λ_2, X_1, X_2, X_3 and X_4 are $\lambda_{1,2} = -\frac{N}{2} \pm \left[\frac{N^2}{4} + \frac{1}{R^2}(1-R)\right]^{0.5}$; $\lambda_3 = 1/R$; $\lambda_4 = 1/R$; $\delta = \lambda_1 - \lambda_2$; $X_1 = e^{(\lambda_1 R\text{NTU}/5)}$; $X_2 = e^{(\lambda_2 R\text{NTU}/5)}$; $X_3 = e^{(\lambda_3 R\text{NTU}/5)}$; $X_4 = e^{(\lambda_4 R\text{NTU}/5)}$;

Note 2: P_2 (P_s) can be found using Equation 2.55 or P_2 (P_s) = P_1R_1, $R_2 = 1/R_1$, $\text{NTU}_2 = R_1\text{NTU}_1$

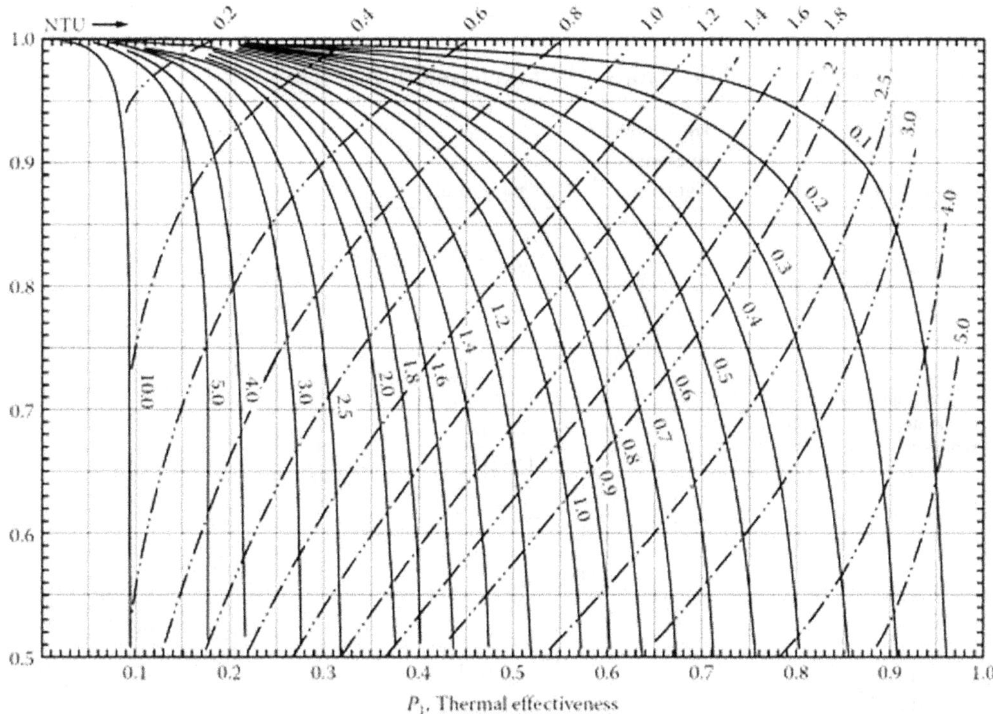

FIGURE 2.32 Thermal effectiveness chart—TEMA E_{1-5} shell. Two parallelflow and three counterflow passes (overall counterflow arrangement); shell fluid mixed, tube fluid mixed between passes. F as a function P for constant R (solid lines) and constant NTU (dashed lines) (as per Equation T18, Table 2.8).

in Table 2.11. The thermal effectiveness charts for J_{1-1}, J_{1-2}, and J_{1-4} are given in Figures 2.37–2.39, respectively.

2.9.6 TEMA *X* Shell

This exchanger is for very-low-pressure applications. For an X_{1-1} exchanger, the thermal effectiveness is the same as single-pass crossflow with both fluids unmixed, that is, the unmixed–unmixed case [1]. An X_{1-2} crossflow shell and tube exchanger is equivalent to Figure 2.40a for both fluids unmixed throughout with overall parallelflow (shell fluid entering at the tube inlet pass end) and Figure 2.40b for overall counterflow (shell fluid entering at the tube exit pass end), respectively.

2.10 THERMAL EFFECTIVENESS OF MULTIPLE HEAT EXCHANGERS

A multiple exchanger assembly consists of two or more exchangers piped or manifolded together into a single assembly with the exchangers arranged either in a parallel, series, or combination of parallel/series arrangement. Multiple exchanger assemblies are used when the customer's heat transfer requirements are too large for either single piece or modular heat exchanger construction. Sometimes, to utilize the allowable pressure drop effectively, multipassing is also employed; this in turn enhances the thermal effectiveness. Multipassing by multiple units is possible with both compact exchangers and shell and tube exchangers. With multipassing, two or more exchangers can be coupled either in an overall parallel or in an overall countercurrent scheme, as shown in Figure 2.41. Multipassing is also possible in a single unit of a compact exchanger (Figure 2.42).

TABLE 2.9
Thermal Effectiveness Relations for *G* Shell (Referred to Tubeside)

Flow arrangement	Equation no./reference/Figure no.	General formula	Value for *R* = 1 and special cases
1–1 *G* shell: tube fluid split into two streams; shell fluid mixed. Stream symmetric	T19 [37] Fig. 2.33	$P=\frac{1}{R}\left[A+B-AB(1+1/R)+AB^2/R\right]$ $A=\frac{1-e^{-\text{NTU}(1+R)/2}}{(1+R)/R}$ $B=\frac{(1-D)}{(1-D/R)}$ $D=\text{e}^{-\text{NTU}(R-1)/2}$	$B=\frac{\text{NTU}}{2+\text{NTU}}$ for $R=1$
1–2 TEMA *G* shell: overall parallelflow arrangement; shell fluid mixed, tube fluid mixed between passes	T20 [18]	$P_1=\frac{(B-\alpha^2)}{R(A-\alpha^2/R+2)}$ $A=\frac{(1-\alpha)^2}{(R-0.5)}$ $B=\frac{4R-\beta(2R-1)}{(2R+1)}$ $\alpha=e^{\frac{-\text{NTU}(2R-1)}{4}}$ $\beta=e^{\frac{-\text{NTU}(2R+1)}{2}}$	For R = 0.5 $P_1=\frac{1+2R\text{NTU}-\beta}{R(4+4R\text{NTU}+R^2\text{NTU}^2)}$ $\beta=e^{(-2R\text{NTU})}$
1–2 TEMA *G* shell: overall counterflow arrangement; shell fluid mixed, tube fluid mixed between passes	T21 [36] Fig. 2.34	$P_1=\frac{(B-\alpha^2)}{R(A+2+B/R)}$ $A=\frac{-(1-\alpha)^2}{(R+0.5)}$ $B=\frac{4R-\beta(2R+1)}{(2R-1)}$ $\alpha=e^{\frac{-\text{NTU}(2R+1)}{4}}$ $\beta=e^{\frac{-\text{NTU}(2R-1)}{2}}$	For R = 0.5 $P_1=\frac{1+2R\text{NTU}-\alpha^2}{R\left[4+4R\text{NTU}-(1-\alpha)^2\right]}$ $\alpha=e^{(-R\text{NTU})}$

Note: P_2 (P_s) can be found using Equation 2.55 or P_2 (P_s) = P_1R_1, $R_2=1/R_1$, $\text{NTU}_2=R_1\text{NTU}_1$

The fundamental formulas for the global effectiveness of multiple units are given next. These equations are not valid for multiple tubeside passes of various TEMA shells. The following idealizations are employed, in addition to those already listed at the beginning of this section:

1. Each pass has the same effectiveness, although any basic flow arrangement may be employed in any pass. If each pass has the same flow arrangement, then NTU is equally distributed between *N* passes, and NTU per pass (NTU_p) is given by NTU/*N*.

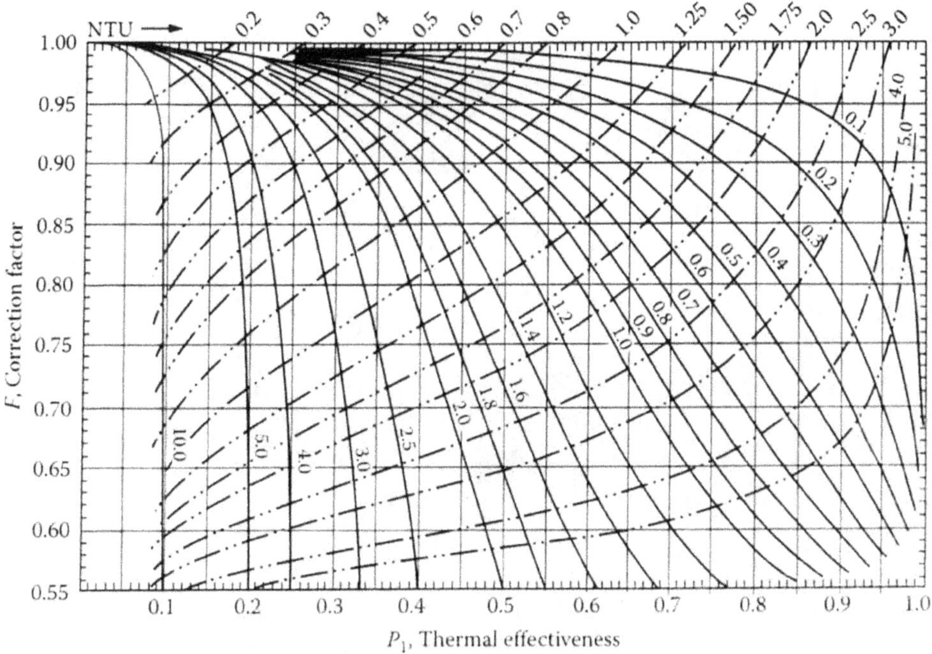

FIGURE 2.33 Thermal effectiveness chart—G_{1-1} shell. Tube fluid split into two streams with shell of mixed fluid. Stream symmetric. F as a function of P for constant R (solid lines) and constant NTU (dashed lines) (as per Equation T19, Table 2.9).

2. The fluid properties are idealized as constant so that C^* or R is the same for each pass. The final results of this analysis are valid regardless of which fluid is being C_{max} or C_{min}.

2.10.1 Two-Pass Exchangers

The expression for the global effectiveness for the parallelflow case is given by Pignotti [43] in the following equation:

$$P_2 = P_A + P_B - P_A P_B (1+R) \tag{2.72}$$

and for the counterflow case by

$$P_2 = \frac{P_A + P_B - P_A P_B (1+R)}{(1 - R P_A P_B)} \tag{2.73}$$

where

P_A and P_B are the thermal effectiveness of individual heat exchanger
R is the capacity rate ratio

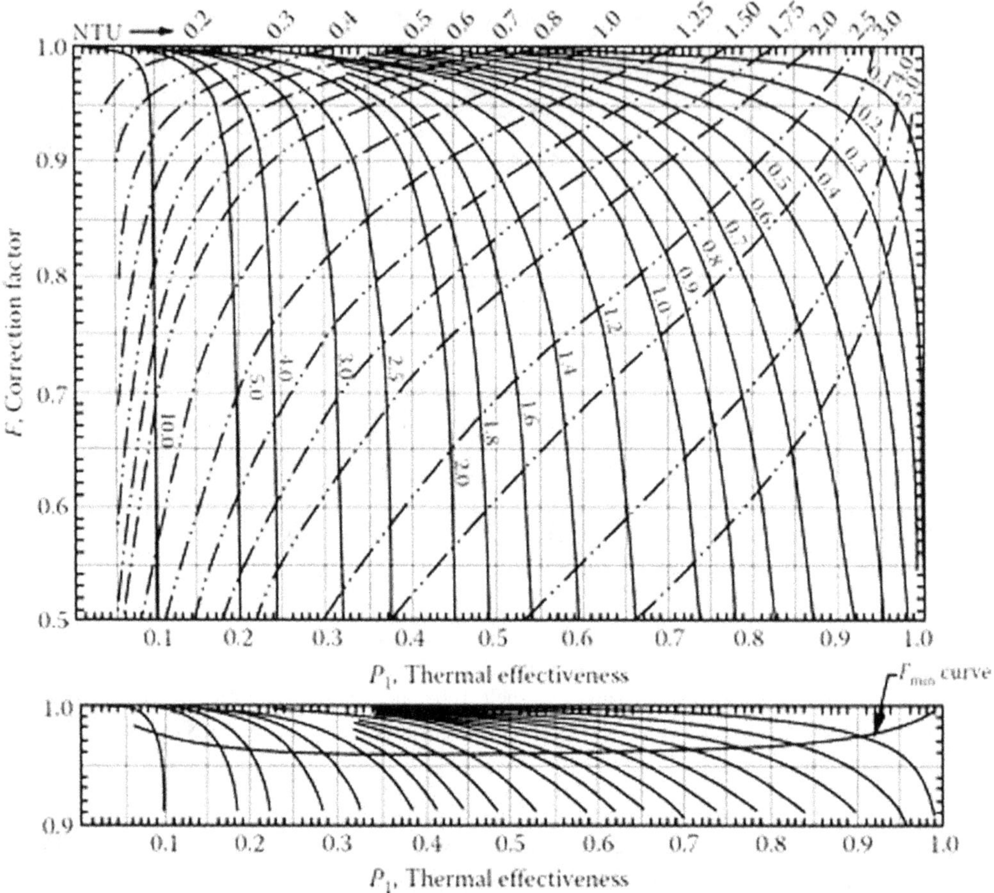

FIGURE 2.34 Thermal effectiveness chart—TEMA G_{1-2} shell. Overall counterflow arrangement; shell fluid mixed, tube fluid mixed between passes. F as a function of P for constant R (solid lines) and constant NTU (dashed lines). F_{min} curve is also shown (as per Equation T21, Table 2.9).

2.10.2 *N*-Pass Exchangers

Counterflow arrangement: The effectiveness P_N of an N-pass assembly, in terms of the effectiveness P_i of each unit, which is equal for all units, has been described by Domingos [44]. For overall countercurrent connection, the effectiveness is given by

$$P_N = \frac{1 - \left[\dfrac{(1-P_iR)}{(1-P_i)}\right]^N}{R - \left[\dfrac{(1-P_iR)}{(1-P_i)}\right]^N} \quad \text{for } R \neq 1 \tag{2.74a}$$

$$= \frac{NP_i}{1+(N-1)P_i} \quad \text{for } R = 1 \tag{2.74b}$$

TABLE 2.10
Thermal Effectiveness Relations for *H* Shell (Referred to Tubeside)

Flow arrangement	Equation no./ reference/ Figure no.	General formula	Value for *R* = 1 and special cases
1–1 *H* shell: tube fluid split into two streams; shell fluid mixed	T22 [37] Fig. 2.35	$P_1 = \frac{1}{R}\{E[1+(1-B/2R)(1-A/2R+AB/R)] - AB(1-B/2R)\}$ $A = \frac{1}{1+1/2R}(1-\exp[-R\,\mathrm{NTU}(1+1/2R)/2]$ $B = (1-D)/(1-D/2R)$ $D = e^{[-R\mathrm{NTU}(1+1/2R)/2]}$ $E = \left(A+B-\frac{AB}{2R}\right)/2$	Same as Equation T22 with $R = 0.5$ $B = \frac{R\mathrm{NTU}}{(2+R\mathrm{NTU})}$
1–2 TEMA *H* shell: overall parallelflow arrangement; shell fluid mixed, tube fluid mixed between passes	T23 [39]	$P_1 = \left[1-\frac{B+4GR}{(1-D)^4}\right]$ $B = (1+H)(1+E)^2$ $G = (1-D)^2(D^2+E^2)+D^2(1+E)^2$ $D = \frac{1-e^{-\alpha}}{1-4R}$ $E = \frac{e^{-\beta}-1}{4R+1}$ $H = \frac{e^{-2\beta}-1}{4R+1}$ $\alpha = \frac{\mathrm{NTU}}{8}(4R-1)$ $\beta = \frac{\mathrm{NTU}}{8}(4R+1)$	For $R = 0.25$ $P_1 = \left[1-\frac{B+4GR}{(1-D)^4}\right]$ $D = -\mathrm{NTU}/8$
1–2 TEMA *H* shell: overall counterflow arrangement; shell fluid mixed, tube fluid mixed between passes	T24 [39] Fig. 2.36	$P_1 = \left[1-\frac{(1-D)^4}{(B-4GR)}\right]$ $B = (1+H)(1+E)^2$ $G = (1-D)^2(D^2+E^2)+D^2(1+E)^2$ $D = \frac{1-e^{-\alpha}}{4R+1}$ $E = \frac{1-e^{-\beta}}{4R-1}$ $H = \frac{1-e^{-2\beta}}{4R-1}$ $\alpha = \frac{\mathrm{NTU}}{8}(4R+1)$ $\beta = \frac{\mathrm{NTU}}{8}(4R-1)$	For $R = 0.25$ $P_1 = \left[1-\frac{(1-D)^4}{(B-G)}\right]$ $B = (1+\alpha)(1+\alpha/2)^2$ $G = (1-D)^2 x$ $(D^2+\alpha^2/4)+$ $D^2(1+\alpha/2)^2$

Note: P_2 (P_s) can be found using Equation 2.55 or P_2 (P_s) $= P_1R_1$, $R_2 = 1/R_1$, $\mathrm{NTU}_2 = R_1\mathrm{NTU}_1$.

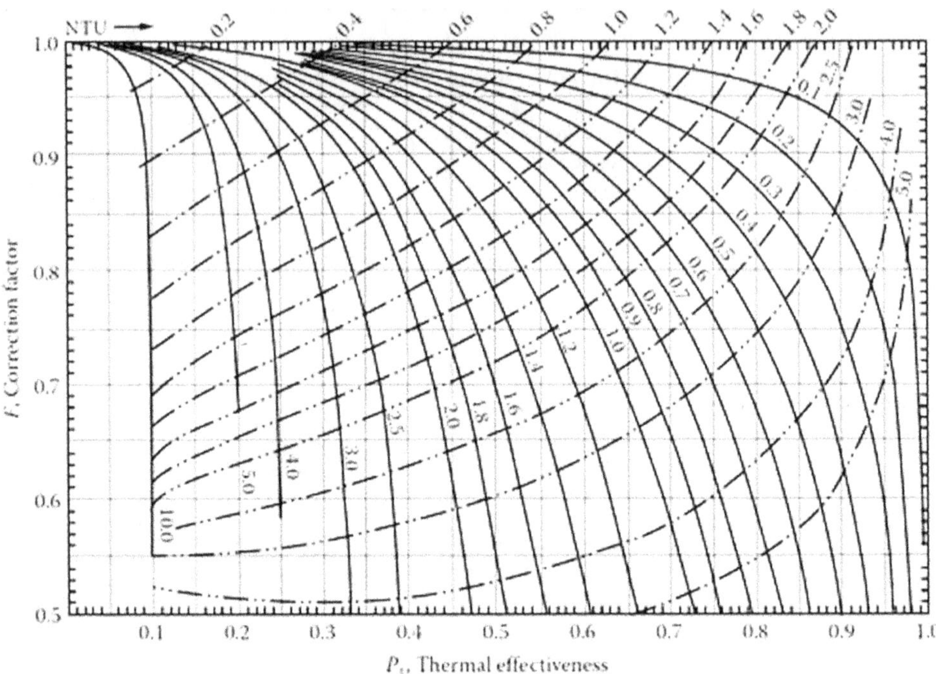

FIGURE 2.35 Thermal effectiveness chart—$H_{1\text{-}1}$ shell. Tube fluid split into two streams. Shell mixed. F as a function of P for constant R (solid lines) and constant NTU (dashed lines) referred to shellside (as per Equation T22, Table 2.10).

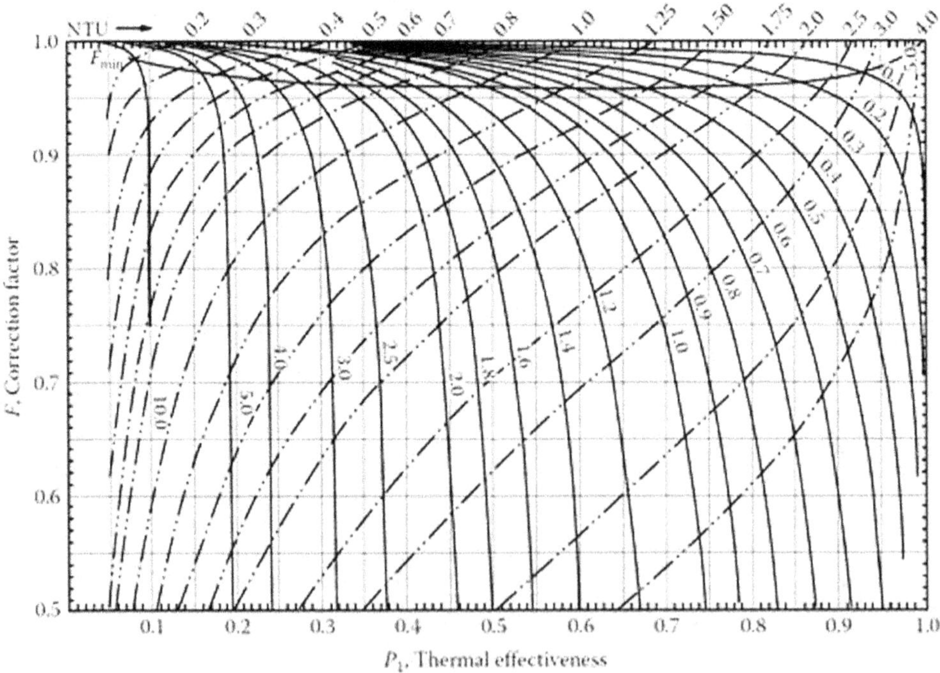

FIGURE 2.36 Thermal effectiveness chart—TEMA $H_{1\text{-}2}$ shell. Overall counterflow arrangement; shell fluid mixed, tube fluid mixed between passes. F as a function of P for constant R (solid lines) and constant NTU (dashed lines). F_{min} curve is also shown (as per Equation T24, Table 2.10).

TABLE 2.11
Thermal Effectiveness Relations for *J* Shell (Referred to Tubeside)

Flow arrangement	Equation no./ reference/ Figure no.	General formula	Values for $R = 1$ and special cases
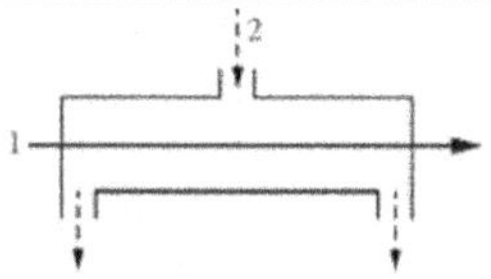 1–1 TEMA *J* shell: shell fluid mixed	T25 [40] Fig. 2.37	$P_1 = 1 - \frac{(2R-1)}{(2R+1)}\left[\frac{2R+\phi^{-(R+0.5)}}{2R-\phi^{-(R-0.5)}}\right]$ $\phi = e^{\mathrm{NTU}}$	For $R = 0.5$, $P_1 = 1 - \frac{1+\phi^{-1}}{2+\mathrm{NTU}}$
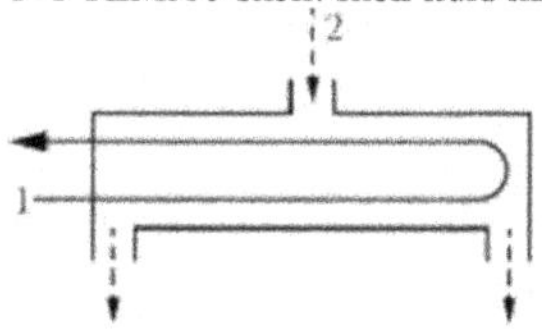 1–2 TEMA *J* shell: shell fluid mixed and tube fluid mixed between passes	T26 [41] Fig. 2.38	$P_1 = \frac{2}{1+2R[1+\lambda A - 2B\lambda]}$ $A = \frac{\phi^{R\lambda}+1}{\phi^{R\lambda}-1}$ $B = \frac{\frac{\phi^{R(1+\lambda)/2}}{\phi^{R\lambda}-1}}{1+\lambda\left(\frac{\phi^{R\lambda}+1}{\phi^{R\lambda}-1}\right)}$ $C = 1 + \frac{\lambda\phi^{R(\lambda-1)/2}}{\phi^{R\lambda}-1}$ $\phi = e^{\mathrm{NTU}}$ $\lambda = \frac{\sqrt{1+4R^2}}{2R}$	Same as Equation T26
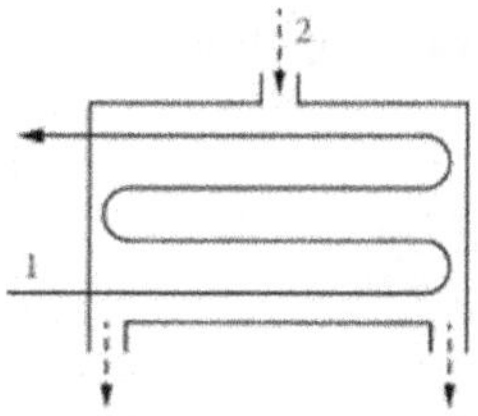 1–4 *J* shell: shell fluid mixed and tube fluid mixed between passes	T27 [41] Fig. 2.39	$P_1 = \frac{4}{1+A+4R[1+\lambda B - 2C\lambda]}$ $A = \frac{2\sqrt{\phi}}{1+\sqrt{\phi}}$ $B = \frac{\phi^{R\lambda}+1}{\phi^{R\lambda}-1}$ $C = \frac{\frac{\phi^{R(1+\lambda)/2}}{\phi^{R\lambda}-1}}{1+\lambda\left(\frac{\phi^{R\lambda}+1}{\phi^{R\lambda}-1}\right)}$ $D = 1 + \frac{\lambda\phi^{R(\lambda-1)/2}}{\phi^{R\lambda}-1}$ $\phi = e^{\mathrm{NTU}}$ $\lambda = \frac{\sqrt{1+16R^2}}{4R}$	Same as Equation T27
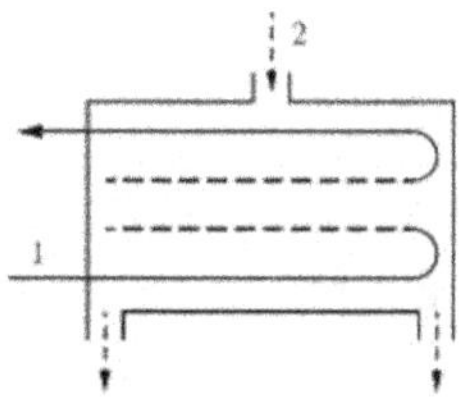 1–*N* *J* shell shell fluid mixed and tube fluid mixed between passes; stream symmetric	T28 [42]	$P = \frac{1}{\frac{R\phi^R}{(\phi^R-1)} + \frac{\phi}{\phi-1} - \frac{1}{\mathrm{NTU}}}$ $\phi = e^{\mathrm{NTU}}$	Same as Equation T28

Note: P_2 (P_s) can be found using Equation 2.55 or P_2 (P_s) $= P_1R_1$, $R_2 = 1/R_1$, $\mathrm{NTU}_2 = R_1\mathrm{NTU}_1$

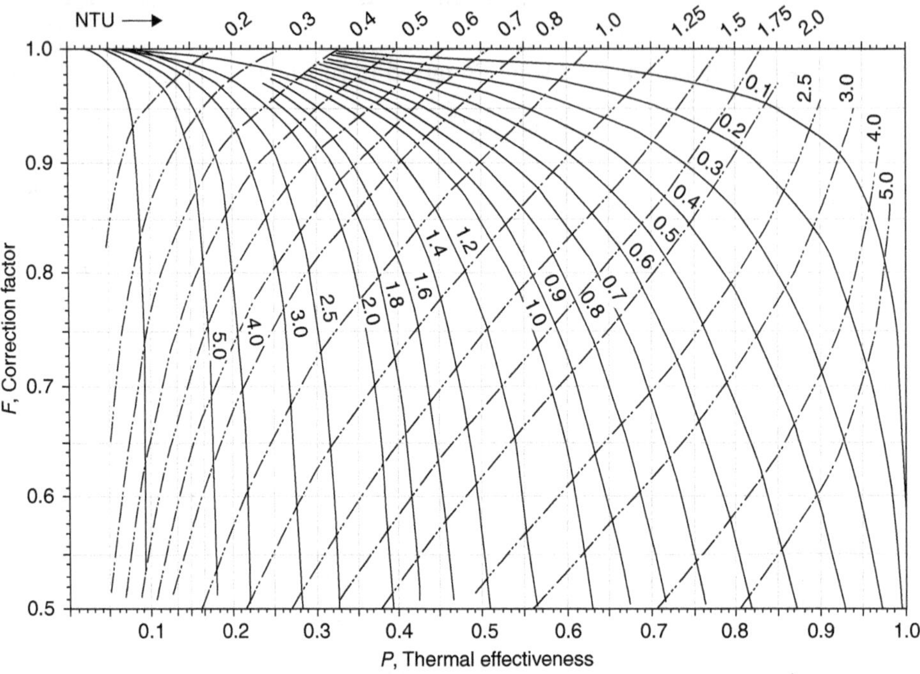

FIGURE 2.37 Thermal effectiveness chart—TEMA J_{1-1} shell. Shell fluid mixed. F as a function of P for constant R (solid lines) and constant NTU (dashed lines) (as per Equation T25, Table 2.11).

where R is the heat capacity rate ratio, which is the same for the individual unit, as for the

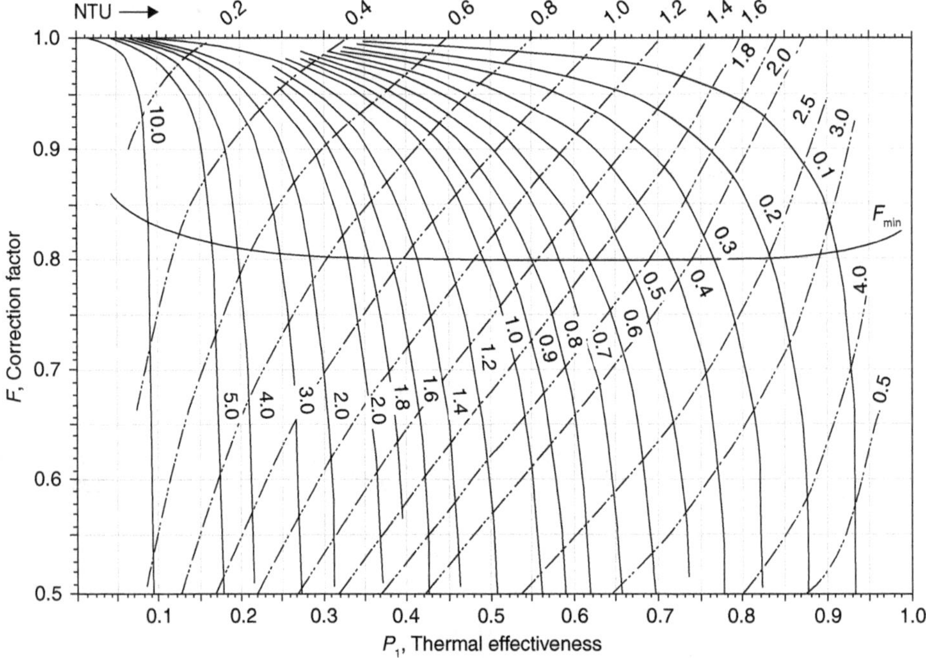

FIGURE 2.38 Thermal effectiveness chart—TEMA J_{1-2} shell. Shell fluid mixed and tube fluid mixed between passes. F as a function of P for constant R (solid lines) and constant NTU (dashed lines), with F_{min} to avoid temperature cross (horizontal bowl-shaped curve) (as per Equation T26, Table 2.11).

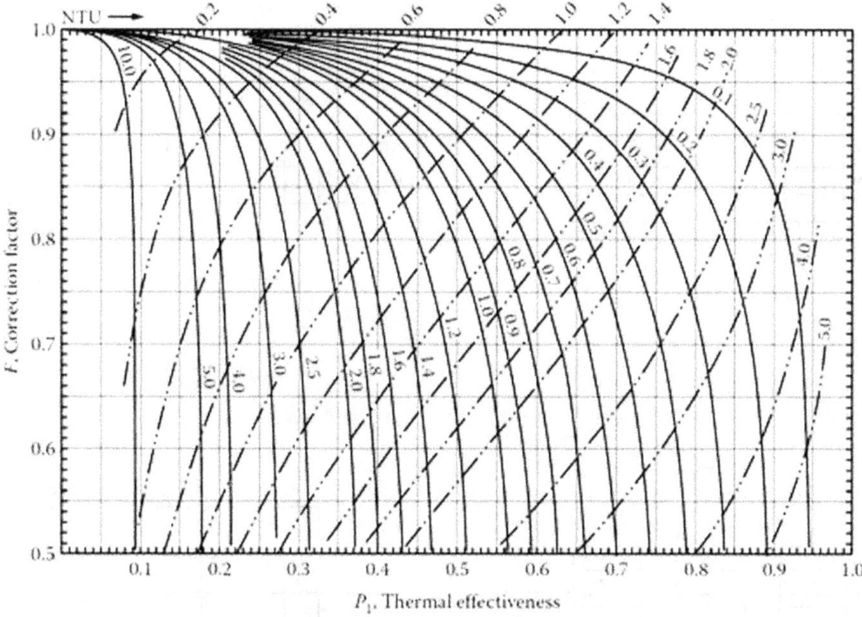

FIGURE 2.39 Thermal effectiveness chart—TEMA $J_{1\text{-}4}$ shell. Shell fluid mixed and tube fluid mixed between passes. F as a function of P for constant R (solid lines) and constant NTU (dashed lines) (as per Equation T27, Table 2.11).

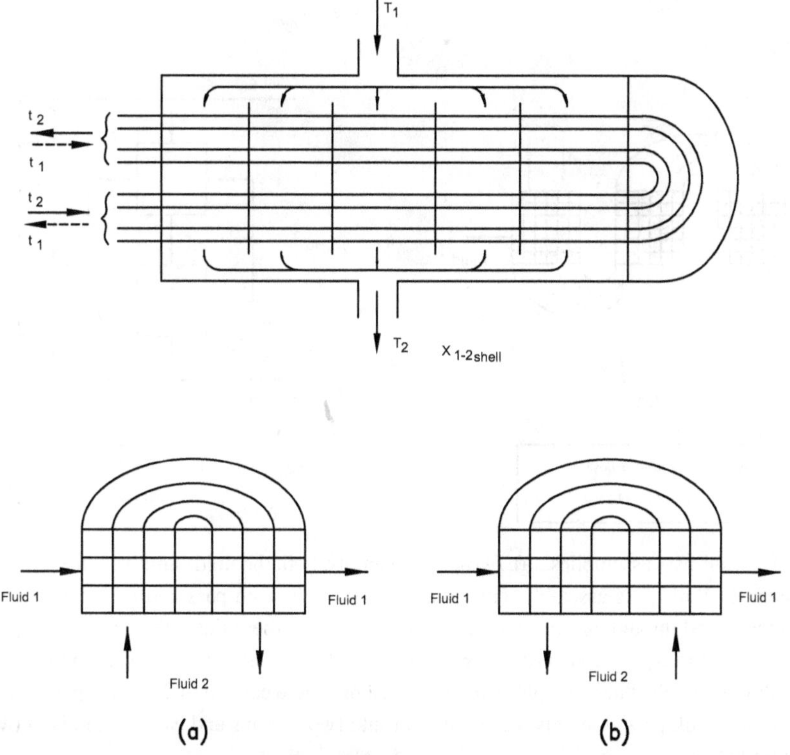

FIGURE 2.40 $X_{1\text{-}2}$ shell arrangement equivalent with two-pass crossflow heat exchanger: (a) parallelflow; (b) counterflow.

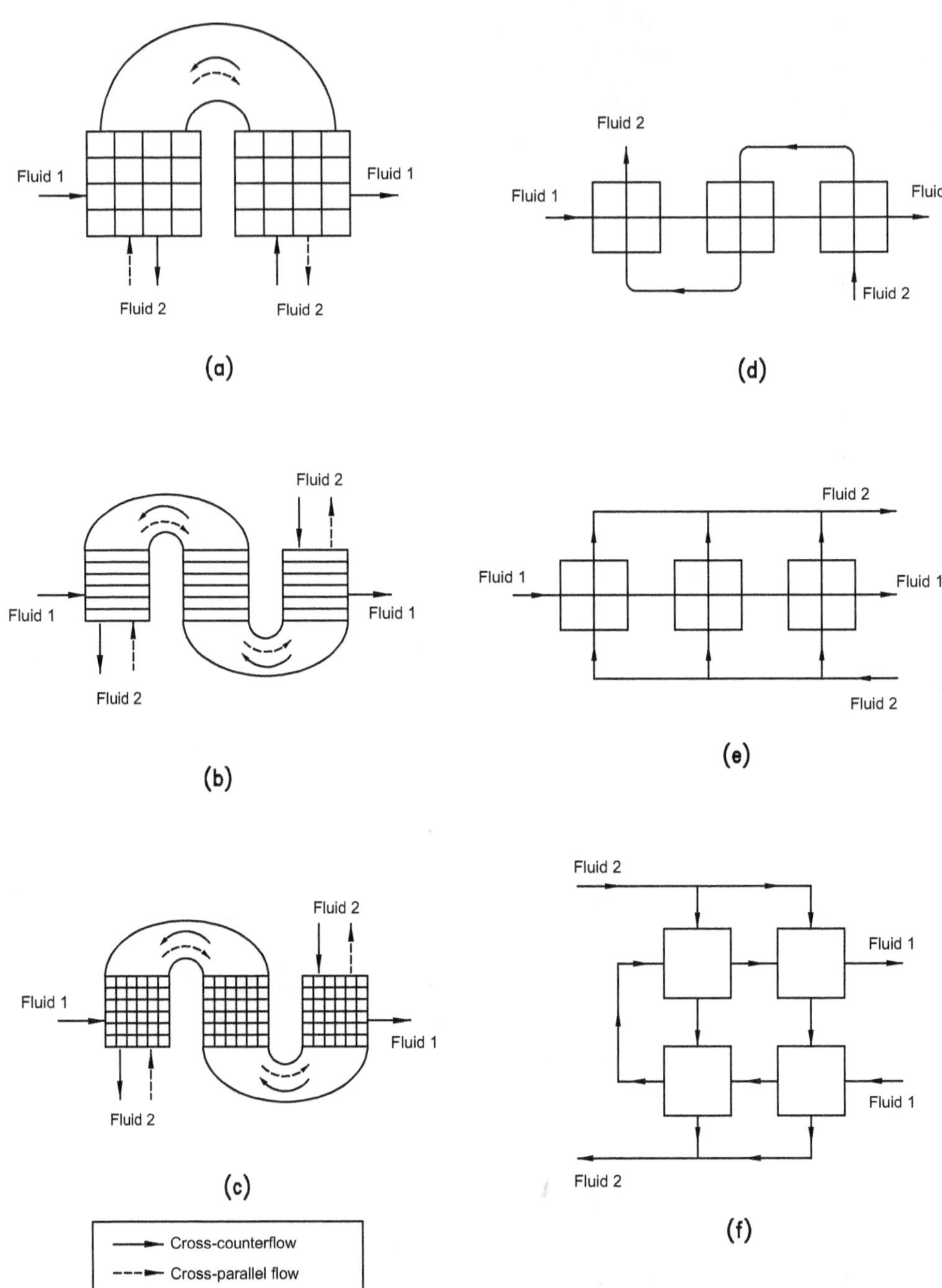

FIGURE 2.41 Multipass assemblies. (a) Two-pass assembly, both the fluids unmixed in each pass and mixed between passes; (b) three-pass assembly, one fluid is unmixed in each pass and mixed between passes, and the other fluid is mixed throughout; (c) three-pass assembly, both the fluids unmixed in each pass and mixed between passes; (d) three-pass counterflow assembly, both the fluids mixed between passes; (e) three-pass parallel-crossflow assembly, fluid 1 is split in to three streams-single pass, fluid 2 in three passes mixed between passes; (f) complex multipass assembly, fluid 2 is split into two streams and each stream is in two passes and fluid 1 is in four passes.

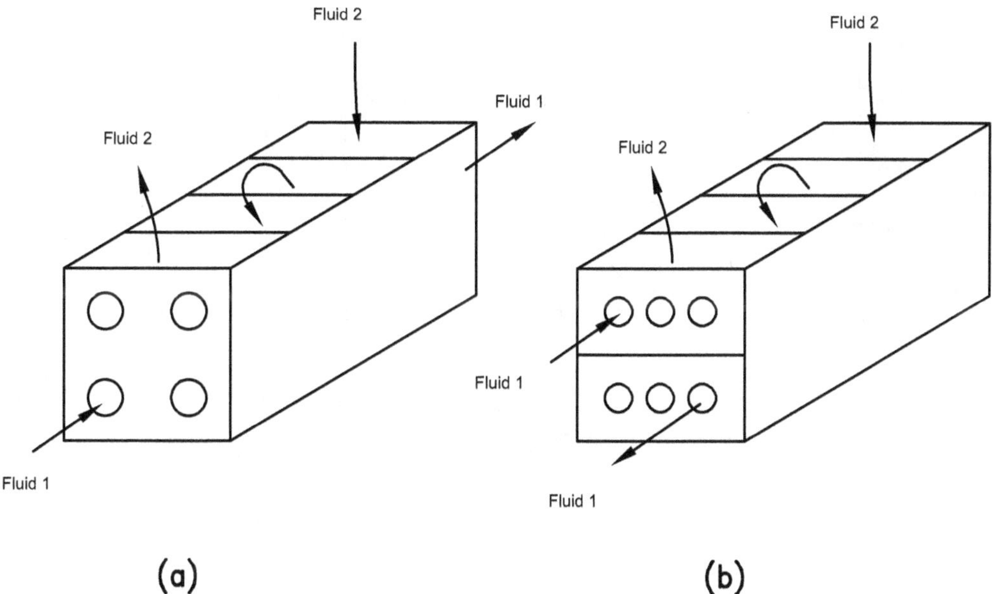

FIGURE 2.42 Multipassing in the same unit.

overall assembly. Alternately, if P_N is known, then the individual component effectiveness can be determined from the following formulas:

$$P_i = \frac{1 - \left[\frac{(1 - P_N R)}{(1 - P_N)}\right]^{1/N}}{R - \left[\frac{(1 - P_N R)}{1 - P_N}\right]^{1/N}} \quad \text{for } R \neq 1 \tag{2.75a}$$

$$= \frac{P_N}{N - P_N (N - 1)} \quad \text{for } R = 1 \tag{2.75b}$$

2.11 MULTIPASS CROSSFLOW EXCHANGERS

Two or more crossflow units can be coupled in two or more passes either in an overall parallelflow or in an overall counterflow arrangement. When both fluids are mixed in the interpass, mixing the resulting flow arrangement can be differentiated only by the flow arrangements in each pass (unmixed–mixed or mixed–mixed). However, if one fluid is unmixed throughout, the order in which the streams enter the next pass must be differentiated. The coupling of the unmixed fluid from one pass to the other pass can be of two types as shown in Figure 2.43 [1]: (1) identical order coupling and (2) inverted order coupling. The coupling is referred to as an identical order if the stream leaving one pass enters the next pass from the same side as in the previous pass, as in Figure 2.43a, whereas a coupling is considered to be in an inverted order if the stream leaving one pass enters the next pass

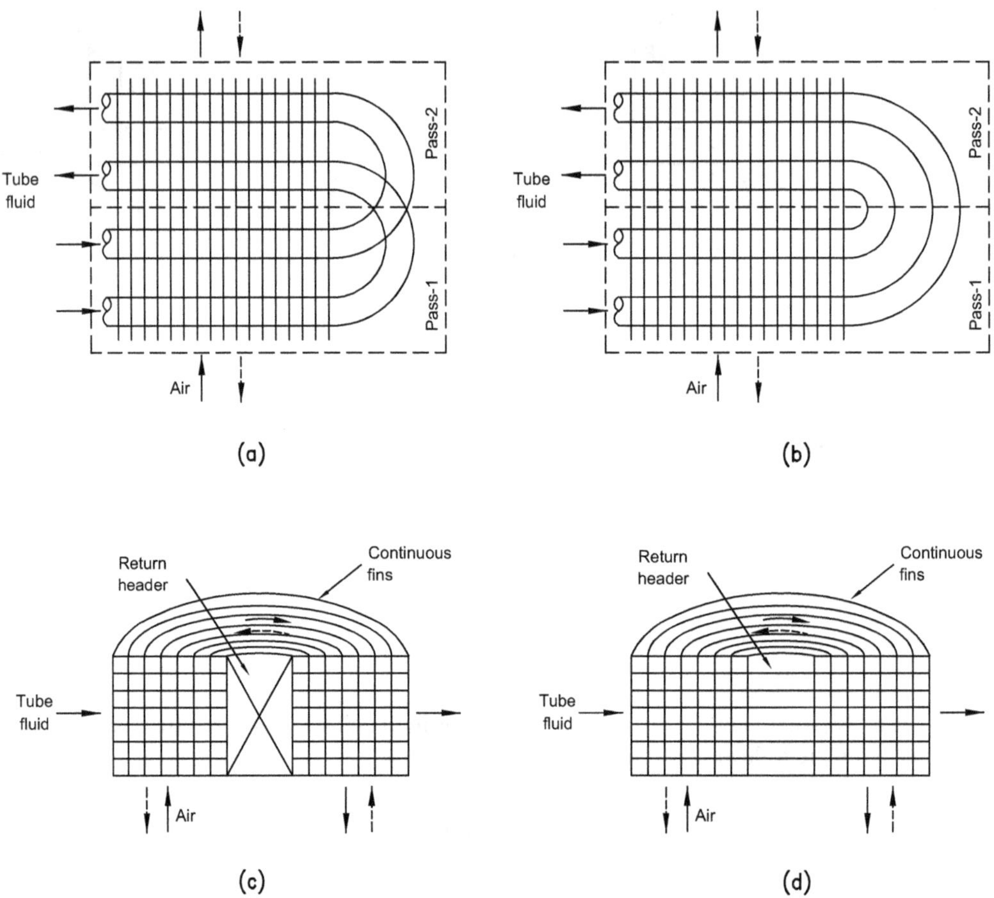

FIGURE 2.43 Two-pass parallel-crossflow and counter-crossflow with both fluids unmixed throughout; coupling of tubefluid between passes: (a) identical order coupling; (b) inverted order coupling. The cases (c) and (d) are symbolic representations of cases (a) and (b), respectively. (Adapted from Shah, R.K. and Mueller, A.C., Heat exchanger basic thermal design methods, in *Handbook of Heat Transfer Applications*, 2nd edn. [W.M. Rohsenow, J.P. Hartnett, and E.N. Ganic, eds.], McGraw-Hill, New York, pp. 4-1–4-77, 1985.)

from the opposite side of the previous pass, as in Figure 2.43b. Inverted-order coupling is very convenient for manufacturing and installation of heat exchangers. Thermal effectiveness must decrease for constant NTU whenever irreversibilities take place, which may be due to mixing between passes, or due to a higher local temperature difference between the fluids, which occurs when the order of the flow arrangement is inverted [28]. Some of the possible cases of two-pass heat exchangers are shown in Figures 2.44a–c. For some cases, the equivalence of thermal effectives is shown in Figures 2.44b and 2.44c. Thermal effectiveness relations for a few cases of two-pass assemblies of heat exchangers arrangement with one fluid unmixed throughout is shown in Table 2.12 and thermal effectiveness relations for a few cases of two-pass assemblies of heat exchangers arrangement with both fluids unmixed in the passes is given in Table 2.13. All these cases were analyzed by Stevens et al. [29], some analytically and others by numerical integration, whereas Baclic [3] solved all possible flow arrangements of two passes units analytically. Multipass crossflow arrangements with complete mixing between passes were analyzed by Domingos [44], and the formulas are also presented in Ref. [2].

TABLE 2.12
Thermal Effectiveness Relations for Two-Pass Assemblies of Heat Exchangers Arrangement with One Fluid Unmixed Throughout

Flow arrangement	Equation no./ reference/ Figure no.	Formula for overall cross-counterflow exchangers
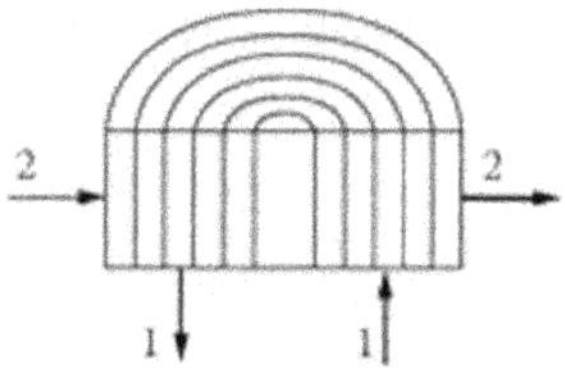 2 2 1 1 Fluid 1 unmixed throughout; inverted order coupling; fluid 2 mixed throughout; I(a) of Figure 2.44a	T29 [28] Fig. 2.51	$P_1 = \frac{1}{R}\left[1 - \frac{1}{\frac{K}{2} + \left(1 - \frac{K}{2}\right)e^{2KR}}\right]$ $K = 1 - e^{(-\text{NTU}/2)}$ Equation for P_2 = Equation T30
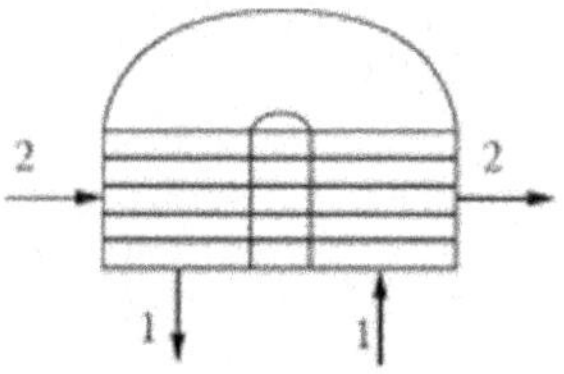 2 2 1 1 Fluid 1 mixed throughout; inverted order coupling; fluid 2 unmixed throughout; I(b) of Figure 2.44a	T30 [28] Fig. 2.52	$P_1 = \left[1 - \frac{1}{\frac{K}{2} + \left(1 - \frac{K}{2}\right)e^{(2K/R)}}\right]$ $K = 1 - e^{(-R\text{NTU}/2)}$ Equation for P_2 = Equation T29
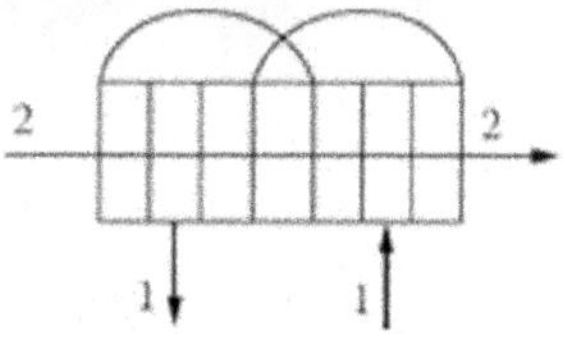 2 2 1 1 Fluid 1 unmixed throughout, fluid 2 mixed throughout; identical order coupling; II(a) of Figure 2.44b	T31 [28] Fig. 2.53	$P_1 = \frac{1}{R}\left[1 - \frac{e^{-KR}}{e^{KR} - K^2 R}\right]$ $K = 1 - e^{(-\text{NTU}/2)}$ Equation for P_2 = Equation T32
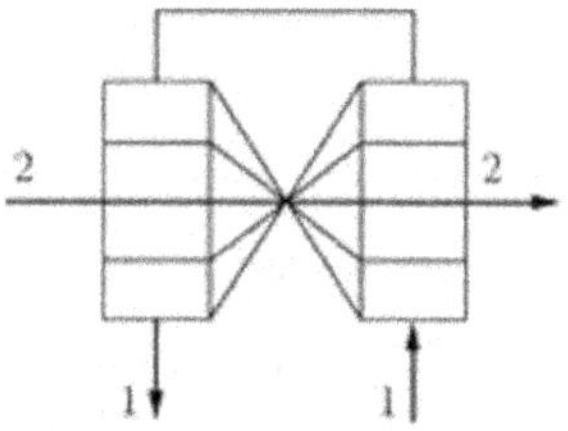 2 2 1 1 Fluid 1 mixed throughout; fluid 2 unmixed throughout, identical order coupling; II(b) of Figure 2.44a	T32 [28] Fig. 2.54	$P_1 = \left[1 - \frac{e^{-K/R}}{e^{K/R} - K^2/R}\right]$ $K = 1 - e^{(-R\text{NTU}/2)}$ Equation for P_2 = Equation T31

Note 1: For an individual case, at $R = 1$, $P_1 = P_2$.

TABLE 2.13
Thermal Effectiveness Relations for Two-Pass Assemblies of Heat Exchangers Arrangement with Both Fluids Unmixed in the Passes

Flow arrangement	Equation no./ reference/ Figure no.	Formula for overall cross-counterflow exchangers
Fluid 1 unmixed throughout; inverted order coupling; fluid 2 unmixed in each pass and mixed between passes; II(a) of Figure 2.44b	T33 [45]	$\varepsilon_1 = \frac{1}{C^*}\left[1 - \frac{\bar{V}_{1/2}^2}{\bar{V}_{1/2} + \bar{\mu}_{1/2}}\right]$ Eqn for ε_2 = Eqn. T34
Fluid 1 unmixed in each pass and mixed between passes; fluid 2 unmixed throughout; inverted order coupling; II(b) of Figure 2.44b	T34 [45] Fig. 2.55	$\varepsilon_1 = 1 - \frac{V_{1/2}^2}{V_{1/2} + \mu_{1/2}}$ Eqn for ε_2 = Eqn T33
Fluid 1 unmixed throughout, identical order coupling; fluid 2 mixed between passes and unmixed in each pass; I(a) of Figure 2.44b	T35 [46] Fig. 2.56	$\varepsilon_1 = \frac{1}{C^*}\left[1 - \frac{\bar{V}_{1/2}^2}{1 + 2\bar{V}_{1/2} - 2\bar{V}_{2/2}}\right]$ Eq for ε_2 = Eqn T36
Fluid 1 mixed between passes and unmixed in each pass; fluid 2 unmixed throughout, identical order; I(b) of Figure 2.44b	T36 [46]	$\varepsilon_1 = 1 - \frac{V_{1/2}^2}{1 + 2V_{1/2} - 2V_{2/2}}$ Eqn for ε_2 =Eqn T35

Note1: For individual case at $C^* = 1$, $\varepsilon_1 = \varepsilon_2$.

Note2: ε_2 can be found using Equation 2.55 or $\varepsilon_2 = \varepsilon_1 R_1$, $R_2 = 1/R_1$, $NTU_2 = R_1 NTU_1$

Terms of the equations:

$$\mu_{1/2} = \frac{2}{C^*\text{NTU}} \sum_{m=0}^{\infty} \sum_{m=0}^{\infty} \frac{(-1)^m (n+m)!}{n!m!} V_{m+2}\left(\frac{C^*\text{NTU}}{2}, \frac{\text{NTU}}{2}\right) V_{n+2}\left(\frac{\text{NTU}}{2}, \frac{C^*\text{NTU}}{2}\right) \quad \text{(A)}$$

Set $C^* = 1/C^*$, NTU = C^*NTU and C^*NTU = NTU in Equation (A), then, Equation (B) is obtained by

$$\bar{\mu}_{1/2} = \frac{2}{\text{NTU}} \sum_{n=0}^{\infty} \sum_{n=0}^{\infty} \frac{(-1)^m (n+m)!}{n!m!} V_{m+2}\left(\frac{\text{NTU}}{2}, \frac{C^*\text{NTU}}{2}\right) V_{n+2}\left(\frac{C^*\text{NTU}}{2}, \frac{\text{NTU}}{2}\right) \quad \text{(B)}$$

in which

TABLE 2.13 (Continued)
Thermal Effectiveness Relations for Two-Pass Assemblies of Heat Exchangers Arrangement with Both Fluids Unmixed in the Passes

Flow arrangement	Equation no./ reference/ Figure no.	Formula for overall cross-counterflow exchangers
$V_0(\xi,\eta)=e^{-(\xi+\eta)}I_0\left(2\sqrt{\xi\eta}\right)$	(C)	
$V_m(\xi,\eta)=e^{-(\xi+\eta)}\sum_{n=m-1}^{\infty}\binom{n}{m-1}\left(\frac{\eta}{\xi}\right)^{n/2}I_n\left(2\sqrt{\xi\eta}\right)\quad m\ge 1$	(D)	
$v(a,b)=e^{-(a+b)}\left[I_0\left(2\sqrt{a\,b}\right)+\sqrt{(b/a)}I_1\left(2\sqrt{ab}\right)-\left(\frac{a}{b}-1\right)\sum_{n=2}^{\infty}\left(\frac{b}{a}\right)^{n/2}I_n\left(2\sqrt{ab}\right)\right]$	(E)	

The expressions for $v_{1/2}$ and $\overline{v}_{1/2}$ are given by

$$v_{1/2}=\exp\left[-(1+C^*)\frac{\text{NTU}}{2}\right]\times\left[I_0\left(\text{NTU}\sqrt{C^*}\right)+\sqrt{C^*}I_1\left(\text{NTU}\sqrt{C^*}\right)-\frac{1-C^*}{C^*}\sum_{n=2}^{\infty}(C^*)^{n/2}I_n\left(\text{NTU}\sqrt{C^*}\right)\right]\quad\text{(F)}$$

$$\overline{v}_{1/2}=\exp\left[-\frac{1+C^*}{C^*}\frac{C^*\text{NTU}}{2}\right]\times\left[I_0\left(\text{NTU}\sqrt{C^*}\right)+\sqrt{\frac{1}{C^*}}I_1\left(\text{NTU}\sqrt{C^*}\right)-(C^*-1)\sum_{n=2}^{\infty}(C^*)^{-n/2}I_n\left(\text{NTU}\sqrt{C^*}\right)\right]\quad\text{(G)}$$

Various forms of Equation (E): $\overline{v}=\overline{v}_{1/1}=v(\text{NTU},\ C^*\text{NTU})$; $\overline{v}_{1/2}=v(\text{NTU}/2,\ C^*\text{NTU}/2)$; $\overline{v}_{2/2}=v(\text{NTU}/2,\ C^*\text{NTU})$; $\overline{v}=\overline{v}_{1/1}=v(C^*\text{NTU},\ \text{NTU})$; $\overline{v}_{1/2}=v(C^*\text{NTU}/2,\ \text{NTU}/2)$; $\overline{v}_{2/2}=v(C^*\text{NTU}/2,\ \text{NTU})$.

Sources: Shah, R.K. and Mueller, A.C., Heat exchanger basic thermal design methods, in *Handbook of Heat Transfer Applications*, 2nd edn., W.M. Rohsenow, J.P. Hartnett, and E.N. Ganic, eds., McGraw-Hill, New York, pp. 4-1–4-77, 1985; Baclic, B.S. and Gvozdenac, D.D., Exact explicit equations for some two–and three-pass cross-flow heat exchanger effectiveness, in *Heat Exchangers; Thermal-Hydraulic Fundamentals and Design*, S. Kakac, A.E. Bergles, and F. Mayinger, eds., Hemisphere, Washington, DC, pp. 481–494, 1981; Baclic, B.S. and Gvozdenac, D.D., NTU relationships for inverted order flow arrangements of two-pass crossflow heat exchangers, in *Regenerative and Recuperative Heat Exchangers*, Vol. 21, R.K. Shah and D.E. Metzger, eds., ASME, New York, pp. 27–41, 1981.

2.11.1 Multipassing with Complete Mixing between Passes

The thermal effectiveness charts are given for the following cases of two- and three-pass cross-counterflow arrangements with complete mixing between passes:

1. Weaker fluid is mixed in all the passes: For the two-pass arrangement (I-a or I-b of Figure 2.44a), the thermal effectiveness chart is given in Figure 2.45, and in Figure 2.46 for the three-pass arrangements.
2. Stronger fluid is mixed in all the passes: For the two-pass arrangement (I-a or I-b of Figure 2.44a), the thermal effectiveness chart is given in Figure 2.47, and in Figure 2.48 for the three-pass arrangements.

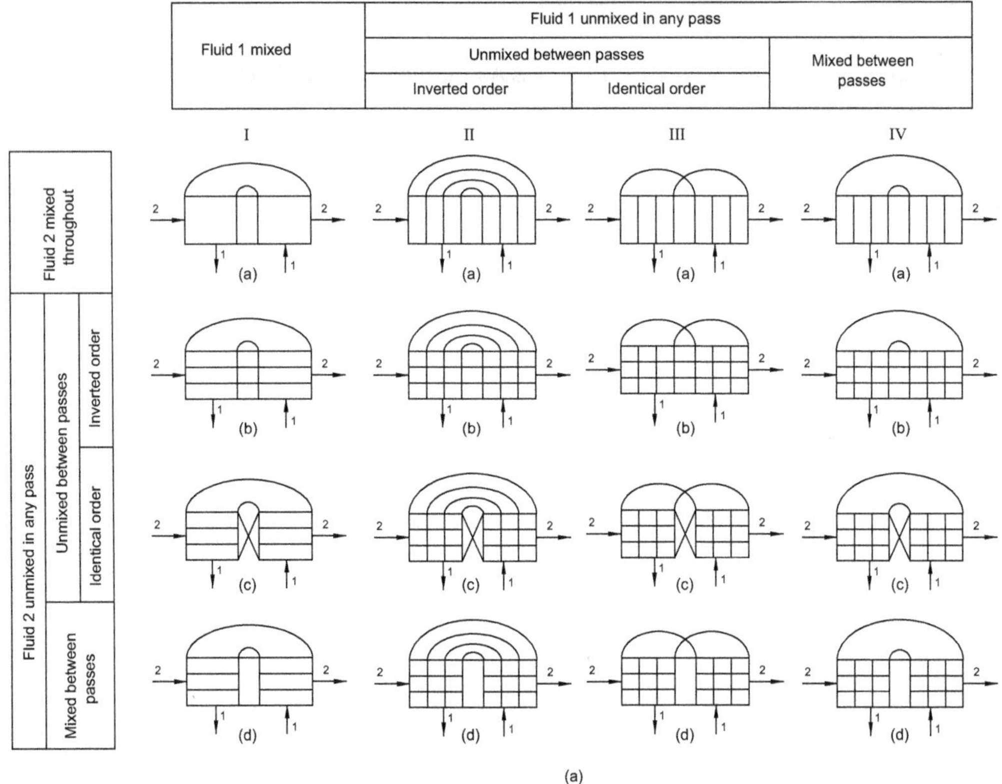

FIGURE 2.44 (a) Flow arrangements for possible cases of two-pass heat exchangers (b) Flow arrangements for few cases of two-pass heat exchangers with one fluid unmixed throught. *Note*: $\varepsilon_{\text{II-a}} > \varepsilon_{\text{III-a}} > \varepsilon_{\text{I-a}}$ (c) Flow arrangements for few cases of two-pass heat exchangers with both the fluids unmixed in each pass *Note*: $\varepsilon_{\text{IV-c}} > \varepsilon_{\text{I-a}} > \varepsilon_{\text{IV-b}} > \varepsilon_{\text{III-a}} > \varepsilon_{\text{II-a}} > \varepsilon_{\text{IV-a}}$.

3. Both the fluids unmixed–unmixed in all the passes: For the two-pass arrangements (IV-b of Figure 2.44b), the thermal effectiveness chart is given in Figure 2.49, and in Figure 2.50 for the three-pass arrangements.

2.11.2 Two Passes with One Fluid Unmixed throughout, Cross-Counterflow Arrangement

Some of the possible cases of two passes with one fluid unmixed throughout with cross-counterflow arrangements are given here, and these cases were solved by Stevens et al. [29]:

1. Two-pass cross-counterflow arrangement: Fluid 1 unmixed throughout, inverted-order coupling; fluid 2 mixed throughout (I-a of Figure 2.44a). Refer to Equation T29 of Table 2.12.
2. Two-pass cross-counterflow arrangement: Fluid 1 mixed throughout; fluid 2 unmixed throughout, inverted-order coupling (I-b of Figure 2.44a). Refer to Equation T30 of Table 2.12.
3. Two-pass cross-counterflow arrangement: Fluid 1 unmixed throughout, identical-order coupling; fluid 2 mixed throughout (II-a of Figure 2.44a). Refer to Equation T31 of Table 2.12.
4. Two-pass cross-counterflow arrangement: Fluid 1 mixed throughout; fluid 2 unmixed throughout, identical-order coupling (II-b of Figure 2.44a). Refer to Equation T32 of Table 2.12.

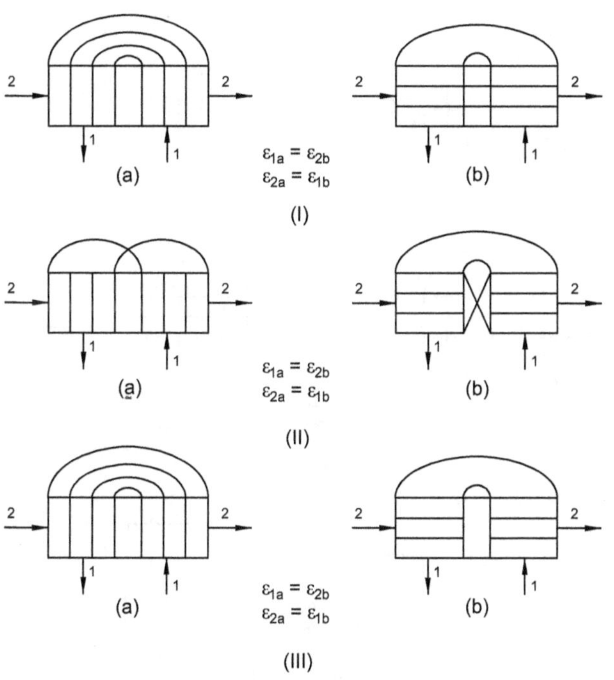

Nature of fluids flow in each pass, between passes and coupling order

Fluid	I		II		III	
	(a)	(b)	(a)	(b)	(a)	(b)
1	Unmixed throughout, inverted order	Mixed throughout	Unmixed in each pass ; identical order	Mixed throughout	Unmixed in each pass, between passes inverted order	Mixed throughout
2	Mixed throughout	Unmixed throughout, inverted order	Mixed throughout	Unmixed in each pass ; identical order	Mixed throughout	Unmixed in each pass, between passes mixed

(b)

FIGURE 2.44 (Continued)

Thermal relation formulas (T29 through T32) for these cases are given in Table 2.12, and thermal effectiveness charts are given in Figures 2.51–2.54, respectively.

Comparison of thermal effectiveness of two-pass crossflow cases:
For the possible cases of two-pass exchangers shown in Figure 2.44a, the thermal reflectiveness comparison is given by

$$\varepsilon_{\text{II-a}} > \varepsilon_{\text{III-a}} > \varepsilon_{\text{I-a}} \text{ of Figure 2.44a.}$$

For NTU = 4 and $C^* = 1$ [29], the thermal effectiveness of cases a, b, and c of Figure 2.44a is less about 5%–7%.

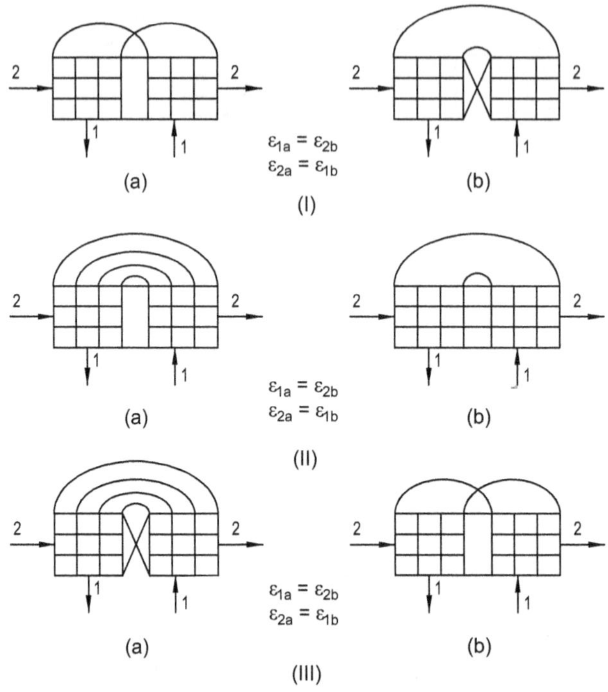

Nature of fluids flow between passes and coupling order ; in each pass unmixed.

Fluid	I		II		III	
	(a)	(b)	(a)	(b)	(a)	(b)
1	Unmixed, identical order	Mixed	Unmixed, inverted order	Mixed	Unmixed, inverted order	Unmixed, identical order
2	Mixed	Unmixed, identical order	Mixed	Unmixed, inverted order	Unmixed, identical order	Mixed

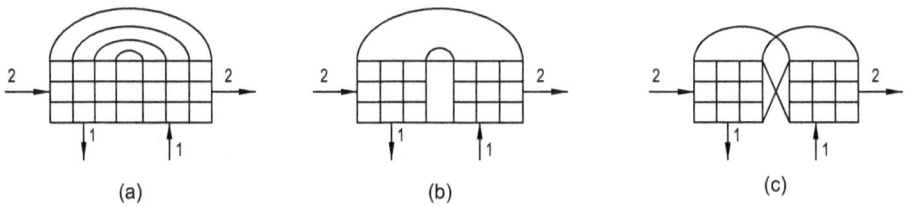

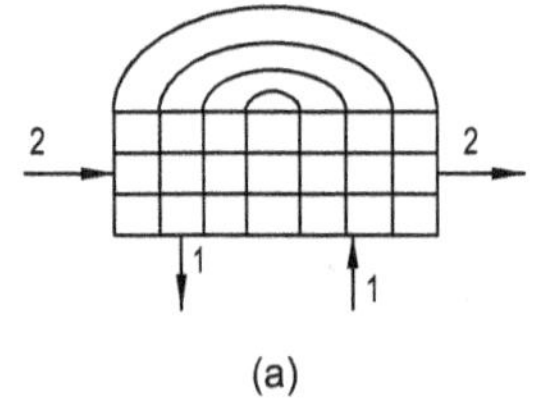

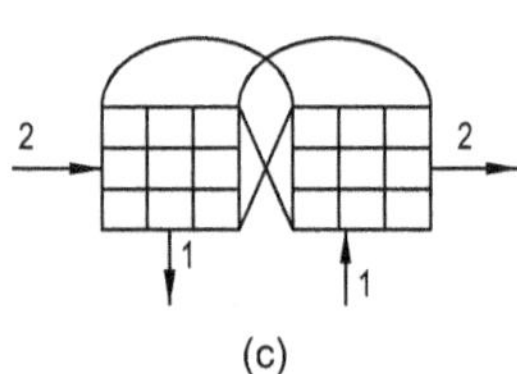

(a) (b) (c)

Maximum thermal effectiveness arrangement

Note :

(a)
1. Both the fluids unmixed throughout.
2. Coupling between passes is inverted order

(b)
1. Both the fluids unmixed in each pass
2. Mixed between passes

(c)
1. Both the fluids unmixed throughout.
2. Coupling between passes is identical order

(c)

FIGURE 2.44 (Continued)

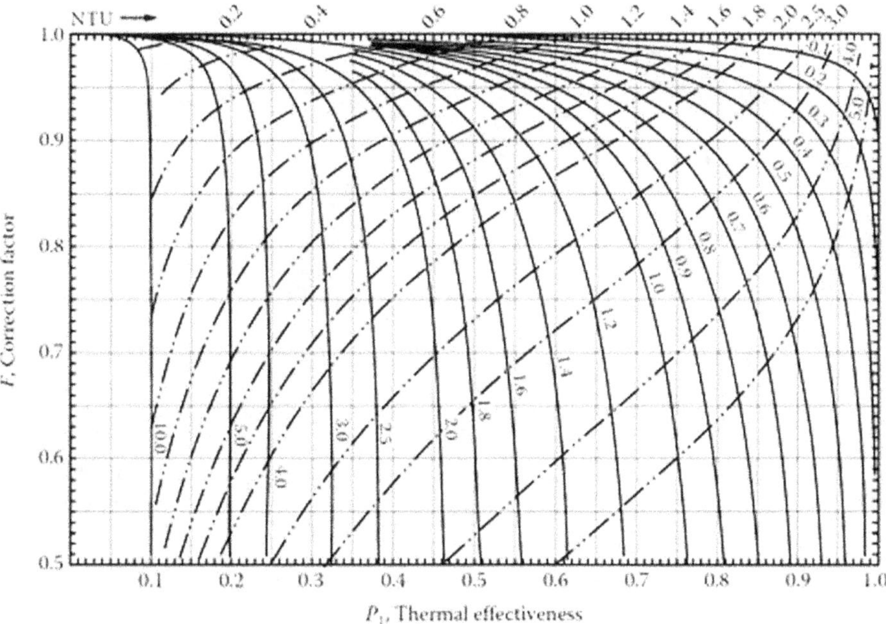

FIGURE 2.45 Thermal effectiveness chart—two-pass cross-counterflow arrangement with complete mixing between passes. Each pass is unmixed–mixed arrangement with weaker fluid mixed (as per Equation T6 of Table 2.4 for single pass; for the flow arrangement shown in III-a or b of Figure 2.44a); *F*–*R*–*P*–NTU chart; *F* as a function of *P* for constant *R* (solid lines) and constant NTU (dashed lines).

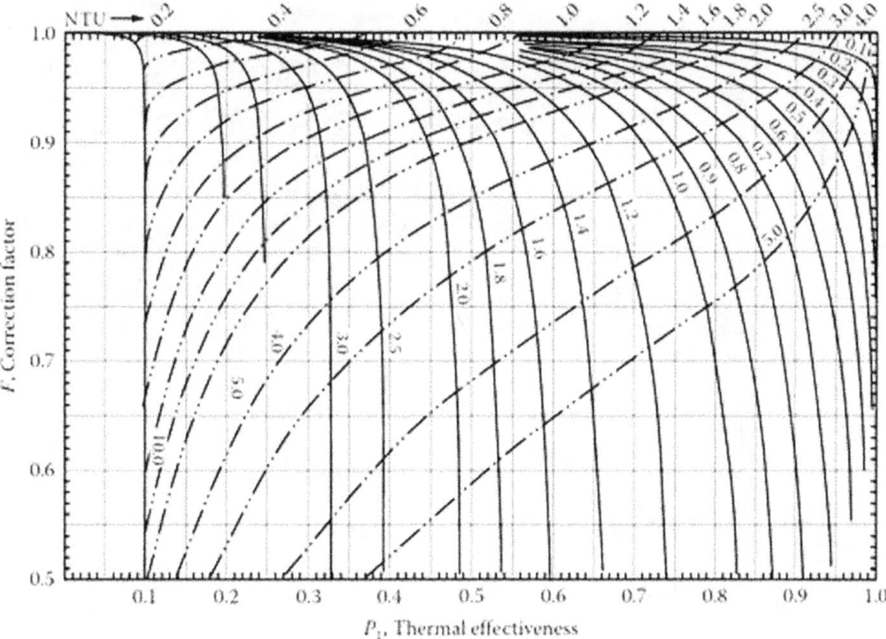

FIGURE 2.46 Thermal effectiveness chart—three-pass cross-counterflow arrangement with complete mixing between passes. Each pass is unmixed–mixed arrangement with weaker fluid mixed (as per Equation T6 of Table 2.4 for single pass; similar to the flow arrangement shown in III-a or b of Figure 2.44a); *F*–*R*–*P*–NTU chart; *F* as a function of *P* for constant *R* (solid lines) and constant NTU (dashed lines).

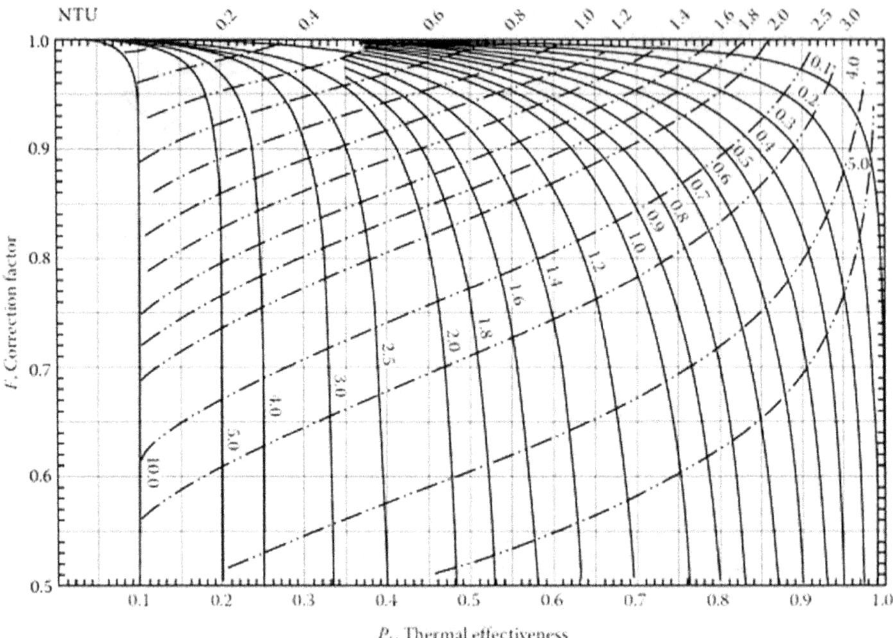

FIGURE 2.47 Thermal effectiveness chart—two-pass cross-counterflow arrangement with complete mixing between passes. Each pass is unmixed–mixed arrangement with stronger fluid mixed (as per Equation T7 of Table 2.4 for single pass; for the flow arrangement shown in III-a or b of Figure 2.44a); *F*–*R*–*P*–NTU chart; *F* as a function of *P* for constant *R* (solid lines) and constant NTU (dashed lines).

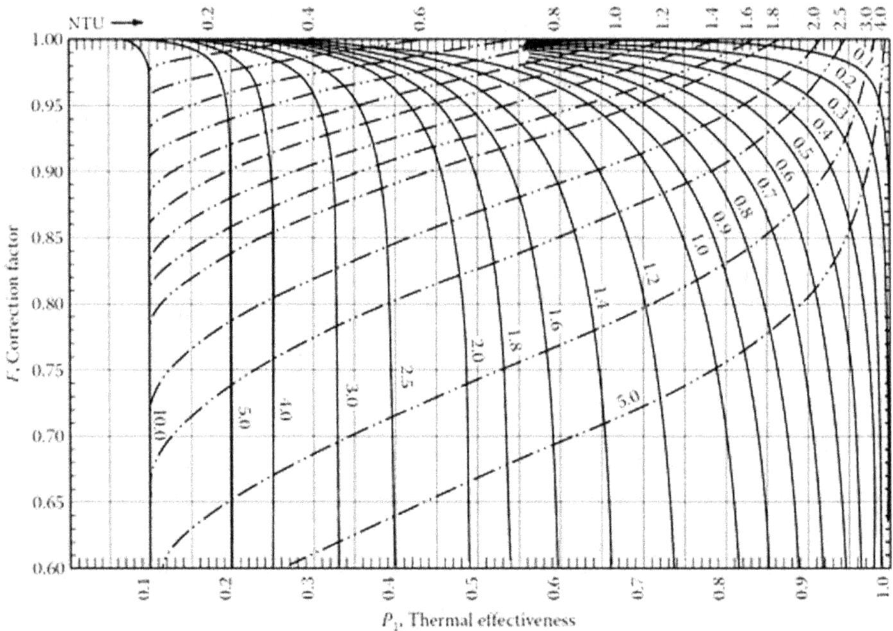

FIGURE 2.48 Thermal effectiveness chart—three-pass cross-counterflow arrangement with complete mixing between passes. Each pass is unmixed–mixed arrangement with stronger fluid mixed (as per Equation T7 of Table 2.4 for single pass; similar to the flow arrangement shown in to III-a or b of Figure 2.44a); *F*–*R*–*P*–NTU chart; *F* as a function of *P* for constant *R* (solid lines) and constant NTU (dashed lines).

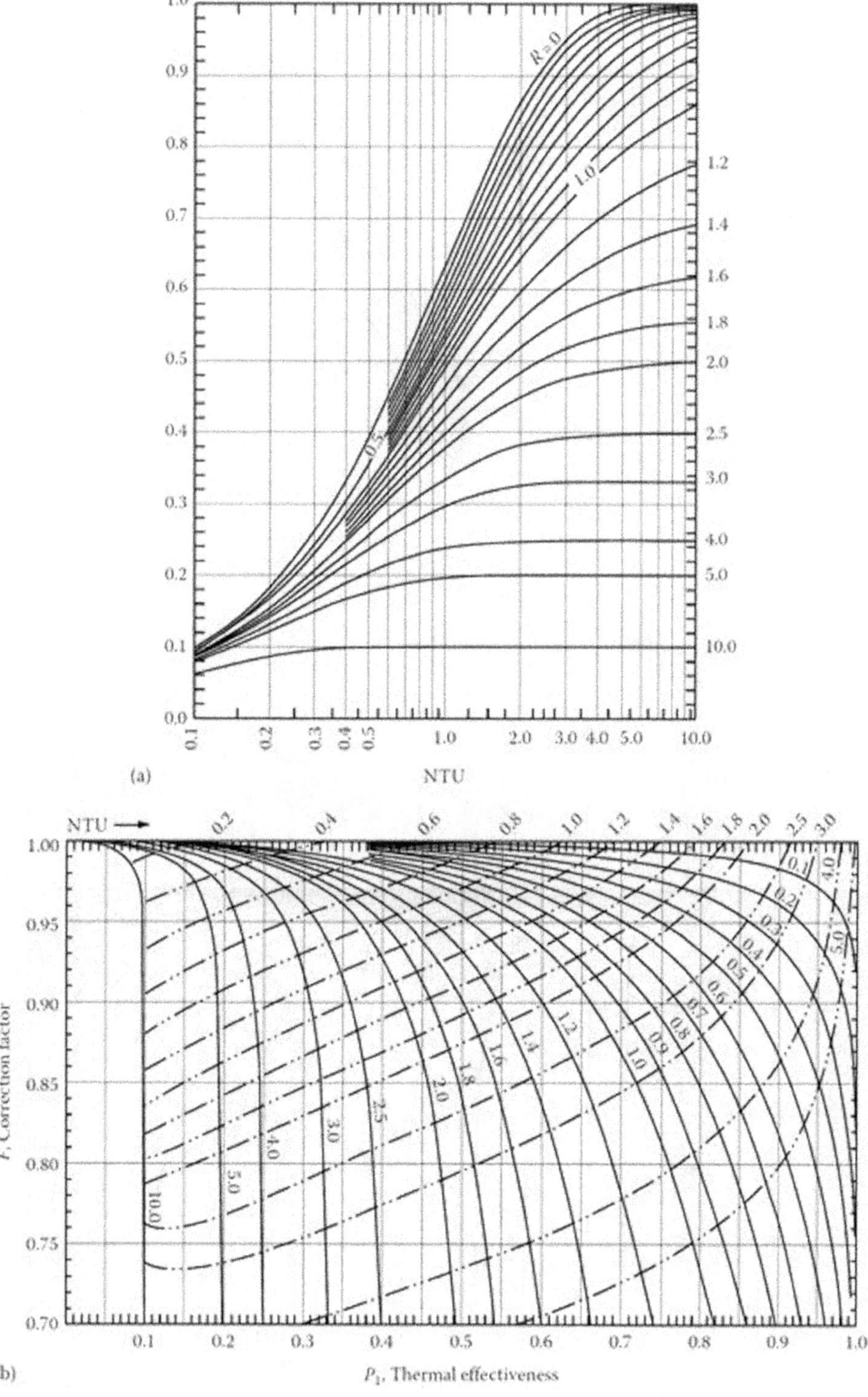

FIGURE 2.49 (a) Thermal effectiveness chart—two-pass cross-counterflow arrangement with complete mixing between passes. In each pass, both the fluids unmixed; *R*–*P*–NTU chart; (b) *F*–*R*–*P*–NTU chart; *F* as a function of *P* for constant *R* (solid lines) and constant NTU (dashed lines) (for the flow arrangement shown in IV-b of Figure 2.44b).

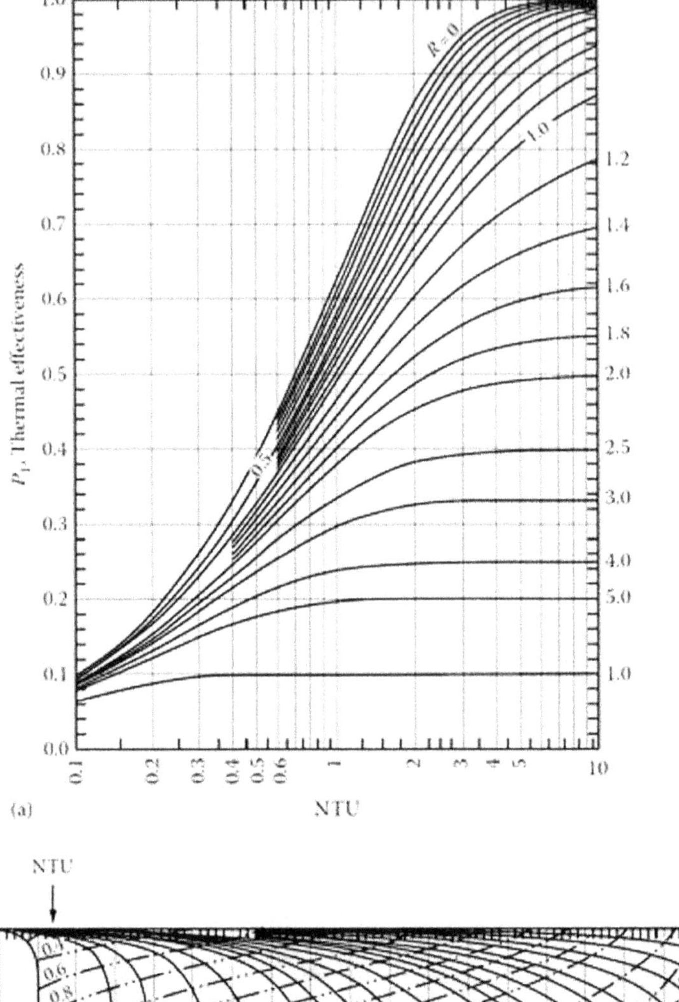

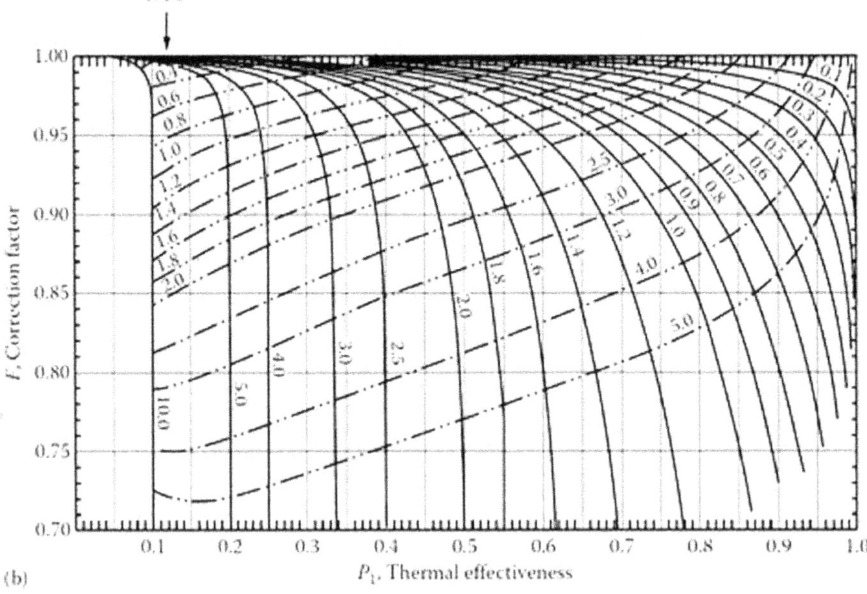

FIGURE 2.50 (a) Thermal effectiveness chart—three-pass cross-counterflow arrangement with complete mixing between passes. In each pass, both the fluids unmixed (for the flow arrangement shown in Figure 2.41c); R–P–NTU chart; (b) F–R–P–NTU chart; F as a function of P for constant R (solid lines) and constant NTU (dashed lines) (for the flow arrangement shown in Figure 2.41c).

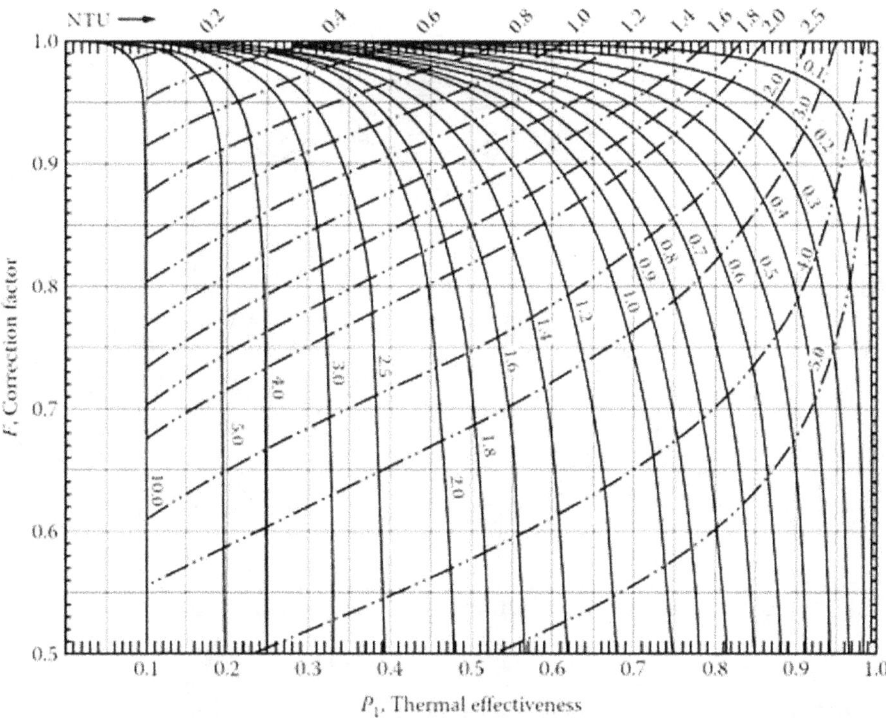

FIGURE 2.51 Thermal effectiveness chart—two-pass cross-counterflow arrangement. Fluid 1 unmixed throughout, inverted order coupling; fluid 2 mixed throughout. F as a function of P for constant R (solid lines) and constant NTU (dashed lines) (as per Equation T29 of Table 2.12, for the configuration shown in I-a of Figure 2.44a).

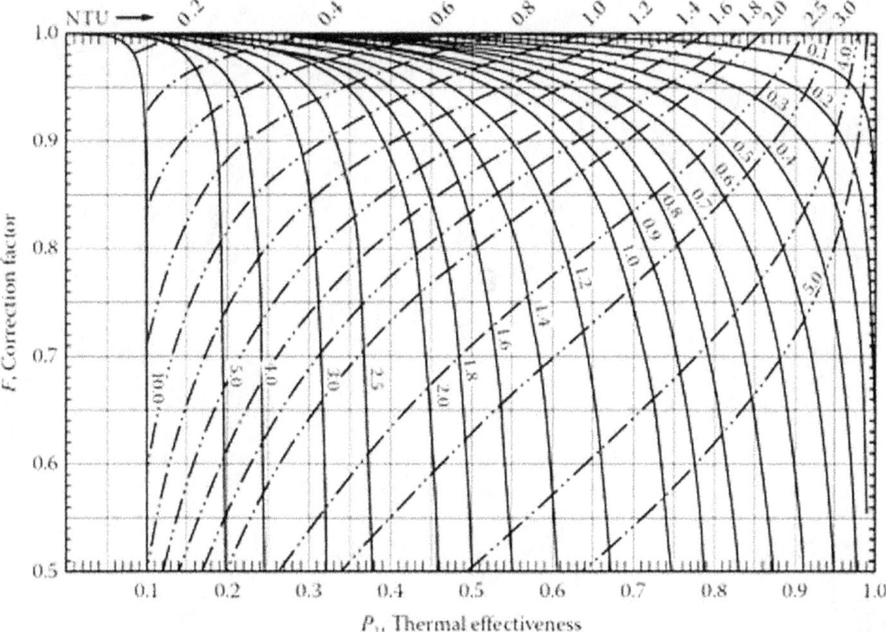

FIGURE 2.52 Thermal effectiveness chart—two-pass cross-counterflow arrangement. Fluid 1 mixed throughout; fluid 2 unmixed throughout, inverted order coupling. F as a function of P for constant R (solid lines) and constant NTU (dashed lines) (as per Equation T30 of Table 2.12, for the configuration shown in I-b of Figure 2.44a).

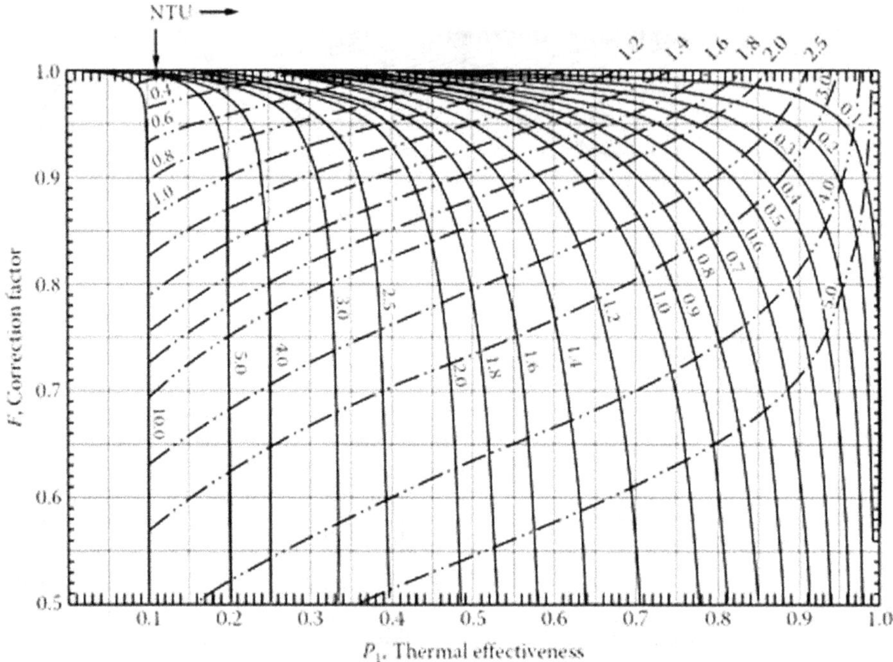

FIGURE 2.53 Thermal effectiveness chart—two-pass cross-counterflow arrangement. Fluid 1 unmixed throughout, identical order coupling; fluid 2 mixed throughout. F as a function of P for constant R (solid lines) and constant NTU (dashed lines) (as per Equation T31 of Table 2.12, for the configuration shown in II-a of Figure 2.44a).

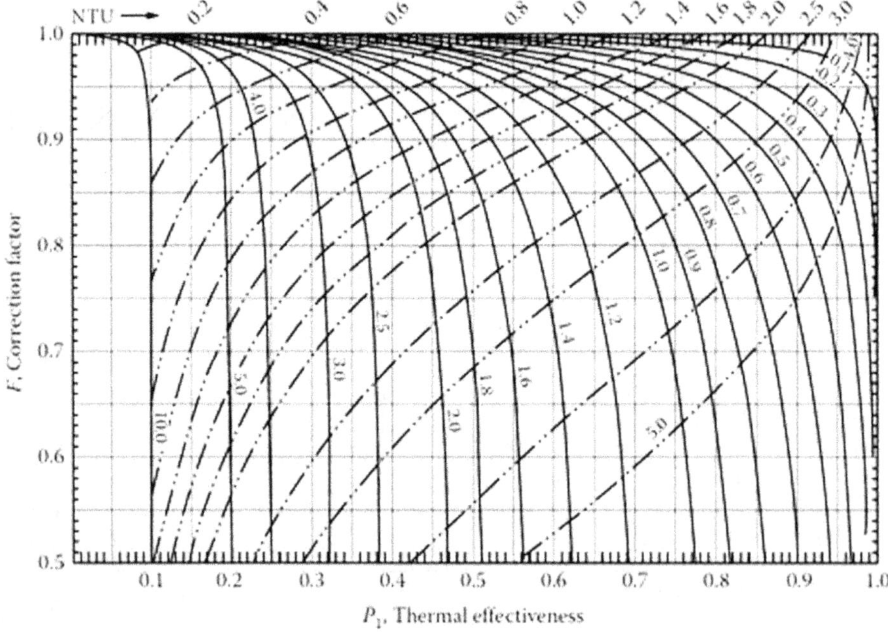

FIGURE 2.54 Thermal effectiveness chart—two-pass cross-counterflow arrangement. Fluid 1 mixed throughout; fluid 2 unmixed throughout, identical order coupling. F as a function of P for constant R (solid lines) and constant NTU (dashed lines) (as per Equation T32 of Table 2.12, for the configuration shown in II-b of Figure 2.44a).

2.11.3 Two Passes with Both Fluids Unmixed–Unmixed in Each Pass and One Fluid Unmixed throughout, Cross-Counterflow Arrangement

Possible cases of two passes with both fluids unmixed–unmixed in each pass and one fluid unmixed throughout with cross-counterflow arrangements are the following:

1. Two-pass cross-counterflow arrangement: fluid 1 unmixed throughout, inverted-order coupling; fluid 2 unmixed in each pass and mixed between passes (II-a of Figure 2.44b). (Refer to Equation T33 of Table 2.13 [45].)
2. Two-pass cross-counterflow arrangement: fluid 1 unmixed in each pass and mixed between passes; fluid 2 unmixed throughout, inverted-order coupling (II-b of Figure 2.44b). (Refer to Equation T34 of Table 2.13 [45].)
3. Two-pass cross-counterflow arrangement: fluid 1 unmixed throughout, identical-order coupling; fluid 2 mixed between passes and unmixed in each pass (I-a of Figure 2.44b). (Refer to Equation T35 of Table 2.13 [46].)
4. Two-pass cross-counterflow arrangement: fluid 1 mixed between passes and unmixed in each pass; fluid 2 unmixed throughout, identical-order coupling (II-b of Figure 2.44b). (Refer to Equation T36 of Table 2.13 [46].)

Thermal relation formulas (T33–T36) for these cases are given in Table 2.13. The thermal effectiveness chart for case 2 is given in Figure 2.55 and for case 3 in Figure 2.56.

Arrangement

1. Fluid 1 unmixed throughout; fluid 2 unmixed throughout. Both fluids coupling in inverted order (IV-a of Figure 2.44b). The thermal effectiveness chart is given in Figure 2.57. This is also equivalent to the TEMA *X*-shell, two passes on the tubeside with counterflow arrangement. For cross-parallelflow, the thermal effectiveness is given by [45]

$$\varepsilon = \mu_{1/2} + \frac{1}{C^*}\bar{\mu}_{1/2} \tag{2.76}$$

   where the terms of the equation were defined earlier. This is also equivalent to the TEMA *X*-shell, two passes on the tubeside with parallelflow arrangement.

2. Fluid 1 unmixed throughout, coupling in identical order; fluid 2 unmixed throughout, coupling in inverted order (III-b of Figure 2.44b).
3. Fluid 1 unmixed throughout, coupling in inverted order; fluid 2 unmixed throughout, coupling in identical order (III-a of Figure 2.44b). The thermal effectiveness chart is shown in Figure 2.58.
4. Fluid 1 unmixed throughout; fluid 2 unmixed throughout. Both fluids coupling in identical order (I-b of Figure 2.44b).

The thermal effectiveness charts shown in Figures 2.57 and 2.58 were arrived at by a numerical method as described by Stevens et al. [29]. For all these four cases, a closed-form solution is given by Baclic [3].

For the flow arrangement of IV-c of Figure 2.44b, *the most efficient case*, the thermal effectiveness for $C^* = 0.5$ is less by 1.8% (approximately) compared to the corresponding counterflow case. The maximum difference is in the NTU range 3.5–4.0, and for other values of NTU, the difference is fast decreasing and approaches zero. Also, for $0.0 < C^* < 0.5$, the difference is decreasing with C^*

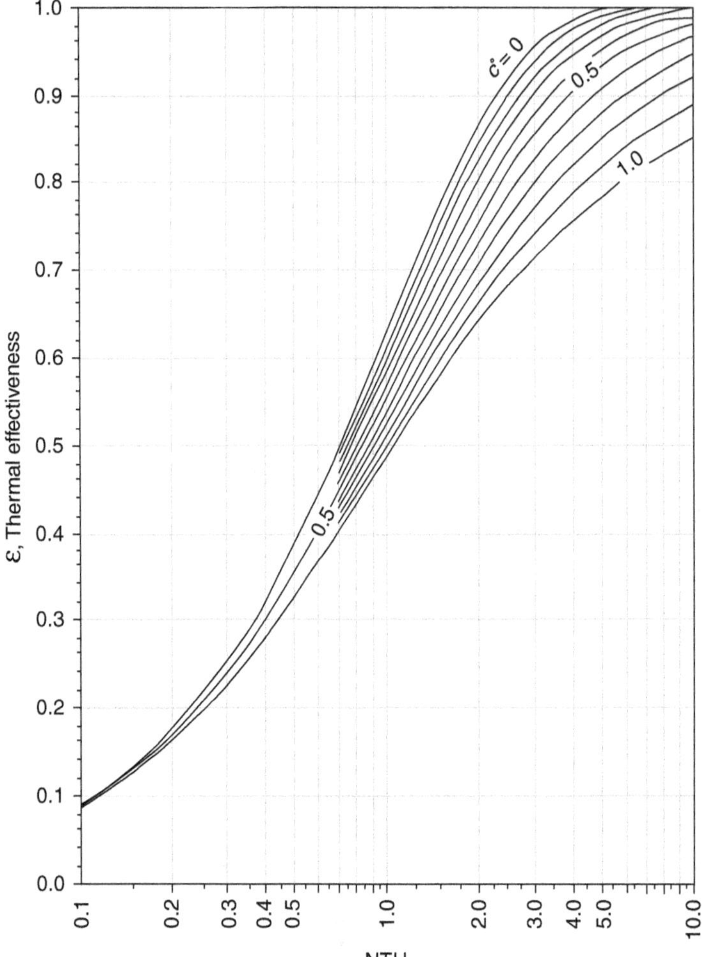

FIGURE 2.55 Thermal effectiveness chart—two-pass cross-counterflow arrangement. Fluid 1 unmixed in each pass mixed between passes; fluid 2 unmixed throughout, inverted order coupling (as per Equation T34 of Table 2.13, for the configuration shown in II-b of Figure 2.44b).

and it is zero for $C^* = 0.0$. For $C^* = 1.0$, the thermal effectiveness is less by 2.9% (approximately) compared to the corresponding counterflow case. The maximum difference is in the NTU range 4–6, and for other values of NTU, the difference is fast decreasing and approaches to zero. Also, for $0.5 < C^* < 1$, the difference is decreasing with decrease in C^*.

Comparison of thermal effectiveness of two-pass crossflow cases:
For the possible cases of two pass exchangers shown in Figure 2.44b, the thermal reflectiveness comparison is given by

$$\varepsilon_{\text{IV-c}} > \varepsilon_{\text{I-a}} > \varepsilon_{\text{IV-b}} > \varepsilon_{\text{III-a}} > \varepsilon_{\text{II-a}} > \varepsilon_{\text{IV-a}}$$

For NTU = 4 and $C^* = 1$ [29],

$$\varepsilon_{\text{IV-c}}/\varepsilon_{\text{IV-a}} = 1.038 \quad \text{and} \quad \varepsilon_{\text{IV-c}}/\varepsilon_{\text{I-a}} = 1.012$$

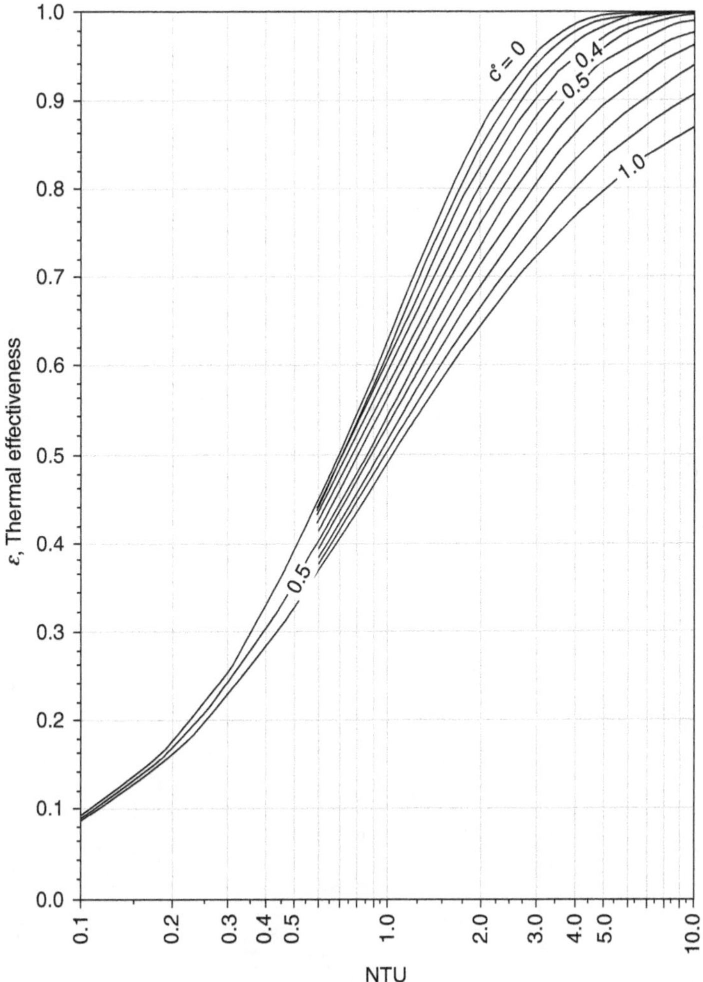

FIGURE 2.56 Thermal effectiveness chart—two-pass cross-counterflow arrangement. Fluid 1 unmixed throughout, identical order coupling; fluid 2 mixed between passes and unmixed in each pass (as per Equation T35 of Table 2.13, for the configuration shown in I-a of Figure 2.44b).

2.12 THERMAL EFFECTIVENESS OF MULTIPLE-PASS SHELL AND TUBE HEAT EXCHANGERS

Since the 1–N exchanger has lower effectiveness, multipassing on the shellside may be employed to approach the counterflow effectiveness. With this concept, the heat exchanger would have M shell passes and N tube passes. Figure 2.59 represents some such multipassing arrangements for $E_{1\text{-}2}$ and $E_{1\text{-}3}$ shells [32]. But multipassing on the shellside decreases the transfer area on the shellside, it is very difficult to fit the partition walls, and the possibility of leakage through the partition plates cannot be ruled out. Therefore, this difficulty is overcome by multiple shells with basic shell arrangements. A configuration with M shell passes, each one with N tube passes (similar to Figure 2.59), is equivalent to a series assembly of M shells each with N tube passes. This is illustrated in Figure 2.60 for $E_{1\text{-}2}$.

Multiple E shells in series, each with two tube passes, $E_{1\text{-}2}$: The thermal effectiveness charts for $2E_{1\text{-}2}$, $3E_{1\text{-}2}$, $4E_{1\text{-}2}$, $5E_{1\text{-}2}$, and $6E_{1\text{-}2}$ are given in Figures 2.61–2.65.

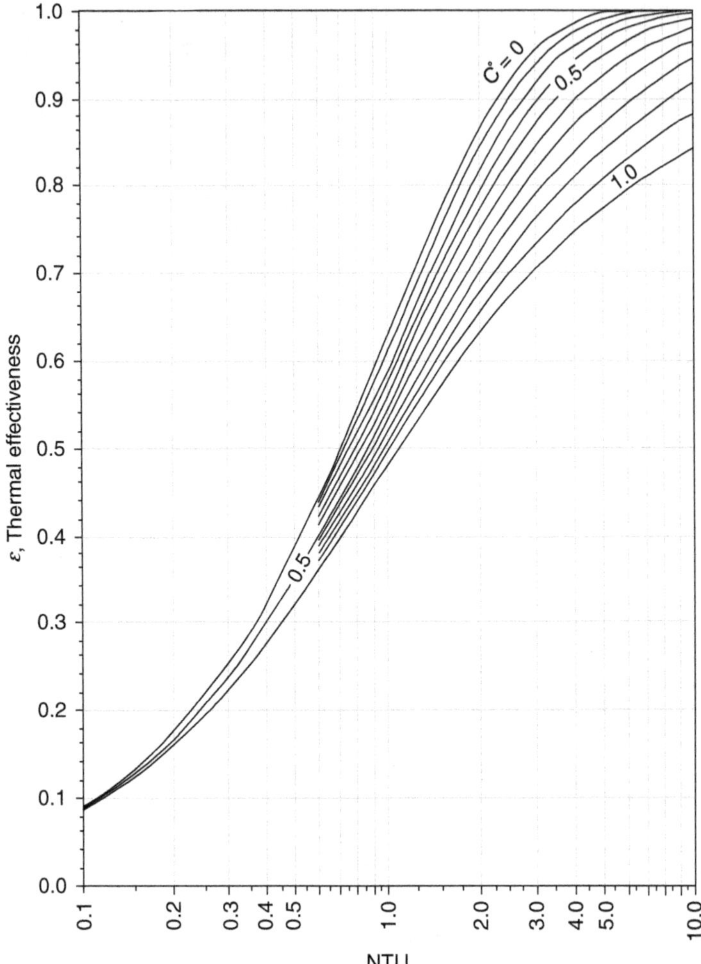

FIGURE 2.57 Thermal effectiveness chart—two-pass cross-counterflow arrangement. Fluid 1 unmixed throughout; fluid 2 unmixed throughout, both the fluids coupling in inverted order (equivalent to TEMA X_{1-2} shell with two passes on the tubeside, counterflow arrangement) (for the configuration shown in IV-a of Figure 2.44b).

2.13 HEAT EXCHANGER THERMAL DESIGN

2.13.1 Fundamentals of Heat Exchanger Design Methodology

Heat exchanger design methodology, shown in Figure 2.66, involves the following major design considerations [47]:

1. Process/design specifications
2. Thermohydraulic design to ensure the required heat exchanger performance and to satisfy the pressure-drop requirements for each stream
3. Flow-induced vibration in the case of shell and tube heat exchanger and individual fin-tube and bare tube bank compact heat exchanger
4. Mechanical design to provide the mechanical integrity required by design codes and operating conditions

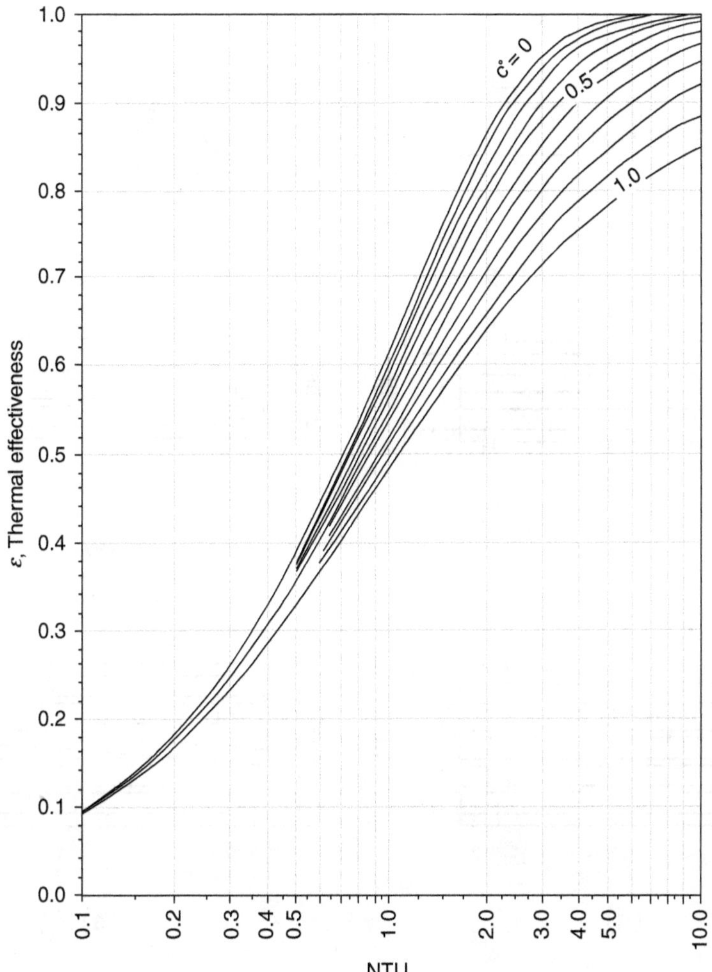

FIGURE 2.58 Thermal effectiveness chart—two-pass cross-counterflow arrangement. Fluid 1 unmixed throughout inverted order; fluid 2 unmixed throughout, coupling in identical order (for the configuration shown in III-a of Figure 2.44b).

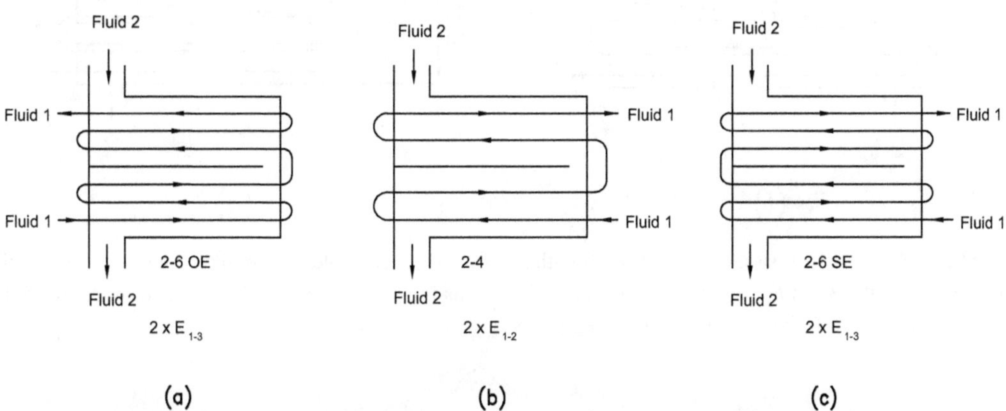

FIGURE 2.59 Shellsidemultipass arrangement for *E* shell. (From Pignotti, A. and Tamborenea, P.I., *Trans. ASME, J. Heat Transfer*, 111, 54, 1988.)

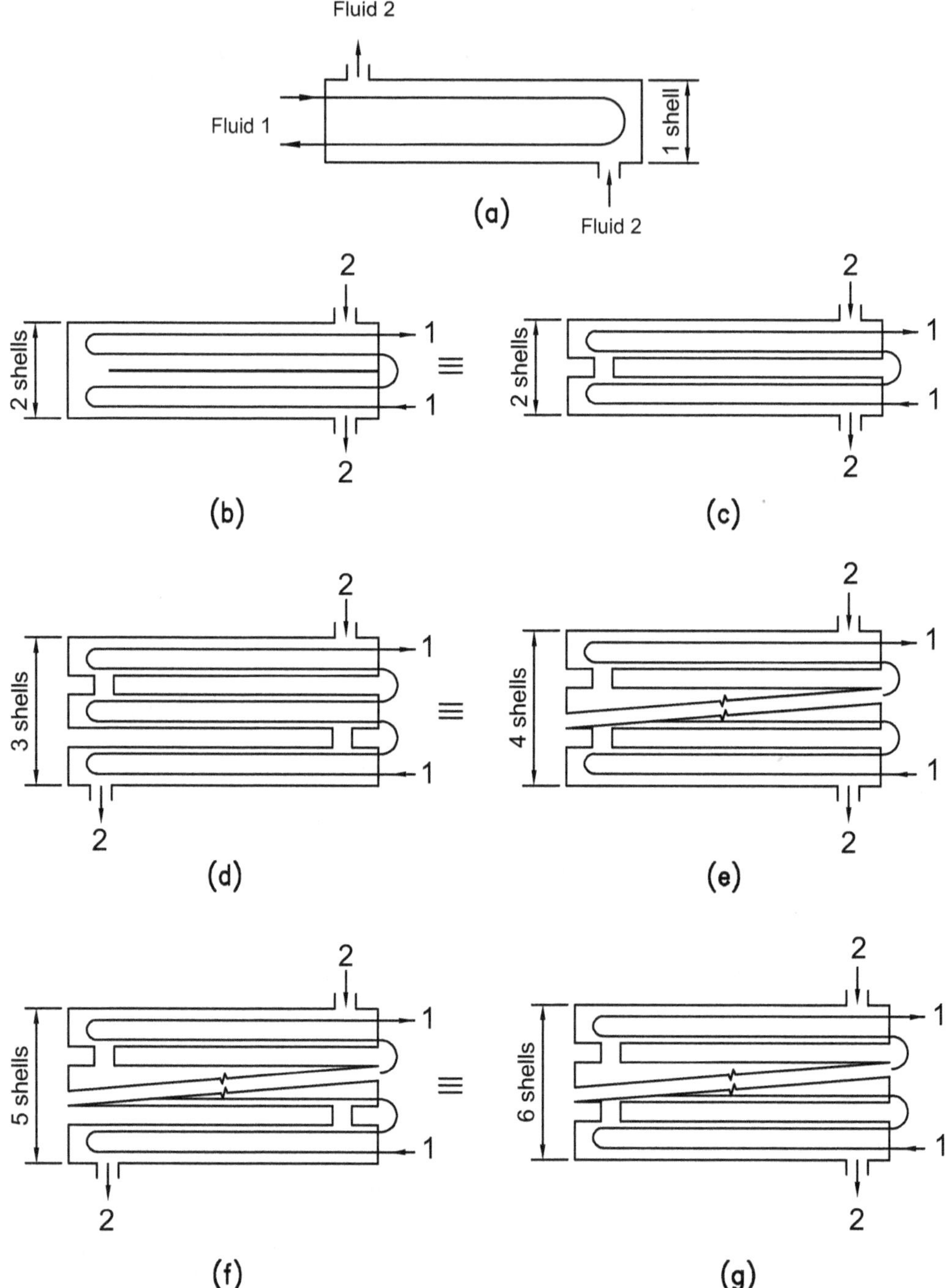

FIGURE 2.60 *M*-shell passes and each shell with *N*-tube pass is equivalent to *M*-shells in series, each with *N*-tube passes. (a) Basic E_{1-2} shell; (b) and (c) two E_{1-2} shells in series; (d) three E_{1-2} shells in series; (e) four E_{1-2} shells in series; (f) five E_{1-2} shells in series; (g) six E_{1-2} shells in series.

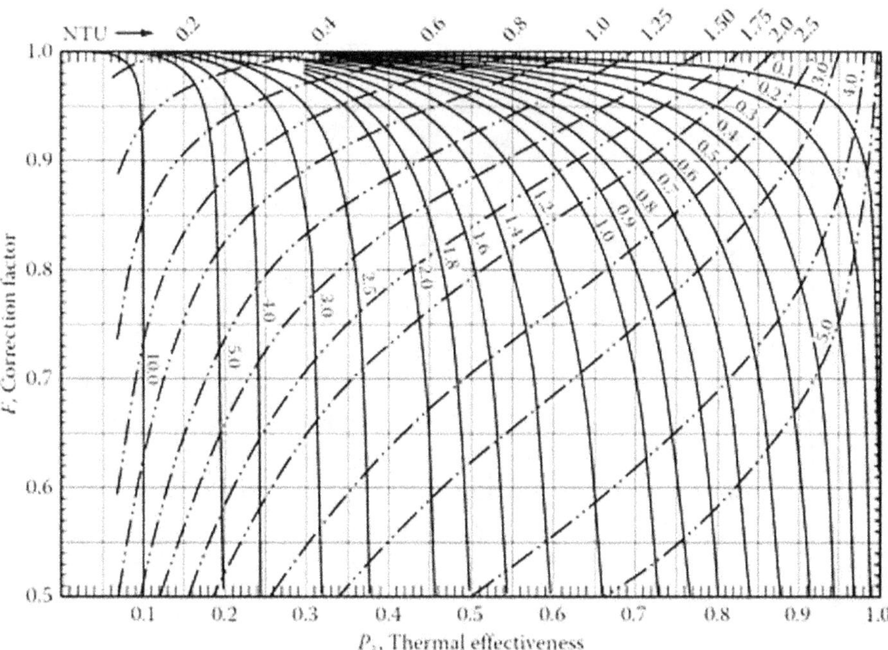

FIGURE 2.61 Thermal effectiveness chart -Two E_{1-2} shells in series; shell fluid mixed, tube fluid mixed between passes with stream symmetric (for the flow arrangement shown in Figure 2.60b). *F* as a function of *P* for constant *R* (solid lines) and constant NTU (dashed lines).

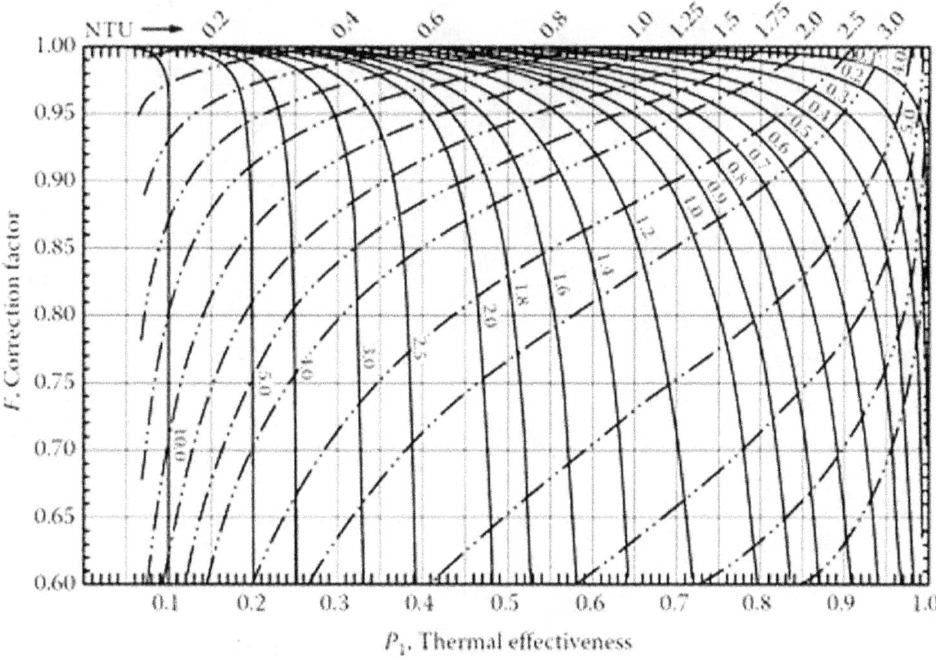

FIGURE 2.62 Thermal effectiveness chart =Three E_{1-2} shells in series; shell fluid mixed, tube fluid mixed between passes with stream symmetric (for the flow arrangement shown in Figure 2.60d). *F* as a function of *P* for constant *R* (solid lines) and constant NTU (dashed lines).

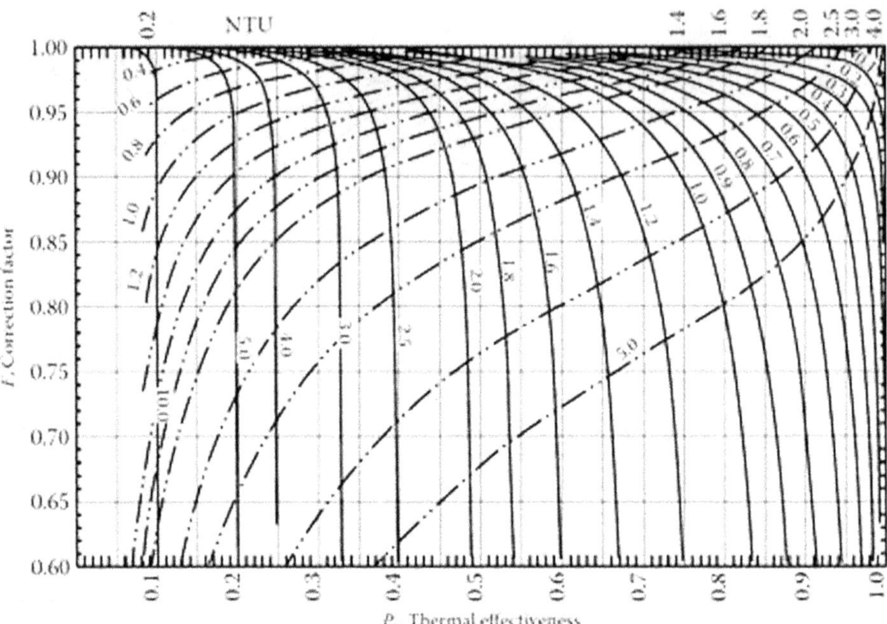

FIGURE 2.63 Thermal effectiveness chart -Four E_{1-2} shells in series; shell fluid mixed, tube fluid mixed between passes with stream symmetric (for the flow arrangement shown in Figure 2.60e). *F* as a function of *P* for constant *R* (solid lines) and constant NTU (dashed lines).

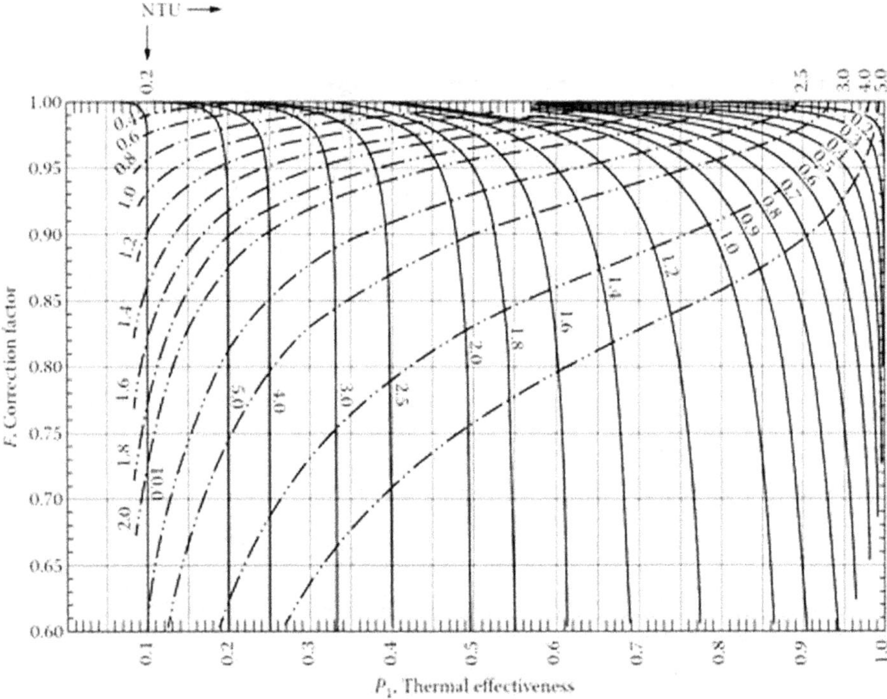

FIGURE 2.64 Thermal effectiveness chart - Five E_{1-2} shells in series; shell fluid mixed, tube fluid mixed between passes with stream symmetric (for the flow arrangement shown in Figure 2.60f). *F* as a function of *P* for constant *R* (solid lines) and constant NTU (dashed lines).

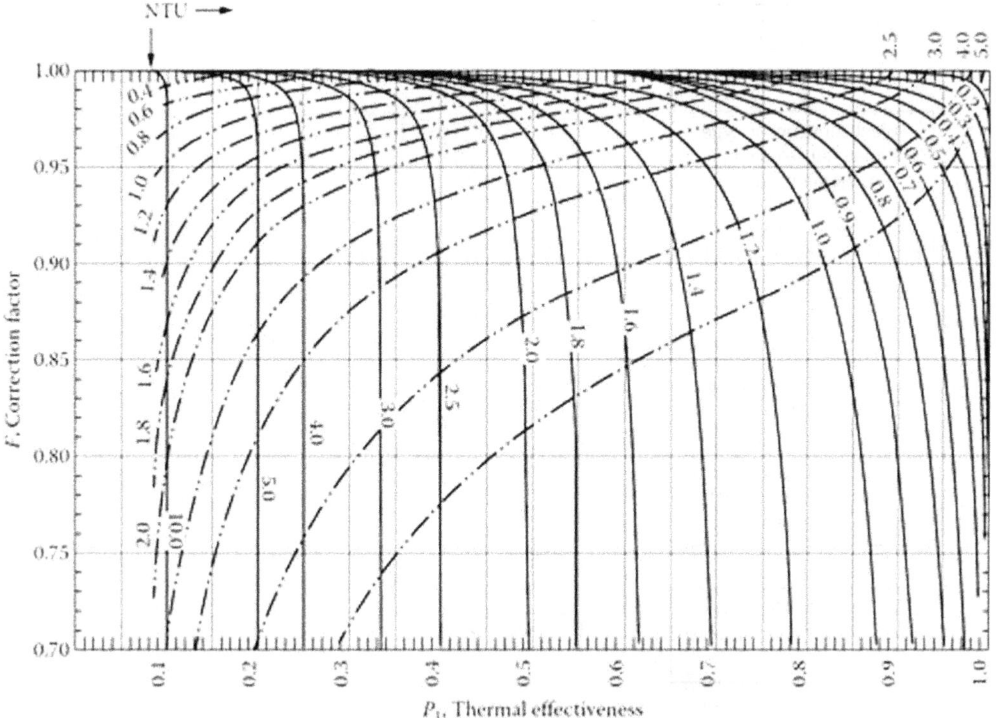

FIGURE 2.65 Thermal effectiveness chart - Six E_{1-2} shells in series; shell fluid mixed, tube fluid mixed between passes with stream symmetric (for the flow arrangement shown in Figure 2.60g). F as a function of P for constant R (solid lines) and constant NTU (dashed lines).

5. Cost and manufacturing considerations
6. Trade-off factors and system-based optimization

Most of these considerations are dependent on each other and should be considered simultaneously to arrive at the optimum exchanger design. Of these design considerations, items 1 and 2 are discussed in this chapter, flow-induced vibration and mechanical design are discussed in Chapter 1 of Volume 3 and Chapter 1 of Volume 2 of the three-volumes set of *Heat Exchanger Design*, respectively, and item 5 was discussed earlier in Chapter 1. The remaining item, trade-off factors and system-based optimization, is not within the scope of this book.

2.13.2 Process/Design Specifications

Process or design specifications include all the necessary information to design and optimize the exchanger for a specific application. It includes the following information [47]: (1) problem specification; (2) exchanger construction type; (3) flow arrangement; (4) construction materials; (5) design considerations like preferred tube sizes, tube layout pattern, maximum shell dimensions, and maintenance consideration; (6) design standard and construction code; (7) safety and protection, high product purity; and (8) special operating considerations such as cycling, upset conditions, etc.

2.13.2.1 Problem Specification

The problem specification involves the process parameters, operating conditions, and environment in which the heat exchanger is going to be operated. Typical details pertaining to problem specification

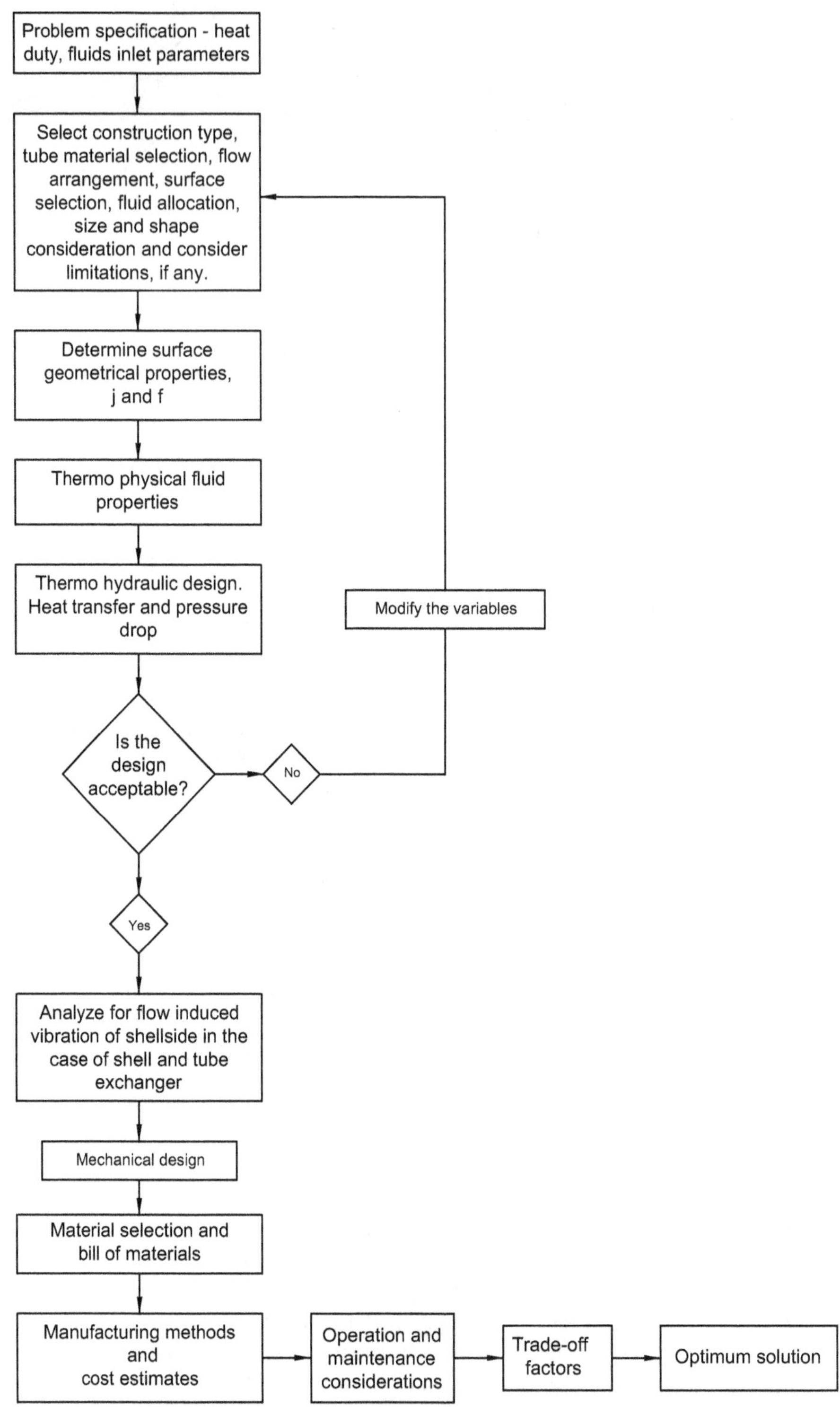

FIGURE 2.66 Heat exchanger design methodology.

include design parameters such as inlet temperatures and pressures; flow rates (including composition for mixtures); vapor quality, heat duty, allowable pressure drops, and fluctuations in the process parameters; overall size, layout, weight, etc.; and corrosiveness and fouling characteristics of fluid. Other factors that must be considered are:

1. Climatic conditions—minimum ambient, frost, snow, hail, and humidity
2. Operating environment—maritime, desert, tropical, seismic, cyclonic, and dust
3. Site layout—proximity to buildings or other cooling equipment, prevailing wind directions, duct allowances, length of pipe runs, and access

2.13.2.2 Exchanger Construction

Based on the problem specifications and experience, the exchanger construction type and flow arrangement are first selected. Selection of the construction type depends upon the following parameters [47]:

1. Fluids (gas, liquids, or condensing/evaporating) used on each side of a two-fluid exchanger
2. Operating pressures and temperatures
3. Fouling
4. Whether leakage or contamination of one fluid to the other is allowed or not
5. Cost and available heat exchanger manufacturing technology

2.13.2.3 Surface Selection

Compact heat exchanger: Factors that influence the surface selection include the operating pressures, fouling, maintenance requirements, erosion, fabricability, cost, etc.

Shell and tube heat exchanger: For shell and tube exchangers, the criteria for selecting core geometry or configurations are the desired heat transfer performance within specified pressure drops, fouling, corrosion, maintenance, repair, cleanability by mechanical means, minimal operational problems (flow-induced vibrations), safety, and cost; additionally, the allocation of fluids on the shellside and the tubeside is an important consideration.

2.13.2.4 Thermohydraulic Design

Heat exchanger thermohydraulic design involves quantitative heat transfer and pressure-drop evaluation or exchanger sizing. Basic thermohydraulic design methods and inputs to these analyses are as follows.

2.13.3 Basic Thermohydraulic Design Methods

As discussed earlier, the ε-NTU, P-NTU, LMTD, or ψ-P method can be used for solving the thermal design problem.

2.13.3.1 Thermophysical Properties

For heat transfer and pressure-drop analysis, the following thermophysical properties of the fluids are needed: dynamic viscosity μ, density, specific heat c_p, surface tension, and thermal conductivity k. For the conduction wall, thermal conductivity is needed.

2.13.3.2 Surface Geometrical Properties

For heat transfer and pressure-drop analyses, at least the following surface geometrical properties are needed on each side of a two-fluid compact heat exchanger:

- Heat transfer area, A, which includes both primary and secondary surface area if any
- Core frontal area, A_{fr}
- Minimum free flow area, A_o
- Hydraulic diameter, D_h
- Flow length, L
- Fin thickness and fin conduction length
- Core volume, V and core dimensions L_1, L_2, and L_3

These quantities are computed from the basic dimensions of the core and heat transfer surface. On the shellside of a shell and tube heat exchanger, various leakage and bypass flow areas are also needed. The procedure to compute these quantities is presented elsewhere.

2.13.3.3 Surface Characteristics

Surface characteristics for heat transfer, j, and flow friction, f, are key inputs for the exchanger heat transfer and pressure-drop analysis, respectively. Experimental results for a variety of exchanger surfaces are presented in Kays and London [48], and correlations in Shah and Bhatti [49]. More references on j and f factors are furnished in Chapter 3.

2.13.4 Design Procedure

The design of heat exchangers requires consideration of the factors just outlined. It is unlikely that any two designers will arrive at exactly the same design for a given set of conditions, as the design process involves many judgments while carrying out the design [50]. Gollin [50] describes the step-by-step design procedure for a heat exchanger. Important steps in the heat exchanger design procedure include the following:

1. Assess the heat transfer mechanisms involved.
2. Select the heat exchanger class.
3. Determine the construction details and select the surface geometry.
4. Determine size and layout parameters keeping in mind the constraints imposed by the purchaser/client.
5. Perform preliminary thermal design known as approximate sizing.
6. Perform the detailed design.
7. Check the design.
8. Optimize the design.
9. Perform the check for flow-induced vibration in the case of shell and tube heat exchanger, individually finned tube, and bare tube bank compact heat exchangers.
10. Perform mechanical design.
11. Estimate cost and finalize the design as per trade considerations.

2.13.5 Heat Exchanger Design Problems

In a broad sense, the design of a new heat exchanger means the selection of exchanger construction type, flow arrangement, tube and fin material, and the physical size of an exchanger to meet the specified heat transfer and pressure-drop requirements. Two most common heat exchanger design problems are the rating and sizing. For an existing exchanger, the performance evaluation problem is referred to as the rating problem. The sizing problem is also referred to as the design problem. Rating and sizing problems are discussed here. For more details on the rating and sizing problems, refer to Refs. [48,51] and Bell [52].

2.13.5.1 Rating

Determination of heat transfer and pressure-drop performance of either an existing exchanger or an already-sized exchanger is referred to as a rating problem. Inputs to the rating problem include [47]: (1) heat exchanger construction details; (2) flow arrangement; (3) overall dimensions; (4) material details; (5) surface geometries and surface characteristics (j and f factors); (6) fluid flow rates; (7) inlet temperatures; and (8) fouling factors. The designer's task is to predict the fluid outlet temperatures, total heat transfer rate, and pressure drop on each side.

2.13.5.2 Rating of a Compact Exchanger

The rating problems for a two-fluid direct-transfer type compact heat exchanger that has gas as a working fluid at least on one side are discussed briefly here, and the detailed rating of a crossflow and counter-crossflow exchanger is described separately. Customarily, the ε-NTU method is employed for compact heat exchangers. Hence, the solution procedure is outlined here using the ε-NTU method. The basic steps involved in the analysis of a rating problem are the determination of

1. Surface geometrical parameters
2. Thermophysical fluid properties
3. Reynolds numbers
4. Surface characteristics, j and f
5. Corrections to the temperature-dependent fluid properties
6. Heat transfer coefficients
7. Fin effectiveness and overall surface effectiveness
8. Thermal resistance due to conduction wall
9. Overall heat transfer coefficient
10. NTU, C^*, and exchanger effectiveness, ε
11. Heat transfer rate, outlet temperatures, and pressure drop on each side

The rating of a compact heat exchanger is discussed in detail in Chapter 3.

2.13.5.3 Rating of a Shell and Tube Exchanger

"Rating" implies that a specific heat exchanger is fairly completely described geometrically (with the possible exception of the length) and the process specifications for the two streams are given. The Bell-Delaware method is a rating method. The basic rating program of the Bell-Delaware method is shown in Figure 2.67, and the method is described in detail in Chapter 4.

2.13.5.4 Sizing

In a sizing or design problem, we determine the physical size (length, width, height, and surface area on each side) of an exchanger. Inputs to the sizing problem are the fluid inlet and outlet temperatures, flow rates, fouling factors, and the pressure drop on each side. The designer's task is to select construction type, flow arrangement, materials, and surface geometry on each side. With the selection of construction types and surface geometries on each side, the problem then reduces to the determination of the core dimensions for the specified heat transfer and pressure-drop performance. However, one can reduce the sizing problem to the rating problem by tentatively specifying the dimensions, then predicting the performance [47]. If the computed results do not agree with the specified values, a new size is assumed and the calculations are repeated.

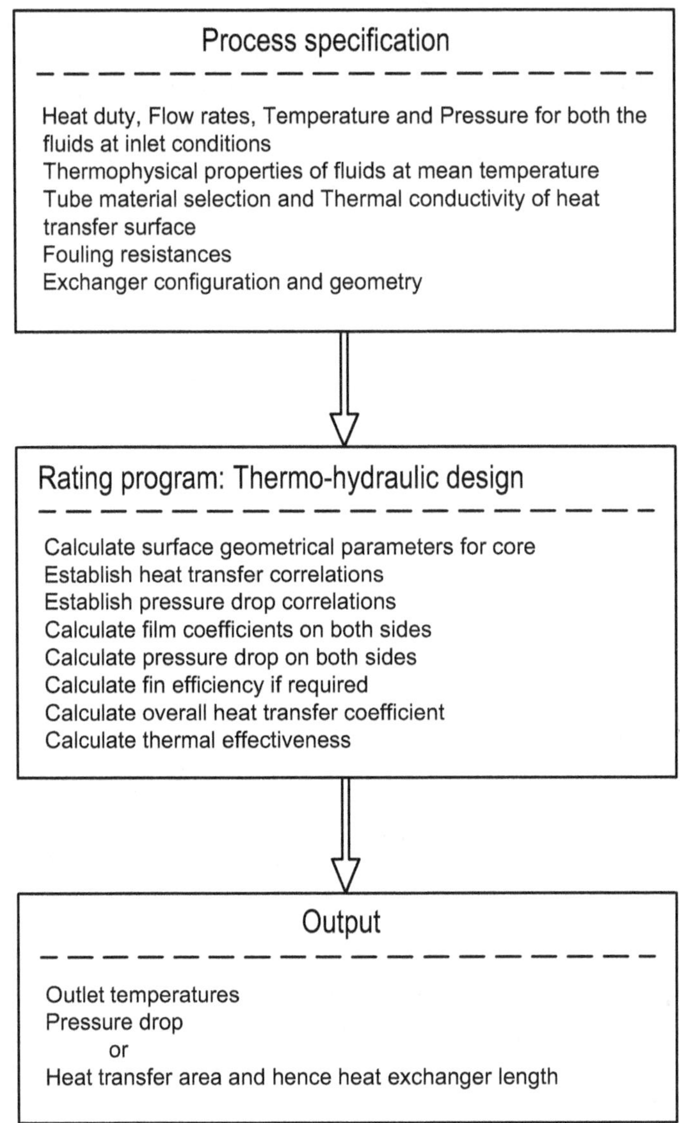

FIGURE 2.67 Rating of shell and tube heat exchanger (Modified from Bell, K. J., Overall design methodology for shell and tube exchangers, in *Heat Transfer Equipment Design* [R. K. Shah, E. C. Subbarao, and R. A. Mashelkar, eds.], Hemisphere, Washington, DC, 1988, pp. 131–144).

2.13.5.5 Size of a Heat Exchanger

For a given heat duty, the size of the heat exchanger is a function of the following parameters:

1. Thermal effectiveness
2. Fluid flow rate
3. Secondary surface area per unit volume
4. Heat transfer surface performance parameters
5. Heat transfer augmentation devices, if any
6. Conductance ratio of the process fluids

The sizing of a compact heat exchanger is discussed in detail in Chapter 3.

2.13.5.6 Sensitivity Analysis

In a sizing problem, sometimes one is interested in determining the sensitivity of certain variables individually. For example, how does the heat transfer vary when changing the fin density in a compact heat exchanger with a secondary surface? In such a case, one inputs a series of values of fin densities at one time, runs the performance (rating) calculations, obtains a series of results, and analyzes them.

2.13.5.7 Sizing of a Compact Heat Exchanger

The principle of compact heat exchanger sizing is discussed in Chapter 3.

2.13.5.8 Sizing of a Shell and Tube Heat Exchanger

Shell and tube heat exchanger design or sizing is based upon: (1) design conditions, that is, fluid flow rates, terminal temperatures, thermophysical fluid properties, and allowable pressure drop; (2) assumptions, heat transfer surface area, overall heat transfer coefficient, or size, length, or number of tubes; and (3) pressure drop across the heat exchanger [53]. The design conditions are fixed by overall plant design and determine the expected performance of the exchanger. Trial-and-error calculations of the film coefficients are used to check the assumptions, which are also checked by an overall heat balance. Finally, the pressure drop is calculated and compared with the allowable values. If the calculated pressure drop is too high, a new set of assumptions is made and rechecked as before.

2.13.6 Heat Exchanger Optimization

The solution to the sizing problem in general is not adequate for the design of a new exchanger, since other constraints in addition to pressure drop are imposed on the design, and the objective of the design is to minimize the weight, volume, and heat transfer surface, and minimum pumping power, pressure drop, or other considerations in addition to meeting the required heat transfer. This is achieved by heat exchanger optimization. Shah et al. [54] reviewed various methods used in the literature for heat exchanger optimization and described numerical nonlinear programming techniques.

2.13.7 Solution to the Rating and Sizing Problem

Now let us discuss the basic steps involved in the solution of the two design problems, the rating and sizing.

2.13.7.1 Rating

The basic steps involved in the solution to the rating problem are as follows [1].

ε-NTU method:

1. Compute C^* and NTU from the input specifications.
2. Determine ε for known NTU, C^*, and the flow arrangement.
3. Compute q from

$$q = \varepsilon C_{\text{min}}\left(t_{\text{h,i}} - t_{\text{c,i}}\right) \tag{2.77}$$

and outlet temperatures from

$$t_{\text{h,o}} = t_{\text{h,i}} - \frac{q}{C_{\text{h}}} \tag{2.78}$$

$$t_{c,o} = t_{c,i} + \frac{q}{C_c} \tag{2.79}$$

LMTD method:

1. Compute R from $R = C_t/C_s$.
2. Assume the outlet temperatures to determine P, or assume P and calculate outlet temperatures. Also calculate LMTD.
3. Determine LMTD correction factor, F.
4. Determine q from $q = UAF$(LMTD).
5. Evaluate the outlet temperatures from known q, C_c, and C_h, and compare with those of step 2.
6. Repeat steps 2–5 until the desired convergence is achieved.

2.13.7.2 Solution to the Sizing Problem

In a sizing problem, U, C_c, C_h, and the terminal temperatures are specified, and the surface area A is to be determined. Or U may be calculated from the specified convective film coefficients and the fouling resistances. The basic steps involved in sizing by the ε-NTU and LMTD methods are as follows [1].

ε-NTU method:

1. Compute ε from the specified inlet and outlet temperatures and calculate C^*.
2. Determine NTU from known ε, C^*, and the flow arrangement.
3. Calculate the required surface area A from $A = (\text{NTU})C_{min}/U$ and from terminal temperatures.

LMTD method:

1. Compute P and R from the specified inlet and outlet temperatures.
2. Determine F from F-P curves for known P, R, and the flow arrangement.
3. Calculate the heat transfer rate q and LMTD.
4. Calculate A from $A = q/[UF(\text{LMTD})]$.

A trial-and-error approach is needed for the solution to the rating problem by the LMTD method.

2.13.8 Pressure-Drop Analysis, Temperature-Dependent Fluid Properties, Performance Failures, Flow Maldistribution, Fouling, and Corrosion

2.13.8.1 Heat Exchanger Pressure-Drop Analysis

The term "pressure drop" refers to the pressure loss that is not recoverable in the circuit. The determination of pressure drop in a heat exchanger is essential for many applications for at least two reasons [1]:

1. The operating cost of a heat exchanger is primarily the cost of the power to run fluid-moving devices such as pumps, fans, and blowers. This pumping power, P_p, is proportional to the exchanger pressure drop as given by

$$P_p = \frac{M\Delta p}{\rho} \tag{2.80}$$

where
M is the mass flow rate
Δp is the pressure drop
ρ is the fluid density

2. The heat transfer rate can be significantly influenced by the saturation temperature change for a condensing/evaporating fluid for a large pressure drop.

The principle of pressure-drop analysis for a heat exchanger is described by Kays [55] and is extended to all types of heat exchangers. In this section, pressure-drop analysis for various types of heat exchangers as per Ref. [2] is discussed.

2.13.8.2 Pressure-Drop Evaluation for Heat Exchangers

The pressure drop associated with a heat exchanger may be considered as having two major components: (1) pressure drop associated with the core or matrix, and (2) pressure drop in inlet and outlet headers, manifolds, nozzles, or ducting due to change in flow area, flow turning, etc. In this section, core pressure drop for extended surface exchangers, regenerators, and tubular exchangers is presented, followed by the pressure drop associated with bends and flow turnings.

2.13.8.3 Pressure Drop through a Heat Exchanger

The pressure drop on any one side consists of pressure losses due to sudden contraction at the core inlet, Δp_{1-2}, core pressure drop, Δp_{2-3}, and the pressure rise due to sudden expansion at the core outlet, Δp_{3-4}. Therefore, the total pressure drop on any one side of the exchanger is given by

$$\Delta p = \Delta p_{1-2} + \Delta p_{2-3} - \Delta p_{3-4} \tag{2.81}$$

Pressure drop for various compact heat exchangers: Pressure drop for various compact heat exchangers on the fin side is given in the following [1,2].

1. Plate-fin heat exchangers

$$\frac{\Delta p}{p_i} = \frac{G^2}{2g_c}\frac{1}{p_i\rho_i}\left[\left(1-\sigma^2+K_c\right)+2\left(\frac{\rho_i}{\rho_0}-1\right)+f\frac{L}{r_h}\rho_i\left(\frac{1}{\rho}\right)_m-\left(1-\sigma^2-K_e\right)\frac{\rho_i}{\rho_0}\right] \tag{2.82}$$

1.1 For simplified formula, refer to Chapter 3, Section 3.7.1.

2a. Tube-fin heat exchangers (individually finned)

$$\frac{\Delta p}{p_i} = \frac{G^2}{2g_c}\frac{1}{p_i\rho_i}\left[2\left(\frac{\rho_i}{\rho_0}-1\right)+f\frac{L}{r_h}\rho_i\left(\frac{1}{\rho}\right)_m\right] \tag{2.83}$$

2b. Tube-fin heat exchangers (continuously finned)

$$\frac{\Delta p}{p_i} = \frac{G^2}{2g_c}\frac{1}{p_i\rho_i}\left[2\left(\frac{\rho_i}{\rho_0}-1\right)+f\frac{4L}{D_h}\rho_i\left(\frac{1}{\rho}\right)_m\right] + \frac{G'^2}{2g_c}\frac{1}{p_i\rho_i}\left[\left(1-\sigma'^2+K_c\right)-\left(1-\sigma'^2-K_e\right)\frac{\rho_i}{\rho_0}\right] \tag{2.84}$$

where

$$G\sigma = G'\sigma' \tag{2.85}$$

3. Regenerator both plate-fin type and randomly staked matrix—same as Equations 2.82 and 2.83.
4. For plate heat exchanger, refer to Chapter 7, Section 7.13.4.3.
5. For shell and tube heat exchanger, refer to Chapter 4, Section 4.31.8.

Generally, the core frictional pressure drop is a dominating term, accounting for about 90% or more of the pressure drop. The entrance and exit losses are important at low values of σ (short flow length) and L (i.e., short core), high values of Reynolds number, and for gases. They are negligible for liquids. The values of K_c and K_e are presented in Ref. [55]. The definitions of terms used in Equations 2.82 through 2.84 are as hereunder:

1. K_c is the contraction loss coefficient. Values of K_c are given in Ref. [55] for four different entrance flow passage geometries.
2. $\left(\frac{1}{\rho}\right)_m$ is defined as

$$\left(\frac{1}{\rho}\right)_m = v_m = \frac{v_i + v_0}{2} = \frac{1}{2}\left(\frac{1}{\rho_i} + \frac{1}{\rho_o}\right) \tag{2.86}$$

where v denotes specific volume, ρ_i denotes density of the fluid at inlet conditions and ρ_o at the outlet conditions. For a perfect gas,

$$\left(\frac{1}{\rho}\right)_m = \frac{R_c T_{lm}}{P_{ave}} \tag{2.87}$$

where
R_c is the gas constant
$P_{ave} = (p_i + p_o)/2$
$T_{lm} = T_m \pm \text{LMTD}$

Here, T_m is the average temperature of the fluid on the other side of the exchanger.

3. K_e is the expansion coefficient. Values of K_e for four different flow passage geometries are presented in Ref. [50].
4. Mass velocity G is given by

$$G = \frac{M}{A_o} \tag{2.88}$$

where A_o is the minimum free flow area.

5. σ' for continuously finned tube-fin heat exchangers (Figure 4.25, Chapter 4 of Volume 1) is given by [56]:

$$\sigma' = \frac{L_3L_1 - L_3\delta N_f L_1}{L_3L_1} \tag{2.89}$$

and

$$G\sigma = G'\sigma' \tag{2.90}$$

2.13.8.4 Shell and Tube Heat Exchangers

The pressure drop on the tubeside is determined from Equation 2.82 with proper values of f, K_c, and K_e. However, in shell and tube exchangers, K_c and K_e for the tube flow are generally neglected since their contribution is small compared to the losses associated with inlet and outlet headers/channels. If U-tubes or hairpins are used in a multipass unit, an additional pressure drop due to a 180° bend needs to be included. The pressure drop associated with such a bend is discussed next.

2.13.8.5 Pressure Drop due to Flow Turning

The pressure drop associated with flow turning, Δp_t, is expressed in the general form as

$$\Delta p_t = K\left(\frac{1}{2}\rho U^2\right) \tag{2.91}$$

where

K is the turning loss coefficient
U is the upstream velocity based on unaffected flow area
ρ is the fluid density at the bulk mean temperature

For a 180° turning, Δp_t is expressed in terms of mass velocity G by

$$\Delta p_t = \frac{4G^2}{2g_c\rho} = 4 \times \text{Velocity head} \tag{2.92}$$

2.13.8.6 Pressure Drop in the Nozzles

Pressure drop in the inlet nozzle, $\Delta p_{n,i}$, in terms of mass velocity through nozzle, G_n is given by

$$\Delta p_{n,i} = \frac{1.0G_n^2}{2g_c\rho} = 1.0 \times \text{Velocity head} \tag{2.93}$$

Pressure drop in the outlet nozzle, $\Delta p_{n,e}$, is given by

$$\Delta p_{n,e} = \frac{0.5G_n^2}{2g_c\rho} = 0.5 \times \text{Velocity head} \tag{2.94}$$

Total pressure drop associated with the inlet and outlet nozzles, Δp_n, is given by

$$\Delta p_n = \frac{1.5G_n^2}{2g_c\rho} = 1.5 \times \text{Velocity head} \tag{2.95}$$

where G_n is the mass velocity through nozzle.

2.13.8.7 Temperature-Dependent Fluid Properties Correction

One of the basic idealizations made in the theoretical solutions for Colburn *j* factor and friction factor *f* is that the fluid properties are constant. Most of the experimental *j* and *f* data obtained involve small temperature differences so that the fluid properties do not vary significantly. For those problems that involve large temperature differences, the constant-property results would deviate substantially and need to be modified [48,51]. Two schemes for such correction commonly used are

1. The property ratio method, which is extensively used for the internal flow problem
2. The reference temperature method, which is most common for the external flow problem

One of the two methods, the property ratio method, is described next. In the property ratio method, all properties are evaluated at the bulk mean temperature, and then all of the variable properties effects are lumped into a function. Shah's approach [51] to temperature-dependent fluid properties correction via the property ratio method is discussed here.

2.13.8.8 Gases

For gases, the viscosity, thermal conductivity, and density are functions of the absolute temperature *T*; they generally increase with temperature. The temperature-dependent property effects for gases in terms of Stanton number St are correlated by the following equations:

$$\frac{Nu}{Nu_{cp}} = \frac{St}{St_{cp}} = \left(\frac{T_w}{T_m}\right)^n \tag{2.96}$$

$$\frac{f}{f_{cp}} = \left(\frac{T_w}{T_m}\right)^m \tag{2.97}$$

where

T_w is the tube wall temperature
T_m is the bulk mean temperature of the fluid

where the subscript cp refers to the constant property variable and all temperatures are absolute. All of the properties in dimensionless groups of Equations 2.95 and 2.96 are evaluated at the bulk mean temperature. The values of the exponents *n* and *m* depend upon the flow regime, namely, laminar or turbulent.

Laminar flow: For laminar flow, the exponents *n* and *m* are given by [57]

$$n = 0.0, \quad m = 1.00 \quad \text{for } 1 < \frac{T_w}{T_m} < 3 \quad (\text{heating}) \tag{2.98}$$

$$n = 0.0, \quad m = 0.81 \quad \text{for } 0.5 < \frac{T_w}{T_m} < 1 \quad (\text{cooling}) \tag{2.99}$$

Turbulent flow:
Gas heating: For gas heating, Sleicher and Rouse [58] recommend the following correlations:

$$Nu = 5 + 0.012\ \text{Re}^{0.83}\left(\text{Pr} + 0.29\right)\left(\frac{T_w}{T_m}\right)^n \tag{2.100}$$

where

$$n = -\left[\log_{10}\left(\frac{T_w}{T_m}\right)\right]^{0.25} + 0.3 \tag{2.101}$$

Equations 2.100 and 2.101 are valid for $0.6 < \text{Pr} < 0.9$, $10^4 < \text{Re} < 10^6$, $1 < T_w/T_m < 5$, $L/D_h > 40$. All the fluid properties in Nu, Re, and Pr are evaluated at the bulk mean temperature. (Nu is Nusslet number, Re is Reynolds number, and Pr is Prandtl number.)

Further, $m = -0.10$ for $1 < T_w/T_m < 2.7.4$.

Gas cooling: Kays and Crawford [59] recommend $n = 0.0$ and $m = -0.10$.

2.13.8.9 Liquids

For liquids, only the viscosity varies greatly with temperature. Thus, the temperature-dependent effects for liquids are correlated by the following equations:

$$\frac{Nu}{Nu_{cp}} = \left(\frac{\mu_w}{\mu_m}\right)^n \tag{2.102}$$

$$\frac{f}{f_{cp}} = \left(\frac{\mu_w}{\mu_m}\right)^m \tag{2.103}$$

where

μ_w is the viscosity of the fluid at tube wall temperature
μ_m is the bulk mean temperature of the fluid

Laminar flow: For laminar flow through a circular tube, the exponents n and m are given by Deissler [60] for heating and by Shannon and Depew [61] for cooling as

$$n = -0.14, \quad m = 0.58 \quad \text{for } \frac{\mu_w}{\mu_m} < 1 \quad (\text{heating}) \tag{2.104}$$

$$n = -0.14, \quad m = +0.54 \quad \text{for } \frac{\mu_w}{\mu_m} > 1 \quad (\text{cooling}) \tag{2.105}$$

Turbulent flow: Petukhov [62] recommends the following values for the exponents n and m:

$$n = -0.11 \quad \text{for } 0.0.8 < \frac{\mu_{\mathrm{w}}}{\mu_{\mathrm{m}}} < 1 \quad (\text{heating}) \tag{2.106}$$

$$n = -0.25 \quad \text{for } 40 > \frac{\mu_{\mathrm{w}}}{\mu_{\mathrm{m}}} > 1 \quad (\text{cooling}) \tag{2.107}$$

which are valid for $2 \le \mathrm{Pr} \le 140$, $10^4 \le \mathrm{Re} \le 1.25 \times 10^5$. Also,

$$m = 0.25 \quad \text{for } 0.35 < \frac{\mu_{\mathrm{w}}}{\mu_{\mathrm{m}}} < 1 \quad (\text{heating}) \tag{2.108}$$

$$m = 0.24 \quad \text{for } 2 > \frac{\mu_{\mathrm{w}}}{\mu_{\mathrm{m}}} > 1 \quad (\text{cooling}) \tag{2.109}$$

which are valid for $1.3 \le \mathrm{Pr} \le 10$, $10^4 \le \mathrm{Re} \le 2.7.3 \times 10^5$.

For liquid heating, Petukhov [62] correlates the variable property friction data in the following forms:

$$\frac{f}{f_{\mathrm{cp}}} = \frac{1}{6}\left(7 - \frac{\mu_{\mathrm{m}}}{\mu_{\mathrm{w}}}\right) \quad \text{for } 0.35 < \frac{\mu_{\mathrm{w}}}{\mu_{\mathrm{m}}} < 1 \tag{2.110}$$

Equation 2.110 is based on the following data: $1.3 \le \mathrm{Pr} \le 10$, $10^4 \le \mathrm{Re} \le 2.7.3 \times 10^5$, and 0.35–2 for $\mu_{\mathrm{w}}/\mu_{\mathrm{m}}$.

2.13.9 Thermal Performance Failures

Satisfactory performance of heat exchangers can be obtained only from units that are properly designed and have built-in quality. Correct installation and preventive maintenance are user responsibilities. The failure of heat exchanger equipment to perform satisfactorily may be caused by one or more factors:

1. Improper thermal design
2. Operating conditions differing from design conditions
3. Uncertainties in the design parameters
4. Excessive fouling
5. Air or gas blanketing
6. Flow maldistribution
7. Deterioration in the geometrical parameters that cause bypassing the main fluids

2.13.10 Maldistribution

A serious deterioration in performance results for a heat exchanger when the flow through the core is not uniformly distributed. This is known as maldistribution. Due to maldistribution, the flow passages nearer to the distributor will be flush with more fluids, whereas the passages further away will have lean flow or starve for want of fluids. Maldistribution causes the following:

1. Thermal performance deterioration
2. Enhanced pressure drop
3. Erosive wear of flow passages with high mass velocity

Typical causes of maldistribution are the following:

1. Poor header design; for example, too small header or distributor located in one end of the core; flow commences from one end of the core and flows perpendicular to the core
2. Blocking of fin passages due to fouling, mechanical damage of core passages
3. Damage of flow passages either at the entrance or at the exit

Maldistribution due to unfavorable nozzle location can be corrected either by relocating the nozzles (Figure 2.68) or using deflecting vanes (for a side-entry nozzle), or baffle plates or perforated plates in the header to direct the fluid uniformly through the tubes [63]. The baffle plate/perforated plate should be installed parallel to the tubesheet.

2.13.11 Fouling

Fouling of the shellside heat transfer surfaces is common. Generally, the effects of fouling are to reduce the leakage of fluid between baffles and tubes by blocking these gaps, and it also results in less leakage paths [64]. Industrial practice is generally to evaluate the heat transfer in the clean condition and apply appropriate fouling resistances to arrive at the heat transfer surface area required in the fouled condition. Pressure drop is lower during commissioning and start-up after cleaning and gradually increases to values those predicted in the fouled condition.

2.13.12 Corrosion Allowance

Corrosion allowances are required for the various heat exchanger components to allow for the material loss due to corrosion in service. Corrosion allowances are specified according to the severity of corrosion and the corrosion resistance property of the material. Corrosion allowance is usually specified by the purchaser and goes hand-in-hand with the material selection.

2.13.13 Computer-Aided Thermal Design

In the present computer era, thermal design is almost exclusively performed by industry using computers. Chenoweth et al. [2], Bell [65], and Palen [66] discuss computer-aided design methods for shell and tube exchangers. Shah [67] discusses in detail the computer-aided thermal design methodology for both compact and shell and tube exchangers. Although there are similarities in the overall structure of the computer programs, the details vary significantly between compact and shell and tube exchangers. In the following sections, we discuss the structure of a computer design method for the thermal design of (1) compact heat exchangers and (2) shell and tube heat exchangers. Salient features of this computer program are discussed.

2.13.13.1 Overall Structure of a Thermal Design Computer Program

The overall structure of a thermal design computer program for a compact heat exchanger consists at a minimum of the following subroutines [67]: (1) input subroutines; (2) geometry subroutine; (3) fluid properties subroutine; (4) surface characteristics subroutine; (5) fin efficiency subroutine;

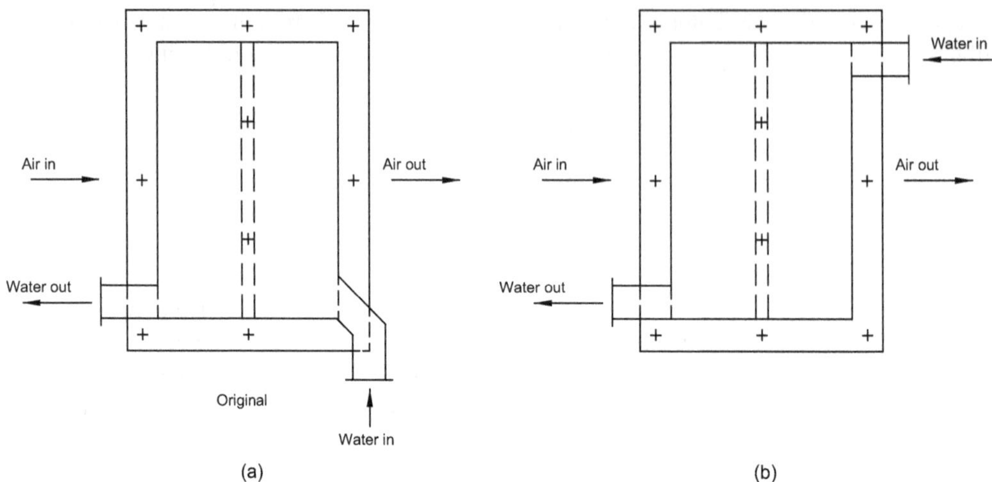

FIGURE 2.68 Maldistribution due to nozzle location with 2 passes on the tubeside corrected by relocating the nozzles.

(6) ε-NTU subroutine; (7) pressure-drop subroutine; (8) rating problem subroutine; (9) sizing problem subroutine; (10) optimization subroutines; and (11) output subroutines.

2.13.13.2 Guidelines on Program Logic

Shah [67] lists several points that should be taken into account in the initial organization and writing of the program:

1. The program should be written in a modular form containing many subroutines rather than one big main program. This allows flexibility in thorough debugging and modification of the program while running the program.
2. Error messages subroutines should be inbuilt that monitor the warning and minor and major errors for the input data and throughout the problem execution.
3. All of the iterative calculations should have a maximum number of iterations specified.

2.13.13.3 Program Structure for a Shell and Tube Exchanger

For a shell and tube exchanger, most of the structures of the subroutines mentioned so far are common but the contents slightly vary as follows:

1. *Geometry subroutine*: This should include the auxiliary calculations on the shellside and a range of geometries including the shell type, number of shells in series, number of shells in parallel, shell diameter, tube length, baffles and baffle cut, various shellside clearances, tube count, and nozzles.
2. Various shellside correction factors calculated for heat transfer and pressure drop.
3. *The thermal effectiveness subroutine*: P-NTU$_t$ relations should be built in for all possible flow arrangements and TEMA shells, including the check for "temperature cross."

2.13.14 Cooperative Research Programs on Heat Exchanger Design

Nowadays, the thermal design of most of the exchangers is carried out using commercially available software such as those developed by cooperative research organizations like Heat Transfer Research Inc. (HTRI) and Heat Transfer and Fluid Flow Services (HTFS). Other software includes

Advanced Pressure Vessel of Computer Engineering, Inc., Blue Springs, MO, Intergraph® PV Elite™, Houston, TX and COMPRESS of CODEWARE, Houston, TX (www.codeware.com). These programs offer design and cost analysis for all primary heat exchanger types and incorporate multiple design codes and standards. However, these programs are application oriented and contain company proprietary data.

2.13.14.1 HTRI

Heat Transfer Research, Inc. (HTRI), Alhambra, CA, is a cooperative research organization whose membership includes many of the leading users of heat exchangers, engineering contractors, and heat exchanger manufacturers. One of the major activities of HTRI is the development of computer programs that enable its members to design and rate heat exchangers. The scope of HTRI software HTRI Xchanger Suite® includes design, rating, and simulation of (a) air-cooled heat exchangers, heat recovery bundles, and air preheaters; (b) hairpin heat exchangers; (c) single- and two-phase shell and tube heat exchangers, including kettle and thermosiphon, reboilers, falling film evaporators, and reflux condensers; (d) plate heat exchangers, and (e) spiral plate heat exchangers; graphical stand-alone tube layout software and flow-induced vibration analysis of individual tubes in a heat exchanger bundle.

2.13.14.2 HTFS

HTFS ASPEN software will simulate, design, and rate the shell and tube heat exchanger, crossflow heat exchanger, plate heat exchanger, etc. The scope of various HTFS ASPEN software includes mechanical design, rating of shell, and tube heat exchangers or basic pressure vessels, design checking and simulation of air-cooled or other crossflow heat exchangers, of multi-stream plate-fin heat exchangers, plate heat exchangers, or brazed plate heat exchangers—it models both single-phase and two-phase streams and multistream plate-fin heat exchanger made of aluminum, stainless steel, and titanium. Simulation covers thermosiphon and crossflow pattern and also detailed layer-by-layer analysis. For example, Aspen Shell and Tube Exchanger generates the design for all major industrial shell and tube heat exchanger equipment types and applications, including single-phase, condensation, and evaporation.

2.13.15 Uncertainties in Thermal Design of Heat Exchangers

A major problem constantly faced by heat exchanger designers is to predict accurately the performance of a given heat exchanger or a system of heat exchangers for a given set of service conditions. The problem is complicated by the fact that uncertainties exist in most of the design parameters and in the design procedures themselves [68]. The design parameters that are used in the basic thermal design calculations of a heat exchanger include process parameters, heat transfer coefficients, tube dimensions (e.g., tube diameter and wall thickness), thermal conductivity of the tube material, and thermophysical properties of the fluids. Nominal or mean values of these parameters are used in the design calculations. However, uncertainties in these parameters prevent us from predicting the exact performance of the unit.

2.13.15.1 Uncertainties in Heat Exchanger Design

Uncertainties in the design parameters are summarized by Al-Zakri et al. [68], Cho [69], and Mahbub Uddin and Bell [70]. Various uncertainties and their reasons are the following:

1. Fluid flow rates, temperatures, pressures, and compositions vary from the design conditions.
2. Temperatures of cooling water and air used to cool process fluids vary with seasonal temperature changes.
3. Physical properties of the process fluids are often poorly known, especially for mixtures.

4. Heat transfer and pressure-drop correlations from which one computes convective heat transfer coefficients and pressure drop have data spreads around the mean values.
5. Manufacturing of heat transfer tubes and other component dimensions influencing thermal performance does not produce precise tube dimensions.
6. Manufacturing tolerances in equipment lead to significant differences in thermohydraulic performance between nominally identical units.
7. Fouling of heat transfer surface is poorly predictable.
8. Miscellaneous factors influence the thermal performance.

2.13.15.2 Uncertainty in Process Conditions

Heat exchangers are designed for a nominal set of operating conditions chosen to represent a relatively ideal state of a system. Process stream flow rates, compositions, and temperatures can all be expected to vary during the lifetime of the exchanger, and seasonal changes in weather constantly change cooling water and air temperatures.

2.13.15.3 Uncertainty in the Physical Properties of the Process Fluids

Correlations for heat transfer coefficients and pressure drops, and the heat transfer rate equation, require values of one or more of the following physical properties: density, specific heat, viscosity, thermal conductivity, latent heat, etc. There are substantial uncertainties in the physical properties of all but a few fluids [71].

2.13.15.4 Flow Nonuniformity

To obtain maximum thermal performance, the flow should be uniform across the entire frontal area of the core. However, the flow may not be uniform due to unfavorable header design, nozzle location, and nonuniform flow passages in the case of tube-fin and plate-fin heat exchangers. For example, automobile radiators are usually designed with the assumptions that the cooling air is uniformly distributed over the radiator core, and the cooling water is uniformly distributed over the radiator tubes [72]. Nonuniform flow distribution adversely affects the thermal performance and also could produce high-velocity regions leading to erosion of flow passages. Flow nonuniformity due to header design and due to unfavorable location of radiator fan of a diesel electric locomotive is discussed next.

Flow nonuniformity due to header design: Normally the frontal area of a compact heat exchanger is very large compared to the dimensions of the nozzles and the core depth. The large frontal area requires turning the streams in a system of headers. The turnings have marked area changes through the headers, leading to nonuniform flow through the core (Figure. 2.69), even though there may be

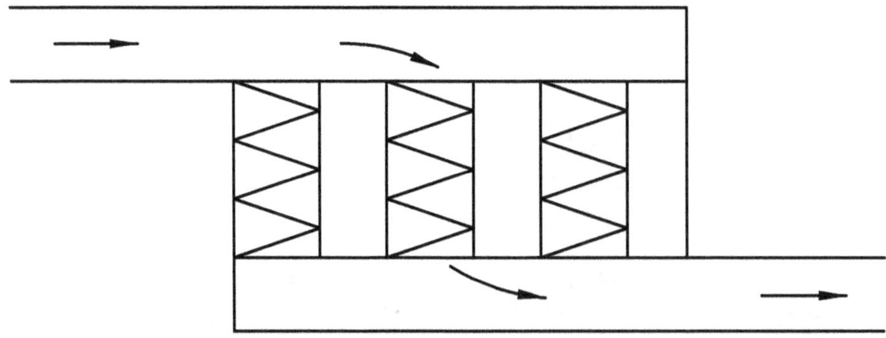

FIGURE 2.69 Nonuniform flow due to flow turning in a header.

uniform flow passages. The thermal performance degradation due to unfavorable nozzle location on the tubeside can be evaluated as discussed by Muller [73]. If the tubeside film coefficient is not controlling the overall thermal resistance, the effect of maldistribution on the overall heat transfer coefficient will be negligible.

Maldistribution due to unfavorable location of radiator fan in a diesel-electric locomotive: Flow nonuniformity on diesel locomotive radiators is usually due to inherent problems on the air side due to unfavorable location of the fan with respect to the radiators. In diesel locomotives, the left and right radiators are mounted vertically; the radiator fan is mounted at the top but at the center of the radiators, which results in nonuniform flow distribution over the radiators. Other reasons for flow nonuniformity are a crowded radiator compartment fitted with lube oil cooler and filter, and radiator fan drives.

2.13.15.5 Nonuniform Flow Passages

For high-surface-density compact heat exchangers such as tube-fin and plate-fin units with parallel fins, the identified nonuniform flow passages are due to [74]: (1) nonuniform fin spacing; (2) recurved fin; and (3) open fin. For mechanically bonded tube-fin cores, there is ample scope for bunching of fins due to deficiency in bullet expansion, thermal fatigue, and cracking of fin collars. Maldistribution due to nonuniform flow passages as shown in Figure 2.70 does not adversely affect the thermal performance in the case of either turbulent flow or highly interrupted laminar flow surfaces like offset strip fins and louvered fins, but can have a marked effect in fully developed laminar flow in continuous cylindrical passage geometries, such as those employed in disk-type rotary regenerators [75]. As a result, the reduction in Colburn j factor is substantial, with a slight reduction in friction factor f. The theoretical analysis of this problem for low Reynolds number laminar flow surfaces is discussed in Refs. [74–76]. To a certain extent, nonuniform flow passages due to fin bunching of mechanically bonded cores or soldered radiator cores can be overcome by combing of fins during maintenance schedules.

2.13.15.6 Uncertainty in the Basic Design Correlations

All of the correlations for the basic fluid flow and heat transfer mechanisms in heat exchangers show scatter when compared to experimental data. Some processes such as nucleate boiling are only very crudely predictable a priori [68].

2.13.15.7 Uncertainty due to Thermodynamically Defined Mixed or Unmixed Flows for Crossflow Heat Exchangers, after Digiovanni and Webb

The thermal effectiveness of crossflow heat exchangers is evaluated for known values of NTU and capacity rate ratio C^*, from the thermodynamically defined fluid mixedness and unmixedness and/or both through the core. The typical patterns of flow mixedness across the core are the following:

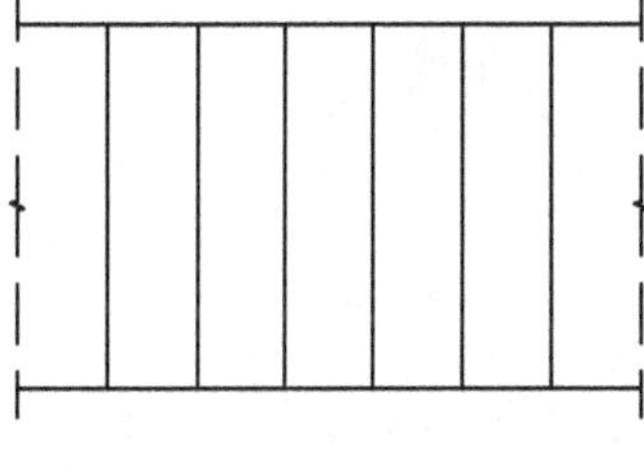

(a) Uniform passages.

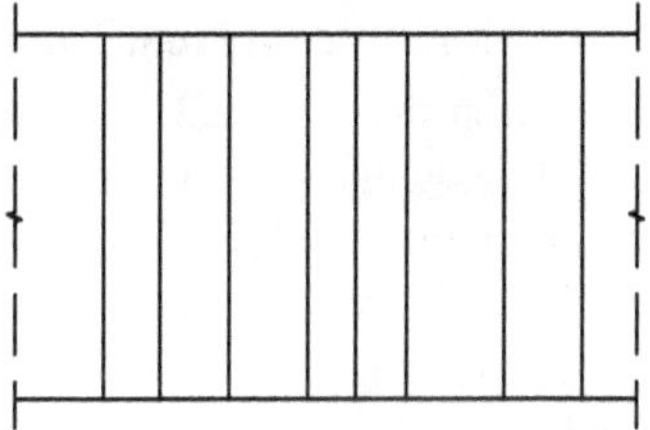

(b) Non-uniform passages.

FIGURE 2.70 Nonuniform flow passages of a compact heat exchanger.

1. Unmixed–unmixed; for example, brazed plate-fin exchanger with plain fins and continuously finned tube compact heat exchanger
2. Unmixed–mixed
3. Mixed–mixed

However, the thermodynamic definitions do not tell whether a specific geometry will provide mixed or unmixed flow. For any values of NTU and C^*, the unmixed–unmixed arrangement yields the highest thermal effectiveness and the mixed–mixed arrangement has the lowest, and the unmixed–mixed case offers intermediate values.

Generally, there is always some mixedness on the air side in the cases of individually finned tube banks and offset strip fins. During the design stage if these cases are assumed to act as a thermodynamically unmixed case, the actual thermal effectiveness will be less than the predicted value. For a tube bank, the level of mixedness depends on the tube layout patterns, pitch ratios, Reynolds number, size of the eddies in the transverse direction, etc.

Correction for partial mixing. Digiovanni and Webb [77] developed a procedure to calculate the effectiveness of crossflow heat exchangers for partially mixed flow. To calculate the correction factor, it is necessary to provide a quantitative definition of the "percent mixing," which is designated as 100y. The partially mixed condition is defined by linear interpolation between the unmixed and the mixed flow effectiveness values. If $y = 0$, the flow stream is unmixed, and $y = 1$ represents the fully mixed stream. There are three possible cases of interest for the partially mixed condition:

1. Partially mixed flow on one stream and unmixed in the other stream
2. One stream partially mixed and the other stream mixed
3. Both streams partially mixed

They presented equations to define y for the three cases of interest. For case 1, y has been defined; for the other two cases refer to Ref. [77].

Case 1: If the measured effectiveness of the partially mixed exchangers is $\varepsilon_{\mathrm{pm,u}}$, then

$$y = \frac{\varepsilon_{\mathrm{u,u}} - \varepsilon_{\mathrm{pm,u}}}{\varepsilon_{\mathrm{u,u}} - \varepsilon_{\mathrm{m,u}}} \tag{2.111}$$

From Equation 2.111, for y% mixedness, $\varepsilon_{\mathrm{pm,u}}$ is given by

$$\varepsilon_{\mathrm{pm,u}} = \varepsilon_{\mathrm{u,u}} - y\left(\varepsilon_{\mathrm{u,u}} - \varepsilon_{\mathrm{m,u}}\right) \tag{2.112}$$

2.13.15.8 Nonuniform Heat Transfer Coefficient

The idealization that the heat transfer coefficient, h, is constant throughout its flow length is not correct. It varies with location, the entrance length effect (due to the boundary layer development), surface temperature, maldistribution, fouling, manufacturing imperfections, fluid physical properties, etc. [78]. Thermal performance degradation due to uncertainties in heat transfer will be felt more in the case of highly viscous fluids and units in which phase change takes place. Temperature effects and thermal entry length effect are significant in laminar flows; the latter effect is generally not significant in turbulent flow except for low Prandtl number fluids. Approximate methods to account for specific variations in heat transfer coefficient for counterflow are analyzed by Colburn [79] and Butterworth [80], for crossflow by Sider-Tate [81] and Roetzel [82], and for 1–2N shell flow by Bowman et al. [83]. These methods are summarized in Refs. [77,78]. The state of the art on nonuniform heat transfer coefficient is reviewed in Ref. [78].

2.13.15.9 Bypass Path on the Air Side of Compact Tube-Fin Exchangers

Owing to the construction requirements, there are various clearances on the shellside of a shell and tube exchanger leading to various bypass paths. Modern procedures for shellside heat transfer calculations do take care of various leakages and bypass streams while evaluating shellside performance. In the case of finned-tube banks, the bypass paths are inevitable between (1) the sidesheet assembly and the outermost tubes, and (2) the core assembly and the housing in which the core is housed. The thermal performance deterioration due to these clearances can be overcome by the following means:

1. Attach seals to the sidesheet assembly to block the clearance between the sidesheet assembly and the outermost tubes
2. Permanently weld strips to the housing (without hindering the fitment of the core), which will block the bypass path between the core assembly and the housing in which the core is housed

These measures are discussed in Chapter 3.

2.13.15.10 Uncertainty in Fouling

The influence of uncertainties inherent in fouling factors is generally greater than that of uncertainties in physical properties, flow rates, or temperatures, such as in heat exchange between dirty aqueous fluids, in which fouling resistances can completely dominate the thermal design [84]. Planned fouling prevention, maintenance, and cleaning schedules make it possible to allow lower resistance values.

2.13.15.11 Miscellaneous Effects

Depending upon the individual design/constructional features, there exist certain miscellaneous effects that add uncertainties to thermal performance. Such effects include stagnant regions, radiation effects, allowance for in-service tube plugging due to leakages that cannot be corrected, influence of brazing/soldering, etc. Compact soldered finned-tube cores and brazed PFHEs will suffer from soldering/brazing-induced surface roughness/partial blockage due to melting of brazing filler metal. This may reduce the j factor slightly but increase the f factor substantially.

2.13.15.12 Determination of Uncertainties

The method usually used by heat exchanger designers is to ignore in the original calculations uncertainties in the design parameters. After the area has been calculated, it is multiplied by a safety factor assigned by the designer to ensure that the exchanger will perform adequately. The safety factor is based mostly upon the designer's experience and judgment and can vary from 15% to 100%. Such methods can unnecessarily add to the cost of the equipment in capital investment [68].

2.13.15.13 Computational Procedures

Buckley [85] was the first to use a statistical approach to the sizing of process equipment. His approach was to first size the unit with nominal values of design parameters. Once the effects of all uncertain parameters on the overall heat transfer area are known, the oversizing can be determined according to the desired confidence level.

Mahbub Uddin and Bell [70] present a Monte Carlo procedure for dealing with the effect of uncertainties on the design, performance, and operation of systems of process heat exchangers.

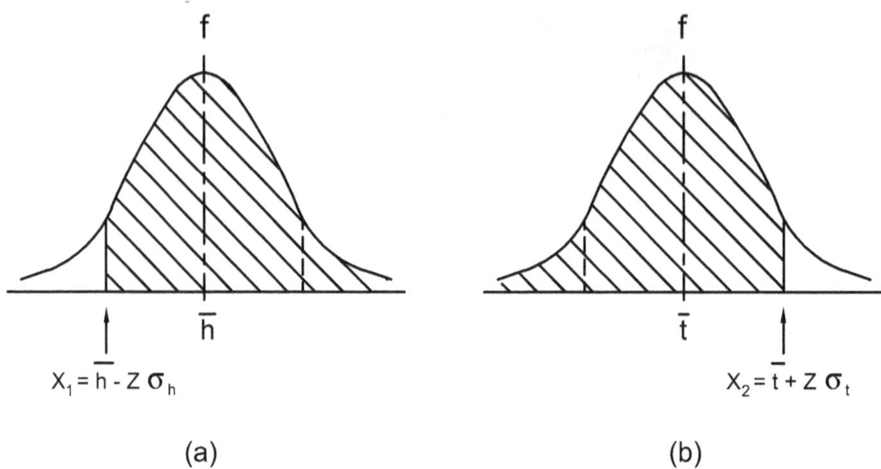

FIGURE 2.71 Normal distribution of (a) heat transfer coefficient, *h*, and (b) tube-wall thickness, *t*.

Cho [69] presents a simple method of assessing the uncertainties in the design of heat exchangers assuming a Gaussian form of data distribution for "uncertain" parameters, which leads to the root sum square (RSS) method of establishing the probability, or confidence level, that the heat exchanger design will meet its performance requirements. This method is briefly discussed next.

Cho's method of uncertainty analysis [69]: The degree of data spread around the mean value may be quantified using the concept of standard deviation, σ. Suppose that the distribution of data points for a certain parameter has a Gaussian or normal distribution. Then the procedure to incorporate the effect of data spread into the design of a heat exchanger is as follows. For example, consider the convective heat transfer coefficient, *h*, which has a data spread as shown in Figure 2.71a. The probability of the value of *h* falling below the mean value is of concern. Therefore, if one takes

$$X_1 = \bar{h} - Z\sigma_h \quad \text{and} \quad X_2 = \infty \tag{2.113}$$

or

$$Z_1 = -Z \quad \text{and} \quad Z_2 = \infty \tag{2.114}$$

then the probability that *h* will attain a value greater than $\bar{h} - Z\sigma_h$ is

$$P\left(X_1 = \bar{h} - Z\sigma_h, X_2 = \infty\right) = \frac{1}{\sqrt{2\pi}} \int_{-z}^{\infty} e^{-Z^2/2} dZ \tag{2.115}$$

where

$$Z = \left(\frac{Z - \bar{X}}{\sigma}\right) \tag{2.116}$$

This probability is known as the cumulative probability or one-sided confidence level and the numerical values are given in Ref. [69]. Here, *Z* indicates the degree of the design confidence level;

for example, if one desires a 90% confidence level, the value of Z becomes 1.282. This means the value of h should be reduced to $\bar{h} - 1.282\sigma_h$ to obtain a 90% level of design confidence. Due to the reduction in the heat transfer coefficient, the surface area is increased to get the desired thermal performance. Equations 2.113, 2.114, and 2.115 are applicable to uncertain parameters whose lower values are of design concern, like heat transfer coefficients and tube-wall thermal conductivity.

Now consider the effect of tube-wall thickness, t, which has a data spread as shown in Figure 2.71b. The value of t greater than the mean value is of concern here. Therefore, if one takes

$$X_1 = -\infty^* \quad \text{and} \quad X_2 = \bar{t} + Z\sigma_t \tag{2.117}$$

or

$$Z_1 = -\infty \quad \text{and} \quad Z_2 = +Z \tag{2.118}$$

then the probability that $\bar{t}$ will attain a value less than $\bar{t} + Z\sigma_t$ is

$$P\left(X_1 = -\infty, X_2 = \bar{t} + Z\sigma_t\right) = \frac{1}{\sqrt{2\pi}} \int_{-z}^{\infty} e^{-Z^2/2} dZ \tag{2.119}$$

Again, Z indicates the degree of the design confidence level. A 90% level of design confidence requires the use of an increased thickness, $\bar{t} + 1.282\sigma_t$ in the sizing calculation. The reduction in heat transfer rate due to increased thickness is to be compensated by an increase in the heat transfer surface area. Equations 2.117, 2.118, and 2.119 are applicable to uncertain parameters whose upper values are of design concern, like tube-wall thickness and fouling resistance.

The effects of data spreads have been discussed for individual parameters. These individual effects usually take place simultaneously, and the combined effect can be assessed for normally distributed parameters, using the RSS method [69]:

$$\Delta X = \sqrt{\sum_{i=1}^{N} \left(\Delta X_i\right)^2} \tag{2.120}$$

2.13.15.14 Additional Surface Area Required due to Uncertainty

The effects of data spread have been discussed for individual parameters. These individual effects usually take place simultaneously and the combined effect is assessed using the RSS method. The total additional surface area ΔA_i required to obtain a certain level of design confidence is calculated from [69]

$$\Delta A = \sqrt{\sum_{i=1}^{N} \left(\Delta A_i\right)^2} \tag{2.121}$$

2.13.15.15 Additional Pressure Drop due to Uncertainty

Equation 2.120 can also be applied to determine the additional pressure drop, $\Delta(\Delta p)$, required to obtain a certain level of design confidence, as given by the following equation [69]:

$$\Delta(\Delta p) = \sqrt{\sum_{i=1}^{N} \left[\Delta(\Delta p)_i\right]^2} \tag{2.122}$$

Nomenclature

A	total heat transfer area, m^2 (ft^2)
A_o	minimum free flow area, m^2 (ft^2)
c_p	specific heat, J/kg °C (Btu/lbm.°F)
C	specific heat of cold fluid, J/kg °C (Btu/lbm.°F)
C_c	specific heat of cold fluid, J/kg °C (Btu/lbm.°F)
C_h	specific heat of hold fluid, J/kg °C (Btu/lbm.°F)
C^*	heat capacity rate ratio = $(mc_p)_{min}/(mc_p)_{max}$
D_h	hydraulic diameter, ($4r_h$), m (ft)
F	LMTD correction factor
f	Fanning friction factor
G	mass velocity, $kg/m^2{\cdot}s$ ($lbm/h{\cdot}ft^2$)
g_c	acceleration due to gravity or proportionality constant 9.81 m/s^2 (32.7.17 ft/s^2) = 1 and dimensionless in SI units
h	convective heat transfer coefficient, $W/m^2{\cdot}°C$ ($Btu/ft^2\ h{\cdot}°F$)
j	Colburn heat transfer factor
K_c	fluid entrance contraction coefficient
K_e	fluid exit expansion coefficient
k	thermal conductivity of the fluid, W/m°C (Btu/h·ft·°F)
L	length in the fluid flow direction, m (ft)
L_1, L_2, L_3	heat exchanger core dimensions of a tube bank, m (ft)
M	mass flow rate of the fluid, kg/s (lbm/h)
m	exponent
NTU	number of transfer units,
Nu	Nusselt number
n	exponent
P_p	fluid pumping power
p	pressure, Pa (lbf/ft^2)
Pr	Prandtl number
Q	total heat duty of the exchanger, W·s (Btu)
q	heat transfer rate, W (Btu/h)
R	heat capacity rate ratio of shell and tube heat exchanger fluids
Re	Reynolds number
r_h	hydraulic radius, m (ft) = $D_h/4$
P	thermal effectiveness of shell and tube heat exchanger
P	fluid friction power
p	pressure, Pa (lbf/ft^2)
St	Stanton number
T	temperature, °C (°F)
t	temperature, °C (°F)
U	upstream fluid velocity, m/s (ft/s)
U	overall heat transfer coefficient, $W/m^2{\cdot}°C$ ($Btu/h{\cdot}fr^2{\cdot}°F$)
Δp	pressure drop, Pa (lbf/ft^2)
ε	thermal effectiveness
μ_m	viscosity of the fluid at bulk mean temperature, kg/m·s or Pa·s (lbm/h·ft)
μ_w	viscosity of the fluid at wall temperature, kg/m·s or Pa·s (lbm/h·ft)
ρ	fluid density, kg/m^3 (lbm/ft^3)
β	area density, m^2/m^3 (ft^2/ft^3)
Δp	pressure drop
η_o	fin efficiency or the overall surface efficiency
σ	the ratio of minimum free flow area to frontal area

Subscripts

avg	average
c	cold
h	hot
cp	constant property
e	exit
i	inlet
lm	log mean
m	mean
n	nozzle
o	outlet
s	shell
t	tube
m,u	mixed, unmixed
pu,u	partially unmixed, unmixed
u,u	unmixed, unmixed
w	wall temperature conditions
1	inlet of the exchanger
2	outlet of the exchanger
min	minimum
max	maxim

ACKNOWLEDGMENT

The author acknowledges Dr. R.K. Shah for providing the thermal relation formulas for various cases of crossflow arrangements and shell and tube heat exchangers.

REFERENCES

1. Shah, R.K., Mueller, A.C. (1985) Heat exchanger basic thermal design methods. In *Handbook of Heat Transfer Applications*, 2nd edn. (W.M. Rohsenow, J.P. Hartnett, and E.N. Ganic, eds.), pp. 4-1–4-77.McGraw-Hill, New York.
2. Shah, R.K. (1983) Heat exchanger basic design methods. In: *Low Reynolds Number Flow Heat Exchangers* (S. Kakac, R.K. Shah, and A.E. Bergles, eds.), pp. 21–71. Hemisphere, Washington, DC.
3. Baclic, B.S. (1990) ε-NTU analysis of complicated flow arrangements. In: *Compact Heat Exchangers—A Festschrift for A. L. London* (R.K. Shah, A.D. Kraus, and D. Metzger, eds.), pp. 31–90. Hemisphere, Washington, DC.

4a. London, A.L., Seban, R.A. (1942) *A Generalization of the Methods of Heat Exchanger Analysis.* Mechanical Engineering Department, Stanford University, Stanford, CA.

4b. London, A.L., Seban, R.A. (1980) A generalization of the methods of heat exchanger analysis. *International Journal of Heat Mass Transfer* 23: 5–16.

5. Shah, R.K., Sekulic, D. (2003) *Fundamentals of Heat Exchanger Design.* Wiley, New York, NY.
6. Kraus, A.D. (2003) Heat exchangers. Chapter 11. In: *Heat Transfer Handbook* (A. Bejan and A.D. Kraus, eds.), pp. 797–911. John Wiley & Sons.
7. Smith, D.M. (1934) Mean temperature difference in crossflow. *Engineering* 138: 479–481, 606–607.
8. Mueller, A.C. (1973) Heat Exchangers, Section 18. In: *Handbook of Heat Transfer* (W. H. Rohsenow and J. P. Hartnett, eds.), pp. 634–793. McGraw-Hill, New York.
9. Singh, K.P. (1981) Some fundamental relationships for tubular heat exchanger thermal performance. *Transactions of ASME, Journal of Heat Transfer* 103: 573–578.
10. Turton, R., Ferguson, C.D., Levenspiel, O. (1984) Performance and design charts for heat exchangers. *Transactions of ASME, Journal of Heat Transfer* 106: 893–895.
11. Bowman, R.A. Mueller, A.C., Nagle, W.M. (1940) Mean temperature difference. *Transactions of ASME* 62: 283–294.

12. Underwood, A.J.V. (1934) The calculation of the mean temperature difference in multipass heat exchangers. *Journal of the Institution of Petroleum Technologists* 20: 145–158.
13. Ten Broeck, H. (1938) Multipass heat exchanger calculations. *Industrial & Engineering Chemistry* 30: 1041–1042.
14. Kays, W.M., London, A.L. (1964) *Compact Heat Exchangers*, 2nd edn. McGraw-Hill, New York.
15. Tubular Exchanger Manufacturers Association (2007) *Standards of Tubular Exchanger Manufacturers Association*, 9th edn. Tubular Exchanger Manufacturers Association, Tarrytown, NY.
16. Schlinder, E.V. editor-in-chief (1983) *Heat Exchanger Design Handbook*, Vol. 1. Hemisphere, Washington, DC.
17. Pignotti, A. (1984) Flow reversibility of heat exchangers. *Transactions of ASME, Journal of Heat Transfer* 106: 361–368.
18. Pignotti, A. (1989) Relation between the thermal effectiveness of overall parallel and counterflow heat exchangers. *Transactions of ASME, Journal of Heat Transfer,* 111: 294–299.
19. Shah, R.K., Sekulic, D.P. (1998) Heat exchangers. In: *Handbook of Heat Transfer Applications* (Rohsenow, W.M., Hartnett, J.P., Ganic, E.N., eds.), 3rd edn., pp. 17.1–17.169, Chapter 17. McGraw-Hill, New York.
20. Nusselt, W. (1930) A new formula for heat transfer in crossflow [in German]. *Tech. Mech. Thermodynamics* 1: 417–422.
21. Mason, J.L. (1955) Heat transfer in crossflow. *Proceedings of the Second U.S. National Congress of Applied Mechanics*, pp. 801–803. Ann Arbor, MI, ASME, New York.
22. Baclic, B.S. (1978) A simplified formula for crossflow heat exchanger effectiveness. *Transactions of ASME, Journal of Heat Transfer* 100: 746–747.
23. Eckert, E.R.G. (1959) *Heat Transfer*, 2nd edn., p. 483. McGraw-Hill, New York.
24. Taborek, J. (1983) F and θ charts for crossflow arrangements. In: *Heat Exchanger Design Handbook*, Vol. 1, Section 1.5.3 (E.V. Schlunder, editor-in-chief), pp. 1.5.3-1–1.5.3–15. Hemisphere, Washington, DC.
25. (a) Pignotti, A., Cordero, G.O. (1983) Mean temperature difference in multipass crossflow. *Transactions of ASME, Journal of Heat Transfer* 105: 584–591; (b) Pignotti, A., Cordero, G.O. (1983) Mean temperature difference charts for air coolers. *Transactions of ASME, Journal of Heat Transfer* 105: 592–597.
26. (a) Pignotti, A. (1986) Analytical techniques for basic thermal design of complex heat exchanger configurations. *Heat Transfer, Eighth International Conference*, San Francisco, CA; (b) Pignotti, A. (1984) Matrix formalism for complex heat exchangers. *Transactions of ASME, Journal of Heat Transfer* 106: 352–360.
27. Schedwill, H. (1968) *ThermischeAuslegung von Kreuzstromwarmeaustauschern*, Fortschr-Ber. VDI-Z Reihe 6 Nr. 19. VDI-Verlag, Dusseldorf, Germany.
28. Nicole, F.J.L. (1972) Mean temperature difference for heat exchanger design. *Council for Scientific and Industrial Research, Special Report Chemistry 223*. Pretoria, South Africa.
29. Stevens, R.A., Fernandez, J., Woolf, J.R. (1957) Mean temperature difference in one, two, three-pass crossflow heat exchangers. *Transactions of ASME* 79: 287–297.
30. Nagle, W.M. (1933) Mean temperature difference in multipass heat exchanger. *Industrial & Engineering Chemistry* 25: 604–609.
31. Baclic, B.S. (1989) 1–2N shell and tube exchanger effectiveness: A simplified Kraus-Kern equation. *Transactions of ASME, Journal of Heat Transfer* 111: 181–182.
32. Pignotti, A., Tamborenea, P.I. (1988) Thermal effectiveness of multiple shell and tube pass TEMA E heat exchangers. *Transactions of ASME, Journal of Heat Transfer* 111: 54–59.
33. Whistler, A.M. (1947) Correction for heat conduction through longitudinal baffle of heat exchanger. *Transactions of ASME* 69: 683–685.
34. Rozenmann, T., Taborek, J. (1972) The effect of leakage through the longitudinal baffle on the performance of two pass shell exchangers. *Heat Transfer–Tulsa, 1972*, AIChE Symposium Series No. 118, 68: 12–20.
35. Taborek, J. (1983) F and θ charts for shell and tube exchangers. In: *Heat Exchanger Design Handbook*, Vol. 1, Section 1.5.2 , pp. 1.5.2-1–1.5.2–15 (E. V. Schlunder, editor-in-chief). Hemisphere, Washington, DC.
36. Schlinder, D.L.. Bates, H.T. (1960) True temperature difference in a 1–2 divided flow heat exchanger, heat transfer—Starrs. *Chemical Engineering Progress Symposium Series* 56: 203–206.

37. Murty, K.N. (1983) Heat transfer characteristics of one- and two-pass split flow heat exchangers. *Heat Transfer Engineering* 4(3–4): 26–34.
38. Singh, K.P., Holtz, M.J. (1979) Generalization of the split flow heat exchanger geometry for enhanced heat transfer, heat transfer—San Diego. *AIChE Symposium Series* 75: 219–226.
39. Ishihara, K., Palen, J.W. (1986) Mean temperature difference correction factor for the TEMA H shell. *Proceedings of the Eighth International Heat Transfer Conference*, San Francisco, CA (C.L. Tien, V.P. Carey, J. K. Ferrel, eds.), pp. 2709–2714. Hemisphere, Washington, DC.
40. Shah, R.K., Pignotti, A. (1989) *Heat exchanger basic design theory and results*. A report prepared under NSF Grant No. INT-8513531, Buffalo, NY. (TEMA G_{1-1} exchanger results are also derived in this report by the chain rule methodology.)
41. Jaw, L. (1964) Temperature relations in shell and tube exchangers having one pass splitflow shells. *Transactions of ASME, Journal of Heat Transfer* 86: 408–416.
42. (a) Gardner, K.A. (1941) Mean temperature difference in multipass exchangers—Correction factors with shell fluid unmixed. *Industrial & Engineering Chemistry* 33 1495–1500; (b) Gardner, K.A. (1941) Mean temperature difference in multipass exchangers. *Industrial & Engineering Chemistry* 33: 1215–1223.
43. Pignotti, A. (1986) Effectiveness of series assemblies of divided-flow heat exchanger. *Transactions of ASME, Journal of Heat Transfer* 108: 141–146.
44. Domingos, J.D. (1969) Analysis of complex assemblies of heat exchangers. *International Journal of Heat Mass Transfer* 12: 537–548.
45. Baclic, B.S., Gvozdenac, D.D. (1981) Exact explicit equations for some two- and three-pass crossflow heat exchanger effectiveness. In: *Heat Exchangers; Thermal-Hydraulic Fundamentals and Design* (S. Kakac, A.E. Bergles, F. Mayinger, eds.), pp. 481–494. Hemisphere, Washington, DC.
46. Baclic, B.S., Gvozdenac, D.D. (1981) NTU relationships for inverted order flow arrangements of two-pass crossflow heat exchangers. In: *Regenerative and Recuperative Heat Exchangers*, Vol. 21 (R.K. Shah, D.E. Metzger, eds.), pp. 27–41. ASME, New York.
47. Shah, R.K. (1988) Heat exchanger design methodology. In: *Heat Transfer Equipment Design* (R.K. Shah, E.C. Subbarao, R.A. Mashelikar, eds.), pp. 17–23. Hemisphere, Washington, DC.
48. Kays, W.M., London, A.L. (1984) *Compact Heat Exchangers*, 3rd edn. McGraw-Hill, New York.
49. Shah, R.K., Bhatti, M.S. (1988) Assessment of correlations for single-phase heat exchangers. In: *Two-Phase Flow Heat Exchangers—Thermal-Hydraulic Fundamentals and Design* (S. Kakac, A.E. Bergles, F.E. Oliveira, eds.), pp. 81–123. Kluwer Academic Publishers, London, UK.
50. Gollin, M. (1984) Heat exchanger design and rating. In: *Handbook of Applied Thermal Design* (E.C. Guyer, ed.), pp. 7-24–7-36. McGraw-Hill, New York.
51. Shah, R.K. (1981) Compact heat exchanger design procedures. In: *Heat Exchangers: Thermal Hydraulic Fundamentals and Design* (S. Kakac, A.E. Bergles, F. Mayinger, eds.), pp. 495–536. Hemisphere, Washington, DC.
52. Bell, K.J. (1988) Delaware method for shellside design. In: *Heat Transfer Equipment Design* (R.K. Shah, E.C. Subbarao, R.A. Mashelkar, eds.), pp. 145–166. Hemisphere, Washington, DC.
53. Cardwell (1950) Optimum tube size for shell and tube type heat exchangers. *Transactions of ASME* 72: 1061–1065.
54. Shah, R.K., Afimiwala, K.A., Mayne, R.W. (1978) Heat exchanger optimization. In: *Heat Transfer*, Vol. 4, pp. 185–191. Hemisphere, Washington, DC.
55. Kays, W.M. (1950) Loss coefficients for abrupt changes in flow cross section with low Reynolds number flow in single and multiple tube system. *Transactions of ASME* 1067–1074.
56. Shah, R.K. (1988) *Personal Communication, dt. 3.1.1998.* Senior Staff Research Scientist, Heat Exchanger Development, DELPHI Harrison Thermal Systems, Lockport, NY.
57. Kays, W.K., Perkins, H.C. (1973) Forces convection, internal flow in ducts. In: *Handbook of Heat Transfer* (W.M. Rohsenow, J.P. Hartnett, eds.). McGraw-Hill, New York.
58. Sleicher, C.A., Rouse, M.W. (1975) A convenient correlation for heat transfer to constant and variable property fluids in turbulent pipe flow. *International Journal of Heat Mass Transfer* 18: 677–683.
59. Kays, W.M., Crawford, M.E. (1980) *Convective Heat and Mass Transfer*, 2nd edn. McGraw-Hill, New York.
60. Deissler, R.G. (1951) *Analytical investigation of fully developed laminar flow in tubes with heat transfer with fluid properties variable along the radius, NACA TN 2410* . NACA, Washington, DC.

61. Shannon, R.L., Depew, C.A. (1969) Forced laminar flow convection in a horizontal tube with variable viscosity and free convection effects. *Transactions of ASME, Journal of Heat Transfer* 91: 251–258.
62. Petukhov, B.S. (1970) Heat transfer and friction in turbulent pipe flow with variable physical properties. In: *Advances in Heat Transfer*, Vol. 6, pp. 503–564. Academic Press, New York.
63. Shah, R.K. (1985) Compact heat exchangers. In: *Handbook of Heat Transfer Applications* (W.M. Rohsenow, J.P. Harnett, E.N. Ganic, eds.), pp. 4-174–4-313. McGraw-Hill, New York.
64. Minton, P.E. (1990) Process heat transfer. *Proceedings of the Ninth International Heat Transfer Conference*, Jerusalem, Paper No. KN-22, 1: 355–363.
65. Chenoweth, J.M., Kistler, R.S. (1976) *Computer program as a tool for heat exchanger rating and design.* ASME Paper No. 76-WA/HT-4.
66. Taborek, J. (1979) Evolution of heat exchanger design techniques. *Heat Transfer Engineering* 1: 15–29.
67. Shah, R.K. (1983) Compact heat exchanger surface selection, optimization, and computer aided thermal design. In: *Low Reynolds Number Flow Heat Exchangers* (S. Kakac, R.K., Shah, A.E. Bergles, eds.), pp. 983–998. Hemisphere, Washington, DC.
68. Al-Zakir, A.S., Bell, K.J. (1981) Estimating performance when uncertainties exist. *Chemical Engineering Progress* 77: 39–49.
69. Cho, S.M. (1986) Uncertainty analysis of heat exchanger thermal hydraulic design. In: *Proceedings of the Thermal/Mechanical Heat Exchanger Design—Karl Gardner Memorial Session*, Vol. 64 (K.P. Singh, S.M. Shenkman, eds.), pp. 33–41. American Society of Mechanical Engineers Winter Meeting, Anaheim, CA, December 7, 1985.
70. Mahbub Uddin, A.K.M., Bell, K. J. (1988) Effect of uncertainties on the design and operation of systems of heat exchangers. In: *Heat Transfer Equipment Design* (R.K. Shah, E.C. Subbarao, R.A. Mashelikar, eds.), pp. 39–47. Hemisphere, Washington, DC.
71. Haseler, L.E., Owen, R.G., Sardesai, R.G. (1984) The sensitivity of heat exchanger calculations to uncertainties in the physical properties of the process fluids. In: *Practical Application of Heat Transfer*, IMechE Conference Publications, 1982–1984, C56/82, pp. 11–20. IMechE, London, UK.
72. Chiou, J.P. (1981) The effect of flow nonuniformity on the sizing of the engine radiator. *800035 SAE* 222–230.
73. Mueller, A.C. (1976) An inquiry of selected topics on heat exchanger design. *Chemical Engineering Progress Symposium Series* 73: 273–287.
74. London, A.L. (1970) Laminar flow gas turbine regenerators—The influence of manufacturing tolerances. *Transactions of ASME, Journal of Engineering Power Series A* 92: 46–56.
75. Shah, R.K., London, A.L. (1980) Effect of nonuniform passages on compact heat exchanger performance. *Transactions of ASME, Journal of Engineering Power* 102: 653–659.
76. Mondt, J.R. (1977) Effects of nonuniform passages on deepfold heat exchanger performance. *Transactions of ASME, Journal of Engineering Power* 99: 657–663.
77. Digiovanni, M.A., Webb, R.L. (1989) Uncertainty in effectiveness for crossflow heat exchangers. *Heat Transfer Engineering* 10: 61–69.
78. Shah, R.K. (1993) Nonuniform heat transfer coefficient for heat exchanger thermal design. In: *Aerospace Heat Exchanger Technology* (R.K. Shah, A. Hashemi, eds.), pp. 417–445. Elsevier Science Publishers, Amsterdam, the Netherlands.
79. Colburn, A.P. (1933) Mean temperature difference and heat transfer coefficient in liquid heat exchangers. *Industrial & Engineering Chemistry* 25: 873–877.
80. Butterworth, D. (1981) Condensers: Thermal-hydraulic design. In: *Heat Exchangers: Thermal-Hydraulic Fundamentals and Design* (S. Kakac, A.E. Bergles, F. Mayinger, eds.), pp. 647–679. Hemisphere/McGraw-Hill, Washington, DC.
81. Sieder, E.N., Tate, G.E. (1936) Heat transfer and pressure drop of liquids in tubes. *Industrial & Engineering Chemistry* 28: 1429–1435.
82. Roetzel, W. (1974) Heat exchanger design with variable transfer coefficient for crossflow and mixed flow arrangements. *International Journal of Heat Mass Transfer* 17: 1037–1049.
83. Bowman, R.A., Muller, A.C., Nagle, W.M. (1940) Mean temperature difference. *Transactions of ASME* 62: 283–294.
84. Larowski, A., Taylor, M.A. (1982) *Systematic Procedures for Selection of Heat Exchangers*, C58/82, pp. 32–56. Institution of Mechanical Engineers, London, UK.
85. Buckley, P.S. (1950) Statistical methods in process design. *Chemical Engineering* September: 112–114.

Bibliography

Bell, K.J. (1988) Overall design methodology for shell and tube exchangers. In: *Heat Transfer Equipment Design* (R.K. Shah, E.C. Subbarao, R.A. Mashelkar, eds.), pp. 131–144. Hemisphere, Washington, DC.

Breber, G. (1988) Computer programs for design of heat exchangers. In: *Heat Transfer Equipment Design* (R.K. Shah, E.C. Subbarao, R.A. Mashelikar, eds.), pp. 167–178. Hemisphere, Washington, DC.

Cho, S.M. (1988) Basic thermal design methods for heat exchangers. In: *Heat Transfer Equipment Design* (R.K. Shah, E.C. Subbarao, R.A. Mashelikar, eds.), pp. 23–38. Hemisphere, Washington, DC.

ESDU International Ltd (1986) *Effectiveness-Ntu Relationships for Design and Performance Evaluation of Two-Stream Heat Exchangers.* Engineering Sciences Data Unit Item No. 86018, July 1986. ESDU International Ltd., McLean, VA.

ESDU International Ltd (1987) *Effectiveness-Ntu Relationships for the Design and Performance Evaluation of Multi-Pass Crossflow Heat Exchangers*. Engineering Sciences Data Unit Item No. 87020, October 1987. ESDU International Ltd., McLean, VA.

Gardner, K.A., Taborek, J. (1977) Mean temperature difference: A reappraisal. *AIChE Journal* 23: 777–786.

Kays, W.M., London, A.L. (1984) *Compact Heat Exchangers*, 3rd edn. McGraw-Hill, New York.

Lord, R.C., Minton, P.E., Sulusser, R.P. (1979) Guide to trouble free heat exchangers. In: *Process Heat Exchange, Chemical Engineering Magazine* (V. Cavaseno, ed.), pp. 60–67. McGraw-Hill, New York.

Palen, J.W. (1986) Design of process heat exchanger by computers—A short history. *Heat Transfer, Eighth International Conference*, pp. 239–248. San Francisco, CA.

Pase, G.K. (1986) Computer programs for heat exchanger design. *Chemical Engineering Progress* 70: 53–56.

3 Compact Heat Exchangers

3.1 CLASSIFICATION AND CONSTRUCTION DETAILS OF COMPACT HEAT EXCHANGERS

Compact heat exchangers are used in a wide variety of applications. Typical among them are the heat exchangers used in automobiles, air conditioning, electronics cooling, waste and process heat recovery, cryogenics, aircraft and spacecraft, ocean thermal energy conservation, solar, and geothermal systems. The need for light-weight, space-saving, and economical heat exchangers has driven the development of compact surfaces.

3.1.1 Characteristics of Compact Heat Exchangers

Specific characteristics of compact heat exchangers are discussed by Shah [1–4] and include the following:

1. Usually with extended surfaces.
2. A high heat transfer surface area per unit volume of the core, usually in excess of 700 m^2/m^3 on at least one of the fluid sides, which usually has gas flow.
3. Small hydraulic diameter.
4. Usually at least one of the fluids is a gas.
5. Fluids must be clean and relatively nonfouling because of small hydraulic diameter (D_h) flow passages and difficulty in cleaning; plain uninterrupted fins are used when "moderate" fouling is expected.
6. The fluid pumping power (i.e., pressure drop) consideration is as important as the heat transfer rate.
7. Operating pressures and temperatures are limited to a certain extent compared to shell and tube exchangers due to thin fins and/or joining of the fins to plates or tubes by brazing, mechanical expansion, etc.
8. The use of highly compact surfaces results in an exchanger with a large frontal area and a short flow length; therefore, the header design of a compact heat exchanger is important for a uniform flow distribution.
9. Fluid contamination is generally not a problem.
10. A variety of surfaces are available having different orders of magnitudes of surface area density.
11. Flexibility in distributing surface area on the hot or cold side as desired by the designer.

DOI: 10.1201/9781003352044-3

3.1.2 Construction Types of Compact Heat Exchangers

Basic construction types of compact heat exchangers are

1. Tube-fin
2. Plate-fin
3. Regenerators

Tube-fin and plate-fin heat exchangers are discussed in this chapter. Regenerators, exclusively used in gas-to-gas applications, could have more compact surface area density compared to plate-fin or tube-fin surfaces. They are covered in detail in a separate chapter. The unique characteristics of compact extended-surface plate-fin and tube-fin exchangers that favor them as compared with the conventional shell and tube exchangers include the following [1,3,4]:

1. Many surfaces are available having different orders of magnitude of surface area density.
2. Flexibility in distributing surface area on the hot and cold sides as warranted by design considerations.
3. Generally substantial cost, weight, or volume savings.

The following factors mainly determine the choice among the three types of compact heat exchangers:

1. Operating pressure and temperature
2. Phases of the fluids dealt with
3. Fouling characteristics of the fluid
4. Allowable pressure drop
5. Strength and ruggedness
6. Restrictions on size and/or weight
7. Acceptable intermixing of the fluids dealt with

3.1.3 Tube-Fin Heat Exchangers

The features of finned-tube are as follows:

Extended Surfaces for Liquids. Extended surfaces used with liquids may be on the inner or outer surface of the tubes. Because liquids have higher heat transfer coefficients than gases, fin efficiency considerations require shorter fins with liquids than with gases.

Internally extended surfaces. Integral internally finned tubes have been developed for tubeside heat transfer augmentation, and they are commercially available in copper and aluminum.

Materials. Finned tubes may be made of two dissimilar metals in order to utilize the advantages of both. Each material has its own advantages and considerations, depending on the end use and effect on performance and efficiency. For example, an aluminum fin can be applied to a stainless steel tube to maximize air cooling. Similar metals are also common—say a copper tube with a copper fin. Copper fin is known for superior heat transfer property but can be susceptible to corrosion. On the other hand, a stainless steel fin has a much lower heat transfer coefficient but is highly resistant to corrosion and has superior tensile strength properties. Materials for the manufacture of fins are limited by the operating temperature of certain applications. For low- to moderate-temperature applications, fins can be made from aluminum, copper, or brass and thus maintain high-fin efficiency.

For high-temperature applications, stainless steel and heat-resistant alloys may be used with a possible reduction in the fin efficiency. Consequently, suitable high-performance surfaces may be selected to offset the reduction in fin efficiency or the fin thickness should be reduced.

Manufacture is by brazing or welding. For low- or moderate-temperature applications, the fins can be mechanically bonded. Low-fin tubes are extruded.

Flow arrangements. Compact exchangers are mostly used as single-pass crossflow or multipass cross-counterflow exchangers.

Applications. Tube-fin surfaces are also known as extended surfaces and extended surface heat exchangers are widely used throughout the industry in a variety of applications. Extended surfaces can be applied to both internal and external tubes and plate surfaces. They are employed when one fluid stream is at a higher pressure and/or has a significantly higher heat transfer coefficient compared to the other fluid stream. For example, in a gas-to-liquid exchanger, the heat transfer coefficient on the liquid side is generally very high compared to the gas side. Fins are used on the gas side to increase the surface area. In a tube-fin exchanger, round and rectangular tubes are most commonly used (although elliptical tubes are also used), and fins are employed either on the outside or on the inside, or on both sides of the tubes, depending upon the application. This chapter discusses external extended surface heat exchangers and internal extended surfaces units are discussed in Chapter 8.

Fins on the outside of the tubes (Figure 3.1) may be categorized as follows:

1. Normal fins on individual tubes (Figure 3.1a), referred to as individually finned tubes.
2. Longitudinal fins on individual tubes (Figure 3.1b), which are generally used in condensing applications and for viscous fluids in double-pipe heat exchangers.
3. Flat or continuous (plain, wavy, or interrupted) external fins on an array of tubes (either circular or flat tube) as shown in Figures 3.1c,d. Since interrupted fins are more prone to fouling, many users prefer plain continuous fins in situations where fouling, particularly by fibrous matter, is a serious problem.

3.1.3.1 Specific Qualitative Considerations for Tube-Fin Surfaces

Specific qualitative considerations for the tube-fin surfaces include the following:

i. Tube-fin exchangers usually have lower compactness than plate-fin units.
ii. A tube-fin exchanger may be designed for a wide range of tube fluid operating pressures, with the other fluid being at low pressure.
iii. Reasonable fouling can be tolerated on the tubeside if the tubes can be cleaned.

3.1.3.2 Applications

Tube-fin exchangers are extensively used as condensers and evaporators in air-conditioning and refrigeration applications, for cooling of water or oil of vehicular or stationary internal combustion engines, and as air-cooled exchangers in the process and power industries. The usual arrangement is that water, oil, or refrigerant flows in the tubes, while air flows across the finned tubes.

3.1.4 Individually Finned Tubes

Individually finned tube geometry is much more rugged than continuous fin geometry, but has lower compactness. The most common individual finned tubes are with plain circular, helical, or annular enhanced fin geometries such as segmented, studded, slotted, or wire-loop fins [5]. Fins are attached

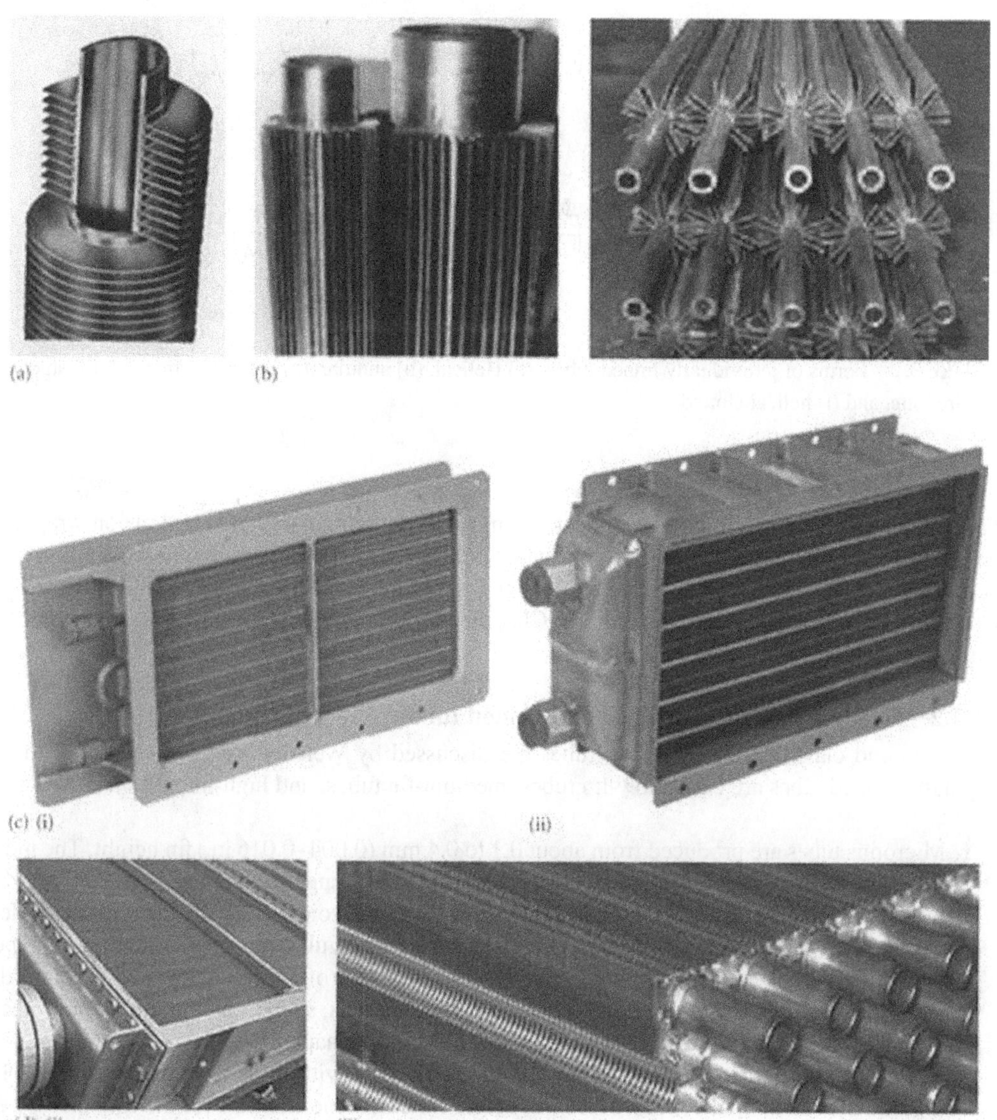

FIGURE 3.1 (a) Aluminum *L* foot tube fin and (b) welded longitudinal fin. (Courtesy of Vulcan Finned Tubes, Tomball, TX.) (c) *Continuous finned* tube heat exchanger—(i) cupronickel tube-fin heat exchanger and (ii) flat-tube heat exchanger to cool hydraulics (ALEX Core). (Courtesy of Lytron Inc., Woburn, MA.) (d) Continuous finned compact cooler. (Courtesy of GEA Heat Exchangers, Catoosa, OK.)

to the tubes by a tight mechanical bond through tension winding, soldering, brazing, or welding. Figure 3.2 shows some of the fin geometries (schematic) that have been used on circular tubes (some of these geometries are discussed as heat transfer enhancement fins in Chapter 8).

Figure 3.3 shows photos of fin-tube geometries, and Figure 3.4 shows circular fin-tube geometrical parameters. Fins are attached to the tubes by a tight mechanical fit, tension winding, adhesive bonding, soldering, brazing, welding, or extrusion or sleeving a liner containing extruded fins on a base metal. Figure 3.5 shows common circular tube-fin classification and manufacturing techniques. Details of various fin geometries are discussed next.

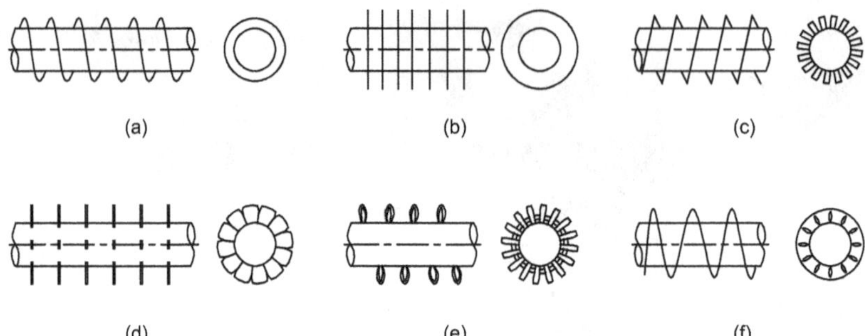

FIGURE 3.2 Forms of individually finned tubes. (a) Helical, (b) annular, (c) helical segmented, (d) studded, (e) wire loop, and (f) helical slotted.

3.1.4.1 Plain Circular Fins

Plain circular fins are the simplest and most common. They are manufactured by tension wrapping the fin strip around a tube, forming a continuous helical fin, or by mounting circular fin disks on the tube. Helically wrapped and extruded fins on circular tubes are frequently used in the process industries and in heat recovery applications. Classification of plain circular fin tubes in terms of fin height is discussed next.

3.1.4.2 Fin Height and Classification of Finned Tubes

Fin height and classification of finned tubes are discussed by Wen-Jei Yan [6]. According to fin height, the finned tubes are called low-fin tubes, medium-fin tubes, and high-fin tubes as follows:

1. Microfins tubes are produced from about 0.1 to 0.4 mm (0.004–0.016 in.) fin height. The most favorable helix angles for heat transfer and pressure drop range from about 7° to 23°, but 18° is most popular. Most microfins have approximately a trapezoidal cross section with a rounded top and rounded corners at the root. Other shapes are triangular, rectangular, and screw type, and fin height is about 0.2–0.3 mm (0.008–0.012 in.). Microfin tubes are available primarily in copper but are also becoming available in other materials, such as aluminum, carbon steel, steel alloys, etc. Microfin tubes are discussed in detail in Chapter 8.
2. In low-fin tubes, fins are extruded from the tube material with fin heights on the order of 1–2 mm. Fins of nominal 1.6 mm (1/16 in.) height have become standard for the low-fin type of duty usually required in shell and tube exchangers. Low-fin tubes are generally manufactured by raising the fin from base tube metal by a fabrication operation (sometimes by an extrusion process). Integral fining ensures the maximum thermal efficiency of the tube since there is no possibility of fins becoming loose (and hence increase in thermal contact resistance) due to corrosion, thermal expansion and contraction, or by mechanical damage in handling.
3. Fins of approximately 3.2 mm (1/8 in.) are referred to as medium fins and are usually employed in shell and tube exchangers.
4. Air-cooled exchangers require high fins, which have a height of 6.4–25 mm (1/4–1.0 in.). Heat recovery systems prefer medium to high fins in tube banks of inline arrangement.

However, Rabas and Taborek [7] classify low-fin tubes as having a fin height less than 6.35 mm (0.25 in.). The major reasons for using low-finned tubes in various types of heat exchangers are as follows [7]:

1. To balance heat transfer resistance
2. To maintain fin bond integrity

FIGURE 3.3 (a) Helical spiral wound crimped finned tube. (b) Annular fin or stamped fin tube. (a and b Courtesy of Fin Tube Products, Inc., Wadsworth, OH.) (c) TURB-XHF (serrated or segmented fin). (d) Stud welded tube. (c and d Courtesy of Fin Tube, LLC, Tulsa, OK; www.fintubellc.com) (e) Wire loop fintube. (TAAM Engineering, Maharastra, India.)

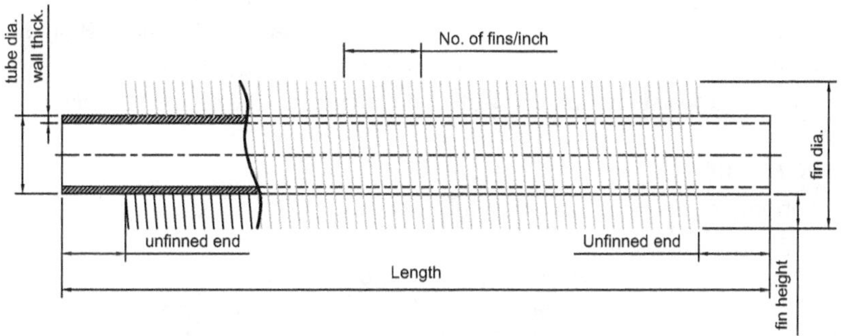

FIGURE 3.4 Individually finned tube details.

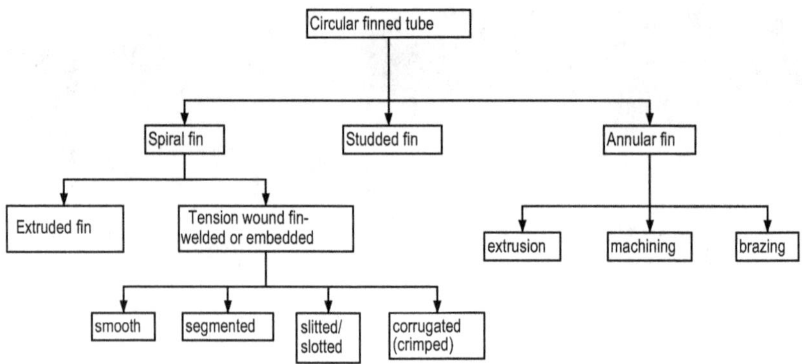

FIGURE 3.5 Tube-fin classification.

3. To reduce fouling
4. To satisfy the tube material specification

3.1.4.3 Enhanced Fin Geometries

Segmented or spine fins: Segmented or spine fins are made by helically winding a continuous strip of metal that has been partially cut into narrow sections. Upon winding, the narrow sections separate and form the narrow strip fins that are connected at the base.

Studded fins: A studded fin is similar to a segmented fin, but individual studs are welded to the tubes.

Slotted fins: Slotted fins have slots in the radial direction; when radially slitted material is wound on a tube, the slits open.

Wire loops around the tube: The wire loops are held in the tube by a tensioned wire within the helix or by soldering. The enhancement characteristic of small-diameter wires is important at low flows where the enhancement of other interrupted fins diminishes.

All of these geometries provide enhancement by the periodic development of thin boundary layers on small-diameter wires or flat strips, followed by their dissipation in the wake region between elements [8].

3.1.4.4 Construction Materials

The materials to be used for the tubes, headers, and water tanks depend on the specific requirements of the application. The most common materials for tubes and fins are copper, aluminum, and steel. For open-circuit installations, the tubes are made from copper, phosphorus-deoxidized copper, aluminum, brass, and copper–nickel–iron alloys, that is, cupronickels, and carbon steel for heat-recovery applications. Fins are mostly from aluminum and copper; for heat-recovery applications they are made from carbon steel. Headers and tube plates are made of carbon steel.

3.1.4.5 Tube Layout

The two basic arrangements of the tubes in a tube bank are staggered and inline. These arrangements are shown in Figure 3.6. The only construction difference between these two is that in the staggered arrangement each alternate row is shifted half a transverse pitch, and the two arrangements differ in flow dynamics. Due to compactness and higher heat transfer, the staggered layouts are mostly used. However, if the air stream is laden with dust, abrasive particles, etc., the inline layout is preferred, being less affected and enabling ease of cleaning.

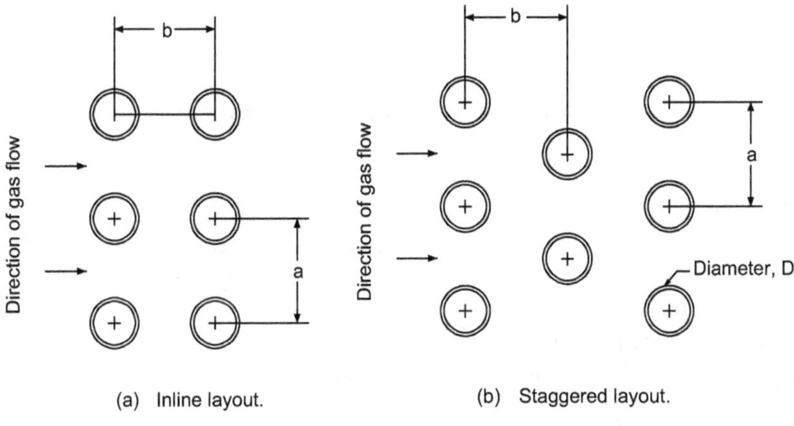

FIGURE 3.6 Bare tube layout: (a) inline and (b) staggered.

3.1.5 Continuous Fins on a Tube Array

This type of tube-fin geometry is most commonly used in (1) air-conditioning and refrigeration exchangers known as coils, in which high-pressure refrigerant is contained on the tubeside; (2) as radiators for internal combustion engines; and (3) for charge air coolers and intercoolers for cooling supercharged engine intake air of diesel locomotives, etc. The tube layout pattern is mostly staggered.

3.1.5.1 Tube: Primary Surface

All tubes feature seamless tubes supplied to ASTM/ASME specifications. Return bends can be supplied of heavier wall tubing to ensure adequately formed thickness in applications where erosion is a concern. Round or flat tubes (rectangular tubes with rounded corners) usually with a staggered tube arrangement are used. Elliptical tubes are also being used. Round tubes are used for high-pressure applications and also when considerable fouling is anticipated. The use of flat tubes is limited to low-pressure applications, such as vehicular radiators. Flat or elliptical tubes, instead of round tubes, are used for increased heat transfer in the tube and reduced pressure drop outside the tubes [2,9]. High-parasitic-form drag is associated with flow normal to the round tubes. In contrast, the flat tubes yield lower pressure loss for flow normal to the tubes due to lower form drag and avoid the low-performance wake region behind the tubes.

3.1.5.2 Fin: Secondary Surface

All coils or continuous fin exchangers feature die formed, flat, or patterned plate-fins of either aluminum or copper. The fins are attached to the tubes by mechanical expansion of the tube, ensuring a permanent fin-to-tube bond. Full fin collars allow for both precise fin spacing and maximum fin-to-tube contact. Fin pattern is optimized to produce highly energy-efficient operation, resistance to airflow, and cleanability.

Flat fin: Flat fin has no corrugation, which provides the lowest possible air pressure drop and lowest fan horsepower.

Star fin: Star fin pattern corrugation around the tubes provides higher heat transfer than the flat fin with a slight increase in air pressure drop.

Wavy fin: Wave fin corrugation across the fin provides the maximum heat transfer rate for a given surface area.

Louvered fin: Heat exchangers with flat tubes and louver fins are widely used due to higher heat transfer and reduced size.

3.1.5.3 Headers

Coils feature either steel or nonferrous manifold headers with threaded pipe connections. Cooling coils are supplied with vents and drains. Header tube holes are drilled. Headers provide proper joint clearances and allow optimum braze metal application. Coils have steel or copper tubesheets and removable steel headers with threaded connections. Header thickness, gasket surface area, and bolt area are designed and tested to provide a leak tight gasket seal.

3.1.5.4 Tube-to-Header Joints

Tube-to-header joints are either roller-expanded, brazed, or soldered.

3.1.5.5 Casings or Tube Frame

The coil incorporates a heavy-duty rectangular structural tube frame, which improves rigidity, squareness, and long-term stability, and is double flanged for coil stacking. Coil casings are either galvanized or stainless steel. Casing supports are provided on various centers for extra support during handling. Coil casings are furnished without mounting holes unless required.

3.1.5.6 Circuiting

Each coil is individually circuited for specific applications in recirculated, flooded, direct expansion, or control pressure receiver refrigeration systems along with water, glycols, or brines.

3.1.5.7 Exchangers for Air Conditioning and Refrigeration

Evaporators and condensers for air conditioning and refrigeration are usually the tube-fin type when air is one of the fluids. These exchangers are referred to as coils (Figure 3.7) when air is one of the fluids. Air-cooled finned coils are made up of a series of finned-tubes carrying the refrigerant. The fans are typically installed on the outside near the coil, on the intake or outlet, or may be attached to the coil itself, as in the case of air-cooled evaporators. The tubes are bent to form a "U" and then placed into the finned coil. Mechanical deformation of the tube against the fins (expansion) creates a joint and thus thermal contact between the tubes and fins. Formation of a mechanical bond requires very little energy compared to the energy required to solder, braze, or weld the fin to the tube. Microchannel heat exchangers of HVAC systems are discussed in detail in Section 3.11 "Microchannel Heat Exchangers."

3.1.5.8 Manufacturing of Tube-Fin Aluminum Heat Exchangers

Two distinct assembly techniques are used for the manufacturing of tube-fin aluminum heat exchangers:

i. Mechanical assembly
ii. Brazing

The primary advantage of mechanically assembled heat exchangers compared to brazed heat exchangers is the lower investment cost. Although the mechanical bond between tube and fin provides good contact, there is still a significant amount of contact resistance that reduces the effective heat conductivity. However, brazed heat exchangers have a better thermal performance

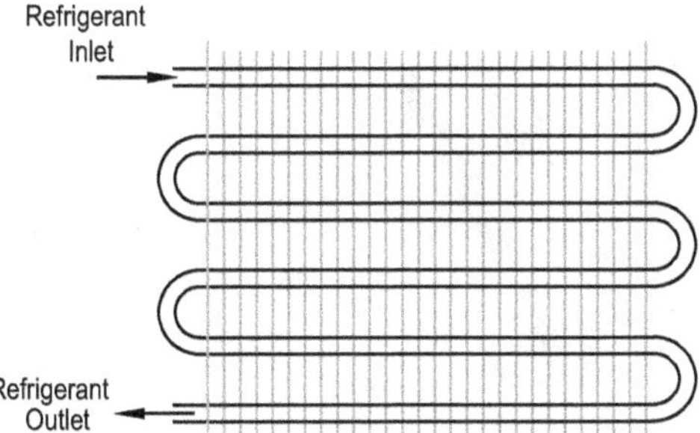

FIGURE 3.7 Heating/cooling coils—schematic.

rating in comparison to mechanically assembled ones. Consequently, brazing is today the dominant assembly method for aluminum heat exchangers. The contact resistance will generally increase during service, further deteriorating heat transfer performance over time.

3.1.5.9 Radiators

Radiators are compact liquid-to-air heat exchangers widely used in automotive vehicles. Major construction types used for radiators include (1) soldered design with flat brass tubes with flat plain or louvered copper fins—tubes and header sheets are made from brass and the water tank from brass or plastics—and (2) brazed aluminum design with flat tube and corrugated fins, usually louvered fins or flat tube continuous fins. These construction types are shown in Figure 3.8. Louvered cores are preferred as they allow a lower fin density than a nonlouvered core of the same area for the same thermal performance. The soldered brass tubes and copper fins were primarily used before the 1980s. Because of their light weight and significantly better durability and operating life, brazed aluminum tubes and aluminum corrugated multilouver fin radiators have almost replaced the copperbrass radiators [2]. Typical fin density ranges from 400 to 1000 fins/m (10–25 fins/in.), and the surface area density (β) ranges from 900 to 1650 m^2/m^3 (275–500 ft^2/ft^3).

3.1.5.10 Effect of Fin Density on Fouling

A major question in the choice of detailed dimensions for a core is the appropriate fin spacing. As the number of fins per inch (or per meter) is increased, the unit can be made progressively more compact. At the same time, it becomes more sensitive to clogging by dirt, soot, airborne dirt, insects, and the like, and more sensitive to nonuniformity in the fin spacing. In practice, it has been found that for most automotive applications, radiators can have 10–12 fins/in. Fourteen fins per inch can be fabricated with little increase in cost per unit surface area, and this closer fin spacing may be justified for installations where space is a constraint and fouling of the surfaces is not a problem [9]. The CuproBraze heat exchanger is discussed in Section 3.13 "CuproBraze Heat Exchanger."

Tube-fin details for air-conditioning and refrigeration applications are furnished by Shah [2]: Evaporator coils for small-capacity systems (less than 20 tons) use 7.9, 9.5, and 12.7 mm (5/16, 3/8, and 1/2 in.) tubes and fin densities in the range of 315–551 fins/m (8–14 fins/in.). For refrigeration applications, lower fin densities (157–315 fins/m or 4–8 fins/in.) are used due to the possibility of frosting and blockage. For condenser coils, tube diameters used are 7.9, 9.5, and 12.7 mm (5/16, 3/8, and 1/2 in.), and fin densities range from 394 to 787 fins/m (10–20 fins/in.) and fin thickness from 0.09 to 0.25 mm (0.0035–0.10 in.).

3.1.5.11 One-Row Radiator

In conventional multirow radiators, the fin array is not in contact with the tubes over the full fin depth. The region of no contact occurs between the tube rows and is illustrated in Figure 3.9. This

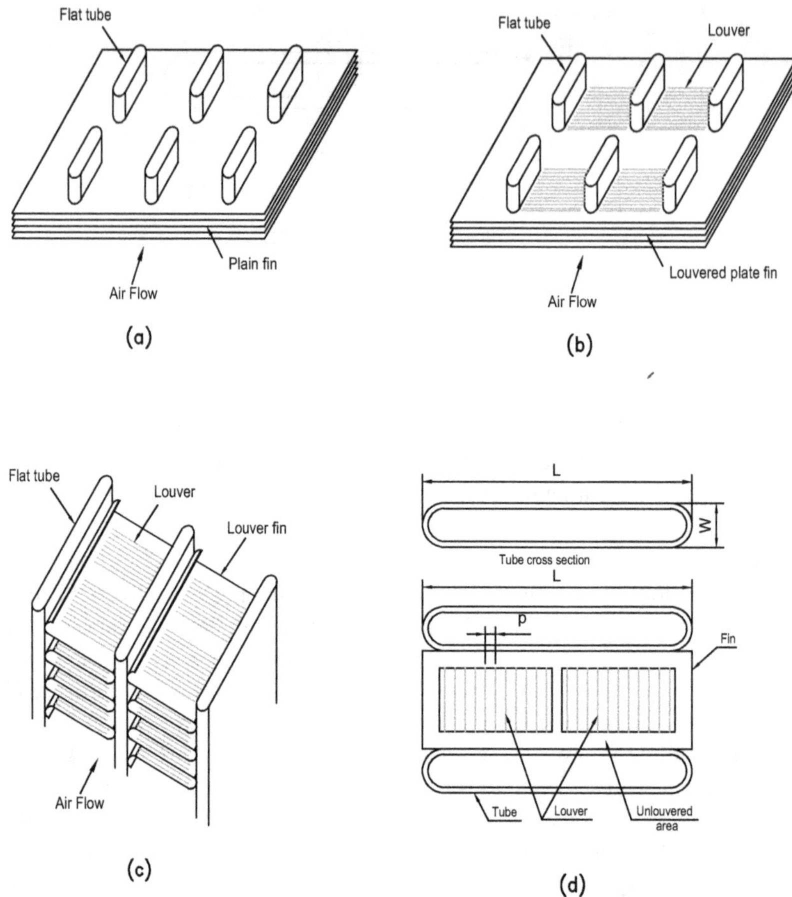

FIGURE 3.8 Tube-fin details of radiator: (a) continuous plain fin-flat tube radiator, (b) continuous louvered fin-flat tube design, (c) single flat tube row louvered fin radiator design, and (d) tube-fin details.

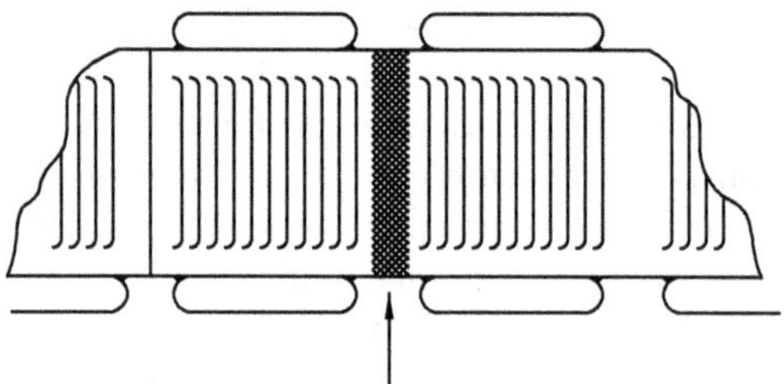

FIGURE 3.9 Region of no contact between flat tube rows of a radiator.

hinders heat conduction from the tubes to the fin region between the tube rows. If the fins are louvered in this region, the conduction path is further impaired. Hence the thermal performance per unit fin depth is low compared to the fin region in contact with the tubes. Remedies for this problem are as follows [10]:

1. Use a one-row core (Figure 3.10) in place of the multirow design. This design eliminates the low-performance fin area between tube rows, and offers either material savings and lower air

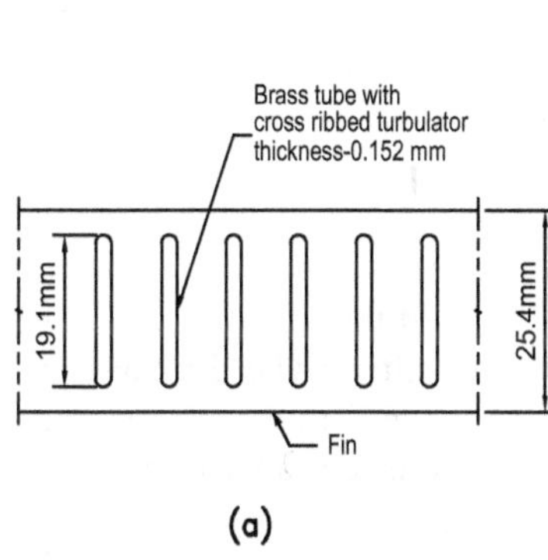

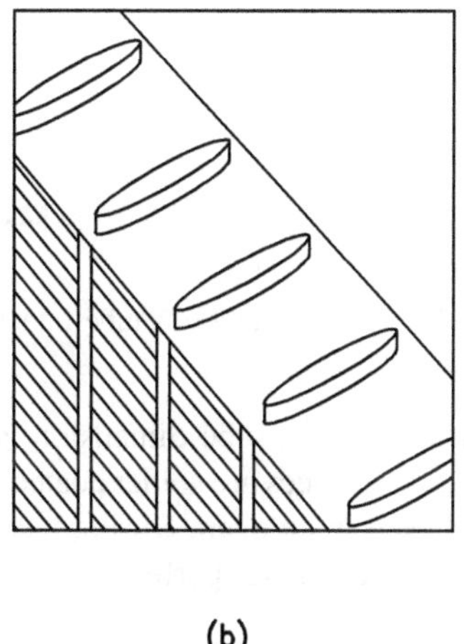

FIGURE 3.10 Single tube row radiator design.

pressure drop, or increased thermal performance. In the United States, manufacturers usually make 19.1 mm (3/4 in.) tubes in 0.152 mm (0.006 in.) thick brass. To prevent the tube wall from becoming concave during core assembly, a heavier wall thickness is usually adopted. Another method followed to prevent the tube wall from becoming concave during assembly is to use cross-ribbed turbulator tubes. The indented cross ribs increase the cross-sectional moment of inertia of the tube wall, increasing its stiffness. The indentations in the walls of turbulator tubes may be beneficial for low-flow-rate applications, where the flow may be laminar. For turbulent flow, the increase in coolant-side heat transfer coefficient will be marginal, because the dominant thermal resistance is typically on the airside, and the ribs or indentations produce large increase in pressure drop.

2. An alternative method of eliminating the ineffective fin area is to eliminate the space between the tube rows. In other words, have the tubes touching in the axial direction. This is referred to as the "tubes-touching" design. To achieve tubes-touching design, the ends of the tubes must be shaped to provide the requirement of minimum ligament between the tubes in the header plate. For the same fin depth in the airflow direction, the tubes-touching design should provide the same thermal performance as a one-row design.

3.1.5.12 Manufacture of Continuous Finned Tube Heat Exchangers

The manufacturing procedure for a continuous finned tube heat exchanger is discussed next. Detailed manufacturing practices are further described in the Chapter 4, "Heat Exchanger Fabrication: Brazing and Soldering," in Volume 2.

Mechanically bonded core: In a rounded hole design, the flat fins are formed with holes larger than the tubes to be inserted. Tubes are assembled in a fixture and fins are inserted. The fin gap (i.e., the fin density) is controlled by the ferrule height (Figure 3.11). After assembly, the tubes are expanded so that metal-to-metal contact is achieved. The tube ends are generally expanded into the header plates. Sometimes, on thinner tube plates, the tubes are brazed. The tubes, fins, and headers together form the core, which is then attached to the side sheet assembly and the header for the tubeside fluid or water tank, as the case may be, and tightly sealed with gaskets.

Soldered core: In a soldered design, which is mostly followed for flat tube-fin exchangers, the radiator cores are built from tinned brass tubes and copper fin sheets. The fins are punched with holes

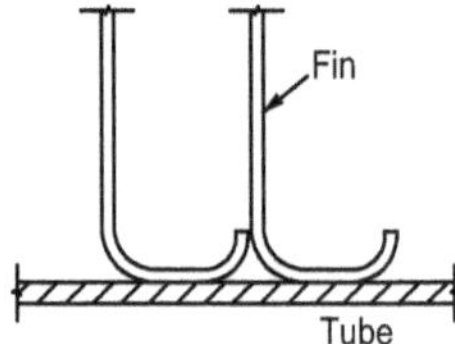

FIGURE 3.11 Ferrule height of a tube-fin heat exchanger.

slightly larger than the tube dimension, and tubes are inserted through the holes. The tube outer surfaces and fin surfaces are solder coated to achieve bonding. After the header plates are assembled, the core is baked in an oven. During baking the solder metal on the tube outer surface and on the fin surface melts and gives perfect metal-to-metal contact. Alternately, a fluxed core assembly is dip soldered.

3.1.6 Surface Selection

Various surfaces are used in compact heat exchanger applications. The proper selection of surface is one of the most important considerations in compact heat exchanger design. The selection criteria for these surfaces are dependent upon (1) the qualitative and (2) the quantitative considerations [11].

3.1.6.1 Qualitative Considerations

The qualitative considerations include the following:

1. Heat transfer requirements
2. Operating temperature and pressure
3. Flow resistance
4. Size or compactness and weight
5. Mechanical integrity
6. Fouling characteristics
7. Availability of surfaces, manufacturing considerations
8. Low capital cost, and low operating cost, especially low pumping cost
9. Designer's experience and judgment
10. Maintenance requirements, reliability, and safety

Heat transfer requirements with low flow resistance to minimize the pumping cost and a compact, low-weight unit are the most important criteria for selecting the surface for a compact heat exchanger. Surface selection influenced by fouling, manufacturing considerations, and cost is discussed next with additional details.

Fouling of compact exchangers: Fouling is one of the major problems in compact heat exchangers, particularly with various fin geometries and fine flow passages that cannot be cleaned mechanically. Other than low-finned tubes, if fouling is a problem, with the understanding of the problem and applying innovative means to prevent/minimize fouling, compact heat exchangers may be used in at least low to moderate fouling applications. Fouling control and methods to reduce fouling are discussed in Chapter 2, Volume 2.

Manufacturing considerations: Most of the heat exchanger manufacturing industries make only a limited number of surfaces due to limited availability of tools and market potential. Select a surface that a company can manufacture, even though theory and analysis can be used to arrive at some high-performance surfaces that may pose problems to manufacture.

Cost: Cost is one of the most important factors in the selection of surfaces, such that the overall heat exchanger is less expensive either in initial cost or in both initial and operating costs. If a plain fin surface can do the job for an application, a plain surface is preferable to expensive complicated surface geometries.

3.1.6.2 Quantitative Considerations

The quantitative considerations include performance comparison of surfaces with some simple "yardsticks." Surface selection by a quantitative method is made by comparing the performances of various heat exchanger surfaces and choosing the best under some specified criteria for a given heat exchanger application. Shah [11] categorizes the surface selection method as follows:

1. Comparisons based on j and f factors
2. Comparison of heat transfer as a function of fluid pumping power
3. Miscellaneous direct comparison methods
4. Performance comparisons with a reference surface

These methods are discussed in Chapter 8.

3.2 GEOMETRICAL RELATIONS: TUBULAR HEAT EXCHANGERS

Geometrical relations are derived for the inline and staggered tube arrangements with bare tubes, individually finned circular tubes, and circular tubes with plain continuous fins. The header dimensions for the tube bank are $L_2 \times L_3$, the core length for flow normal to the tube bank is L_2, and the no-flow dimension is L_3. The lateral pitch is given by P_t, and longitudinal pitch by P_l. Flow is idealized as being normal to the tube bank on the outside. Surface geometrical relations for tube-fin heat exchangers are derived from first principles by Shah [12,13], and for individually finned tubes by Idem et al. [14]. The surface geometrical relations given here are based on Shah [13].

3.2.1 Tube Inside Surface

Tubes have inside diameter d_i, length between headers L_1, header thickness T_h, and total length including header plates thickness as $L_1 + 2T_h$. The total number of tubes, N_t, is given by

$$N_t = \frac{L_2 L_3}{P_t P_l} \tag{3.1}$$

The geometrical properties of interest for analysis are:

$$\text{Total heat transfer area } A = \pi d_i L_1 N_t \tag{3.2}$$

$$\text{Minimum free flow area } A_o = \left(\frac{\pi}{4}\right) d_i^2 N_t \tag{3.3}$$

$$\text{Hydraulic diameter } D_h = d_i$$

$$\text{Tube length for heat transfer} = L_1$$

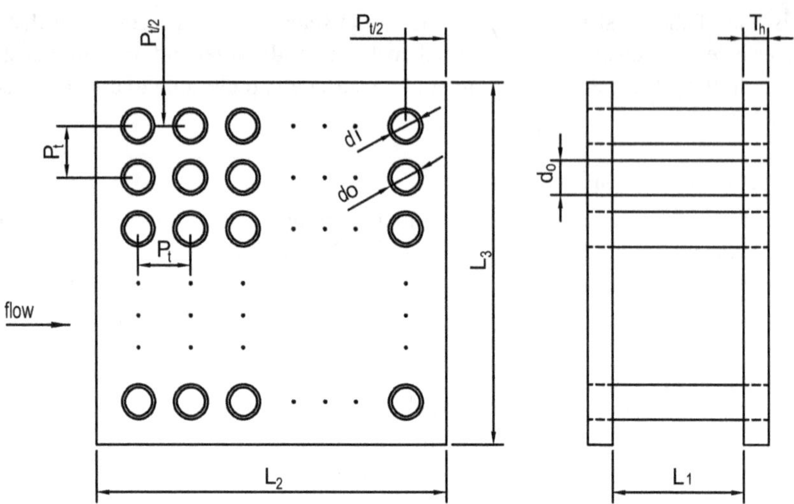

FIGURE 3.12 Inline tube bank geometry (bare tube).

$$\text{Tube length for pressure drop} = L_1 + 2T_h$$

The geometrical properties given by Equations 3.2 and 3.3 also hold good for other types of arrangements discussed next.

3.2.2 Tube Outside Surface

3.2.2.1 Bare Tube Bank

Inline arrangement: The geometrical properties of the inline tube bank as shown in Figure 3.12 are summarized here. The number of tubes in one row (in the flow direction) is

$$N_1 = \frac{L_3}{P_t} \tag{3.4}$$

The total heat transfer area consists of the area associated with the tube outside surface given by

$$A = \pi d_o L_1 N_t \tag{3.5}$$

The minimum free flow area $A_{o,}$ frontal area A_{fr}, the ratio of free flow area to frontal area σ, and hydraulic diameter D_h are given by

$$A_o = \left(P_t - d_o\right) N_t' L_1 \tag{3.6}$$

$$A_{Fr} = L_1 L_3 \tag{3.7}$$

$$\sigma = \frac{\left(P_t - d_o\right)}{P_t} \tag{3.8}$$

$$D_h = \frac{4A_o L_2}{A} \tag{3.9}$$

$$\begin{aligned} &\textit{Flow length for pressure drop calculation} = L_2 \\ &\textit{Heat exchanger volume } V = L_1 L_2 L_3 \end{aligned} \tag{3.10}$$

It must be emphasized that the foregoing definition of the hydraulic diameter is used by Kays and London [15] for tube banks. However, many correlations use the tube outside diameter or fin root diameter as the characteristic dimension in the heat transfer and pressure-drop correlations. Therefore, it is essential to find out first the specific definition of the characteristic length before using a particular correlation.

The geometrical properties given by Equations 3.6–3.10 also hold good for other types of arrangements discussed later.

Staggered arrangement: A staggered tube bank layout and its unit cell are shown in Figure 3.13. The total number of tubes with half tubes in the alternate row is given by Equation 3.1. If the half tubes are eliminated from the alternate intermediate tube rows, then the number of tubes in the first row will be L_3/X_t, and in the second row $L_3/X_t - 1$. The minimum free flow area occurs either at the front or through the diagonals. The minimum free flow area is then given by [12,13]

$$A_o = \left[\left(\frac{L_3}{P_t} - 1\right)z + \left(P_t - d_o\right)\right]L_1 \tag{3.11a}$$

where

$$z = 2x \quad \text{if } 2x < 2y \tag{3.11b}$$

$$z = 2y \quad \text{if } 2y < 2x \tag{3.11c}$$

in which $2x$ and y are defined as

$$2x = \left(P_t - d_o\right) \tag{3.12a}$$

$$y = \left[\left(\frac{P_t}{2}\right)^2 + \left(P_l\right)^2\right]^{0.5} - d_o \tag{3.12b}$$

In Equation 3.11a, the last term $(P_t - d_o)L_1$ corresponds to the free flow area between the last tube at each end in the first row and the exchanger wall.

3.2.2.2 Circular Fins on Circular Tubes

The basic core geometry of an idealized single-pass crossflow exchanger for an inline arrangement is shown in Figure 3.14. The total number of tubes in this exchanger is given by Equation 3.1. It is idealized that the root of a circular fin has an effective diameter d_r and the fin tip has a diameter d_f. Depending upon the manufacturing techniques, d_r may be the tube outside diameter or tube outside

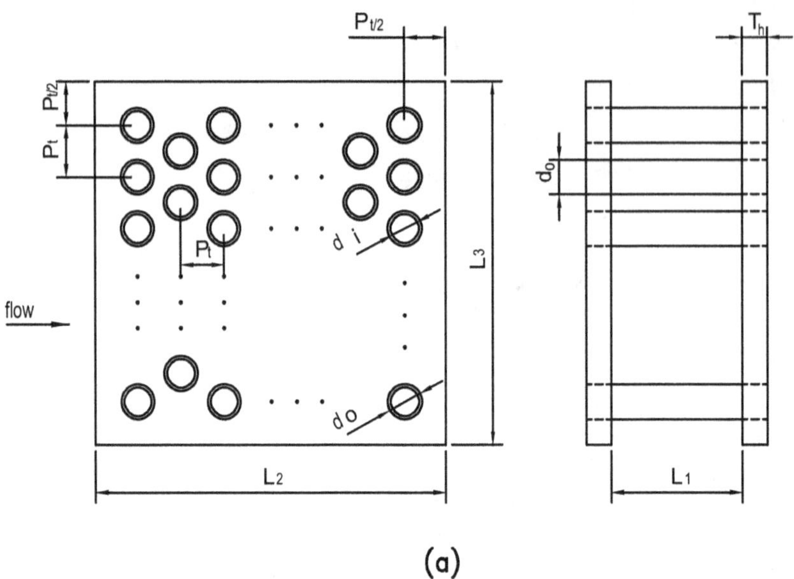

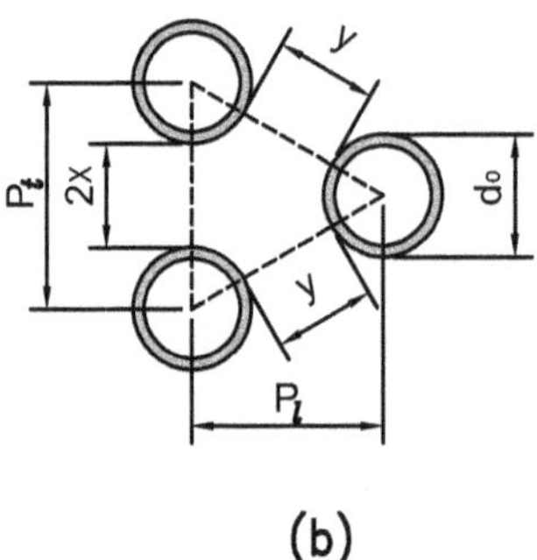

FIGURE 3.13 Staggered tube bank (bare tube): (a) layout and (b) unit cell.

diameter plus two fin collar thickness. The total heat transfer area consists of the area associated with the exposed tubes (primary area) and fins (secondary area). The primary area, A_p, is the tube surface area minus the area blocked by the fins. This is given by

$$A_p = \pi d_r \left(L_1 - t_f N_f L_1\right) N_t \tag{3.13}$$

where

t_f is the fin thickness
N_f is the number of fins per unit length
The fin surface area, A_f, is given by

$$A_f = \left[\frac{2\pi\left(d_f^2 - d_r^2\right)}{4} + \pi d_f t_f \right] N_f L_1 N_t \tag{3.14}$$

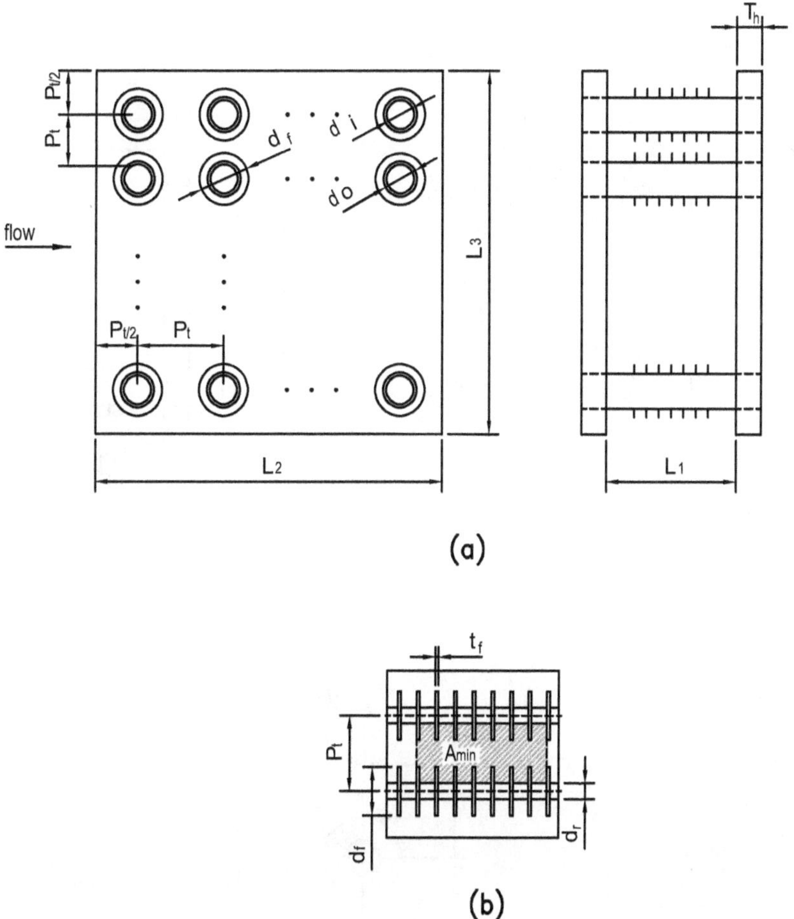

FIGURE 3.14 Individually finned circular fin tube. (a) Inline layout and (b) diagram showing frontal area (dotted lines) and free flow area (hatched area).

and therefore, the total heat transfer surface area A is equal to $A_p + A_f$, given by

$$A = \pi d_r\left(L_1 - t_f N_f L_1\right)N_t + \left[\frac{2\pi\left(d_f^2 - d_r^2\right)}{4} + \pi d_f t_f\right]N_f L_1 N_t \tag{3.15}$$

Minimum free flow area: Inline arrangement: The minimum free flow area for the inline arrangement is that area for a tube bank, Equation 3.5, minus the area blocked by the fins:

$$A_o = \left[\left(P_t - d_r\right)L_1 - \left(d_f - d_r\right)t_f N_f L_1\right]\left(\frac{L_3}{P_t}\right) \tag{3.16}$$

Minimum free flow: Staggered arrangement: The basic core geometry of an idealized single-pass crossflow exchanger for staggered arrangement is given in Figure 3.15. The total number of tubes in this exchanger is given by Equation 3.1. For the staggered tube arrangement, the minimum free flow

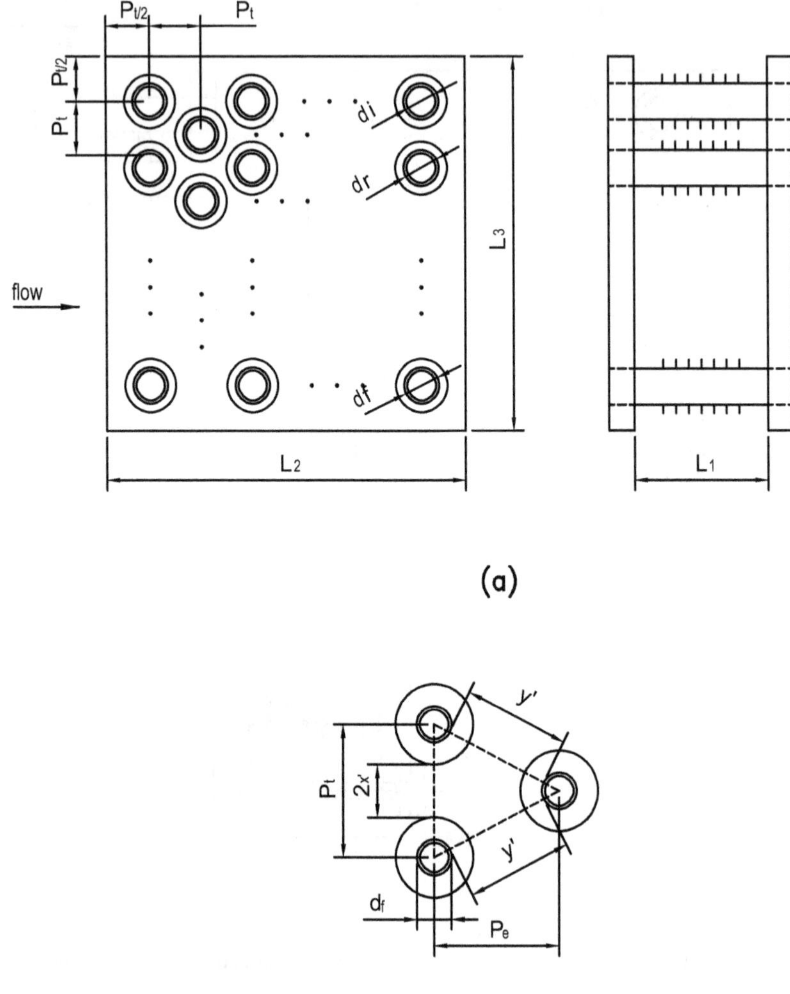

FIGURE 3.15 Individually finned tube staggered arrangement. (a) Geometry and (b) unit cell. (After Shah, R.K., Compact heat exchangers, in *Handbook of Heat Transfer Applications*, W.M. Rohsenow, J.P. Harnett, and E.N. Ganic, eds., McGraw-Hill, New York, 1985, pp. 4-176–4-312.)

area could occur either through the transverse pitch or through the diagonals similar to those of the unit cell shown in Figure 3.15b. The minimum free flow area is given by [12,16]

$$A_o = \left[\left(\frac{L_3}{P_t} - 1\right)z' + \left(P_t - d_r\right) - \left(d_f - d_r\right)t_f N_f\right]L_1 \tag{3.17}$$

where

$$z' = 2x' \quad \textit{if } 2x' < 2y' \tag{3.18a}$$

$$z' = 2y' \quad \text{if } 2y' < 2x' \tag{3.18b}$$

in which $2x'$ and y' are given by

$$2x' = (P_t - d_r) - (d_f - d_r)t_f N_f \tag{3.19a}$$

$$y' = \left[\left(\frac{P_t}{2}\right)^2 + (P_l)^2\right]^{0.5} - d_r - (d_f - d_r)t_f N_f \tag{3.19b}$$

Dimensions $2x'$ and y' cannot be depicted in the figure.

3.2.2.3 Continuous Plain Fins on Circular Tubes

The tube-fin arrangement is shown in Figure 3.16. The total heat transfer area consists of the area associated with the exposed tubes (primary area) and the fins (secondary area). The primary area is the same as that given by Equation 3.5, minus the area blocked by the fins. Therefore, A_p is given by [12,17]

$$A_p = \pi d_r (L_1 - t_f N_f L_1) N_t \tag{3.20}$$

The total fin surface area, A_f, consists of (1) the fin surface area, and (2) the area with leading and trailing edges, given by

$$A_f = 2\left[(L_2 L_3 - \pi d_r^2 N_t)/4\right] N_f L_1 + 2L_3 t_f N_f L_1 \tag{3.21}$$

and

$$\text{Total heat transfer surface area } A = A_p + A_f \tag{3.22}$$

Minimum free flow area for the inline arrangement: The minimum free flow area for an inline arrangement is that area for a tube, Equation 3.5, minus the area blocked by the fins [12,17]

$$A_o = \left[(P_t - d_r)L_1 - (d_f - d_r)t_f N_f L_1\right]\left(\frac{L_3}{P_t}\right) \tag{3.23}$$

Minimum free flow area for the staggered arrangement: For the staggered tube arrangement as shown in Figure 3.16b, the minimum free flow area could occur either through the transverse pitch or through the diagonals. The minimum free flow area is given by [12,13] as follows:

$$A_o = \left\{(L_3 / P_t - 1)z'' + \left[(P_t - d_r) - (P_t - d_o)t_f N_f\right]\right\} L_1 \tag{3.24}$$

In Equation 3.24, z'' is defined as

$$z'' = 2x'' \quad \text{if } 2x'' < 2y'' \tag{3.25a}$$

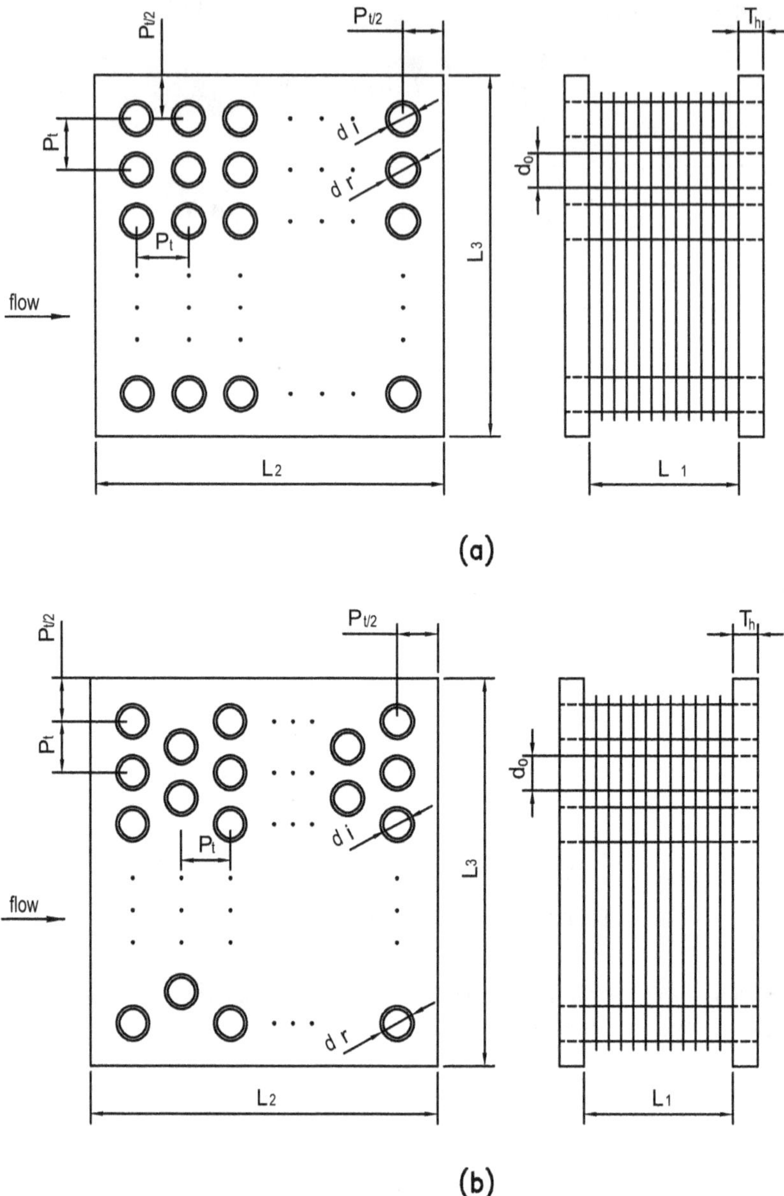

FIGURE 3.16 Continuous flat fin-tube heat exchanger: (a) inline layout and (b) staggered layout.

$$z'' = 2y'' \quad \text{if } 2y'' < 2x'' \tag{3.25b}$$

where $2x''$ and y'' are given by

$$2x'' = \left(P_t - d_r\right) - \left(P_l - d_r\right)t_f N_f \tag{3.26a}$$

$$y'' = \left[\left(\frac{P_t}{2}\right)^2 + \left(P_l\right)^2\right]^{0.5} - d_r - \left(P_t - d_r\right)t_f N_f \tag{3.26b}$$

Minimum free flow area for pressure drop: For the determination of entrance and exit pressure losses, the area contraction and expansion ratio σ' is needed at the leading and trailing fin edges. It is given by [12,17] as follows:

$$\sigma' = \frac{L_3 L_1 - L_3 t_f N_f L_1}{L_3 L_1} \tag{3.27}$$

3.3 FACTORS INFLUENCING TUBE-FIN HEAT EXCHANGER PERFORMANCE

Important factors that influence the heat transfer performance of tube-fin heat exchangers include the following:

1. Tube layout (staggered vs. inline arrangement)
2. Equilateral layout versus equivelocity layout
3. Number of rows
4. Tube pitch
5. Finned tube variables (fin height, spacing, thickness)
6. Fin tubes with surface modifications
7. Side leakage
8. Boundary-layer disturbances and characteristic flow length
9. Contact resistance in finned tube heat exchangers
10. Induced draft versus forced draft

Taborek [7] and Johnson and Rabas [18] discuss the effect of various tube-fin parameters on thermohydraulic performance.

3.3.1 Tube Layout

The two basic arrangements of the tubes in a tube bank are (1) staggered arrangement with equilateral triangular pitch and (2) inline arrangement. The extent of transverse flow in the inline layout is small in comparison to the staggered arrangement. The most commonly used is the staggered layout. There are very few applications of low-finned tubes with the inline arrangement. It may be used when airside fouling problems are encountered. For identical operating conditions, the inline arrangement has low heat transfer, of the order of 70%, and pressure drop is also of this order.

3.3.2 Equilateral Layout versus Layout

In the conventional equilateral staggered layout, the area available for airflow in the transverse direction is half of that available in the diagonal direction; therefore, energy is lost in accelerating and decelerating the air stream as it flows through the bank. In the equivelocity layout, these two areas are made equal by adjusting the transverse (P_t) and longitudinal (P_l) pitches, thus minimizing these losses. In general, for applications where horsepower is evaluated at a high premium and a reduced pressure drop is attractive, even at the expense of some loss of thermal performance, an expanded pitch layout type should be strongly considered.

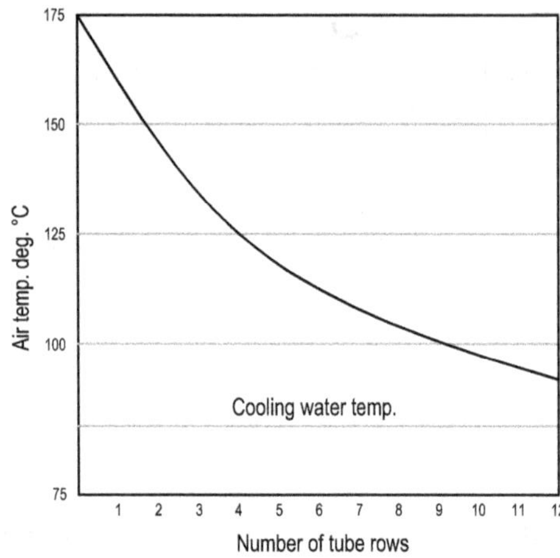

FIGURE 3.17 Temperature drop through a multi-tube row heat exchanger—schematic (for a typical case).

3.3.3 Number of Tube Rows

The addition of tube rows in the crossflow direction is the common way of increasing the heat transfer area when the frontal area of the tube bank is fixed. Such an addition of tube rows increases the flow length for the air and may affect the fluid dynamics with a possible effect on heat transfer and pressure drop. Pressure drop through the tube bundle increases with the number of tube rows. The temperature drop through the tube bundle also increases with the number of tube rows, though the effect is very large with the first few tube rows and for a very large number of tube rows the effect is small. Temperature drop through a multirow tube bank is shown schematically in Figure 3.17. The row effect on heat transfer coefficient is discussed by Johnson [18]. A basic difference exists between the row effects for the inline and the staggered arrangements. For an inline arrangement, the heat transfer coefficient decreases with an increase in the number of rows. Brauer [19] shows that the one-row tube bank heat transfer coefficient is approximately 50%–80% higher than the average for four rows, whereas for a staggered arrangement the four-row tube bank has approximately 30% higher heat transfer coefficient (average) than a one-row tube bank.

3.3.4 Tube Pitch

Tube pitch is commonly used in heat-exchanger design to affect the compactness of the core and flow conditions. The hydraulic diameter increases with the pitch. Based on comparable air velocities, the pitch has no effect on the heat transfer coefficient. However, increasing the pitch strongly affects the pressure drop, as it decreases the pressure drop through the bundle [18].

3.3.5 Tube-Fin Variables

The tube-fin variables that affect the thermal performance include fin height, fin spacing, and thickness and thermal conductivity of the fin material. The surface area of the tube bank can be increased by employing fin tubes with high (longer) fins and denser spacing. However, increasing the fin height will decrease the fin effectiveness, and may lower heat transfer compared with the short fins. Denser fins may lower the heat transfer because of laminarization of the flow conditions between the fins [18]. Fin efficiency increases with a reduction in fin height and an increase in fin thickness. Thermal conductivity is usually not a problem in selecting the material of a primary surface, but it is important in extended surfaces because of the relatively long and narrow heat flow path, which affects their heat transfer performance [6].

3.3.5.1 Fin Height and Fin Pitch

The fin height and fin pitch that can be used effectively are fundamentally related to the boundary-layer thickness [6]. Wen-Jei Yang [6] states the rule of thumb: (1) For large thermal boundary-layer thickness, the heat transfer coefficient, *h*, is low and thus high fins are employed; (2) low fins are employed in the case of a thin boundary layer where *h* is high; and (3) if the fin spacing is too small, the main stream cannot penetrate the spaces between the fins, resulting in no increase in the thermal boundary area and an excessive pressure drop.

3.3.6 Finned Tubes with Surface Modifications

The term "surface modification" refers to fin tubes that have extended surfaces that are mechanically deformed to various shapes by crimping, corrugating, cutting, slotting, or serrating the fins; for continuous plate-fins, surface interruptions constitute surface modification [18]. Any surface modification will increase the heat transfer and pressure drop as compared with the identical smooth fin. Common surface modifications for circular fins are discussed further in Chapter 8.

3.3.7 Side Leakage

The heat transfer and pressure-drop performances of a tube bank are affected by the leakage flow between the outermost tube and the side walls. This is equivalent to the bundle-to-shell bypass stream (C-stream) leakage in a shell and tube exchanger. The effect of bypass flow can be minimized by attaching side seals to the exchanger in every even row as shown in Figure 3.18. Clearance between the core and the housing can be minimized by welding metal strips to the housing or attaching metal strips to the header as shown in Figure 3.18b.

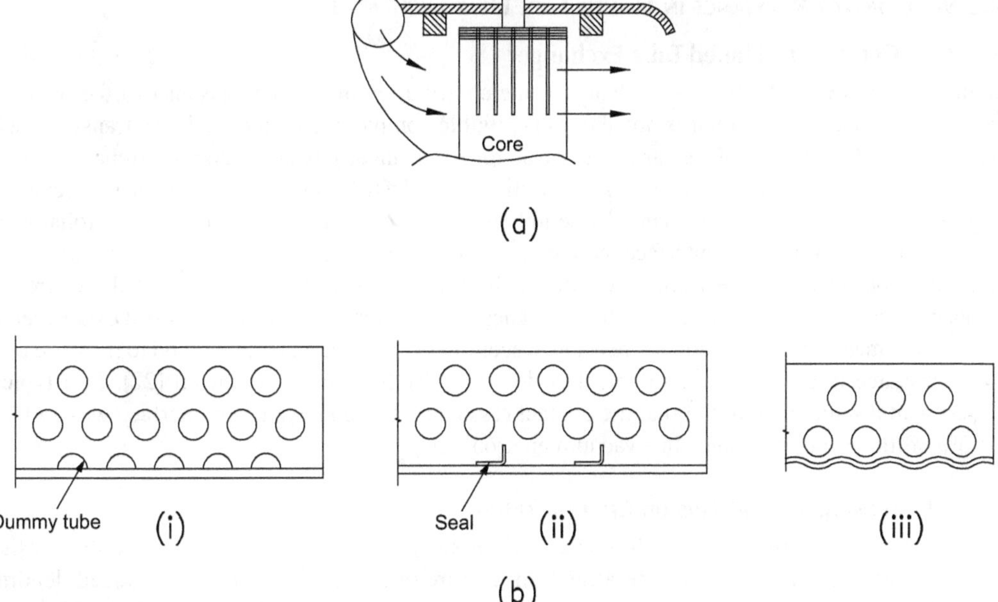

FIGURE 3.18 Sealing strips to arrest bypass flow: (a) between the core and the housing and (b) between the tubes and side sheets.

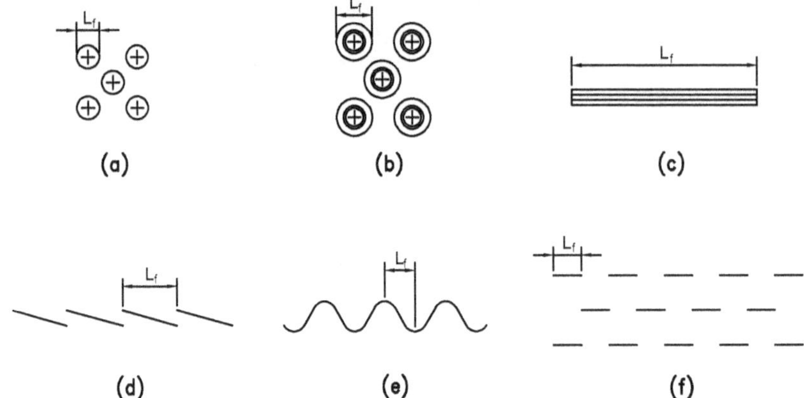

FIGURE 3.19 Characteristic flow length between major boundarylayerdisturbance for various heat exchanger surfaces. (Adapted from LaHaye, P.G. et al., *Trans. ASME, J. Heat Transfer*, 96, 511, 1973.)

3.3.8 Boundary-Layer Disturbances and Characteristic Flow Length

Periodic interruption of the boundary layer is achieved by breaking the total flow length into many characteristic lengths by means of OSFs, louvered fins, wavy fins, etc. [20]. The characteristic flow length, L_f, is illustrated in Figure 3.19. The characteristic flow length and the hydraulic diameter, D_h, are responsible for the numerical merits of a surface geometry. The representative length of finned heat exchangers as parameters to represent their performance suggests that it would be possible to enhance their performance by shortening the representative length of the fin and using finer segmentation than the regions of louvered or OSFs [21]. The application of the pin finned surface can be considered one of the best candidates.

3.3.9 Contact Resistance in Finned Tube Heat Exchangers

3.3.9.1 Continuous Finned Tube Exchanger

In the design of finned tube heat exchangers, the contribution of contact resistance at the interface between the tube and fin collar is not always negligible compared with the total heat transfer resistance (Figure 3.20) [21]. This is particularly true for mechanically bonded tube-fin exchangers used in low-pressure applications such as air-conditioning industries and air coolers for supercharged engine intake air. In the heat exchanger assembly process, the initial interference, the microhardness of the softer material at the interface, contact area, and contact pressure are parameters that determine the contact resistance [22]. Selection of the proper bullet size can optimize the degree of contact at the interface. Over the last three decades, many researchers have developed experimental methods to measure thermal contact resistance accurately. An experimental method to measure contact resistance in a vacuum environment is described by Nho and Yovanovich [22]. In a typical experimental setup, they found that the contact resistance at the tube and fin interface was 17.6%–31.5% of the overall resistance in a vacuum environment.

3.3.9.2 Tension-Wound Fins on Circular Tubes

For a round tube with aluminum fins wrapped on steel tubes under tension, the contact resistance increases quickly with the operating temperature or the tension would be reduced, leading to fin loosening, due to the higher thermal expansion of aluminum. The operating temperature is limited to temperatures of 125°C or less, whereas the "L" fin accommodates temperatures between 150–170°C

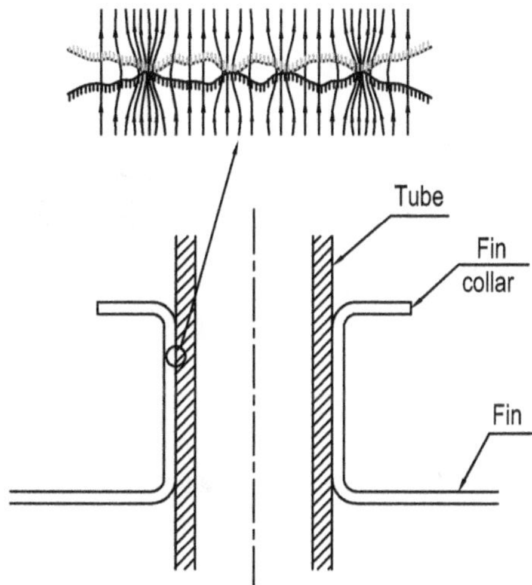

FIGURE 3.20 Metal-to-metal contact between the tube and the fin.

3.3.9.3 Wire Finned Tube

A wire loop tube consisting of a series of elongated wire loops enhancing the surface of the tube is spirally wound on to the tube wall and held in position with a binding wire at the base of the loops. The loops and binding wire are then soft-soldered to the tube wall to give a metallic bond between the wire loops and the tube. The wire loop secondary surface gives excellent heat transfer performance due to its ability to promote turbulence in the fluid passing over it. Temperatures of up to 250°C can be applied to this type of finned tube.

3.3.9.4 Integral Finned Tube

With an integral finned tube, there is no concern at the fin bond about heat transfer resistance and the integrity of tubing, regardless of temperature and pressure levels. However, only certain fin geometries can be obtained with an extrusion process.

3.3.10 Induced Draft versus Forced Draft

The heat transfer and pressure drop depend on the method used to generate the flow on the tube bank, namely, induced draft or forced draft. When a fluid enters the tube bank, the level of turbulence can change with penetration into the bundle. The tube bundle heat transfer coefficient and pressure-drop performance vary for each additional row for a shallow exchanger, and become independent of the number of rows only for the deep bundles with a very large number of tube rows [7]. Common practice is to apply a row correction multiplier to the deep bundle correlations (ideal bundle) to account for these variations.

3.3.11 Round Tube versus Flat Tube

Conventional soldered copper–brass radiators uses a flat tube. Though round tubes provide a strong joint at the header plates, because of their low surface-to-volume ratios, round tubes are inefficient at transferring heat from the tubeside fluid. The surface-to-volume ratio can be dramatically increased using flat tubes. The air flow pattern around a round tube and flat tube is shown in Figure 3.21.

Young Touchstone, USA/UK, offers its customers the best of both types with its patented FLAT-ROUND technology. It uses brass tubes that are flat in the core and round at the headers.

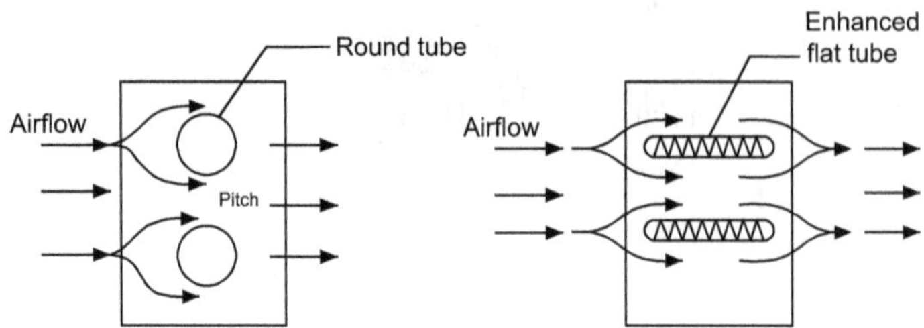

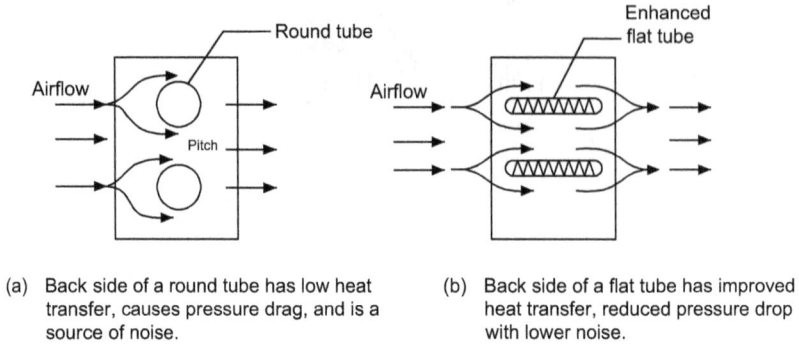

(a) Back side of a round tube has low heat transfer, causes pressure drag, and is a source of noise.

(b) Back side of a flat tube has improved heat transfer, reduced pressure drop with lower noise.

FIGURE 3.21 Air flow pattern around a round tube and flat tube. (Adapted from NOCOLOK® Flux Brazed Aluminum Heat Exchangers for the Refrigeration and Air Conditioning Industry, A Solvay Special Chemicals Aluminum Heat Exchangers, pp. 1–7.)

This FLAT-ROUND design combines the superior airflow and heat transfer of flat tubes with reliable tube-to-header mechanical bonding for exceptional durability. Combining the proven FLAT-ROUND technology with CuproBraze technology results in the toughest, most efficient cooling systems in the industry.

3.3.12 Surface Coating

Surface modification is an effective way to enhance the efficiency of heat exchangers such as higher heat transfer, corrosion resistance, etc. Surface modification by forming a microporous coated layer can greatly enhance the boiling heat transfer and thus achieve a high performance. For example, in aluminum finned tube air conditioner heat exchangers, during the refrigeration process, water molecules condense into droplets adhering to the surface of the fins, increasing the wind resistance of the heat exchanger, resulting in a decrease of the heat transfer efficiency, and hindering the continuous condensation of water molecules. To solve the above problems, manufacturers resort to coating desiccants on the fins and changing the physical or chemical properties of the heat exchanger fin surface to form wettability surfaces, including hydrophobic and hydrophilic surfaces.

3.3.12.1 Hydrophobic, Hydrophilic, and Nano Coating

Hydrophobicity is a property of a substance that repels water. This means it lacks affinity for water, and tends to repel or not to absorb water. Hydrophilic refers to having a strong affinity for water. Materials with a special affinity for water and those it spreads across, maximizing contact, are known as hydrophilic; those that naturally repel water, causing droplets to form, are known as hydrophobic. Both classes of materials can have a significant impact on the performance of heat transfer surfaces.

Hydrophobic coating. The epoxy-based hydrophobic coating effectively inhibits dust and bacterial accumulation.

Hydrophilic coating. The hydrophilic coating improves airflow by reducing the thickness of condensing water layers and avoids carry-over of condensed water at high air velocities.

Nano coatings improve heat transfer capabilities, resist corrosion of salt water and dust, and protect against organic solvents and chemicals, reduce dirt accumulation, etc.

3.3.13 Finned Tube-Fin Aluminum Heat Exchangers of Automobiles

3.3.13.1 Applications of Aluminum Tube-Fin Heat Exchangers

In automobiles, aluminum heat exchangers are used in the following main application categories [23]:

i. Engine water cooling (radiator)
ii. Oil cooling (oil of the engine main lubrication circuit, the manual and automatic transmission, the power steering, etc.)
iii. Condensers and evaporators for air-conditioning systems
iv. Heaters
v. Charge air cooling (water cooled, air cooled unit is BAHX).

Some of the above heat exchangers are shown in Figure 3.22.

3.3.13.2 Advantages of Aluminum in the Design of Heat Exchangers

Aluminum offers a number of advantageous material characteristics for heat exchangers [23]:

i. Significant potential for lightweight design
ii. Highly automated, reliable manufacturing process (brazing)
iii. High thermal conductivity, also when joined by brazing
iv. Excellent corrosion resistance
v. Good formability
vi. Adequate strength to resist temperature and pressure cycles
vii. Easy recyclability, i.e. an environmentally friendly solution
viii. Commercial availability of a wide range of aluminum alloys and product forms to meet different design options.

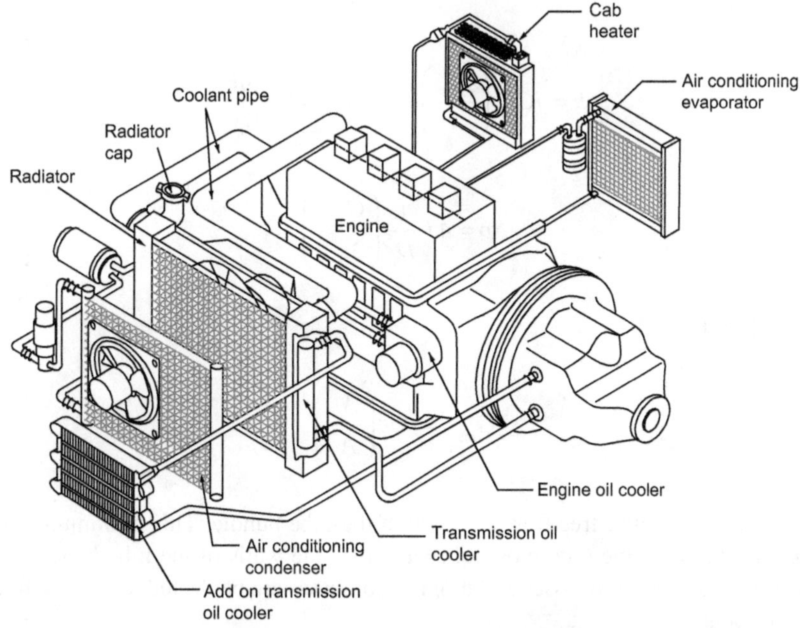

FIGURE 3.22 Aluminum finned tube heat exchangers of automobiles. (Adapted and modified from NOCOLOK® flux brazing process Solvay Fluor GmbH, Hannover, Germany, pp. 1–7.)

3.4 THERMOHYDRAULIC FUNDAMENTALS OF FINNED TUBE HEAT EXCHANGERS

It is common practice to use the Colburn j factor to characterize the heat transfer performance and the Fanning friction factor f to characterize the pressure drop of the tube banks as functions of Reynolds number. This section provides empirical correlations for j and f factors. With these correlations, the performance of geometrically similar heat exchangers can be predicted, within the parameter range of the correlations.

3.4.1 Heat Transfer and Friction Factor Correlations for Crossflow over Finned Tube Banks

Over the last 50 years, many investigations have been carried out to determine the heat transfer and pressure drop over the tube banks. A standard reference for the heat transfer and fraction data of PFHE surfaces is the book by Kays and London [15]. This book presents j and f versus Reynolds number plots for tube banks, tube-fin heat exchangers, and 52 different plate-fin surface geometries. Some of these data were published in *Transactions of the ASME*. Shah and Bhatti [24] summarize important theoretical solutions and correlations for simple geometries that are common in compact heat exchangers. Most of the experimental data have been obtained with air as the test fluid. The j factor and fanning f factor are defined rather consistently in the literature by the following equations:

$$j = \frac{h(\mathrm{Pr})^{2/3}}{Gc_\mathrm{p}} \tag{3.28a}$$

$$= \frac{h}{Gc_\mathrm{p}}\left(\frac{\mu c_\mathrm{p}}{k}\right)^{2/3} \tag{3.28b}$$

$$h = jGc_\mathrm{p}\left(\frac{\mu c_\mathrm{p}}{k}\right)^{-2/3} \quad \text{or} \quad h = \frac{\mathrm{Nu}k}{D_h} \tag{3.28c}$$

$$\Delta p = 4f\frac{L}{D_\mathrm{h}}\left(\frac{G_\mathrm{max}^2}{2g_\mathrm{c}\rho}\right) \tag{3.29}$$

where G_max is given by

$$G_\mathrm{max} = \rho U_\mathrm{f}\left(\frac{A_\mathrm{fr}}{A_\mathrm{o}}\right) \tag{3.30}$$

and is based on the minimum free flow area within the tube bundle. This minimum free flow area includes the gaps between the outermost tubes and the side walls of the tube bundle. In Equation 3.29, contraction and expansion losses and momentum losses are neglected; only flow friction effect is taken into account.

The Reynolds number is defined as Re

$$\mathrm{Re} = \frac{Gd_\mathrm{r}}{\mu} \quad \text{or} \quad \mathrm{Re} = \frac{Gd_\mathrm{o}}{\mu} \quad \text{or} \quad \mathrm{Re} = \frac{GD_\mathrm{h}}{\mu} \tag{3.31}$$

where

d_r is the fin root diameter or the tube outside diameter, which depends on the tube manufacturing techniques

D_h is the hydraulic diameter

The j factor is related to Stanton number and Prandtl number by

$$j = \mathrm{St}(\mathrm{Pr})^{2/3} \tag{3.32}$$

where the Stanton number is given by

$$\mathrm{St} = \frac{\mathrm{Nu}}{\mathrm{Re}(\mathrm{Pr})} \tag{3.33}$$

From Equations 3.32 and 3.33, the expression for j is given by

$$j = \frac{\mathrm{Nu}}{\mathrm{Re}(\mathrm{Pr})^{1/3}} \tag{3.34}$$

3.4.2 The *j* and *f* Factors

References for the j and f factors include Thome [25], Y.G. Park and A. M. Jacobi [26], Giedt [27], and others covered later.

3.4.2.1 Bare Tube Bank

Experimental heat transfer and fluid friction data for six staggered circular tube patterns and one inline arrangement are given in Ref. [28]. The results of Ref. [28] have been presented in Kays and London [15].

The Giedt [27] correlation states that the thermal performance of a bare tube bank is given by

$$j = \frac{0.376}{\mathrm{Re}^{0.4}} \tag{3.35}$$

a. Inline Tube Bank

For a 10-row-deep inline tube bank, the correlation for j is given by [29]

$$j = \frac{0.33}{\mathrm{Re}^{0.4}} \tag{3.36}$$

This correlation is valid for tube diameter from 1/4 to 2 in., and Reynolds number range 100–80,000.

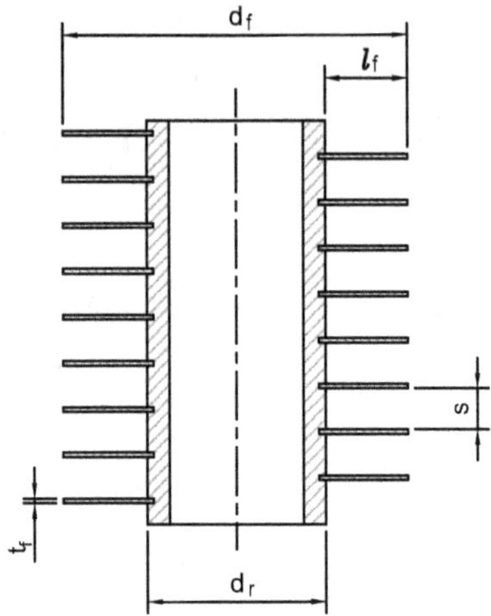

FIGURE 3.23 Tube-fin details.

3.4.2.2 Circular Tube-Fin Arrangement

The basic geometry is shown in Figure 3.23.

Experimental heat transfer and fluid friction data for three circular finned tube surfaces are given in London et al. [30]. The results of Ref. [30] have been presented in Kays and London [15].

1. Briggs and Young Correlations [31]
   For low-fin tube banks, the correlation based on experimental heat transfer data is

$$\mathrm{Nu} = 0.1507\,\mathrm{Re}_\mathrm{d}^{0.667}\,\mathrm{Pr}^{1/3}\left(\frac{s}{l_\mathrm{f}}\right)^{0.164}\left(\frac{s}{t_\mathrm{f}}\right)^{0.075} \tag{3.37}$$

   where $[s = (l - N_f t_f)/N_f]$ with a standard deviation of 3.1%. This is applicable for a circular finned tube with low-fin height and high-fin density, Re_d (based on tube outer diameter) = 1000–20,000, $N_r \geq 6$, and equilateral triangular pitch.

   For high-fin tube banks, the correlation based on experimental heat transfer data is

$$\mathrm{Nu} = 0.1378\mathrm{Re}_\mathrm{d}^{0.718}\mathrm{Pr}^{1/3}\left(\frac{s}{l_\mathrm{f}}\right)^{0.296} \tag{3.38}$$

   with a standard deviation of 5.1%.

   For all tube banks, the correlation based on regression analysis is

$$\mathrm{Nu} = 0.134\mathrm{Re}_\mathrm{d}^{0.681}\mathrm{Pr}^{1/3}\left(\frac{s}{l_\mathrm{f}}\right)^{0.200}\left(\frac{s}{t_\mathrm{f}}\right)^{0.1134} \tag{3.39}$$

$$j = 0.134\mathrm{Re}_\mathrm{d}^{-0.319}\left(\frac{s}{l_\mathrm{f}}\right)^{0.200}\left(\frac{s}{t_\mathrm{f}}\right)^{0.1134} \tag{3.40}$$

   with a standard deviation of 5.1%.

2. Robinson and Briggs Correlations [32]
For equilateral triangular pitch with high-finned tubes,

$$f = 9.465\mathrm{Re}_{\mathrm{d}}^{-0.316}\left(\frac{P_t}{d_\mathrm{o}}\right)^{-0.927} \tag{3.41}$$

with a standard deviation of 7.8%. This is applicable for the following parameter definitions:
Re_d = 2000–50,000
d_f = 1.562–2.750
P_t = 1.687–3.50

For an isosceles triangular layout,

$$f = 9.465\mathrm{Re}_{\mathrm{d}}^{-0.316}\left(\frac{P_\mathrm{t}}{d_\mathrm{r}}\right)^{-0.927}\left(\frac{P_\mathrm{t}}{P_1}\right)^{0.515} \tag{3.42}$$

This is applicable for d_o (mm) = 18.6–40.9, $l_\mathrm{f}/d_\mathrm{o}$ = 0.35–0.56, l_f/s = 3.5–5.3, σ= 13.3 -17.4, $P_\mathrm{t}/d_\mathrm{o}$ = 1.8–3.6, P_1/d_o = 1.8–3.6, $N_\mathrm{r} \geq 6$.

3. Rabas, Eckels, and Sabatino Correlation [33]

For tube banks arranged on an equilateral triangular pitch with low-finned tubes,

$$\begin{aligned} j &= 0.292Re_{\mathrm{d}}^{-m}\left(\frac{s}{d_\mathrm{f}}\right)^{1.116}\left(\frac{s}{l_\mathrm{f}}\right)^{0.257}\left(\frac{t_\mathrm{f}}{s}\right)^{0.666} \\ &\quad \times\left(\frac{d_\mathrm{f}}{d_\mathrm{r}}\right)^{0.473}\left(\frac{d_\mathrm{f}}{t_\mathrm{f}}\right)^{0.7717}\alpha_\mathrm{h}\alpha_\mathrm{n} \end{aligned} \tag{3.43a}$$

$$m = 0.415 - 0.0346\ln\left(\frac{d_\mathrm{f}}{s}\right) \tag{3.43b}$$

$$\begin{aligned} f &= 3.805\,Re_{\mathrm{d}}^{-0.234}\left(\frac{s}{d_\mathrm{f}}\right)^{0.251}\left(\frac{l_\mathrm{f}}{s}\right)^{0.759} \\ &\quad \times\left(\frac{d_\mathrm{r}}{d_\mathrm{f}}\right)^{0.729}\left(\frac{d_\mathrm{r}}{P_\mathrm{t}}\right)^{0.709}\left(\frac{P_\mathrm{t}}{P_1}\right)^{0.379} \end{aligned} \tag{3.44}$$

where α_h and α_n are a temperature-dependent fluid properties correction factor and a row correction factor, respectively. The row correction factor has been covered earlier. Rabas and Taborek [7] recommend the following correlations for α_h:

$$\alpha_\mathrm{h} = \left(\frac{T_\mathrm{b}+273}{T_\mathrm{w}+273}\right)^{0.25} \tag{3.45a}$$

$$\alpha_{\mathrm{h}} = \left(\frac{Pr(T_{\mathrm{b}})}{Pr(T_{\mathrm{w}})} \right)^{0.26} \tag{3.45b}$$

Equation 3.45a is from Weierman [34], and Equation 3.45b is from ESDU [35]. However, the Prandtl number format is recommended by Rabas and Taborek [7] for liquids and there is no fluid property correction term for the friction factor f. These are applicable for $l_{\mathrm{f}} < 6.35$ mm, $1000 < \mathrm{Re_d} < 25{,}000$, 3.76 mm $\leq d \leq$ 31.75 mm, $246 \leq N_{\mathrm{f}} \leq 1181$ fins/m, 15.08 mm $\leq P_{\mathrm{t}} \leq$ 111 mm, 10.32 mm $\leq P_{\mathrm{l}} \leq$ 96.11 mm, $P_{\mathrm{l}} \leq P_{\mathrm{t}}$, $d_{\mathrm{f}}/s < 40$, and $N_{\mathrm{r}} > 6$.

4. ESDU Correlation [35]

$$j = 0.183\ \mathrm{Re_d^{+0.3}}\mathrm{Pr}^{0.027} \left(\frac{s}{l_{\mathrm{f}}} \right)^{0.36} \left(\frac{l_{\mathrm{f}}}{d_{\mathrm{f}}} \right)^{0.11} \left(\frac{P_{\mathrm{t}}}{d_{\mathrm{f}}} \right)^{0.06} \alpha_{\mathrm{h}}\alpha_{\mathrm{n}} \tag{3.46a}$$

$$f = 4.71\ \mathrm{Re_d^{(-0.286)}} \left(\frac{X_{\mathrm{t}}}{d_{\mathrm{r}}} - 1 \right)^{(-0.36)} \left(\frac{l_{\mathrm{f}}}{s} \right)^{0.51} \left(\frac{X_{\mathrm{t}} - d_{\mathrm{r}}}{X_{\mathrm{l}} - d_{\mathrm{r}}} \right)^{0.536} \tag{3.46b}$$

This is applicable for $1000 < \mathrm{Re_d} < 80{,}000$ or 100,000 and $N_{\mathrm{t}} > 10$.

5. Ganguly et al. [36] Correlation

$$j = 0.255\mathrm{Re_d^{-03}} \left(\frac{d_{\mathrm{f}}}{s} \right)^{-0.3} \tag{3.47}$$

$$f = F_{\mathrm{s}} f_{\mathrm{p}} \tag{3.48a}$$

where

$$F_{\mathrm{s}} = 2.5 + \frac{3}{\pi} \tan^{-1}\left[0.5\left(A/A_{\mathrm{p}} - 5 \right) \right] \tag{3.48b}$$

$$f_{\mathrm{p}} = 0.25\beta\left(\mathrm{Re_d^{-0.25}} \right) \tag{3.48c}$$

$$\beta = 2.5 + 1.2\left(X_{\mathrm{t}} - 0.85 \right)^{-1.06} + 0.4\left(\frac{X_{\mathrm{l}}}{X_t} - 1 \right)^3 - 0.01\left(\frac{X_t}{X_{\mathrm{l}}} - 1 \right)^3 \tag{3.48d}$$

For both j and f factors, 95% of the data correlated within 15% and 100% of the data correlated within 20%. This is applicable for $l_{\mathrm{f}} \leq 6.35$ mm, $800 \leq \mathrm{Re_d} \leq 800{,}000$, $20 \leq \theta \leq 40$, $X_{\mathrm{t}} \leq 4$, and $N_{\mathrm{r}} \geq 3$.

6. Elmahdy and Biggs Correlation [37]

For a multirow staggered circular or continuous fin-tube arrangement,

$$j = C_1\mathrm{Re_h^{C_2}} \tag{3.49a}$$

where

$$C_1 = 0.159\left(\frac{t_f}{l_f}\right)^{0.141}\left(\frac{D_h}{t_f}\right)^{0.065} \tag{3.49b}$$

$$C_2 = -0.323\left(\frac{t_f}{l_f}\right)^{0.049}\left(\frac{s}{t_f}\right)^{0.077} \tag{3.49c}$$

This correlation is also valid for circular fin-tube arrangement. The experimentally determined j factor is 3%–5% lower than that predicted analytically.

This correlation is applicable for the following parameter: $Re_d = 200–2000$, $D_h/t_f = 3.00–33.0$, $t_f/l_f = 0.01–0.45$, $d_f/P_1 = 0.87–1.27$, $s/t_f = 2.00–25.0$, $d_f/P_t = 0.76–1.40$, $d/d_f = 0.37–0.85$, $\sigma = 0.35–0.62$.

3.4.3 Continuous Fin on Circular Tube

3.4.3.1 Gray and Webb Correlation [38]

Four-row array:

$$j_4 = 0.14\ \text{Re}_d^{-0.328}\left(\frac{P_t}{P_1}\right)^{-0.502}\left(\frac{s}{d_o}\right)^{0.0312} \tag{3.50}$$

For this form, 89% of data were correlated within ±10%.

Up to three rows:

$$\frac{j_N}{j_4} = 0.991\left[2.24\ \text{Re}_d^{-0.092}\left(\frac{N_r}{4}\right)^{-0.031}\right]^{0.607(4-N_r)} \tag{3.51}$$

This is used for $N_r = 1–3$ (for $N_r > 4$, the row effect is negligible). The standard deviation is −4% to +8%.

Pressure drop correlation: Gray and Webb correlation is based on a superposition model that was initially proposed by Rich [39]. The basic model is written as

$$\Delta_p = \Delta p_f + \Delta p_t \tag{3.52}$$

where

Δp_t, is the pressure-drop component due to the drag force on the tubes
Δp_f is the pressure-drop component due to the friction factor on the fins

Accordingly, the expressions for Δp_t and Δp_f are given by

$$\Delta p_f = f_f \frac{A_f}{A_c}\frac{G^2}{2g_c\rho} \tag{3.53a}$$

$$\Delta p_t = f_f \frac{A_t}{A_{c,t}} \frac{G^2}{2g_c\rho} \tag{3.53b}$$

$$f = f_f \frac{A_f}{A} + f_t\left(1 - \frac{A_f}{A}\right)\left(1 - \frac{t_f}{S}\right) \tag{3.54a}$$

$$f_f = 0.508\, Re_d^{-0.521}\left(\frac{P_t}{d_o}\right)^{1.318} \tag{3.54b}$$

Several correlations are available for the friction factor of flow normal to tube banks [28,40].

Parameters are N_r= 1–8 or more, Re_d = 500–24,700, P_t/d_o = 1.97–2.55, P_l/d_o = 1.70–2.58, S/d_o= 0.08–0.64, $\mathrm{Re} = Gd_o/\mu$.

3.4.3.2 McQuistion Correlation [41]

Continuous-fin tube heat exchanger with four rows:

$$j_4 = 0.0014 + 0.2618\, \mathrm{Re}_d^{-0.4}\left(\frac{A}{A_t}\right)^{-0.15} \tag{3.55}$$

where A/A_t is the ratio of the total surface area to the area of the bone tube. Here, 90% of data were correlated within ±10%.

Less than four rows:

$$\frac{j_N}{j_4} = \frac{1.0 - 128N_r\, \mathrm{Re}_L^{-1.2}}{1.0 - 5120\, \mathrm{Re}_L^{-1.2}} \tag{3.56}$$

for N_r = 1–3 with a standard deviation of –9% to +18%.

$$f = 4.904\times 10^{-3} + 1.382\alpha^2 \tag{3.57a}$$

$$\alpha = \mathrm{Re}_d^{-0.25}\left(\frac{R}{R^*}\right)^{0.25}\left[\frac{(X_a - 2R)N_f}{4(1 - n_f t_f)}\right]^{-0.4}\left(\frac{P_t}{2R^*} - 1\right)^{-0.5} \tag{3.57b}$$

$$\left(\frac{R^*}{R}\right) = \frac{A/A_t}{(\mathrm{P}_t - 2R)n_f + 1} \tag{3.57c}$$

$$\frac{A}{A_t} = \frac{4}{\pi}\frac{P_l}{D_h}\frac{P_t}{d_o}\sigma \tag{3.57d}$$

$$R = \frac{d_o}{2}$$

Parameters are

$$P_l, P_t = 1-2 \text{ in.}$$

$$\alpha = 0.08 - 0.24$$

$$d_o = \frac{3}{8} - \frac{5}{8} \text{ in.}$$

$$\text{Tube spacing} = 1-2 \text{ in.}$$

$$N_f = 4 - 14 \text{ fin/in.}$$

$$t_f = 0.006 - 0.010 \text{ in.}$$

3.4.4 Continuous Fin on Flat Tube Array

Performance figures for three geometries are given in Kays and London [15].

New data are presented for two new types of rippled fin by Maltson et al. [42]. For these types of layouts, fin length for calculating fin efficiency is half the spanwise length of fin between tubes.

Kroger [43] presented performance characteristics for a six rows deep (3 × 2) core geometry given in Figure 3.24, whose *j* and *f* factors are given by

$$j = \frac{0.174}{\text{Re}_h^{0.387}}, \quad f = \frac{0.3778}{\text{Re}_h^{0.3565}} \tag{3.58}$$

These equations were also found to correlate with the data for having eight rows deep (4 × 2).

3.5 PLATE-FIN HEAT EXCHANGERS

A plate-fin heat exchanger (PFHE) consists of a block of alternating layers of corrugated fins. The layers are separated from each other by special parting sheets and sealed along the edges by means of side bars, and are provided with inlet and outlet ports for the streams. The stacked assembly is brazed in a fluxless vacuum furnace or NOCOLOK® flux brazing to become a rigid core. To complete the heat exchanger, headers with nozzles are welded to it.

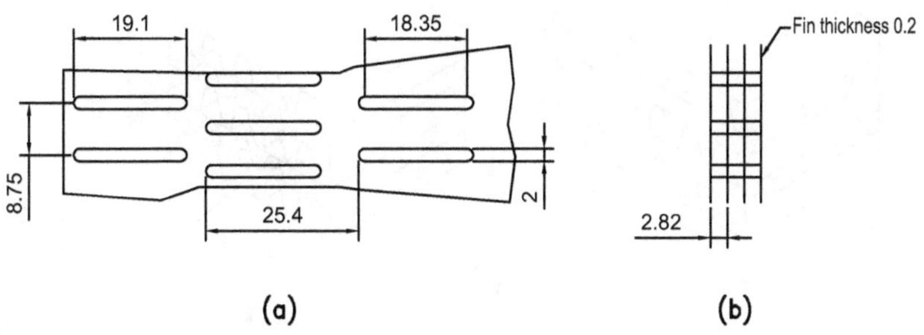

FIGURE 3.24 Geometrical details of a flat tube-fin radiator to calculate *j* and *f* factors as per Equation 3.58 (Parameters are in mm.)

Two important sources on PFHE and brazed aluminum plate-fin heat exchanger are ALPEMA Standard [44] and *The Plate-Fin Heat Exchangers Guide to Their Specification and Use* [45]. The definition for PFHE given in Ref. [45] is as follows: Plate-fin heat exchangers (PFHE) are a form of a compact heat exchanger consisting of a stack of alternate flat plates called "parting sheets" and fin corrugations, brazed together as a block. The heat exchanger is normally specified by its outside dimensions in the following order: Width (W) × Stack Height (H) × Length (L).

The basic elements of a PFHE and flow arrangements are shown in Figure 3.25 and a brazed unit is Figure 3.26. Fluid streams flow along the passages made by the corrugations between the parting sheets. The corrugations serve as both secondary heat transfer surfaces and mechanical supports for the internal pressures between layers. In liquid or phase-change (to gas) applications, the parting sheets may be replaced by flat tubes on the liquid or phase-change side. Further, PFHEs are widely used in various industrial applications because of their compactness. For many years, PFHEs have been widely used for gas separation in cryogenic applications and for aircraft cooling duties. In aerospace applications weight saving is of paramount importance. In the following sections, first the salient features of PFHE are covered followed by specific features of brazed aluminum PFHEs as per ALPEMA Standard [44].

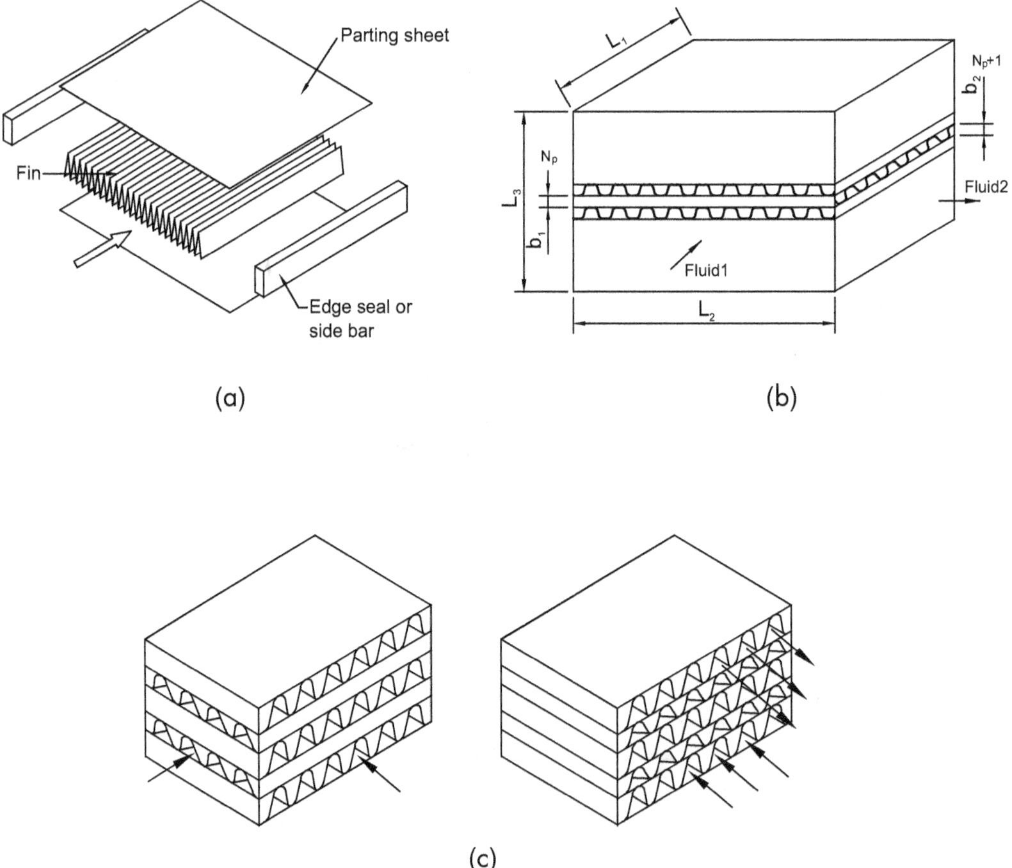

FIGURE 3.25 Plate-fin heat exchanger: (a) basic elements, (b) PFHE, (c) crossflow and counterflow and (d) counterflow arrangement arrangements(Photo courtesy of Fives Cryo, Golbey, France.)

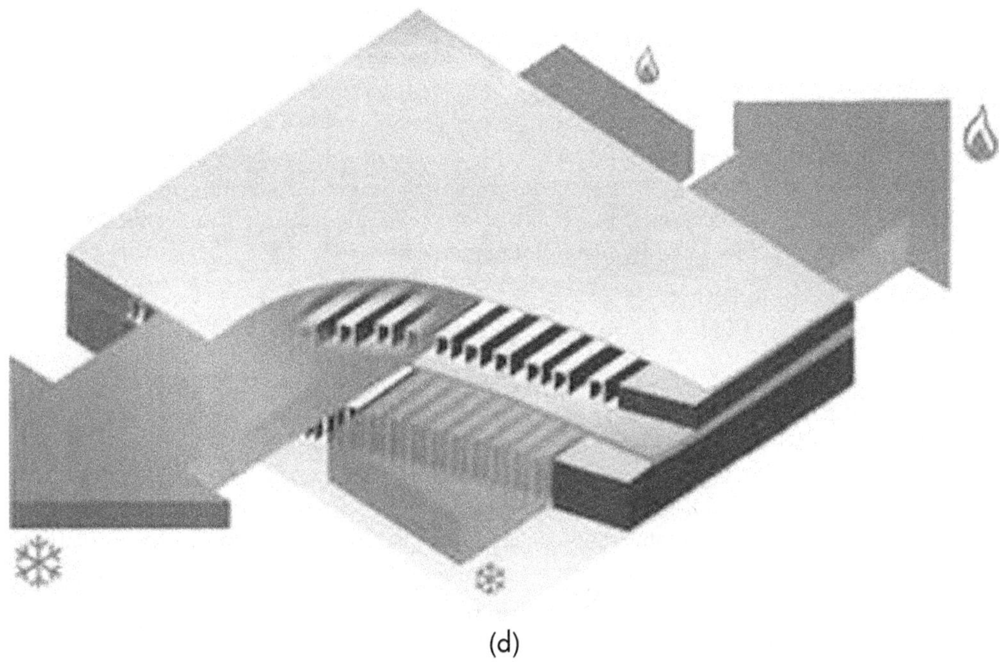

(d)

FIGURE 3.25 (Continued)

3.5.1 PFHE: Essential Features

The salient features of PFHE as discussed by Shah include the following [1,2]:

1. Plate-fin surfaces are commonly used in gas-to-gas exchanger applications. They offer high area densities (up to about 6000 m^2/m^3 or 1800 ft^2/ft^3).
2. The passage height on each side could be easily varied. Different fins (such as [a] rectangular or triangular fin either plain or with louver or perforation, [b] offset strip fin (OSF), and [c] wavy fin) can be used between plates for different applications.
3. Plate-fin exchangers are generally designed for low-pressure applications, which operating pressures limited to about 1000 kPag (150 psig).
4. The maximum operating temperatures are limited by the type of fin-to-plate bonding and the materials employed. Plate-fin exchangers have been designed from low cryogenic operating temperatures (all-aluminum PFHE) to about 800°C (1500°F) (made of heat-resistant alloys).
5. Fluid contamination (mixing) is generally not a problem since there is practically zero fluid leakage from one side to the other of the exchanger.

3.5.2 Application for Nonfouling Service

The PFHE is limited in application to relatively clean streams because of its small flow passages. It is generally not designed for applications involving heavy fouling since there is no easy method of cleaning the exchanger. If there is a risk of fouling, use wavy fins and avoid serrated fins. Upstream strainers should be employed where there is any doubt about solids in the feed. Measures to overcome fouling problems with PFHE are described by Shah [2].

3.5.3 Size

Maximum size is limited by the brazing furnace dimensions, and by the furnace lifting capacity. The heating characteristics of denser blocks also impose limitations on the maximum size which can be brazed. Typical maximum dimensions for low-pressure aluminum PFHEs are 1.2 m × 1.2 m in cross section × 6.2 m along the direction of flow [1,2,45], and 1.0 m × 1.0 m × 1.5 m for non-aluminum PFHEs [45]. Higher pressure units, being heavier per volume, make both handling and brazing more difficult, thus reducing the economic maximum size. Duties that call for larger units are met by welding together several blocks, or by manifolding pipework.

3.5.4 Advantages of PFHEs

The principal advantages of PFHEs over other forms of heat exchangers are summarized in Ref. [44] and include the following:

1. Plate-fin heat exchangers, in general, are superior in thermal performance to those of the other types of heat exchangers employing extended surfaces.
2. PFHE can achieve temperature approaches as low as 1°C between single-phase streams and 3°C between multiphase streams. Typically, overall mean temperature differences of 3°C–6°C are employed in aluminum PFHE applications [45].
3. With their high surface compactness, ability to handle multiple streams, and with aluminum's highly desirable low-temperature properties, brazed aluminum plate-fins are an obvious choice for cryogenic applications.
4. Very high thermal effectiveness can be achieved; for cryogenic applications, effectiveness of the order of 95% and above is common.
5. Provided the streams are reasonably clean, PFHEs can be used to exchange heat in most processes, for the wide range of stream compositions and pressure/temperature envelopes [45].
6. Large heat transfer surface per unit volume is possible.
7. Low weight per unit heat transfer.
8. Possibility of heat exchange between many process streams.
9. Provided the correct materials are selected, the PFHE can be specified for temperatures ranging from near absolute zero to more than 800°C, and for pressures up to at least 140 bar [1]. According to Shah [1], usually PFHEs do not involve both high temperature and high pressure together.
10. PFHE offers about 25 times more surface area per equipment weight than the shell and tube heat exchanger [45].

3.5.5 Limitations of PFHEs

Limitations other than fouling and size include the following [1]:

1. With a high-effectiveness heat exchanger and/or large frontal area, flow maldistribution becomes important.
2. Due to short transient times, a careful design of controls is required for startup compared with shell and tube exchangers.
3. Flow oscillations could be a problem.

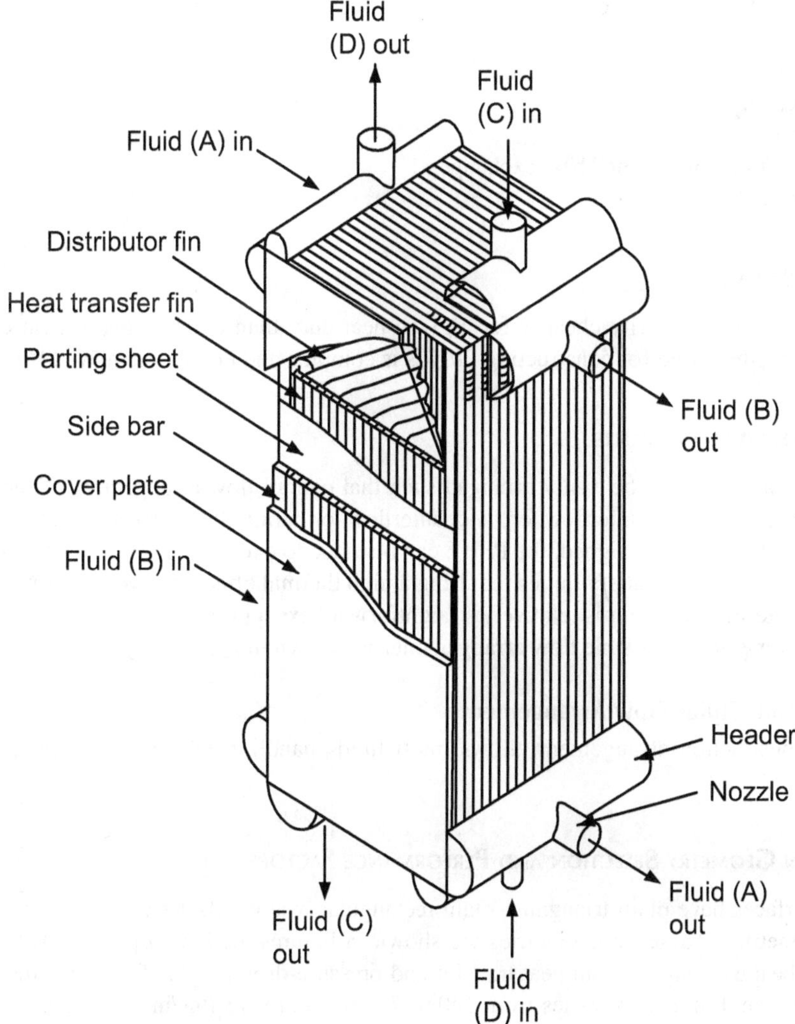

FIGURE 3.26 Brazed aluminum plate-fin heat exchanger structure. (*Note*: A, B, C, D are different fluids.) (From Linde AG, Engineering Division. With permission.)

3.5.6 Applications

Current areas of application include [45] the following:

1. Cryogenics/air separation
2. Petrochemical production
3. Syngas production
4. Aerospace
5. Land transport (automotive, locomotive)
6. Oil and natural gas processing

PFHEs are used in all modes of heat duty, including the following [45]:

1. Heat exchange between gases, liquids, or both
2. Condensing
3. Boiling
4. Sublimation ("reversing" heat exchangers)
5. Heat or cold storage

3.5.7 Economics

The PFHE is not necessarily cheaper for a given heat duty than other forms of heat exchangers, because the method used for constructing PFHEs is complex and energy-intensive [45].

3.5.8 Flow Arrangements

The fins on each side can be easily arranged such that overall flow arrangement of the two fluids can result in crossflow, counterflow, cross-counterflow, or parallelflow, though parallelflow is generally not used. Properly designed, the PFHE can be made to exchange heat in perfect counterflow, which permits PFHEs to satisfy duties requiring a high thermal effectiveness. The construction of a multifluid plate-fin exchanger is relatively straightforward except for the inlet and outlet headers for each fluid. Some of the possible flow arrangements are shown in Figure 3.27.

3.5.8.1 Multi Fluids Flow Arrangements

The core construction arrangement allows multi fluids handling in the same unit, as shown in Figure 3.28.

3.5.9 Fin Geometry Selection and Performance Factors

Plate-fin surfaces have plain triangular, plain rectangular, wavy, offset strip, louver, perforated, or pin fin geometries. These fin geometries are shown in Figures 3.29–3.33 [8]. PFHEs are superior to tube-fin heat exchangers from heat transfer and pressure-drop points of view for the given total packaging space. Fin density varies from 120 to 700 fins/m (3–18 fins/in.), thickness from 0.05 to 0.25 mm (0.002–0.01 in.), and fin heights may range from 2 to 25 mm (0.08–1.0 in.) [2].

The herringbone and serrated fins provide the greatest surface area and the highest heat transfer performance. They are particularly suitable for applications involving close temperature approaches. Where there are critical pressure drop requirements, the plain and perforated fins can be used.

Fins, also known as a secondary surface, provide an extended heat transfer surface. They provide a connecting structure between the parting sheets, thereby creating the essential structural and pressure-holding integrity of the heat exchanger.

All except for the plain surfaces have features in their geometry to enhance heat transfer film coefficients by thinning/disruption of the hypothetical laminar film. Thermal hydraulic characteristics ("*j*" and "*f*" factors) of fins are influenced by fin density, fin thickness, fin height, perforation fraction (if perforated), wave angle and wavelength (if herringbone/wavy), and serration length (if serrated)

3.5.9.1 Plain Fin

Plain fin corrugation is the simplest type of fining. These surfaces are straight fins that are uninterrupted (uncut) in the fluid flow direction. Triangular and rectangular passages are more common. When the fins are straight along the flow length, the boundary layers tend to be thick, resulting in

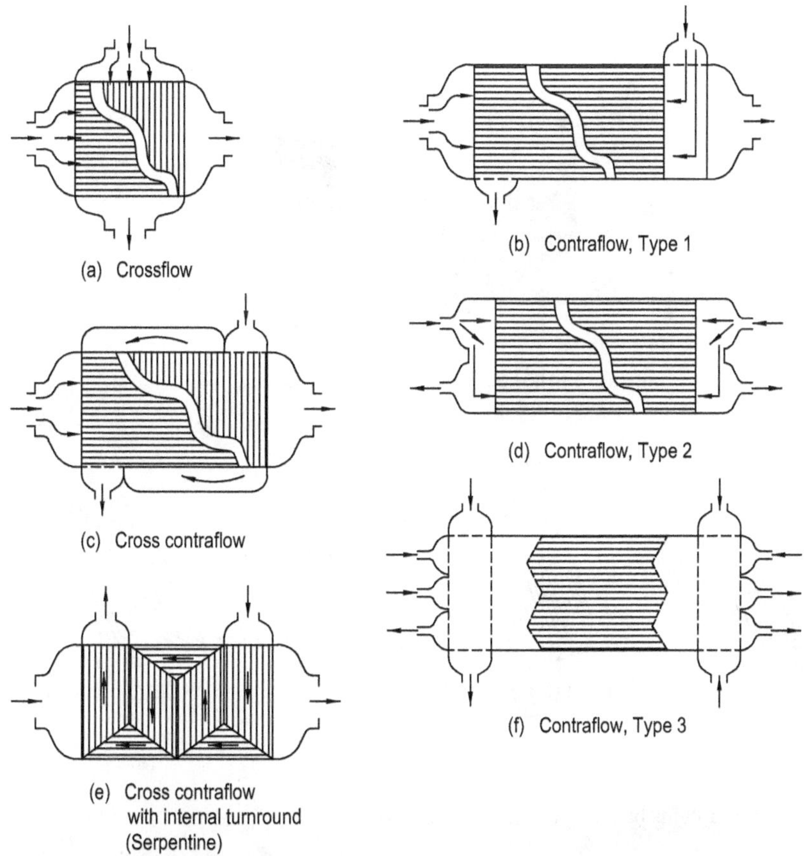

FIGURE 3.27 Fluid flow arrangement through PFHE. (*Note*: contraflow is to be read as counterflow.) (Courtesy of Chart Industries Inc., Garfield Heights, OH.)

lower values of the heat transfer coefficient. When they are wavy or off-set strip fin along the flow length, the boundary layers are thinner or the growth of boundary layers is disrupted periodically, respectively, resulting in a higher heat transfer coefficient.

Plain fin surfaces have pressure-drop and heat transfer characteristics similar to flow through small-bore tubes, i.e., relatively low pressure drop and heat transfer, but a high ratio of heat transfer to pressure drop [45]. For a flow channel with rectangular or triangular cross section, under turbulent flow conditions, standard equations for turbulent flow in circular tubes may be used to calculate j and f, provided the Reynolds number (Re) is based on the hydraulic diameter, D_h. If the Re based on hydraulic diameter is less than 2000, one may use theoretical laminar flow solutions for j and f.

Plain fins are preferred for very low Reynolds number applications and in applications where the pressure drop is very critical. Condensing duties require minimal pressure drop or else the heat release curve can significantly be altered and the overall duty may not be met. Therefore, for condensation, plain fins are normally specified.

3.5.9.2 Plain-Perforated Fin

Perforated fins have either round or rectangular perforations, having the size and longitudinal and transverse spacings as major perforation variables. The flow passages are either triangular or rectangular. A metal strip is first perforated, then corrugated. Perforated areas vary from about 5% to

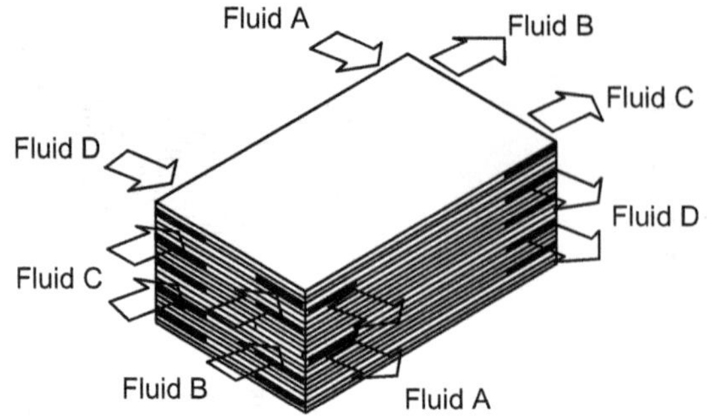

(a) PFHE heat exchanger with 4 fluids.

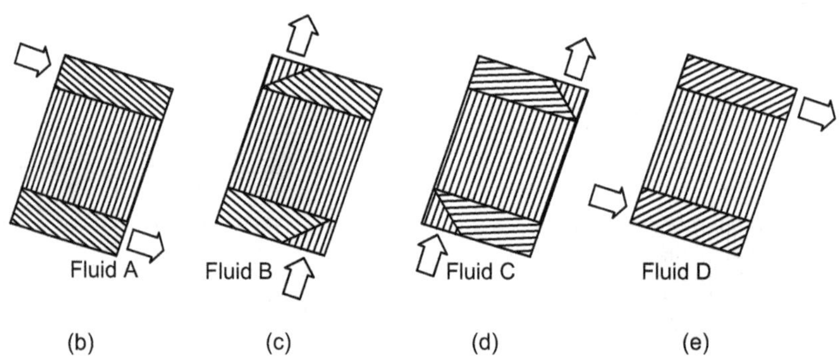

FIGURE 3.28 Multi fluids handling capability of a PFHE.

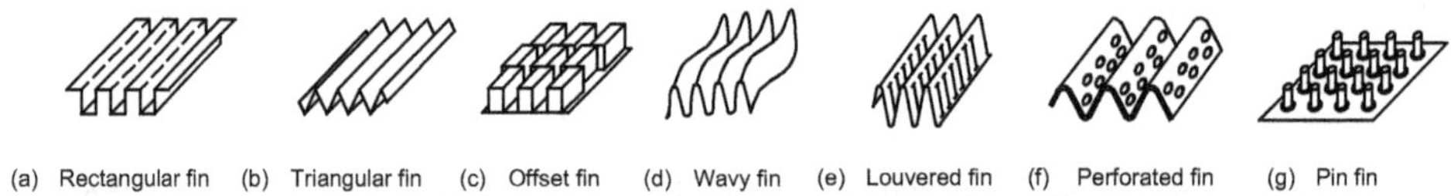

FIGURE 3.29 PFHE fin geometries.

25% of the sheet area [45]. When the corrugations are laid across the flow, the stream is forced to pass through these small holes. Salient performance features include:

1. The perforated holes promote turbulence, which increases the local heat transfer coefficient compared to plain fins, but as the percentage of perforated holes increase, the loss of heat transfer surface offsets this advantage [45].
2. The perforated surface is prone to flow-induced noise and vibration.

Applications: Perforated fins are now used in only a limited number of applications [17].
Typical applications include the following:

1. "Turbulators" in oil coolers.
2. In boiling applications to maintain a wetted surface and minimize depositions/concentrations [45]. Perforated corrugations permit interchannel fluid migration, which evens out surges and vibration and avoids localized concentrations or depositions.

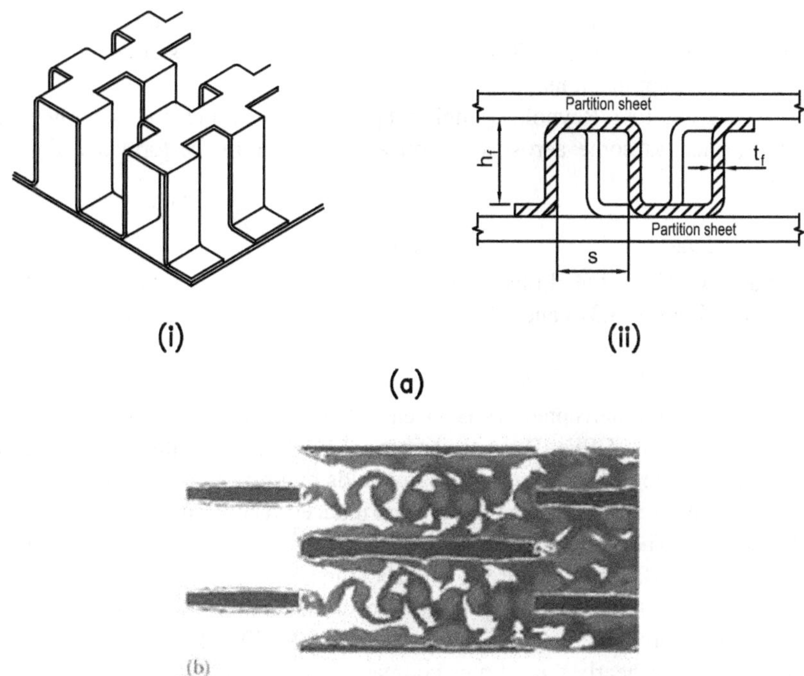

FIGURE 3.30 (a) Offset strip fin—schematic. (b) Flow past offset strip fin—formation of turbulence behind the fin. (Photo courtesy of Fives Cryo, Golbey, France.)

3. Perforated fins were once used in two-phase cryogenic air separation exchangers, but now they have been replaced by OSFs.

Shah [46] provides a very detailed evaluation of the perforated fin geometries.

3.5.9.3 Offset Strip Fin

The offset strip fin (OSF) is the commonly used geometry in PFHEs. The fin has a rectangular cross section; it is cut into small strips of length *l*, and every alternate strip is offset by about 50% of the fin pitch in the transverse direction, as schematically shown in Figure 3.30a. Fin spacing, fin height, fin thickness, and strip length in the flow direction are the major variables of OSFs.

OSF geometry is characterized by high heat transfer area per unit volume, and high heat transfer coefficients. The heat transfer mechanism in OSFs is described by Joshi and Webb [47] and is as follows. The heat transfer enhancement is obtained by periodic growth of laminar boundary layers on the fin length, and their dissipation in the fin wakes, as shown in Figure 3.30b. This enhancement is accompanied by an increase in pressure drop because of increased friction factor. A form drag force, due to the finite thickness of the fins, also contributes to the pressure drop. Fluid interchange between channels is possible. OSFs are used in the approximate Reynolds number range of 500–10,000 [17]. Salient features pertaining to thermal performance include the following [45]:

1. Commonly used in air separation plants where high thermal effectiveness at low mass velocities is required [45].
2. The heat transfer performance of OSFs is increased by a factor of about 1.5–4 over plain fins of similar geometry, but at the expense of higher pressure drop [17].

3. At high Reynolds numbers, the *j* factor decreases, while the friction factor remains constant because of the high form drag. Therefore, offset fins are used less frequently for very high Reynolds number applications.
4. They are used at low Reynolds number applications calling for accurate performance predictions, such as some aerospace applications; other fin performance data are not as repeatable.

3.5.9.4 Serrated Fins

These are similar to OSFs. Serrated fins are formed by simultaneously folding and cutting alternative sections of fins (Figures 3.31a and 3.32d). These fins are also known as lanced fins.

3.5.9.5 Herringbone or Wavy Fins

Since wavy fins have noninterrupted walls in each flow channel, they are less likely to catch particulates and foul than are OSFs. Their performance is competitive with that of the OSFs, but the friction factor continues to fall with increasing Reynolds numbers [45]. The waveform in the flow direction provides effective interruptions to flow and induces very complex flows. A wavy fin stack is shown in Figure 3.34. The augmentation is due to Goertler vortices, which form as the fluid passes over the concave wave surfaces. These are counterrotating vortices, which produce a corkscrew-like pattern and probable local flow separation that will occur on the downstream side of the convex surface [8,17]. In the low-turbulence regime (Re of about 6000–8000), the wall corrugations increase the heat transfer by about nearly three times compared with the smooth wall channel. Therefore, wavy fins are often a better choice at the higher Reynolds numbers typical of the hydrocarbon

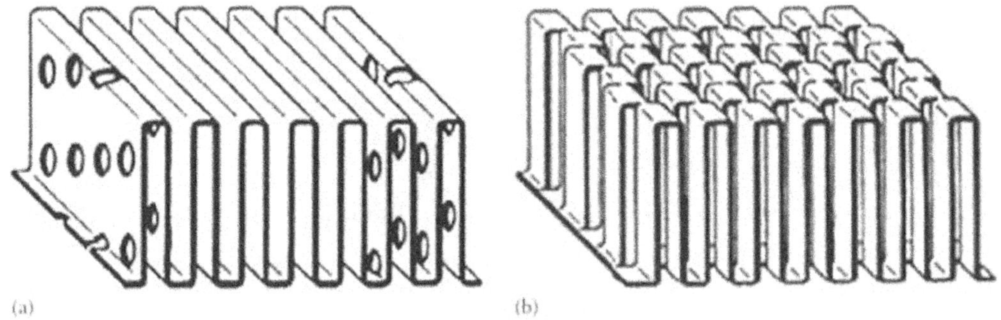

FIGURE 3.31 Brazed aluminum plate-fin heat exchanger fin profiles—(a) perforated fin and (b) offset strip fin (schematic). (From Linde AG, Engineering Division. With permission.)

FIGURE 3.32 Brazed aluminum plate-fin heat exchanger fin profiles—(a) serrated fins (b) OSF and (c) perforated fins. (Courtesy of Fives Cryo, Golbey, France.)

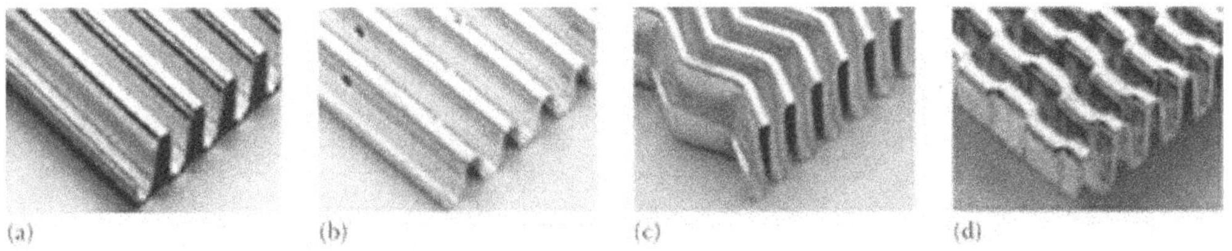

FIGURE 3.33 Brazed aluminum plate-fin heat exchanger fin profiles: (a) plain fin, (b) perforated fin, (c) herringbone fin, and (d) serrated fin. (Courtesy of Chart Industries Inc., Garfield Heights, OH.)

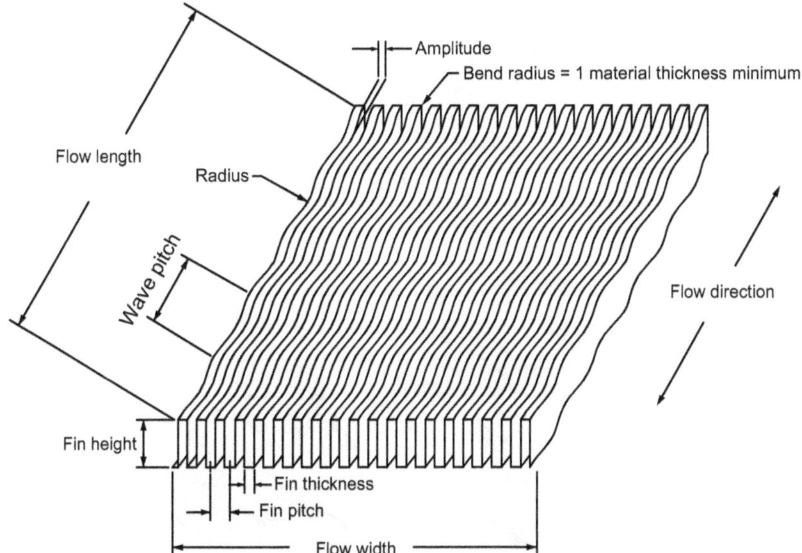

FIGURE 3.34 Wavy fin (Adapted from https://www.robfin.com/ruffled-herringbone-folded-fins/).

industry; the smooth surface allows the friction factor to fall with increasing Reynolds number. Kays and London [15] provide curves of j and f versus Reynolds number for two wavy-fin geometries.

3.5.9.6 Louver Fins

The louvers are essentially formed by cutting the sheet metal of the fin at intervals and by rotating the strips of metal thus formed out of the plane of the fin. Louvers can be made in many different geometries, some of which are shown in Figure 3.35. Note that the parallel louver fin and OSF both have small strips aligned parallel to the flow. The louvered fin geometry bears a similarity to the offset strip (Figure 3.30a). Rather than offsetting slit strips, the entire slit fin is rotated 20°–60° relative to the airflow direction. As such, they are similar in principle to the OSF. Louvered fins enhance heat transfer by providing multiple flat plate leading edges with their associated high values of heat transfer coefficient.

The operating Reynolds number range is 100–5000, depending upon the type of louver geometry employed [17].

Applications: Their attractive thermal performance characteristics in terms of compactness, light weight, and low pumping power for a given heat transfer duty make louver fins a potential candidate for aerospace applications [48]. They are recognized as very effective heat transfer surfaces, leading to compact solutions to cooling problems. Louvered fin surfaces are now used widely for

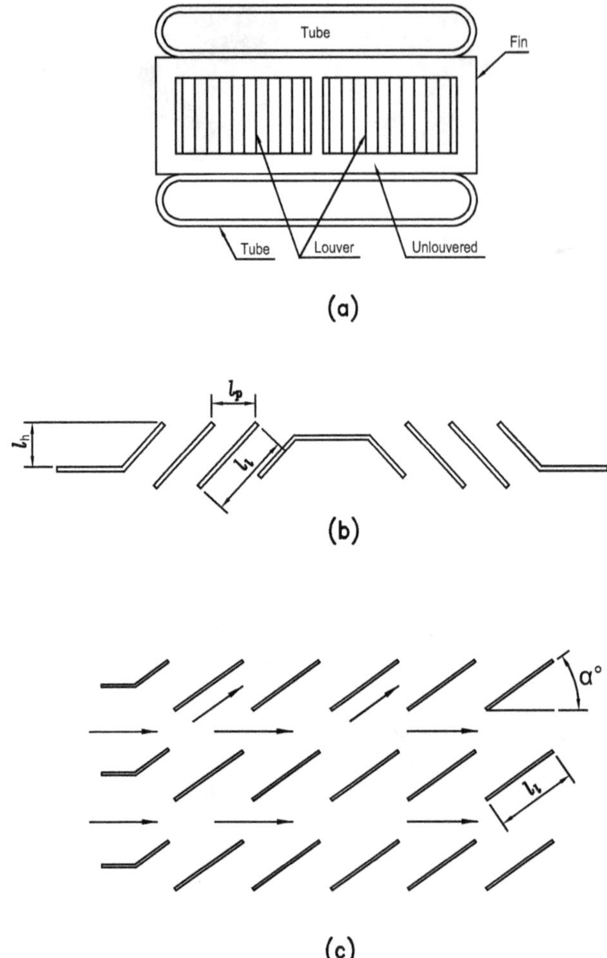

FIGURE 3.35 Louvered fin details. (a) Tube-louver fin arrangement, (b) louver fin geometry, and (c) fluid flow through louver. (*Note*: l_h is louver height, l_p is louver pitch, l_l is louver length and α is louver angle.)

engine cooling equipment, air-conditioning heat exchangers (evaporators and condensers), and aircraft oil and air coolers. They share this quality with the OSF surfaces [16]. The louvered surface is the standard geometry for automotive radiators. For the same strip width, the louvered fin geometry provides heat transfer coefficients comparable to OSFs.

3.5.9.7 Pin Fins

An array of wall-attached cylinders (e.g., rods, wires) mounted perpendicular to the wall is known as pin fins. They can be manufactured at a very high speed continuously from a thin wire. After the wire is formed into rectangular passages, the top and bottom horizontal wire portions are flattened for brazing or soldering with the plates. Pins can be of circular or elliptical shapes. The enhancement mechanism due to a pin array is the same as that of the OSF, namely, repeated boundary-layer growth and wake dissipation [8]. The potential application for pin fins is at very low Reynolds number (Re < 500) for which the pressure drop is of no major concern. Present applications for pin fins include [17] (1) electronic cooling devices with generally free convective flows over the pin fins and (2) to cool turbine blades.

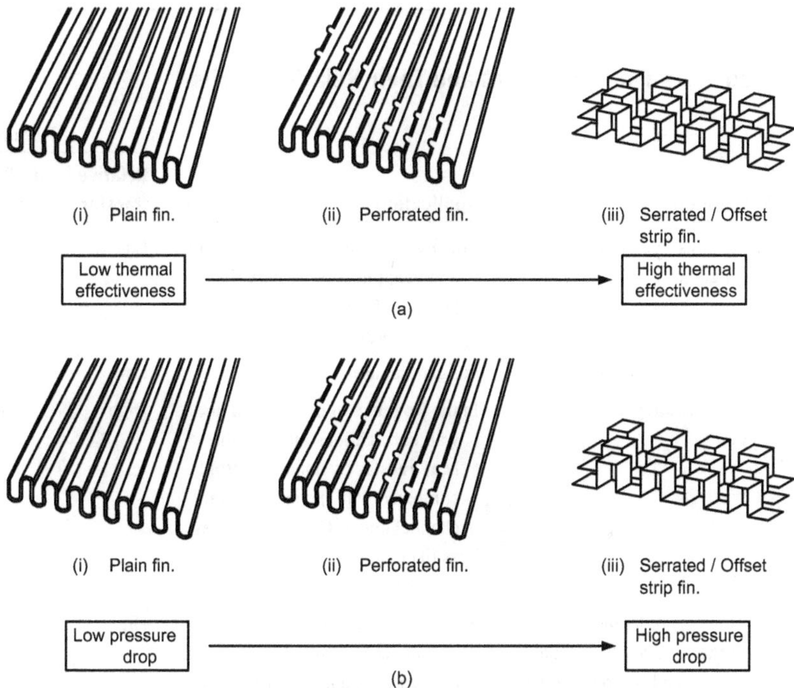

FIGURE 3.36 Comparison of performance of commonly used plate-fin geometries.

Limitations of pin fins:

1. The surface compactness achieved by the pin fin geometry is much lower than that achieved by the strip fin or louver fin surfaces.
2. Due to vortex shedding behind the circular pins, noise and flow-induced vibrations can take place.

For some fin geometries, Figure 3.34 shows comparison of performances in respect of thermal effectiveness and pressure drop and Table 3.1 shows commonly used plate-fin geometries and their relative merits.

3.5.9.8 Fin Corrugation Code

A commonly encountered code for fin corrugation is 350S1808 [45]. The first three digits give the fin height in thousandths of an inch (350 = 0.35 in.). The letter gives the type of corrugation (S = serrated). Other types are P = plain, R = perforated, H = herringbone. The next two digits give the fins per inch (18 = 18 fpi). The following two digits give the fin thickness in thousandths of an inch (08 = 0.008 in.). The time following the oblique stroke (if any) is used only for a perforated fin and it is the porosity, σ.

3.5.10 Corrugation Selection

Corrugations, edge bars, and parting sheets are chosen primarily to contain the appropriate pressure. Corrugation selection depends on factors such as compactness, pressure, mechanical integrity, manufacture, performance, velocity limits, and fouling [45].

TABLE 3.1
Commonly Used Plate-Fin Geometries and Their Relative Merits

Corrugation	Description	Application	Features: Relative heat transfer	Features: Relative pressure drop
Plain	Straight fins (rectangular or triangular)	Low Reynolds number applications and in applications where the pressure drop is very critical, e.g., condensation	Lowest	Lowest
Perforated	Straight fin with small holes	For general use	Low	Low
Herringbone or wavy fin	Smooth but wavy, about 10 mm pitch	In the Re range of 6000–8000, the wall corrugations increase the heat transfer by about threefold compared with the smooth wall channel due to Goertler vortices. Less likely to catch particulates and foul than are OSFs	High	High
Louvered fin	The louvers are formed by cutting the sheet metal of the fin at intervals and by rotating the strips of metal thus formed out of the plane of the fin	Radiators, air conditioning heat exchangers (evaporators and condensers), and aircraft oil and air coolers	Highest	Highest
OSF	Straight but offset by half a pitch (usually about every 3–4 mm)	Air separation plants and low Reynolds number applications calling for accurate performance predictions, e.g., aerospace applications	Highest	Highest

3.5.11 Construction Materials

PFHEs are made in a variety of materials to suit a wide range of process streams, temperatures, and pressures. Material specifications should comply with the appropriate section of the ASME or to other applicable codes.

3.5.11.1 Aluminum

Aluminum is preferred for cryogenic duties, because of its relatively high thermal conductivity, strength at low temperatures, and low cost. For cryogenic services, aluminum alloy 3003 is generally used for the parting sheets, fins, and edge bars that form the rectangular PFHE block. Headers and nozzles are made from aluminum alloy 3003, 5154, 5083, 5086, or 5454 [45]. Above ambient temperature, with an increase in temperature, most aluminum alloys rapidly lose their strength. For land-based transport vehicles, aluminum PFHEs are used up to 170°C–180°C to cool supercharged engine intake air.

3.5.11.2 Other Metals

Stainless steels and most nickel alloys are used for PFHEs, particularly for high-temperature services. Stainless steels have poor thermal conductivity, but their higher strength allows thinner parting plates and fins than with aluminum, which offsets some of the reduction in heat transfer.

3.5.12 Mechanical Design

The PFHE is a pressure vessel. It is required to be designed and constructed in accordance with a recognized Pressure Vessel Code. The mechanical design of a PFHE can be divided into the conventional pressure vessel area of header tanks, nozzle, nozzle reinforcements, lifting devices, pipe loads, etc., and the less familiar area of the block itself. Details on mechanical design of vessel components are given in Chapter 1 of Volume 2.

3.5.13 Manufacture, Inspection, and Quality Control

PFHEs are manufactured by brazing. Heat exchanger manufacture by brazing is described in Chapter 4 of Volume 2. The manufacturer is responsible for the mechanical design and for the thermal design where the latter is not specified by the purchaser. Salient features of quality control and inspection of PFHE are detailed in Refs. [17,45].

3.5.14 Brazed Aluminum Plate-Fin Heat Exchanger (BAHX)

Though most of the details on PFHE were covered earlier, specific details on aluminum PFHE are discussed here. Over the past 40 years, brazed aluminum plate-fin heat exchangers (BAHXs) have become the preferred type of exchanger for a variety of applications, mainly in heating and cooling of liquids and gases, and condensing and boiling with single- and multicomponent streams. BAHXs are capable of handling a wide variety of noncorrosive streams, up to a pressure of 100 bar (1450 psi) and from cryogenic temperature of −269°C to 204°C (−452°F to 400°F). BAHXs can handle several streams in the same exchanger.

A brazed aluminum PFHE is shown in Figure 3.37 and collections of different shapes/configuration of PFHEs are shown in Figures 3.38 and 3.39.

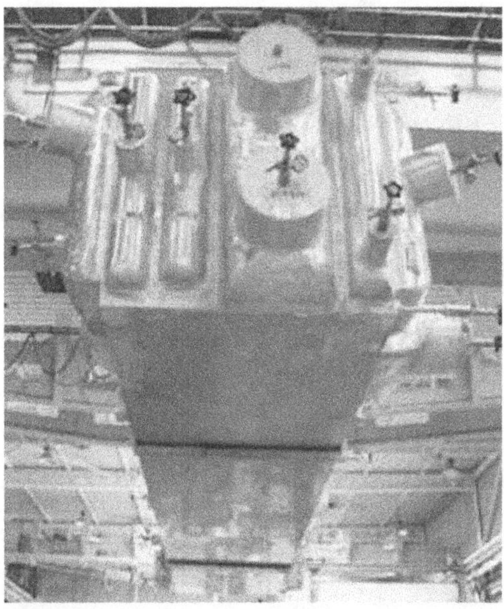

FIGURE 3.37 Brazed aluminum plate-fin heat exchanger. (Courtesy of Fives Cryo, Golbey, France.)

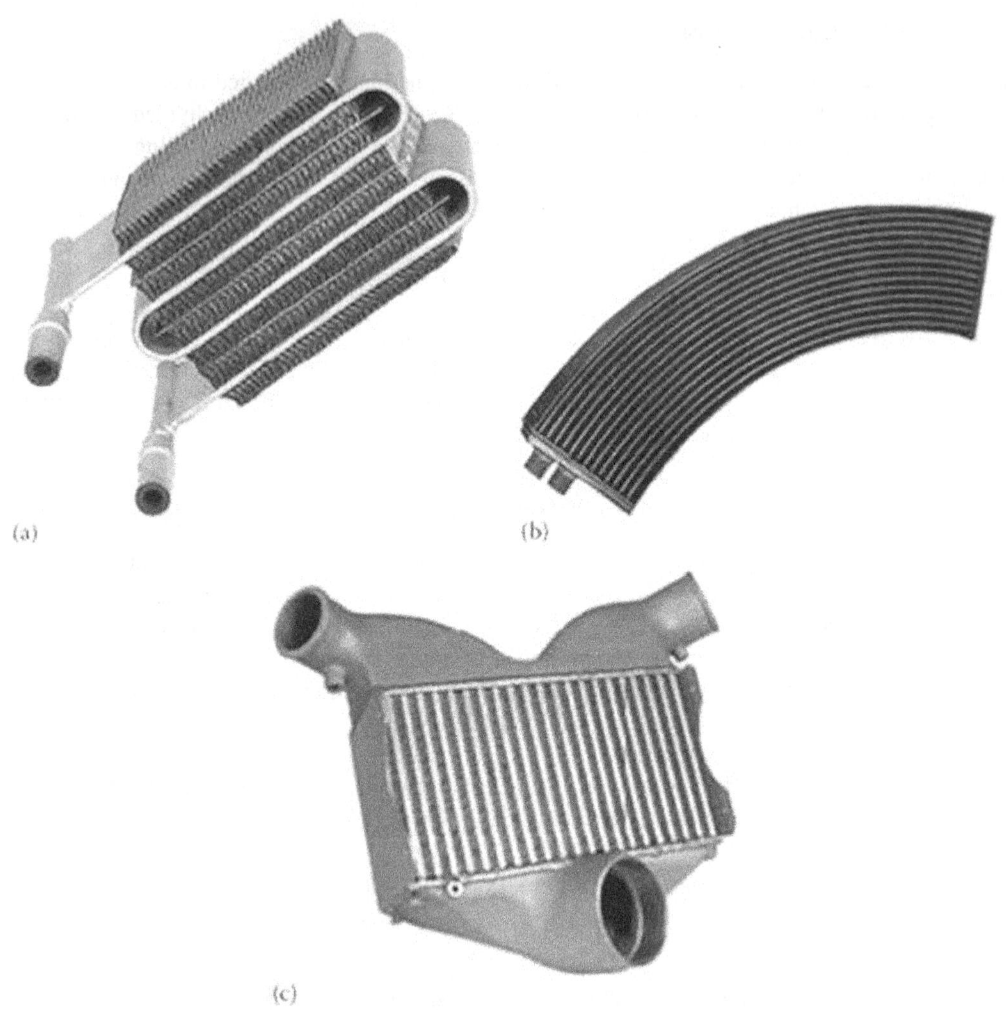

FIGURE 3.38 Varieties of brazed aluminum PFHEs configuration—(a) heat exchanger with single continuous piece of aluminum flat tube, (b) flat tube heat exchanger with curved tubes, and (c) plate-fin heat exchanger for air-to-air application. (Courtesy of Lytron Inc., Woburn, MA.)

3.5.14.1 ALPEMA Standard

Brazed aluminum PFHEs are designed and manufactured as per the guidelines of ALPEMA Standard, 3rd edition, 2010 [44]. Some of the members of ALPEMA Standards are (in alphabetical order)

1. Chart Energy and Chemicals Inc., 2191 Ward Avenue, La Crosse, WI
2. FivesCryo, 25, Rue du Fort BP 8788194 Golbey, Cedex, France
3. Kobe Steel Ltd., Energy System Center, Takasago Equipment Plant, 2-3-1, Shinhama, Arai-cho, Takasago-Shi, Hyogo-Ken, 676-8670, Japan
4. LINDE AG, Engineering Division, Works Schalchen, D-83342 Tacherting, Germany
5. Sumitomo Precision Products Co Ltd., Thermal Energy Systems Engineering Department, 1-10 Fuso-cho, Amagasaki, Hyogo Pref. 660-0891, Japan

Another PFHE manufacturer, among others, includes Lytron Inc., 55 Dragon Court, Woburn, MA.

FIGURE 3.39 Varieties of brazed aluminum plate-fin heat exchangers. (Courtesy of Sumitomo Precision Products Co. Ltd., Thermal Energy Systems Engineering Department, Japan.)

3.5.14.2 Applications

PFHEs can be used for a wide range of applications, especially for low-temperature services and treatment of clean fluids. Such applications include petrochemical plants, gas treatment plants, natural gas liquefaction plants, air separation plants, and helium liquefaction plants, among others.

3.5.14.3 Heat Exchanger Core

Each heat exchanger is built by stacking layers of fins separated by parting sheets and sealed along the edges with side bars. The matrix assembly is brazed in a vacuum/controlled atmosphere brazing (CAB) furnace to form an integral, rigid heat exchanger block. Headers and supports welded onto the brazed matrix complete the unit. Design variations in the configuration of the heat exchanger matrix can accommodate an almost unlimited range of flow options, including counterflow, crossflow, parallelflow, multipass, and multistream formats. The structure and components of brazed aluminum heat exchangers (BAHXs) are shown in Figure 3.26. The PFHE core inner details, nozzles, and inside flow pattern are shown in Figure 3.40.

- Fin height: 3.8–12 mm
- Fin thickness: 0.15–0.61 mm
- Fin pitch: 1.15–3.5 mm (i.e., fin density is 22–6 fpi)
- Surface area density, area/volume: 1500 m^2/m^3
- Design pressure up to a maximum of 110 bar (1600 psig)
- The size is available up to (W × H × L)—1.5 × 3.0 × 8.2 M

A heat exchanger's size, number of layers, type of fins, stacking arrangement, and stream circuiting will vary depending on the application requirements.

3.5.14.4 Flow Arrangement

A simple crossflow layout is generally suitable for low to moderate duties. It is used extensively for gas/liquid applications and is especially effective when handling a low-pressure gas stream on one side of the heat exchanger. For heavier duty tasks, where the mean effective temperature difference in crossflow may be significantly reduced, the counterflow pattern offers an efficient solution. The higher levels of efficiency achieved by counterflow units are essential to most low-temperature applications.

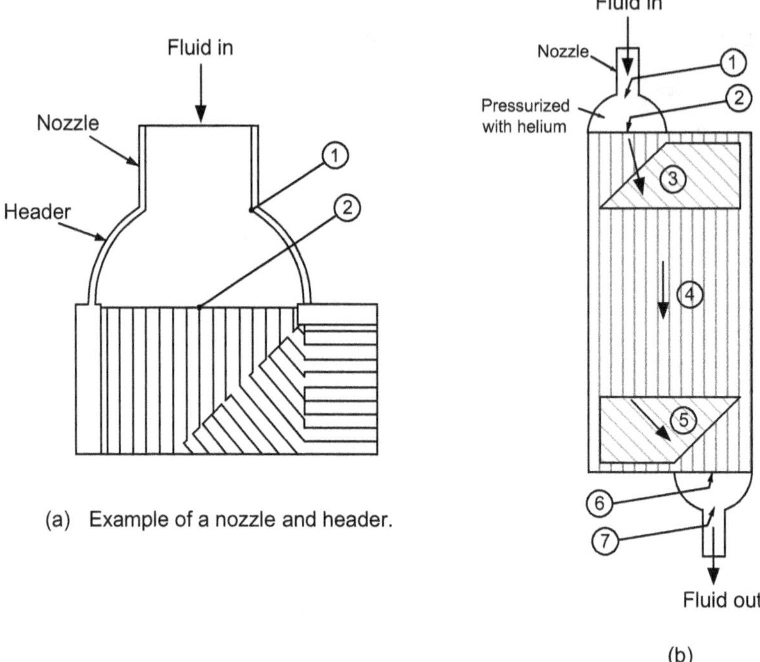

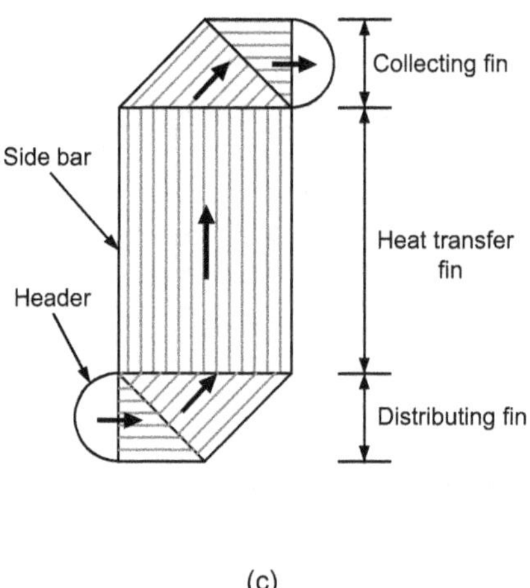

(c)

FIGURE 3.40 PFHE core inner details, (a) nozzles and (b and c) inside fluid flow pattern.

3.5.14.5 Core Details

Distributor fins: Distributor fins distribute the fluid between the port and the heat transfer fins. The distributor fin used adjacent to a port is called a port fin. The distributor fin used between a port fin and a heat transfer fin is called a turning fin. The typical distributor fin thickness is 0.45 mm (0.016 in.).

Parting sheets: Parting sheets (sometimes referred to as separator sheets) contain the fluids within individual layers in the exchanger and also serve as the primary heat transfer surface. Their selection is mainly based on design pressures. Parting sheets are normally clad on both sides with a brazing alloy. However, unclad parting sheets are available, where brazing is performed using brazing foils (filler material). Standard parting sheet thicknesses typically vary between 0.8 and 2.0 mm. Typical parting sheet thicknesses are 1.0, 2.0, and 3.0 mm (0.040, 0.080, and 0.125 in.).

Cap sheet or outside sheets: Cap sheets serve as the outside parting sheets. Standard cap sheet thicknesses are typically 5 or 6 mm. However, thicknesses from 2 to 10 mm are also used for special applications.

Side and end bars: Side and end bars enclose individual layers and form the protective perimeter of the exchanger. Solid extruded forms are used. Side bar heights are the same as the fin heights, and width is selected by the manufacturer according to the design pressure and typically varies between 10 and 25 mm.

Permissible temperature differences between adjacent streams: Aluminum PFHEs are produced by brazing. To maintain the thermal stresses within the acceptable limits for the material used, the maximum permissible temperature difference between streams shall be about 50°C.

Support angles: Support angles are typically 90° extruded aluminum angles welded to the exchanger bar face for the purpose of supporting or securing an exchanger in its installed position. Other support configurations, such as pedestal bases, are also available.

Lifting lugs: Lifting lugs (aluminum) are lift attachment points strategically welded to the exchanger block assembly for the specific purpose of lifting the exchanger into its installed position.

Header/nozzle configurations: Streams to and from the heat exchanger enter and leave by means of various header/nozzle configurations. Nozzles are the pipe sections used to connect the heat exchanger headers to the customer piping. Headers are the half cylinders which provide for the distribution of fluid from the nozzles to or from the ports of each appropriate layer within the heat exchanger.

Ports: Ports are the openings in either the side bar or the end bar, located under the headers, through which the fluids enter or leave individual layers.

Construction materials: Typical construction materials for the BAHX components are hereunder.

Cap sheets—AA 3003	Headers and nozzles, flanges—AA 5083/5454
Parting sheets—AA 3003	Support angles—AA 6061-T6
Side and end bars—AA 3003	Lifting lugs—AA 5083
Heat transfer fins—AA 3003	
Distributor fins—AA 3003	

Life: A design life similar to tubular heat exchangers can be expected from them.

Construction code: The design, construction, and testing of BAHXs are governed by the existing national rules applying to pressure vessels.

3.5.14.6 Aluminum PFHE Varieties in use

a. *Modular exchanger assembly*: A modular exchanger assembly consists of two or more exchanger blocks, welded together prior to attaching the headers, to form a single-piece exchanger. This form of construction is used when the customer's heat exchange requirements exceed the maximum block size which can be furnace brazed. Modular construction eliminates the need for costly piping to interconnect separate, individual exchangers.

b. *Cold box*: An economic alternative to separate installation on site is the assembly of various cryogenic plant components into a steel containment (cold box). Interconnecting piping, vessels, valves, and instrumentation are included in this packaged unit to form, after filling with insulation material, a ready-to-operate unit.

c. *Block-in-shell*: One or more heat exchanger blocks installed in a shell instead of a tube bundle. This is a highly efficient and economical alternative to standard shell and tube heat exchangers. The benefits include tighter approach temperatures, smaller in size and weight, lower capital costs, and lower operating costs. A typical block-in-shell is shown in Figures 3.41(a) and (b).

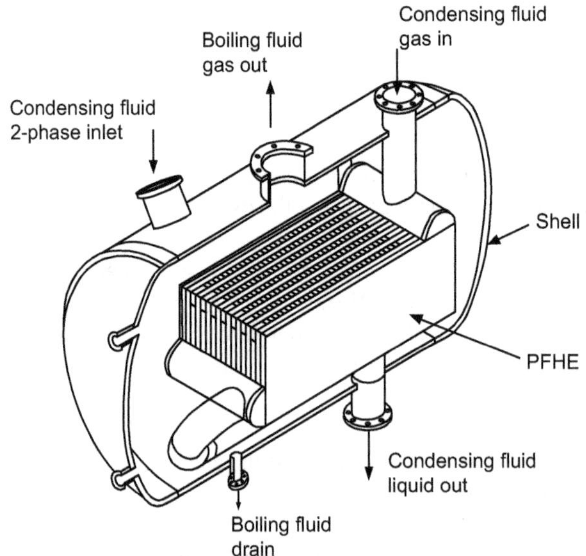

FIGURE 3.41 (a) Schematic of a block-in-shell PFHE and (b) Brazed aluminum heat exchanger: block-in-shell. (From Linde AG, Engineering Division. With permission.)

d. *Transition joint*: A transition joint is a bimetallic coupling used to make the transition from aluminum to stainless steel piping. Transition joints are available in various configurations.

e. *Water bath vaporizers*: Water bath vaporizers are used for the vaporization of all liquefied air gases, carbon dioxides, and hydrocarbons in single or multistream operations, including on request the pressure build-up vaporizer. The Linde water bath vaporizer consists of a water vessel in which a coil-wound tube bundle is submerged, optimized to withstand quick start-ups and temperature changes. The water bath is heated by direct steam injection or hot water circulation.

f. *Air-heated vaporizers*: For example, Linde air-heated vaporizers are heat exchangers for evaporating and superheating of cryogenic fluids, such as oxygen, nitrogen, argon, hydrogen, and carbon dioxide. A new generation of all-aluminum vaporizers ensures maximum air circulation due to optimized fin and vaporizer geometrics.

g. *Modular BAHX assembly*: A modular BAHX assembly consists of two or more individually brazed BAHX blocks that are welded together prior to attaching the headers to form a single-piece BAHX. This form of construction is used when the customer's heat exchange requirements exceed the maximum block size that can be furnace brazed. Modular construction eliminates the need for costly piping to interconnect separate, individual BAHX.

h. *Multiple BAHX assembly*: A multiple BAHX assembly, often referred to as a "battery," consists of two or more BAHX piped or manifolded together into a single assembly with the individual BAHX arranged either in a parallel, series, or combination parallel series arrangement. Multiple BAHX assemblies are used when the customer's heat transfer requirements are too large for either single-piece or modular BAHX construction.

i. *Core-in-Kettle® assembly.* A Core-in-Kettle® assembly consists of a cylindrical pressure vessel, usually carbon steel or stainless steel, which contains and supports one or more BAHX and associated piping including transition joints. In operation, one fluid is piped through the headered stream of the BAHX and the other partially fills the vessel and communicates with the open (unheadered) stream of the BAHX. Core-in-Kettle heat exchangers are designed to replace shell-and-tube heat exchangers with the direct benefits of lower installation costs, reduced operating costs, less replacement time, and reduced horsepower requirements. The Core-in-Kettle design is capable of achieving tight temperature approaches down to 2°F, thereby increasing plant capacity and reducing horsepower requirements.

3.5.14.7 Rough Estimation of the Core Volume

To obtain a quick indication of the heat exchanger volume required for a certain duty, the following simple relation as per ALPEMA standard may be used [44]:

$$V = \frac{Q/\text{MTD}}{\Gamma} \tag{3.59}$$

where

- V is the required volume of heat exchanger (without headers)
- Q is the overall heat duty
- MTD is the mean temperature difference between streams
- Γ is the coefficient: 100,000 for hydrocarbon applications; 50,000 for air separation applications

The values of 100,000 and 50,000 represent the product, UAd, assuming an overall heat transfer coefficient of 200 W/m^2 and 100 W/m^2 K, respectively, and a mean geometric heat transfer surface density of 500 m^2/m^3.

The weight of a complete heat exchanger may be assumed to be 1000 kg per unit core volume (m^3). This value varies in practice between 650 and 1500 kg/m^3.

3.5.14.8 Provisions for Thermal Expansion and Contraction

Provisions for thermal expansion and contraction of the heat exchanger in the horizontal plane at the support location must be provided. The expected thermal movement should be calculated in both horizontal directions by the following equations:

$$\Delta L = 12.6\times 10^{-6}\times L\times \Delta T \text{ in.} \quad (\text{in FPS units}) \tag{3.60}$$

$$\Delta L = 97.1\times 10^{-6}\times L\times \Delta T \text{ m} \quad (\text{in SI units}) \tag{3.61}$$

where

L is the distance in m/in. between extreme bolts in the direction under consideration

ΔT is the change in temperature in °C(°F) at the support location from the installed (ambient) temperature to the coldest possible operating temperature

ΔL is the expected thermal movement in m/in. which will result from this calculation

If the expected thermal movement in both directions is 12.7 mm (0.5″) or less, the bolt hole diameters in the aluminum support angles should be oversized by adding the maximum expected thermal movement to the bolt diameter. If the expected thermal movement exceeds 12.7 mm (0.5 in.) in one of the horizontal directions, a slotted hole should be used with a slot length equal to the bolt diameter plus the maximum expected thermal movement.

3.5.14.9 Mechanical Design of Brazed Aluminum Plate-Fin Heat Exchangers

The design of a brazed aluminum PFHEs is the result of the mechanical strength analysis of

i. The plate-fin structure under pressure
ii. The influence of headers on the plate-fin structure
iii. The header/nozzle assembly

Pressure vessel codes generally do not contain formulae for the fins. The calculation methods used by manufacturers have been approved by the applicable code authority. The calculated stresses are compared with the maximum allowable stress as per the code.

3.5.14.10 Codes

BAHXs are normally designed and manufactured in accordance with Section VIII, Division I of the ASME Pressure Vessel Code, carry the "U" stamp, and are registered with the National Board of Boiler and Pressure Vessel Inspectors. Manufacturers can manufacture to other international standards, such as PED European Pressure Equipment Directive, AD-Merkblatter, Australian Standard, British Standard, Chinese Standard, CODAP, Japanese Industrial Standards, Raccolta/VSR, Stoomwezen Grondslagen, Swedish Pressure Vessel Code, Russian Gost, etc.

Associated piping is normally designed and manufactured in accordance with the ASME B31.3 Piping Code. The ASME pressure vessel and piping code boundaries are indicated on the drawing. Heat exchangers and piping are sometimes designed and manufactured to other national codes. The governing national code is specified on the drawing and exchanger nameplate.

3.5.14.11 Construction Materials

Typical materials for use in the construction of brazed aluminum PFHEs are (1) core matrix (fins, plates, side bars) is aluminum alloy AA 3003 and (2) headers/nozzles are aluminum alloy AA 5083 and their mean metal temperature limitations as per ASME code are alloy AA 3003 is −269/+204°C and that of AA 5083 is −269/+65°C.

3.5.14.12 Manufacture

After forming, cutting, and cleaning of the single parts, the block is stacked and prepared for high vacuum brazing at about 600°C or NOCOLOK flux brazing. For welding of headers and nozzles to the core, proven welding procedures, such as GTAW and GMAW, are used.

3.5.14.13 Quality Assurance Program and Third-Party Inspection

Quality management is an essential part of all manufacturers' strategy. The companies are certified to various national and international standards. Acceptance inspection is carried out by the firm's own specialists and further by experts from various international inspection organizations/agencies, such as TÜV, Lloyd's Register, Stoomwezen, Bureau Veritas, Det Norske Veritas, etc.

3.5.14.14 Testing of BAHX

Extensive test procedures are carried out on the completely assembled heat exchangers. These include radiographic and dye penetrant testing of weld seams, pressure test, and helium leak tests and, if requested, flow tests. A BAHX is first structurally tested. For hydrostatic test methods, each stream is pressurized, with the other streams at zero pressure, to 1.3 times the design pressure as per the requirements of ASME. The unit remains pressurized for the stipulated period as per the code. Pneumatic test methods can be used with the pressure level being adjusted to 1.1 times design pressure but exceeding a maximum of 150 psi. Leak testing is performed after the structural test has been successfully performed. Internal leak testing is performed by pressurizing each stream individually and monitoring small nozzle valves on the unpressurized streams using a soap film solution. External leak testing is performed using soap solution on the exchanger external joints. For critical applications, helium vacuum leak testing may also be used to validate the leak tightness of an exchanger. The leak testing arrangement is shown in Figure 3.42. This method consists of pressurizing the system with a high-pressure gas, usually dry air or nitrogen. Then the part is isolated from the gas supply and, after a stabilizing period, its internal pressure is monitored over time. From the amount of pressure loss and the length of time for the loss to occur, the leak rate can be calculated. If the pressure remains the same, that component is leak-free.

3.5.14.15 Guarantees

Manufacturers shall offer a thermal performance guarantee and a mechanical guarantee. Details shall be agreed upon by the purchaser and the manufacturer.

3.5.14.16 ALEX: Aluminum Fin-Steel Tube System

ALEX is the trade name of a BAHX developed independently by Kobe Steel in 1963 for use as the regenerator in air separation plants. The core tube is steel with aluminum coated in a flat shape with parallel sides and smooth fins having a reduced tendency for fouling. Kobe Steel is one of the members of ALPEMA standards and is the top user of ALEX for cold box for LNG applications, core-in-drum type heat exchangers, etc. A small air-cooled condenser ALEX

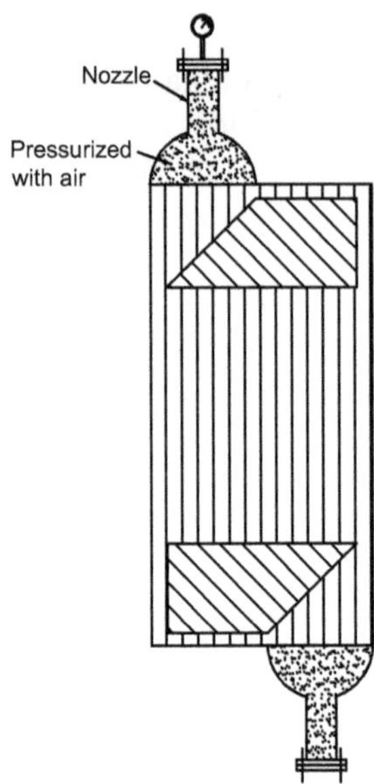

FIGURE 3.42 PFHE Leak testing arrangement. (*Note*: A PFHE without cover sheet is shown. But actual testing will be performed in a fully completed unit only.)

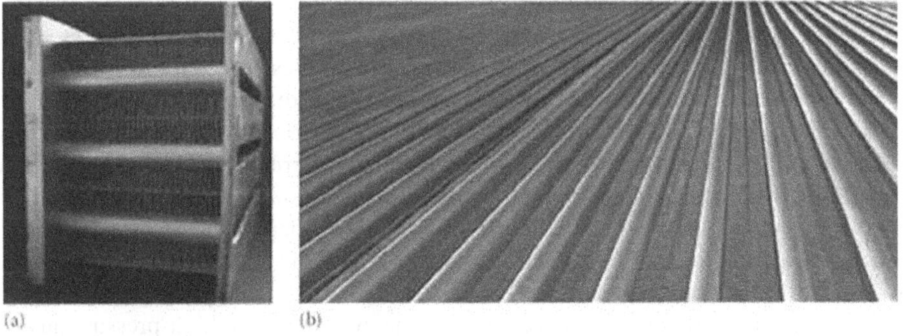

FIGURE 3.43 Aluminum Exchanger (ALEX) core—(a) mini core and (b) aircooled heat exchanger with single row type with steel tubes and ALEX core. (Courtesy of GEA Power Cooling, Inc., Lakewood, CO.)

core is shown in Figure 3.43a, and an air-cooled condenser made of ALEX core is shown in Figure 3.43b.

3.5.15 Comparison of Salient Features of Plate-Fin Heat Exchangers and Coil-Wound Heat Exchanger

Both the brazed aluminum PFHEs and coiled tube heat exchangers find application in liquefication processes. The salient features of these two types of heat exchangers are compared in Table 3.2 for the intended service.

3.5.16 Heat Exchanger Specification Sheet for Plate-Fin Heat Exchanger

A heat exchanger specification sheet for PFHE is shown in Figure 3.44.

TABLE 3.2
Comparison of Salient Features of Plate-Fin Heat Exchangers and Coil-Wound Heat Exchangers

	Plate-fin heat exchangers	Coil-wound heat exchangers
Characteristics	Extremely compact, multiple streams, single- and two-phase streams	Compact, multiple streams, single- and two-phase streams
Fluid	To be very clean	To be clean
Flow types	Counterflow, crossflow	Cross counterflow
Heating surface compactness	300–1400 m^2/m^3	20–300 m^2/m^3
Materials	Aluminum	Aluminum, stainless steel (SS), carbon steel (CS), special alloys
Temperature	−269°C to +65°C (150°F)	All
Pressure	Up to 115 bar (1660 psi)	Up to 250 bar (3625 psi)
Applications	Cryogenic plants, noncorrosive fluids. Very limited installation space	Also for corrosive fluids and thermal shocks

Source: Linde AG, Engineering Division, Germany. With permission.

3.5.17 Operation and Maintenance

3.5.17.1 Field Testing and Repair

On-site repair of plate-fin heat exchangers depends on the type of damage. Seal welding of leakages or blocking of damaged passages may apply. It is recommended that any repair be carried out only by the manufacturer or a recognized specialist repair team. All repairs should be accompanied by thermal and hydraulic reviews.

Maximum working pressures and temperatures are always specified on the manufacturer's nameplate. These values should not be exceeded during field testing or operation. Since it is extremely difficult to dry brazed aluminum heat exchangers in the field, only a clean dry gas should be used for leak testing. Internal leaks in a brazed aluminum heat exchanger are generally indicated by a change of purity in any of the fluid streams.

External leaks can be determined by sight, smell, audible sounds of leaking fluid, external gas monitoring equipment, or localized cold spots appearing on the external insulation. An air-soap test is effective for locating external leaks. An air test with soap applied to nozzle connections or a nitrogen-freon test can be used to identify the streams involved in an internal or cross pass leak. Internal and external leaks usually can be repaired by blocking layers, making localized external welds, etc.

3.5.17.2 Leak Test

In order to ascertain the absence of a leak from one chamber toward any other chamber or into the atmosphere, a leak test is necessary. The two methods normally followed are (i) air test and (ii) helium test for external leak test and inter-stream leak test. A schematic of a leak test is shown in Figure 3.42. Qualified manufacturers' representatives are usually required to establish the exact location of an internal leak and to make any repairs. Details of field testing and repair can be found in Refs. [43,48].

3.5.17.3 The Lifetime of a Plate-Fin Heat Exchanger

Under steady-state conditions the lifetime of a plate-fin heat exchanger is comparable to that of other heat exchange equipment types. Situations that can reduce the BAHX lifetime include operation

Fives Cryo
Design and Sales Department
25 bis, rue du Fort - B.P.87
88194 Golbey Cedex - FRANCE
Tel : +33 (0)3 29 68 00 01

fives cryogenie

HEAT EXCHANGER SPECIFICATION SHEET

CUSTOMER :	PROJECT :	N.C ORDER No. :
ITEM No. :	LOCATION :	CUST. JOB.No. :
	PLANT SERVICE :	CONSTRUCTION CODE :

01	Fluid								
02	Total Flow rate	kg/h							
03	Vapor Flow rate In	kg/h							
04	Vapor Flow rate Out	kg/h							
05	-MCL WGT. In Out								
06	Liquid Flow rate In	kg/h							
07	Liquid Flow rate Out	kg/h							
08	-MCL WGT. In Out								
09	Temperature In	°C							
10	Temperature Out	°C							
11	Dew Point / Bubble Point								
12	Operating Pressure	bar a							
13	Allowable Pressure Drop	mbar							
14	Total Heat Transferred	MW							
15	Corrected Mtd (Global)	°C							
16	Fouling Factor	Km/W							
17	Design Temperature	°C	-196 / +65						
18	Design Pressure	bar g							
19	Hydraulic Test Pressure	bar g							
20	Number of Assemblies :		Width :	mm	Type of Heat Exchanger :				
21	No. of Cores/Assembly :		Height :	mm	Total No. of Layers / Core:				
22	Total Number of Cores :		Length :	mm	Parting Sheets (Ext.6.00 mm):		mm		
23	No. of Passages / Core								
24	Effective Passage Width	mm							
25	Types of Fins								
26	Fin Height	mm							
27	Fin Thickness	mm							
28	Number of Fins Per Inch	FPI							
29	Effective Passage Length	mm							
30	Total Heat Transfer Area	m^2							
31	Total Free Flow Area	cm^2							
32	Calculated Pressure Drop	mbar							
33	Header Size In Out	mm							
34	Nozzle Size In Out	mm							
35	Submanifold Size In Out	mm							
36	Manifold Size In Out	mm							
37	Connections In Out	Type							
38	Aluminium Ansi Flange	Class							
39	Transition Joints								
40									
41	Notes :								

0			
Rev.	Date	Issued By	Approved By

FIGURE 3.44 PFHE specification sheet N°. (Courtesy of Fives Cryo, Golbey, France.)

outside recommended operation conditions, even occasionally, and repairs. Great care should be taken in performance monitoring and leak detection under these conditions.

3.5.17.4 Monitoring the Operating Status of Exchangers

ALPEMA [44] recommends pressure and temperature measurement devices on every stream, preferably at the inlet and outlet of each stream in the heat exchanger. Flow measurements are desirable when practical. Flow rates can often be estimated from a material balance. A user should evaluate what instrumentation is appropriate and cost effective.

3.5.17.5 Mercury-Tolerant Construction

Under certain process conditions, liquid elemental mercury can have severe detrimental effects on unprotected BAHX. Chart was the first manufacturer to identify and address the conditions where mercury contamination in the feed gas could be harmful to aluminum heat exchangers. Chart's proprietary mercury-tolerant solution combines multiple design, material selection, and fabrication features [49].

3.5.17.6 Asset Integrity Management (AIM)

Asset Integrity Management (AIM) is a proactive preventative maintenance program that enables plant stakeholders to maximize the reliability and integrity of their assets. It combines multiple activities as given below [50]:

1. Site visit by skilled personnel to evaluate the current health of equipment fleets
2. Historic analysis of plant distributed control system (DCS) to identify any operating patterns outside the manufacturer's guidelines
3. Continued analysis to mitigate against future non-ideal operating practices
4. Corrective action recommendations
5. Process optimization recommendations to support operational challenges
6. Future event planning support, such as shutdowns and debottlenecking
7. Operator training and education
8. 24/7 emergency support

3.5.17.7 Lifecycle OEM Service and Support

Annual Service Plans. The Lifecycle Service Program will mitigate the risk of downtime and the associated lost production and revenue [51]:

- i. Assess current condition of the BAHX inventory
- ii. Evaluate current operations versus BAHX recommended operating guidelines
- iii. Plant process optimization recommendations to support operational challenges
- iv. Mitigate current and future risk of failure
- v. Support customers in managing BAHX inventory fleet
- vi. Plans for necessary cleaning and repairs and future planned core replacements
- vii. Detailed analysis of operational and shutdown data
- viii. Annual site inspection visits that include operator training
- ix. Emergency assistance to sites located overseas

3.5.18 Surface Geometrical Relations for PFHE

3.5.18.1 Surface Geometrical Parameters: General

The following are some of the basic relationships between the surface and core geometries on one side of the compact heat exchangers.

a. Hydraulic Diameter, D_h

The generalized relation for hydraulic diameter is given by

$$D_h = \frac{4A_o L}{A} \tag{3.62}$$

where
- A is the total heat transfer area
- A_o is the free flow area
- L is the flow length

The hydraulic radius r_h is given by $D_h/4$.

b. Surface Area Density α and σ

For a compact heat exchanger with a secondary surface on one side, it is customary to designate the ratio of the total heat transfer surface area on one side of the exchanger to total volume of the exchanger by α, and the ratio of the free flow area to the frontal area by σ. Thus,

$$\alpha_1 = \frac{A_1}{V} \quad \text{and} \quad \alpha_2 = \frac{A_2}{V} \tag{3.63}$$

$$\sigma_1 = \frac{A_{0.1}}{A_{fr,1}} = \left(\frac{AD_h/4}{A_{fr}L}\right)_1 = \frac{(AD_h/4)_1}{V} \tag{3.64a}$$

$$\sigma_2 = \frac{A_{0.2}}{A_{fr,2}} = \left(\frac{AD_h/4}{A_{fr}L}\right)_2 = \frac{(AD_h/4)_2}{V} \tag{3.64b}$$

From the definition of hydraulic diameter D_h, the relation between hydraulic diameter, surface area density α, and σ is given by

$$\left(\frac{D_h}{4L}\right)_1 = \left(\frac{A_o}{A}\right)_1 = \left(\frac{\sigma}{L\alpha}\right)_1 \tag{3.65a}$$

$$\left(\frac{D_h}{4L}\right)_2 = \left(\frac{A_o}{A}\right)_2 = \left(\frac{\sigma}{L\alpha}\right)_2 \tag{3.65b}$$

and

$$D_{h,1} = 4\frac{\sigma_1}{\alpha_1}, \quad D_{h,2} = 4\frac{\sigma_2}{\alpha_2} \tag{3.66}$$

3.5.19 Surface Geometrical Properties of PFHE

The following geometrical properties on each side are needed for the heat transfer and pressure drop analysis of a PFHE [44]:

1. The primary and secondary surface area (if any), A_p and A_s, respectively
2. Minimum free flow area, A_o
3. Frontal area, A_{fr}
4. Hydraulic diameter, D_h
5. Flow length, L
6. Fin dimensions—thickness and length (height)

3.5.19.1 Heat Transfer Area

The total heat transfer area consists of the sum of the primary and secondary heat transfer areas and their calculation method is given later. The primary heat transfer surface within the heat exchanger consists of the bare parting sheet and the fin base directly brazed to the parting sheet and the secondary heat transfer surface is provided by the fins. The cross-sectional view of the fin and parting sheet is shown in Figure 3.45. The effectiveness of the secondary surface to transfer heat is given by the fin efficiency. The per unit area of each parting sheet is given by [44]

a. Primary surface area, A_p is given by $2(1 - nt_f)$ (3.67)
b. Secondary surface, A_s is given by $2n\ (l_f - 1)$ (3.68)

where
n is the fin density (m^{-1}), i.e. number of fins per unit length
t_f is the fin thickness (m)
l_f is the fin height (m)

The effective heat transfer surface area for a passage can be estimated by the following equation:

$$A = A_p + \eta\phi A_s \tag{3.69}$$

Single blanking and fin efficiency, η is given by

$$\eta = \frac{\tan h(\beta/2)}{\beta/2} \tag{3.70}$$

where
A_p is the primary heat transfer surface of a stream (refer to Figure 3.45)
A_s is the secondary heat transfer surface of a stream (refer to Figure 3.45)
η is the passage fin efficiency for single banking
l_f is the fin height
t_f is the fin thickness
n is the fin density

$$\beta = h\left(\frac{2\alpha}{k_f t_f}\right)^{0.5} \tag{3.71}$$

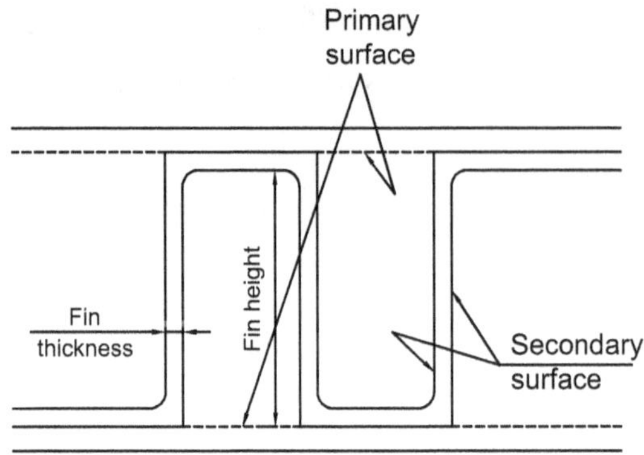

FIGURE 3.45 Primary and secondary heat transfer area for a unit flow passage.

α is the effective heat transfer coefficient of a stream
ϕ is the unperforated fraction (1 – ratio of perforated area to the total fin area)/100
k_f is the thermal conductivity of fin material
h is the heat transfer coefficient

The thermal length of a brazed aluminum PFHE is defined as the effective length of the finned region between, but not including, the distributors.

The information not specified directly is computed based on the specified surface geometries and core dimensions. Consider a PFHE as shown in Figure 3.25b in which L_1 and L_2 are the flow lengths, N_p and $N_p + 1$ are the number of flow passages on sides 1 and 2, respectively, and V is the total core volume. The following formulas define some of the geometrical properties on any one side of the PFHE. These geometrical relations are derived in Refs. [15,52,53].

The volume between plates on each side is given by

$$V_{p,1} = L_1L_2\left(b_1N_p\right) \quad V_{p,2} = L_1L_2b_2\left(N_p + 1\right) \tag{3.72}$$

Heat transfer area on each side is given by

$$A_1 = \beta_1V_{p,1} \quad A_2 = \beta_2V_{p,2} \tag{3.73}$$

where β_1 and β_2 are the surface area density on each side per unit volume between plates.

The ratio of minimum free flow area to frontal area (σ) on side 1 is given by

$$\sigma_1 = \frac{A_{0,1}}{A_{fr,1}} = \frac{A_{0,1}L_1}{A_{fr,1}L_1} = \frac{A_1D_{h,1}/4}{V} = \frac{V_{p,1}\beta_1D_{h,1}/4}{V} \tag{3.74a}$$

$$= \frac{b_1N_p\beta_1D_{h,1}/4}{b_1N_p + b_2\left(N_p + 1\right) + 2a\left(N_p + 1\right)} \approx \frac{b_1\beta_1D_{h,1}/4}{b_1 + b_2 + 2a} \tag{3.74b}$$

$$\sigma_2 = \frac{b_2\left(N_p + 1\right)\beta_2D_{h,2}/4}{b_1N_p + b_2\left(N_p + 1\right) + 2a\left(N_p + 1\right)} \approx \frac{b_2\beta_2D_{h,2}/4}{b_1 + b_2 + 2a} \tag{3.74c}$$

where a is the thickness of parting plates. The ratio is given similarly for side 2. Here, the last approximate equality is for the case when N_p is very large or the numbers of passages on the two sides are the same.

The heat transfer surface area on one side divided by the total volume V of the exchanger, designated as α_1, is

$$\alpha_1 = \frac{A_1}{V} = \frac{4\sigma_1}{D_{h,1}} = \frac{b_1\beta_1}{b_1 + b_2 + 2a} \tag{3.75a}$$

Similarly, α_2 is given by

$$\alpha_2 = \frac{A_2}{V} = \frac{4\sigma_2}{D_{h,2}} = \frac{b_2\beta_2}{b_1 + b_2 + 2a} \tag{3.75b}$$

3.5.20 Correlations for *j* and *f* Factors of Plate-Fin Heat Exchangers

3.5.20.1 Offset Strip Fin Heat Exchanger

Wieting correlations [54] consist of power-law curve fits of the j and f values for 22 heat exchanger surface geometries over two Reynolds number ranges: $\mathrm{Re}_h \le 1000$, which is primarily laminar, and $\mathrm{Re}_h \ge 2000$, which is primarily turbulent.

For $\mathrm{Re}_h \le 1000$,

$$j = 0.483\left(\frac{l_f}{D_h}\right)^{-0.162} \xi^{-0.184}\, \mathrm{Re}_h^{-0.536} \tag{3.76}$$

$$f = 7.661\left(\frac{l_f}{D_h}\right)^{-0.384} \xi^{-0.092}\, \mathrm{Re}_h^{-0.712} \tag{3.77}$$

For $\mathrm{Re}_h \ge 2000$,

$$j = 0.242\left(\frac{l_f}{D_h}\right)^{-0.322} \left(\frac{t_f}{D_h}\right)^{0.089} \mathrm{Re}_h^{-0.368} \tag{3.78}$$

$$f = 1.136\left(\frac{l_f}{D_h}\right)^{-0.781} \left(\frac{t_f}{D_h}\right)^{0.534} \mathrm{Re}_h^{-0.198} \tag{3.79}$$

where Re_h is the Reynolds number based on hydraulic diameter D_h given by

$$D_h = \frac{2sh_f}{s + h_f} \tag{3.80}$$

and α is the aspect ratio, s/h_f (refer to Figure 3.46).

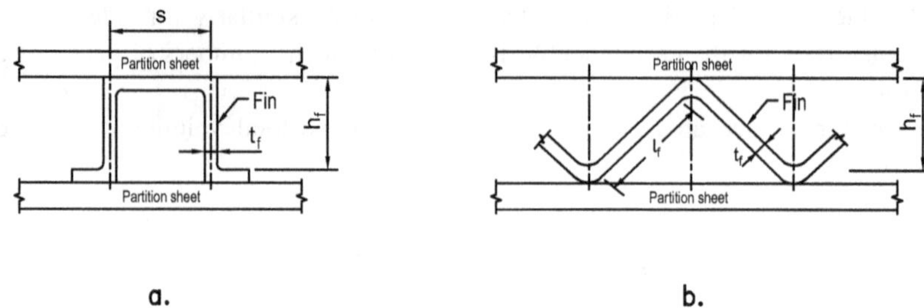

FIGURE 3.46 Fin length (a) Offset strip fin(rectangular fin) and (b) triangular fin.

Joshi and Webb [47] developed a more sophisticated theoretical model to predict the characteristics of j and f versus Reynolds number for the OSF. This model preferably includes all geometrical factors of the array (s/d, t_f/l_f, t_f/s) and heat transfer to the base surface area to which the fins are attached.

Manglik and Bergles [55] reviewed the available j and f data and developed improved correlations for j and f, with a single predictive equation representing the data continuously from laminar to turbulent flow. Their correlations for j and f are as follows:

$$j = 0.6522\,\mathrm{Re}_\mathrm{h}^{-0.5403}\xi^{-0.1541}\delta^{0.1499}\eta^{-0.0678} \times\left\{1+5.269\times10^{-5}\,\mathrm{Re}_\mathrm{h}^{1.340}\xi^{0.504}\delta^{0.456}\eta^{-1.055}\right\}^{0.1} \tag{3.81}$$

and

$$f = 9.6243\,\mathrm{Re}_\mathrm{h}^{-0.7422}\xi^{-0.1856}\delta^{0.3053}\eta^{0.2659} \times\left\{1+7.6669\times10^{-8}\,\mathrm{Re}_\mathrm{h}^{4.429}\xi^{0.92}\delta^{3.767}\eta^{0.236}\right\}^{0.1} \tag{3.82}$$

where ξ is the aspect ratio = s/h_s; $\delta = t_f/l_f$; $\eta = t_f/s$; and Re_h is based on the hydraulic diameter D_h given by the following equation:

$$D_\mathrm{h} = \frac{4sh_f l_f}{2\left(sl_f + h_f l_f + t_f h_f\right) + t_f s} \tag{3.83}$$

These equations correlate with the experimental data for the 18 cores within ±20%.

3.5.20.2 Louvered Fin

Based on tests of 32 louvered-fin geometries, Davenport [16] developed multiple regression correlations for j and f versus Re as follows:

$$j = 0.249\mathrm{Re}_\mathrm{lp}^{-0.42} l_\mathrm{h}^{0.33} H_1^{0.26}\left(\frac{l_1}{H_1}\right)^{1.1} \quad 300 < \mathrm{Re} < 4000 \tag{3.84}$$

$$f = 5.47\mathrm{Re}_\mathrm{lp}^{-0.72} l_\mathrm{h}^{0.37} H_1^{0.23} l_\mathrm{p}^{0.2}\left(\frac{l_1}{H_1}\right)^{0.89} \quad 70 < \mathrm{Re} < 1000 \tag{3.85a}$$

$$f = 0.494\text{Re}_{\text{lp}}^{-0.39} H_{\text{l}}^{0.46} \left(\frac{l_h}{H_{\text{l}}}\right)^{0.33} \left(\frac{l_{\text{l}}}{H_{\text{l}}}\right)^{1.1} \quad 1000 < \text{Re} < 4000 \tag{3.86}$$

where

Re_{1p} is the Reynolds number based on louver pitch
l_{h} is the louver height
l_{l} is the louver length
l_{p} is the louver pitch
H_l is the fin height

Here, 95% of the data could be correlated within ±6% for heat transfer, and within ±10% for the friction factor. The required dimensions are in millimeters. Figure 3.34 defines the terms in the equations. The characteristic dimension in the Reynolds number is louver pitch, l_{p}, and not D_{h}.

3.5.20.3 Pin Fin Heat Exchangers

For the heat transfer characteristics of pin-finned surfaces, refer to Kays [56], Norris et al. [57], and Kanzaka et al. [58], among others. The results of Ref. [56] were presented in Kays and London [15]. The works of Kays [56] and Norris [57] are limited to the heat transfer characteristics of pin fins having their representative length comparable to those of louvered or offset fins currently used [58]. To materialize a heat exchanger that has thermal performance superior to those of offset or louvered fins requires a more finely segmented surface of less than 1 mm at maximum. Kanzaka et al. [58] studied the heat transfer and fluid dynamic characteristics of pin-finned surfaces of various profiles, namely, circular, square, and rectangular, and both inline and staggered arrangements.

Staggered arrangements: For their test element, the expression for the Nusselt number is

$$\text{Nu}_\phi = 0.662\ \text{Re}_\phi^{0.5}\text{Pr}^{1/3} \tag{3.87}$$

Inline arrangements: The expression for the Nusselt number is

$$\text{Nu}_\phi = 0.440\ \text{Re}_\phi^{0.5}\text{Pr}^{1/3} \tag{3.88}$$

where ϕ is the characteristic length. The expressions for Nusselt number and $d\phi$ are

$$\text{Nu}_\phi = \frac{hd_\phi}{k} \quad \text{Re}_\phi = \frac{Gd_\phi}{\mu} \tag{3.89}$$

$$d_\phi = \frac{\pi d_{\text{p}}}{2} \quad \text{for circular fins} \tag{3.90a}$$

$$d_\phi = \frac{a_{\text{p}} + b_{\text{p}}}{2} \quad \text{for rectangular fins} \tag{3.90b}$$

where

a_{p} and b_{p} are the rectangular pin fin dimensions (i.e., cross section is $a_{\text{p}} \times b_{\text{p}}$)
d_{p} is the circular fin diameter

3.6 FIN EFFICIENCY

Fins are primarily used to increase the surface area and consequently to enhance the total heat transfer rate. Both conduction through the fin cross section and convection from the surface area take place. Hence, the fin surface temperature is generally lower than the base (prime surface) temperature if the fin convects heat to the fluid. Due to this, the fin transfers less heat than if it had been at the base temperature. This is described by the fin temperature effectiveness or fin efficiency η_f. The fin temperature effectiveness is defined as the ratio of the actual heat transfer, q, through the fin to that which would be obtained, q_i, if the entire fin were at the base metal temperature:

$$\eta_f = \frac{q}{q_i} \tag{3.91}$$

The fin efficiency for plate-fin surfaces in heat exchanger design can be determined from Gardner [59], Kern and Kraus [60], Scott and Goldschmidt [61], Schmidt [62], Zabronsky [63], Lin and Sparrow [64], Shah [16,65], and others. The fin efficiency of a plain rectangular profile cross section is discussed in this section.

3.6.1 Fin Length for Some Plate-Fin Heat Exchanger Fin Configurations

The two most commonly used plate-fin geometries are rectangular and triangular passages, shown in Figure 3.46. From a review of Figure 3.46, the fin length l_f for a rectangular passage is

$$l_f = \frac{h_f - t_f}{2} \approx \frac{h_f}{2} \tag{3.92}$$

Note that no effect of the fin inclination is taken into account in Equation 3.92 for the triangular fin. Numerous fin geometries and corresponding fin lengths for various flow passage configurations are presented by Shah [53].

3.6.2 Circular Fin

Kern and Kraus [60] *formula for rectangular-profile circular fin*: For circular fins (Figure 3.47), the fin efficiency is given by

$$\eta_f = \frac{2r_o}{m\left(r_e^2 - r_o^2\right)}\left[\frac{I_1(mr_e)K_1(mr_o) - K_1(mr_e)I_1(mr_o)}{I_o(mr_o)K_1(mr_e) + I_1(mr_e)K_o(mr_o)}\right] \tag{3.93}$$

where

$m = (2h/k_f t_f)^{0.5}$

I_n is the modified Bessel function of the first kind (nth order)

K_n is the modified Bessel function of the second kind (nth order)

k_f is the fin material thermal conductivity

r_e is the fin radius

r_o is the tube outer radius

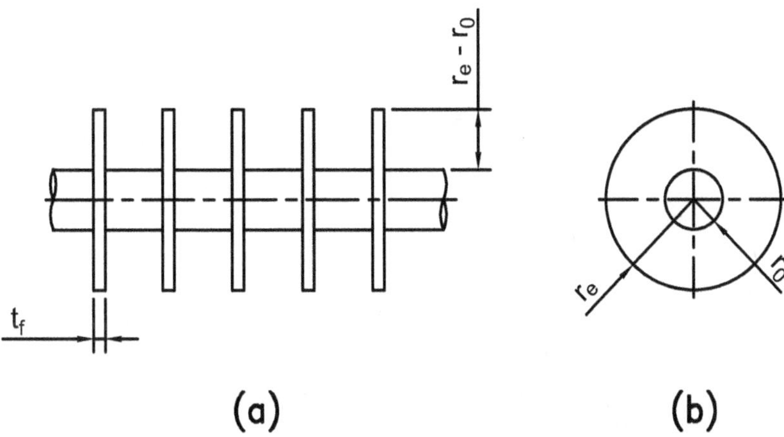

FIGURE 3.47 Circular fin details to calculate fin efficiency as per Equations 3.93, 3.97 and 3.98a and 3.98b.

t_f is the fin thickness
h is the heat transfer coefficient

The fin efficiency expressed by Equation 3.93 does not lend itself to comparison with the efficiencies of fins of other radial profiles but can be adjusted by expressing the efficiency in terms of the radius ratio r_o/r_e and a parameter ϕ:

$$\rho = \frac{r_o}{r_e} \tag{3.94}$$

$$\phi = \left(r_e - r_o\right)^{3/2}\left(\frac{2h}{k_f A_p}\right)^{0.5} \tag{3.95}$$

where A_p is the profile area of the fin

$$A_p = t_f\left(r_e - r_o\right) \tag{3.96}$$

Substituting the values of ρ and ϕ in Equation 3.93, the resulting formula for fin efficiency is given by

$$\eta_f = \frac{2\rho}{\phi(1+\rho)}\left[\frac{I_1\left[\phi/(1-\rho)\right]K_1\left[\rho\phi/(1-\rho)\right] - K_1\left[\phi/(1-\rho)\right]I_1\left[\rho\phi/(1-\rho)\right]}{I_0\left[\rho\phi/(1-\rho)\right]K_1\left[\phi/(1-\rho)\right] + I_1\left[\phi/(1-\rho)\right]K_0\left[\rho\phi/(1-\rho)\right]}\right] \tag{3.97}$$

Scott and Goldschmidt: An excellent approximation of Equation 3.97 has been provided by Scott and Goldschmidt [61] as follows:

$$\eta_f = x\left(ml_e\right)^{-y} \quad \text{for } \phi > 0.6 + 2.257\left(r^*\right)^{-0.445} \tag{3.98a}$$

$$\eta_{\mathrm{f}} = \frac{(\tanh \phi)}{\phi} \quad \text{for } \phi \le 0.6 + 2.257\left(r^*\right)^{-0.445} \tag{3.98b}$$

where

$$x_1 = \left(r^*\right)^{-0.246} \tag{3.99}$$

$$\phi = ml_e \left(r^*\right)^{\exp\left(0.13ml_e - 1.3863\right)} \tag{3.100}$$

$$y_i = 0.9107 + 0.0893\, r^* \quad \text{for } r^* \le 2 \tag{3.101a}$$

$$= 0.9706 + 0.17125 \ln r^* \quad \text{for } r^* > 2 \tag{3.101b}$$

$$r^* = \frac{r_{\mathrm{e}}}{r_{\mathrm{o}}}, \tag{3.102}$$

$$l_{\mathrm{e}} = l_{\mathrm{f}} + \frac{t_{\mathrm{f}}}{2} \tag{3.103}$$

Schmidt method: The Schmidt [62] formula for the efficiency of a plane circular fin is given by

$$\eta = \frac{\tanh ml^*}{ml^*} \tag{3.104}$$

where l^* is given by

$$l^* = \left(r_{\mathrm{e}} - r_{\mathrm{o}}\right)\left[1 + \frac{t_{\mathrm{f}}}{2\left(r_{\mathrm{e}} - r_{\mathrm{o}}\right)}\right]\left[1 + 0.35 \ln(\rho)\right] \tag{3.105}$$

in which $\rho = r_{\mathrm{e}}/r_{\mathrm{o}}$. It was shown that for $0.5 \le \rho \le 1$ and $1 \le \rho \le 8$, the error using Equation 3.104 is less than 1% of the exact value of the fin efficiency. Because the fin efficiency is very close to unity for typical low-finned tube geometries, Rabas and Taborek [7] do not recommend any corrections to this method to account for departure from a rectangular profile and for the nonuniformity of the heat transfer coefficient over the height of the fin.

3.6.3 Plain Continuous Fin on Circular Tubes

A recurring arrangement of extended surface is that of a single sheet of metal pierced by round tubes in either a square or equilateral triangular array. For polygonal fins, the methods suggested in Refs. [16,62–64] can be used.

Schmidt method: Schmidt [62] extended the circular fin efficiency formula expressed in Equation 3.104 to rectangular and hexagonal layout fin surfaces as shown in Figure 3.48. The extension was

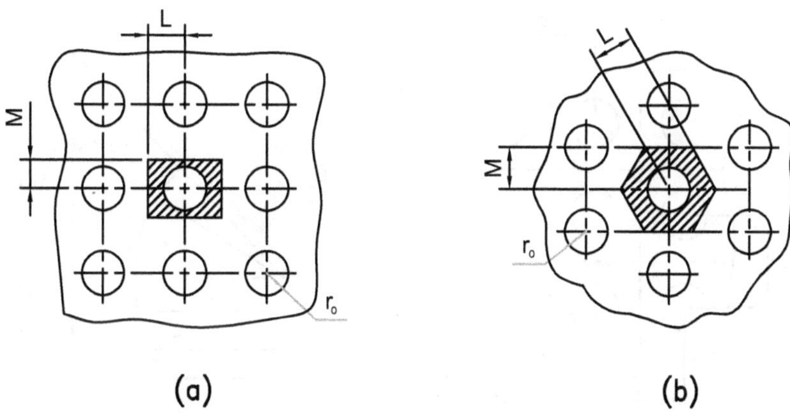

FIGURE 3.48 Geometrical details of continuous fin-tube heat exchanger layouts. (a) Rectangular tube array and (b) hexagonal tube array.

made by analyzing these surfaces on the basis of equivalent circular fins. The expression for the ratio of the equivalent fin radius to the tube radius, ρ_e, is expressed as follows:

For rectangular fins,

$$\rho_e = 1.28\lambda_1 (\beta_1 - 0.2)^{0.5} \tag{3.106}$$

For hexagonal fins,

$$\rho_e = 1.27\lambda_1 (\beta_1 - 0.3)^{0.5} \tag{3.107}$$

where

$$\lambda_1 = \frac{M}{r_o} \tag{3.108}$$

$$\beta_1 = \frac{L}{M} \tag{3.109}$$

and where L and M are defined in Figure 3.48. Of these two, L must always be the larger dimension. The fin efficiencies calculated in this way always fall within the known upper and lower limits defined by the inner and outer circumscribing circles and are very close to that of a circular fin of equal area. The accuracy of this method is especially good for square or regular hexagonal fins, where $L = M$.

Sparrow and Lin method: The Sparrow and Lin method [64] considered both the square fin and hexagonal fin formed by tubes on equilateral triangular pitch as shown in Figure 3.49. For the fins shown in Figure 3.50, the fin efficiency is obtained from the numerical temperature solution and is conveniently represented in terms of a fictitious edge radius r_e^* that corresponds to a radial fin having the same surface area as the square and hexagonal fins under consideration. This fictitious edge radius for the square fin is given by

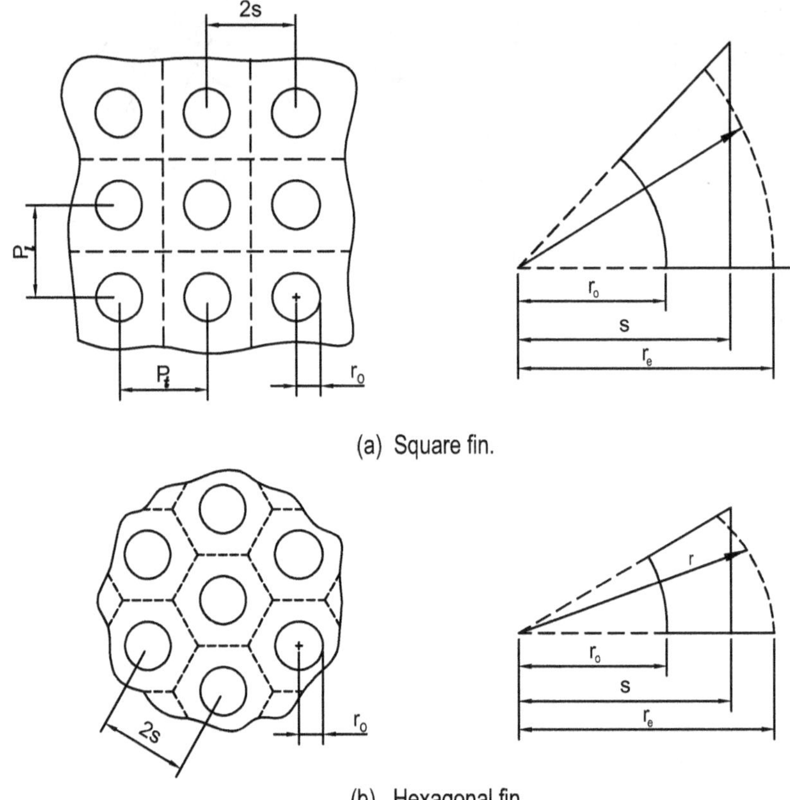

FIGURE 3.49 Fictitious edge radius for unit fin surface of continuous fin with circular tube. (a) Square layout (i.e., square fin) and (b) staggered layout (60°) (i.e., hexagonal fin).

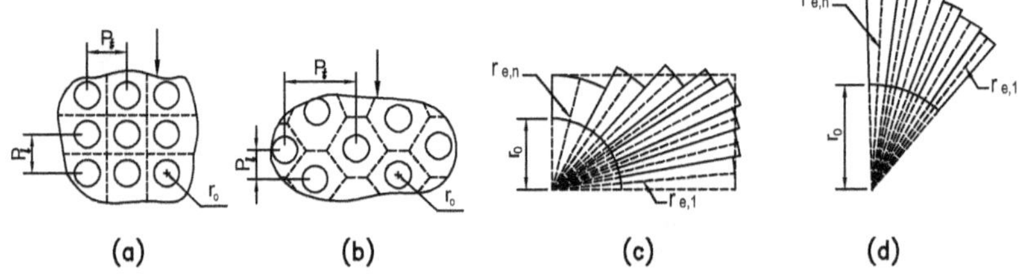

FIGURE 3.50 Fin sectors to determine fin efficiency of continuous finned-tube heat exchanger. (a) Inline layout, (b) staggered layout, (c) fin sectors for inline layout, and (d) fin sectors for staggered layout.

$$r_e^* = \left(\frac{2}{\sqrt{\pi}}\right)s \tag{3.110}$$

and for the hexagonal fin

$$r_e^* = \left(\frac{2\sqrt{3}}{\pi}\right)^{1/2} s \tag{3.111}$$

The Sparrow and Lin method satisfies exactly the isothermal boundary condition at the fin base and fulfills approximately to any desired accuracy the adiabatic boundary condition at the fin edge.

Sector method: The fin efficiency of the polygonal fin is obtained by an approximate method referred to as the "sector method" [53]. The flat fin is broken down into n sectors bounded by idealized adiabatic planes as indicated by the dotted lines for the inline and staggered tube arrangements in Figure 3.48. The radius of each circular sector is determined by the length of a construction line originating from the tube center and passing through the midpoint of each line segment. The fin efficiency of each sector is then determined by use of Equations 3.97, 3.98, or 3.104. The fin efficiency for the entire surface can then be determined by the summation as [53]

$$\eta_f = \frac{\sum_{i=1}^{n} \eta_{f,i} A_{f,i}}{\sum_{i=1}^{n} A_{f,i}} \tag{3.112}$$

where $A_{f,i}$ is the area of each fin sector. The approximation improves as n becomes very large, but for practical purposes, only a few segments will suffice to provide η_f within a desired accuracy of 0.1%. The fin efficiency calculated by the sector method is lower than that for the actual flat fin, whereas the equivalent annulus method yields η_f values that are higher than those by the sector method for highly rectangular fin geometry around the tubes [53].

3.6.4 Basic Heat Transfer Relations for PFHE based on Ref. [44]

The required surface area of a brazed aluminum plate-fin heat exchanger can be obtained from [44]:

$$UA_r = \frac{Q}{MTD} \tag{3.113}$$

where

- U is overall heat transfer coefficient between streams (W/m^2 K)
- A_r is required overall effective heat transfer surface area (m^2)
- Q is heat to be transferred (W)
- MTD is mean temperature difference between composite or combined streams (K)

Equation (3.113) becomes

$$UA_r = \Sigma \frac{Q_i}{LMTD_i} \tag{3.114}$$

where i refers to a particular stream.

Overall Effective Heat Transfer Surface of Exchanger

The overall effective heat transfer surface can be estimated from Equation (3.115). The thermal resistance of the parting sheet between the two streams can usually be ignored primarily because it is made from thin aluminum sheet.

$$\frac{1}{UA_r} = \frac{1}{\Sigma(h_0 A)_{wi}} + \frac{1}{\Sigma(h_0 A)_{ci}} \tag{3.115}$$

where
h_0 is effective heat transfer coefficient of a stream (W/m^{2}K)
A is effective heat transfer surface of a passage or layers of a stream (m^2)
A_d is designed (or estimated) overall effective heat transfer surface (m^2)
Suffix wi, ci is warm or cold stream i

3.6.4.1 Effective Heat Transfer Coefficient of Each Stream

The heat transfer coefficient of each stream can be estimated from Equation (3.116).

$$h = \frac{j\,G_m C_p}{Pr^{2/3}} \tag{3.116}$$

where
h is heat transfer coefficient of a stream (W/m^{2}K)
j is Colburn factor for a finned passage
G_m is mass flux of a stream (kg/m^{2}s)
C_p is specific heat capacity of a stream at constant pressure (J/kg K)
Pr is Prandtl number of a stream ($C_p\,\mu$/k)
k is thermal conductivity of a stream (W/m K)
μ is dynamic viscosity of a stream (Ns/m^2)

The effective heat transfer coefficient of each stream, α_0, can be estimated from Equation (3.117) which takes the fouling resistance into account.

$$\frac{1}{h_o} = \frac{1}{h} + R_F \tag{3.117}$$

where
R_F is fouling resistance of a stream (m^{2}K/W)
Equation (3.116) can be used for single phase streams, i.e., all vapor or all liquid flow.

3.7 COMPONENTS OF PRESSURE LOSS FOR A PFHE

The individual pressure losses within a heat exchanger typically consist of [44]

1. Expansion loss into the inlet header
2. Contraction loss at the entry to the core
3. Loss across the inlet distributor
4. Loss across the heat transfer length
5. Loss across the outlet distributor
6. Expansion loss into the outlet header
7. Contraction loss into the outlet nozzle

3.7.1 Single-Phase Pressure Loss

The frictional pressure loss across a plate-fin passage and at any associated entry, exit, and turning losses can be expressed by [44]

$$\Delta P = 4f\left(\frac{L}{D_h}\right)\left(\frac{G_m^2}{2\rho}\right) + K\left(\frac{G_m^2}{2\rho}\right) \tag{3.118}$$

where

f is the fanning friction factor
L is the passage length
D_h is the hydraulic diameter of passage
G_m is the mass velocity (mass flux) of stream
ρ is the density of a stream
K is the expansion, contraction, or turning loss coefficient
ΔP is the overall pressure drop

3.8 RATING AND SIZING OF A COMPACT EXCHANGER

In a rating problem, the geometry and size of the heat exchanger are fully specified. Incoming fluid flow rates and temperatures are known. It is used to calculate the thermal effectiveness and quantity of heat transferred and pressure drop of each stream. This is a quite straightforward calculation, with one exception. Because the exit stream temperatures are not known, the average temperatures at which the fluid properties are evaluated are not known.

In a sizing problem, the heat exchange requirement is specified and the designer must calculate the heat exchanger size. Normally, pressure drop limits are given for each fluid stream. Since the entering flow rates, temperatures, and pressures are given, and the thermal or heat duty (or leaving temperatures) is specified, the thermal effectiveness ε and NTU are directly calculable. A sizing problem is considerably more complex than a rating problem. For a sizing calculation, the calculation result cannot be completed by a single, step-by-step rational decision path. A number of decisions must be made prior to making the thermal performance calculations. These include the selection of the following [53,66]:

1. Heat exchanger fluid flow arrangement, e.g., counterflow, cross flow.
2. Heat exchanger materials for tube as it is influenced by fluid pressure, temperature, and corrosion potential and fin material from heat transfer, resistance to atmospheric corrosion, fabrication methods, etc.
3. Fin geometry and fin thickness.
4. Fouling considerations which influence the type of surface geometry and fin spacing that may be selected.
5. Heat exchanger frontal area. This key decision establishes the Reynolds number for each flow stream. The pressure drops are directly dependent on this decision.

3.8.1 Rating of a Compact Exchanger

The rating problems for a two-fluid direct transfer type compact crossflow and counterflow exchanger with gas as a working fluid at least on one side are discussed in this section. The surface employed on the gas side of this exchanger has a high surface area density and low hydraulic diameter. Customarily, the ε-NTU method is employed for compact heat exchangers. Hence, the solution procedure is outlined using the ε-NTU method. The rating procedure given here is as in Refs. [53,67,68].

3.8.1.1 Rating of Single-Pass Counterflow and Crossflow Exchangers

The basic steps of the rating problem are shown in Figure 3.51 and involve the determination of the following parameters.

1. Surface geometrical properties. These include
   A_o minimum free flow area
   A heat transfer surface area (total of primary surface area, A_p, and secondary surface area, A_f)

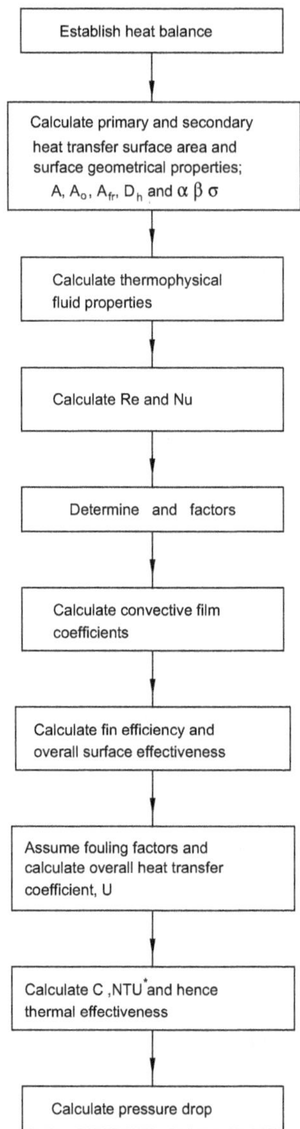

FIGURE 3.51 Flowchart for rating of a compact heat exchanger. (*Note*: Wherever applicable the calculation is to be done for both fluid 1 and fluid 2.)

L flow length
D_h hydraulic diameterβ heat transfer surface area densityσ ratio of minimum free flow area to frontal areal_f fin lengtht_f fin thicknessAlso included are specialized dimensions used for heat transfer and pressure-drop correlations.

2. Fluid physical properties. Determine the thermophysical properties at bulk mean temperature for each fluid, namely, hot and cold fluids. The properties needed for the rating problem are μ, c_p, k, and Pr. Since the outlet temperatures are not known for the rating problem, they are guessed initially. Unless it is known from past experience, assume an exchanger effectiveness as 60%–75% for most single-pass crossflow exchangers, and 80%–85% for single-pass counterflow and two-pass cross-counterflow exchangers. For the assumed effectiveness, calculate the fluid outlet temperatures by

$$t_{h,o} = t_{h,j} - \varepsilon \frac{C_{min}\left(t_{h,i} - t_{c,i}\right)}{C_h} \quad (3.119a)$$

$$t_{c,o} = t_{c,i} + \varepsilon \frac{C_{min}}{C_c}\left(t_{h,i} - t_{c,i}\right) \quad (3.119b)$$

Initially, assume $C_c/C_h \approx M_c/M_h$ for a gas-to-gas exchangers, or $C_c/C_h = (Mc_p)_c/(Mc_p)_h$ for a gas-to-liquid exchanger with approximate values of c_p. For exchangers with $C^* \geq 0.5$ (usually gas-to-gas exchangers), the bulk mean temperature on each side will be the arithmetic mean of the inlet and outlet temperatures on each side. For exchangers with $C^* < 0.5$ (usually gas-to-liquid exchangers), the bulk mean temperature for C_{max} as the hot fluid is $t_{h,m} = (t_{h,i} + t_{h,o})/2$, $t_{c,m} = t_{h,m}$ − LMTD and for C_{max} as the cold fluid $t_{c,m} = (t_{c,i} + t_{c,o})/2$, $t_{h,m} = t_{c,m}$ + LMTD [53,66].

Once the bulk mean temperatures are obtained on each side, obtain the fluid properties from thermophysical property tables or from standard thermal engineering books.

3. Reynolds numbers. Calculate the Reynolds number and/or any other pertinent dimensionless groups needed to determine *j* or Nu and *f* of heat transfer surfaces on each side of the exchanger.
4. Compute *j* or Nu and *f* factors.
5. Correct Nu (or *j*) and *f* for variable fluid property effects in the second and subsequent iterations.
6. From Nu or *j*, determine the heat transfer coefficient for both the fluid streams.

$$h = \frac{\mathrm{Nu}k}{D_h} \quad (3.120)$$

or

$$h = jGC_p Pr^{-2/3} \quad (3.121)$$

7. Determine the fin efficiency η_f and the overall surface efficiency η_o given by

$$\eta_f = \frac{\tanh ml}{ml} \quad (3.122)$$

where l is the fin length.

$$\eta_o = 1 - \frac{A_f}{A}\left(1 - \eta_f\right) \quad (3.123)$$

8. Overall conductance. Calculate the wall thermal resistance, R_w. Knowing the fouling resistances $R_{f,h}$ and $R_{f,c}$ on the hot and cold fluid sides, respectively, calculate the overall thermal conductance UA from

$$\frac{1}{UA} = \frac{1}{(\eta_o hA)_h} + \frac{R_{f,h}}{(\eta_o A)_h} + R_w + \frac{1}{(\eta_o hA)_c} + \frac{R_{f,c}}{(\eta_o A)_c} \tag{3.124}$$

$$= R_h + R_1 + R_w + R_2 + R_c \tag{3.125}$$

9. Calculate NTU, C^*, and exchanger effectiveness, ε. If the thermal effectiveness is above 80%, correct for wall longitudinal conduction effect.
10. Compute the outlet temperature from Equations 3.119a and 3.119b. If these outlet temperatures differ significantly from those assumed in step 2, use these outlet temperatures in step 2 and continue iterating steps 2–9 until the assumed and computed outlet temperatures converge within the desired degree of accuracy. For a gas-to-gas exchanger, probably one or two iterations will be sufficient.
11. Compute the heat duty from

$$q = \varepsilon C_{min}\left(t_{h,i} - t_{c,i}\right) \tag{3.126}$$

12. Calculate the core pressure drop. The friction factor f on each side is corrected for the variable fluid properties as discussed in Chapter 2. The wall temperature T_w is computed from

$$T_{w,h} = t_{m,h} - q\left(R_h + R_1\right) \tag{3.127}$$

$$T_{w,c} = t_{m,c} - q\left(R_c + R_2\right) \tag{3.128}$$

3.8.1.2 Shah's Method for rating of Multipass Counterflow and Crossflow Heat Exchangers

The rating procedure for multipass crossflow exchangers with fluids mixed between passes is described by Shah [66]. The solution procedure for the rating problem for the two-pass crossflow exchangers with flows unmixed in the passes but mixed between passes is also very similar to the sizing of a single-pass crossflow. Only some of the calculations on the two pass side need to be modified, and only those points are summarized here.

1. In order to compute fluid bulk mean temperature and thermophysical properties of fluids, first guess the overall thermal effectiveness, ε_N. Assume it to be 80%–85% unless it is known from past experience. Assume that the NTU per pass is the same. The individual pass effectiveness ε_p is related to overall effectiveness ε_N for the case of fluids mixed between passes. Compute approximate values of C_c and C_h since we don't know yet the accurate values of the specific heats.
2. Determine the intermediate and outlet temperatures by solving the following individual pass effectiveness and overall effectiveness equations:

$$\varepsilon_{p1} = \frac{C_c\left(t_{c,o} - t_{c,int}\right)}{C_{min}\left(t_{h,i} - t_{c,int}\right)} \tag{3.129}$$

$$\varepsilon_{p2} = \frac{C_c\left(t_{c,int} - t_{c,i}\right)}{C_{min}\left(t_{h,int} - t_{c,i}\right)} \tag{3.130}$$

$$\varepsilon = \frac{C_c\left(t_{c,o} - t_{c,i}\right)}{C_{min}\left(t_{h,i} - t_{c,i}\right)} \tag{3.131}$$

In the foregoing three equations, there are three unknowns: $t_{c,o}$, $t_{c,int}$, and $t_{h,int}$. Hence, they can be evaluated exactly and then from the overall energy balance, $t_{h,o}$ is calculated.

$$C_c\left(t_{c,o} - t_{c,i}\right) = C_h\left(t_{h,i} - t_{h,o}\right) \tag{3.132}$$

Since we know all terminal temperatures for each pass, we can determine the bulk mean temperature for each fluid in each pass and subsequently the fluid properties separately for each pass.

3.8.2 Sizing of a Compact Heat Exchanger

The basis of sizing involves coupling of heat transfer and flow friction in the derivation of the core mass velocity G on each side of a two-fluid exchanger. Subsequently, the sizing problem is carried out in a manner similar to the rating problem.

3.8.2.1 Core Mass Velocity Equation[1]

The dominant term in the expression for the pressure drop is the core friction term. The entrance and exit effects are generally relatively small and are of the opposite sign. Similarly, the flow acceleration term is relatively small in most heat exchangers, being generally less than 10% of the core friction term [53], so their elimination is usually warranted in a first approximation. With these approximations and $L/r_h = A/A_o$, the expression for pressure drop may be written after rearrangement as

$$\frac{G^2}{2g_c p_i \rho_i} = \frac{\Delta p}{p_i}\frac{A_o}{A}\frac{1}{\rho_i (1/\rho)_m}\frac{1}{f} \tag{3.133}$$

In the absence of fouling resistances, the overall NTU is related to NTU_h, NTU_c, and the wall resistance, R_w, by

$$\frac{1}{NTU} = \frac{1}{NTU_h\left(C_h/C_{min}\right)} + C_{min}R_w + \frac{1}{NTU_c\left(C_c/C_{min}\right)} \tag{3.134}$$

The wall resistance term is generally small and hence neglected in the first-approximation calculation. Hence, the number of transfer units on one side of interest (either hot or cold), designated as NTU_1, may be estimated from the known overall NTU as given next.

If both fluids are gases, one can start with the estimate that the design is "balanced" by a selection of the hot and cold side surfaces so that $R_h = R_c = R_o/2$, that is, $NTU_h = NTU_c = 2NTU$. Then

$$NTU_1 = NTU_2 = 2NTU \tag{3.135}$$

For a gas–liquid heat exchanger, one might estimate

$$NTU_{gas\ side} = 1.1(NTU) \tag{3.136}$$

The term NTU_1 is related to the Colburn factor j on side 1 by

$$\mathrm{NTU}_1 = \left(\frac{\eta_o hA}{MC_p}\right)_1 = \left(\eta_o \frac{hA}{GC_p A_o}\right)_1 = \left(\eta_o jPr^{-2/3}\frac{A}{A_o}\right)_1 \tag{3.137}$$

Eliminating (A/A_o) from Equations 3.133 and 3.137, and simplifying the expression, we get the core mass velocity G for one side:

$$G = \left[\left(\frac{2g_c \eta_o}{(1/\rho)_m Pr^{2/3}}\right)\left(\frac{\Delta p}{\mathrm{NTU}}\right)\left(\frac{j}{f}\right)\right]^{0.5} \tag{3.138}$$

The feature that makes this equation so useful is that the ratio j/f is a relatively flat function of the Reynolds number. Thus, one can readily estimate an accurate magnitude of j/f based on a "ballpark" estimate of Re. If there are no fins, overall surface efficiency $\eta_o = 1$. For a "good design," the fin geometry is chosen such that η_o is in the range 70%–90%. Therefore, $\eta_o = 0.8$ is suggested as a first approximation to determine G from Equation 3.138.

3.8.2.2 Procedure for sizing a Compact Heat Exchanger

The procedure for sizing any of the compact heat exchangers (Figure 3.52) is almost inevitably an iterative one and thus lends itself very conveniently to computer calculations. Kays and London [15] illustrate the procedure for sizing a crossflow heat exchanger. To illustrate this, a simple crossflow heat exchanger is considered, as shown in Figure 3.32(b). It is assumed that each fluid is unmixed throughout. The two fluids are designated by h and c. The problem is to determine the three dimensions L_1, L_2, and L_3. The first step after choosing the two surfaces is to assemble the geometric characteristics of the surface—D_h, σ, α, β, A_f/A, and A_w/A for the hot side and cold fluids side. A first estimate of G_h and G_c can be made using Equation 3.138. After the determination of mass velocity, carry out the calculations similar to a rating problem. The sequence of parameters to be calculated after determination of mass velocity includes Re, η_f, η_o, j, f, h, UA, A_o, A_f, flow length, K_c, K_e, and $(\Delta p/p)$, both on the hot side and cold side. The pressure drop results are then compared with those specified for the design, and the procedure is repeated until the two pressure drops are within the specified value.

3.9 OPTIMIZATION OF PLATE-FIN EXCHANGERS AND CONSTRAINTS ON WEIGHT MINIMIZATION AFTER [69]

Minimizing the material volume, or weight, of the PFHE core does not necessarily represent the optimum solution, even for weight-sensitive aerospace applications; size and shape can be important design considerations [69]. Minimum weight core could be longer than is desirable because of the low fin efficiency implied. If the fin is too thin, its efficiency will be low, and to compensate both core length and flow area will be high, giving an excessively high weight. On the other hand, a very thick fin giving high fin efficiency will yield a low core length but with low porosity (free flow area) and weight. Minimizing the components' thickness can also contribute to low weight. However, there are limitations in reducing components beyond certain limits as follows:

The thicknesses of fin material and separating plates have lower limits set by pressure-retaining capability.

Plate-fin cores are usually made from sheet stock of a fixed range of thickness; rolling finstock to a special "optimum" thickness could be uneconomic.

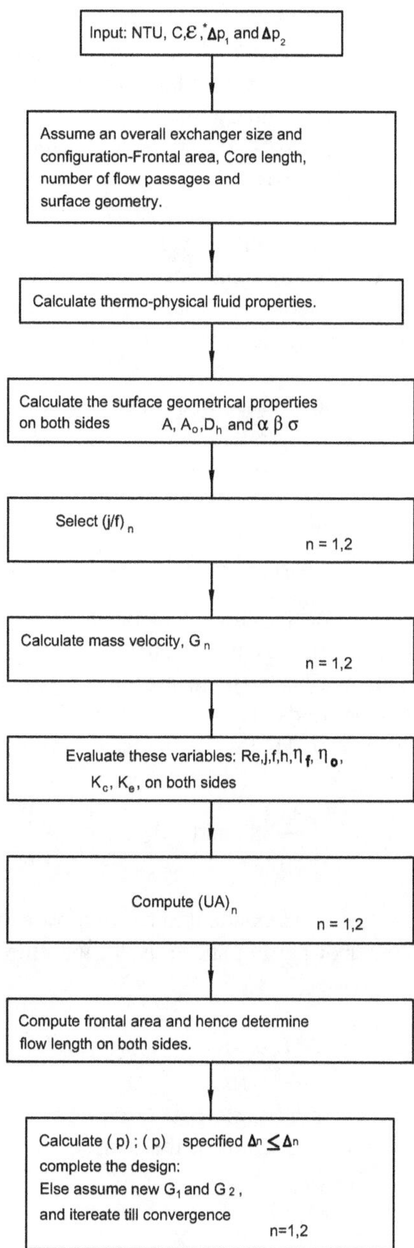

FIGURE 3.52 Flowchart for sizing of a compact heat exchanger. (*Note*: 1, 2 are both sides fluids designation.)

It may not be possible to form fins of sufficient dimensional accuracy if they are too thin; if they can be made, they might deform unsatisfactorily on brazing.

There may also be lower thickness limits set by erosion and corrosion problems.

3.9.1 Effect of Longitudinal Heat Conduction on Thermal Effectiveness

The ε-NTU and LMTD methods discussed are based on the idealization of zero longitudinal heat conduction in the wall in the fluid flow direction. If a temperature gradient is established in the wall in the fluid flow direction, heat transfer by conduction takes place from the hotter to the colder region of the wall in the longitudinal direction. This longitudinal conduction in the wall flattens the temperature distributions, reduces the mean outlet temperature of the cold fluid, increases the mean outlet temperature of the hot fluid, and thus reduces the thermal effectiveness [66,70]. The reduction in the effectiveness at a specified NTU may be quite significant for a compact exchanger designed for high thermal effectiveness (above ~80%) and a short flow length, L. The shell and tube exchangers and plate exchangers are usually designed for an effectiveness of 60% or less per pass. Therefore, the influence of heat conduction in the wall in the flow direction is negligible for these effectiveness levels. The presence of longitudinal heat conduction in the wall is incorporated into the thermal effectiveness formula by an additional parameter, λ, referred to as the longitudinal conduction parameter. Another parameter that influences the longitudinal wall heat conduction is the convection conduction ratio, $\eta_o hA$. Thus, in the presence of longitudinal heat conduction in the wall, the exchanger thermal effectiveness is expressed in a functional form [70]:

$$\varepsilon = \phi\left\{\text{NTU}, C^*, \eta_o hA, \lambda\right\} \tag{3.139}$$

3.9.2 Longitudinal Conduction Influence on Various Flow Arrangements

The influence of longitudinal conduction in the case of parallelflow is not discussed because high-performance exchangers are not designed as for parallelflow arrangements. For high-performance exchangers, the effect is discussed in Chapter 6. In a crossflow arrangement, the wall temperature distributions are different in the two perpendicular flow directions. This results in a two-dimensional longitudinal conduction effect, and is accounted for by the cold-side and hot-side parameters defined as [70]:

$$\lambda_c = \left(\frac{k_w A_k}{LC}\right)_c \quad \text{and} \quad \lambda_h = \left(\frac{k_w A_k}{LC}\right)_h \tag{3.140}$$

Note that the flow lengths L_c and L_h are independent in a crossflow exchanger. If the separating wall thickness is a, and the total number of parting sheets is N_p, then the transfer area for longitudinal conduction is given by [70] as follows:

$$A_{k,c} = 2N_p L_n a \quad \text{and} \quad A_{k,h} = 2N_p L_c a \tag{3.141}$$

Chiou [71] analyzed the problem for a single-pass crossflow heat exchanger with both fluid unmixed and tabulated values for the reduction in thermal effectiveness due to longitudinal wall heat conduction.

3.9.3 Comparison of Thermal Performance of Compact Heat Exchangers

Among the three types of compact heat exchangers, viz., tube-fin, continuous fin, and PFHE, PFHE has maximum heat transfer capacity compared to the other two types of heat exchangers. Figure 3.53 shows a comparison of the thermal performances of PFHEs, flat tube heat exchangers (oil coolers), and tube-fin heat exchangers. Performance is shown as Q/ITD, the heat load divided by the difference in incoming temperature of the liquid and air.

3.10 AIR-COOLED HEAT EXCHANGER (ACHE)

Air-cooled heat exchangers (ACHEs), or "fin-fans," are usually composed of rectangular tube bundles containing several rows of tubes on a triangular pitch. ACHEs are alternate heat rejection

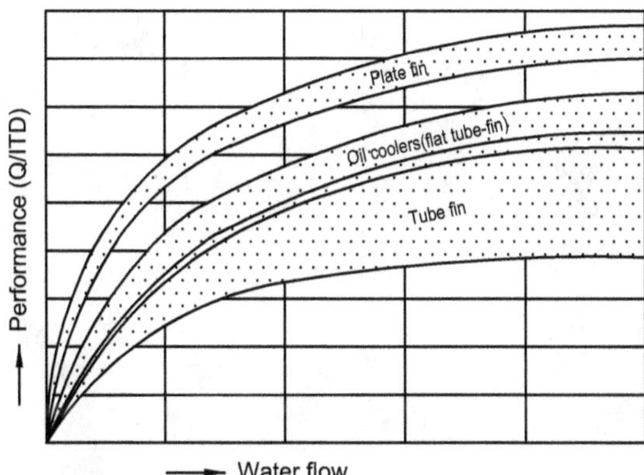

FIGURE 3.53 Chart for comparison of thermal performance of PFHEs, flat tube heat exchangers (oil coolers), and tube-fin heat exchangers performance is shown as Q/ITD, the heat load divided by the difference in incoming temperature of the liquid and air. (Courtesy of Lytron Inc., Woburn, MA.)

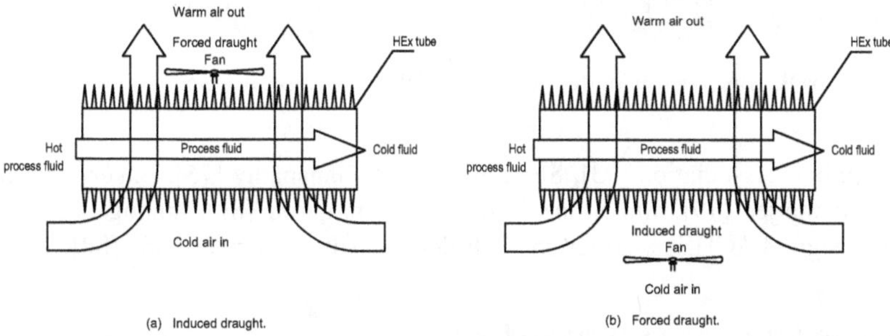

FIGURE 3.54 Principle of ACHE.

devices that are used frequently in place of the conventional water-cooled shell and tube heat exchanger or plate and frame heat exchanger to cool a process fluid. ACHEs have been successfully and economically used in liquid cooling for compressor engine and jacket water and other recirculating systems, petroleum fractions, oils, etc. They can be used in all climates. In an ACHE, hot process fluid enters one end of the ACHE and flows through tubes, while ambient air flows over and between the tubes, which typically have externally finned surfaces. The process heat is transferred to the air, which cools the process fluid, and the heated air is discharged into the atmosphere. There are two basic types of ACHEs in use:

i. Forced draft—the fan is located below the process bundle and air is forced through the tubes
ii. Induced draft—the fan is located above the process bundle and air is pulled, or induced, through the tubes.

   Additionally, there are natural draft air-cooled heat exchangers—Natural draft air-cooled exchangers operate the way their name suggests: there are no fans to push or pull air through the tube bundle. Instead, a chimney above the tube bundle creates the draft that drives air through the tube bundle.

The principle of ACHE, both forced draft and induced draft, is shown in Figure 3.54, and a forced draft ACHE unit is shown in Figure 3.55. For details on ACHE, refer to Ref. [25], the Basics of Air-Cooled Heat Exchangers, Hudson Products Corporation [72], Air-Cooled Exchangers, Gas

FIGURE 3.55 A bank of ACHEs. (Courtesy of GEA Iberica S.A., Vizcaya, Spain.)

Processors Suppliers Association [73], Steve Boes [74], Ganapathy [75], Baker [76], Shipes [77], Brown [78], Mukherjee [79], Jim Stone [80], and Giammaruti [81], among others. One of the standards for design of ACHEs is American Petroleum Institute Standard 661 [82].

3.10.1 Forced Draft versus Induced Draft

One of the design criteria affecting the performance of ACHEs is the type of draft: forced draft or induced draft. Both these arrangements are discussed below [72,75,83].

3.10.1.1 Forced Draft

The majority of air-cooled exchangers are of forced draft construction. Among other reasons, the accessibility of fans, actuators, and drives is much better for maintenance and there is thus a strong preference for this arrangement. Forced draft shall be selected for critical and condensing duties where the difference between the design product outlet temperature and the design air inlet temperature is 15°C or higher. A common problem with forced draft coolers is accidental warm air recirculation. Forced draft fans have the advantage of handling cold air entering the exchanger, requiring smaller volumes of air and less horsepower. They are best adapted for cold-climate operation with warm air recirculation. Also, forced draft fans afford a higher heat transfer coefficient relative to induced draft, because forced draft fans cause turbulence across the tube bundle [75].

Forced Draft Disadvantages

i. Low natural draft capability on fan failure due to small stack effect.
ii. Less uniform distribution of air over the bundle.
iii. Increased possibility of hot air recirculation, resulting from low discharge velocity from the bundles, high intake velocity to the fan ring, and no stack.
iv. Complete exposure of the finned tubes to sun, rain, and hail, which results in poor process control and stability.

3.10.1.2 Induced Draft

Compared with the forced draft design, induced draft design has the following advantages:

i. Better distribution of air across the section.
ii. Less possibility of the hot air recirculating around to the intake of the sections. The hot air is discharged upward at approximately 2.5 times the velocity of intake, or about 1500 ft/min.
iii. Less effect of sun, rain, and hail, since 60% of the face area of the section is covered.
iv. In the event of fan failure, the natural draft stack effect is much greater with induced draft.
v. Better process control and stability because the plenum covers 60% of the bundle face area, reducing the effects of sun, rain, and hail.
vi. Increased capacity in the fan-off or fan-failure condition, since the natural draft stack effect is much greater.

The disadvantages of induced draft design are:

1. Difficult to remove bundles for maintenance.
2. High-temperature service limited due to the effect of hot air on the fans.
3. Possibly higher horsepower requirements if the effluent air is very hot.
4. Effluent air temperature should be limited to 200–220°F to prevent damage to fan blades, bearings, belts, or other mechanical equipment in the hot air stream.
5. Fans are less accessible for maintenance, and maintenance may have to be done in the hot air generated by natural convection.
6. Plenums must be removed to replace bundles.

Recommendations include:

a. Induced draft units should be used whenever hot-air recirculation is a potentially critical problem.
b. Forced draft units should be used whenever the design requires pour point protection, or winterization.

3.10.2 Construction of ACHE

Assembly details of a ACHE are shown in Figure 3.56 and its construction details and components are as follows [72,73,82]:

1. One or more tube bundles of heat transfer surface. Major elements of tube bundle assembly except heat exchanger core are shown in Figure 3.57.
2. An air pumping device, such as an axial flow fan or blower driven by an electric motor and power transmission device such as belt (up to 50 hp) or gear (above 50 hp) to mechanically rotate the air-moving device.
3. Bay. One or more tube bundles, serviced by two or more electric-motor-driven fans—most have a speed reducer consisting of either a belt drive (e.g., V-belt or cog belt) or a right-angle gear drive, including the structure, plenum, and other attendant equipment which is called a bay in an air-cooled heat exchanger. Refer to Figure 3.58 for typical bay arrangements.
4. Unless it is a natural draft application, the moving of air across the tube bundle may be either forced draft (fan is located below the tube bundle and air is forced through the fin tubes) or induced draft (fan is located above the bundle and air is induced or pulled through the fin tubes).

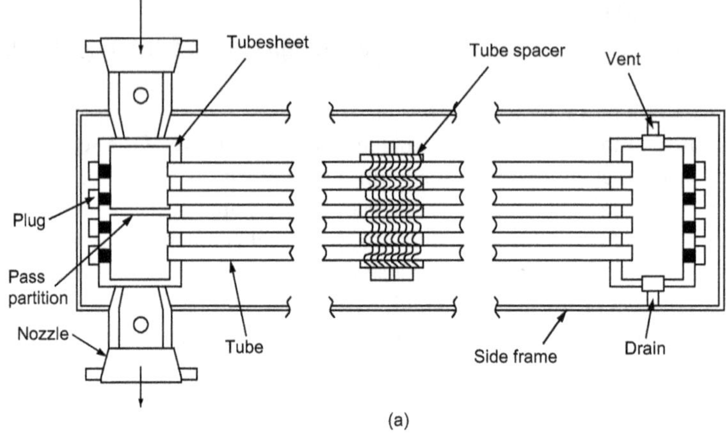

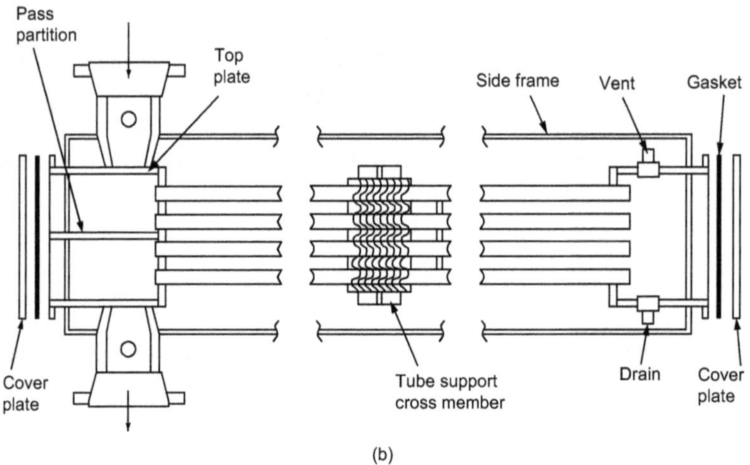

FIGURE 3.56 Assembly and Construction details of a ACHE.

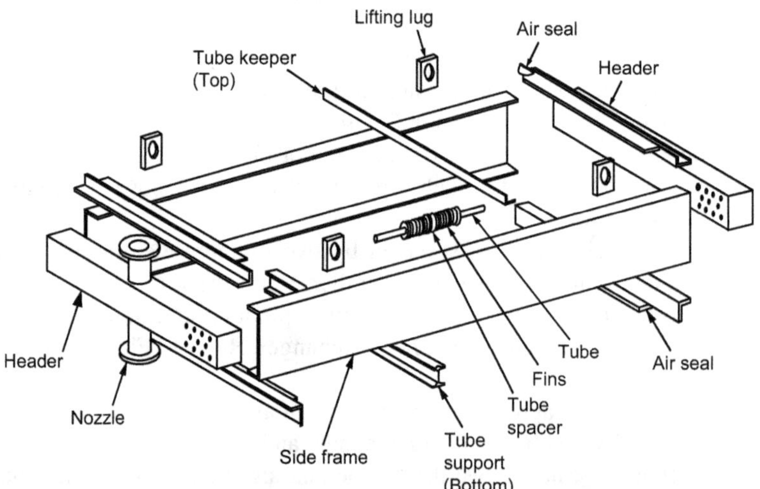

FIGURE 3.57 Major elements of tube bundle assembly except heat exchanger core.

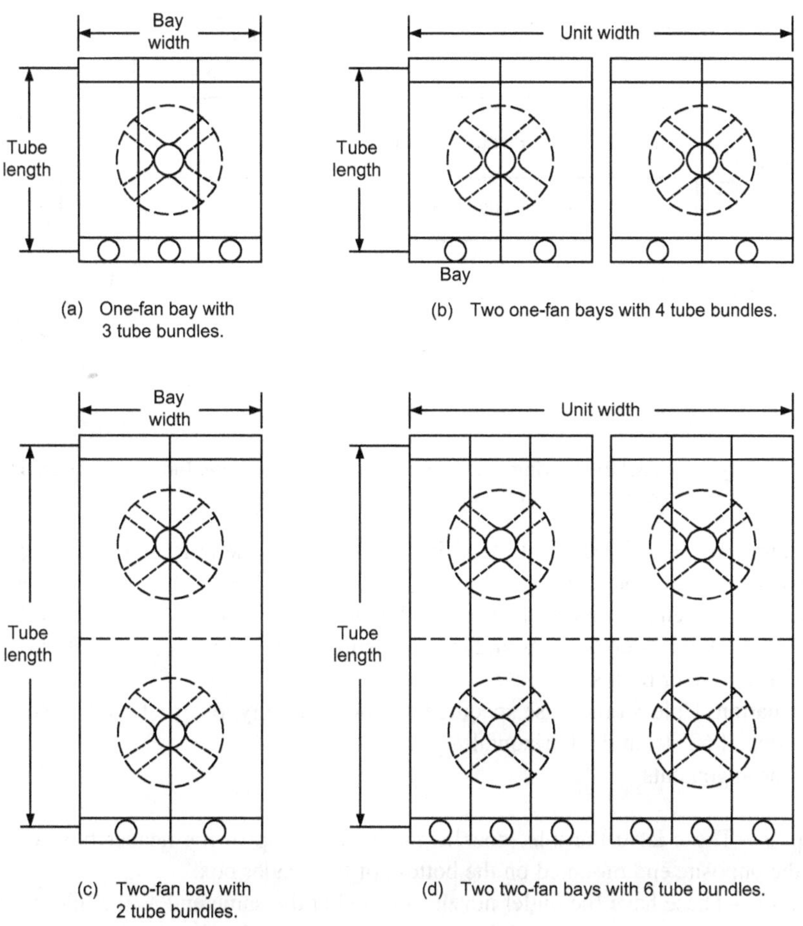

FIGURE 3.58 ACHE bay arrangement.

5. A support structure high enough to allow air to enter beneath the ACHE at a reasonable flow rate.
6. Optional features like (i) header and fan maintenance walkways with ladders to climb (Figure 3.59); (ii) louvers for process outlet temperature control; (iii) recirculation ducts and chambers for protection against freezing or solidification of high pour point fluids in cold weather; (iv) variable pitch fan hub or frequency drive for temperature control; and (v) a plenum between the bundle(s) and the air-moving device.
7. Anti-rotation devices should be installed on fans to prevent belt damage when fans are restarted after being out of service. These also provide a safety measure for personnel performing maintenance.
8. Plenum. The air plenum is an enclosure that provides for the smooth flow of air between the fan and bundle.
9. Humidification sections or air washers: If the geographic location is such that the relative humidity is low most of the year, a humidification section could be installed below the unit. This, in effect, moisturizes the inlet air down to its wet bulb temperature which could be 5–11°C cooler than ambient. However, care should be taken to insure that air entering the tube bundle is dry.

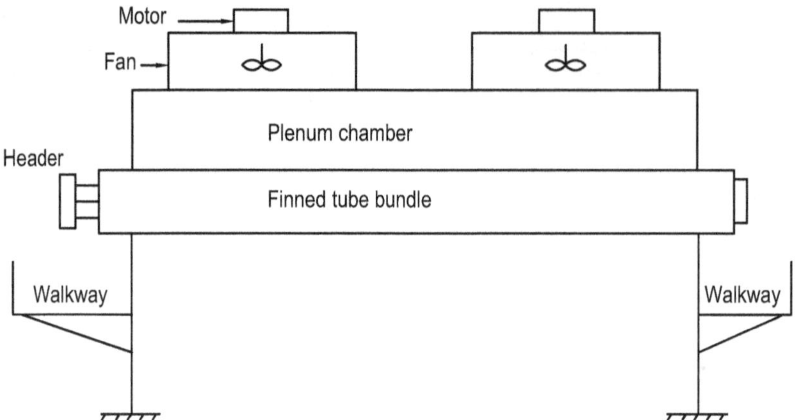

FIGURE 3.59 Layout of ACHE showing tube bank with walkways. (*Note*: Induced draft design.)

10. Winterized unit. A forced-draft ACHE is outfitted with one or more methods, e.g., air outlet louvers, fans equipped with variable frequency drives (VFDs), hot air recirculation systems to control the temperature of the process fluid leaving the exchanger. This type of unit is typically found in colder climates and in hotter climates for process fluids with high viscosities and/or high pour points.
11. Orientation. Unless otherwise specified, the horizontal type is preferred. A-frames are usually sloped 60° from the horizontal.
12. Pass arrangements

Single pass—These have the inlet nozzles mounted on top of the header box, with the outlet nozzles at the opposite end mounted on the bottom of the header box.

Double pass—These have the outlet nozzles located at the same end as the inlet nozzles, so for additional surface area, more passes can be added, or additional units can be installed and located side by side. Figure 3.60 shows tubeside pass arrangements.

13. Structure

The structure consists of columns, braces, and cross beams that support the exchanger at a sufficient elevation above grade to allow the necessary volume of air to enter below at an approach velocity low enough to allow unimpeded fan performance and to prevent unwanted recirculation of hot air. To conserve ground space in oil refineries and chemical plants, ACHEs are usually mounted above, and supported by, pipe racks with other equipment occupying the space underneath the pipe rack. ACHE structures are designed for appropriate wind, snow, seismic, piping loads, and dead and live loads.

14. Drive

Fin-fans are typically driven by gears or belts (either V-belts or synchronous belts) powered by electric motors. A common configuration includes an electric motor with a V-belt drive. In recent years, however, synchronous (timing) belt drives have been replacing V-belt drives to increase efficiency.

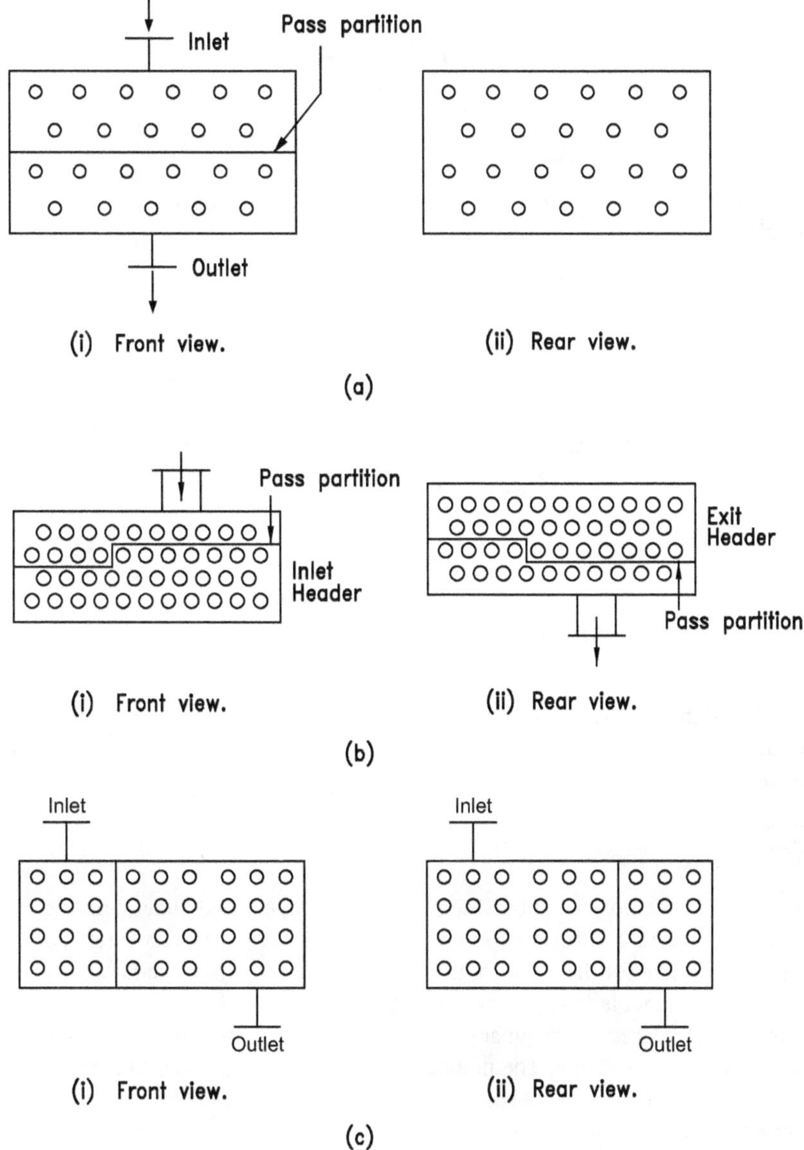

FIGURE 3.60 ACHE: (a) inlet or front header and (b) and (c) exit or rear headers with three passes arrangement on the tubeside.

15. Fans

Number of Fans. At least two fans shall be provided for each bay. Any deviation from this requirement will need the prior approval of the company.

16. Plenums, dispersion angle, and fan coverage

The fan coverage is the ratio of the fan area to the bundle face area. The higher this ratio, the better the fan coverage. The API minimum is 40% with a 45° maximum dispersion angle from the fan ring to the middle of the tube bundle at the middle of the sides or the middle of the ends of each

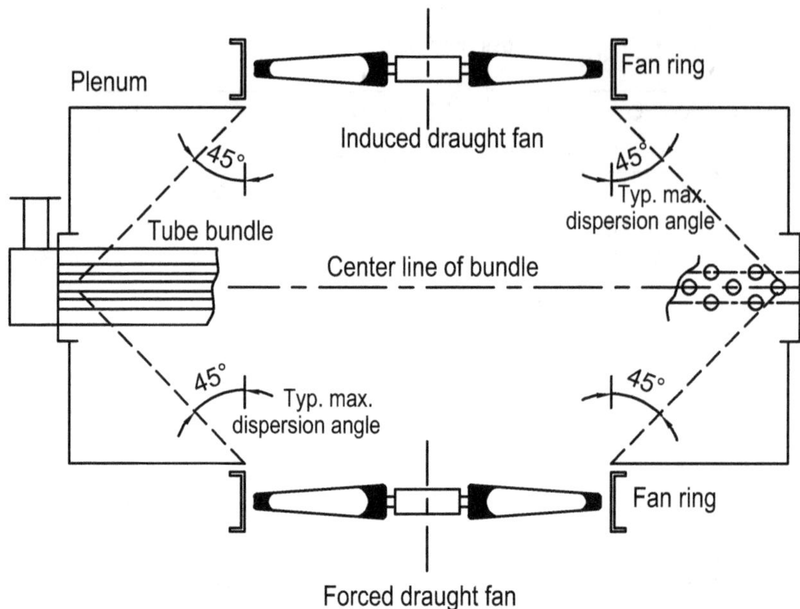

FIGURE 3.61 Condition for dispersion angle of the ACHE fan.

fan chamber. More fan coverage or a lower dispersion angle can improve the air distribution. The condition for dispersion angle of the fan is shown in Figure 3.61. Each fan shall be located such that its dispersion angle shall not exceed 45° at the bundle centerline [82].

3.10.3 ACHE Core

A tube bundle is an assembly of tubes, headers, side frames, and tube supports, as shown in Figure 3.55. Usually the tube surface exposed to the passage of air has an extended surface in the form of fins to compensate for the low heat transfer rate of air at atmospheric pressure and at a low enough velocity for reasonable fan power consumption.

The tubes are usually placed in triangular pitch and in horizontal rows three to eight high. A four-row tube stack is a very common design. The number of passes on the tubeside (the process fluid side) is made possible by internal baffling of the header boxes. For horizontal bundles, vertical baffles result in a side-by-side crossflow arrangement, whereas horizontal baffles result in a counter-crossflow arrangement relative to airflow, as shown in Figure 3.60. Inlet and outlet nozzles may or may not be on the same header box. Tubes are attached to the tubesheet by roller expansion (i.e., tube rolling-in) or welding.

Fin-fans are typically driven by gears or belts (either V-belts or synchronous belts) powered by electric motors. A common configuration includes an electric motor with a V-belt drive. In recent years, however, synchronous (timing) belt drives have been replacing V-belt drives to increase efficiency. The fan blades may be of aluminum, plastic, or, in the case of corrosive atmospheres, stainless steel. The drive can be an electric motor with gears or V-belts. Before discussing the construction details and design aspects, the factors that favor air cooling over water cooling and forced draft vs induced draft are discussed.

3.10.4 American Petroleum Institute Standard API 661/ISO 13706

American Petroleum Institute Standard 661/ISO 13706 [82] is meant for ACHEs for general refinery service. The scope of the standard includes minimum requirements for design, materials selection, fabrication, inspection, testing, and preparation for shipment of refinery process ACHEs. The standard specifies that the pressure parts shall be designed in accordance with ASME Code, Section VIII, Division 1 [84].

3.10.5 Air versus Water Cooling

Two primary methods of process cooling are (1) water cooling and (2) air cooling. The choice between air or water as coolant depends on many factors such as: (1) cooler location; (2) space for cooling system; (3) effect of weather; (4) design pressure and temperature; (5) danger of contamination; (6) fouling, cleaning, and maintenance; and (7) capital costs. Environmental concerns such as shortage of makeup water, blowdown disposal, and thermal pollution have become an additional factor in cooling system selection. In moderate climates air cooling will usually be the best choice for minimum process temperatures above 65°C, and water cooling for minimum process temperatures below 50°C. Between these temperatures a detailed economic analysis would be necessary to decide the best coolant.

3.10.5.1 Air Cooling

Since air is a universal coolant, there are numerous applications where economic and operating advantages are favorable to air-cooled heat transfer equipment. However, applications are limited to cases where the ambient air dry bulb temperature is below the desired cooling or condensing temperature. Air cooling is increasing in use, particularly where water is in short supply. Factors that favor air cooling include the following:

1. Air is available free in abundant quantity with no preparation costs.
2. Water is corrosive and requires treatment to control both scaling and deposition of dirt, whereas air is mostly noncorrosive. Therefore, material selection is governed by process fluids routed through the tubeside.
3. Mechanical design problems are eased with ACHEs since the process fluid is always on the tubeside.
4. Danger of process fluid contamination does not arise.
5. Airside fouling can be periodically cleaned by air blowing, and chemical cleaning can be carried out either during half-yearly or yearly attention. Water-cooled systems may require frequent cleaning.
6. Maintenance costs for ACHEs are about 20%–30% of those for water-cooled systems [75].
7. Air cooling eliminates the environmental problems like heating up of lakes, rivers, etc. due to discharge of cooling water, blowdown, and washout.

Air cooling has the following disadvantages:

1. ACHEs require large surfaces because of their low heat transfer coefficient on the airside and the low specific heat of air. Water coolers require much less heat transfer surface.
2. ACHEs cannot be located next to large obstructions to avoid air recirculation.
3. Because of air's low specific heat, and dependence on the dry-bulb temperature, air cannot usually cool a process fluid to low temperatures. Water can usually cool a process fluid from 10°F to 5°F lower than air, and recycled water can be cooled to near the wet-bulb temperature of the site in a cooling tower [75].
4. The seasonal variation in air temperatures can affect performance and make temperature control more difficult. Low winter temperatures may cause process fluids to freeze.
5. ACHEs are affected by hailstorms and may be affected by cyclonic winds.

6. Noise is a factor with ACHEs. Low-noise fans can reduce this problem but at the cost of fan efficiency and higher energy costs.
7. Special controls may be needed for cold weather protection.
8. The process fluid cannot be cooled to the same low temperature as cooling water.

3.10.6 Tube Bundle Construction

The prime tube is usually round and of any metal suitable for the process fluid, with due consideration being given to corrosion, pressure, and temperature limitations. Fins can be formed from aluminum, copper, or steel, among other materials. The most prevalent fin material is aluminum because it has good thermal properties, it is lightweight, and it is easy to fabricate. Steel is typically used for high-temperature applications and copper for special severe-corrosion applications.

3.10.6.1 Tube Bundle Fin Geometry

The tube-fin geometries, viz., attachment of fins to the tubes, include extruded, embedded, and welded—single-footed or double-footed (Figure 3.62). These fins can be solid or serrated, tension wound or welded, and of a variety of metals. Fins are attached to the tubes in a number of ways:

1. Extruded—an extrusion process in which the fins are extruded from the wall of an aluminum tube that is integrally bonded to the base tube for the full length.
2. Embedded—helically wrapping a strip of aluminum to embed it in a pre-cut helical groove and then peening back the edges of the groove against the base of the fin to tightly secure it.
3. *Wrap-on (L-Base or L-Foot)*—wrapping on an aluminum strip that is footed at the base as it is wrapped on the tube.

Figure 3.62 shows a cutaway view of these finned tubes.

For higher process temperatures, most customers prefer either embedded (400°C; 750°F) or extruded fins (310°C; 590°F). Extruded fins are made by putting an aluminum sleeve (sometimes called a muff) over the tube, then passing the tube through a machine which has rollers which squish the aluminum to form fins. The process is similar to a thread-rolling. This results in a fin tube which has extremely good metal-to-metal contact between the tube and fin sleeve. Extruded fins are often used in coastal regions or on offshore platforms for this reason. Extruded fins typically have the maximum ability to transfer heat due to the absence of contact resistance at the fin and tube base. High-finned (extruded) monometallic fin-tubes and bimetallic fin-tubes are shown in Figure 3.63. Serrations or slitting the fin tips help to increase the airside heat transfer coefficient. However, this improvement is accompanied by higher airside pressure drop and hence will require higher motor power and, depending on the physical location of the ACHE, serrated fins will have a greater propensity for fouling. To check the adequacy of bonding between the fin and tube, manufacturers subject

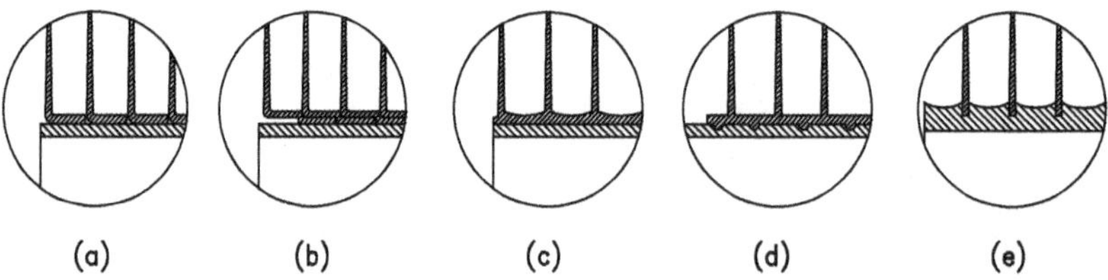

FIGURE 3.62 Tube-fin joint types based on fin joining method with tube (a) welded single L footed, (b) welded double L footed, (c) extruded, (d) serrated, and (e) embedded.

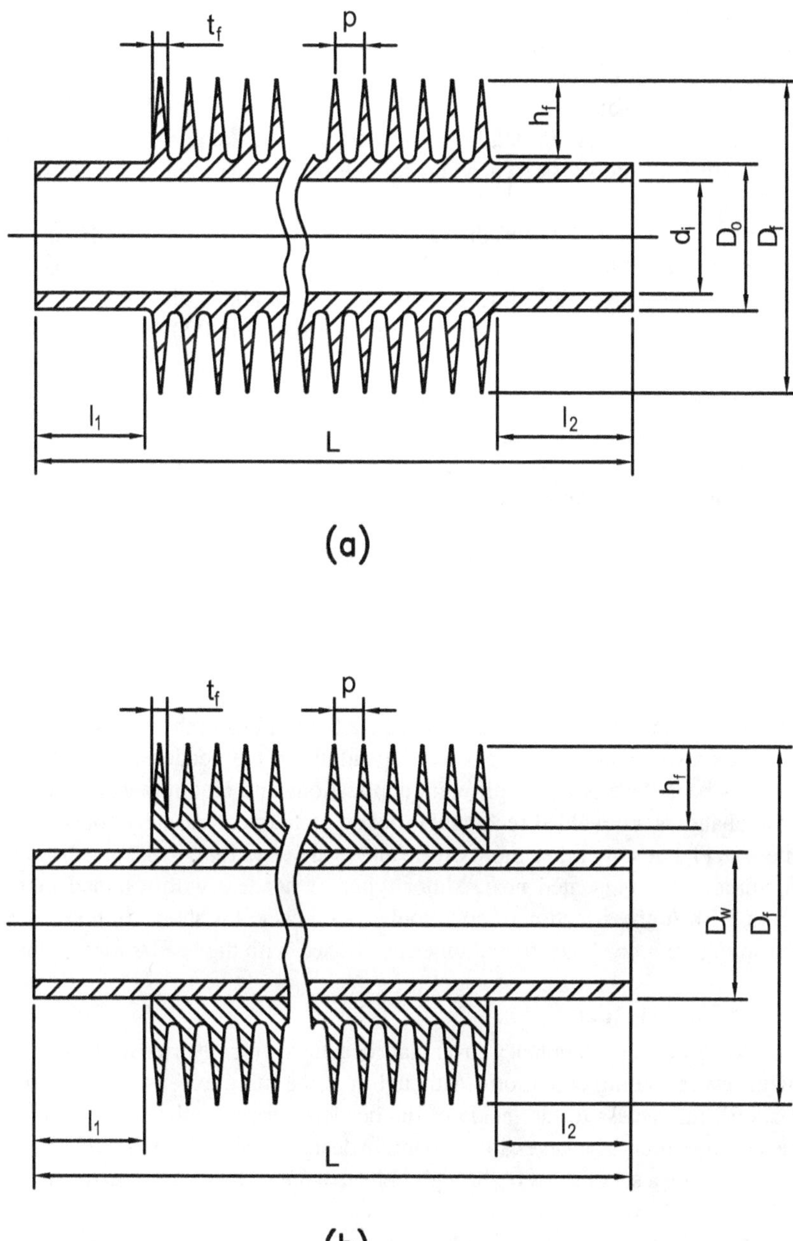

FIGURE 3.63 High finned tube—(a) extruded monometallic fin tube and (b) bimetallic fin tube. *Note*: Sections 1 and 2 are unfinned sections. d_i is tube inner diameter, D_o or D_w is tube outer diameter, D_f is fin diameter, L is tube length and h_f is fin height, p is fin pitch and t_f is fin thickness.

the fin to a tensile test by pulling the fin using a spring balance. Other tests involve such aspects as fin spacing, diameter, and thickness.

The fin most commonly used for ACHEs is aluminum, tension wound, and footed. These are used where the process temperatures are below about 177°C (350°F). The API 661 specification calls for cast zinc bands at the ends of the tubes to prevent the fins from unwrapping. It is cheaper but susceptible to damage due to thermal upsets and airside corrosion.

The embedded fins have the highest temperature capabilities. Embedded fins have less contact resistance to heat transfer than do tension-wound fins, but corrosion is likely to be induced at the groove between fin and tube. Additionally, embedded fin tubes require tubes that are nominally one gauge heaver in wall thickness to account for the mechanical attachment of the fin into the groove [75].

The tubes are usually 1 in. (25.4 mm) diameter; fin density varies from 7 to 16 fins/in. (276–630 fins/m), fin height from 3/8 to 5/8 in. (9.53–15.88 mm), and fin thickness from 0.012 to 0.02 in. (0.3–0.51 mm). The tubes are arranged in standard bundles ranging from 4 to 40 ft (1.22–12.20 m) long and from 4 to 20 ft (1.22–6.10 m) wide, but usually limited to 9.90–10.8 ft (3.2–3.5 m).

3.10.6.2 Fin Materials

Aluminum fins are used because of high thermal conductivity, light weight, and formability, but their use is limited to low-temperature applications. Either SO_2 containing or marine atmospheres may corrode aluminum fins, so carbon steel might be preferred for such conditions. As the base tube is usually carbon or low-alloy steel, galvanic corrosion at the fin/tube junction is expected. Hence, finned tubes of carbon steel can be hot-dip galvanized to protect against galvanic corrosion and also to provide a metallic bond between the fin and the bare tube [68]. Carbon steel tubes with carbon steel or alloy steel fins are used in boiler and fired heater applications.

3.10.6.3 Header

Headers are the boxes at the tube ends which distribute the fluid from the piping to the tubes. Tube-to-header joints are either welded or expanded depending on the header design. A wide range of header designs has been developed for different applications and requirements. Almost all headers on air-cooled exchangers are welded rectangular boxes. Headers are usually constructed of carbon steel or stainless steel, but sometimes more exotic alloys are used for corrosion resistance. Standard header configurations are presented next. Major types of headers with a finned tube bundle are shown in Figure 3.64. A cross section of commonly used headers is shown in Figures 3.65 and the tubular header in Figure 3.66 shows finned tubes positioned with the header for welding.

3.10.6.4 Cover Plate Header

Either with removable cover or a bonnet, a cover plate header is typically used in chemical applications or services with severe fouling conditions. Although it is the more expensive alternative, the cover plate design allows full access to the inside of the headers for inspection and cleaning. The cover plate design is usually limited, however, to a maximum design pressure of 350 psig. The cover plate is attached to the header by a set of studs or through-bolts to a flange around the perimeter of the header.

3.10.6.5 Cover Plate Header (with Stud Bolts)

The cover plate header with stud bolts features top or bottom nozzles to allow cover removal without disassembly of the piping (Figure 3.64a). Nozzles may alternatively be integrated in the cover. The header is fit for a maximum pressure of 50 bar depending on the fluid temperature and type of sealing. A nonstandard header design is required for higher pressures. Tubes may be weld connected or expanded.

3.10.6.6 Welded Header (D-Type)

The welded header is mainly used for clean products or vacuum pressure conditions. The tubes are welded into tube sheets to which the bonnet-type header with the necessary nozzles is welded (Figure 3.64b).

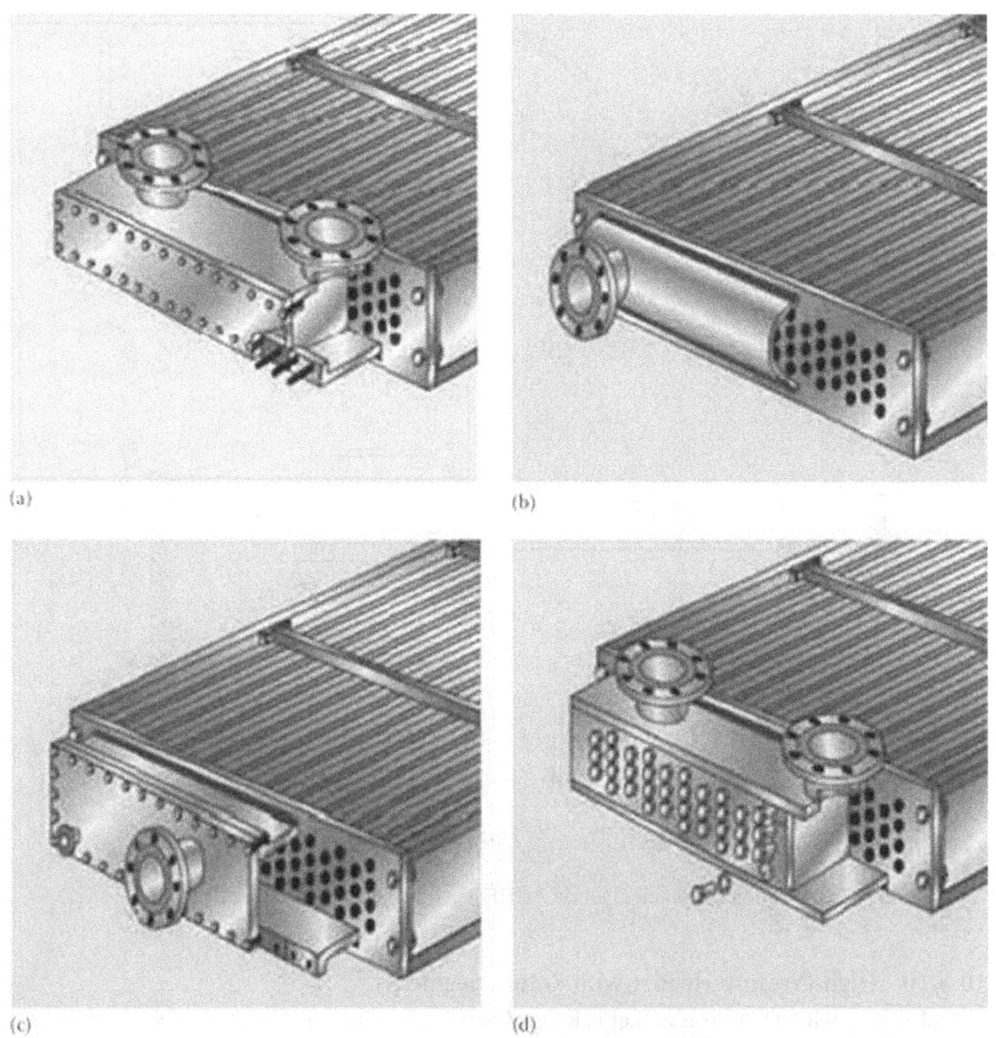

FIGURE 3.64 ACHE header types—(a) cover plate header with stud bolts, (b) welded header (D type), (c) cover plate header with through-bolts, and (d) plug type header. (Courtesy of GEA Luftkühler GmbH, Bochum, Germany.)

3.10.6.7 Cover Plate Header (with Through-Bolts)

The header features a removable cover plate with through-bolts for easy inspection and tube cleaning (Figure 3.64c). It is normally sufficient to remove a pipe bend for inspection, since the removal of the bend allows examination of part of the tube array. Tubes may be weld connected or expanded.

3.10.6.8 Plug Header

A vast majority of the headers are of the plug type. Plug header construction uses a welded box which allows partial access to tubes by means of shoulder plugs opposite the tubes. The design consists of a shoulder plug opposite each tube which allows access for inspection and cleaning of individual tubes (Figure 3.64d).

3.10.6.9 Pipe Manifold or Billet Header

Pipe manifold headers are common for all pressures, including full vacuum. Billet headers, machined from a solid piece of material, are used in extremely high-pressure (>10,000 psig) applications.

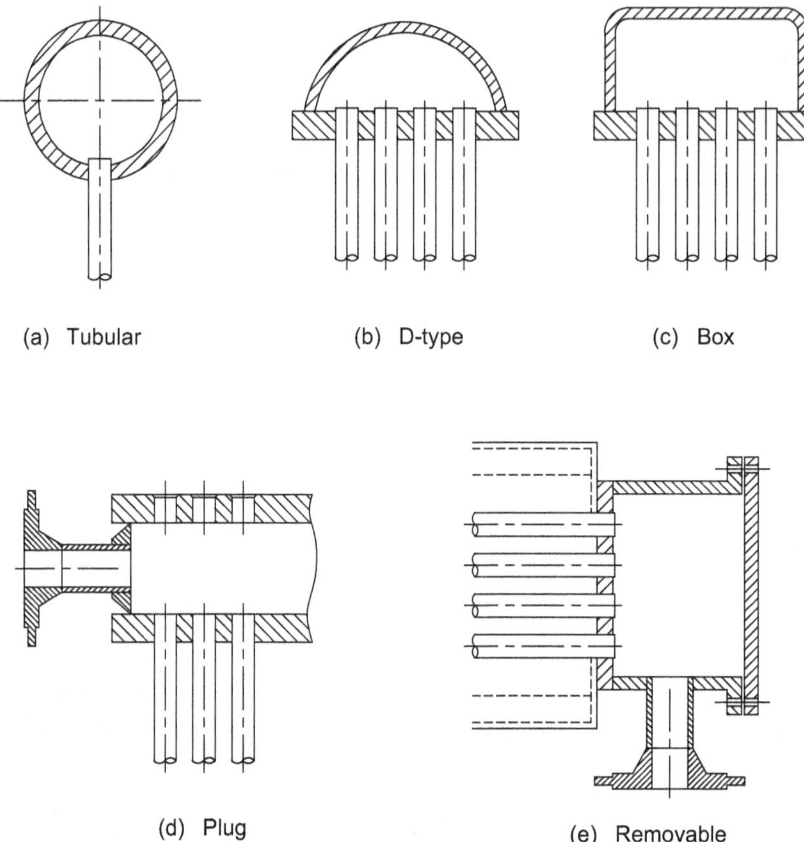

FIGURE 3.65 Cross section of commonly used ACHE headers.

3.10.6.10 High-Pressure Header with Return Segments

A threaded plug with a soft iron gasket is located opposite each tube. The plug may be removed for further tube expansion or for tube cleaning.

Number of passes on the tubeside: On the tubeside, ACHEs can be designed with up to 10 or more passes. But normally, the number of passes adopted is up to four. Figure 3.60 shows front and rear headers with single-pass and three-pass arrangements on the tubeside. Figure 3.67 shows a four-pass rectangular tubeplate header.

Number of tube rows: A minimum of three rows is used in the tube bank. The usual maximum limit is eight rows, though occasionally up to 12 rows are used.

3.10.6.11 Orientation of Tube Bundle

ACHE bundles can be installed either vertically or horizontally. Various orientations of ACHEs are shown in Figure 3.68. The most common orientation is in the horizontal plane. A considerable reduction in ground area can be made if bundles are vertically mounted, but the performance of the unit is greatly influenced by the prevailing wind direction. In general, the use of vertically mounted bundles is confined to small, packaged units. A compromise, which requires about half the ground area of the horizontal unit, is the A- or V-frame unit. In this type, two bundles sloped at 45°–60° from the horizontal are joined by their headers at the top or bottom to form the sloping side of the

FIGURE 3.66 Finned tubes positioned with the header for welding. (Courtesy of Polysoude S.A.S, Cedex, France.)

FIGURE 3.67 Four pass rectangular tube plate of an energy recuperator. (From Thermofin, www.thermofin.net)

A (i.e., roof type) or V, respectively. The A-frame type with forced draft fans below (Figure 3.69) is the more common and is used in steam-condensing applications.

3.10.6.12 Fan Ring Design

The fan ring design varies among manufacturers. The common configurations for fan rings are [83] the following: (1) eased inlet; (2) taper fan ring; (3) straight inlet; (4) flanged inlet; and (5) channel rings, which they are illustrated in Figure 3.70. Due to the cost of manufacture, the eased or taper fan rings are generally not utilized in standard products; however, they are available when required for special applications.

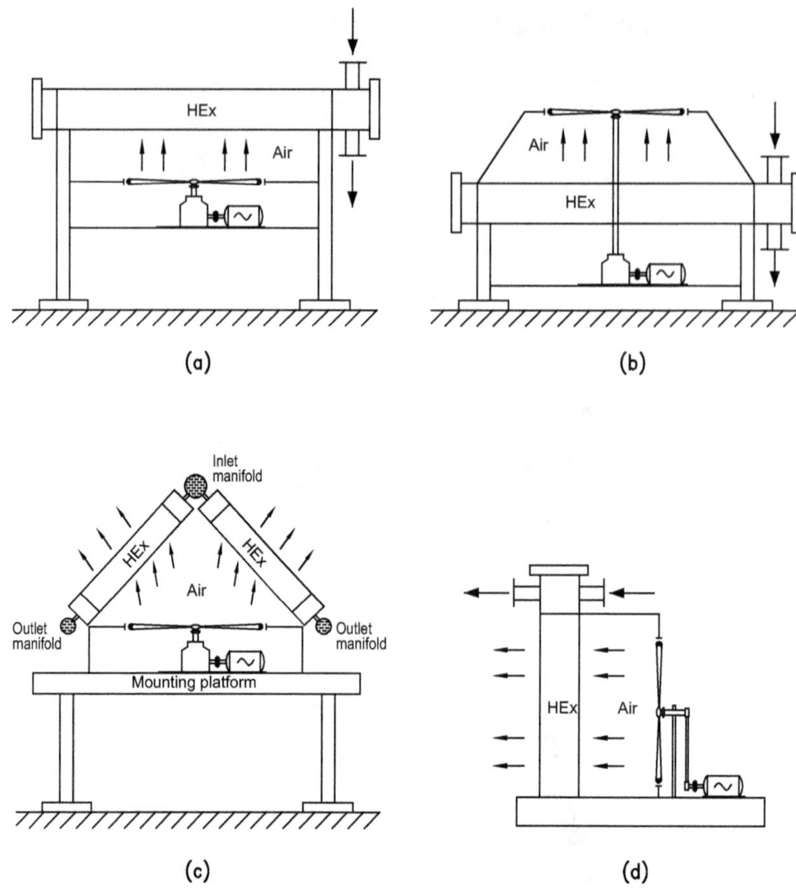

FIGURE 3.68 Orientation of ACHE tube bundle—(a) horizontal, forced draft, (b) horizontal, induced draft, (c) vertical, and (d) A-frame.

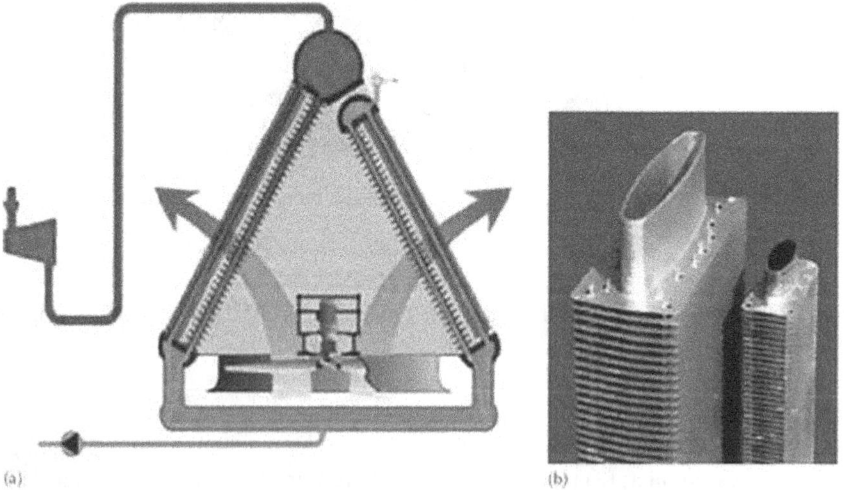

FIGURE 3.69 Air-cooled condensers (A-type orientation)—(a) two tube rows design with hot dip galvanized elliptical fin tubes with rectangular fins and (b) hot dip galvanized elliptical fin tube. (Courtesy of GEA Iberica S.A., Vizcaya, Spain.)

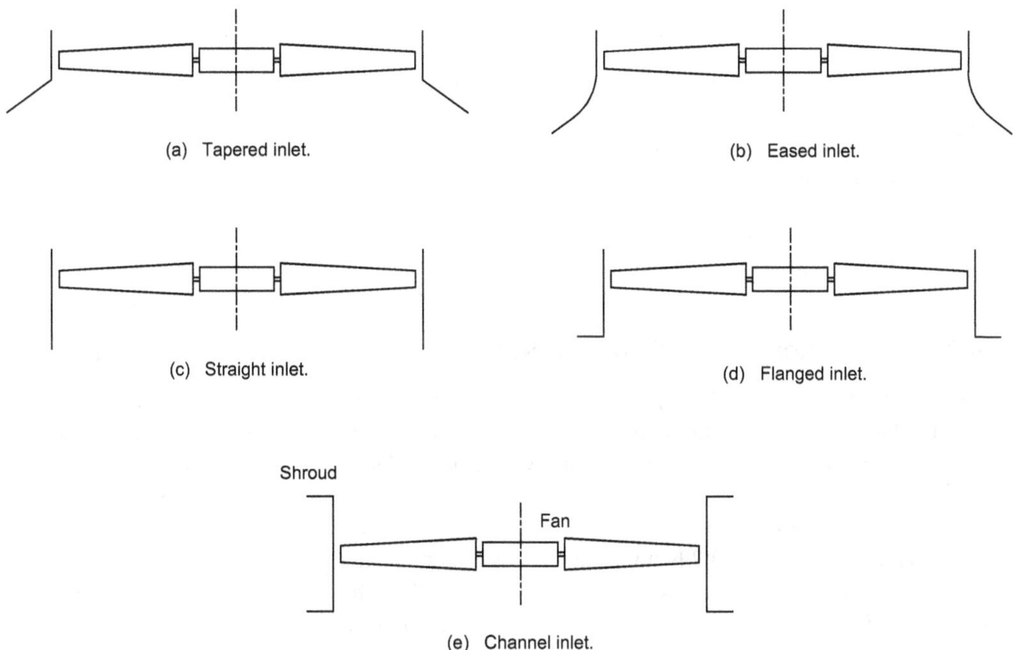

FIGURE 3.70 Fan ring configurations.

Tapered or eased rings: The tapered inlet and eased rings both allow for a more uniform exit of the air from the fan ring. Most fan design programs will indicate slightly less airside pressure drop and less horsepower required for this configuration. In addition, these fan rings allow for better air dispersion since the air is directed when it leaves the ring. In most ACHEs, the cost of producing this configuration outweighs the increased savings in horsepower, or in airflow efficiency.

Straight, flanged inlet or channel rings: These are the most common fan rings utilized by manufacturers. This ring is easily produced, and provides good air movement if close tip clearance between the ring and the fan is maintained. The depth of this ring will vary with the fan selected.

3.10.7 Operation and Maintenance

The reliability and thermal performance of any ACHE depends on how well its mechanical components are maintained periodically [85]:

1. Check belt tension on a regular basis, usually every six weeks at a minimum for ACHEs that are in continuous service. Check belts for wear, and sprockets for tooth wear or cracking.
2. Inspect fans at least annually by following these procedures:
   - i. Visually inspect each blade and the hub, looking for cracking, rubbing, or excessive wear.
   - ii. Check all bolting hardware for proper torque, especially the blade-clamping bolts.
   - iii. Check blades for proper pitch as per the ACHE specification sheet. (Be sure to use the fan manufacturer's blade angle procedure, as this can vary from one manufacturer to another.)
   - iv. Check the blade tracking to be sure that all blades are riding in the same plane in the fan ring.
   - v. Check blade tip clearances to be sure that they are all within recommended limits.

3.10.8 Suggestions to Improve Reliability and the Performance of ACHE

Suggestions to improve reliability and optimize the performance of ACHE by following these steps:

1. Perform routine maintenance.
2. Optimize performance
   a. Determine original design conditions/parameters/performance.
   b. Repair back to original condition.
   c. Perform upgrades.

3.10.9 Optimization of ACHE Performance

If air-cooled heat exchangers (ACHEs) are not meeting the required cooling demands, they need to be examined in the following areas and the necessary data collected. This will allow the end user to make an informed decision on options to improve the particular ACHE situation [86]:

i. Are the fans leaking air?
ii. What is the actual pitch? Is it the same as originally designed?
iii. What is the actual horsepower (hp) being used? Is there further potential in the motor?
iv. What kind of drives are installed—V-belt, timing belt, or gearbox?
v. What is the "tip clearance"?
vi. Are "inlet bells" installed?
vii. Are "seal discs" or "air seals" installed? If not, is the ACHE design lacking seal discs, or have they been removed and not reinstalled?
viii. Are the fin-tubes clean?
ix. Is the fan running at the design speed shown on the original specification sheet? Is the fan running in the correct direction, and are the blades pitched with the leading edge down?
x. Are these fans delivering the airflow that the original specification sheet suggests?
xi. Is this original design sufficient to meet your requirements now and in the future?

Ref. [86] addresses the above issues.

3.10.9.1 Methods of Optimizing Thermal Performance

Optimizing thermal performance is discussed in Ref. [74]. Here are the steps to bring the exchanger to the desired condition [74]:

Step 1—Determine the original design performance of the ACHE.
Step 2—Inspect the exchanger assemblies and subassemblies as listed below and make repairs to return it to its original condition:

Tube bundle—frame, tube supports and tube keepers, contoured fin supports or wiggle strips, air seals, headers, and tubes.

Fans—blade pitch, blade tracking, tip clearance, seal disc, inlet bell, fan speed, vibration or pulsing.

Mechanical components—belts, sprockets, alignment, and motors.

Controls—louvers, actuators, steam coil, variable-frequency drives, and auto-variable hub.

Once the mechanical inspection is completed, steps should be taken as necessary to bring the ACHE back to the as-built condition. In some instances, simple bundle cleaning and fan adjustments, repairs, and/or modifications can bring the unit back to near-design performance.

Step 3—Perform upgrades.

If bringing the ACHE back to its original condition does not provide enough cooling duty, you still have several options to improve and optimize ACHE. Typical steps that can be taken to optimize ACHE, in order of increasing cost are hereunder:

Fans and mechanical components
In most applications, one of the quickest ways to get more cooling duty out of an ACHE is to increase airflow from the fans by increasing the fan blade pitch, increase the fan speed, reset tip clearances, install inlet bells, install a fan seal disc, and install high-efficiency fans.

Tube bundle
Clean the finned-tube bundle and clean inside the tubes.

Retube the bundle—Overall, a 10–50% increase in ACHE duty can be achieved by retubing the bundle, depending on the level of finned tube deterioration.

Replace the bundle—Severe finned-tube corrosion is sometimes an indication that the headers and bundle frame may have reached the end of their useful life. It is sometimes more expedient to replace the entire bundle with a new one. As with retubing, 10–50% increases in ACHE duty can be achieved.

3.10.10 Operation and Control of ACHEs

In addition to the fact that the process flow rate, composition, and inlet temperature of the fluid may vary from the design conditions, the ambient air temperature varies throughout a 24-hour day and from day to day. Since air coolers are designed for maximum conditions, some form of control is necessary when overcooling of the process fluid is detrimental, or when saving fan power is desired. Although control could be accomplished using bypassing of process fluid, this is rarely done, and the usual method is air flow control. Varying air flow can be accomplished by [72]:

1. Adjustable louvers on top of bundles.
2. Two-speed fan motors.
3. Fan shut-off in sequence for multifan units.
4. AUTO-VARIABLE® fans.
5. Variable frequency for fan motor control.

3.10.10.1 Common Operating Problems for Air-Coolers

Air-cooled heat exchangers (ACHEs) experience operating problems not encountered in other types of heat exchangers. Some of the more common operating problems with air-coolers to be addressed to increase the heat transfer performance of the air-cooler are hereunder[87].

3.10.10.2 Reduced Air Flow Rate due to Dirty Tube Bundles

The most efficient way to determine if an air-cooler is dirty and experiencing reduced air flow is to develop an air flow profile using an anemometer. Air flow measurements are a non-intrusive way to determine when the tube bundle of an air-cooler needs to be cleaned.

Reverse Flow
There are two areas where reverse flow is most prevalent. The most common one is at the tip of the fan blade where it meets the plenum housing. This gap should be approximately 3/8" but not greater

than 3/4" as per API 660. A less common form of reverse flow in ACHE occurs if there is a gap in the area above the motor or hub. Air will loop back through the center of the fan blade and be caught in a recycle. This problem will again be obvious if an air flow profile is taken.

Blade Pitch

A common problem with air-coolers is improper blade pitch angle. It is always best to refer to the manufacturer's specifications to set the optimum blade pitch angle. Generating air flow profiles can help narrow in on the optimum blade pitch angle.

Motor Current

A related problem that is often encountered is motor not running near the full load amps (FLA). To optimize peak air flow and heat transfer, fan motors should operate near their FLA set point. It is preferred to have the %FLA at or above 85.

Mechanical Integrity

The overall condition of an ACHE can greatly reduce its ability to transfer heat. The following list contains common areas where air-coolers can experience minor problems that are relatively easy to fix [87]:

1. Louvers. Missing or inoperable louvers are a common problem. At full open, the louvers should be at least 50–60% open to allow unimpeded air to travel through the tube bundle.
2. Plenum. The plenum should be inspected periodically to confirm that no panels are missing and that no large holes exist.
3. Summer water spray on tube bundle. Although a fairly common practice, it is not recommended to spray water on tube bundles to provide temporary additional heat transfer capacity during hot summer days. This practice leads to a reduction in performance over time and if the bundle is sprayed directly, tube-to-fin bonding, fouling, and corrosion problems, and in time the tube bundle needs to be replaced.
4. Bent or crushed tube fins. A common problem is bent or crushed tube fins. In this case, a comb-type device can be used to rake through and lift the fins back into a position perpendicular to the tubes.

3.10.10.3 Airside Fouling of ACHEs

Airside fouling of heat exchangers depends on the number of factors on the heat exchanger (system) side and the airside. In HVAC systems, fin and tube heat exchangers are extensively used for residential, commercial, and industrial applications. For the fin and tube exchangers, fouling is dependent on the following factors: (i) fin type; (ii) number of tube rows; (iii) heat exchanger type; (iv) fin spacing; (v) refrigerant type; and (vi) filters, if any.

Cleaning

Internal tube cleaning is performed by automated high-pressure water-jetting techniques such as tube lancing equipment. Chemical cleaning can be used to remove sticky pollutants that present challenges to hydro-jetting applications [88].

a. *Dry cleaning.* Use soda blasting to clean induced and forced draft ACHE. For induced draft ACHE, this method can be carried out fully online. For forced draft ACHE, it is required for the fan to be stopped.
b. *Foam cleaning.* Foam cleaning allows the active ingredient to penetrate into the pollution as it clings to the fins. The foam with the dislodged pollution is then washed down with water.
c. *Mechanized hydro-jetting.* For the cleaning of A-frame and forced draft ACHE, use multiple nozzle jetting heads, which move along a rail by chain and air-motor.

3.10.11 Problems with Heat Exchangers in Low-Temperature Environments

ACHEs are designed to perform at ambient temperatures ranging from −60°F to 130°F. In extremely cold environments, overcooling of the process fluid may cause freezing. This may lead to tube burst, and hence freeze protection is required to prevent plugging or damage to the tubes. Proper design of ACHEs for cold climate service requires a well-balanced consideration of fluid flow, heat transfer, structural design, air movement, wind effects, temperature control in cold climate, and economics [77].

3.10.11.1 Temperature Control

Several methods are used to control the performance of ACHEs to meet variations in weather and process requirements. The current practice for ACHE design in cold climates for problem services includes: (1) recirculation of hot air through the tube bundle; (2) steam coils provided to preheat air for start-up conditions—these may be mounted at the cooler base to warm up the inlet air, and according to the recommendation of API 661, this steam coil should be separate from the main cooler bundle; (3) control devices that enable part of the process fluid to bypass the unit; (4) for some units, particularly viscous oil coolers, concurrent design is appropriate, where the hot fluid enters nearest the cold ambient wall, and the outlet where wall temperature is critical is in the warm air stream [89]; and (5) in certain cases, simply controlling the process outlet temperature with fan switching, two-speed fan drives, variable-speed motor automatic louvers, variable pitch fans, or autovariable-pitch fans is sufficient. Each case must be considered on its merits to decide on the best method of control.

3.10.11.2 Extreme Temperature Controls

For extreme cases of temperature control, such as prevention of freezing in cold climates in winter, or prevention of solidification of high pour-point or high melting point materials, more sophisticated designs are available. Extreme case controls include [72]:

1. Internal Recirculation. By using one fixed-pitch fan blowing upward and one AUTO-VARIABLE pitch fan, which is capable of negative pitch and thus of blowing air downward, it is possible to temper the air to the coldest portion of the tubes and thus prevent freezing.
2. External Recirculation. This is a more positive way of tempering coolant air, but is practical only with forced draft units.
3. Co-current Flow. For high pour-point streams, it is often advisable to ensure a high tube wall temperature by arranging the flow co-currently, so that the high inlet temperature process fluid is in contact with the coldest air.
4. Use Auxiliary Heating Coils. Steam or glycol heating coils are placed directly below bundles. Closing a louver on top of a bundle will allow the heating coil to warm the bundle or keep it warm in freezing weather, so that on start-up or shut-down the material in the bundle will not freeze or solidify.

3.10.11.3 Recirculation ACHE

There are situations where a minimum tubewall temperature must be maintained thought the year. Hence a special arrangement is needed to maintain a constant air temperature (the design ambient temperature) of the fan delivered to the tube bundle at all times, and such an arrangement is shown in Figure 3.71. Figure 3.71 shows the louvers arrangement to maintain constant air temperature delivered by a fan throughout the year [79].

3.10.12 Design of ACHEs for Viscous Liquids

Film coefficients for laminar flow inside tubes are very low and of the same order of magnitude as film coefficients for air flowing over the outside of bare tubes. Therefore, there is generally no

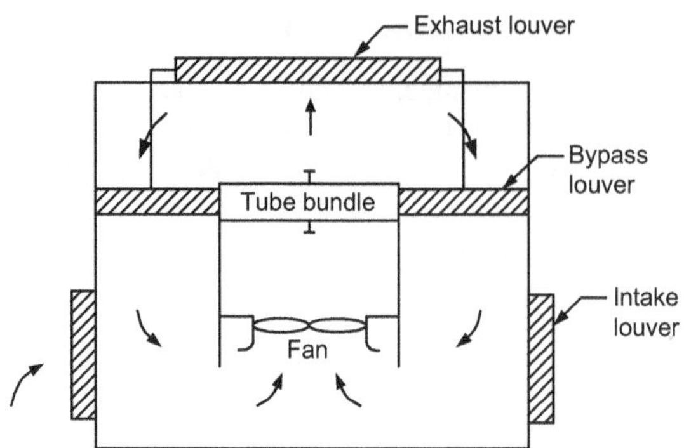

FIGURE 3.71 Louvers arrangement to maintain constant inlet air temperature throughout the year. (Adapted from Mukherjee, R. (1997) Effectively design air cooled heat exchangers. *Chemical Engineering Progress*, pp. 26–47.)

advantage in using fins on the airside to increase the overall heat transfer rate since the inside laminar flow coefficient will be controlling.

3.10.13 Windmilling or Free-Spinning of ACHE

Free-spinning (or "windmilling") occurs when ACHE fans are powered off, becoming free to rotate backwards. In this condition, fans not only pose a safety hazard to maintenance workers, but they also place a strain on motor system components when they power back up. There is a simple, economical solution to this problem. It's called an anti-rotation device, and it makes ACHE fans safer to work on while reducing wear and tear on fan system components. It's a simple, mechanical device that bolts onto the large fan pulley with no additional modifications required. Also, if the system is built so that more than one fan shares the same housing, the fans can windmill backward when one or more fans are turned off. This can be very destructive to the drive system once this fan gets a signal to go forward. It can be controlled by VFDs.

3.10.13.1 Anti-Rotation Devices

Anti-rotation devices like the Gates Draftguard unit provide an economical solution to the two major problems created by windmilling ACHE fans [90,91]. From a safety standpoint, they secure fan drives and prevent them from rotating freely when not receiving power, allowing maintenance technicians access to the fan cage without risk of injury. Secondly, they prevent hard starts by allowing fan drives to power up from a neutral, standstill position, minimizing damage to drive components caused by shock loading. A Gates Draftguard™ anti-rotation device is shown in Figure 3.72. The device was a one-way clutch system comprised of a shaft bushing, clutch adaptor, clutch housing, clutch bearing with seal, clutch shaft, bottom plate, and fixed surface. The clutch adaptor also included a locking plate and pin.

3.10.13.2 V-Belt versus Synchronous Belt Drive

An ACHE V-belt drive, even when properly installed and maintained, is 93% efficient or less. V-belt drive efficiency is further compromised by hard starts when ACHE drives are windmilling. Many users of ACHE systems have converted to synchronous belt drives for greater efficiency, which translates to higher energy savings. Synchronous, or toothed, belts run on toothed sprockets. While VFDs may ease wear and tear on motor system components, they are costly, require additional wiring, and do not address the safety issue caused by windmilling ACHE fans.

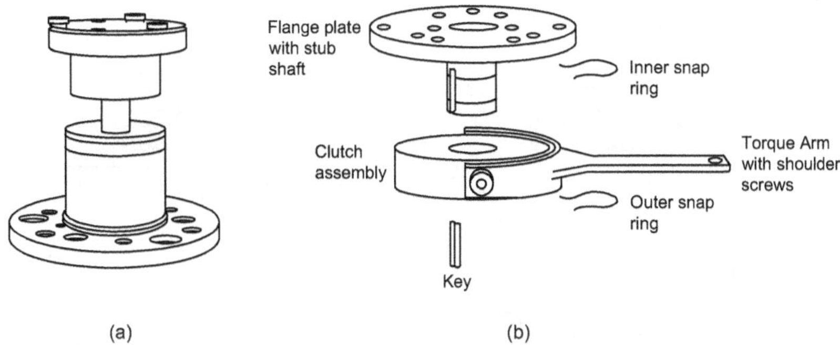

FIGURE 3.72 Draft guard assembly

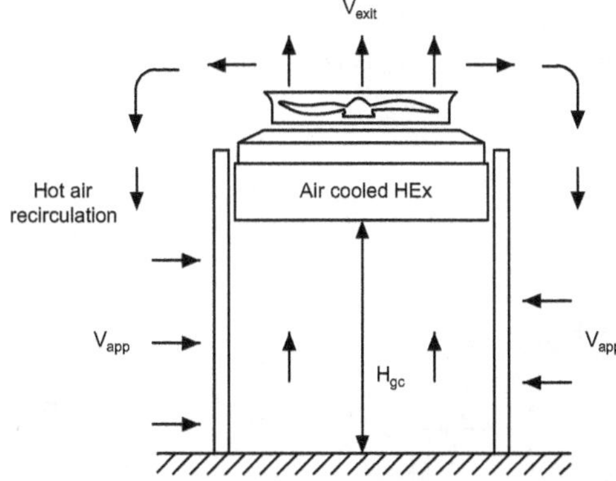

FIGURE 3.73 Air recirculation (Illustration). (Adapted from www.calgavin.com/articles/heat-exchanger-problems-part-3.)

3.10.13.3 Preventing Hard Starts with Variable Frequency Drives

One solution to avoiding shock load when starting up a windmilling ACHE fan is to install a VFD. A VFD regulates motor speed electronically by changing the frequency of the alternating current supplied to the motor.

3.10.14 Hot Air Recirculation

Recirculation of hot plume air occurs in most forced draft ACHEs. The resultant increase in inlet cooling air reduces the effectiveness of the heat exchanger. To reduce the amount of recirculation, wind walls may be erected along the periphery of the heat exchanger. This problem has been analyzed experimentally and numerically by Kroger [92]. For design aspects of hot air circulation, refer to Gunter et al. [93].

3.10.14.1 Why Does Hot Air Recirculation Happen?

The hot air leaving the cooler is traveling vertically upwards and is less dense than the surrounding air (and therefore buoyant). Recirculation requires some of this hot air to travel downwards in order to re-enter the cooler, as shown in Figure 3.73. For this to occur, the air pressure at the sides of the cooler must be less than the air pressure at the top of the cooler. This pressure difference is the driving force for recirculation. Air approaching the sides of the cooler (V_{app}) is somewhat higher than the velocity of the air leaving the top of the cooler (V_{exit}). To avoid hot air recirculation, the air approach velocity should be less than the air exit velocity [94].

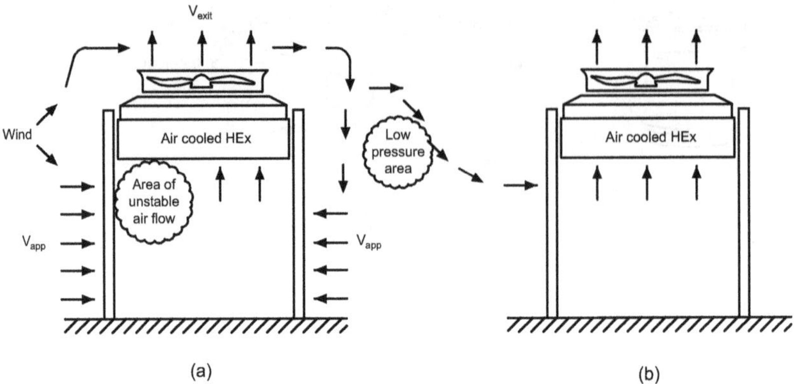

FIGURE 3.74 Recirculation due to Wind effect. (Adapted from www.calgavin.com/articles/heat-exchanger-problems-part-3 and JGC and PETRONAS Collaborate to Improve Efficiency of Existing LNG Plants Using AI and IoT | News Releases 2018 | JGC HOLDINGS CORPORATION.)

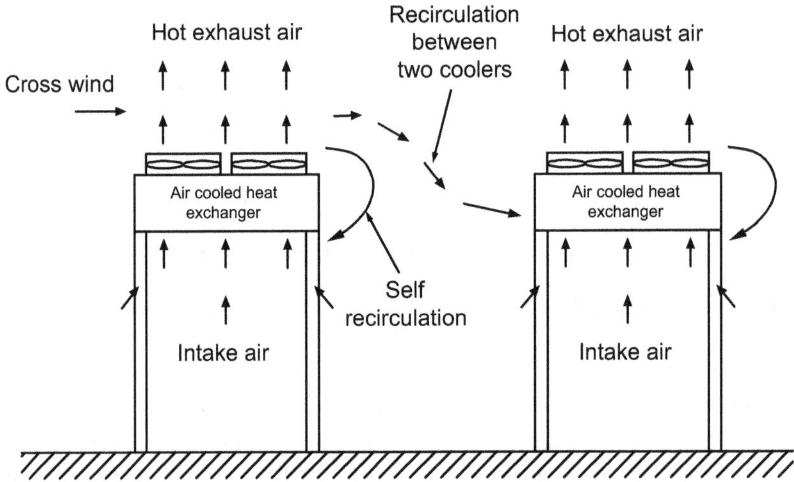

FIGURE 3.75 Hot air travels from the exit of one cooler to the inlet of another. (Adapted from www.jgc.com/en/business/ tech-innovation/tech-journal/pdf/jgc-tj_03–02(2014).pdf)

Wind Effects

Cross winds can have other damaging effects on air-cooled exchangers. For example, the fans on the upwind side of the cooler can be affected by unstable airflows caused by the wind interacting with the cooler structure. Wind effect is shown in Figure 3.74 [95].

Plant Layouts for Multiple Air Cooler Installations

As shown in Figure 3.75, hot air can travel from the exit of one cooler to the inlet of another. It is therefore important that the overall plant layout minimizes the risk of such hot air migration to vulnerable coolers. This requires consideration of the orientation of coolers to the prevailing summer wind direction, relative cooler positions and elevations, mixing of forced and induced draft coolers, and the influence of local topology. The presence of other sources of hot air (e.g., engine exhausts) should also be considered [96].

An example of a restricted approach area is where a cooler is positioned in the middle of a large bank of air coolers, as shown in Figure 3.76 [93]. Compared to a stand-alone cooler, such

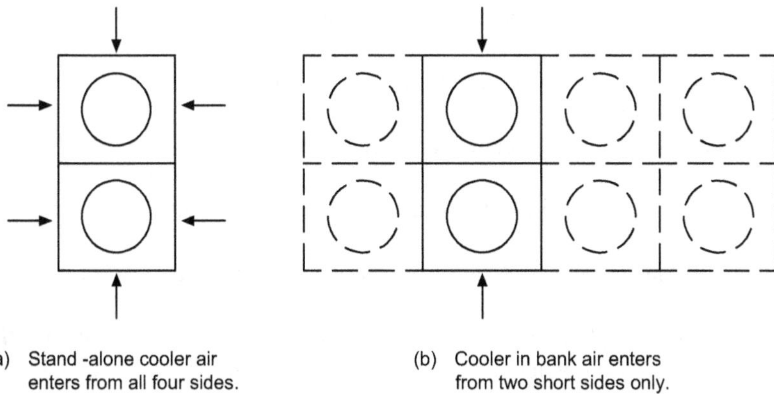

(a) Stand -alone cooler air enters from all four sides.
(b) Cooler in bank air enters from two short sides only.

FIGURE 3.76 Restricted approach for air entry. (Adapted from www.calgavin.com/articles/heat-exchanger-problems-part-3).

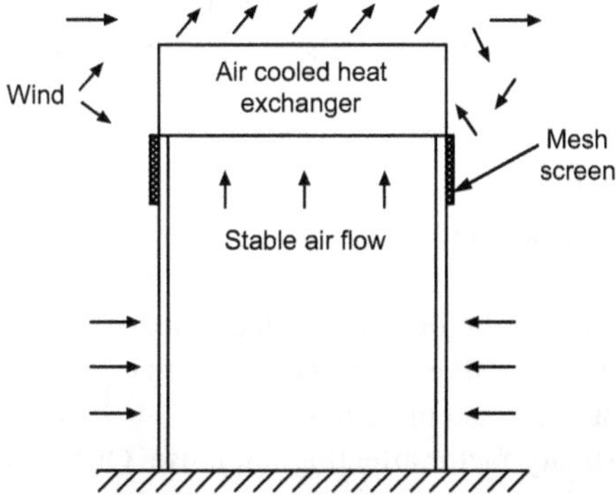

FIGURE 3.77 Installation of mesh screen. (Adapted from www.calgavin.com/articles/heat-exchanger-problems-part-3).

an arrangement significantly reduces the air approach area and therefore leads to relatively high approach velocities.

Designing to Avoid Hot Air Recirculation Problems

Use mesh screens combined with increased ground clearance to protect large forced draft air coolers from the undesirable effects of cross winds. Such screens are mounted on the sides of the air cooler structure and provide stable air conditions in the fan entrance area, as shown in Figure 3.77. The screens also ensure greater physical separation between the exit and approach air streams, such that the risk of recirculation is reduced [94].

3.10.14.2 Avoiding Hot Air Recirculation

Problems associated with hot air recirculation are the direct effect of poor exchanger design and location. Minimum allowable distances between air coolers and other process equipment should be considered.

i. Using induced draft fans which force the air away from the bundle.
ii. Baffles and/or a stack on top of the bundle for a forced draft unit (or fan on an induced draft unit) will also direct the air away from the bundle.

iii. A-frame, V-frame, and vertical bundle arrangements should not be used if recirculation is a potential problem.
iv. Units should preferably be placed in the open and at least 23–30 m from any large building or obstruction to normal wind flow.
v. Units should not be located near heat sources.

HTRI-proposed solutions to eliminate hot air recirculation include [97]:

i. Allowing all fans to be at the same vertical elevation when adding the new ACHEs to the existing units
ii. Implementing solid plate seals to close off short lateral gaps between the ACHEs
iii. Adding wind walls and wind screens around the outer periphery of the ACHEs.

3.10.15 Fan Noise

Air-cooled heat exchangers are a source of plant noise. Air-cooled heat exchangers have four main noise sources: fans, drives, motors, and structural vibration. ACHE noise is mostly generated by fan blade vortex shedding and air turbulence. Other contributors are the speed reducer (high torque drives or gears) and the motor [98,99].

3.10.15.1 Fan Noise Control

Methods to control fan noise are hereunder:

i. Fan noise is the greatest contributor to air-cooled heat exchanger noise. Fan noise is a function of fan tip speed, input power, diameter, and pitch angle, plus factors for tip clearance, inlet flow conditions, motor noise and drive noise.
ii. Blade design is the primary factor affecting fan noise. One of the most effective methods of air cooler noise control is to utilize low-noise fan blades.
iii. Inlet flow conditions, tip clearance, and recirculation at the fan hub can also affect fan noise.
iv. Inlet bells should be smooth and rounded, and obstacles in the air stream should be minimized to promote nonturbulent flow into the fan.

Noise due to air flow is moderated with a baffled ACHE design as shown in Figure 3.78 [98].

3.10.15.2 Drive Noise

Drive noise can contribute 1–3 dBA to the overall noise of the air-cooled heat exchanger. Gears emit noise at the gear mesh frequency, the frequency at which the teeth of the gear sprockets come together. In general, double reduction gears are quieter than single reduction gears, and worm gears are quieter than both of these other designs.

3.10.15.3 Motor Noise

The effect of motor noise is much less than fan noise. Motors can affect the noise level of the air-cooled heat exchanger by 1 dBA. Motor noise can be reduced by using a premium efficiency motor instead of a standard efficiency motor.

3.10.15.4 Noise from Structural Vibration

Periodic forces from rotating equipment can drive resonances in the structure of the air-cooled heat exchanger. Vibrations in panels can produce low-frequency noise. The problem can and should be eliminated in the design stage by analyzing the structure for its resonant frequency and making sure the fan and motor do not operate near this frequency.

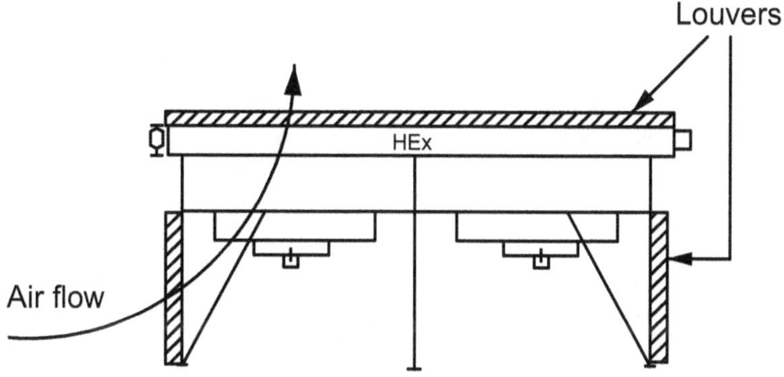

(a) ACHE - Conventional design.

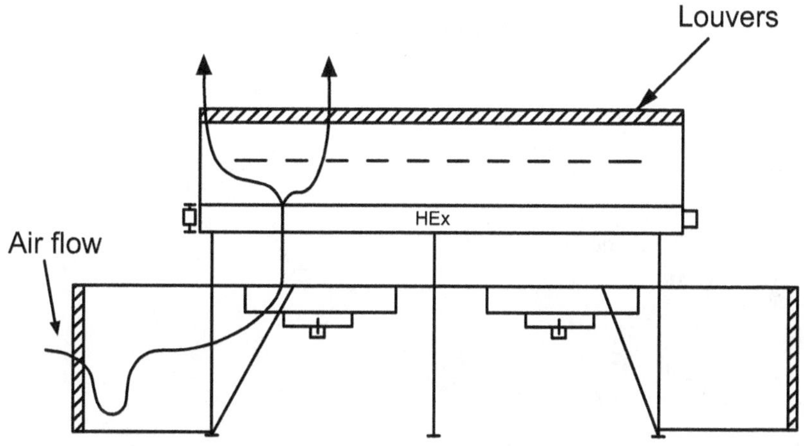

(b) ACHE - Baffled design.

FIGURE 3.78 Noise due to air flow is controlled by a baffled air inlet of an ACHE. Adapted from Pinkerton, A., Chapple, S. (1993) *The design of Quiet Air-Cooled Heat Exchangers, A Paper for The Energy Resources Conservation Board*, November 9, 1993, Calgary, Alberta pp. 1–20. Hudson Products Corporation, Houston, Texas.)

Noise due to air flow is controlled by installing baffled air inlet to an ACHE, as shown in Figure 3.78 [98].

3.10.16 Design of ACHE

3.10.16.1 Design Variables

Modern ACHEs are designed thermally by computers, which are capable of examining all design variables to produce the optimum unit. In view of the increased size and cost of these units in large plants and the competitive nature of the industry, improved and more sophisticated designs are essential to satisfy the needs of the industry and society [92]. Important design variables pertaining to the tube bundle include:

Tube (diameter, length and wall thickness, number of tube rows)
Fins (height, spacing, thickness)

TABLE 3.3
Software Program Structure for Design of Air-Cooled Heat Exchanger

Input parameters	Heat duty, process parameters, fouling resistances, allowable pressure drop, minimum ambient temperature
Code	ASME Code, Section VIII, Div. 1
Standard	API 661
Header types	Plug, studded cover, flanged confined cover, flanged full face cover, bonnet, U-tube, pipe (dished), header gasket type
Tube parameters	Diameter (internal, outer), length and wall thickness, and material
Bundle parameters	Size (length, width, and depth)
	Tubes per row, tube rows per pass, passes per bundle, bundle connection, and bay connection
Tube layout pattern	Triangular, rotated triangular, square, diamond
Fin size	High fin (fin height, spacing/fin density, thickness, diameter of fin)
Fin configuration	Circular, tension wound, others
Fin attachment	Extruded, L-type weld, L-type tension, embedded, sleeve, metal coated, etc.
Nozzle details	Number of inlet and outlet nozzles, inlet and outlet nozzle diameter
Type of flange	Slip-on, weld neck, lap joint, ring type joint, long weld neck
Bundle orientation	Horizontal, vertical or inclined
Fan	Horse power rating, size, number of blades, rpm, fan pitch control, noise level, and type of drive
Draft type	Forced, induced draft
Others	Recirculation, operation of louvers, controls for winter operation

Space (length, width, and depth)
Number of tube rows
Number of passes
Face area
Horsepower availability
Plot area

Designers usually optimize criteria such as [78]:

Capital versus running cost
Forced versus induced draft fan
Type and spacing of fins
Number of tube rows
Fan design and noise level

Some of the governing factors in the design of the ACHE and hence the thermal design program structure is shown in Table 3.3.

Software for ACHE Design
ACHE design software is made to demonstrate the thermal design and sizing calculations of air-cooled heat exchangers. The software will design horizontal induced draft or forced draft units for liquid and gas services. Software features and details of calculations are discussed in Ref. [100].

3.10.16.2 Air Velocity

If the airflow rate is known, the tentative air velocity may be chosen, which establishes the cooler face area. In air-cooled heat exchanges, face velocity is usually in the range of 1.5–4 m/s [100]. Face velocity is based on the gross cross-sectional area for airflow (face area), as though the tubes were absent.

3.10.16.3 Airside Pressure Losses, Fan Power, and Noise

The total airside pressure loss is the sum of pressure loss through the finned-tube bundle and the pressure loss due to the fan and plenum. Bundle pressure loss is calculated from the makers' test data or correlations.

3.10.16.4 Capital versus Running Costs

A total cost evaluation usually consists of the following four elements [72,79]:

1. Project equipment cost
2. Field installation cost
3. Electrical distribution cost
4. Operating cost

Consider the preferred drive and cost of horsepower, and the payback time for balancing capital investment for adding surface area, against operating costs of fan horsepower.

3.10.16.5 Design Air Temperature

The following data are needed for realistic estimates of the design air temperature [75]:

An annual temperature probability curve
Typical daily temperature curves
Duration frequency curves for the occurrence of the maximum dry-bulb temperature

3.10.16.6 Design Tips

A list of points to help to create the most economical design of ACHE for process industries as suggested by GEA Heat Exchangers/GEA Rainey Corporation, Catoosa, OK, are as follows:

a. Maximize tube length while maintaining 40% or more fan coverage.
b. Design air cooler with a 1:3 ratio (width to core length ratio) which helps reduce the header size, the most expensive portion of an air cooler, and maintain proper fan coverage.
c. Minimize tube rows to increase heat transfer effectiveness of area, minimize header thickness. Typically between four and six tube rows (maximum).
d. Try to use 25.4 mm (1 in.) tube diameter, depending on service. Even high-viscosity services that appear to benefit from larger diameter tubes can typically be designed cheaper with more 25.4 mm (1 in.) diameter tubes. Minimum thickness of tubeplate shall be 20 mm for carbon and low-alloy steel tubes and 15 mm for high-alloy steel or other material [82].
e. Utilize allowable pressure drop. This allows more passes in the bundle reducing the cooler size.

Other design tips are as follows [101]:

i. A general rule of thumb for the airside face velocity through the coils is as follows:
   Three row coil 800–850 fpm, four row coil 500–700 fpm, five row coil 450–600 fpm, and six row coil 350–500 fpm.
   On new construction, good design practice would normally restrict the number of tube rows to three. This allows for some modification, if needed later, to allow for higher heat load applications. Normally, on gas compressor applications, the air is at such a high temperature after the four rows that additional cooling from additional rows is minimal.
ii. A minimum fan to coil face area of 40%.
iii. Air dispersion angle of 45° should not be exceeded, without compensating for this in the design.

iv. At least two fans per bay for each ACHE should be provided in order to maintain operability during breakdown/maintenance of one fan. Fans should be operated in the mid-range of the fan performance, this should be applied to tip speed, ability to handle the static pressure, and blade angle.
v. The fan tip speed shall not exceed the maximum value specified by the fan manufacturer for the selected fan type. Fan tip speed shall not exceed 60 m/s (12,000 ft/min) unless approved by the purchaser. The radial clearance between the fan tip and the fan orifice ring shall be as specified by the design.
vi. More surface area is always better than more airflow.
vii. Units should not be located near heat sources. Direct drive (fan shaft directly coupled or through gear box to electric motor) is preferred over V-belt drive.

3.10.17 Thermal Design of Air-Cooled Heat Exchangers

In thermal design of an ACHE, there are more parameters to be considered than for shell and tube exchangers. ACHEs are subject to a wide variety of constantly changing climatic conditions which pose problems of control not encountered with shell and tube exchangers. Designers must achieve an economic balance between the cost of electrical power for the fans and the initial capital expenditure for the equipment. A decision must be made as to what ambient air temperature should be used for the design. Since the number of tube rows, the face area, the air face velocity, and the geometry of the surface can all be varied, it is possible to generate many solutions to a given thermal problem. However, there is obviously an optimum solution in terms of capital and operating costs.

1. Thermal design correlations for "*j*" and "*f*" factors.
i. The tubeside heat transfer and pressure drop are calculated in the same way as for shell and tube heat exchangers.
ii. For the airside heat transfer rate a number of calculation methods are available, including correlations by Ref. [25], Briggs and Young [31], and ESDU [35].
iii. The fin efficiency is in the range 0.8–0.9 for fin types and dimensions generally used in ACHEs.
iv. For predicting the airside pressure loss across the finned tube bank—those most commonly used correlations are by Robinson and Briggs [32] and ESDU [35].
v. Typical values of overall heat transfer coefficient for various fluids are given in ESDU [35] and these may be used to obtain approximate sizes. This item also describes the *C-value method* of comparing costs for various heat exchanger types.

3.10.17.1 Air-Cooled Heat Exchanger Design Procedure

Preliminary sizing by Brown's method: Once the inlet temperature is known, a reliable first approximation of the cooler design may be obtained. There are several short-cut manual methods available in the literature. One such method is that of Brown [78]. Brown considers his method will establish a size within 25% of optimum. The method can be stated by the following simple steps.

First, an overall heat transfer coefficient, U is assumed, depending on the process fluid and its temperature range. Approximate overall heat transfer coefficients for ACHEs are tabulated in Ref. [78].

Second, the air temperature rise $(t_2 - t_1)$ is calculated by the following empirical formula:

$$t_2 - t_1 = 0.005\ U\left(\frac{T_1 + T_2}{2} - t_1\right) \tag{3.142}$$

where T_1 and T_2 are hot process fluid terminal temperatures.

Third, the estimate is based on bare tubes, with a layout and fan horsepower estimated from that, so as to avoid the complexity of fin types. Approximate bare tube surface versus unit sizes are tabulated in Ref. [78].

Preliminary Sizing

The procedure to follow in ACHE design includes the following steps:

1. Specify process data and identify site data.
2. Under normal operating conditions, air outlet temperatures should not exceed 60°C with fans in operation and 80°C with free convection on the airside.
3. Assume the layout of the tube bundle (from preliminary sizing), fin geometry, and air temperature rise.
4. Calculate film coefficients and overall heat transfer coefficient, mean temperature difference and correction *F*, and surface area. Check this surface against the assumed layout.
5. If the required surface matches with the assumed layout, calculate the tubeside pressure drop and verify with the specified value.
6. If the surface area and tubeside pressure drop are verified, calculate the airside pressure drop and fan horsepower.

Final Sizing [78]

1. Select a value for the overall heat transfer coefficient.
2. Choose a bare tube size and material to satisfy the process conditions. Normal base tube sizes are 19.05 mm or 25.4 mm outside diameter for liquid cooling or condensing under pressure or 38.1 mm to 50.8 mm for condensing under vacuum.
3. Pick a design value for the inlet air temperature. This is usually the 95% dry bulb temperature (i.e., the maximum daily temperature does not exceed this value 95% of the time). Choose an outlet air temperature, typically 55°C. Due to climate change, in many tropical countries, in summer the ambient temperature itself attains 50°C.
4. Calculate the mean temperature difference (MTD). Be sure to use the proper F correction factor for crossflow with the number of tube passes likely to be employed.
5. Calculate A_o, the bare tube heat transfer surface area required.
6. Calculate the air flow rate from the heat balance equation. Calculate the volumetric air flow rate, in m^3/s, at the inlet conditions.
7. Using a face velocity of 3 m/s for air, compute the face area of the bundle.
8. Choose the number of tube rows and tube length and pitch to satisfy both the bundle face area and the required heat transfer area.
9. Choose the number of tube passes to obtain a process side velocity in a reasonable range (1.0–2.0 m/s for liquid cooling).
10. Calculate fan power: Power = BHP/η_{Motor}.

3.10.17.2 Air-Cooled Heat Exchanger Data/Specification Sheet

The specification defines the minimum requirements for the thermal and mechanical design, supply of materials, fabrication, inspection, testing, guarantee, and delivery of an air-cooled heat exchanger, to be installed on site. An ACHE data/specification sheet is shown in Figure 3.79. It should be submitted to the vendor along with the job "specific requirements" as follows [76]:

1. Location of the plant and elevation above sea level
2. Location of the unit in the plant
3. Temperature variation due to weather conditions

Air Cooled Heat Exchanger Specification Sheet	
Customer :	Equip. Tag No :
Plant Location :	Manufacturer :
Service :	Size :
Job Number :	Surface area per bundle :
Model No :	Type : Forced or Induced Draft
Heat duty :	Orientation : A - or V - frame

Process Data					
Fluid Circulated :					
	In	Out		In	Out
Fluid quantity; total			Molecular Weight, Vapor		
Liquid			Molecular Weight, Non condensable		
Vapor			Viscosity		
Non condensable			Thermal conductivity		
Steam / water			Latent heat		
Temperature			Inlet pressure (Specify gauge or absolute)		
Dew Point			Outlet Pressure		
Freezing Point - Bubble point			Maximum Allowable Pressure Drop		
Pour Point			Calculated pressure drop		
Density			Fouling resistance (min.)		
Specify heat					

Air Side Data	
Air quantity	Altitude
Air quantity / fan	Temperature in
Static Fouling resistance : Pressure :	Temperature out

Mass Velocity	Face Velocity	Min.ambient temperature
Design Pressure	Test Pressure	Design Temperature
Tube bundle	Headers	Tube details
Size	Type	Material
No./bay	Material	ASTM : Seamless / Welded
Arrangement :	Passes :	O.D./Min.Thk.
Bundles in Parallel	Plug design	Tubes / bundle
Bays in Parallel	Material	Tube pitch
Bundle Frame	Gasket Material	Fin Type
Miscellaneous	Corrosion Allow	Fin Material
Structure	Inlet nozzle(s)	Fin O.D.
Surface Prep: Galvanised / Painted	Outlet nozzle(s)	Fin Thickness
Louvers Auto/Manual	Nozzle Rating/Type	Fin density (Fins/inch):
Standards / Code		API-661/ASME Code

FIGURE 3.79 ACHE specification sheet.

Mechanical Drive Equipment		
Fan-Constant-or Variable pitch	Drive	Speed Reducer
Mfg/Model	Type : Belt/Gear	Type
No./bay	No./bay	No./bay
HP / fan	HP/Drive motor	Model
Fan Diameter / RPM	RPM	HP rating
No.blades	Enclosure	Speed ratio
Pitch : Adjustable or Auto	Volt / Ph / Cycles	Manufacturer
Blade / Hub Material	Manufacturer	Coupling Model
Vibration Switch	Louvers	Control : Open / Closed
Noise : dB	Motor-Constant- or Variable freq.	Fan - Install Inlet bells - and Fan seal disc.
Remarks : 1. Recirculation ducts and chambers. 2. Support structure height. 3. Maintenance walkways.		

Air Cooled Heat Exchanger Specification Sheet

FIGURE 3.79 (Continued)

4. Seismic and wind loads
5. Variations in operation
6. Process control instrumentation
7. Fire protection
8. Pumping power. The fan power (P_p) required is given by

$$P_p = \frac{M\Delta p}{\rho} \text{ or} \tag{3.143a}$$

$$P_p = \frac{GA_{min}\Delta p}{\rho} \tag{3.143b}$$

9. Special surface finish required if any

3.10.18 ACHE Fabrication

Supplier's Responsibility

The supplier shall be responsible for the complete design, construction, testing, and process and mechanical guarantees of the equipment, including full compliance with all applicable design codes and standards,

3.10.18.1 Factory Acceptance Test (FAT)

Prior to the FAT, the equipment/systems shall have been completely tested by the supplier as follows [102]:

i. All piping and tubing shall have been adequately tested for leaks.
ii. All electrical circuits shall have been actuated to check their function.
iii. All nameplates shall have been checked for correct spelling and size of letters.

A FAT shall consist of, as a minimum, a full functional test of the equipment/system. Before shipment to the fabrication site, satisfactory functioning and performance of the system must be demonstrated to the satisfaction of the buyer.

3.10.18.2 Hydrostatic (Pressure) Testing

All pressure-retaining equipment and parts (i.e. tubes, headers, etc.) shall be hydrostatically pressure tested, performed by the supplier in accordance with the applicable code and ACHE specification.

3.10.18.3 Functional Testing

The supplier shall supply test procedure(s) for functional testing of the complete equipment/systems over their full functional ranges. Test procedure(s) shall be established to demonstrate that the equipment/systems and all associated components will function satisfactorily in service.

3.10.18.4 Preparation for Dispatch

The supplier's "scope" shall include, as a minimum, the following:

1. Tube bundles complete
2. Fans with fan rings and supports complete
3. General4. Surface preparation and coating as per coating specification

3.10.19 Modular Air-Cooled Heat Exchanger

Alfa Laval's ACHE model G is a modular air-cooled heat exchanger that provides reliable and cost-effective cooling of natural gas, water, and/or synthetic oils, cooling duties that can be found around a natural gas compressor. A flexible modular design allows the cooling units to be easily customized to the needs and options within ACHE model G's design limits. The ACHE model G is fast and easy to [103]:

i. Size and buy—pre-configured modules in flexible configurations
ii. Manufacture and deliver—pre-engineered blocks in compact design for rapid transportation
iii. Erect and install—bolt down the modules, plug in the power cables, connect the flanges, and start

Installation and commissioning

i. Install and fix module on the supporting structure
ii. Connect motor to the supply and check belt
iii. Match module process flanges to piping

3.10.20 Holtec's HI-MAX™ Fin Tube

All Stainless Steel All Welded HI-MAX fin tubes (Figure 3.80) eliminate the thermal stresses associated with aluminized carbon steel tubes and aluminum fins (used in peer systems), thus extending the performance life of the air-cooled condenser. The stainless steel bundles also provide superior internal and external corrosion protection, which mitigates two significant problems with today's ACHEs: (1) flow-accelerated corrosion and (2) iron carryover into the condensate. The

FIGURE 3.80 *HI-MAX*™ all stainless steel all welded fin tube. (Courtesy of Holtec International, Marlton, NJ.)

HI-MAX fin-to-tube bond is made by laser welding and hence stronger in construction. HI-MAX does not have a galvanic coating to combat corrosion.

3.10.21 Performance Control of ACHEs

In addition to the fact that the process flow rate, composition, and inlet temperature of the fluid may vary from the design conditions, the ambient air temperature varies throughout a 24-h day and from day to day. Since ACHEs are designed for maximum conditions, some form of control is necessary when overcooling of the process fluid is detrimental, or when saving fan power is desired. Although control could be accomplished using by-passing of process fluid, this is rarely done, and the usual method is airflow control [72].

3.10.21.1 Varying Airflow

Varying airflow can be accomplished by [72]:

i. Adjustable louvers on top of the bundles
ii. Two-speed fan motors
iii. Fan shut-off in sequence for multifan units
iv. AUTO-VARIABLE® fans
v. Variable frequency fan motor control

3.10.22 Airside Performance Evaluation

For airside performance evaluation, the following minimum information are to be established:

i. Air flow rate
ii. Motor power consumption
iii. Static pressure losses
iv. Inlet/outlet air temperatures
vi. Fan efficiency

Refer to Ref. [104] for details on airside performance evaluation.

3.10.22.1 Air-Cooled Heat Exchangers Performance Testing—ASME PTC 30–1991 (R2021)

This code provides uniform methods and procedures for testing the thermodynamic and mechanical performance of air-cooled heat exchangers, and for calculating adjustments to the test results to design conditions for comparison with the guarantee [105]. This code recommends methods for obtaining data, measurements, observations, and samples to determine the following: (a) physical dimensions; (b) air flow rate; (c) airside pressure differential; (d) fan driver power; (e) sound level; (f) atmospheric; (g) environmental effects; (h) wind velocity; (i) air temperature; (j) entering air temperature; (k) exit air temperature; (l) process fluid temperatures; (m) process fluid pressures; (n) process fluid flow rate; (o) composition of process fluid; (p) percent capability; and (q) process fluid pressure drop.

Worked Example 3.1

Consider a heat exchanger that is comprised of one row of four, annularly finned tubes, as shown in Figure 3.80, i.e., an annular fin-tube heat exchanger. Hot air flows over the outside surfaces of the tubes, and water flows through the insides of the tubes, as a single pass. The inside and outside radii of the tube are r_i and r_o, respectively, and the tubes have a length of L and thermal conductivity of k and for fin k_f W/(mK). The outer fin radius is R, its thickness is δ, the fin pitch is P_f fins/m. Water enters at $T_{t,i}$ ºC and flows at a total mass flow rate of m_w kg/s. The air flows with a velocity of V_{fr} m/s at the frontal face of the heat exchanger and enters with a temperature of $T_{a,i}$ ºC. For this heat exchanger operating under these steady-state conditions, find the total rate of heat transfer using both the LMTD and ε-NTU approaches. Note that "a" refers to airside and "t" the tubeside (Thome, 2004).

Solution by LMTD method

In addition to energy balances, the heat transfer can be characterized by a rate equation in terms of a log mean temperature difference (LMTD)

$$Q = UAF\Delta T_{LMTD}$$

with

$$\Delta T_{LMTD} = \left[\left\{ \left(T_{a,i} - T_{t,o}\right) - (T_{a,o} - T_{t,i}) \right\} / \log \left\{ \frac{T_{a,i} - T_{t,o}}{(T_{a,o} - T_{t,i}} \right\} \right]$$

$$Q = \rho_{a,i} A_{fr} V_{fr} C_a \left(T_{a,i} - T_{a,o}\right)$$

In this method, the rate equation and energy balance on the water and air streams takes the following form:

$$Q = m_a C_a (T_{a,i} - T_{a,o}) \text{ and } Q = m_t C_t (T_{t,o} - T_{t,i})$$

where the subscripts "a" and "t" indicate airside and tubeside, and "i" and "o" indicate the inlet and outlet, respectively.

The equations set for the solution by LMTD method is shown in Table 3.4 and by ε-NTU method is shown in Table 3.5.

TABLE 3.4
Equation Set for an LMTD Solution for an Annular Fin-Tube ACHE Example

Description, airside	Equation
Reynolds number, Re_a	$2(m_a/4)(\pi r_i \mu_a)$
Heat transfer coefficient, h_a	$(0.023\, Re_a^{0.8} Pr_a^{0.4})k_a/(2r_i)$ (based on Dittus-Boelter)
R_{Ta}	$1/(h_c 8\pi r_i L)$
Core frontal area, A_{fr}	$4\,(2R)\,L$
Total number of fins, N_f	$4\,(P_f L)$
Total fin area, A_f	$N_f\left[2\pi(R^2 - r_O^2) + 2\pi R\delta\right]$
Total outside tube surface area, A_{Th}	$4(2\pi r_O L) - N_f(2\pi r_O \delta) + A_f$
Minimum air flow area, A_{min}	$A_{fr} - 4\,(2r_O L) - N_f\, 2(R - r_O)\delta$
Airside hydraulic diameter, $D_{h,a}$	$4\,A_{min}(2R)/\,A_{Th}$
Air mass flow rate, m_a	$\rho_{a,i} V_{fr} A_{fr}$
Airside Reynolds number, Re_a	$m_a D_{h,a}/(A_{min}\mu_a)$
Airside j factor, j_a	$0.0265\ Re_a^{-0.22}$ (*Kearney and Jacobi,1995)
Airside h eat transfer coefficient, h_a	$j_a m_a C_a/(A_{min} Pr_a^{2/3})$
Parameter to find fin efficiency, m	$\sqrt{2R^2 h_a/(k_f \delta)}$

Fin efficiency, $$\eta_f = \frac{2Rr_o}{m(R^2 - r_o^2)}\left[I_1(m)K_1\left(\frac{mr_o}{R}\right) - K_1(m)I_1\left(\frac{mr_o}{R}\right)\right]\left[I_0\left(\frac{mr_o}{R}\right)K_1(m) + I_1(m)K_1\left(\frac{mr_o}{R}\right)\right]^{-1}$$

Overall surface heat transfer efficiency, η_0	$1 - A_f(1-\eta_f)/\,A_{Th}$
Airside thermal resistance, $R_{T,a}$	$1/(h_a \eta_0 A_{T,a})$
Waterside thermal resistance, $R_{T,t}$	$1/(h_t \eta_0 A_{T,t})$
Tube wall thermal resistance, $R_{T,w}$	$\ln(R/r_o)/(8\,\pi L k)$
UA	$(R_{T,a} + R_{T,t} + R_{T,w})^{-1}$
Airside heat transfer rate, Q_a	$m_a C_a\,(T_{a,i} - T_{a,o})$
Waterside fluid heat transfer rate, Q_t	$m_t C_t\,(T_{t,o} - T_{t,i})$ Note: $Q_a = Q_t$

$$Q = UA\ F\left[\left\{(T_{a,i} - T_{t,o}) - (T_{a,o} - T_{t,i})\right\}/\log\left\{\frac{T_{a,i} - T_{t,o}}{(T_{a,o} - T_{t,i})}\right\}\right]$$

(Assume F= 1 and iterate)

Source: Thome, J.R. (Ed.) (2004). *Heat to Air Cooled Heat Exchangers, Chapter 6, Engineering Data Book III*. Wolverine Tube, Inc., Decatur, AL, pp. 6.1–6.40.

Notes: *Kearney, S.P., Jacobi, A.M. (1995) Effects of gull-wing baffles on the performance of a single-row, annularly finned tube heat exchanger. *International Journal of HVAC&R Research, ASHRAE* 1(4): 257–272.

Nomenclature for the worked example only.

C is specific heat

D_h is hydraulic diameter, $D_h = 4A_{min}\ L/\,A_T$

j is Colburn factor, j=Nu/(Re $Pr^{1/3}$)

k is thermal conductivity

L is airside flow depth, the distance from inlet to outlet face

TABLE 3.5
Equation Set for an $\varepsilon - NTU$ for an Annular Fin-Tube ACHE Example

Q =	$\varepsilon(m_a C_a)(T_{a,i} - T_{t,i})$ (if AIR stream is with minimum mC)
NTU =	$UA/(m_a C_a)$
$\varepsilon = \frac{m_t C_t}{m_a C_a}\left[1 - \exp\left\{-\frac{m_a C_a}{m_t C_t}(1 - e^{-NTU})\right\}\right]$	

m is mass flow rate, or fin efficiency parameter
N is number of tube rows in the air-flow direction
N_f is total number of fins
NTU is number of transfer units, $NTU = \text{UA}/(\text{mC})_{min}$
R is thermal resistance
T is temperature
UA is overall heat transfer conductance

Greek symbols
μ is dynamic viscosity
ρ is density

Other subscripts
a is airside
t is tubeside
f is fin
i is in
o is out
Min is minimum
T is tube
t is tube side
w is wall or water

3.11 MICROCHANNEL HEAT EXCHANGERS (MCHES)

MCHEs are fin and tube heat exchangers where one fluid, usually refrigerant or water, flows through tubes or enclosed channels either plain or enhanced, while air flows crosscurrent through the connected fins. The hydraulic diameter of the channels is less than 1 mm. A typical aluminum MCHE has three major components: header, multiport microchannel tube, and louvered fin. These components are joined together by controlled atmosphere brazing. Aluminum microchannel heat exchangers have advantages such as high thermal performance and significant weight reduction compared to conventional copper heat transfer coils. The development of extruded microchannel tubes for refrigerant passage has significantly improved the thermal performance of heat exchangers. A typical aluminum microchannel heat exchange is illustrated in Figure 3.82 [23] and its assembly of components are shown in Figure 3.83 [106]. These components are joined together by brazing. Using a single material for construction produces a consistent rate of heat transfer. The high temperatures used in brazing produce a strong metallurgical alloy at the joins. These methods reduce resistance to heat transfer. Figure 3.84 shows fluid flow details.

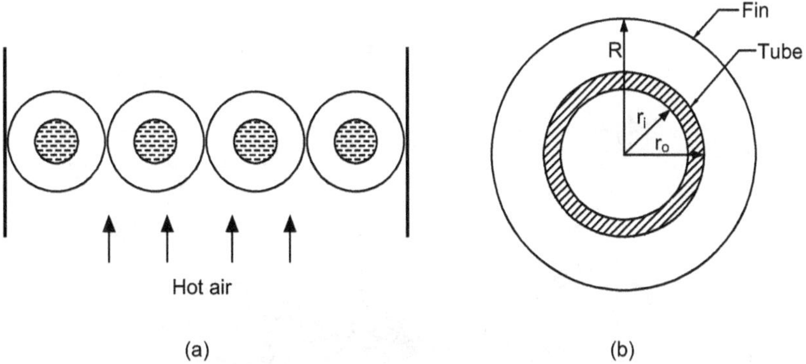

FIGURE 3.81 One row of four annular fin-tube heat exchanger (of worked example).

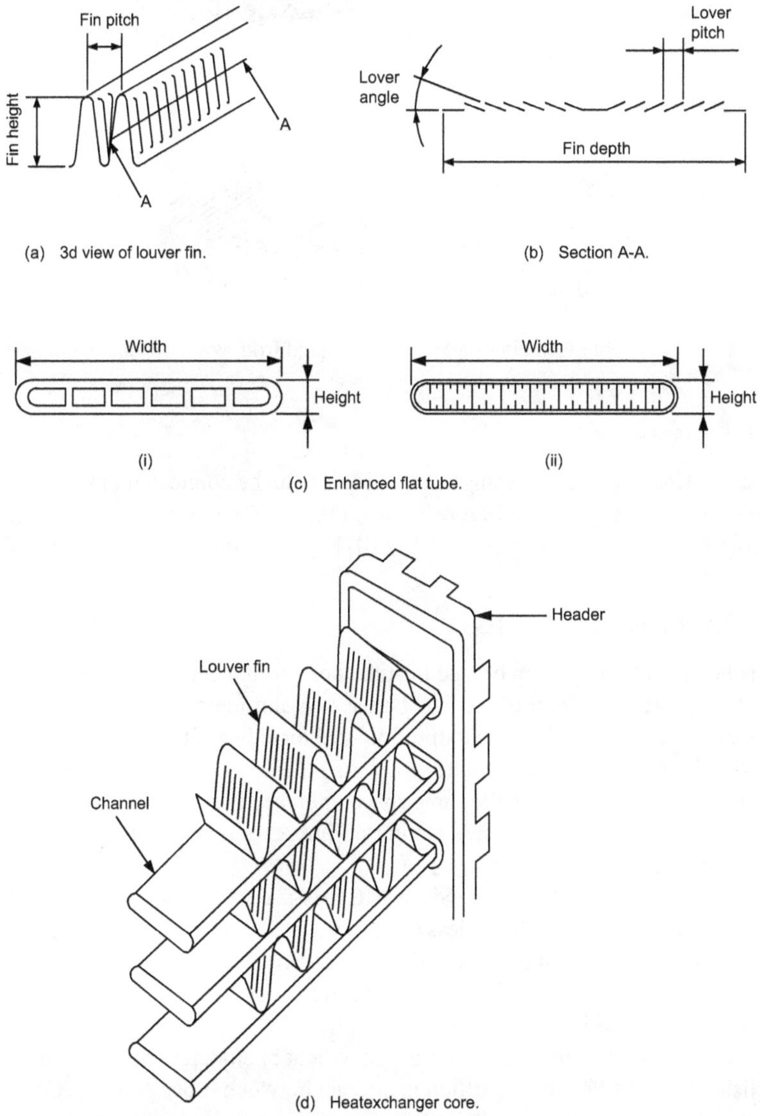

(a) 3d view of louver fin.

(b) Section A-A.

(i)

(ii)

(c) Enhanced flat tube.

(d) Heatexchanger core.

FIGURE 3.82 A louvered fin microchannel heat exchanger.

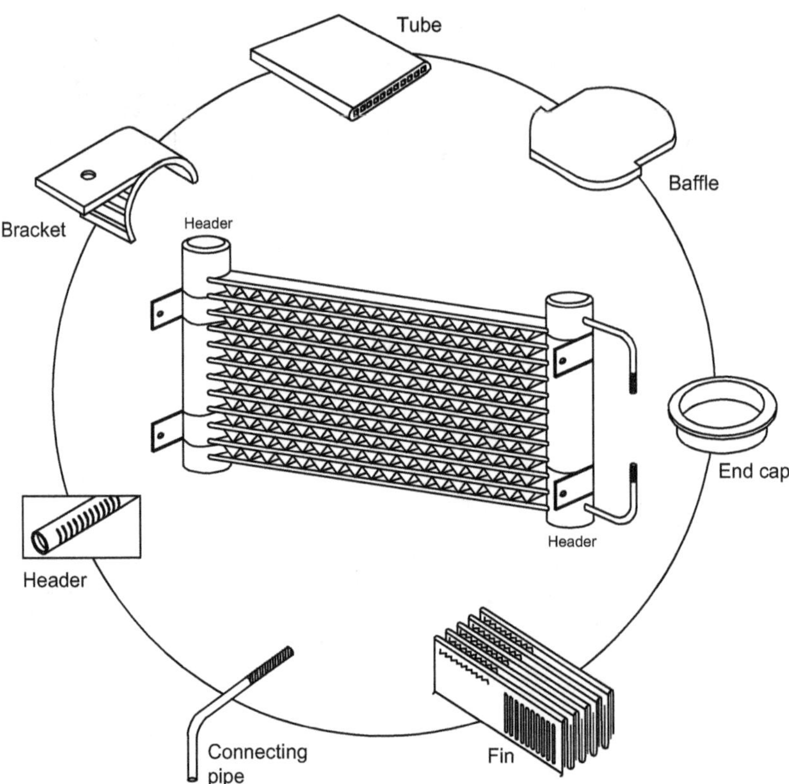

FIGURE 3.83 Components of assembly of a MCHE. (Adopted from www.yit-th.com/major-components).

3.11.1 Pass Arrangement

A few passes can be incorporated in a single unit, or units can be connected in series. A unit with two units connected in series is shown in Figure 3.85(a) [107] and a two-pass arrangement in the same unit can be arranged as shown in Figure 3.85(b) [107]. A three-pass unit is shown in Figure 3.86.

3.11.2 Benefits of Microchannels

A microchannel coil is all-aluminum brazed fin-tube heat exchanger constructed to reduce size, use less refrigerant, be more durable, transfer heat better, and eliminate formicary corrosion. The small hydraulic diameter of the channels leads to multiple technical benefits. The primary benefit in using MCHEs in HVAC is efficiency. Many smaller conduits provide a greater surface area for refrigerant-to-wall contact than larger tubes do in the same space. Increased surface area improves the efficiency of heat transfer, with some manufacturers stating an improvement of 20–40% over traditional fin and tube heat exchangers.

MCHEs require 30% less refrigerant. Since MCHEs are so much more efficient, they can be smaller (up to 30%) and weigh less (60% less) than comparable heat exchangers. Their size and efficiency mean that smaller fans can be used, with lower overall energy use of the system. An added benefit to smaller fans is less noise. MCHEs allow more flexibility in layout and design without worrying about noise complaints [107].

The traditional way of manufacturing finned tube heat exchangers by mechanical tube expansion has the disadvantage of a lack of sufficient contact between tubes and fins. On the other hand, the brazing process that metallurgically bonds fins and tubes in microchannel coils eliminates

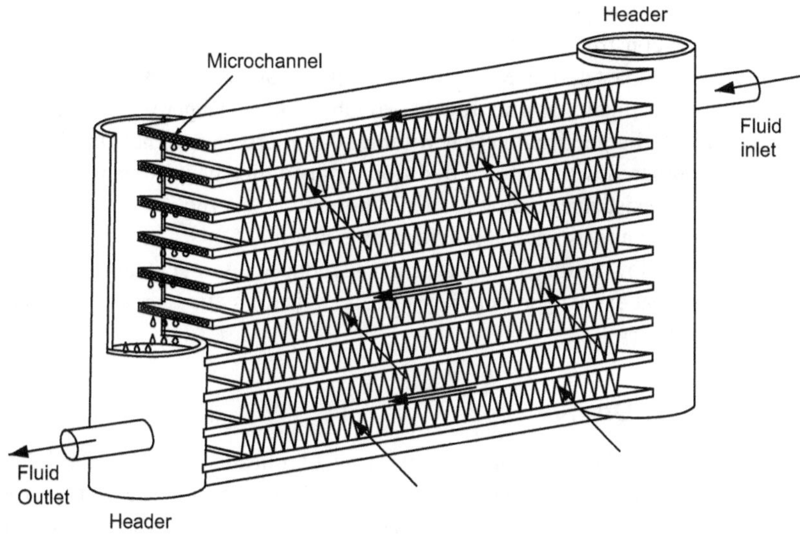

FIGURE 3.84 Components of assembly of a microchannel heat exchanger.

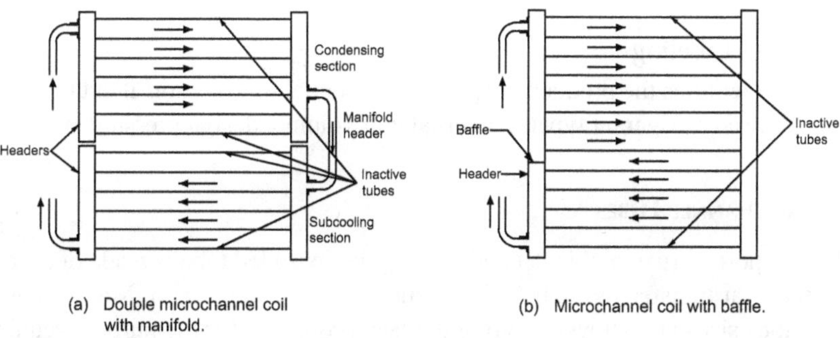

FIGURE 3.85 Two passes arrangements (a) Two MCHEs connected in series and (b) Two passes arrangement in the same unit.

the drawback of contact resistance. These microchannel coils are less than two inches thick, allowing for easy removal of any debris that may be caught within the coil. This is not so with fin/tube coils, which are often much thicker, with staggered tube patterns using corrugated fins which make debris removal difficult, if not impossible, in some circumstances. The durability of microchannel coils also allows for pressure washing (using a broad spray pattern). Microchannel condensers and evaporators outperform traditional finned tube heat exchangers in these aspects [108–110]:

i. Lower raw material cost (aluminum vs. copper)
ii. Highly automated, reliable manufacturing process
iii. Higher thermal conductivity (integral brazing vs. mechanical expanding, larger heat transfer surface, higher louvered fin thermal performance)
iv. Potential for compact and lightweight design
v. Perfect corrosion resistance
vi. Lower airside pressure resistance since the flow length is less
vii. Easy recyclability

viii. The geometry of louvered fins ensures a higher heat transfer coefficient than with the plain fins used in the construction of finned tube heat exchangers
ix. Smaller sizes, compact design. Microchannel heat exchangers with 32-mm tube width provide almost equal performance as a six-row finned tube heat exchanger with the same face area
x. Low refrigerant charge. The internal volume of a microchannel heat exchanger is significantly less compared with that of a traditional finned tube heat exchanger
xi. High corrosion resistance since it is an all-aluminum construction
xii. Increased corrosion resistance is provided by e-coating
xiii. Low weight. The low weight, compact size, and reduced internal volume of an all-aluminum microchannel heat exchanger were key factors in its dramatic increase in popularity in the automobile industry

3.11.3 Structure and Design

The core of a microchannel heat exchanger consists of two headers, namely the inlet and outlet headers, flat multiport microchannel tubes and fins, brazed together with the non-corrosive fluxes. Heat transfer is maximized by using louvered fins between the flat tubes. All the above elements are made of aluminum alloys, distinct in their composition and clads. The core assembly also consists of refrigerant inlet and outlet manifold, refrigerant connections, reinforcement frame/case, and mountings.

3.11.3.1 Product Labeling

The product label identifies the product and provides essential information about the product and its use, including allowed refrigerant type(s), internal coil volume, design pressure, and temperature.

3.11.4 Microchannel Tubes

Also called multi-port extrusion, this flat and rectangular extruded tube is made of several channels that increase the heat transfer through a higher surface per volume ratio. Microchannel tubes are available in various sizes and alloys, ensuring the best properties for the purpose required, and can be delivered with a coating, ready for assembly and oven brazing.

Since microchannel tubes have sub-millimeter diameters, it is important to avoid particles entering the heat exchanger. For this purpose, it is recommended to install a mesh strainer with a sieve size of 0.25 mm or less. Refer to possible multiport flat tubes which find applications are shown in Figure 1.25(i) [23,111].

3.11.5 Applications of Microscale Heat Exchangers

These are currently used in automotive, commercial and residential heating/cooling, aircrafts, manufacturing, cooling electronics applications, etc.

HVAC Coils. Microchannel heat exchangers are suitable for use as condensing coils, DX evaporators, heat pump coils, and water coils for cooling and heating applications [112].

3.11.6 Comparison of Conventional Finned Tube HEx versus MCHE

The favorable features of MCHE compared to conventional finned tube heat exchangers are given below [113]:

i. Less weight and reduced thickness—50% less weight with the same capacity
ii. Less refrigerant charge—50% less internal volume

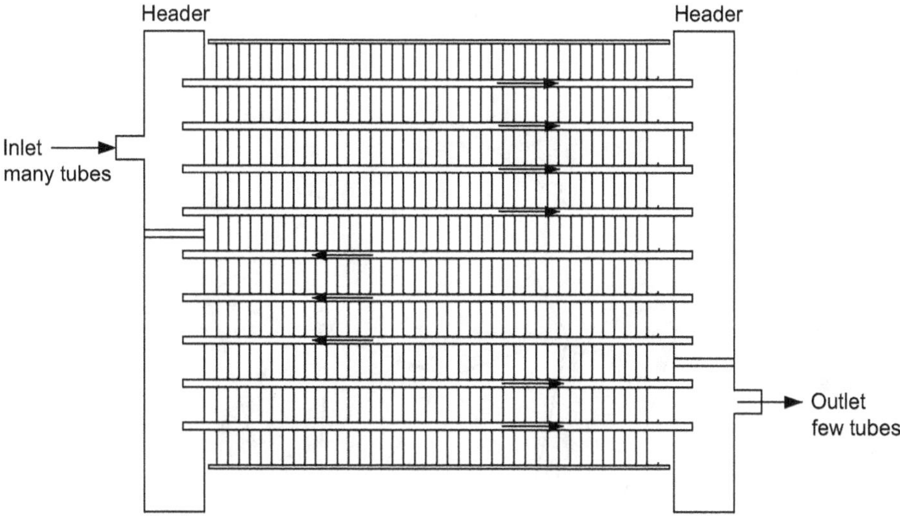

FIGURE 3.86 A three pass unit.

iii. Energy consumption—40% less energy consumptions thanks to lower pressure drops on the airside
iv. Sound power level—10% less sound power level thanks to a reduced thickness of the core compared to a ¾-row tube and fin coil

The above features are shown in Figure 3.87 for comparison.

3.11.7 Disadvantages of Microscale Heat Exchangers

i. One of the main disadvantages of microchannel heat exchangers is the high pressure loss that is associated with a small hydraulic diameter (http://me1065.wikidot.com/micro-scale-heat-exchangers).
ii. Condensate can interfere with efficiency by blocking air flow through fins. Another issue is refrigerant flow distribution.

3.11.8 Microchannel Heat Exchanger Market-Drivers

The microchannel heat exchangers industry is majorly driven by the growing applications of HVAC systems in air conditioning and refrigeration as they offer enhanced efficiency, robust design, and compactness. The demand for microchannel heat exchangers is majorly influenced by the growing applications for the automotive sector. The top-10 microchannel heat exchanger companies include Danfoss Industries Pvt. Ltd., Kaltra Innovativetechnik GmbH, Sanhua Holding Group Co., MAHLE GmbH, Modine Manufacturing, API Heat Transfer Inc., Climetal S.L., Hanon Systems, DENSO Corporation, and Modine Manufacturing Company [114].

3.11.9 Cleaning Procedures

Regular coil maintenance, including annual cleaning, enhances the unit's operating efficiency by minimizing compressor head pressure and amperage draw. Cleaning with cleansers or detergents is strongly discouraged due to the all-aluminum construction; clean water should prove sufficient.

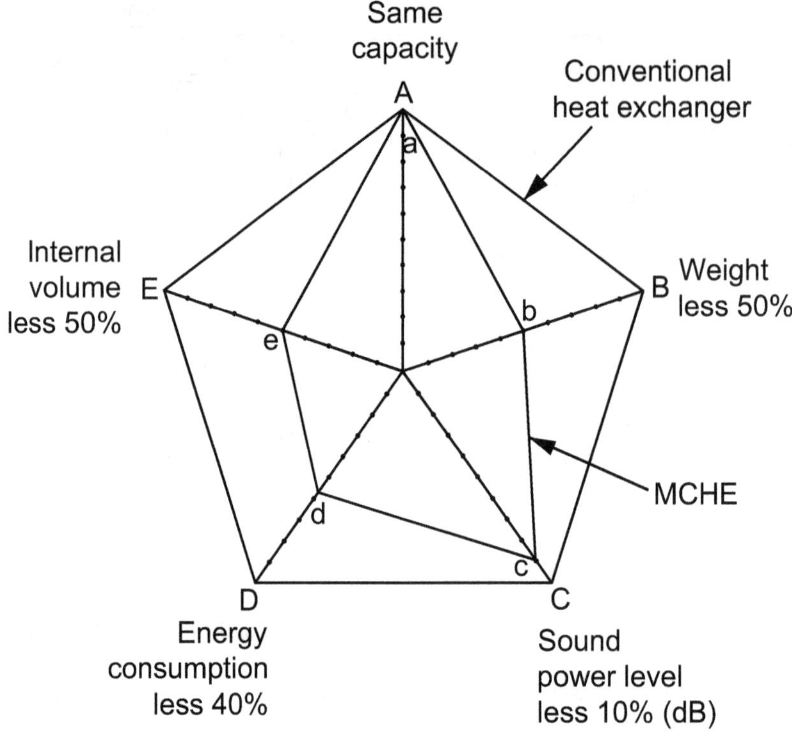

FIGURE 3.87 Comparison of features of conventional finned-tube HEx vs MCHE, based on Ref.(113).

Microchannel coils can be more susceptible to corrosion if the cleanser or detergent used is not thoroughly washed or rinsed off. Any breach in the tubes can result in refrigerant leaks.

3.11.10 Repair: Microchannel Welding

Microchannel coils are easily cleaned and can be field repaired using a simple alloy-based solder. Making these repairs requires a delicate touch and the ability to weld small surfaces. When welding or brazing a micro-channel, it is important to use low temperatures and either a propane torch or MAPP gas. The flame needs to be moved constantly to prevent the channels from melting, which would make any problems worse.

3.11.11 Corrosion in all-Aluminum Microchannel Coil

3.11.11.1 Forms of Corrosion

Aluminum is affected by the following forms of corrosion: atmospheric corrosion, uniform corrosion, galvanic corrosion, pitting corrosion, and erosion corrosion.

3.11.11.2 Corrosion Prevention

To prevent corrosion, proper aluminum alloy selection, surface treatment, i.e., surface coating, and field maintenance are necessary. There are several corrosion prevention options and methods that can be adopted for all-aluminum heat exchangers. These include the application of protective coatings and regular surface cleaning.

3.11.11.3 Materials Selection

Material properties are crucial for heat exchanger durability and corrosion resistance, especially when it comes to operating in aggressive atmospheres like highly polluted industrial and urban areas, coastal zones, and other corrosive environments. Heat exchangers of microchannel coil technology utilize several aluminum alloys with a metallic coating after brazing. To achieve the highest product performances, manufacturers use aluminum alloys and clads of series 3xxx, 4xxx, 7xxx, as well as strong long-life alloys of series 9xxx.

Cooling water system pH control: To prevent corrosion of the aluminum which may be caused by an oxide layer dissolving at extreme pHs (in both acidic and basic environments), the typical pH range for microchannel heat exchangers is 6.0–8.5. Kaltra, Germany, one of the leading MCHE manufacturers, recommends a well-proven mechanism of corrosion inhibition of aluminum alloys by sodium molybdate.

3.11.11.4 Closed-Loop Systems

In most HVAC applications, refrigerant circuits are multi-metal, and the points of contact between dissimilar metals are subject to galvanic corrosion. One method of preventing galvanic corrosion is the elimination of direct bi-metallic contacts. For closed-loop chilled- water systems, it is recommended that pH-neutral, molybdate-based multi-metal corrosion inhibitors be added to the cooling medium which is compatible with glycols and suitable for all water qualities.

3.11.11.5 Protective Coatings

For working in corrosive environments such as sea coasts, industrial zones, and highly polluted areas, microchannel coils may require additional protection. Electrocoating is a process that uses an electrical current to deposit an organic coating from a paint bath onto a heat exchanger body.

3.11.11.6 Sacrificial Zinc Coatings

Applying a zinc layer on top of an aluminum alloy protects the core of the tube by providing a preferred path for corrosion to spread.

3.11.11.7 Trivalent Chromium Process Coating

Trivalent chromium process (TCP) conversion coating is a type of conversion coating used to passivate aluminum alloys as a corrosion inhibitor.

3.11.11.8 Pre-coated Aluminum Fin/Copper Tube Coils

Pre-coated aluminum fin/copper tube coils have a durable coating applied to the fin. This design offers protection in mildly corrosive coastal environments, but is not recommended in severe industrial or coastal environments [115].

3.11.11.9 Corrosion of Mechanically Bonded Copper Tube/Copper Fin Coils

Typically, a copper wavy fin is mechanically bonded to the standard copper tube. Protective isolators are installed between the coil assembly and side sheet metal coil support pan to further protect the coil assembly from galvanic corrosion. Uncoated copper coils are not suitable for dense urban, polluted coastal applications, industrial applications, or industrial marine applications since many pollutants attack copper, putting both the fin and the tube at risk. E-coated MCHE coils should be considered for such applications [115].

3.11.11.10 E-Coating Process

Electrocoating of aluminum fin and copper fin coils is a multi-step process that ensures coils are properly coated, cured, and protected from environmental attack. E-coated coils provide superior

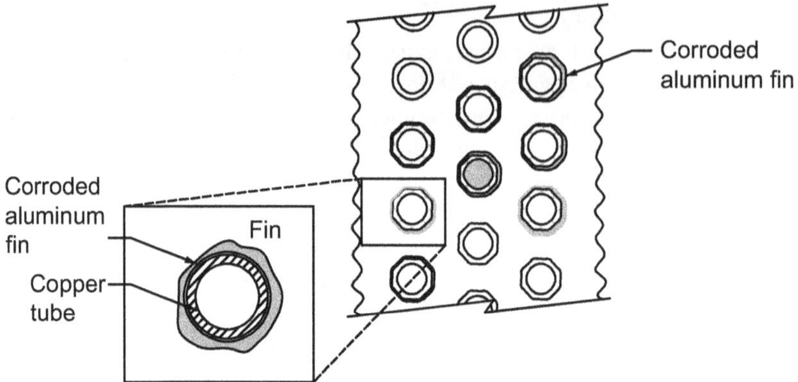

FIGURE 3.88 Illustration of galvanic corrosion of aluminium fin- copper tube heat exchanger based on Ref(115).

protection in the most severe environments. Finally, a UV-protective topcoat is applied to shield the finish from ultraviolet degradation.

Galvanic corrosion of an aluminum fin/copper tube coil is shown in Figure 3.88, and Figure 3.89 [115] shows a precoated coil assembly.

3.11.11.11 Condensation Management

Condensation management is essential to ensuring the efficient function of heat exchangers and critical for the correct operation of evaporators. Hydrophilic topcoats, surfaces tending not to adsorb water or be wetted by water, are effective in environments of excessive condensation, as they enable water to flow easily off the heat exchanger surface by virtue of lowered surface tension, thereby enhancing heat transfer performance in wet and frosty conditions. At the same time, hydrophilic coatings protect heat exchangers from the corrosive effect of water and exhibit fine performance in providing protection in salty environments. A coating is considered hydrophilic in the case that the contact angle of a water droplet's edge with the surface is less than 90°. In sum, the use of a hydrophilic polymer [116]:

i. Improves condensate drainage
ii. Increases performance in wet conditions
iii. Provides antimicrobial protection
iv. Enhances the corrosion resistance of the heat exchanger, contributing to its extended lifetime

3.12 CHILLER

3.12.1 Working Principle

A chiller is a heat-transferring device that uses mechanical refrigeration to remove heat from a continuous flow of process liquid and reject it to the environment, thus lowering the temperature of the process liquid. The chiller is connected to the process liquid system through an evaporator: a heat exchanger in which heat captured by the liquid flow is transferred to the flow of refrigerant fluid, and as the heat transfer takes place, the refrigerant evaporates, changing from a low-pressure liquid into vapor, while the temperature of the process liquid is reduced. Then, refrigerant gas flows to the compressor, which raises the pressure and temperature of the discharged gas which in turn is cooled in the condenser where the refrigerant changes back from a vapor to a liquid state. The latent heat is rejected to the environment in either water or ambient air, depending on the type of condenser

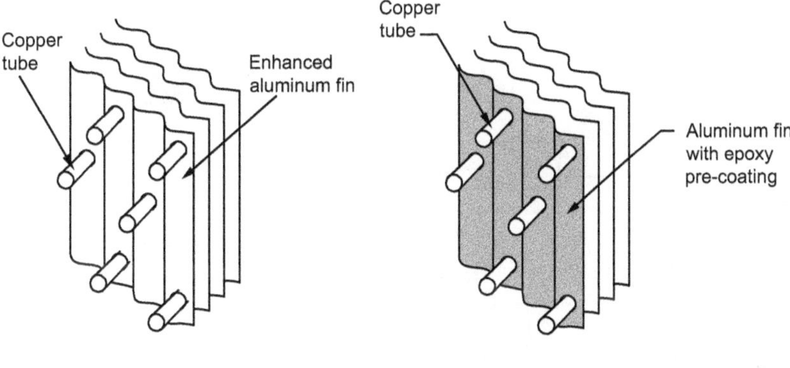

(a) Standard coil construction.

(b) Pre-coated coil assembly.

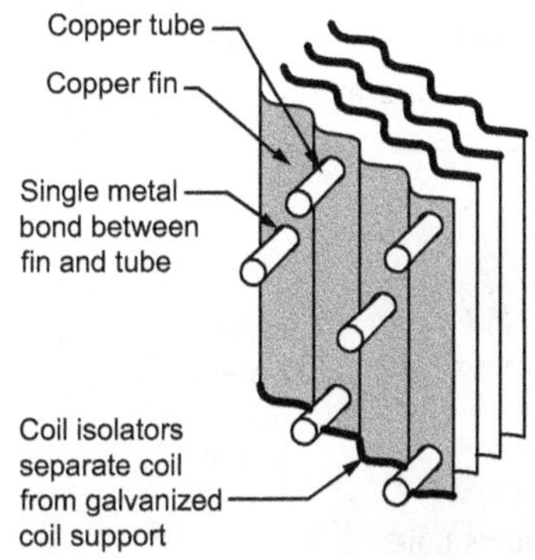

(c) Copper tube - Copper fin coil assembly.

FIGURE 3.89 Precoated heat exchanger coil based on Ref(115).

used in the system: air-cooled or water-cooled. Chillers with air-cooled condensers utilize air-to-refrigerant heat exchangers, while water-cooled chillers reject the heat via a cooling tower or dry cooler [117]. A chiller is shown for water cooling or a process fluid cooling in Figure 3.90.

3.12.2 Air-cooled versus Water-Cooled Chillers

An air-cooled condenser uses ambient air for cooling and condensing the refrigerant gas back to a liquid. It usually is located inside the chiller or remotely, but ultimately it dissipates the heat to the atmosphere with the airstream passing through the condenser coil. A water-cooled condenser repetitively cycles the water between the cooling tower (or dry cooler) and the condenser. Hot refrigerant gas enters the condenser and transfers its heat into the water which is transported up to the cooling tower (or dry cooler) and rejected to the atmosphere.

3.12.3 Pros and Cons

Air-cooled chillers fit better for cooling plants of up to 2 MW capacity, for cold or mild climates, and for regions where water resources are limited. Water-cooled chillers are perfectly suitable for large cooling plants, e.g., for district cooling.

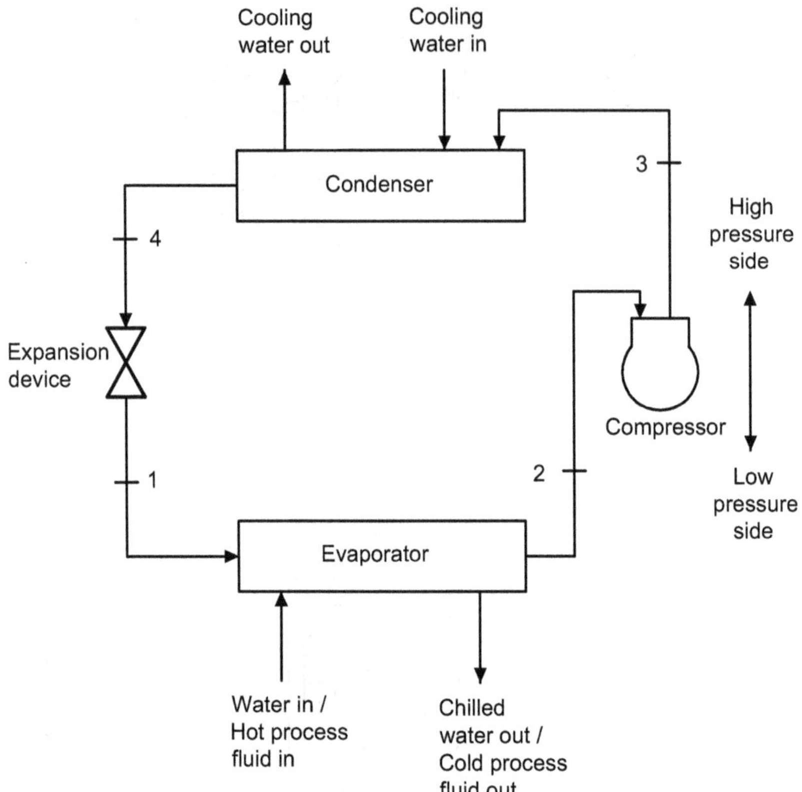

FIGURE 3.90 A chiller for cooling a process fluid.

3.12.4 Aspects of the System Design

Though there is a great deal of variety for installation of both air-cooled and water-cooled chillers, some general installation rules apply and can help to guide the chiller choice [117].

3.12.4.1 Installation Areas

Air-cooled chillers need fresh make-up air for condensers to extract heat from the refrigerant and almost invariably installed outdoors, while all water-cooled chillers intended for indoor installation should be in properly fitted mechanical rooms. Outdoor setup requires electrical control panels suitable for appropriate environmental conditions.

The units with water-cooled condensers are compact compared to air-cooled chillers, but require large outdoor areas for installing cooling towers or dry coolers and space for circulation pumps and other system components.

3.12.4.2 Water Consumption

Water-cooling systems equipped with cooling towers consume significant amounts of water due to evaporation and the need for periodic water removal and refreshing, so air-cooled chillers should be the preferential choice for water conservation, especially in regions experiencing water scarcity. High water consumption is a core issue of water-cooled cooling systems and a primary concern for customers.

3.12.4.3 Noise Output

Air-cooled chillers produce higher noise levels (outdoor) via condenser fans.

3.13 CUPROBRAZE HEAT EXCHANGER

The CuproBraze® brazing process is a technique for manufacturing heat exchangers, particularly automotive radiators. Developed by the International Copper Association, CuproBraze technology enables the manufacture of light, strong, efficient, and compact heat exchangers, including radiators, oil coolers, heater cores, charge air coolers, and condensers, from high strength and high conductivity copper and copper alloys. These materials and technology offer several benefits over existing systems, including a 10% cost saving over conventional aluminum heat exchangers while offering better energy efficiency. Other features that favor CuproBraze radiators include better corrosion resistance, repairability, and higher shock- and vibration-withstanding capacity. Manufacturing processes are now being applied globally in the manufacture of advanced heat exchangers using the new brazing process, known as CuproBraze. CuproBraze technology is flexible and scalable. The International Copper Association licenses CuproBraze technology free of charge to heat exchanger manufacturers.

3.14 MICROGROOVE COPPER TUBE HEAT EXCHANGER

Internally grooved copper tubes also known as "microfin tubes" are a small-diameter coil technology for modern air conditioning and refrigeration systems. Grooved coils facilitate more efficient heat transfer than smooth coils. Small-diameter coils have better rates of heat transfer than conventional-sized condenser and evaporator coils with round copper tubes and aluminum or copper fins that have been the standard in the HVAC industry for many years. Small-diameter coils can withstand higher pressures required by the new generation of environmentally friendlier refrigerants. They have lower material costs because they require less fin and coil materials. Also, they enable the design of smaller and lighter high-efficiency air conditioners and refrigerators because the evaporator and condenser coils are smaller and lighter [118,119]. A microgroove copper tube heat exchanger coil is shown in Figure 3.91.

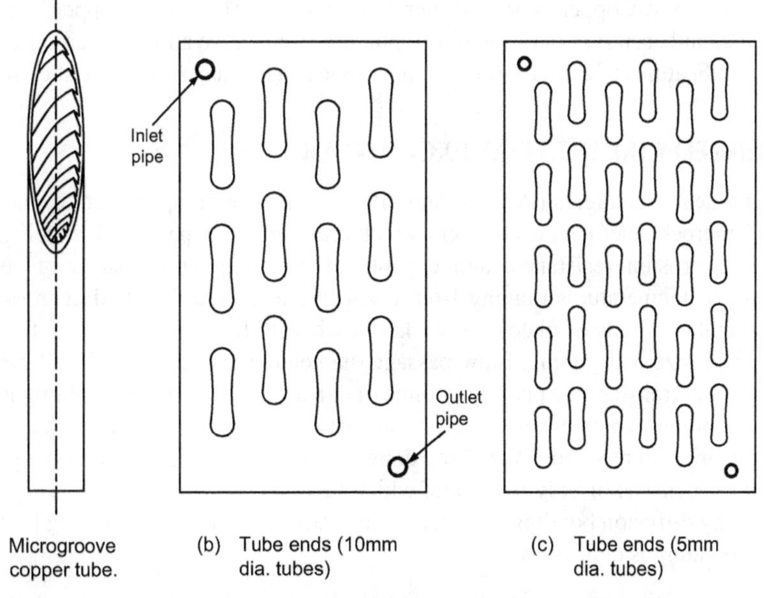

FIGURE 3.91 MicroGroove copper tube heat exchanger coil.

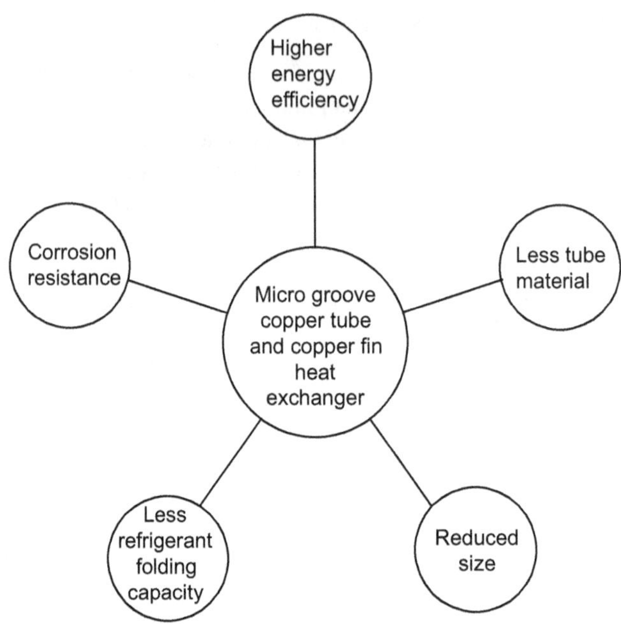

FIGURE 3.92 Features and advantages of microGroove copper tube heat exchanger coil. (Adapted from https://microgroove.net/).

3.14.1 Features

All-copper coils maintain their heat transfer efficiency over long periods of time. The smaller-diameter inner-grooved copper tube heat exchanger not only has a better heat transfer coefficient than the larger-diameter tubes but also is more energy efficient and consumes less material, thereby making it more cost effective. These heat exchangers have shown a reduction in materials (up to 15%) and refrigerant charge (up to 15%) in the system besides improvements in performance and energy efficiency. Copper is an antimicrobial material. The use of copper coils to inhibit the growth of fungi and bacteria is a recent development in innovative air conditioning and refrigeration products [119]. Features of a microgroove copper tube heat exchanger coil are shown in Figure 3.92.

3.15 PRINTED CIRCUIT HEAT EXCHANGER

Printed circuit heat exchangers (PCHEs) and diffusion bonded compact heat exchangers (DCHEs) are kinds of microchannel heat exchangers which are highly compact, with excellent heat transfer performance, corrosion resistance, and capable of operating at pressures of several hundred atmospheres and temperatures ranging from cryogenic to several hundred degrees centigrade. In a DCHE, several hundreds of plates are stacked, each with flow passages. Significant features are the flow passage size and joining. Flow passage diameters are each several millimeters in size, to ensure a large heat transfer area per unit volume of at least 1000 m^2/m^3. The joining is accomplished by diffusion bonding, which covers high-pressure resistance up to 100 MPa. The flow passages of a DCHE are fabricated by chemical etching done on the material plates, rather than by fin forming. Thus each layer consists of only one plate, which facilitates assembly by stacking. The joining is accomplished by diffusion bonding, which can offer stronger joints than brazing [120–123].

Diffusion bonding is defined as

> a method for joining comprising the steps of closely sticking base materials together and pressing these materials against each other at a temperature not exceeding their melting

points, while suppressing their plastic deformation to a minimum, so as to cause the diffusion of atoms at the joining interface to complete the bonding.

"Diffusion-bonding" creates a core with no joints, welds, or other potential points of failure. The resulting unit combines exceptional strength and integrity with high efficiency and performance. The plate material depends on the application. Several materials, including stainless steel, nickelbase alloy, and titanium, have so far been evaluated for their flow-passage fabrication and bonding performance.

By incorporating diffusion bonding with microchannel technology, OEMs like Heatric manufacture units that are up to 85% smaller and lighter than traditional technologies such as shell and tube heat exchangers. This reduction in size can lead to significant savings in structural costs due to elimination of excess pipework, frames, and associated equipment. Heatric's PCHEs are proven technology, having worked in the upstream hydrocarbon processing, petrochemical, and refining industries for many years. A PCHE is shown in Figure 3.93 [123].

3.15.1 Trade Names

i. The Diffusion-bonded Compact Heat Exchanger (DCHE) is a trademark of KOBELCO, Japan.
ii. Printed Circuit Heat Exchanger (PCHE) is the trademark of HEATRIC, a division of Meggitt.
iii. The Marbond heat exchanger is a trademark of Chart Heat Exchangers Company.

3.15.2 Features of PCHE

PCHEs are all-welded, there is no braze material employed in construction, and no gaskets are required. Hence the potential for leakage and fluid-incompatibility is reduced. In fact, the high level of constructional integrity renders PCHEs exceptionally well suited to critical high-pressure applications, such as high-pressure gas exchangers on offshore platforms. Significant features are the low passage size and joining. Flow passage diameters are each several millimeters in size, to ensure a large heat transfer area per unit volume of at least 1000 m^2/m^3.

3.15.3 Materials of Manufacture

Materials for manufacture include stainless steel 304 and 316, Duplex (S31803), Titanium (Grade 2), 6 Moly (NO8367), Alloy 617, etc.

3.15.4 Industry Applications of PCHE

Industrial applications of PCHE include [124]:

i. Heatric's diffusion-bonded heat exchangers are utilized in a wide range of industrial applications including the oil and gas industry, power generation, chemical processing, and others.
ii. Industrial gases. PCHEs act as a key enabling technology for many air-processing systems and can significantly enhance effectiveness while reducing the overall footprint of the equipment.

3.15.5 The Benefits of Heatric's Diffusion-Bonded Heat Exchangers

The benefits of high-integrity DB PCHE include [125]:

i. Compact footprint and reduced weight for structural and space cost savings
ii. The widest operating parameters and performance of any exchanger, including high efficiency and effectiveness due to the unique flow-plate construction

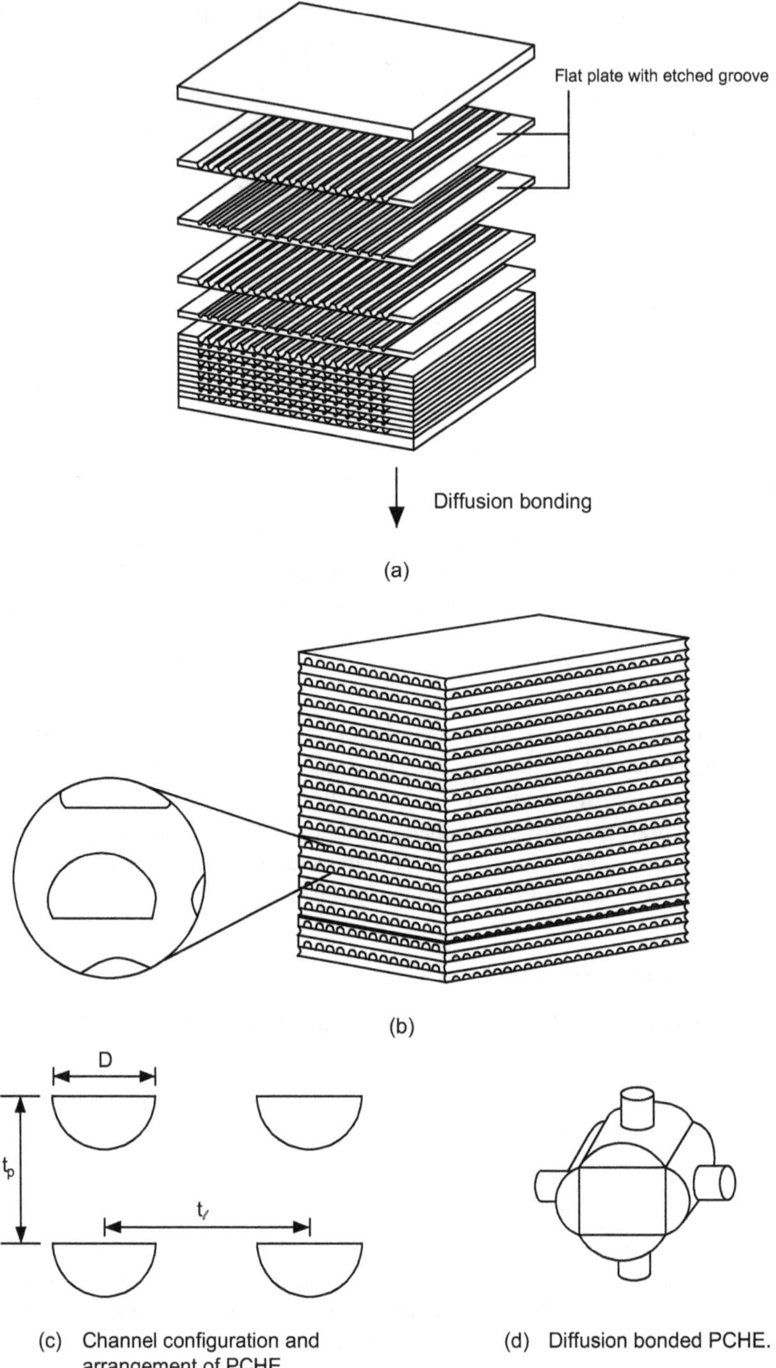

FIGURE 3.93 PCHE(Schematic) (Adapted from www.vpei.com/diffusion-bonded-microchannel-heat-exchangers/).

iii. Further operational savings through reduced fluid flows
iv. Multi-process integration into a single unit for capital and operational cost savings
v. Increased safety in operation due to their high integrity

3.15.6 Thermal Design

Detailed thermal design of PCHEs is supported by proprietary design software developed by the manufacturers which allows for infinite geometric variations to passage arrangements in design optimization. Few flow passage arrangements inside a PCHE are shown in Figure 3.94 [126,127]. Variations to passage geometry have negligible production cost impact since the only tooling required for each variation is a photographic transparency for the photo-chemical machining process. Ports for fluid entry and exit are shown in Figure 3.95.

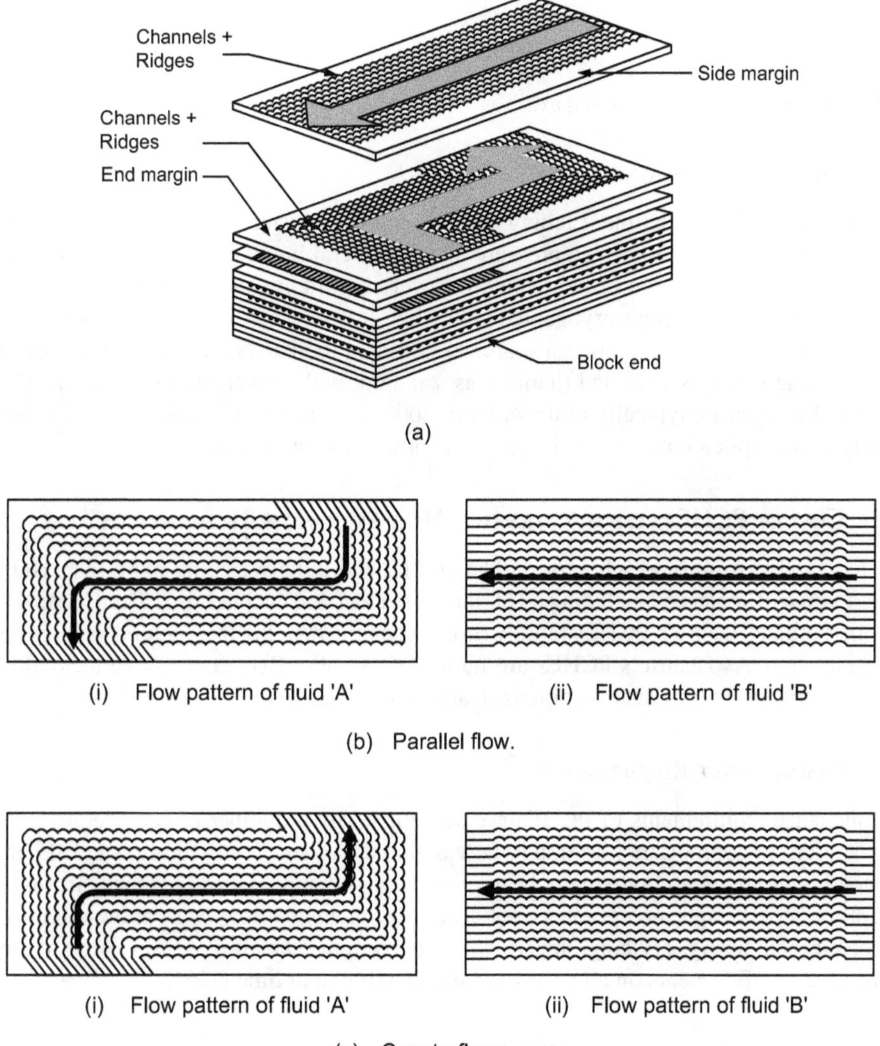

FIGURE 3.94 Examples of flow passages pattern inside a PCHE(Schematic). (Adapted from *Innovative Compact Heat Exchangers*. Proceedings of ICAPP '10 San Diego, CA, USA, June 13–17, 2010 Paper 10300, pp. 218–226.)

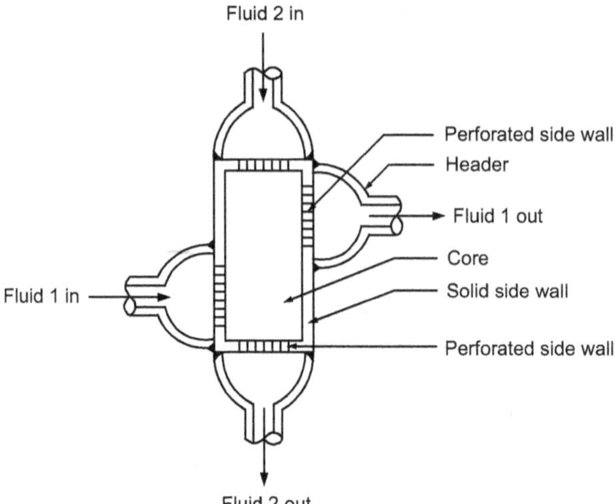

FIGURE 3.95 Ports for fluids entry and exit.

3.15.7 Mechanical Design

The mechanical design is flexible. Etching patterns can be adjusted to provide high-pressure containment where required—the design pressure may be several hundred atmospheres. The diffusion bonded/all-welded construction is compatible with very high-temperature operation, and the use of austenitic stainless steels allows cryogenic application. It is worthy of note that vibration is absent from PCHEs as this can be an important source of failure in shell and tube exchangers. Construction materials include stainless steel and titanium as standard, with nickel and nickel alloys also commonly used. Passages are typically of the order of 2000 microns in a semi-circular cross section for reasonably clean applications, although there is no absolute limit on passage size.

3.15.8 Cost of PCHEs versus Alternative Heat Exchanger Technologies

Heatric has produced units weighing less than 100 kg and units weighing more than 100 tonnes, utilizing a range of materials including 316L stainless steel and Inconel 617. The CAPEX cost for systems that utilize Heatric's PCHEs rapidly drops when the benefits in structural savings are taken into consideration. As Heatric's PCHEs are up to 85% smaller than competitive technologies the reduction in structural costs, such as pipework and skid volume, can be substantial.

3.15.9 Maintenance Requirements

The maintenance requirements of PCHEs are determined by a number of considerations, such as the duty and use of appropriate filtration (where required). When Heatric designs PCHEs the operational conditions are incorporated into the approach and accommodations are made to ensure maximum uptime. The design incorporates easy-access *maintenance nozzles* to allow all areas of the core to be cleaned during servicing, thus restoring maximum efficiency. Also, Heatric strainers are recommended as OE fitments on all units to ensure maximum uptime [128].

3.15.10 Process Filtration for PCHEs

Heatric recommends that strainers are fitted on all PCHEs as a preventative maintenance action. The strainer requirements are dictated by a number of variables such as fluid viscosity, particulate size, operational pressure, operational flow rate, and system configuration/environment.

3.15.11 Cleaning of PCHEs

The requirements for PCHE cleaning are determined by a number of factors, including the operating conditions and maintenance strategy. Use of the correct filtration method can massively reduce the need for cleaning. Heatric has developed an effective PCHE cleaning method using ultra high-pressure (UHP) water jetting.

3.15.12 Fabrication of a Diffusion-Bonded PCHE

Diffusion bonding is a key step in producing a DCHE. In a DCHE, several hundred plates are stacked, each having flow passages. The stacked plates must be homogeneously bonded together and finally, headers and nozzles are welded to the core in order to direct the fluids to the appropriate sets of passages. Welded and diffusion-bonded PCHEs employ no gaskets or braze material, resulting in superior integrity compared to other technologies that may use gaskets or brazing as part of their construction. Here is a general overview of fabricating a diffusion-bonded heat exchanger [123,129–131]:

1. Thermal design of the heat exchanger
2. Etch plates or channels with the flow pattern design
3. Clean the channels
4. Assemble the channels in a counter-flow configuration
5. Diffusion bonding
6. Machine the diffusion-bonded core to size
7. Welding is required for manifolds, flanges, etc.
8. Inspection and testing (hydrostatic test and leak test)
9. Packing and shipping

The diffusion bonding and manufacture of a PCHE is shown in Figure 3.96.

3.15.13 The Marbond Heat Exchanger

The Marbond heat exchanger, manufactured by Chart Heat Exchangers, is formed of slotted flat plates, that is, plates which have been chemically etched through. The plate pack is then diffusion-bonded together [132]. A representative form of the surface with channels for flow is shown schematically in Figure 3.97. Passage forms are very similar to those of a plate-fin (PFHE) surface. As with the PCHE, the range of constructional materials is only limited by their ability to be diffusion-bonded.

3.16 3D Printing or Additive Manufacturing of Heat Exchangers

3.16.1 Additive Manufacturing

Additive manufacturing (AM) is the industrial production name for 3D printing, a computer-controlled process that creates three-dimensional objects by depositing materials, usually in layers and with minimal post-processing. Using computer-aided design (CAD) or 3D object scanners, additive manufacturing can create precise geometric shapes. These are built layer by layer, as with a 3D printing process, which is in contrast to traditional manufacturing that often requires machining or other techniques to remove surplus material. Ideal for creating complex geometric shapes, additive manufacturing is suited for making bespoke parts and prototypes, while the computer-aided design allows for any design changes to be effected quickly and efficiently. Producing heat exchangers

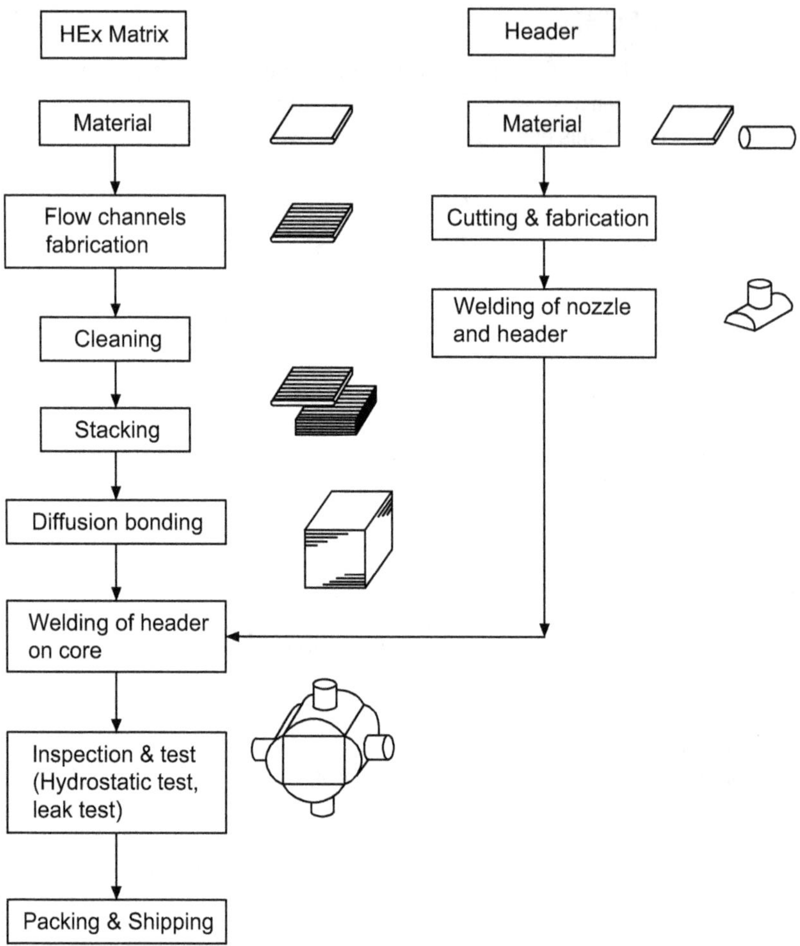

FIGURE 3.96 Diffusion bonding and manufacture of a PCHE (Adapted from www.kobelco.co.jp/english/products/ecmachinery/dche/).

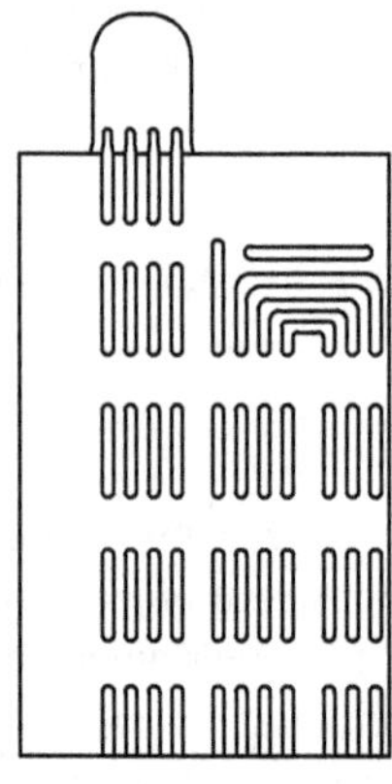

FIGURE 3.97 Marbond heat exchanger etched plate showing flow paths. (Adapted from Hesselgreaves, J.E., Law, R., Reay, D. (2017) *Chapter 2 Industrial Compact Exchangers, Compact Heat Exchangers*, pp. 35–89, http://dx.doi.org/10.1016/B978–0-08-100305–3.00002–1, Elsevier Ltd.)

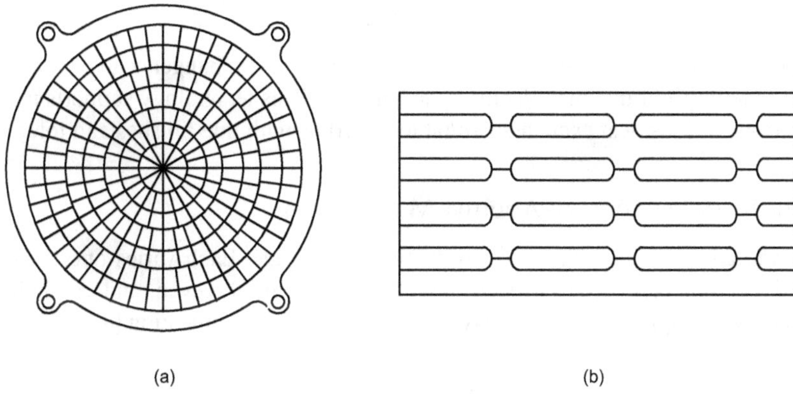

FIGURE 3.98 A heat exchanger geometry produced by an AM process.

using conventional techniques is often a complex and time-consuming task, requiring multiple steps such as forming and welding. Furthermore, there is an increasing demand to make heat exchangers more compact and efficient to improve on ever-growing performance requirements [133–135].

Additive manufacturing enables manufacturing of seamless, leak-proof, and efficient heat exchangers regardless of the type of heat exchanger. Very complex designs can be realized, and the ratio between the heat exchanging surfaces to the total volume of the heat exchanger can be increased significantly. The increased performance allows miniaturization of the heat exchanger. This versatile manufacturing technique is widely used in industries including aerospace, automotive, and medical areas. Figure 3.98 shows a heat exchanger geometry produced by an AM process [135,136].

3.16.2 Advantages

The adoption of AM can generate advantages in various areas. For example, fast prototyping, the acceleration of R&D, and the timely delivery of spare parts. Key benefits of AM include tool-less manufacturing, increased geometric freedom in part design, no or less sub-assemblies, non-physical inventory, rapid spare part production, etc. As far as heat exchangers are concerned, the benefit of using AM is in providing a maximum heat transfer surface in a smaller space. The use of AM in situations that require lightweight, space-saving, and complex components enables an improved flow performance to be achieved.

An efficient heat-exchanger requires a large surface-to-volume ratio. This can be achieved, for example, by the preparation of fine channels in the body of the heat-exchanger, which allows fast heat transfer with the heat-exchange fluid. Additive manufacturing provides the freedom to design complex geometries. Furthermore, production by additive manufacturing can reduce the consumption of raw materials by up to 75%, as conventional machining is avoided, thus also reducing the environmental impact [137].

3.16.3 Materials

AM can produce heat exchangers from a wide range of materials ranging from aluminum alloys to high-temperature alloys such as Inconel 718 and Inconel 625. Other materials, such as copper and copper alloys, are also used for additively manufactured heat exchangers for their high conductivity, which makes them ideal for heat transfer applications. Multi-material heat exchangers are also possible.

3.16.4 Additive Manufacturing of Heat Exchangers

After designed systems are validated by CFD simulations, heat exchangers can be manufactured by selective laser melting additive manufacturing technology. This technology supports manufacturing components from various heat exchanger suitable materials with fine finishes and tolerances.

3.16.5 The Seven Processes of Additive Manufacturing

There are seven additive manufacturing production techniques. Each varies due to the materials, layering, and machine technology needed. EWI specializes in all seven, and can help your team identify, design, and implement the process that is right for your application [138,139]:

1. Powder bed fusion
2. Directed energy deposition
3. Binder jetting
4. Sheet lamination
5. Material extrusion
6. Material jetting
7. Vat photopolymerization

3.16.5.1 Laser Metal Deposition

Laser metal deposition (LMD) is an additive manufacturing process in which a laser beam forms a melt pool on a metallic substrate, into which powder is fed. The powder melts to form a deposit that is fusion-bonded to the substrate. The required geometry is built up in this way, layer by layer. Both the laser and nozzle from which the powder is delivered are manipulated using a gantry system or robotic arm.

3.16.5.2 SLM

Selective laser melting (SLM) is an additive manufacturing process used to build 3D metal objects using high-power laser beams. The SLM technique uses a ytterbium fiber laser to fully melt a pre-deposited layer of a single-component metallic powder onto a substrate according to a computer-generated pattern. By successive powder deposition of these layers at 50 micron thickness, fully dense, micro-crossflow heat exchangers have been built. The construction chamber is filled with an inert gas to prevent oxidation of the metal throughout the construction phase [140]. 3D printing of an object by SLM process is shown in Figure 3.99.

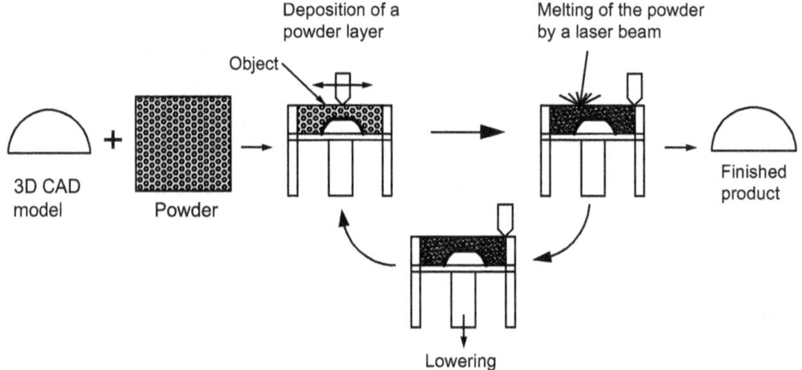

FIGURE 3.99 3D printing of an object by SLM process. (Adapted from Shah, R.K., Ishizuka, M., Rudy, T.M., Wadekar, V.V. (2005) *Proceedings of Fifth International Conference on Enhanced, Compact and Ultra-Compact Heat Exchangers: Science, Engineering and Technology*. Whistler, British Columbia, Canada.)

3.16.6 Why Use Additive Manufacturing for the Manufacture of heat exchangers?

The design of a heat exchanger is often a balance between maximizing the surface area of the part while also minimizing the pressure drop through the heat exchanger. While larger industrial heat exchangers are still manufactured by conventional manufacturing techniques, metal AM is finding more and more applications in smaller, more compact designs. Metal additive manufacturing is preferred for the manufacture of heat exchangers for the following reasons [141–146]:

i. Heat exchangers as a single piece. AM has the advantage of being able to produce the core and manifolding as a single part. In other words, no extra steps of forming, assembly, brazing, or welding, so the entire process is faster and monolithic heat exchangers are significantly lighter weight.
ii. Smaller and more efficient. AM can create the unique shapes and internal features only possible with AM and it facilitates the miniaturization of these components. These heat exchanges with much more compact shapes can fit tight space requirements.
iii. Higher internal surface. Extremely thin-walled internal core lattices increase the surface area and efficiency of heat exchangers. Metal additive manufacturing processes, such as laser powder bed fusion, have the ability to produce walls of 0.1 millimeters thickness or less.
iv. Reduced weight and size. 3D printing allows for making the device lighter and smaller, but with the same or even better performance.
v. New shapes and internal features made possible with 3D printing facilitate the miniaturization of heat exchangers. As a result, 3D-printed heat exchangers have a much more compact shape that fits tight space requirements.
vi. Simplified production. Traditionally, the production of heat exchangers involves multiple steps, including forming, brazing, and welding. When using 3D printing to produce a heat exchanger directly, all these operations can be eliminated, thus streamlining the production process.
vii. Better quality. Furthermore, a 3D-printed heat exchanger is built in one operation so there are no seams or joints that could develop leaks. Due to more straightforward production, process variability is lower and overall quality is expected to be much higher.

3.16.7 Benefits

Additive manufacturing provides the freedom to design complex geometries. Furthermore, production by additive manufacturing can reduce the consumption of raw materials by up to 75%, as conventional machining is avoided, thus also reducing the environmental impact. Additive manufacturing of heat exchangers gives the following benefits [147]:

1. Lightweight and space efficient
2. Parts without joint
3. Less lead time in reaching the market
4. Excellent thermal management

With the above benefits, AM of heat exchangers is a good application in electronics, chemical, aerospace, automotive, chemical, power plants, gas turbines, reactors, etc.

3.16.8 AM Process Capabilities and Advantages of 3D-Printing Heat Exchangers

1. One of the main goals of AM is to reduce a part's size. Compared to a traditional part, it is possible to reduce the weight by between 20% and 50%, and in some cases more, and consequently also the dimensions. In some applications not only lightness but also compactness becomes important.

2. A heat exchanger fits inside a system where there are a vast multitude of components. With AM, the heat exchanger is designed and 3D printed to be installed exactly in the available space and has exact connections.
3. AM is well known for its applications in the aerospace, automotive, and motorsport industries. Its use in the oil & gas and chemical plant industries is growing, but it remains minimally utilized compared to traditional manufacturing methods.
4. The internal structure of the heat exchanger can be designed using the unique shapes/lattice structures available with additive manufacturing, such as lattice structure.
5. Higher internal surface. Extremely thin-walled internal core lattices increase the surface area and efficiency of heat exchangers.
6. The 3D-printed internal structures form the channels for the liquids to flow through, in the process promoting turbulence, which in turn improves heat transfer. Turbulence also helps to prevent fouling in the exchanger and keeps the channels clear.
7. AM provides a tool-less process; the manufacturing constraints and tooling costs typical for subtractive and multi-part assembly operations are eliminated. In addition, cost does not scale with complexity with AM.
8. Design freedom enables the implementation of nonlinear, tapered, and optimized variable geometries in extended structures like fins, vanes, blades, capillary wicking structures, heat pipes, and conformal internal passages.
9. AM can create any size or shape, and unique shapes and internal features only possible with AM. These heat exchanges with much more compact shapes can fit tight space requirements.

3.16.9 AM Design Constraints

AM design constraints include [148]:

i. Aspect ratio
ii. Bridging distance
iii. Maximum hole size
iv. Overhang angle
v. Minimum wall thickness
vi. Moving part tolerances

3.16.10 Fundamental Considerations or Requirements to be Met for Adopting AM

1. Heat exchangers are extremely well suited to AM. However, a deep understanding and combination of Design for AM (DfAM), fundamental principles of heat transfer/fluid mechanics, thermo-fluid simulation, and AM process know how are required to achieve competitive results [149].
2. Importance of surface area density

This remains the first order for performance of heat exchangers. Aim to maximize the surface area that is packaged into a given volume without compromising/increasing the weight of the part. Increase the surface area through the use of lattice structures, pin fins, and microchannels.

3. Large CAD and build data

The importance of surface area density means that AM heat exchangers inherently contain large and dense arrays of complex features. This naturally results in very large native CAD files, which in turn results in very large STL and AM build files.

4. Understanding the implications of defects is critical instead of trying to eliminate them. In depth and robust development programs, process monitoring, computed tomography, performance, and durability/fatigue testing are integral and continue to be performed during a typical product development and are important.
5. The biggest challenge is often at the verification and testing and to ensure that all the powder has been removed from the internal channels and all the internal walls have been created perfectly.

3.16.11 Technology Accelerators for Mainstream Heat Exchanger Production

Technology accelerators for mainstream heat exchanger production include [149]:

1. Increased productivity rates of AM platforms
2. Improved detail resolution—a 2–3-fold reduction in the minimum achievable wall thickness on medium- to large-format AM systems
3. New materials
4. Standards and certification—the design and implementation of specific standards can only help and assist in the qualification and validation of AM heat exchangers

Nomenclature

A	total heat transfer area, m^2 (ft^2)
A_f	fin surface area, m^2 (ft^2)
A_{fr}	frontal area, m^2 (ft^2)
A_o	minimum free flow area, m^2 (ft^2)
A_p	primary heat transfer surface area, m^2 (ft^2)
A_s	secondary heat transfer surface area, m^2 (ft^2)
C, c_p	specific heat, J/kg °C (Btu/lbm.°F)
C_c	specific heat of cold fluid, J/kg °C (Btu/lbm.°F)
C_h	specific heat of hold fluid, J/kg °C (Btu/lbm.°F)
C^*	heat capacity rate ratio for a compact exchanger= $(mc_p)_{min}/(mc_p)_{max}$
D_h	hydraulic diameter, $(4r_h)$, m (ft)
d_i	inside diameter of the tueb, m (ft)
d_f	fin diameter of circular fin, m (ft)
d_o	outside diameter of the tube, m (ft)
d_r	effective fin diameter at the collar, m (ft)
F	LMTD correction factor
f	fanning friction factor
f_f	friction factor on the fin surfaces
f_t	friction factor due to drag force on the tubes
G, G_m	mass velocity, $kg/m^2 \cdot s$ ($lbm/h \cdot ft^2$)
G_{max}	mass velocity based on free flow area, $kg/m^2 \cdot s$ ($lbm/h \cdot ft^2$)
g_c	acceleration due to gravity or proportionality constant 9.81 m/s^2 (32.17 ft/s^2) = 1 and dimensionless in SI units
h	convective heat transfer coefficient, $W/m^2 \cdot °C$ ($Btu/ft^2\ h \cdot °F$)
j	Colburn heat transfer factor
j_N	Colburn heat transfer factor for N number of tube rows
K	fluid entrance contraction or exit expansion coefficient
k	thermal conductivity of the fluid, W/m°C (Btu/h·ft·°F)
k_f	thermal conductivity of the fin material, W/m°C (Btu/h·ft·°F)
L	length in the fluid flow direction, m (ft)
L_f	characteristic fin length, m (ft)
l_f	fin length, m (ft)

L_1, L_2, L_3	heat exchanger core dimensions of a tube bank, m (ft)
M	mass flow rate of the fluid, kg/s (lbm/h)
N_1	number of tubes in one row
N_t	total number of tubes
N_f	fin density of a tubular exchanger
N_p	number of flow passages on side 1 of PFHE
$N_p + 1$	Number of flow passages on side 2 of PFHE
NTU	number of transfer units
Nu	Nusselt number
N_r	number of tube rows,
n	fin density of PFHE
P_p	fluid pumping power
P_l	longitudinal pitch, m^2 (ft^2)
P_T, P_t	lateral pitch, m^2 (ft^2)
p	pressure, Pa (lbf/ft^2)
Pr	Prandtl number
Q	total heat duty of the exchanger, W·s (Btu)
q	heat transfer rate, W (Btu/h)
R	heat capacity rate ratio of shell and tube heat exchanger fluids
Re	Reynolds number
Re_d	Reynolds number based on tube outer diameter
Re_h	Reynolds number based on hydraulic diameter
r_h	hydraulic radius, m (ft) = $D_h/4$
St	Stanton number
s	gap between two fins, m (ft)
t_f	thickness, m (ft)
T	temperature, °C (°F)
t	temperature, °C (°F)
T_h	header plate thickness, m (ft)
U_f	upstream fluid velocity on the fin side, m/s (ft/s)
U	overall heat transfer coefficient, W/m^2·°C (Btu/h·fr^2·°F)
V_P	volume between plates on one side, m^3(ft^3)
α	surface area density of PFHE, m^2/m^3 (ft^2/ft^3)
α	the effective heat transfer coefficient of a stream (Eq. 3.71), W/m^2·°C (Btu/ft^2 h·°F)
Δp	pressure drop, Pa (lbf/ft^2)
ε	thermal effectiveness
μ	viscosity of the fluid, kg/m·s or Pa·s (lbm/h·ft)
ρ	fluid density, kg/m^3 (lbm/ft^3)
β	surface area density, m^2/m^3 (ft^2/ft^3)
ηf	fin efficiency
η_o	overall surface efficiency
σ	the ratio of minimum free flow area to frontal area

Subscripts

ave	average
b	bulk mean temperature
c	cold
h	hot
e	exit
i	inlet
lm	log mean
m	mean

n	nozzle
o	outlet
w	wall
1	side 1 or inlet of the exchanger
2	side 2 or outlet of the exchanger
min	minimum
max	maximum

NOTE

1 After Shah, R.K., Compact heat exchanger design procedures. In: *Heat Exchangers: Thermal-Hydraulic Fundamentals and Design*, Kakac, S., Bergles, A.E., and Mayinger, F., eds., Hemisphere, Washington, DC, pp. 495–536, 1981 also Ref. [53].

REFERENCES

1. Shah, R.K., Robertson, J.M. (1993) Compact heat exchangers for the process industry. :n *Energy Efficiency in Process Technology*, P.A. Pilavachi (Ed.), pp. 565–580. Elsevier Applied Science, London, UK
2. Shah, R.K. (1991) Compact heat exchanger technology, and applications. In: *Heat Exchange Engineering, Vol. 2, Compact Heat Exchangers: Techniques of Size Reduction*, E.A. Foumeny, P.J. Heggs (Eds.), pp. 1–29. Ellis Horwood, New York.
3. Shah, R.K. (1983) Classification of heat exchangers. In: *Low Reynolds Number Flow Heat Exchangers*, S. Kakac, R.K. Shah, A.E. Bergles (Eds.), pp. 9–13. Hemisphere, Washington, DC.
4. Shah, R.K. (1981) Classification of heat exchangers. In: *Heat Exchangers: Thermal-Hydraulic Fundamentals and Design*, S. Kakac, A.E. Bergles, F. Mayinger (Eds.), pp. 9–46. Hemisphere, Washington, DC.
5. Webb, R.L. (1987) Enhancement of single phase heat transfer. In: *Handbook of Single Phase Convective Heat Transfer*, S. Kakac, R.K. Shah, W. Aung (Eds.), Chapter 17, pp. 17.1–17.62. John Wiley & Sons, Inc., New York.
6. Yang, W.-J. (1993) High performance heat transfer surfaces: Single phase flows. In: *Heat Transfer in Energy Problems*, pp. 109–116.
7. Rabas, T.J., Taborek, J. (1987) Survey of turbulent forced convection heat transfer and pressure drop characteristics of low finned tube banks in crossflow. *Heat Transfer Engineering* 8: 49–61.
8. Webb, R.L. (1983) Enhancement for extended surface geometries used in air cooled heat exchangers. In: *Low Reynolds Number Flow Heat Exchangers*, S. Kakac, R.K. Shah, A.E. Bergles (Eds.), pp. 721–733. Hemisphere, Washington, DC.
9. Fraas, A.P., Ozisik, M.N. (1965) *Heat Exchanger Design*. John Wiley & Sons, New York.
10. Webb, R.L., Farrell, P.A. (1990) *Improved thermal and mechanical design of copper/brass radiators*, SAE, 900724, pp. 737–748. SAE, Warrendale, PA, 1990.
11. Shah, R.K. (1978) Compact heat exchanger surface selection methods. In: *Heat Transfer*, Volume 4, pp. 193–199. Hemisphere, Washington, DC.
12. Shah, R.K. (1993) Surface geometrical properties—Tube-fin heat exchangers. Class notes for course no. MEA 522, pp. 413–418. Spring 1993.
13. Shah, R.K. (1985) Compact heat exchangers. In: *Handbook of Heat Transfer Applications*, W.M. Rohsenow, J.P. Harnett, E.N. Ganic (Eds.), pp. 4-176–4-312. McGraw-Hill, New York.

14a. Idem, S.A., Jung, C., Gonzalez, G.J., Goldschmidt, V.W. (1987) Performance of air-to-water copper finned tube heat exchangers at moderately low airside Reynolds numbers, including the effects of baffles. *International Journal of Heat and Mass Transfer* 30: 1733–1741.

14b. Idem, S.A., Jacobi, A.M., Goldschmidt, V.W. (1990) Heat transfer characterization of a finned tube heat exchanger (with and without condensation). *Transactions of ASME, Journal of Heat Transfer* 112: 64–70.

14c. Hedderich, P.C., Kelleher, M.D., Vanderplasts, G.N. (1982) Design and optimization of air cooled heat exchangers. *Transactions of ASME, Journal of Heat Transfer* 104: 683–690.
15. Kays, W.M., London, A.L. (1983) *Compact Heat Exchangers*, 3rd edn. McGraw-Hill, New York.
16. Davenport, C.J. (1983) Correlations for heat transfer and flow friction characteristics of louvred fin. *Heat Transfer—Seattle, AIChE Symposium Series* 79(225): 19–27.
17. Shah, R.K. (1983) Compact heat exchanger surface selection, optimization, and computer aided thermal design. In: *Low Reynolds Number Flow Heat Exchangers*, S. Kakac, R.K. Shah, A.E. Bergles (Eds.), pp. 983–998. Hemisphere, Washington, DC.
18. Johnson, B.M. (1986) The performance of extended surface heat exchangers. In: *Handbook of Heat and Mass Transfer, Vol. 1, Heat Transfer Operation*, N.P. Cheremisinoff (Ed.), pp. 767–805. Gulf Publishing Company, Houston, TX.
19. Brauer, H. (1961) *Kaltetchnik*, 13 Jehrgang, Heft 8, pp. 274–279.
20. LaHaye, P.G., Neugebuer, F.J., Sakhuja, R.K. (1974) A generalized prediction of heat transfer surfaces. *Transactions of ASME, Journal of Heat Transfer* 96: 511–517.
21. Fletcher, L.S. (1988) Recent developments in contact conductance heat transfer. *Transactions of ASME, Journal of Heat Transfer* 110: 1059–1070.
22. Nho, K.M., Yovanovich, M.M. (1990) Effect of oxide layers on measured and theoretical contact conductances in finned tube heat exchangers. In: *Compact Heat Exchangers*, R.K. Shah, A.D. Kraus, D. Metzger (Eds.), pp. 397–419. Hemisphere, Washington, DC.
23. European Aluminium Association (2011) *Applications—Power train—Heat exchangers, the aluminium automotive manual*, Version 2011, pp. 1–46. European Aluminium Association (auto@eaa.be).
24. Shah, R.K., Bhatti, M.S. (1998) Assessment of correlations for single-phase heat exchangers. In: *Two-Phase Flow Heat Exchangers—Thermal-Hydraulic Fundamentals and Design*, S. Kakac, A.E. Bergles, F.E. Oliveira (Eds.), pp. 81–122. Kluwer Academic Publishers, London, UK.
25. Thome, J.R. (2004) Chapter 6, Heat transfer to air cooled heat exchangers. In: *Engineering Data Book III*, pp. 6.1–6.40. Wolverine Tube, Inc..
26. Park, Y.G., Jacobi, A.M. (2001) *Air-Side Performance Characteristics of Round- and Flat-Tube Heat Exchangers: A Literature Review, Analysis and Comparison*, pp. 1–111. ACRCCR-36, May 2001.
27. Giedt, W.H. (1957) *Principles of Engineering Heat Transfer*. Van Nostrand, New York.
28. Kays, W.M., London, A.L., Shah, R.K. (1954) Heat transfer and friction characteristics of gas flow normal to tube banks—Use of a transient test technique. *Transactions of ASME* 76: 387–396.
29. Gram, A.J., Mackey, C.O., Monroe, E.S. (1958) Convection heat transfer and pressure drop of air flowing across inline tube banks II—Correlation of data for ten row deep tube banks. *Transactions of ASME* 80: 25–35.
30. London, A.L., Kays, W.M., Johnson, D.W. (1952) Heat transfer and flow-friction characteristics of some compact heat exchanger surfaces. *Transactions of ASME* 74: 1167–1178.
31. Briggs, D.E., Young, E.H. (1963) Convection heat transfer and pressure drop of air flowing across triangular pitch banks of finned tubes. *Chemical Engineering Progress Symposium Series* 59: 1–10.
32. Robinson, K.K., Briggs, D.E. (1966) Pressure drop of air flowing across triangular pitch banks of finned tubes. *Chemical Engineering Progress Symposium Series* 62: 177–184.
33. Rabas, T.J., Eckels, P.W., Sabatino, R.A. (1981) The effect of fin density on the heat transfer and pressure drop performance of low finned tube banks. *Chemical Engineering Communications* 10: 127–147.
34. Weierman, C. (1977) Pressure drop data for heavy duty finned tubes. *Chemical Engineering Progress* 73: 69–72.
35. ESDU (1973) *Engineering sciences data unit convective heat transfer during crossflow of fluid over plain tube banks*. ESDU, Item number 73031.
36. Ganguli, A., Yilmaz, S.B. (1987) New heat transfer and pressure drop correlations for cross flow over low finned tube banks. *AIChE Symposium Series Heat Transfer—Pittsburg* 83: 9–19.
37. Elmahdy, A.H., Biggs, R.C. (1978) *Finned tube heat exchanger: Correlation of dry surface heat transfer data*, pp. 262–273. ASHRAE, No. 2544.
38. Gray, D.L., Webb, R.L. (1986) Heat transfer and friction correlations for plate fin and tube heat exchangers having plain fins. *Proceedings of the Eighth International Heat Transfer Conference*, Volume 6, pp. 2745–2750. San Francisco, CA.
39. Rich, D.G. (1973) The effect of fin spacing on the heat transfer and friction performance of multirow, smooth plate fin tube heat exchanger. *ASHRAE Transactions* 79: 137–145.

40. Zukauskas, A.A. (1989) *High Performance Single-Phase Heat Exchangers*, J. Karni (Ed., English edition). Hemisphere, Washington, DC.
41a. McQuiston, F.C. (1981) Finned tube heat exchangers—State of the art for the air side. *Transactions of ASHRAE* 87: 1077–1085.
41b. McQuiston, F.C. (1978) Heat, mass and momentum transfer data for five plate fin tube heat transfer surfaces. *Transactions of ASHRAE* 84: 266–281.
42. Maltson, J.D., Wilcock, D., Davenport, C.J. (1989) Comparative performance of rippled fin plate fin and tube heat exchangers. *Transactions of ASME, Journal of Heat Transfer* 111: 21–28.
43. Kroger, D.G. (1985) *Radiator characterisation and optimization*, pp. 2.984–2.990. SAE 840380.
44. ALPEMA Standard (2010) *The Brazed Aluminium Plate-Fin Heat Exchanger Manufacturer's Association*, 3rd edn. Didcot, Oxon, UK.
45. Taylor, M.A. (Ed.) (1980) *Plate-Fin Heat Exchangers, Guide to Their Specification and Use*. HTFS (Harwell Laboratory), Oxon, UK.
46a. Shah, R.K. (1975) *Perforated heat exchanger surfaces part 1—Flow phenomena, noise and vibration characteristics*. ASME Paper No. 75-WA/HT-8.
46b. Shah, R.K. (1975) *Perforated heat exchanger surfaces part 2—Heat transfer and flow friction characteristics*. ASME Paper No. 75-WA/HT-9.
47. Joshi, H.M., Webb, R.L. (1987) Heat transfer and friction in the offset strip-fin heat exchanger. *International Journal of Heat and Mass Transfer* 30: 69–84.
48. Cowell, T.A., Heikal, M.R., Achaichia, A. (1993) Flow and heat transfer in compact louvred fin surfaces. In: *Aerospace Heat Exchanger Technology*, R.K. Shah, A. Hashemi (Eds.), pp. 549–560. Elsevier Science Publishers, Amsterdam, the Netherlands.
49. (2021) *Installation, operation, and maintenance manual for Chart Brazed Aluminum Heat Exchangers (BAHX) and Core-in-Kettle® Assemblies*, pp. 1–50.
50. Chart Industries, Inc. (2020) *Brazed Aluminium Heat Exchangers*, pp. 1–26. Chart Industries, Inc., Wisconsin, WI.
51. https://files.chartindustries.com/ChartLifecycleBAHXServices.pdf
52. Kays, W.M., London, A.L. (1950) Heat transfer and flow friction characteristics of some compact heat exchanger surfaces, Part 2—Design data for thirteen surfaces. *Transactions of ASME* 72: 1087–1097.
53. Shah, R.K. (1981) Compact heat exchanger design procedures. In: *Heat Exchangers: Thermal-Hydraulic Fundamentals and Design*, S. Kakac, A.E. Bergles, F. Mayinger (Eds.), pp. 465–536. Hemisphere, Washington, DC.
54. Weiting, A.R. (1975) Empirical correlations for heat transfer and flow friction characteristics of rectangular offset-fin plate-fin heat exchangers. *Transactions of ASME, Journal of Heat Transfer* 97: 488–490.
55. Manglik, R.M., Bergles, A.E. (1990) The thermal hydraulic design of the rectangular offset strip-fin compact heat exchanger. In: *Compact Heat Exchangers—A Festschrift for A. L. London*, R.K. Shah, A.D. Kraus, D. Metzger (Eds.), pp. 123–149. Hemisphere, Washington, DC.
56. Kays, W.M. (1955) Pin-fin heat exchanger surfaces. *Transactions of ASME* May: 471–483.
57. Norris, R.H., Spofford, W.A. (1942) High-performance fins for heat transfer. *Transactions of ASME* 64: 489–496.
58. Kanzaka, M., Iwabuchi, M., Aoki, Y., Ueda, S. (1989) Study on heat transfer characteristics of pin finned plate type heat exchangers. *Heat Transfer—Philadelphia* 85: 306–311.
59. Gardner, K.A. (1945) Efficiency of extended surface. *Transactions of ASME* 67: 621–632.
60. Kern, D.Q., Kraus, A.D. (1972) *Extended Surface Heat Transfer*. McGraw-Hill, New York.
61. Scott, T.C., Goldschmidt, I. (1979) Accurate, simple expressions for the fin efficiency of single and composite extended surfaces. *Advances in Enhanced Heat Transfer, ASME Symposium* 100–122: 79–85.
62. Schmidt, T.E. (1949) Heat transfer calculations for extended surfaces. *Journal of ASHRAE* April: 351–357.
63. Zabronsky, H. (1955) Temperature distribution and efficiency of a heat exchanger using square fins on round tubes. *ASME Journal of Applied Mechanics* 22: 119–122.
64. Sparrow, E.M., Lin, S.H. (1964) Shorter communication. *International Journal of Heat and Mass Transfer* 7: 951–953.

65. Shah, R.K. (1971) Temperature effectiveness of multiple sandwich rectangular plate-fin surfaces. *Transactions of ASME, Journal of Heat Transfer* 91: 471–473.
66. Shah, R.K. (1988) Plate-fin and tube-fin heat exchanger design procedures. In: *Heat Transfer Equipment Design*, R.K. Shah, D.C. Subbarao, R.M. Mashelekar (Eds.), pp. 255–266. Hemisphere, Washington, DC.
67. Webb, R.L. *Specification of Rating and Sizing Problems.* DOI 10.1615/hedhme.a.000303
68. *Procedures for the Thermal Sizing Problem.* Heat Exchanger Design Handbook Multimedia Edition, DOI 10.1615/hedhme.a.000306; https://hedhme.com/content_map/?link_id=17251&article_id=306
69. Hesselgravcs, J.E. (1993) Optimizing size and weight of plate fin heat exchangers. In: *Aerospace Heat Exchanger Technology*, R.K. Shah, A. Hashemi (Eds.), pp. 391–399. Elsevier Science Publishers, Amsterdam, the Netherlands.
70. Shah, R.K., Mueller, A.C. (1985) Heat exchanger basic thermal design methods. In: *Handbook of Heat Transfer Applications*, W.M. Rohsenow, J.P. Hartnett, E.N. Ganic (Eds.), 2nd edn., pp. 4-1–4-77. McGraw-Hill, New York.
71. Chiou, J.P. (1980) The advancement of compact heat exchanger theory considering the effects of longitudinal heat conduction and flow nonuniformity. In: *Symposium on Compact Heat Exchanges—History, Technological Advancement and Mechanical Design Problems*, Volume. 10, R.K. Shah, C.F. McDonald, C.P. Howard (Eds.), pp. 101–121. ASME, New York.
72. Hudson Products Corporation (2010) *The Basics of Air-Cooled Heat Exchangers*, pp. 1–16. Hudson Products Corporation, A Subsidiary of Hudson Products Holdings, Inc., Texas. www.hudsonproducts.com
73. Gas Processors Suppliers Association Section 10, *Air-Cooled Exchangers*, pp. 10–19. Gas Processors Suppliers Association, GPSA.
74. Boes, S. (2017) *Improve Air-Cooled Heat Exchanger Performance, Back to Basics*, pp. 1–7. Hudson Products Corp. www.aiche.org/cep
75. Ganapathy, V. (1979) Design of aircooled exchangers—Process design criteria. In: *Process Heat Exchange, Chemical Engineering Magazine*, V. Cavaseno (Ed.), pp. 418–425. McGraw-Hill, New York.
76. Baker, W.J. (1980) Selecting and specifying air cooled heat exchangers. *Hydrocarbon Process* May: 173–177.
77. Shipes, K.V. (1974) Air cooled exchangers in cold climates. *Chemical Engineering Progress* 70: 53–58.
78a. Brown, R. (1979) Design of aircooled exchangers—A procedure for preliminary estimates. In: *Process Heat Exchange, Chemical Engineering Magazine*, V. Cavaseno (Ed.), pp. 414–417. McGraw-Hill, New York.
78b, Brown, R. (1978) *Chemical Engineering* 85: 108–111.
79. Mukherjee, R. (1997) Effectively design air cooled heat exchangers. *Chemical Engineering Progress*, pp. 26–47.
80. Stone, J. *Air-Cooled Heat Exchangers—General Information*, pp. 1–4. Stone Process Equipment Co., Akron, OH, www.stoneprocess.com/acooler.htm
81. Giammaruti, R. (2010) Performance improvement to existing air-cooled heat exchangers, Paper No. TP04–13, Presented at the *Cooling Technology Institute Annual Conference*, Houston, TX, February 2–11, 2003. August 2010, pp. 1–16.
82. American Petroleum Institute (2013) *API Standard 661, Petroleum, Petrochemical, and Natural Gas Industries—Air-cooled Heat Exchangers*, 7th Edition. American Petroleum Institute, Washington, DC.
83. KLM Technology (2011) *Process design of air cooled heat exchangers (air coolers) (project standards and specifications)*, pp. 1–19. KLM Technology, Johor Bahru, Malaysia.
84. American Society of Mechanical Engineers (2021) *ASME Boiler and Pressure Vessel Code, Section VIII, Division 1—Pressure Vessels*. American Society of Mechanical Engineers, New York.
85. www.aiche.org/resources/publications/cep/2017/january/improve-air-cooled-heat-exchanger-performance. American Institute of Chemical Engineers.
86. Agius, N. (2006) Hydrocarbon Processing, pp. 89–94. Gulf Publishing Company,.
87. www.hcheattransfer.com/ache.html
88. https://cr3.group/solutions/finfan-optimisation/

89. North, C.D.R. (1980) Air coolers. In: *Developments in Heat Exchanger Technology-1*, D. Chisholm (Ed.), pp. 155–177. Applied Science Publishers, London, UK.
90. Gates Draftguard™ Anti-Rotation Device. Gates Australia Pty Ltd, Dandenong South, Victoria, pp. 1–5.
91. Gates Corporation (2017) *Solving windmilling problems on belt ache fan systems how to improve worker safety and reduce maintenance*, pp. 1–5. Gates Corporation.
92. Kroger, D.G. (1993) Experimental heat transfer, fluid mechanics and thermodynamics in the development of large air cooled heat exchangers. In: *Experimental Heat Transfer, Fluid Mechanics and Thermodynamics*, M.D. Kelleher, et al. (Eds.), pp. 135–141. Elsevier Science Publishers, Amsterdam, the Netherlands.
93. Gunter, A.Y., Shipes, K.V. (1971) *Hot Air Recirculation by Air Coolers*. Twelfth National Heat Transfer Conference, AIChE-ASME, August 1971, Tulsa, Oklahoma. Preprinted for the conference by American Institute of Chemical Engineers, New York, pp. 1–24.
94. www.calgavin.com/articles/heat-exchanger-problems-part-3
95. www.jgc.com/en/business/tech-innovation/tech-journal/pdf/jgc-tj_03–02(2014).pdf
96. Kubota, K. (2014) Hot Air Recirculation Phenomenon in an Air Cooled LNG Plant. *JGC* 3(2): 1–8.
97. www.htri.net/case-study-reducing-hot-air-recirculation-around-air-cooled-heat-exchangers
98. Pinkerton, A., Chapple, S. (1993) *The design of Quiet Air-Cooled Heat Exchangers, A Paper for The Energy Resources Conservation Board,* November 9, 1993, Calgary, Alberta pp. 1–20. Hudson Products Corporation, Houston, Texas.
99. Gunter, A.Y., Shipes, K.V. (1972) *Noise Control of Air-Cooled Heat Paper for a Presentation at a Session on Noise—Results of Various Approaches to Control.* Hudson Products Corporation, Houston, Texas. Presented at the 37th Mid year, May 10, 1972, Meeting of the American Petroleum Institute's Division of Refining in the Waldorf-Astoria, New York.
100. www.webbusterz.com/products/air-cooled-heat-exchanger-design
101. Amercool Manufacturing Inc. *Basics of Air Cooled Heat Exchangers*, Amercool Manufacturing Inc., Tulsa, OK, www.amercool.com
102. Neptune Energy, *Netherlands B.V. Specification 407 rev. 3*, pp. 1–20.
103. ACHE Model G, PEP00262EN, pp. 1–4. Alfalaval.com
104. American Institute of Chemical Engineers (1978) *Air-cooled heat exchangers—A guide to performance evaluation*. AIChE Equipment Testing Procedure. American Institute of Chemical Engineers.
105. ASME. (1991) *Performance Test Code 30—Air Cooled Heat Exchangers, PTC 30*. ASME, New York, NY.
106. www.yit-th.com/major-components
107. www.therma.com/microchannel-heat-exchangers/
108. www.kaltra.com/single-post/2018/04/09/microchannel-heat-exchangers
109. www.kaltra.com/single-post/2016/09/26/microchannel-technology-for-hvac-and-process-cooling
110. General Service Bulletin (2011) *Microchannel Coil Servicing Guidelines*, pp. 1–13. Trane Unitary Light and Large Commercial Units
111. www.hydro.com/en/aluminium/products/precision-tubes/inner-grooved-tubes/
112. www.kaltra.com/single-post/2016/09/02/microchannel-coil-fix-or-replace
113. www.thermokey.com/en/products/coils/microchannel-liquid-coolers/7
114. www.marketsandmarkets.com/Market-Reports/microchannel-heat-exchanger-market-32760657.html
115. *Selection Guide: Environmental Corrosion Protection Condenser Coils and Cooling/Heating Coils for Commercial Products*. December 2012, Carrier Corporation Syracuse, New York, pp. 1–15.
116. https://refindustry.com/news/market-news/hydrophilic-topcoat-for-microchannel-for-kaltra-heat-exchangers/
117. www.kaltra.com/single-post/2019/08/12/air-cooled-chillers-vs-water-cooled-chillers
118. https://microgroove.net/sites/default/files/webinar_iv_slide_show_final.pdf
119. https://microgroove.net/
120. Southall, D., Le Pierres, R., Dewson, S.J. (2008) *Design Considerations for Compact Heat Exchangers*, Paper 8009, Proceedings of ICAPP '08, Anaheim, CA USA, June 8–12, 2008. Heatric Division of Meggitt (UK) Ltd., England.

121a. www.kobelco.co.jp/english/products/ecmachinery/dche/
121b. www.kobelco.co.jp/english/products/ecmachinery/dche/files/Microchannel_dche_smcr_brochure.pdf
122. Hesselgreaves, J.E., Law, R., Reay, D. (2017) *Chapter 2 Industrial Compact Exchangers, Compact Heat Exchangers*, http://dx.doi.org/10.1016/B978–0-08-100305–3.00002–1, pp. 35–89. Elsevier Ltd.
123. www.vpei.com/diffusion-bonded-microchannel-heat-exchangers/
124. www.heatric.com/heat-exchangers/industry-applications/
125. www.heatric.com/heat-exchangers/benefits-heatric-exchangers/
126. Southall, D., Le Pierres, R., Dewson, S.J. (2008) *Design Considerations for Compact Heat Exchangers, Paper 8009*, Proceedings of ICAPP '08, Anaheim, CA USA, June 8–12, 2008. Heatric Division of Meggitt (UK) Ltd., , England.
127. *Innovative Compact Heat Exchangers*. Proceedings of ICAPP '10 San Diego, CA, USA, June 13–17, 2010 Paper 10300, pp. 218–226.
128. www.heatric.com/home/heatric-services/heat-exchanger-cleaning/
129. Tsopanos, S., Sutcliffe, C.J., Owen, I. (2005) The manufacture of micro cross-flow heat exchangers by selective laser melting. *Proceedings of Fifth International Conference on Enhanced, Compact and Ultra-Compact Heat Exchangers: Science, Engineering and Technology*, CHE 2005—53, R.K. Shah, M. Ishizuka, T.M. Rudy, V.V. Wadekar (Eds.). Engineering Conferences International, Hoboken, NJ, USA, September 2005.
130. www.kobelco.co.jp/english/products/ecmachinery/dche/
131. Miwa, Y., Noishiki, K., Suzuki, T., Takatsuki, K. (2013) *Manufacturing Technology of Diffusion-bonded Compact Heat Exchanger (DCHE)*, pp. 51–56. Kobelco Technology Review No. 32.
132. Hesselgreaves, J.E., Law, R., Reay, D. (2017) *Chapter 2 Industrial Compact Exchangers, Compact Heat Exchangers*, pp. 35–89, http://dx.doi.org/10.1016/B978–0-08-100305–3.00002–1, Elsevier Ltd.
133. www.twi-global.com/technical-knowledge/faqs/what-is-additive-manufacturing#AdditiveManufacturingTechnologies
134. www.twi-global.com/what-we-do/research-and-technology/technologies/additive-manufacturing
135. www.confluxtechnology.com/article/the-status-quo-of-additive-manufacturing-for-heat-exchangers-and-where-to-next
136. Tirelli, V. (2019) Advantages of 3D-printing heat exchangers. *Heat Exchanger World*, pp. 46–49.
136.1 https://aidro.it/wp-content/uploads/2021/03/hxw-46–48-additive-manufac-1.pdf
137. https://coolinnov.eu/additive-manufacturing
138. https://ewi.org/capabilities/additive-manufacturing/the-seven-processes-of-additive-manufacturing/
139. www.twi-global.com/technical-knowledge/faqs/what-is-additive-manufacturing
140. Shah, R.K., Ishizuka, M., Rudy, T.M., Wadekar, V.V. (2005) *Proceedings of Fifth International Conference on Enhanced, Compact and Ultra-Compact Heat Exchangers: Science, Engineering and Technology*. Whistler, British Columbia, Canada.
141. Stratasys Direct, Inc. (2017) *Advancing thermal management with additive manufacturing*. Stratasys Direct, Inc.
142. www.additivalab.com/ways-to-design-a-more-efficient-heat-exchanger-with-metal-am/
143. https://all3dp.com/1/better-heat-exchangers-with-additive-manufacturing/#additively-manufactured-heat-exchanger-gallery
144. https://amfg.ai/2019/07/17/3d-printing-for-heat-exchangers-application-spotlight/
145. https://amfg.ai/2019/07/17/3d-printing-for-heat-exchangers-application-spotlight/
146. https://gen3d.com/news-and-articles/video-am-in-10-additive-manufacturing-and-heat-exchangers/
147. https://3dincredible.com/additive-manufacturing-of-heat-exchangers-the-best-solution-for-thermal-management/
148. www.qats.com/Product/Heat-Sinks/BGA-Heat-Sink--High-Aspect-Ratio-Ext/Standard-Pin-Fin/ATS054054016-PF039/2494.aspx
149. www.confluxtechnology.com/article/the-status-quo-of-additive-manufacturing-for-heat-exchangers-and-where-to-next

BIBLIOGRAPHY

Agarwal, R.S., Ramaswamy, M., Srivastava, V.K. (1991) Simulation of air-cooled condensers. Journal of Energy, Heat Mass Transfer (India) 13: 145–164.

Bell, K.J. (1990) Application of plate-fin heat exchangers in the process industries. In: Compact Heat Exchangers (R.K. Shah, A.D. Kraus, D. Metzger, eds.), pp. 591–602. Hemisphere, Washington, DC.

Bergles, A.E., Blumenkrantz, A.R., Taborek, J. (1974) Performance evaluation criteria for enhanced heat transfer surfaces. In: Heat Transfer, Vol. II, pp. 239–243. JSME, Tokyo, Japan.

Fletcher, L.S. (1993) Experimental techniques for thermal contact resistance measurements. In: Experimental Heat Transfer, Fluid Mechanics and Thermodynamics (M.D. Kelleher et al., eds.), pp. 195–206. Elsevier Science Publishers, Amsterdam, the Netherlands.

Ganguli, A., Tung, S.S., Taborek, J. (1985) Parametric study of air cooled heat exchanger finned tube geometry. Heat Transfer—Denver 81: 122–128.

Glass, J. (1979) Design of aircooled exchangers—Specifying and rating fans. In: Process Heat Exchange, Chemical Engineering Magazine (V. Cavaseno, ed.), pp. 426–430. McGraw-Hill, New York.

GPSA (1987) Gas Processors Suppliers Association Engineering Data Book, Vol. 1, 10th edn. GPSA.

Huang, L.J., Shah, R.K. (1992) Assessment of calculation methods for efficiency of straight fins of rectangular profile. International Journal of Heat and Fluid Flow 13: 282–293.

Kern, M.J., Wallner, R. (1986) Water Cooled Charge Air Coolers for Heavy Diesel Engines, C116/86, pp. 261–268. IMechE, London, UK.

London, A.L., Shah, R.K. (1968) Offset rectangular plate-fin surfaces—Heat transfer and flow friction characteristics. Transactions of ASME, Journal of Engineering and Power 90A: 218–228.

McQuiston, F.C. (1971) Heat transfer and flow friction data for fin-tube surfaces. Transactions of ASME, Journal of Heat Transfer 93C: 249–250.

McQuiston, F.C., Tree, D.R. (1972) Optimum space envelopes of the finned tube heat transfer surface. ASHRAE 78: 144–152.

Mueller, A.C. (1988) Air cooled heat exchangers. In: Heat Transfer Equipment Design (R.K. Shah, E.C. Subbarao, R.A. Mashelikar, eds.), pp. 179–190. Hemisphere, Washington, DC.

Nir, A. (1991) Heat transfer and friction factor correlations for crossflow over staggered finned tube banks. Heat Transfer Engineering 12: 43–58.

Rich, D.G. (1966) The efficiency and thermal resistance of annular and rectangular fins. Proceedings of the 3rd International Heat Transfer Conference 111: 281–289.

Shah, R.K., London, A.L. (1971) Influence of brazing on very compact heat exchanger surfaces. ASME Paper No. 71-HT-29.

Shah, R.K., Webb, R.L. (1983) Compact and extended heat exchangers. In: Heat Exchangers: Theory and Practice (J. Taborek, G.F. Hewitt, N. Afgan, eds.), pp. 425–468. Hemisphere/McGraw-Hill, Washington, DC, 1983.

Soland, J.G., Mack, W.M. Jr., Rohsenow, W.M. (1978) Performance ranking of plate-fin heat exchanger surfaces. Transactions of ASME, Journal of Heat Transfer 100: 514–519.

Sunden, B., Svantesson, J. (1990) Thermal hydraulic performance of new multilouvered fins. Heat Transfer—Jerusalem, Paper No. 14-HX-16, pp. 91–96.

Thome, J.R. (Ed.) (2004). *Heat to Air Cooled Heat Exchangers, Chapter 6, Engineering Data Book III.* Wolverine Tube, Inc., Decatur, AL, pp. 6.1–6.40.

Taborek, J. (1985) Bond resistance and design temperatures for high finned tubes—A reappraisal. Proceedings of the ASME, Thermal/Mechanical Heal Exchanger Design, Karl Gardner Memorial Session (K.P. Singh and S. M. Shenkman, eds.), Vol. 118, pp. 49–57.

Ward, D.J., Young, E.H. (1959) Heat transfer and pressure drop of air in forced convection across triangular pitch banks of finned tubes. Chemical Engineering Progress Symposium Series 55: 37.

Ward, D.J., Young, E.H. (1963) Heat transfer and pressure drop of air in forced convection across triangular pitch banks of finned tubes. Chemical Engineering Progress Symposium Series 59: 37–44.

Webb, R.L. (1980) Air side heat transfer in finned tube heat exchangers. Heat Transfer Engineering 1: 33–49.

Webb, R.L. (1983) Compact heat exchangers. In: Heat Exchanger Design Handbook (E. U. Schlunder, editor-in-chief), Vol. 3, Section 3.9. Hemisphere, Washington, DC.

Zukauskas, A. (1972) Heat transfer from tubes in cross flow. In: Advances in Heat Transfer (J.P. Hartnett and T. Irvine, Jr., eds.), Vol. 8, pp. 93–160. Academic Press, New York.

4 Shell and Tube Heat Exchanger Design

4.1 INTRODUCTION

The most commonly used heat exchanger is the shell and tube type. This is the "workhorse" of industrial process heat transfer. It has many applications in the power generation, petroleum refinery, chemical, and process industries. They are used as oil coolers, condensers, feedwater heaters, etc. Other types of heat exchangers are used when it is economical to do so. Though the application of other types of heat exchangers is increasing, the shell and tube heat exchanger will maintain its popularity for a long time, largely because of its versatility [1].

As its name implies, this type of heat exchanger consists of a shell (a large pressure vessel) with a bundle of tubes inside it. One fluid runs through the tubes, and another fluid flows over the tubes (through the shell) to transfer heat between the two fluids. The set of tubes is called a tube bundle. Typically, the ends of each tube are terminated through holes in a tubesheet. The tubes are mechanically rolled or welded into the tubesheet face. Tubes may be straight or bent in the shape of a U, called U-tubes.

This type of heat exchanger consists of a shell with an internal tube bundle that is typically supported by tubesheets and intermittent tube support plates known as baffles. The tube bundle may be composed of several types of tubes supported by baffles. The tube pitch (center-to-center distance of adjoining tubes) is typically a minimum of 1.25 times the tube outer diameter or larger. Typically, the ends of each tube are terminated through holes in a tubesheet.

4.2 CLASSIFICATION OF SHELL AND TUBE HEAT EXCHANGERS

Heat exchangers have been developed with different approaches to meet different requirements such as to differential thermal expansion of tubes, ease of maintenance and servicing, etc. Three principal types of heat exchangers—(1) fixed tubesheet exchangers; (2) U-tube exchangers; and (3) floating head exchangers—satisfy these design requirements. Brief construction details are hereunder:

1. Fixed Tubesheet Heat Exchanger. A heat exchanger with two stationary tubesheets, each attached to the shell and channel. The heat exchanger contains a bundle of straight tubes connecting both tubesheets.
2. U-Tube Heat Exchanger. A heat exchanger with one stationary tubesheet attached to the shell and channel. The heat exchanger contains a bundle of U-tubes attached to the tubesheet.
3. Floating Tubesheet Heat Exchanger. A heat exchanger with one stationary tubesheet attached to the shell and channel, and one floating tubesheet that can move axially. The heat exchanger contains a bundle of straight tubes connecting both tubesheets.

DOI: 10.1201/9781003352044-4

4.3 CONSTRUCTION DETAILS FOR SHELL AND TUBE EXCHANGERS

The following major components perform a function mainly related to pressure and fluid containment. Their design is carried out in accordance with the relevant pressure vessel code and standard.

i. Shell
ii. Dished heads and flat heads
iii. Tubesheets
iv. Bolted flanged joint
v. Expansion joint
vi. Nozzle
vii. Vessel support
viii. Other pressure and nonpressure parts

The expansion joint is an important component in the case of a fixed tubesheet exchanger for certain design conditions. Other components include tie-rods and spacers, impingement plates, sealing strips, etc. The selection criteria for a proper combination of these components are dependent upon the operating pressures, temperatures, thermal stresses, corrosion characteristics of fluids, fouling, cleanability, and cost. A large number of geometrical variables are associated with each component, and they are discussed in detail in this chapter. A cut section of a STHE is shown in Figure 4.1 and its major components are shown in Figure 4.2.

4.4 TEMA CLASSIFICATION OF HEAT EXCHANGERS BASED ON SERVICE CONDITION

TEMA has set up mechanical standards for three classes of shell and tube heat exchangers: *R, C*, and *B* [2]. Class *R* heat exchangers specify the design and fabrication of unfired shell and tube heat exchangers for the generally severe requirements of petroleum and related processing applications, class *C* for the generally moderate requirements of commercial and general process applications, and class *B* for chemical process service. Salient features of TEMA standards are discussed in Chapters 1 and 4 of Volume 2.

4.5 DESIGN STANDARDS

4.5.1 TEMA Standard

TEMA standards [2] are followed in most countries for the design of shell and tube heat exchangers. Each section is identified by an uppercase letter symbol, which precedes the paragraph numbers of

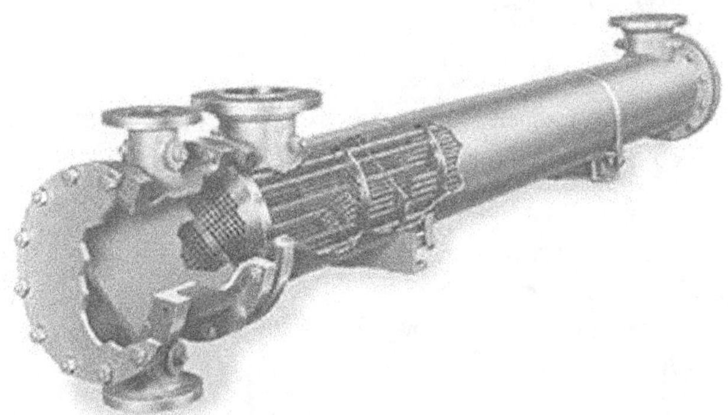

FIGURE 4.1 Cut section of a shell and tube heat exchanger. (Courtesy of API Heat Transfer Inc., Buffalo, NY.)

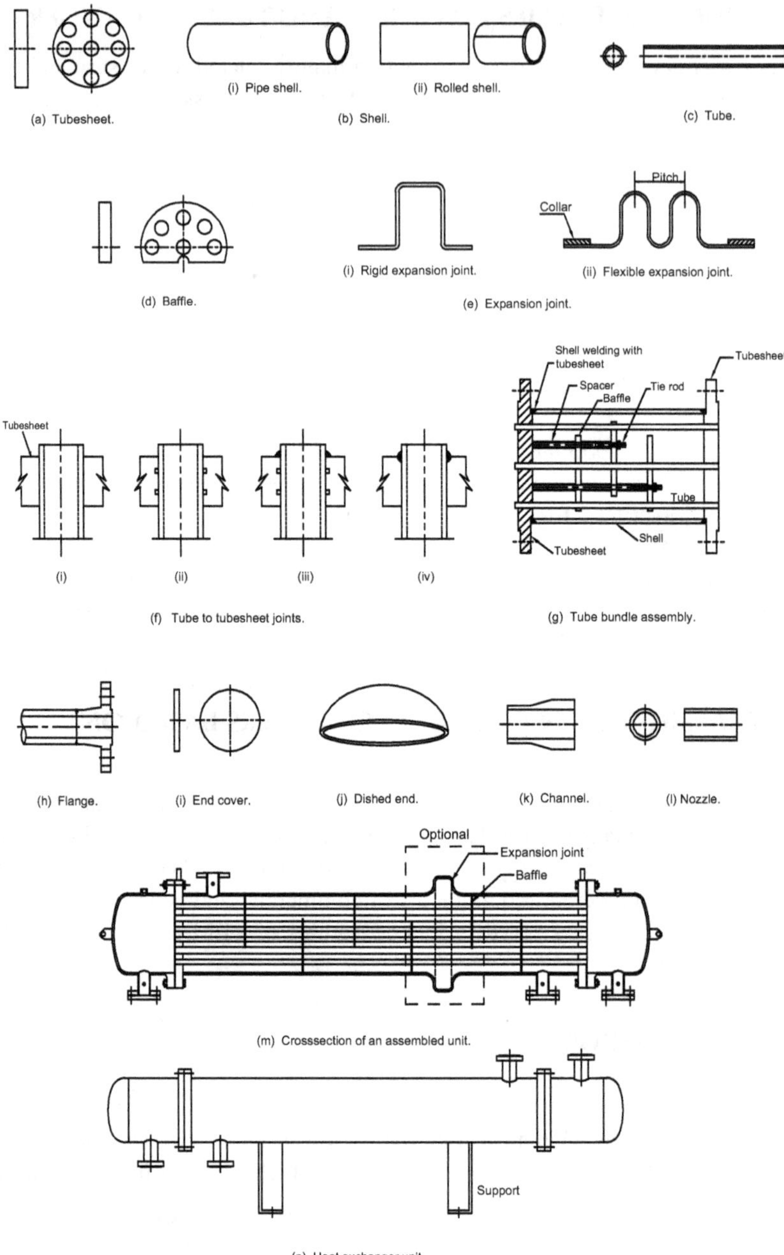

FIGURE 4.2 Major components of a shell and tube heat exchanger. (Note – Item (e) flexible membrane type is used for piping applications.)

the section and identifies the subject matter. Also, TEMA classes R, C, and B have been combined into one section titled class RCB.

The scope of TEMA standards [2] is shown in Table 4.1 and some of the major differences between TEMA classes *R*, *C*, and *B* are briefly given in Table 4.2.

TABLE 4.1
Scope of TEMA Standards [2]

Parameter	Limit
Inside diameter	100 in. (2540 mm)
Nominal diameter × pressure	100, 000 psi (17.5 × 10^6 kPa
Design Pressure	3,000 psi (20,684 kPa)
Shell wall thickness	3 in. (76 mm)
Stud diameters (approx.)	4 in. (102 mm)
Construction code	ASME Section VIII, Div. 1
Pressure source	Indirect (unfired units only)

TEMA Standards Contents

Section	Symbol	Paragraph
1	N	Nomenclature
2	F	Fabrication tolerances
3	G	General fabrication and performance information
4	E	Installation, operation, and maintenance
5	RCB	Mechanical standards TEMA class RCB heat exchangers
6	V	Flow-induced vibration
7	T	Thermal relations
8	P	Physical properties of fluids
9	D	General information
10	RGP	Recommended good practice
Appendix	A	Tubesheets

4.5.2 TEMA Standards Scope and General Requirements (Section 5, RCB-1.1.1)

The TEMA mechanical standards are applicable to unfired shell and tube heat exchangers with inside diameters not exceeding 100 in. (2540 mm), a maximum product of nominal diameter (in.) and design pressure (psi) of 100,000 psi (20,684 kPa), or a maximum design pressure of 3,000 psi. The intent of these parameters is to limit the maximum shell wall thickness to approximately 3 in. (76 mm) and the maximum stud diameter to approximately 4 in. (102 mm).

4.5.3 Scope of TEMA Standards

The scope of TEMA standards is given in Table 4.1. Each section is identified by an uppercase letter symbol, which precedes the paragraph numbers of the section and identifies the subject matter.

4.5.4 Construction Code

According to Section RCB-1.1.3, the construction of heat exchangers shall comply with the ASME boiler and pressure vessel code (BPVC), Section VIII, Div. 1.

4.6 OTHER STANDARDS USED FOR DESIGN OF HEAT EXCHANGERS

Some design standards used for the mechanical design of heat exchangers include the following: TEMA, HEI, and API. TEMA standards founded in 1939, the Tubular Exchanger Manufacturers Association, Inc., is a group of leading manufacturers of shell and tube heat exchangers who have pioneered the research and development of heat exchangers for over 60 years. TEMA standards are followed in most countries of the world for the design of shell and tube heat exchangers.

4.6.1 Heat Exchange Institute Standards

The HEI, Cleveland, Ohio, is an association of manufacturers of heat transfer equipment used in power generation. The association promotes improved designs by developing equipment design standards. It publishes standards for tubular heat exchangers used in power generation. Such exchangers include surface condensers, feedwater heaters, and other power plant heat exchangers. Among these standards are:

i. Standards for Direct Contact Barometric and Low Level Condensers, 9th Edition, 2014
ii. Standards for Shell and Tube Heat Exchangers, 5th Edition, 2013
iii. Standards for Steam Surface Condensers, 12th Edition, 2017
iv. Standards for Steam Jet Vacuum Systems, 7th Edition, 2017
v. Standards and Typical Specifications for Tray Type Deaerators, 10th Edition, 2016
vi. Standards for Closed Feedwater Heaters, 9th Edition, 2015
vii. Standards for Air Cooled Condensers, 2nd Edition, 2016

4.6.2 ASME Standards

i. Steam Surface Condenser, ASME PTC 12.2-2010(2020)
This performance test code establishes equipment performance metrics with the philosophy of promoting testing.
ii. Closed Feedwater Heaters, ASME PTC 12.1-2020

This code applies to all horizontal and vertical heaters except those with partial-pass full-length drain cooling zones. The heater design is based on a specific operating condition that includes flow, temperature, and pressure. This specific condition constitutes the design point that is found on the manufacturer's feedwater heater specification sheet.

4.6.3 API 660-2015, Shell and Tube Heat Exchangers

This standard specifies requirements and gives recommendations for the mechanical design, material selection, fabrication, inspection, testing, and preparation for shipment of shell and tube heat exchangers for the petroleum, petrochemical, and natural gas industries. This standard is applicable to the following types of shell and tube heat exchangers: heaters, condensers, coolers, and reboilers. This standard is not applicable to vacuum-operated steam surface condensers and feed-water heaters.

4.6.4 ISO 16812-2019 Petroleum, Petrochemical and Natural Gas Industries—Shell and Tube Heat Exchangers

This supplements API Std 660, 9th edition (2015), the requirements of which are applicable with the exceptions specified in this document and applicable to the shell and tube heat exchangers like heaters, condensers, coolers, and reboilers. It is not applicable to vacuum-operated steam surface condensers and feed-water heaters.

4.6.5 PIP VESSM001-2017

Specification for Small Pressure Vessels and Heat Exchangers with Limited Design Conditions

4.6.6 Codes

Many pressure vessel codes including ASME Boiler and Pressure Vessel Code Sec VIII Div 1 discuss STHE design.

4.7 DESIGN CODES

A code is a system of regulations or a systematic book of law often given statutory force by state or legislative bodies. Codes are intended to set forth engineering requirements deemed necessary for safe design and construction of pressure vessels, piping, etc., e.g., ASME Code for Boilers and Pressure Vessels.

4.7.1 ASME Codes

Among the codes, the ASME code [3] for the construction of boilers and pressure vessels including heat exchangers is the most widely used and is the referred-to code around the world today. ASME code establishes minimum rules of safety governing the design, fabrication, inspection, and testing of boilers, pressure vessels, and nuclear power plant components.

4.7.2 ASME Code Sec VIII Div 1 Part UHX Rules for Shell and Tube Heat Exchangers [3]

(a) The rules in Part UHX cover the minimum requirements for design, fabrication, and inspection of shell and tube heat exchangers.
(b) The rules in Part UHX cover the common types of shell and tube heat exchangers and their elements but are not intended to limit the configurations or details to those illustrated or otherwise described therein. Designs that differ from those covered in this Part shall be in accordance with U-2(g).

4.7.3 Other Codes

Apart from ASME code, many other codes are issued by various countries such as CODAP, PED European Pressure Equipment Directive, BS EN 13445, PD 5500., AD Merkblatter, etc. Generally, heat exchanger standards quote certain codes to be followed for construction of the heat exchangers.

4.8 TEMA SYSTEM FOR DESCRIBING HEAT EXCHANGER TYPES

The major components of a shell and tube heat exchanger are the front head, shell section, and rear head. Each of these components is available in a number of standard designs. In TEMA standards, they are identified by an alphabetic character. A heat exchanger unit is designated using the designations of front head, shell, and rear head. It consists of three alphabetic characters, such as AES, AKT, AJW, BEM, AEP, and CFU. Seven major types of shells, five types of front heads, and eight types of rear heads known as heat exchanger nomenclature as per TEMA are shown in Figure 4.3. In addition to these types, special types of shells and heads are also available depending upon the applications and customer needs. Figures 4.4a–f illustrate various types of heat exchangers.

The nomenclature of heat exchanger components is as follows: (1) stationary head—channel; (2) stationary head—bonnet; (3) stationary head flange—channel or bonnet; (4) channel cover; (5) stationary head nozzle; (6) stationary; (7) tubes; (8) shell; (9) shell cover; (10) shell flange—stationary head end; (11) shell flange—rear head end; (12) shell nozzle; (13) shell cover flange;

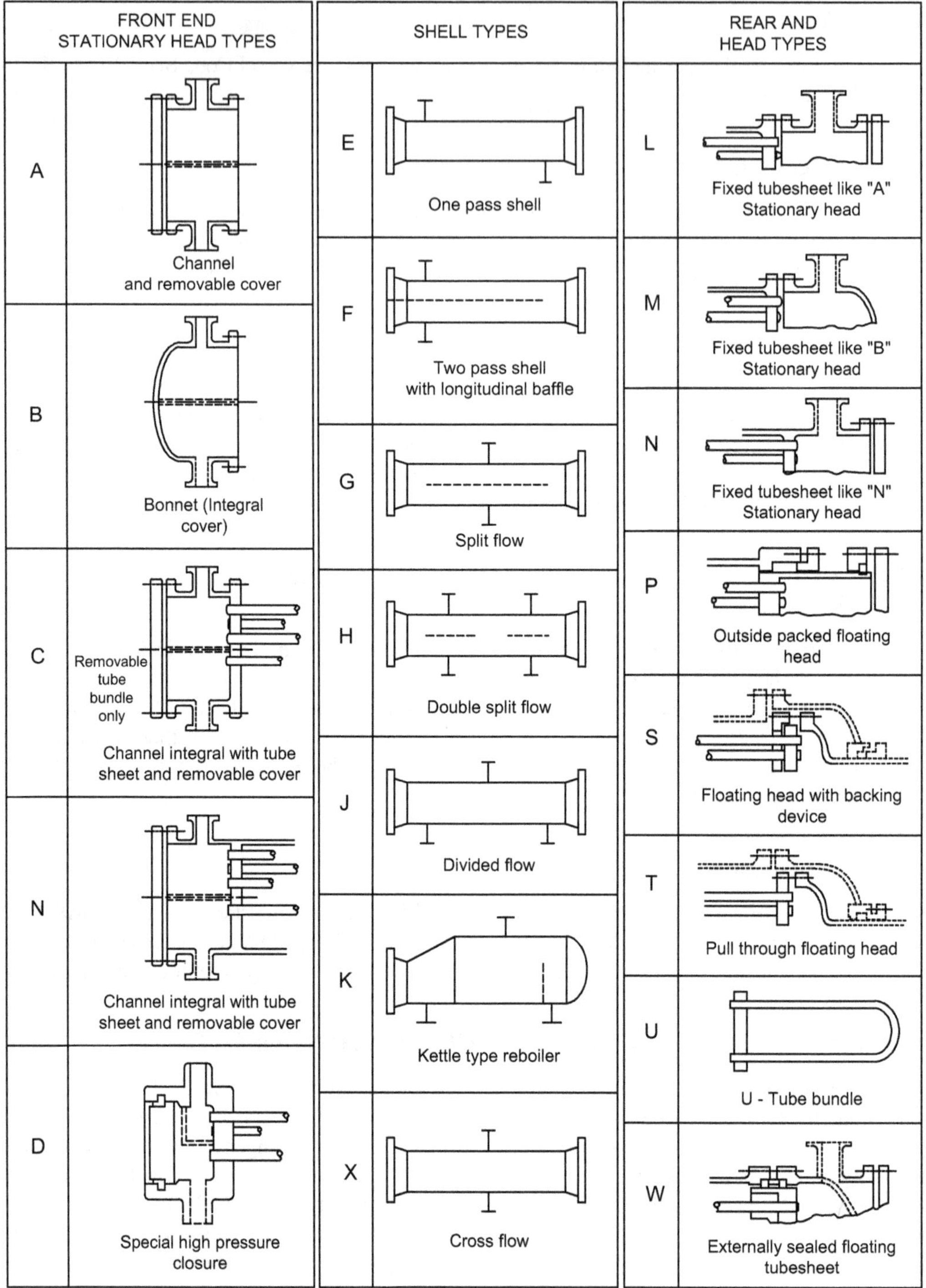

FIGURE 4.3 Nomenclature for heat exchanger components. (© Standards of Tubular Exchanger Manufacturers Association, Inc., 19th edition, 2019.)

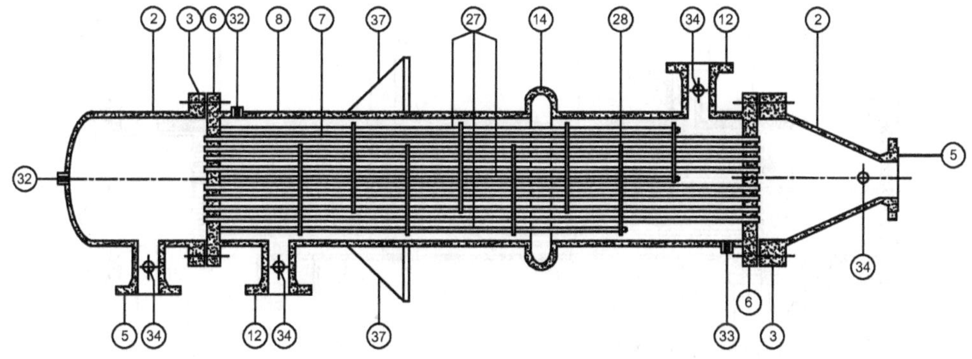

(a) BEM - Fixed tubesheet HEx

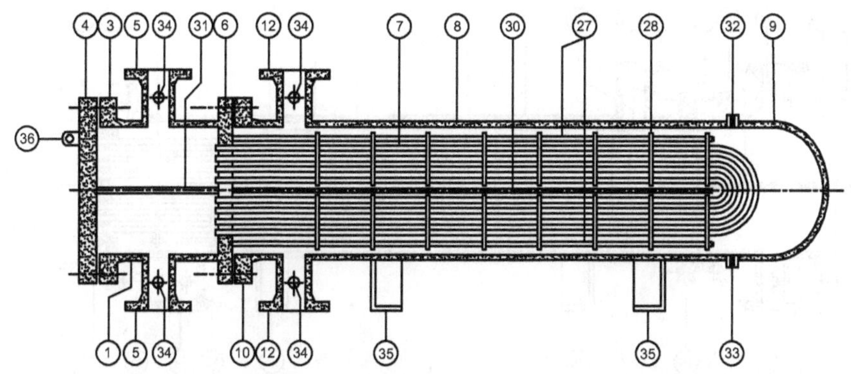

(b) CFU - U tube HEx

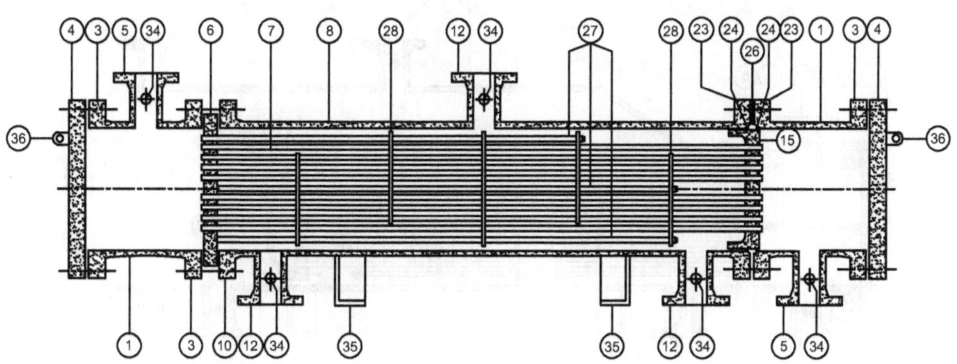

(c) AJW - Divided flow externally sealed floating head type.

FIGURE 4.4 Heat exchanger types as per TEMA designation.

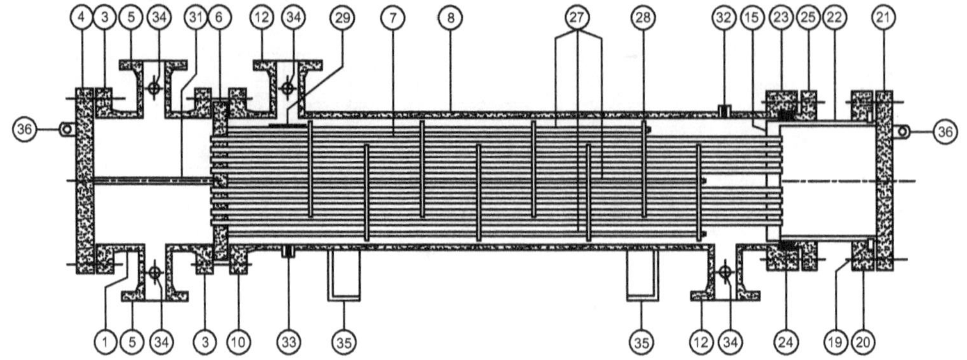

(d) AEP - Pull through floating head type.

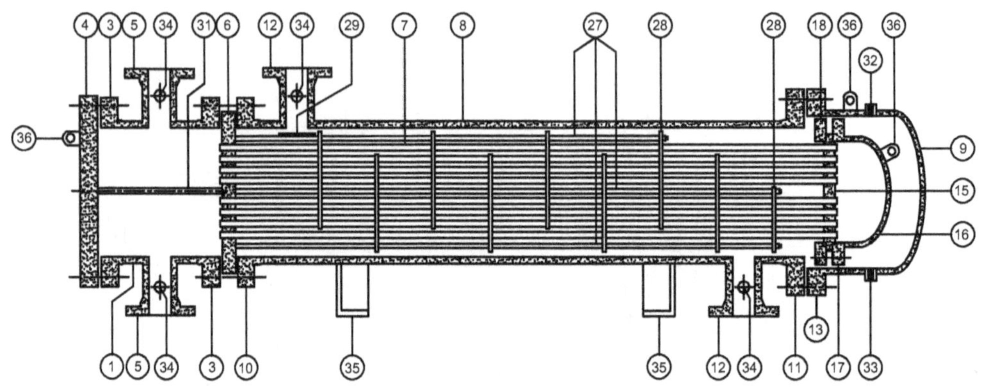

(e) AES - Internal split ring floating head type.

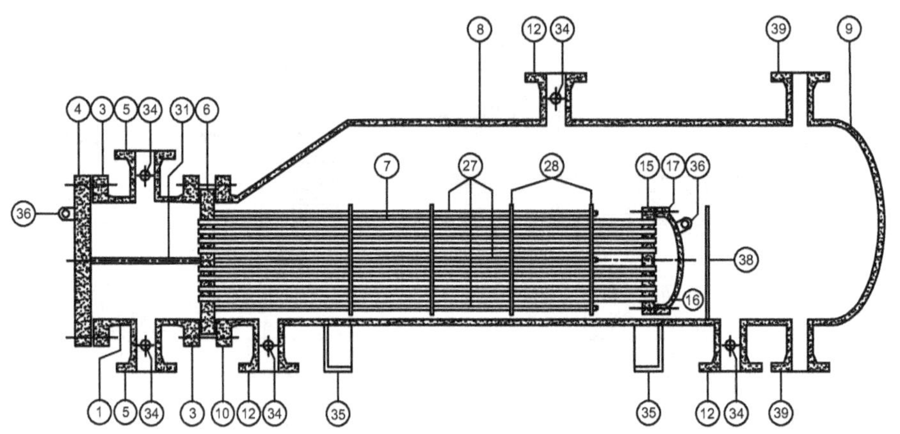

(f) AKT - Kettle type reboiler.

FIGURE 4.4 (Continued)

(14) expansion joint; (15) floating; (16) floating head cover; (17) floating head cover flange; (18) floating head backing device; (19) split shear ring; (20) slip-on backing flange; (21) floating head cover—external; (22) floating skirt; (23) packing box; (24) packing; (25) packing gland; (26) lantern ring; (27) tie-rods and spacers; (28) transverse baffles or support plates; (29) impingement plate; (30) longitudinal baffle; (31) pass partition; (32) vent connection; (33) drain connection; (34) instrument connection; (35) support saddle; (36) lifting lug; (37) support bracket; (38) weir; (39) liquid level connection, and (40) floating head support. (© Standards of Tubular Exchanger Manufacturers Association, Inc., 10th edn., 2019.)

4.8.1 Fixed Tubesheet Exchangers

This is the most popular type of shell and tube heat exchanger. These units are constructed with the tubesheets integral with the shell. The fixed tubesheet heat exchanger uses straight tubes secured at both ends into tubesheets, which are firmly welded to the shell. Hence, gasketed joints are minimized in this type, and thereby the least amount of maintenance is required. They may be used with *A*-, *B*-, or *N*-type front heads and *L*-, *M*-, or *N*-type rear heads. A BEM-type fixed tubesheet heat exchanger is shown in Figure 4.4a. Fixed tubesheet heat exchangers are used where [4]:

1. It is desired to minimize the number of joints.
2. Temperature conditions do not represent a problem for thermal stress.
3. The shellside fluid is clean and tube bundle removal is not required.
4. High degree of protection against contamination of streams.
5. Double tubesheet is possible.
6. Mechanical tubeside cleaning is possible.

Provision is to be made to accommodate the differential thermal expansion of the shell and the tubes when the thermal expansion is excessive. Fixed tubesheet exchangers can be designed with removable channel covers, "bonnet"-type channels, integral tubesheets on both sides, and tubesheets extended as shell flanges. They can be designed as counterflow one-to-one exchangers (i.e., single pass on the tubeside and on the shellside) or with multipasses on the tubeside. Where three or five passes have been specified for the tubeside, the inlet and outlet connections will be on opposite sides. For fixed tubesheet exchangers, a temperature analysis must be made considering all phases of operation (i.e., start-up, normal, upset, abnormal) to determine if thermal stress is a problem and how to relieve it [4]. Figure 4.5 shows important dimensions of a fixed tubesheet heat exchanger.

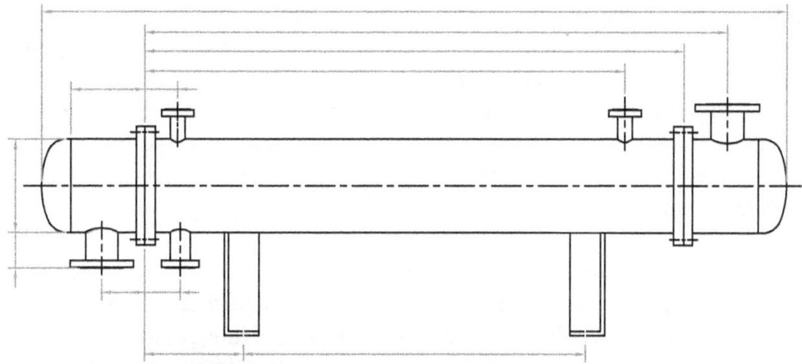

FIGURE 4.5 Important dimensions of a fixed tubesheet heat exchanger.

4.8.2 U-Tube Exchangers

In this type of construction, tube bundle and individual tubes are free to expand and the tube bundle is removable. A U-tube exchanger, CFU type, is shown in Figure 4.4d. U-tube exchangers can be used for the following services [4]:

1. Clean fluid on the tubeside
2. Extreme HP on one side
3. Temperature conditions requiring thermal relief by expansion
4. For H_2 service in extreme pressures, utilizing an all-welded construction with a nonremovable bundle
5. To allow the shell inlet nozzle to be located beyond the bundle
6. High degree of protection against contamination of streams
7. Double tubesheet is possible

4.8.2.1 Shortcomings of U-Tube Exchangers

Some of the drawbacks associated with U-tube exchangers are the following [4]:

1. Mechanical cleaning from inside tubes is difficult; chemical cleaning is possible.
2. Flow-induced vibration can also be a problem in the U-bend region for the tubes in the outermost row because of long unsupported span.
3. Interior tubes are difficult to replace, often requiring the removal of outer layers or plugging the leaking tube.
4. Erosion damage to U-bends is seen with high tubeside velocity.
5. Tubeside pressure drop will be on the higher side due to flow reversals in the U-end.

U-tube exchangers may be used with *A*-, *B*-, *C*-, *N*-, or *D*-type front heads.

4.8.3 Floating Head Exchangers

The floating head exchanger consists of a stationary tubesheet and one floating tubesheet that is free to accommodate the thermal expansion of the tube bundle. Floating head exchangers with TEMA designations AES, AEP, and AJW are shown in Figures 4.4a, c, and f, respectively. They may be used with *A*-, *B*-, or *C*-type front heads. There are four basic types of floating head exchangers. They are discussed next.

1. Floating head, outside packed floating head: The floating head (*P* head), outside packed stuffing box heat exchanger uses the outer skirt of the floating tubesheet as part of the floating head. The packed stuffing box seals the shellside fluid while allowing the floating head to move. The tube bundle is removable. Maintenance is also very easy since all bolting is from the outside only. With this floating head, any leak (from either the shellside or the tubeside) at the gaskets is to the outside, and there is no possibility of contamination of fluids. Since the bundle-to-shell clearance is large (about 1.5 in. or 38 mm), sealing strips are usually required. The earlier types are recommended for LP, low-temperature, nonhazardous fluids.

2. Floating head, externally sealed floating tubesheet: The floating head (*W* head), externally sealed floating tubesheet, or outside packed lantern ring heat exchanger uses a lantern ring around the floating tubesheet to seal the two fluids as the floating tubesheet moves back and forth. The lantern ring is packed on both sides and is provided with vent or weep holes so that leakage through either should be to the outside. The number of tube passes is limited to one or two. The tube bundle is removable. This is the lowest cost of floating head design and can be used with type *A*, *B*, or *C* front heads. This type is recommended for LP, low-temperature, and nonhazardous fluids.

3. Floating head, pull-through head: In the floating head (*T* head), pull-through head exchanger, a separate head or cover is bolted to the floating tubesheet within the shell. In this design, the tube bundle can be removed without dismantling the joints at the floating end. Due to the floating head bonnet flange and bolt circle, many tubes are omitted from the tube bundle at the tube bundle periphery, and hence it accommodates the smallest number of tubes for a given shell diameter. This results in the largest bundle-to-shell clearance or a significant bundle-to-shell bypass stream *C*. To overcome the reduction in thermal performance, sealing devices are normally required, and the shell diameter is somewhat increased to accommodate the required surface area size. An ideal application for the *T* head design is as the kettle reboiler, in which there is ample space on the shellside and the flow bypass stream *C* is of no concern.

4. Floating head with backing device: In the floating head (*S* head) with backing device, the floating head cover (instead of being bolted directly to the floating tubesheet as in the pull-through type) is bolted to a split backing ring. The shell cover over the floating head has a diameter larger than the shell. As a result, the bundle to shell clearance is reasonable, and sealing strips are generally not required. The tube bundle is not removable. Both ends of the heat exchanger must be disassembled for cleaning and maintenance. This type is recommended for HP, nonhazardous process fluids.

4.9 DIFFERENTIAL THERMAL EXPANSION

Means should be identified to accommodate the thermal expansion or contraction between the shell and the tube bundle due to high mean metal temperature differentials between the shell and the tube bundle. This is particularly true for fixed tubesheet exchangers. Differential thermal expansion is overcome by the following means in shell and tube heat exchangers:

U-tube design: The U-bend design allows each tube to expand and contract independently.

In a *fixed tubesheet exchanger*, the tubesheets are welded to the shell. If large temperature differences exist between the shell and tube side fluids, it may be necessary to incorporate an expansion bellows in the shell, to eliminate excessive thermal stresses caused by differential expansion between shell and tubes . Such bellows are often a source of weakness and prone for failure in operation. In circumstances where the consequences of failure are particularly grave U-Tube or Floating Header units are normally used [5].

Floating head designs: The floating head exchangers solve the expansion problem by having one stationary tubesheet and one floating tubesheet that is free to accommodate the thermal expansion of the tube bundle.

4.10 COMPONENTS OF STHE

4.10.1 Shell

Heat exchanger shells are manufactured in a large range of standard sizes, materials, and thicknesses. Shell shape is generally circular, however the steam surface condenser of power plants may be rectangular. Smaller sizes are usually fabricated from standard-size pipes. Larger sizes are fabricated from plate by rolling. The cost of the shell is much more than the cost of the tubes; hence, the designer should try to accommodate the required heat transfer surface in one shell. It has been found that a more economical heat exchanger can usually be designed by using a small-diameter shell and the maximum shell length permitted by such practical factors as plant layout, installation, servicing, etc. [6]. Up to six shorter shells in series is common, and this arrangement results in countercurrent flow close to performance as if one long single shell design was used. Nominal shell diameters and shell thicknesses are furnished in TEMA Tables R-3.13 and CB-3.13 [2]. Roundness and consistent shell inner diameter are necessary to minimize the space between the baffle edge and the shell, as excessive space allows fluid bypass and reduced performance.

4.10.2 Tubes

Tubes of circular cross section are exclusively used in exchangers. Since the desired heat transfer in the exchanger takes place across the tube surface, the selection of tube geometrical variables is important from the performance point of view [7]. Important tube geometrical variables include tube outside diameter, tube wall thickness, tube pitch, and tube layout patterns. The tube layout pattern is shown in Figure 4.6. Tubes should be able to withstand the following:

1. Operating temperature and pressure on both sides
2. Thermal stresses due to the differential thermal expansion between the shell and the tube bundle
3. Corrosive nature of both the shellside and tubeside fluids

There are two types of tubes: straight tubes and U-tubes. The tubes are further classified as

1. Plain tubes
2. Finned tubes
3. Duplex or bimetallic tubes
4. Enhanced surface tubes

Extended or enhanced surface tubes are used when one fluid has a substantially lower heat transfer coefficient than the other fluid. Doubly enhanced tubes with enhancement both on thc inside and outside of thc tubes are available that can reduce the size and cost of the exchanger. Extended surfaces of finned tube types can provide two to four times as much heat transfer area on the outside as the corresponding bare tube, and this enhanced area helps to offset a lower outside heat transfer coefficient. More recent developments include corrugated tube, which has both inside and outside heat transfer enhancement, a finned tube, which has integral inside turbulators as well as extended outside surface, and tubing, which has outside surfaces designed to promote nucleate boiling [8].

4.10.2.1 Tube Diameter

Tube size is specified by outside diameter and wall thickness. From the heat transfer point of view, smaller diameter tubes yield higher heat transfer coefficients and result in a compact exchanger. However, larger diameter tubes are easier to clean, more rugged, and are necessary when the allowable tubeside pressure drop is small. Almost all heat exchanger tubes fall within the range of $\frac{1}{4}$ in. (6.35 mm) to 2 in. (4.8 mm) outside diameter. TEMA tube sizes in terms of outside diameter are $\frac{1}{4}, \frac{3}{8}, \frac{1}{2}, \frac{5}{8}, \frac{3}{4}, \frac{7}{9}$, 1, 1.25, 1.5, and 2 in. (6.35, 9.53, 12.70, 14.88, 19.05, 22.23, 24.40, 31.75, 38.10, and 50.80 mm). Standard tube sizes and gauges for various metals are given in TEMA Table RCB-2.21. These sizes give the best performance and are most economical in many applications. Most popular are the $\frac{3}{8}$-in. and $\frac{3}{4}$-in. sizes, and these sizes give the best all-round performance and are most economical in most applications [6]. Use $\frac{1}{4}$ in. (6.35 mm) diameter tubes for clean fluids. For

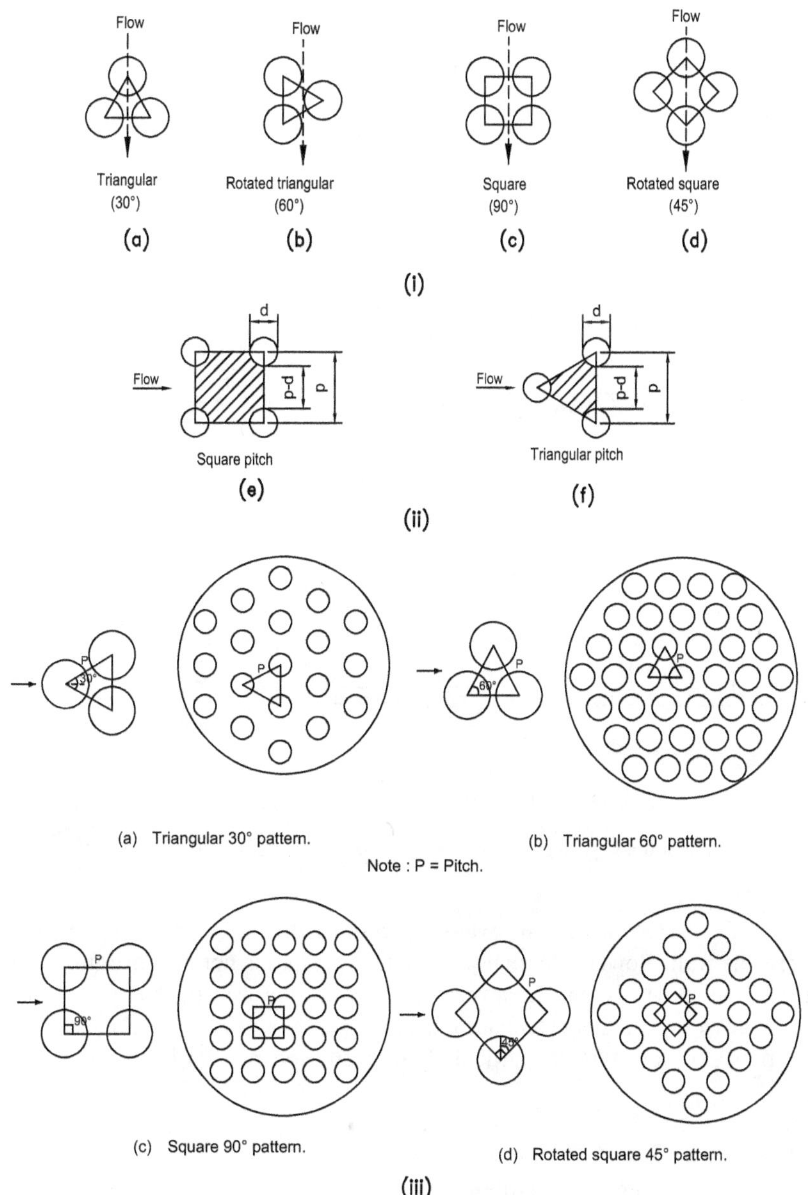

FIGURE 4.6 (i–iii) Tube layout patterns.

mechanical cleaning, the smallest practical size is $\frac{3}{4}$ in. (19.05 mm). Tubes of diameter 1 in. are normally used when fouling is expected because smaller ones are not suitable for mechanical cleaning, and falling film exchangers and vaporizers generally are supplied with 1.5- and 2-in. tubes [9].

4.10.2.2 Tube Wall Thickness

The tube wall thickness is generally identified by the Birmingham wire gauge. Standard tube sizes and tube wall thicknesses in inches are presented in TEMA Table RCB-2.21. Tube wall thickness

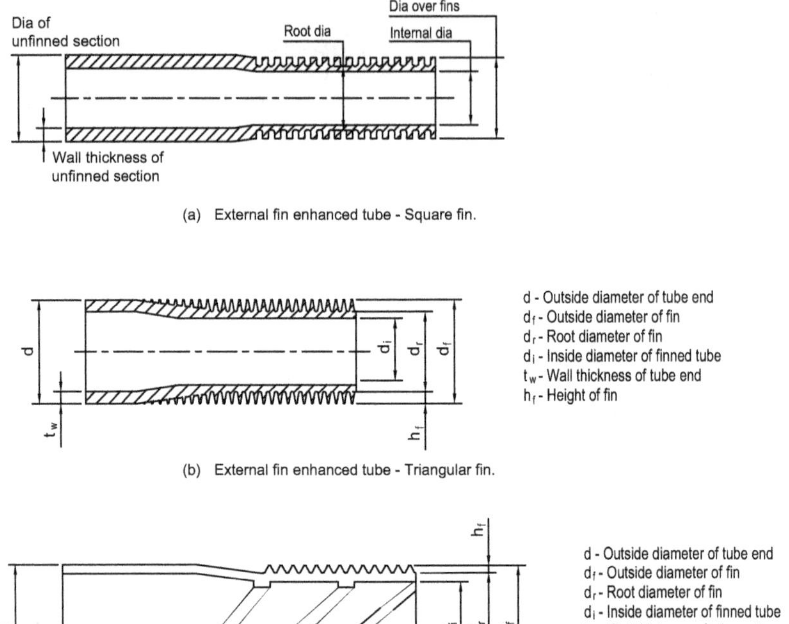

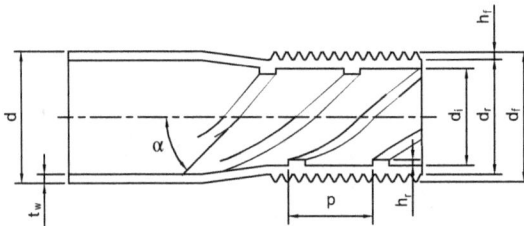

FIGURE 4.7 Low-finned tubes.

must be checked against the internal and external pressures separately, or the maximum pressure differential across the wall. However, in many cases, the pressure is not the governing factor in determining the wall thickness. Except when pressure governs, the wall thickness is selected on these bases [10]: (1) providing an adequate margin against corrosion; (2) fretting and wear due to flow-induced vibration; (3) axial strength, particularly in fixed exchangers; (4) standardized dimensions; and (5) cost.

4.10.2.3 Low-Finned Tubes

Shell and tube exchangers employ low-finned tubes (Figure 4.7) to increase the surface area on the shellside when the shellside heat transfer coefficient is low compared to the tubeside coefficient—e.g., when shellside fluid is highly viscous liquids, gases, or condensing vapors. The low-finned tubes are generally helical or annular fins on individual tubes.

Fin tubes for a shell and tube exchanger are generally "low-fin" type with fin height slightly less than $\frac{1}{16}$ in. (1.59 mm). The most common fin density range is 19–40 fins/in. (748–1575 fins/m). The surface area of such a fin tube is about 2.5–3.5 times that of a bare tube [7]. The finned tube has bare ends having conventional diameters of bare tubing; the diameter of the fin is either slightly lower than or the same as the diameter of the bare ends, depending upon the manufacturer. In addition to the geometrical variables associated with bare tubes, the additional geometrical dimensions associated with a fin tube are root diameter, fin height, and fin pitch.

4.10.2.4 Tube Length

For a given surface area, the most economical exchanger is possible with a small shell diameter and long tubes, consistent with the space and availability of handling facilities at site and in the fabricator's shop [10]. Therefore, minimum restrictions on length should be observed. However, for offshore applications, long exchangers, especially with removable bundles, are often very difficult to install and maintain economically because of space limitations [11]. In this case, shorter and larger shells are preferred despite their higher price per unit heat transfer surface. Standard lengths as per TEMA standard RCB-2.1 are 96, 120, 144, 196, and 240 in. Other lengths may be used.

4.10.2.5 Means of Fabricating Tubes

Tubing used for heat exchanger service may be either welded or seamless. The welded tube is rolled into a cylindrical shape from strip material and is welded automatically by a precise joining process. A seamless tube may be extruded or hot pierced and drawn. Copper and copper alloys are available only as seamless products, whereas most commercial metals are offered in both welded and seamless versions.

4.10.2.6 Duplex or Bimetallic Tubes

Duplex or bimetallic tubes are available to meet the specific process problem pertaining to either the shellside or the tubeside. For example, if the tube material is compatible with the shellside fluid, but not compatible with the tubeside fluid, a bimetallic tube allows it to satisfy both corrosive conditions.

4.10.2.7 Number of Tubes

The number of tubes depends upon the fluid flow rate and the available pressure drop. The number of tubes is selected such that the tubeside velocity for water and similar liquids ranges between 3–8 ft/s (0.9–2.4 m/s) and the shellside velocity between 2–5 ft/s (0.6–1.5 m/s) [7]. The lower velocity limit is desired to limit fouling; the higher velocity is limited to avoid erosion–corrosion on the tubeside, and impingement attack and flow-induced vibration on the shellside. When sand, silt, and particulates are present, the velocity is kept high enough to prevent settling down.

4.10.2.8 Tube Count

To design a shell and tube exchanger, one must know the total number of tubes that can fit into the shell of a given inside diameter. This is known as the tube count. Factors on which the tube count depends are discussed in Refs. [10], [12], and [13]. Such factors include the following:

- Shell diameter
- Outside diameter of the tubes
- Tube pitch
- Tube layout pattern—square, triangular, rotated square, or rotated triangular
- Clearance between the shell inside diameter and the tube bundle diameter
- Type of exchanger, i.e., fixed tubesheet, floating head, or U-tube
- Number of tubeside passes
- Tie-rods and sealing devices that block space
- Type of channel baffle, i.e., ribbon, pie shape, vertical, etc. [13]

The conventional method of obtaining the tube count by plotting the layout and counting the tubes (thus the tube count) is cumbersome, time-consuming, and prone to error. Tables of tube count are available in Refs. [12,14] and others, which often cover only certain standard combinations of pitch,

FIGURE 4.8 U-tube.

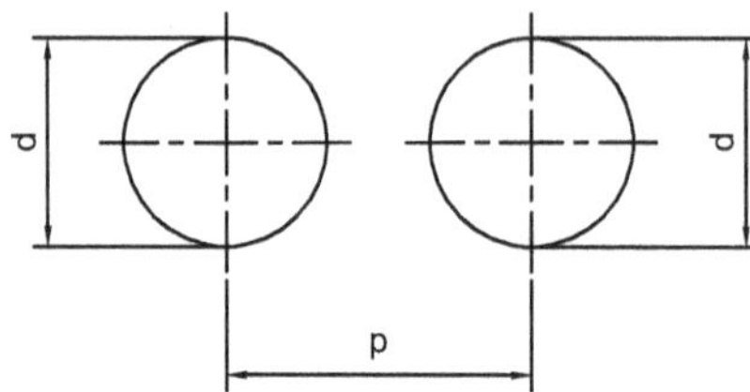

FIGURE 4.9 Tube pitch and ligament width ($p - d$).

tube diameter, and layout parameters. A mathematical approach using number theory is suggested by Phadke [12] to predict the tube count and presented tube count for various combinations of tube layout parameters. His method eliminates the disadvantages of drawing the tube layout pattern and can accommodate any configuration.

4.10.2.9 U-Tube

U-Tube U-Bend Requirements as per TEMA. When U-bends are formed, it is normal for the tube wall at the outer radius to thin. As per TEMA section RCB-2.33, for a U-tube shown in Figure 4.8, the minimum tube wall thickness in the bent portion before bending shall be [2]

$$t_o = t_1 \left(1 + \frac{d_o}{CR_b} \right) \tag{4.1}$$

where

- C is the thinning constant
- t_o is the original tube wall thickness
- t_1 is the minimum tube wall thickness calculated by code rules for a straight tube subjected to the same pressure and metal temperature
- d_o is the tube outer diameter
- R_b is the mean radius of bend

4.10.3 Tube Arrangement

4.10.3.1 Tube Pitch

The selection of tube pitch is a compromise between a close pitch for increased shellside heat transfer and surface compactness, and a larger pitch for decreased shellside pressure drop and fouling, and ease in cleaning. In most shell and tube exchangers, the minimum ratio of tube pitch to tube outside diameter (pitch ratio) is 1.24. The minimum value is restricted to 1.25 because the ligament (a ligament is the portion of material between two neighboring tube holes) may become too weak for proper rolling of the tubes into the tubesheet. The ligament width is defined as the tube pitch minus the tube hole diameter; this is shown in Figure 4.9.

4.10.3.2 Tube Layout

Tube layout arrangements (Figure 4.6) are designed so as to include as many tubes as possible within the shell to achieve maximum heat transfer area. Sometimes a layout is selected that also permits access to the tubes for cleaning as required by process conditions. Four standard types of tube layout patterns are triangular (30°), rotated triangular (60°), square (90°), and rotated square (45°). (Note that the tube layout angle is defined in relation to the flow direction and is not related to the horizontal or vertical reference line arrangement, and that the 30°, 60°, and 45° arrangements are "staggered," and 90° is "in-line.") For identical tube pitch and flow rates, the tube layouts in decreasing order of shellside heat transfer coefficient and pressure drop are 30°, 45°, 60°, and 90°. Thus, the 90° layout will have the lowest heat transfer coefficient and pressure drop. The selection of the tube layout pattern depends on the following parameters, which influence the shellside performance and hence the overall performance:

1. Compactness
2. Heat transfer
3. Pressure drop
4. Accessibility for mechanical cleaning
5. Phase change if any on the shellside

4.10.3.3 Triangular and Rotated Triangular Arrangements

Triangular and rotated triangular layouts (30° and 60°) provide a compact arrangement, better shellside heat transfer coefficients, and stronger tubesheets for a specified shellside flow area. For a given tube pitch/outside diameter ratio, about 15% more tubes can be accommodated within a given shell diameter using these layouts [10].

4.10.3.4 Square and Rotated Square Arrangements

When mechanical cleaning is necessary on the shellside, 45° and 90° layouts must be used with a minimum gap between tubes of 6.35 mm. There is no theoretical limit to tube outer diameter for mechanical cleaning, but the 6.35 mm clearance limits the tubes to a minimum of $\frac{5}{8}$ or $\frac{3}{4}$ in. outer diameter in practice [11]. The square pitch is generally not used in the fixed design because of no need of mechanical cleaning on the shellside. These layout patterns offer lower pressure drops and lower heat transfer coefficients than triangular pitch. The 45° layout is preferred for single-phase laminar flow or fouling service, and for condensing fluid on the shellside. Shah [7] suggests a square layout for the following applications:

1. If the pressure drop is a constraint on the shellside, the 90° layout is used for turbulent flow, since in turbulent flow, 90° has superior heat transfer rate and less pressure drop.
2. For reboilers, a square layout will be preferred for stability reasons. The 90° layout provides vapor escape lanes.

4.11 BAFFLES

Baffles must generally be employed on the shellside to support the tubes, to maintain the tube spacing, and to direct the shellside fluid across or along the tube bundle in a specified manner. There are a number of different types of baffles, and these may be installed in different ways to provide the flow pattern required for a given application.

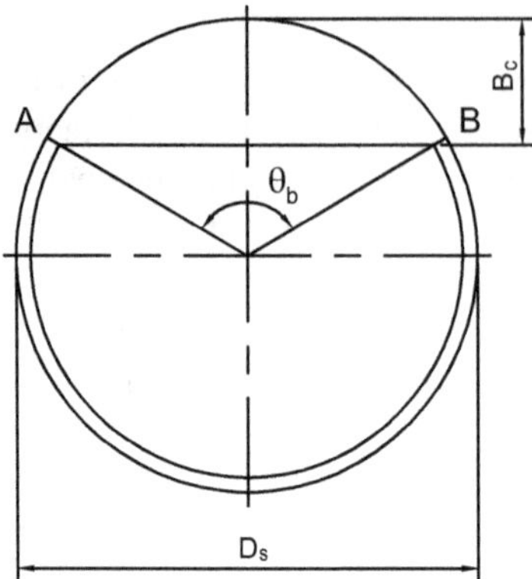

FIGURE 4.10 Baffle cut.

4.11.1 Classification of Baffles

Baffles are either normal or parallel to the tubes. Accordingly, baffles may be classified as transverse or longitudinal. The transverse baffles direct the shellside fluid into the tube bundle at approximately right angles to the tubes and increase the turbulence of the shell fluid. Every shell and tube exchanger has transverse baffles except the X and K shells, which have only support plates. The longitudinal baffles are used to control the direction of the shellside flow. For example, F, G, and H shells have longitudinal baffles. In the F shell, an overall counterflow is achieved.

4.11.2 Transverse Baffles

Transverse baffles are of two types: (1) plate baffles and (2) rod baffles. The three types of plate baffles are (1) segmental; (2) disk and doughnut; and (3) orifice baffles.

4.11.2.1 Segmental Baffles

The segmental baffle is a circular disk (with baffle holes) having a segment removed. Predominantly, a large number of shell and tube exchangers employ segmental baffles. This cutting is denoted as the baffle cut, and it is commonly expressed as a percentage of the shell inside diameter, as shown in Figure 4.10. Here the percent baffle cut is the height, *H*, given as a percentage of the shell inside diameter, D_s. The segmental baffle is also referred to as a single segmental baffle. The heat transfer and pressure drop of crossflow bundles are greatly affected by the baffle cut. The baffle cuts vary from 20% to 49% with the most common being 20–25%, and the optimum baffle cut is generally 20%, as it affords the highest heat transfer for a given pressure drop. Baffle cuts smaller than 20% can result in high-pressure (HP) drop. As the baffle cut increases beyond 20%, the flow pattern deviates more and more from crossflow [9] and can result in stagnant regions or areas with lower flow velocities; both of these reduce the thermal effectiveness of the bundle [1].

Baffle spacing: The practical range of single segmental baffle spacing is $\frac{1}{5}-1$ shell diameter [1], though the optimum could be 40–50% [1]. TEMA Table RCB-4.52 [2] provides maximum baffle spacing for various tube outer diameters, tube materials, and the corresponding maximum allowable temperature limit. The baffles are generally spaced between the nozzles. The inlet and outlet baffle spacings are generally larger than the "central" baffle spacing to accommodate the nozzles,

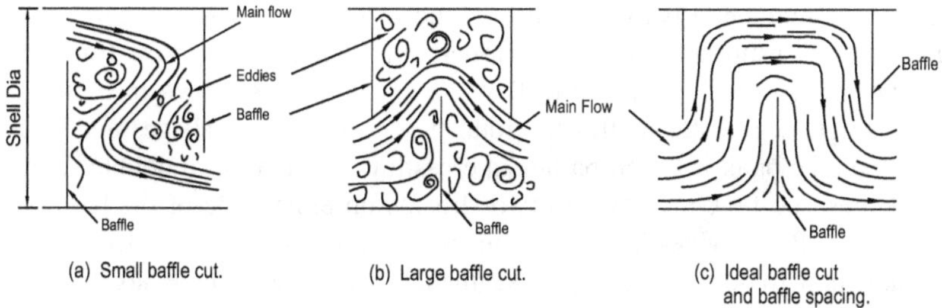

FIGURE 4.11 Shellside flow distribution influenced by baffle cut.

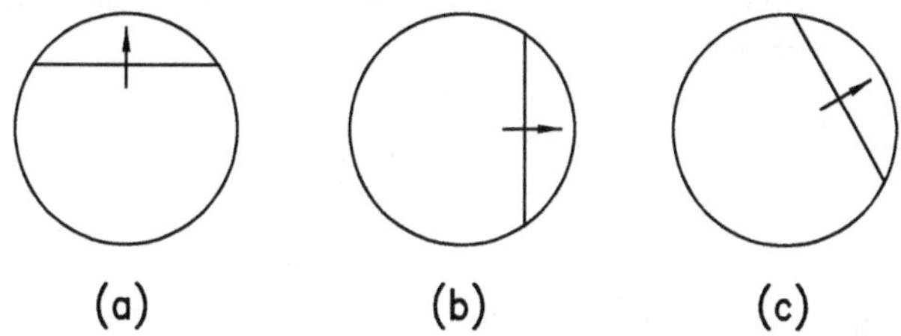

FIGURE 4.12 Baffle cut orientation: (a) horizontal; (b) vertical; and (c) rotated.

since the nozzle dimensions frequently require that the nozzle should be located far enough from the tubesheets.

Baffle thickness: TEMA Tables R-4.41 and CB-4.41 [2] provide the minimum thicknesses of transverse baffles applying to all materials for various shell diameters and plate spacings.

Shellside flow distribution: Segmental baffles have a tendency to poor flow distribution if the spacing or baffle cut ratio is not in correct proportion, as shown in Figure 4.11 [15]. Too low or too high a ratio results in maldistribution and produces inefficient heat transfer and also favors fouling. For low-pressure (LP)-drop designs, choose baffles that ensure a more uniform flow such as multisegmental, disk and doughnuts, and rod baffles.

Orientation of baffles: Alternate segmental baffles are arranged at 180° to each other, which cause shellside flow to approach crossflow through the bundle and axial flow in the baffle window zone. All segmental baffles have horizontal baffle cuts as shown in Figure 4.12a. Unless the shellside fluid is condensed, the horizontal baffle cut should be used for single-phase application, to reduce accumulation of deposits on the bottom of the shell and to prevent stratification of the shellside fluid [9]. The direction of the baffle cut is selected as vertical (Figure 4.12b) for the following shellside applications [10]: (1) for condensation, to allow the condensate to flow freely to the outlet without covering an excessive amount of tubes [4,7]; (2) for boiling or condensing fluids, to promote more uniform flow; and (3) for solids entrained in liquid (to provide least interference for the solids to fall out).

4.11.2.2 Double Segmental and Multiple Segmental Baffles

Various multi-segmental baffles can be used to reduce baffle spacing or to reduce crossflow because of pressure limitations. The multi-segmental baffles are characterized by large open areas and some allow the fluid to flow nearly parallel to the tubes, offering a much lower pressure drop [16].

Double and triple segmental baffle layouts are shown in Figure 4.13a, Figure 4.13b shows the design requirement to ensure adequate tube support [8], idealized fluid flow fraction through the baffles is shown in Figure 4.13c, and fluid flow stream pattern is shown in Figure 4.13d. In an exchanger with single segmental baffles the total flow, except for leakages and bypass streams, passes through the tube bank between baffles in crossflow, whereas with double segmental baffles barring the leakages, the flow divides into two streams on either side of the baffle, and in triple segmental baffles, the flow divides into three streams, as shown in Figures 4.13c and 4.13d. Due to this, heat exchangers with double or multiple segmental baffles can handle larger fluid flows on the shellside [7]. The flow on the shellside is split into two or more streams as per the number of baffle segments, namely, double, triple, multiple, etc.; hence, the danger of shellside flow-induced vibration is minimal.

Double segmental and triple segmental baffle heat exchanger units under fabrication are shown in Figures 4.14a and b, respectively.

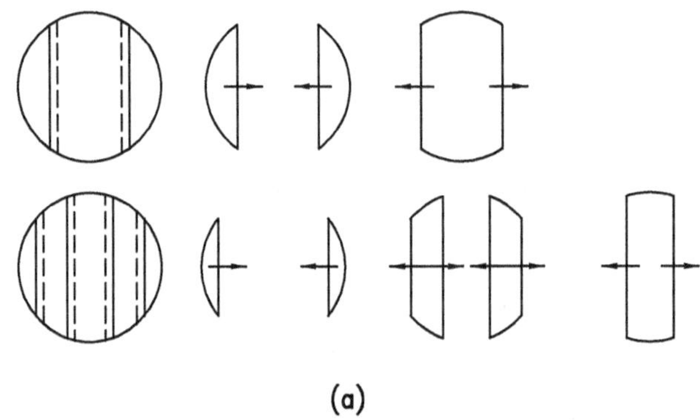

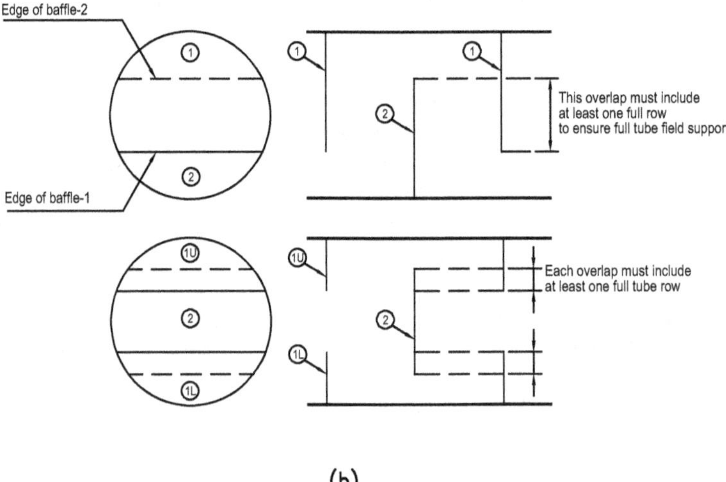

FIGURE 4.13 Segmental baffles layout. (a) Double and triple segmental baffles with end view flow pattern, (b) design requirement to ensure adequate tube support with double and triple segmental baffles. (c) idealized flow fraction with segmental baffles, and (d) stream flow distribution with segmental baffles.

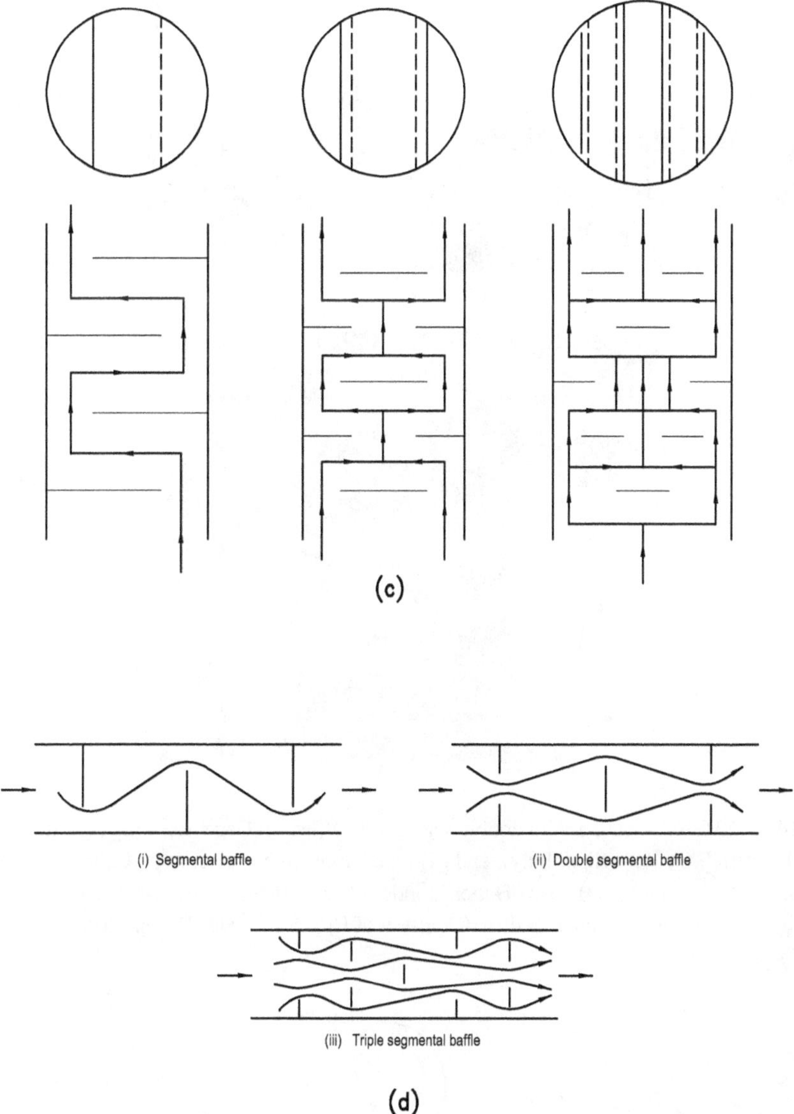

FIGURE 4.13 (Continued)

4.11.2.3 Window Baffles

These are considered when crossflow is not practical because of pressure-drop limitations. Window baffles allow the fluid to flow parallel to the tubes, offering much lower pressure drop [9].

4.11.3 Disk and Doughnut Baffle

The disk and doughnut baffle is made up of alternate "disks" and "doughnut" baffles, as shown in Figure 4.15. Disk and doughnut baffle heat exchangers are primarily used in nuclear heat exchangers [7]. This baffle design provides a lower pressure drop compared to a single segmental baffle for the same unsupported tube span and eliminates the tube bundle to shell bypass stream. Thermal design of disk and doughnut baffles is discussed in Annexure Section 4.A.3.

FIGURE 4.14 (a) Heat exchanger with double segmental baffles. (i) Cage assembly (Courtesy of Brembana Costruzioni Industriali S.p.A., Milan, Italy), and (ii) tube bundle under assembly. (Courtesy of Peerless Mfg. Co. of Dallas, TX, Makers of Alco and Bos-Hatten brands of heat exchangers.) (b) Shell and tube heat exchanger tube bundle cage with triple segmental baffles. (Courtesy of Heat Exchanger Design, Inc., Indianapolis, IN.)

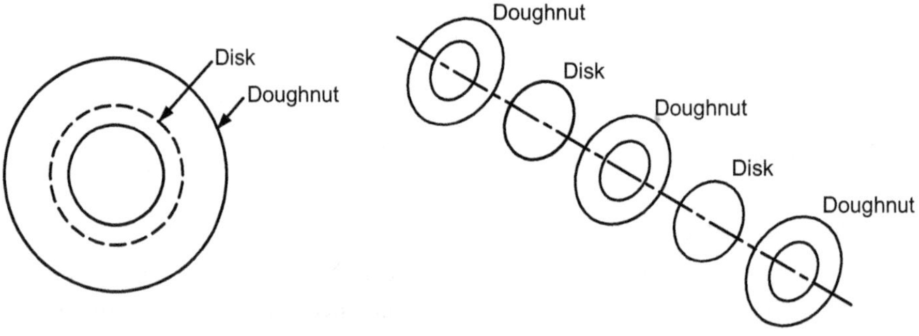

FIGURE 4.15 Disk and doughnut baffles.

4.11.4 Orifice Baffle

In an orifice baffle, the tube-to-baffle hole clearance is large so that it acts as an orifice for the shellside flow (Figure 4.16). These baffles do not provide support to tubes, and, due to fouling, the annular orifices plug easily and cannot be cleaned. This baffle design is rarely used.

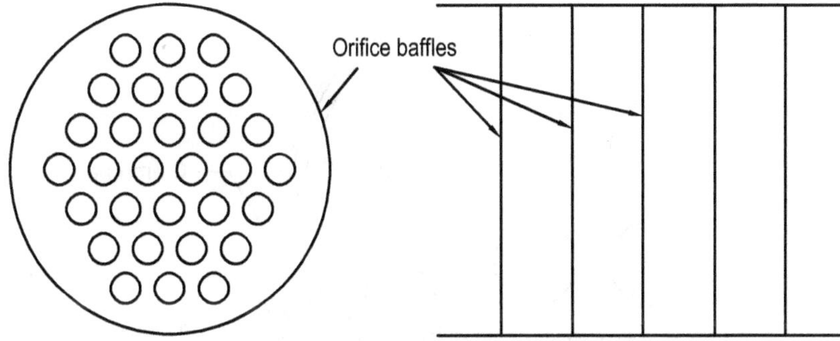

FIGURE 4.16 Orifice baffle.

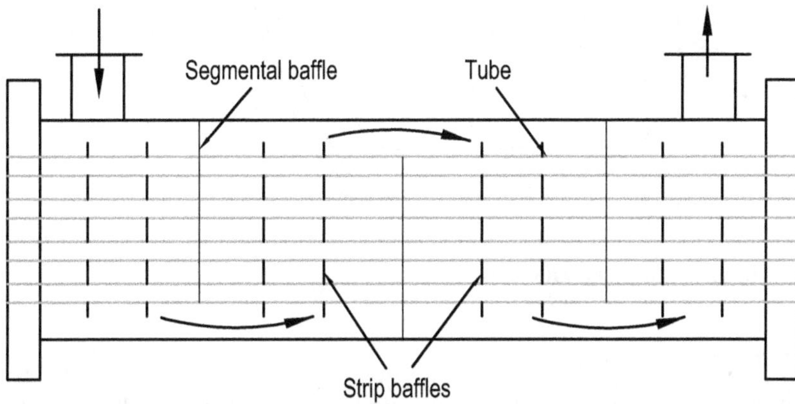

FIGURE 4.17 NTIW design.

4.11.5 No Tubes in Window

The baffle cut area, or baffle window region, is generally filled with tubes. Since the tubes in the window zone are supported at a distance of two times the central baffle spacing, they are most susceptible to vibration. To eliminate the susceptibility of tube vibrations, the tubes in the window zone are removed and therefore all tubes pass through all baffles. Additional support plates are introduced between the main baffles to reduce the unsupported span of the tubes as shown in Figure 4.17, thus providing an increase in the natural frequency of the tubes. The resultant design is referred to as the segmental baffle with no-tubes-in-window (NTIW) design. NTIW design has the following characteristics [17]:

1. Pressure drop about one-third that of single segmental baffle design
2. Uniform shellside flow pattern resembling that of an ideal tube bank, which offers high shellside heat transfer coefficient and low fouling tendency
3. The baffle cut and the number of tubes removed vary from 15% to 25%
4. Very LP drop in the window and correspondingly lower bypass and leakage streams

Since the loss in heat transfer surface is considerable in an NTIW design, this can be minimized by having small baffle cuts and possibly by an increase in the shellside fluid velocity or larger shell diameter to contain the same number of tubes. It may be noted that the design of tubesheet of NTIW does not fall within the TEMA standard.

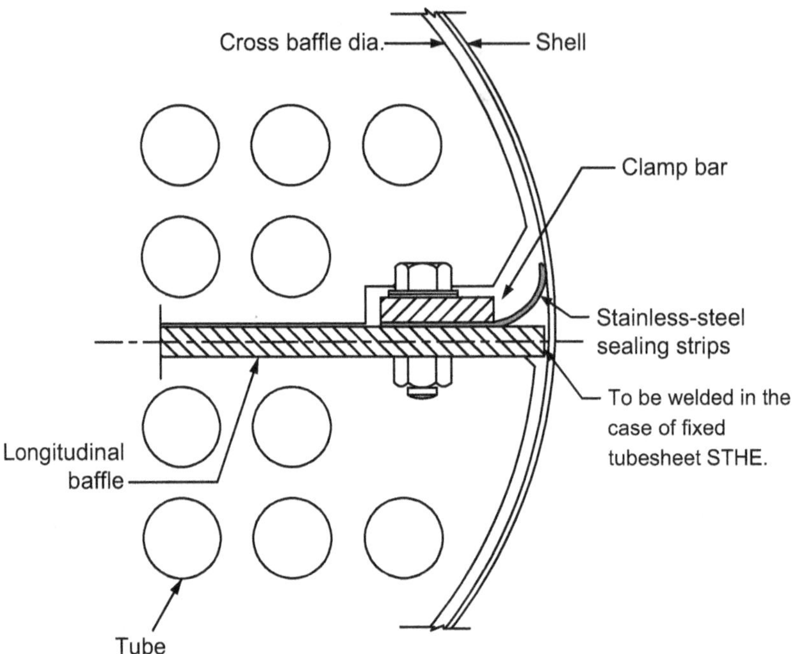

FIGURE 4.18 Longitudinal baffle with seals.

4.11.6 Longitudinal Baffles

Longitudinal baffles divide the shell into two or more sections, providing multipass on the shellside. For example, a single longitudinal baffle from one tubesheet to just short of the other tubesheet produces an F shell, that is, a shell with two shell passes. Longitudinal baffles are also employed to produce G and H shells. It will be apparent that in the case of removable tube bundles, the longitudinal baffle will be a part of the tube bundle and can therefore not be fixed to the shell. In order to prevent bypassing of the shellside fluid from the first pass to the second pass along the edges of the longitudinal baffle in an F shell, flexible strips are employed at both ends of the longitudinal baffle, all along the length of the shell, as shown in Figure 4.18. However, in the case of fixed-tubesheet heat exchangers, since the tube bundle cannot be removed from the shell, the longitudinal baffle can be fixed to the shell by welding. However, this type should not be used unless the baffle is welded to the shell and tubesheet. Nevertheless, several sealing devices have been tried to seal the baffle and the shell, but none has been very effective [9]. Figure 4.18 shows longitudinal baffle with seals.

Gupta [18] lists some sealing devices that are used to seal the baffle and shell. They are the following:

- Sealing strips or multiplex arrangement
- Packing arrangement
- Slide-in or tongue-and-groove arrangement

If the baffle is not welded, bypassing occurs from one side to the other, which adversely affects the heat transfer coefficient and makes its accurate prediction rather difficult. Hence, it is better to weld than to prefer this design. Common methods to weld the longitudinal baffle to the shellside are shown in Figure 4.19 [19]. When multipass shells are required, it is economical to use a

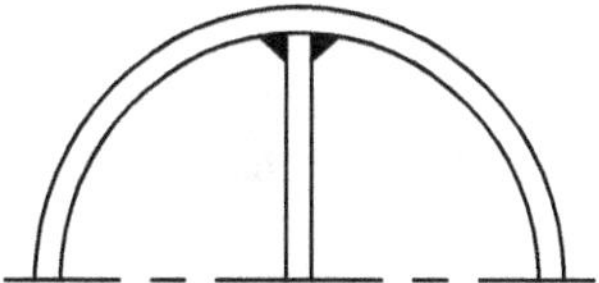

FIGURE 4.19 Longitudinal baffle and shell weld joint.

separate shell, unless the shell diameter is large enough to easily weld a longitudinal baffle to the shell [9].

4.11.6.1 TEMA Guidelines for Longitudinal Baffle

R-4.4.2 Longitudinal Baffles

Longitudinal Baffles with Leaf Seals

Longitudinal baffles with leaf (or other type) seals shall not be less than 1/4" (6.4 mm) nominal metal thickness.

Welded-in Longitudinal Baffles

The thickness of longitudinal baffles that are welded to the shell cylinder shall not be less than the thicker of 1/4" (6.4 mm) or the thickness calculated using the following formula:

$$\mathrm{tt} = b\sqrt{\frac{\Delta p.B}{1.5S}} \tag{4.2}$$

where
t is the minimum baffle plate thickness, in. (mm)
B is the value as shown in Table RCB-9.1.3.2
Δp is the maximum pressure drop across baffle, psi (kPa)
S is the code allowable stress in tension, at design temperature, psi (kPa)
b is the plate dimension, in. (mm), refer Table RCB-9.1.3.2
a is the plate dimension, in. (mm), refer Table RCB-9.1.3.2

The longitudinal baffle shall be considered fixed along the two sides where it is welded to the shell cylinder. It shall be considered simply supported along the sides where it is supported by the tubesheet groove or transverse baffle.

4.11.7 Rod Baffles

The Phillips RODbaffle design uses alternate sets of rod grids instead of plate baffles, enabling the tubes to be supported at shorter intervals without resulting in a large pressure drop. Flow-induced vibration is virtually eliminated by this design. The flow is essentially parallel to tube axis; as a result of the longitudinal flow, it has LP drop to heat transfer conversion characteristics. The tube layout is usually 45° or 90°. The design of the rod baffle heat exchanger is covered separately in this chapter.

4.11.8 Nest Baffles and Egg-Crate Tube Support

NEST™ baffle: This is a patented design intended to overcome the danger of flow-induced vibration of tubes. In this design, each tube rests in a V-shaped cradle and is supported at line segments (Figure 4.20a). These elements are preformed to the desired tube pitch and ligament sizte. The flow is parallel to the

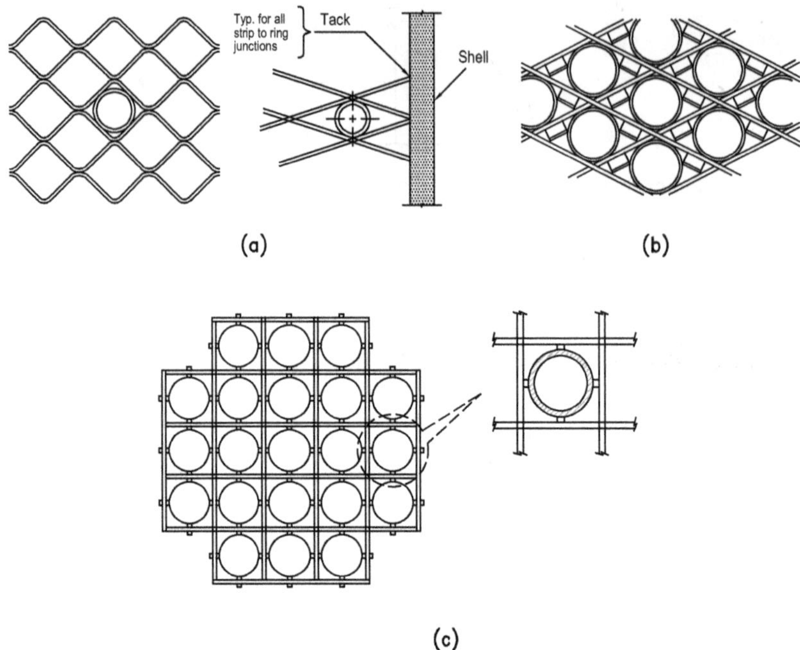

FIGURE 4.20 Special types of plate baffles. (a) NEST™, (b) EGG-CRATE baffle support, and (c) patented baffles with a square tube layout.

tube bundle, and hence the vibration problem is greatly reduced [20]. It is claimed that the pressure drop is lower for the same amount of heat transfer compared to a segmentally baffled exchanger.

EGG-CRATE-GRID™ support (Figure 4.20b) is a simple and economical support for heat exchanger tubes, which can droop or collapse under stress and elevated temperature. This support is fabricated from commercial flat strip material, typically of stainless steel; strip ends are tack welded to the heat exchanger shell and tubes are welded to strips at specified intervals. This design eliminates conventional the tube bundle assembly method, which requires tube insertion through drilled holes.

Figure 4.20c shows patented baffles similar to the EGG-CRATE baffle but with a square tube layout.

4.11.9 Low-Pressure Drop Baffles

Another type of non-segmental baffle made of rods (Figure 4.21) was developed by Koch Heat Transfer, USA, and Holtec International, USA, to ensure LP drop on the shellside since the shellside flow is longitudinal in nature. This baffle type is different from the Phillips RODbaffle heat exchanger.

4.11.10 Grimmas Baffle

The Grimmas baffle is a patented version of the plate baffle, which ensures an axial flow and improves heat transfer [21]. The design is shown in Figure 4.22.

4.11.11 Wavy Bar Baffle

The wavy bar baffle design, as shown in Figure 4.23, is similar to the Grimmas baffle but made of a wavy bar. This design ensures axial flow and improves heat transfer and ensures LP drop on the shellside.

FIGURE 4.21 Low pressure drop baffles. (Courtesy of Holtec International, Marlton, NJ.)

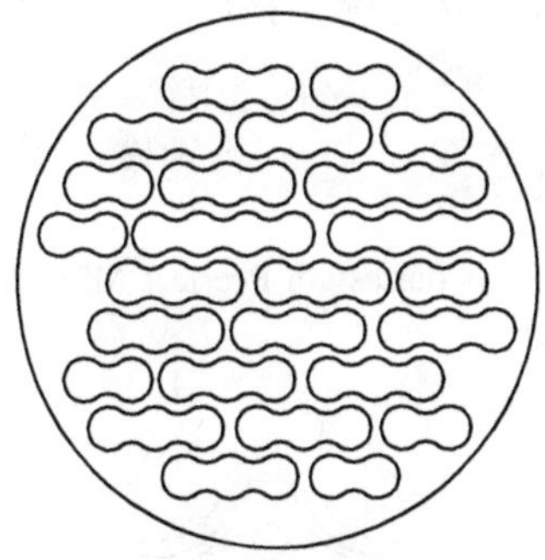

FIGURE 4.22 Grimmas baffle.

4.11.12 Baffles for Steam Generator Tube Support

Figure 4.24 shows five types of baffles—four with plate baffles and one with lattice bars similar to the EGG-CRATE baffle used in steam generator applications [22].

4.12 TUBESHEET

A tubesheet is an important component of a heat exchanger. It is the principal barrier between the shellside and tubeside fluid pressures. The cost of drilling and reaming the tube holes as well as the overall cost of the tubesheet of a given dimension will have a direct bearing on the heat exchanger cost. Additionally, the proper design of a tubesheet is important for safe and reliable operation of the heat exchanger. A tubesheet assembled in a STHE is shown in Figure 4.25. The tubesheet design procedure for fixed, floating head, and U-tubesheet procedure as per ASME Code Sec VIII Div 1 and TEMA procedure is non-mandatory.

4.12.1 Tubesheet Design as per ASME Code Sec VIII Div 1

Part UHX Rules for Shell and Tube Heat Exchangers UHX-1

a. Rules for U-tube heat exchangers are covered in UHX-12.
b. Rules for fixed tubesheet heat exchangers are covered in UHX-13.
c. Rules for floating tubesheet heat exchangers are covered in UHX-14.

FIGURE 4.23 Wavy bar baffle design. (Courtesy of Peerless Mfg. Co. of Dallas, TX, Makers of Alco and Bos-Hatten brands of heat exchangers.)

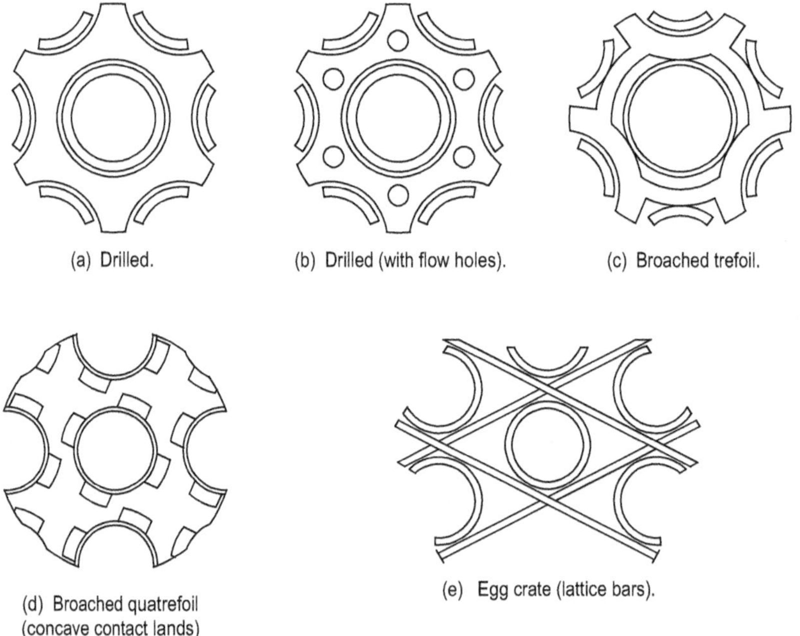

FIGURE 4.24 Types of baffle plates for a steam generator. (a) Drilled, (b), drilled (with flow holes), (c) broached trefoil, (d) broached quatrefoil (concave contact lands), and (e) egg crate (lattice bars).

FIGURE 4.25 Titanium tubesheet assembled in a STHE. (Courtesy of TITAN Metal Fabricators Inc, Camarillo, CA.)

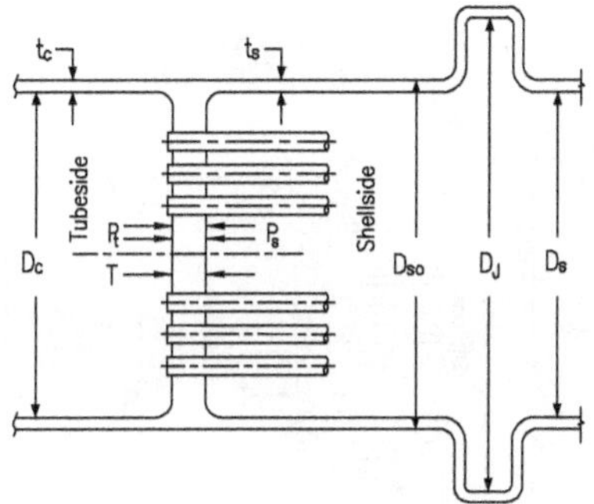

a. Tubesheet integral with shell and channel.

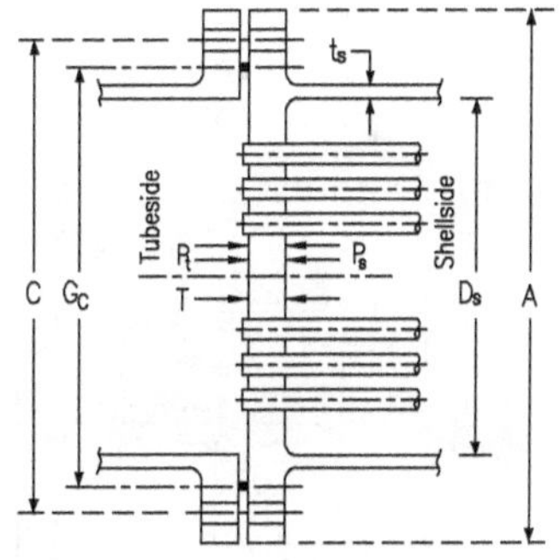

b. Tubesheet integral with shell and gasketed with channel, extended as a flange.

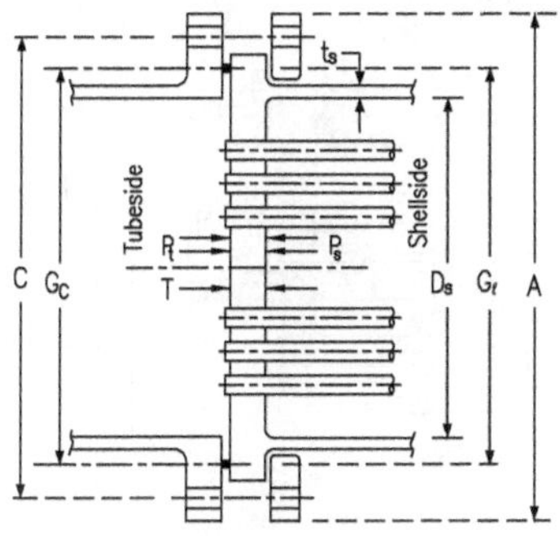

c. Tubesheet integral with shell and gasketed with channel, not extended as a flange.

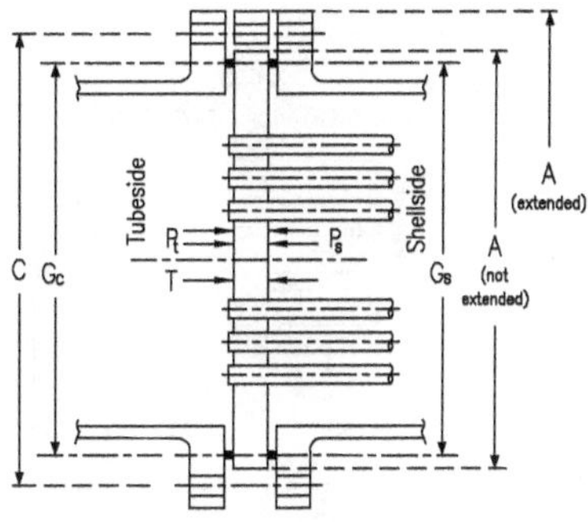

d. Tubesheet gasketed with shell and channel.

FIGURE 4.26 Tubesheet connection with the shell and channel of a fixed tubesheets heat exchanger as per ASME Code Sec VIII dIV 1.

4.12.2 Tubesheet and Its Connection with the Shell and Channel

A tubesheet is an important component of a heat exchanger. It is the principal barrier between the shellside and tubeside fluids. Proper design of a tubesheet is important for the safety and reliability of a heat exchanger. Tubesheets of surface condensers are rectangular shaped. Tubesheets are connected to the shell and the channels either by welds (integral) or with bolts (gasketed joints), or with a combination of these.

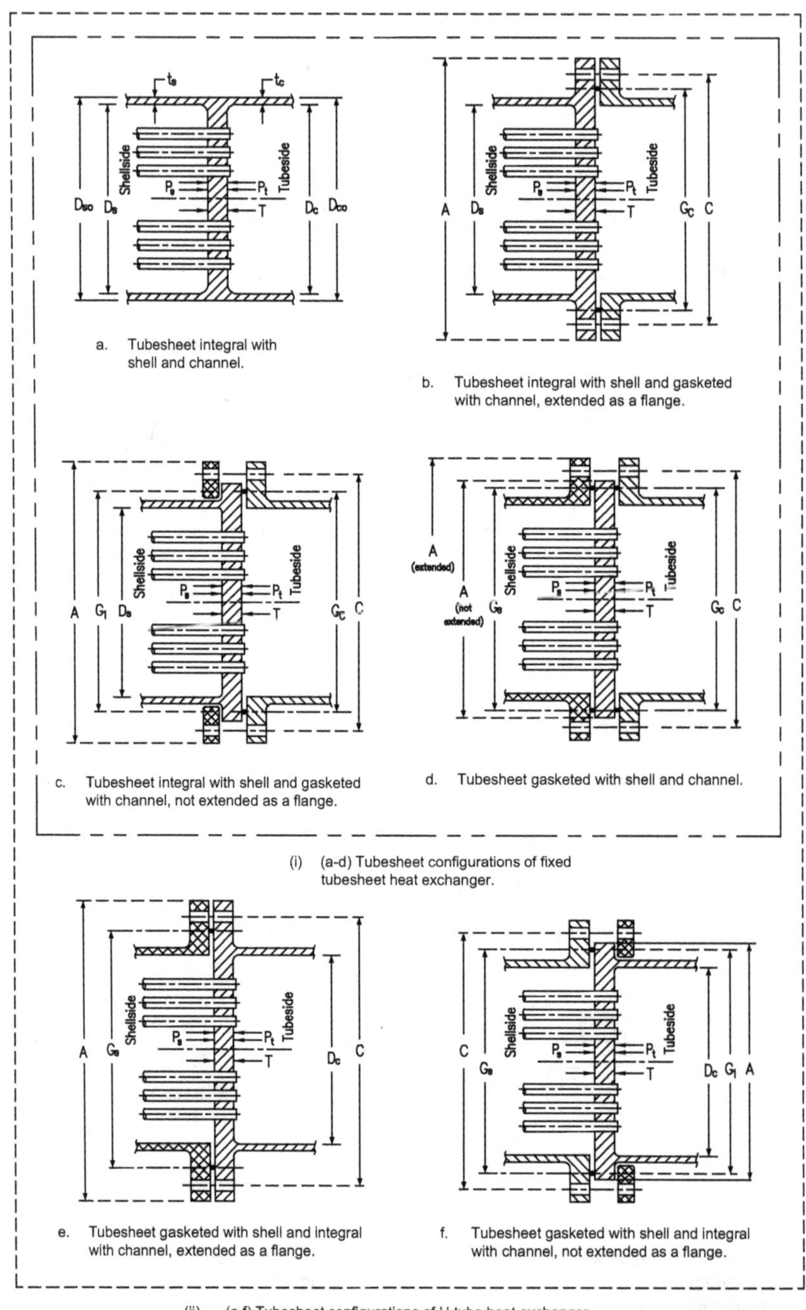

Shellside
Tubeside
a. Tubesheet integral with shell and channel.
b. Tubesheet integral with shell and gasketed with channel, extended as a flange.
c. Tubesheet integral with shell and gasketed with channel, not extended as a flange.
d. Tubesheet gasketed with shell and channel.
A (extended)
A (not extended)
(i) (a-d) Tubesheet configurations of fixed tubesheet heat exchanger.
e. Tubesheet gasketed with shell and integral with channel, extended as a flange.
f. Tubesheet gasketed with shell and integral with channel, not extended as a flange.
(ii) (a-f) Tubesheet configurations of U-tube heat exchanger.

FIGURE 4.27 Tubesheet connection with the shell and channel of a U-tube and fixed tubesheet of Floating head heat exchanger as per ASME Code Sec VIII dIV 1.

4.12.2.1 Tubesheet Connection with the Shell and Channel

Tubesheets are mostly flat circular plates with a uniform pattern of drilled holes. Tubesheets of surface condensers are rectangular in shape. The tubesheet is connected to the shell and the channel either by welding (integral) or bolts (gasketed joints), or a combination thereof. As defined by ASME Code Sec VIII, Div 1, there are four categories for fixed tubesheet heat exchangers and six categories for U-tube and floating head heat exchangers, as shown in Figure 4.26 and Figure 4.27, respectively. Floating head constructions are shown in Figure 4.28.

4.12.3 Clad and Faced Tubesheets

Metal cladding is specifically used to describe a metal outer layer—cladding—that is bonded to a core metal material. A primary purpose of cladding vessels is to make them corrosion-resistant while enhancing their mechanical strength. It is well known that various grades of austenitic SSs of types 304, 304L, 308, 316, and 347, nickel, Monel, Inconel, cupronickel, aluminum, copper, zirconium,

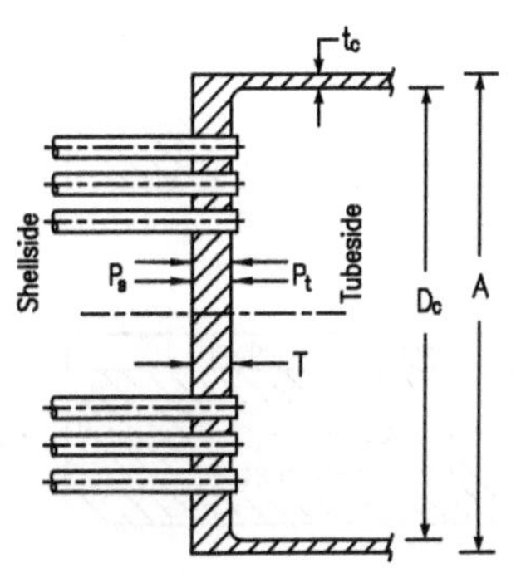

(a) Tubesheet integral with shell and channel.

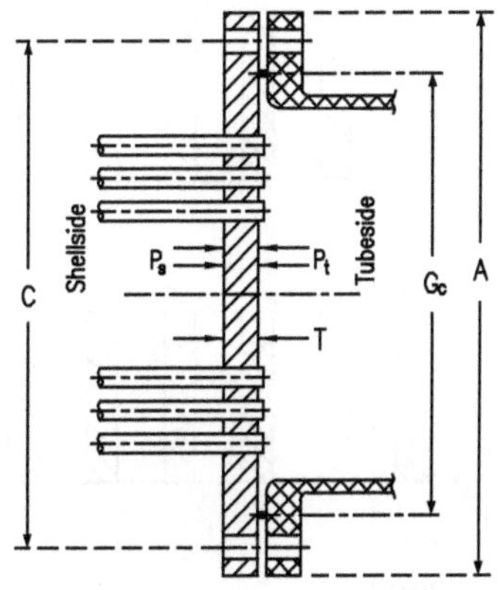

(b) Tubesheet gasketed, extended as a flange.

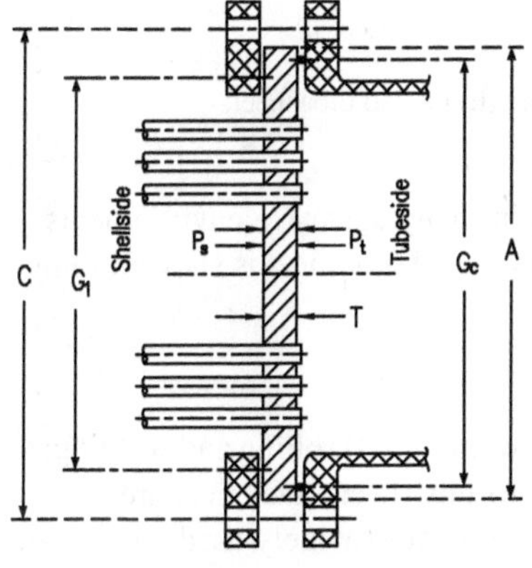

(c) Tubesheet gasketed, not extended as a flange.

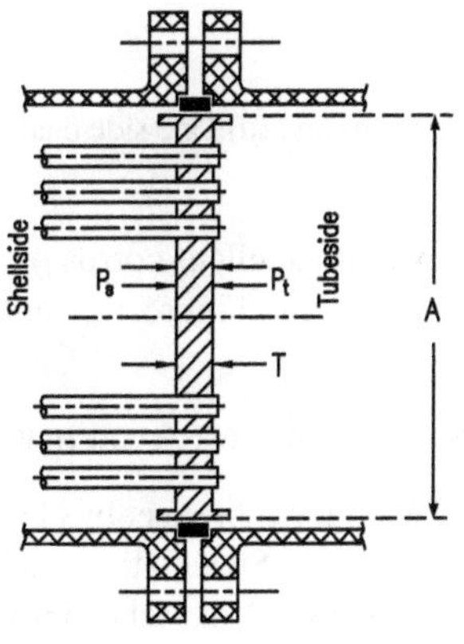

(d) Tubesheet internally sealed.

FIGURE 4.28 Construction details of floating head of floating head heat exchanger as per ASME Code Sec VIII dIV 1.

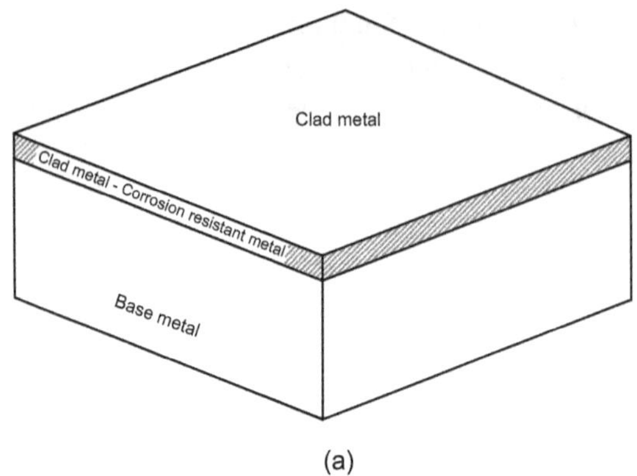

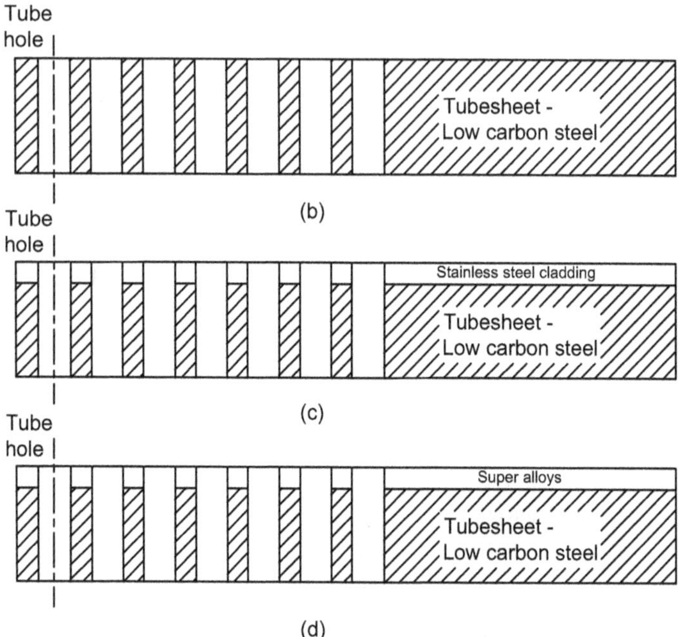

FIGURE 4.29 (a), (c) and (d) One side clad tubesheet and (b) unclad tubesheet.

titanium, etc., exhibit excellent corrosion resistance in many corrosive environments. A one-sided clad tubesheet is shown in Figure 4.29 and a both-sided clad tubesheet is shown in Figure 4.30.

4.12.4 Tube-to-Tubesheet Attachment

Tubes are attached to the tubesheet by (1) rolling; (2) welding; (3) rolling and welding; (4) explosive welding; and (5) brazing. Schematic sketches of tube-to-tubesheet attachment are given in Chapter 4 of Volume 2. Expansion of the tubes into the tubesheet is most widely used and is satisfactory for many services. However, when stresses are higher, or where pressures are such that significant leakage could occur, or where contamination between fluids is not permitted, the tubes are welded to the tubesheet. Explosion welding can be used instead of conventional welding where there is incompatibility between tube and tubesheet materials and for tube plugging under hazardous conditions. Figure 4.31 shows a schematic of a few tube-to-tubesheet connections [23].

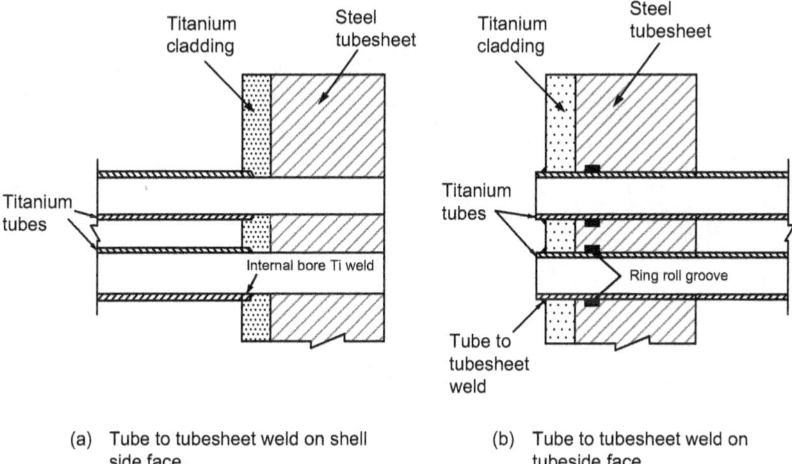

FIGURE 4.30 Both sides titanate cladded tubesheet with tube-to-tubesheet joints.

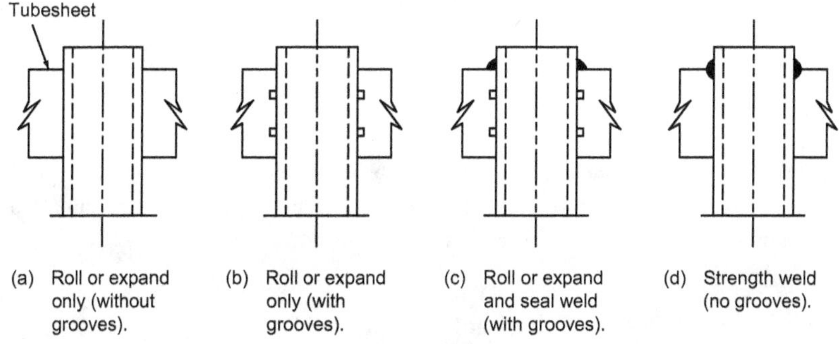

FIGURE 4.31 Tube-to-tubesheet joint types. (Adopted from Harry W. Ebert, Improving the reliability of tube-to-tubesheet joint, *Welding Journal*, September 2000, pp. 47–50.)

4.12.5 Double Tubesheets

Double-tubesheet heat exchangers are used for applications wherein mixing of tubeside and shellside fluid must be avoided to maintain product purity or mixing if one of the fluids is toxic in nature. Another example is a power plant surface condenser. In this application, water is used as a cooling medium. The cooling water (raw water) can be sea, river, tank, or pond water with a lot of contaminants, and if mixing of cooling water and condensate takes place, it leads to unacceptable chemistry of boiler feed water. Because the condensate is pumped back to the boiler, the cooling water mixing with condenser water can lead to many problems on the boiler side. To overcome this possibility, the provision of double-tubesheet construction is preferred for power plant condensers.

4.12.5.1 Types of Double-Tubesheet Designs

Two designs of double tubesheets are available: (1) the conventional double tubesheet design, which consists of two individual tubesheets at each end of the tubes, and (2) the integral double-tubesheet design [24].

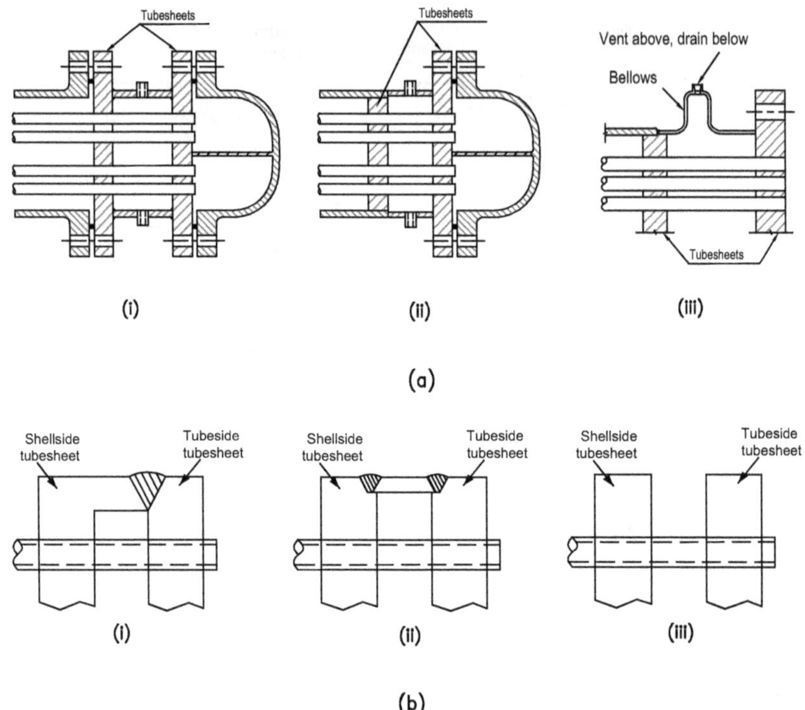

FIGURE 4.32 Double tubesheet shell and tube heat exchanger—schematic. (a) (i) Removable tube bundle with light gauge shroud between two tubesheets, (ii) fixed tubesheets with light gauge shroud, and (iii) fixed tubesheet with bellows and (b) double tubesheets connection methods.

4.12.5.2 Conventional Double-Tubesheet Design

In a conventional double-tubesheet design, the tubesheets are installed with a small space between them. The space is usually open to the atmosphere. Sometimes a thin strip is welded to avoid ingress of dust and dirt, or an expansion joint is welded with a vent at the top and a drain at the bottom. These patterns are shown schematically in Figure 4.32, and double-tubesheet heat exchangers are shown in Figures 4.33 and 4.34. While selecting material for double-tubesheet design, the outer should be compatible with the tubeside fluid and the inner should be compatible with the shellside fluid. The most important consideration is the differential radial expansion of the two tubesheets, which will stress the tubes. The double tubesheet can be installed only in the U-tube, fixed tubesheet, and floating head, outside packed stuffing box exchangers. It is not feasible to use the double tubesheet in heat exchanger types such as [6] (1) floating head, pull-through bundle; (2) floating head with split backing ring; and (3) floating head, outside packed lantern ring exchangers. An expression for the space between the tubesheet pairs, Δx is given by [24]

$$\Delta x = \frac{\sqrt{1.5 E_{\mathrm{T}} \mathrm{d}_{\mathrm{o}} \delta}}{S_{\mathrm{t}}} \tag{4.3}$$

where δ is the free differential in-plane movement in inches (mm) given by

$$\delta = 0.5 D_{otl} \left[\alpha_{\mathrm{h}} \left(T_{\mathrm{h}} - T_{\mathrm{amb}} \right) - \alpha_{\mathrm{c}} \left(T_{\mathrm{c}} - T_{\mathrm{amb}} \right) \right] \tag{4.4}$$

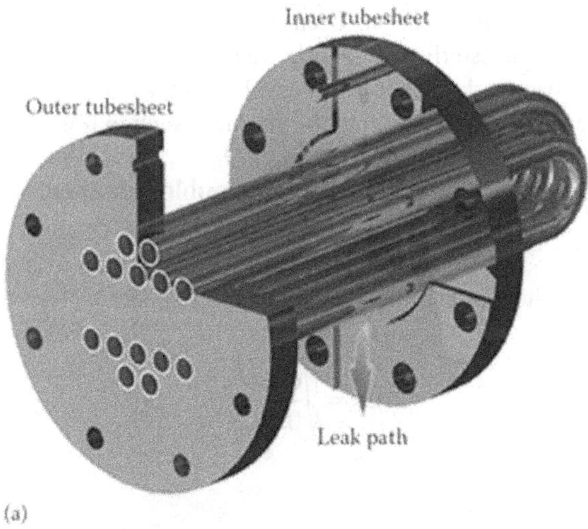

FIGURE 4.33 Double tubesheet shell and tube heat exchanger. (a) Schematic (b) heat exchanger.(Courtesy of Allegheny Bradford Corporation, Bradford, PA.) and (c) Heat exchanger unit. (Courtesy of GEA Heat Exchangers Ltd, Brazil/GEA do Brasil Intercambiadores Ltd, Brazil.)

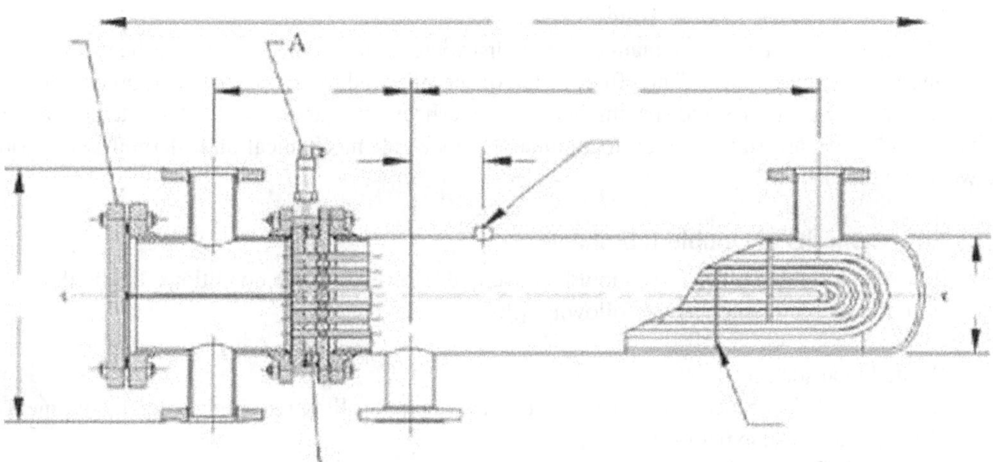

FIGURE 4.34 Double tubesheet shell and tube heat exchanger with real-time leak indicator (A). (Courtesy of Alstrom Corporation, Bronx, NY.)

and where

- d_o is the tube outer diameter, inches (mm)
- E_T is Young's modulus for the tube, psi (Pa)
- D_{otl} is the tube outer limit, inches (mm)
- T_{amb} is the ambient (assembly) temperature, °F (°C)
- T_c is the temperature of the colder tubesheet, °F (°C)
- T_h is the temperature of the hotter tubesheet, °F (°C)
- α_c is the coefficient of thermal expansion of the colder tubesheet between the assembly temperature and temperature T_c, in./(in. °F) [mm/(mm °C)]
- α_h is the coefficient of thermal expansion of the hotter tubesheet between the assembly temperature and temperature T_h, in./(in. °F) [mm/(mm °C)]
- S_t is the allowable stress in the tubes, psi (Pa)

The provision of a double tubesheet has been mandatory for UK power-station condensers in recent years to eliminate any possibility of cooling water entering the steam space of the condenser [25]. The tubesheet interspace is drained to a low-level vessel, which is maintained at the condenser absolute pressure by means of a connection to the air pump suction line.

4.12.5.3 TEMA Guidelines for Spacing between Double Tubesheets

RCB-7.1.2.6.2 Minimum Spacing Between Tubesheets

The minimum spacing Δx, in. (mm), between tubesheets required to avoid overstress of tubes resulting from differential thermal growth of individual tubesheets is given by:

$$\Delta_x = \sqrt{\frac{d_0 \Delta r E_T}{0.27\ Y_T}} \tag{4.5}$$

where

d_o is tube outer diameter, in. (mm)
E_T is the Young's modulus of tube material, psi(kPa),
Y_T is the yield strength of tube material at maximum metal temperature, psi (kPa),
Δr is the differential radial expansion between adjacent tubesheets, in. (mm),

RCB-7.1.2.5 Connected Double Tubesheets

The tubesheets are connected in a manner which distributes axial load between tubesheets by means of an interconnecting cylinder. The effect of the differential radial growth between tubesheets is a major factor in tube stresses and spacing between tubesheets. It is assumed that the interconnecting cylinder and tubes are rigid enough to mutually transfer all mechanical and thermal axial loads between the tubesheets.

4.12.5.4 Demerits of Double Tubesheets

If at all practical, double tubesheets should be avoided. However, some conditions dictate their use along with the problems such as the following [26]:

1. Wasted tube surface
2. Increased fabrication cost due to additional drilling and rolling requiring special equipment
3. Differential radial expansion
4. Differential longitudinal expansion
5. All conditions of start-up, shutdown, and steam out (in the case of condenser) must be considered since they will generally be more severe on the double-tubesheet section than in the operating condition

4.13 TUBE BUNDLE

A tube bundle is an assembly of tubes, baffles, tubesheets, spacers and tie-rods, and longitudinal baffles, if any. Spacers and tie-rods are required for maintaining the space between baffles. Refer to TEMA for details on the number of spacers and tie-rods. A straight tube bundle and a U-tube bindle are shown in Figure 4.35.

4.13.1 Spacers, Tie-Rods, and Sealing Devices

The tube bundle is held together and the baffles are located in their correct positions by a number of tie-rods and spacers. All the rods have spacer tubes fitted over them, each spacer

FIGURE 4.35(1) Shell and tube heat exchanger tube bundle assembly. (a) Straight tube bundle (Courtesy of Vermeer Eemhaven B.V., Rotterdam, the Netherlands) and (b) U-tube bundle. (Courtesy of Holtec International, Marlton, NJ.)

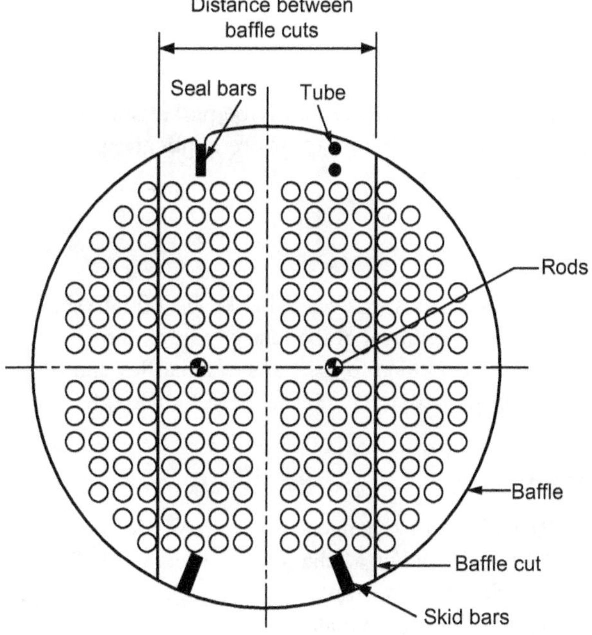

FIGURE 4.35(2) Shell and tube heat exchanger tube bundle assembly cross section showing sealing devices/strips and skid bars.

being tube or pipe with an inside diameter greater than that of the tie-rod diameter, and a length equal to the required baffle spacing. The tie-rods are screwed into the stationary tubesheet and extend the length of the bundle up to the last baffle, where they are secured by locknuts (refer to Figure 4.35(a)). Between baffles, tie-rods have spacers fitted over them. Tie-rods and spacers may also be used as a sealing device to block bypass paths due to pass partition lanes or the

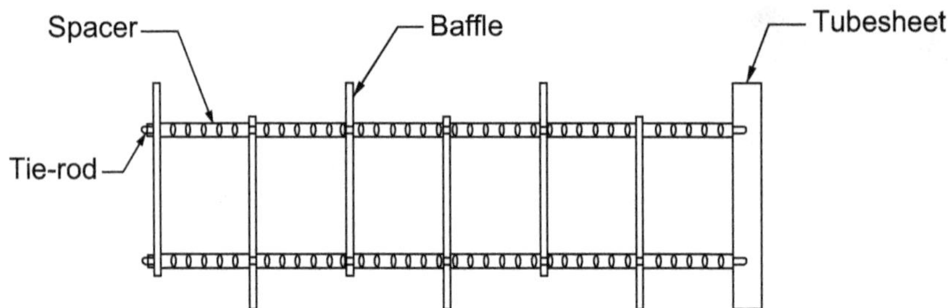

FIGURE 4.36 Baffle cage assembly.

clearance between the shell and the tube bundle. Undesirable shell side leakage paths, such as the gaps due to pass partitions or gaps between bundle and shell, shall be blocked by tie-rods and spacer and sealing devices, shown in Figure 4.35(b) and also in Figure 4.70 and Figure 4.71. A baffle cage assembly showing the tie-rods and spacers is shown schematically in Figure 4.36 and also see Figure 4.2(g).

4.13.2 Outer Tube Limit

The outer tube limit (OTL) is the diameter of the largest circle, drawn around the tubesheet center, beyond which no tube may encroach. The OTL is shown schematically in Figure 4.37.

4.13.3 Bundle Weight

The maximum bundle weight that can conveniently be pulled should be specified and allow for the buildup of fouling and scaling deposits. Offshore applications are particularly sensitive to weight.

4.13.4 Sliding Bar

All removable tube bundles whose mass exceeds 5,450 kg (12,000 lb) shall be provided with a continuous sliding surface to facilitate removal of the tube bundle as shown in Figure 4.35(2). If sliding bars are provided, they should be welded to the transverse baffles and the support plate to form a continuous sliding surface. For more details, refer to TEMA standard.

4.13.5 Tubesheet Pulling Eyes

As per TEMA Section RCB-7.5, in exchangers with removable tube bundles having a nominal diameter exceeding 12" (305 mm) and/or a tube length exceeding 96” (2,438 mm), the stationary tubesheet shall be provided with two tapped holes in its face for pulling eyes. These holes shall be protected in service by plugs of compatible material. Provision for means of pulling may have to be modified or waived for special construction, such as clad tubesheets or manufacturer’s standard, by agreement between the manufacturer and the purchaser.

4.14 TUBESIDE PASSES

The simplest flow pattern through the tubes is for the fluid to enter at one end and exit at the other. This is a single-pass tube arrangement. To improve the heat transfer rate, higher velocities are preferred. This is achieved by increasing the number of tubeside passes. Tubeside multiple passes are

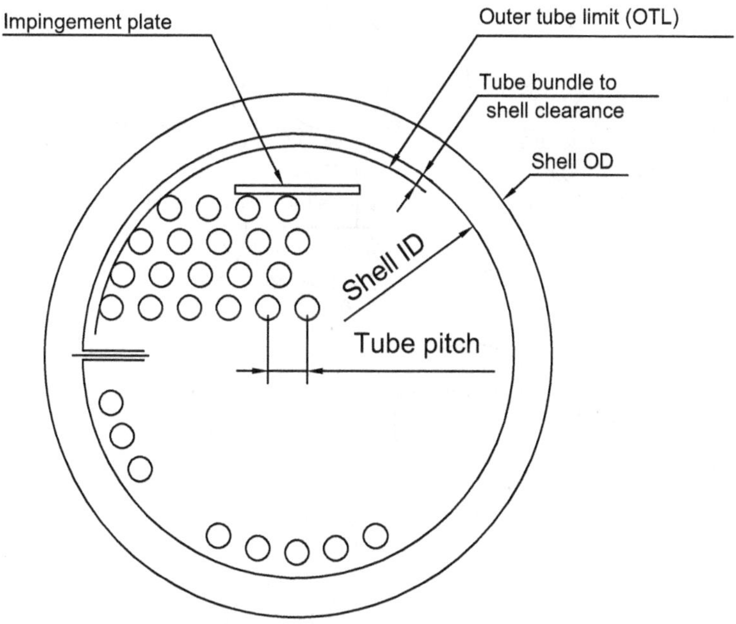

FIGURE 4.37 Definition of OTL.

normally designed to provide a roughly equal number of tubes in each pass to ensure an even fluid velocity and pressure drop throughout the bundle. In a two-tube-pass arrangement, the fluid flows through only half of the total tubes, so that the Reynolds number is high. Increasing the Reynolds number results in increased turbulence and Nusselt number and finally an increase in the overall heat transfer coefficient.

4.14.1 Number of Tube Passes

The number of tubeside passes generally ranges from one to eight. The standard design has one, two, or four tube passes. The practical upper limit is 16. Partitions built into heads known as partition plates or pass ribs control tubeside passes. This is shown schematically in Figure 4.38. The pass partitions may be straight or wavy rib design. The maximum number of tubeside passes is limited by workers' abilities to fit the pass partitions into the available space and the bolting and flange design to avoid interpass leakages on the tubeside. In multipass designs, an even number of passes is generally used; odd numbers of passes are uncommon and may result in mechanical and thermal problems in fabrication and operation. The number of tube passes depends upon the available pressure drop, since a higher velocity in the tube results in a higher heat transfer coefficient, at the cost of increased pressure drop.

Figures 4.39 and 4.40 show tube bundle assemblies with multipasses on the tubeside.

4.15 END CHANNEL AND CHANNEL COVER

Tubeside enclosures are known as end channels or bonnets. They are typically fabricated or cast. They are attached to the tubesheet by bolting with a gasket between the two metal surfaces. Cast heads are typically used up to 250 mm (14 in.) and are made from cast iron, steel, bronze, or stainless steel. They typically have pipe thread connection. Fabricated heads can be made in a wide variety of configurations. They can have a metal cover that allows servicing the tubes without disturbing

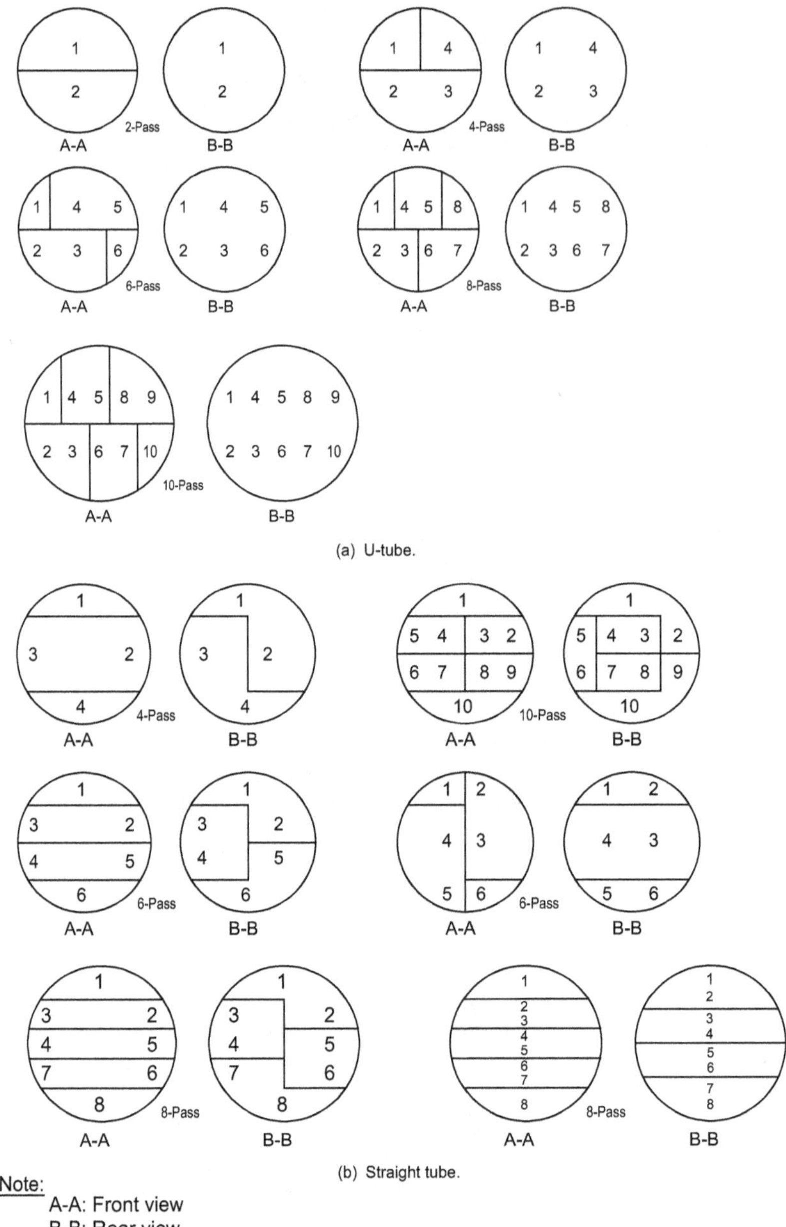

FIGURE 4.38 Typical tubeside partitions for multipass arrangement. (a) U-tube and (b) straight tubes. *Note:* A–A: Front view and B–B: Rear view.

the shell or tube piping. Heads can have axially or tangentially oriented nozzles, which are typically ANSI flanges. Flanged nozzles or threaded bonnets are used in smaller exchangers.

Channel covers: These are round plates that tend to be bolted to the channel flanges. Figure 4.41 show a heat exchanger with bonnet and channel cover and Figures 4.42 and 4.43 show pass partitions in front and rear channels.

FIGURE 4.39 Multipass titanium tube bundle assemblies. (Courtesy of TITAN Metal Fabricators Inc, Camarillo, CA.)

FIGURE 4.40 Eight-pass shell and tube heat exchanger tube bundle showing a few top rows of tubes not incorporated to avoid erosion due to high impingement velocity. (Courtesy of Riggins Company, Hampton, VA.)

FIGURE 4.41 U-tube shell and tube heat exchangers showing bonnet and channel cover plate. (Courtesy of Brembana Costruzioni Industriali S.p.A., Milan, Italy.)

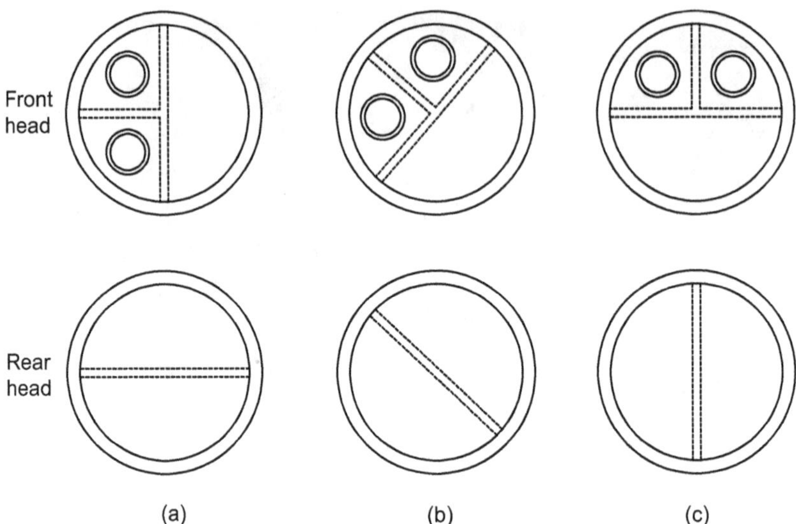

FIGURE 4.42 Four-pass partitions in front channel. (a) Pass partition plate is vertical, (b) pass partition plate is inclined, and (c) Pass partition is horizontal.

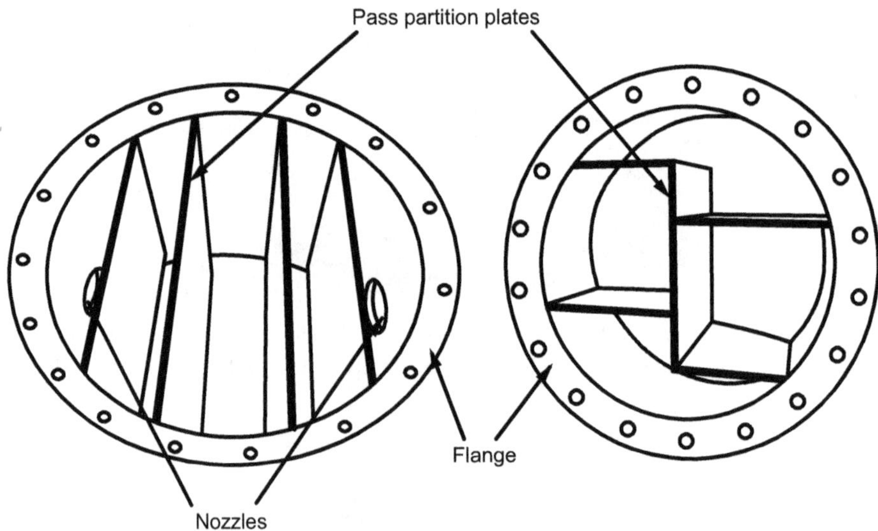

FIGURE 4.43 Six-pass partitions in front channel and rear channel. (Courtesy of Festival City Fabricators, div. of CSTI, Stratford, Ontario, Canada.)

Pass divider or ribs: This is required in an exchanger with two or more tubeside passes. The divider is required in both bonnets and channels (Figures 4.43). Front and rear head pass ribs and gaskets are matched. Pass ribs in cast heads are integrally cast and then machined to sizes; in fabricated heads, they are welded in place. On the tubesheet, the pass rib(s) requires either removing tubes to allow a straight pass rib or machining the pass rib with curves around the tubes, which is more costly to manufacture. Where a full bundle tube count is required to satisfy the thermal requirements, the pass rib approach may lead to considering the next larger shell diameter.

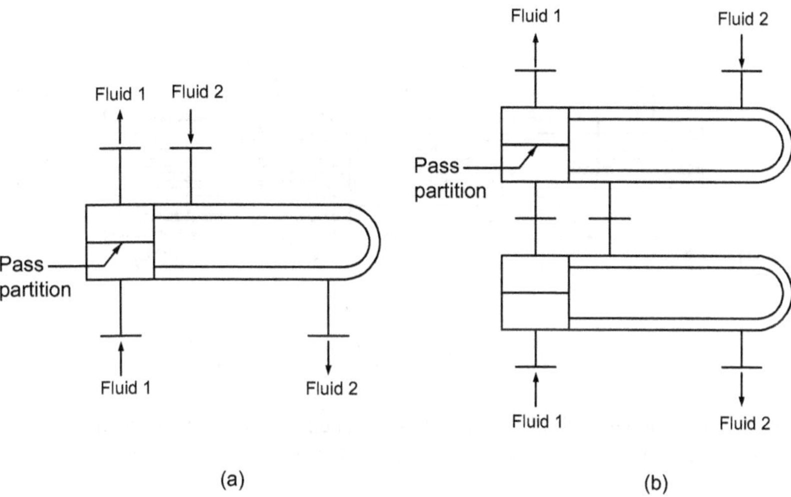

FIGURE 4.44 Stack of two shells in series(Schematic).

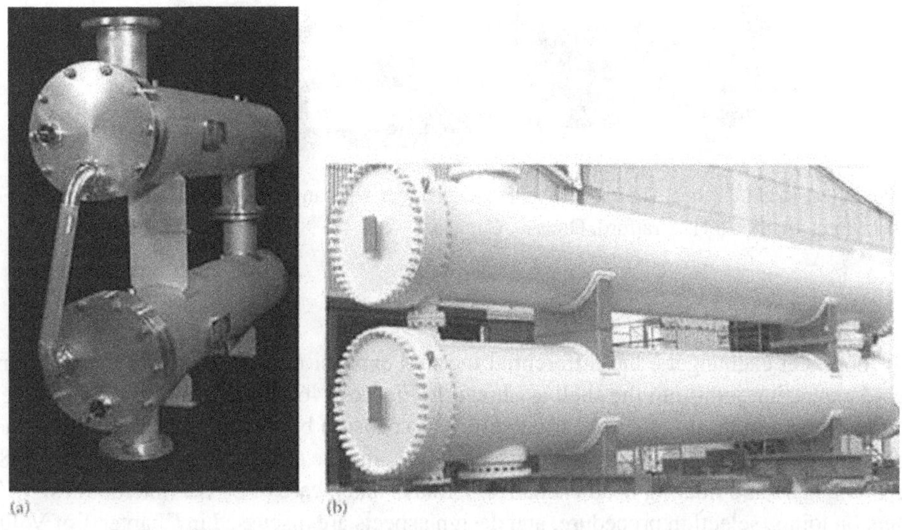

FIGURE 4.45 (a and b) Stack of two shells in series/shell and tube heat exchangers stack. (Photo a: Courtesy of Allegheny Bradford Corporation, Bradford, PA and photo b: Courtesy of Brembana Costruzioni Industriali S.p.A., Milan, Italy.)

4.16 SHELLSIDE PASSES AND SHELL STACKING

For exchangers requiring high effectiveness, multipassing is the only alternative. Shellside passes could be made with the use of longitudinal baffles. However, multipassing on the shell with longitudinal baffles will reduce the flow area per pass compared to a single pass on the shellside. This drawback is overcome by shells in series or stacking one above the other, as shown in Figures 4.44 and 4.45, which is also equivalent to multipassing on the shellside. For the case of the overall direction of two fluids in counterflow, as the number of shellside passes is increased to infinity (practically above four), its effectiveness approaches that of a pure counterflow exchanger. In heat recovery trains and some other applications, up to six shells in series are commonly used.

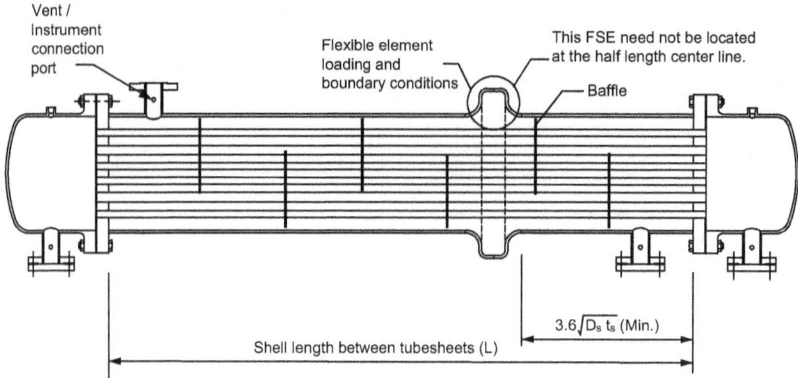

FIGURE 4.46 Cross-sectional view of a fixed tubesheet shell and tube heat exchanger with an expansion joint (schematic). (Adopted from ASME Code Sec VIII Div 1[3].)

FIGURE 4.47 Fixed tubesheet shell and tube heat exchanger with an expansion joint. (Courtesy of Festival City Fabricators, div. of CSTI, Stratford, Ontario, Canada.)

4.17 EXPANSION JOINT

In fixed tubesheet exchangers, the differential thermal expansion problem is overcome by incorporating an expansion joint into the shell as shown in Figure 4.46. A heat exchanger with an expansion joint is shown in Figure 4.47. Alternatively, an expansion bellow fitted on one end of the tube bundle, as shown in Figure 4.48, can be devised to overcome the thermal expansion problem [8]. For U-tube exchangers and floating head exchangers, this is taken care of by the inherent design. Types of expansion joints, selection procedure, and design aspects are discussed in Chapter 1 of Volume 2.

4.18 DRAINS AND VENTS

All exchangers need to be drained and vented; therefore, care should be taken to properly locate and size drains and vents. Additional openings may be required for instruments such as pressure gauges and thermocouples.

4.19 NOZZLES AND IMPINGEMENT PROTECTION

Nozzles are used to convey fluids into and out of the exchanger. These nozzles are pipes of constant cross section welded to the shell and the channels. The nozzles must be sized with the understanding that the tube bundle will partially block the opening. Whenever a high-velocity fluid is entering the shell, some type of impingement protection is required to avoid tube erosion and vibration. As per TEMA standard, impingement protection is mandatory if the inlet fluid is a gas with abrasive particles and in a heat exchanger that receives liquid–vapor mixtures or vapor with entrained

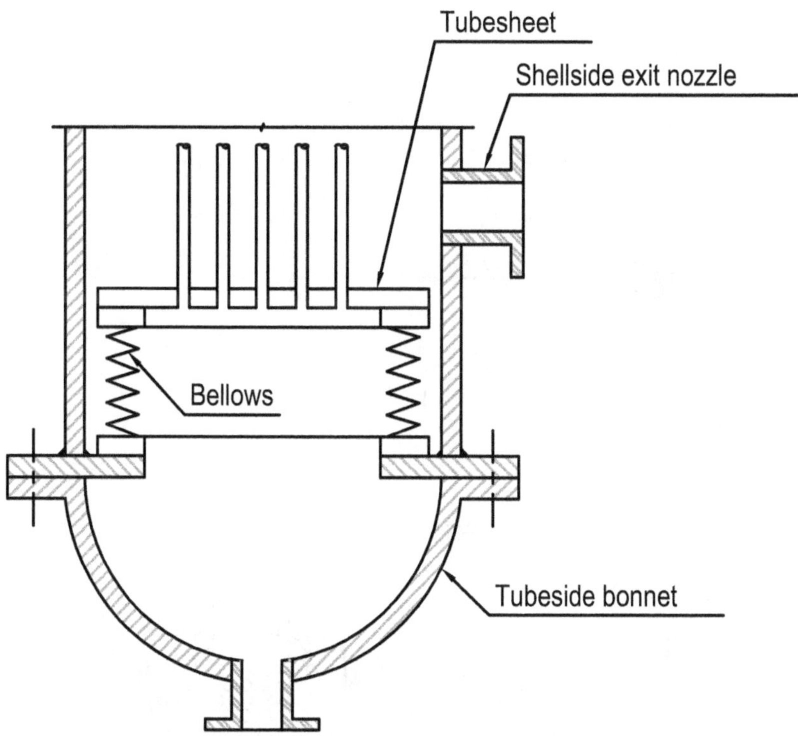

FIGURE 4.48 Shell and tube heat exchanger with an expansion bellow on one end of the tube bundle to overcome thermal expansion problem. (Adapted from Bell, K.J. and Mueller, A.C. *Wolverine Heat Transfer Data Book II*, Wolverine Division of UOP Inc., Decatur, AL, 1984.)

liquid where the erosive effects are intense. For single-phase fluid, TEMA standards define when the installation of impingement protection is required, depending on the kinetic energy of the fluid entering the shell. For liquids with abrasive particles, impingement protection is required if the inlet mass velocity ρV^2 exceeds 744 kg/m^2, and for clean single-phase fluids, impingement protection is required if the inlet mass velocity ρV^2 exceeds 2,230 kg/m^2. Forms of impingement protection include (1) impingement plate; (2) impingement rods; and (3) annular distributors. Figures 4.49–4.51 show impingement plate protection.

4.19.1 Minimum Nozzle Size

Gollin [27] presents a methodology for determining the minimum inside diameter of a nozzle for various types of fluid entering or leaving the unit. The actual size of nozzle used will depend on the pressure, material, corrosion allowance, and pipe schedule.

The minimum inside diameter, $d_{\min}$, is calculated from

$$d_{\min} = \sqrt[4]{\frac{M^2}{3.54 \times 10^6 \times \Delta p_v \rho}} \tag{4.6}$$

where

M is the mass flow rate of fluid through nozzle (lb/h)
ρ is the density of fluid (lb/ft^3)
Δp_v is the velocity head loss through the nozzle (psi)

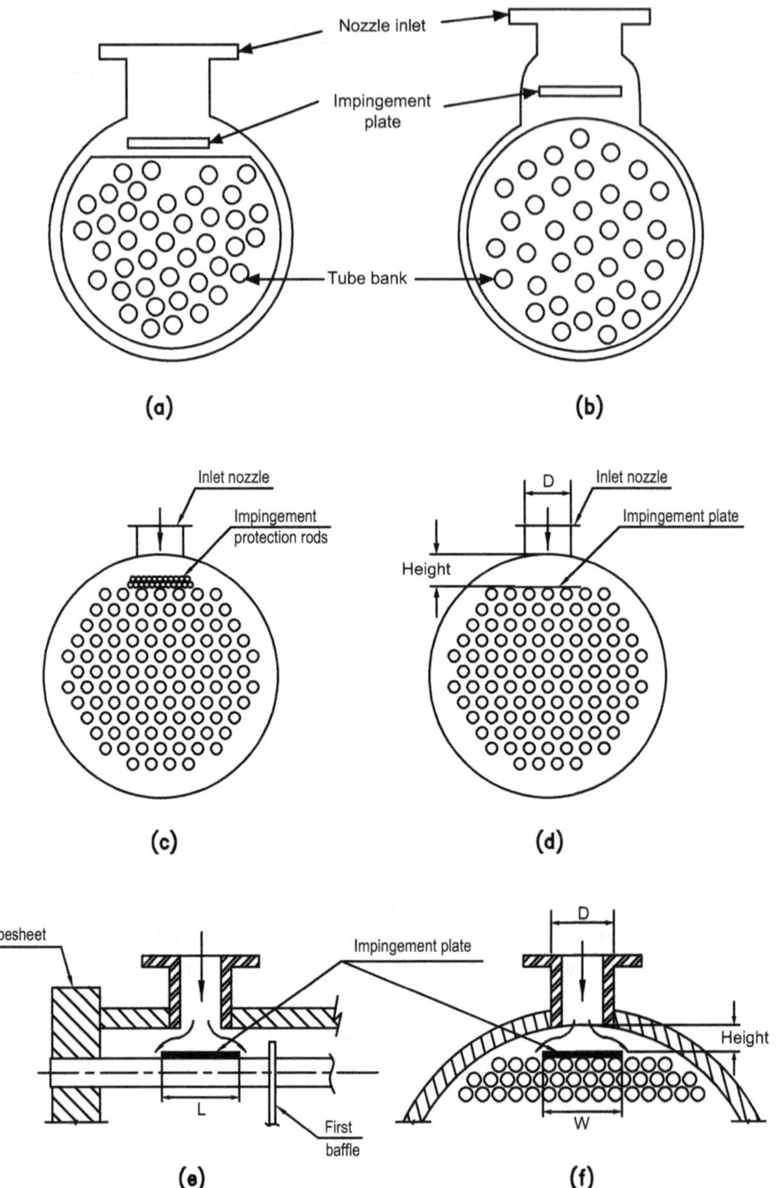

FIGURE 4.49 (a–f) Impingement protection arrangement- Rod and plate arrangement.

4.19.2 Distributor Belt

RGP-RCB-4.6.1 Distributor Belt

A distributor belt is an annular space on the outer diameter of the shell which provides impingement protection, reduced bundle ρV^2, reduced shellside pressure drop, improved flow distribution at the inlet and/or outlet nozzle, and that can reduce flow-induced tube vibration issues. Also, by putting the annular space external to the shell, the full shell diameter may be utilized for the tube field distributor belts which can be designed to incorporate flexible shell elements (RCB-8 FSE) and, as such, serve dual design roles. Since the distributor belt is a high point and low point in a horizontal shell, vent and drain connections should be provided. Figure 4.52 shows a heat exchanger with a distributor arrangement.

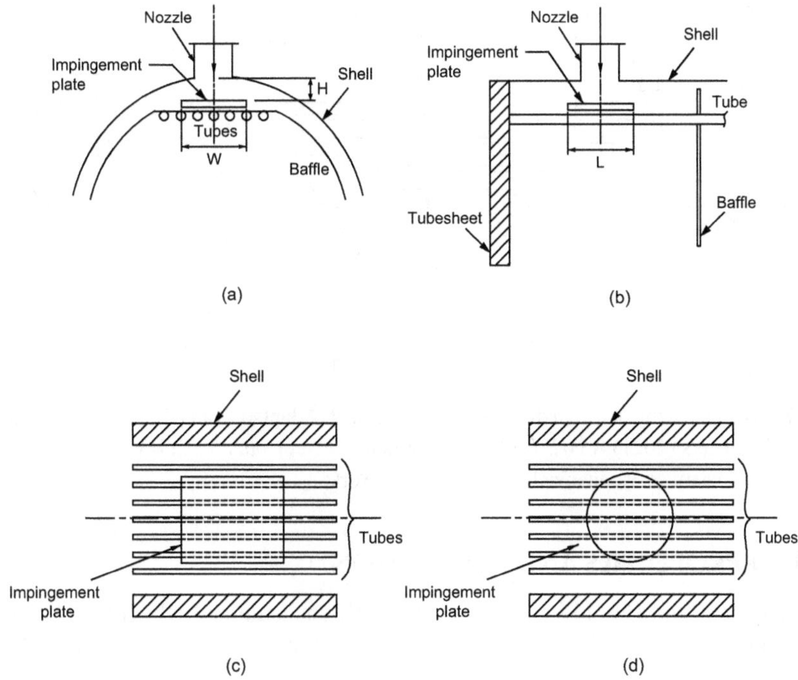

FIGURE 4.50 Impingement protection arrangement –plate type.

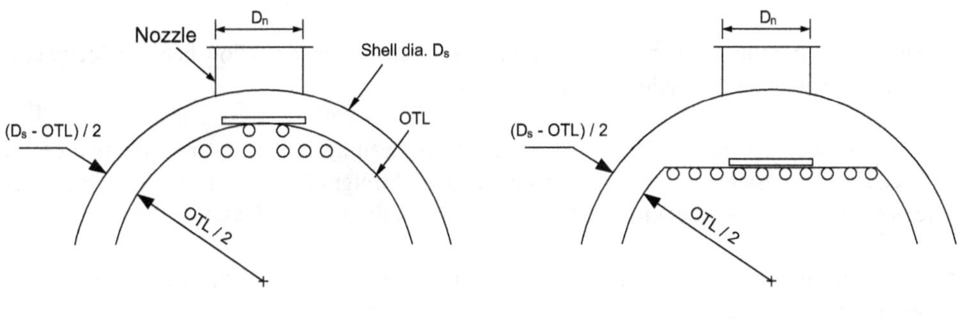

(a) Impingement plate - Full layout.

(b) Impingement plate - Partial layout.

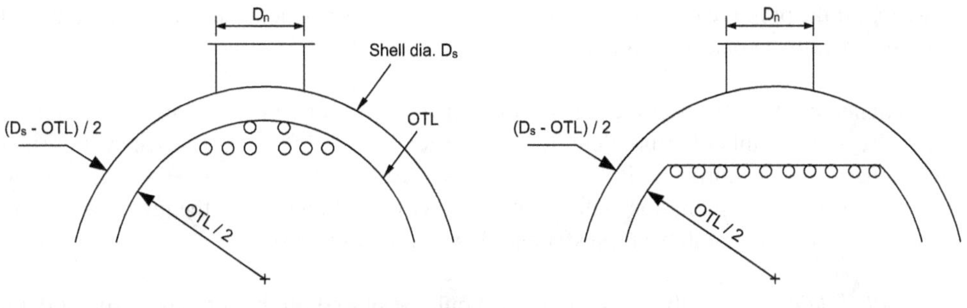

(c) No impingement plate - Full layout.

(d) No impingement plate - Partial layout.

FIGURE 4.51 (a, b) Tube bundle with impingement protection arrangement and (c) and (d) without protection.

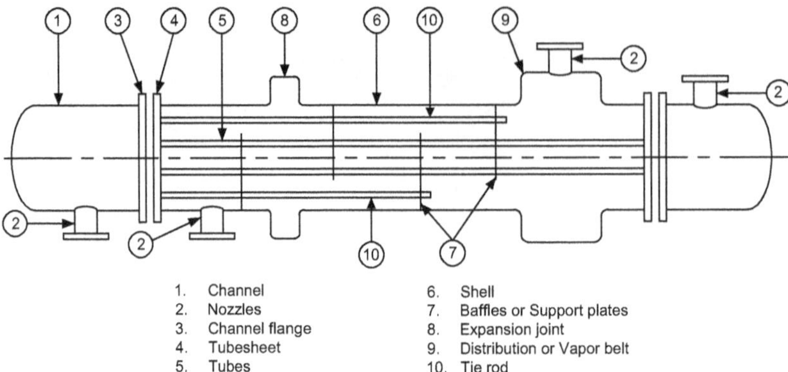

FIGURE 4.52 Heat exchanger with a distributor arrangement. Adapted from ASME Code. American Society of Mechanical Engineers (2021) *ASME Boiler and Pressure Vessel Code, Section VIII, Division 1— Pressure Vessels*. American Society of Mechanical Engineers, New York.

4.20 FLUID PROPERTIES AND ALLOCATION

To determine which fluid should be routed through the shellside and which fluid on the tubeside, consider the following factors. These factors are discussed in detail in Refs. [8, 28, 29].

Corrosion: Fewer corrosion-resistant alloys or clad components are needed if the corrosive fluid is placed on the tubeside.

Fouling: This can be minimized by placing the fouling fluid in the tubes to allow better velocity control; increased velocities tend to reduce fouling.

Cleanability: The shellside is difficult to clean; chemical cleaning is usually not effective on the shellside because of bypassing and requires the cleaner fluid. Straight tubes can be physically cleaned without removing the tube bundle; chemical cleaning can usually be done better on the tubeside.

Temperature: For high-temperature services requiring expensive alloy materials, fewer alloy components are needed when the hot fluid is placed on the tubeside.

Pressure: Placing an HP fluid in the tubes will require fewer costly HP components and the shell thickness will be less.

Pressure drop: If the pressure drop of one fluid is critical and must be accurately predicted, then that fluid should generally be placed on the tubeside.

Viscosity: Higher heat transfer rates are generally obtained by placing a viscous fluid on the shellside. The critical Reynolds number for turbulent flow in the shell is about 200; hence, when the flow in the tubes is laminar, it may be turbulent if the same fluid is placed on the shellside. However, if the flow is still laminar when in the shell, it is better to place the viscous fluid only on the tubeside since it is somewhat easier to predict both heat transfer and flow distribution [29].

Toxic and lethal fluids: Generally, the toxic fluid should be placed on the tubeside, using a double tubesheet to minimize the possibility of leakage. Construction code requirements for lethal service must be followed.

Flow rate: Placing the fluid with the lower flow rate on the shellside usually results in a more economical design and a design safe from flow-induced vibration. Turbulence exists on the shellside at much lower velocities than on the tubeside.

4.21 SHELL AND TUBE HEAT EXCHANGER SELECTION

The fixed tubesheet exchanger is usually the cheapest. If, however, an expansion joint has to be used, then a U-tube exchanger may prove cheaper. If the bundle has to be removable, a U-tube will be the cheapest. For guidelines on selection of shell and tube heat exchanger refer [5, 30]

4.22 TEMA SHELL TYPES

Seven types of shells are standardized by TEMA [2]. They are as follows:

1. *E* One-pass shell
2. *F* Two-pass shell with longitudinal baffle
3. *G* Split flow
4. *H* Double split flow
5. *J* Divided flow
6. *K* Kettle type reboiler
7. *X* Crossflow

Shellside flow for *E, F, G, H,* and *J* shells are shown schematically in Figure 4.53. A brief description of each type is provided next.

4.22.1 TEMA *E* SHELL

The *E* shell is the most common due to its cheapness, simplicity, and ease of manufacture. It has one shell pass, with the shellside fluid entry and exit nozzles attached at the two opposite ends of the shell. The tubeside may have a single pass or multiple passes. The tubes are supported by transverse baffles. This shell is the most common for single-phase shell fluid applications. Multiple passes on the tubeside reduce the exchanger effectiveness or LMTD correction factor *F* over a single-pass arrangement. If *F* is too low, two *E* shells in series may be used to increase the effective

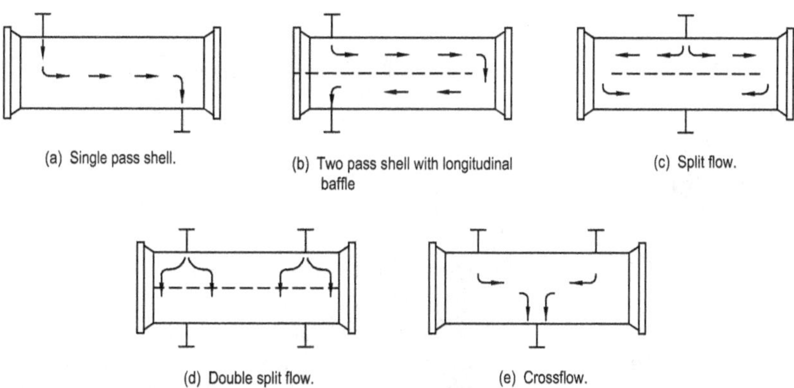

FIGURE 4.53 TEMA shell types and shellside flow distribution pattern. (a) Single-pass shell, (b) two-pass shell with longitudinal baffle, (c) split flow, (d) double split flow, and (e) crossflow.

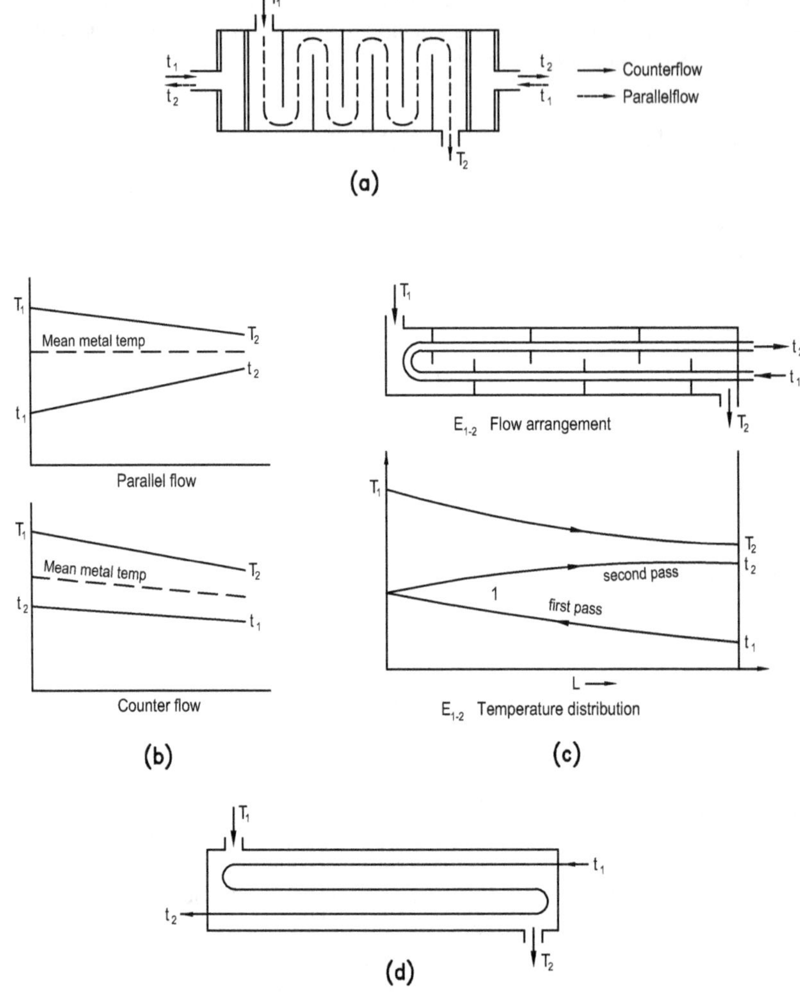

FIGURE 4.54 TEMA *E* shell flow arrangement and temperature distribution. (a) $E_{1\text{-}1}$—parallelflow and counterflow arrangement; (b) $E_{1\text{-}1}$—temperature distribution and (c) $E_{1\text{-}2}$—flow arrangement and temperature distribution.

temperature difference and thermal effectiveness. Possible flow arrangements are $E_{1\text{-}1}$, $E_{1\text{-}2}$, E_{13}, and $E_{1\text{-}N}$ (Figure 4.54).

4.22.2 TEMA *F* Shell

The *F* shell with two passes on the shellside is commonly used with two passes on the tubeside as shown in Figure 4.55 so that the flow arrangement is countercurrent, resulting in an *F* factor of 1.0. This is achieved by the use of an *E* shell having a longitudinal baffle on the shellside. The entry and exit nozzles are located at the same end. The amount of heat transferred is more than that in an *E* shell but at the cost of an increased pressure drop (compared to *E* shell, the shellside velocity is two times higher and pressure drop is eight times higher). Although ideally this is a desirable flow arrangement, it is rarely used because of many problems associated with the shellside longitudinal baffle. They include the following:

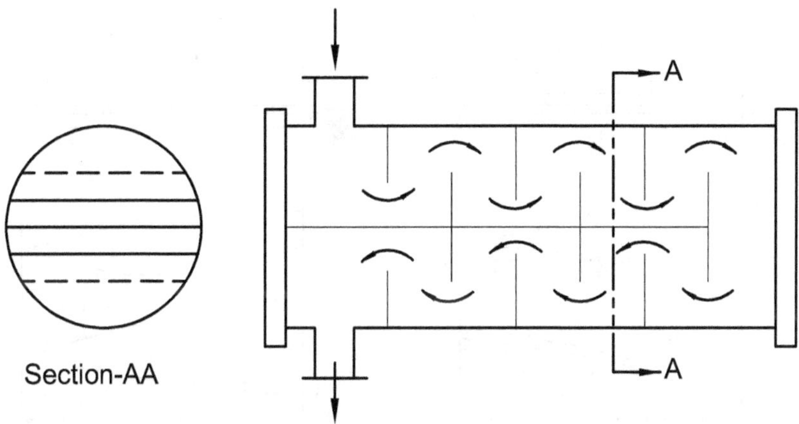

FIGURE 4.55 TEMA *F* shell.

1. There will be a conduction heat transfer through the longitudinal baffle because of the temperature gradient between the shellside passes.
2. If the longitudinal baffle is not continuously welded to the shell, or if seals are not provided effectively between the longitudinal baffle and the shell, there will be fluid leakage from the HP to LP side.

Both these factors will reduce the mean temperature difference and the exchanger effectiveness more than the gain by achieving the pure counterflow arrangement. The welded baffle construction has a drawback: It does not permit the withdrawal of the tube bundle for inspection or cleaning. Hence, if one needs to increase the mean temperature difference, multiple shells in series are preferred over the *F* shell.

4.22.3 TEMA *G, H* Shells

TEMA *G* and *H* shell designs are most suitable for phase change applications where the bypass around the longitudinal plate and counterflow is less important than even flow distribution. In this type of shell, the longitudinal plate offers better flow distribution for vapor streams and helps to flush out noncondensables. They are frequently specified for use in horizontal thermosiphon reboilers and total condensers.

4.22.4 TEMA *G* Shell or Split Flow Exchanger

In this exchanger, there is one central inlet and one central outlet nozzle with a longitudinal baffle. The shell fluid enters at the center of the exchanger and divides into two streams. Hence, it is also known as a split flow unit. Possible flow arrangements ($G_{1\text{-}1}$, $G_{1\text{-}2}$, and $G_{1\text{-}4}$) are shown in Figure 4.56. *G* shells are quite popular with heat exchanger designers for several reasons. One important reason is their ability to produce "temperature correction factors" comparable to those in an *F* shell with only a fraction of the shellside pressure loss (same as *E* shell) in the latter type [31].

4.22.5 TEMA *H* Shell or Double Split Flow Exchanger

This is similar to the *G* shell, but with two inlet nozzles and two outlet nozzles and two horizontal baffles resulting in a double split flow unit. It is used when the available pressure drop is very limited.

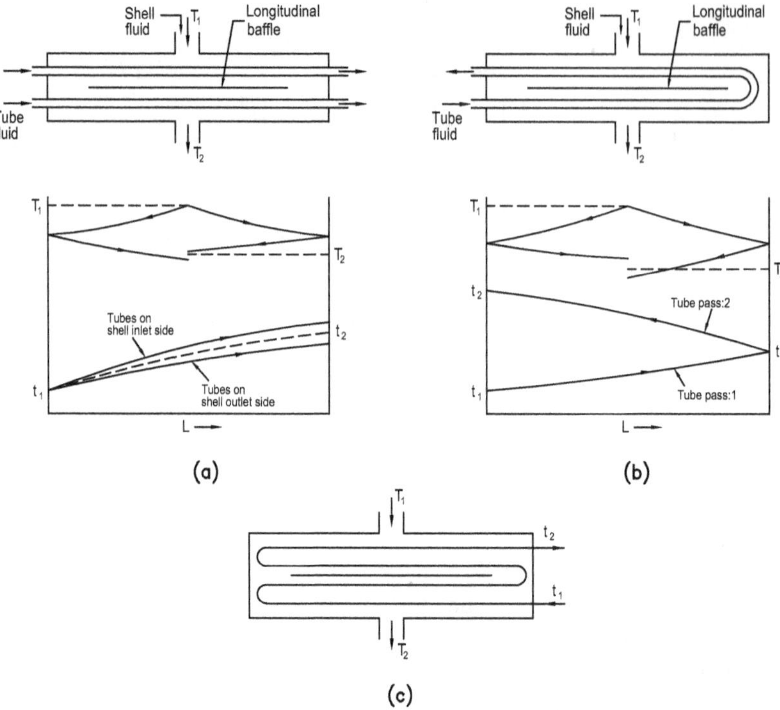

FIGURE 4.56 *G* shell flow arrangement and temperature distribution. (a) $G_{1\text{-}1}$ and (b) $G_{1\text{-}2}$.

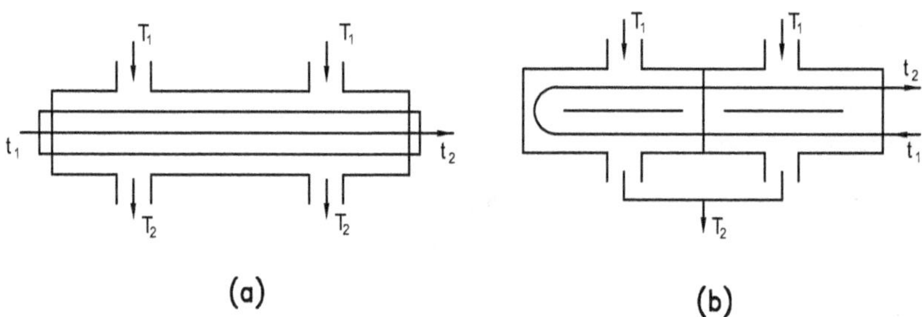

FIGURE 4.57 *H* shell flow arrangement. (a) $H_{1\text{-}1}$ and (b) $H_{1\text{-}2}$.

Possible pass arrangements are $H_{1\text{-}1}$ and $H_{1\text{-}2}$ (Figure 4.57). The *H* shell approaches the crossflow arrangement of the *X* shell, and it usually has low shellside pressure drop compared to *E, F*, and *G* shells [7], i.e., compared to *E* shell, the shellside velocity is one-half and pressure drop is one-eighth.

4.22.6 TEMA *J* Shell or Divided Flow Exchanger

The divided flow *J* shell has two inlets and one outlet or one inlet and two outlet nozzles (i.e., a single nozzle at the midpoint of shell and two nozzles near the tube ends). With a single inlet nozzle at the middle, the shell fluid enters at the center of the exchanger and divides into two streams. These streams flow in longitudinal directions along the exchanger length and exit from two nozzles, one at each end of the exchanger. The possible pass arrangements are one shell pass, and one, two, four, *N* (even), or infinite tube passes. *J* shell pass arrangements with temperature distribution are shown in Figure 4.58 [32]. With a TEMA *J* shell, the shellside velocity will be one-half that of the TEMA

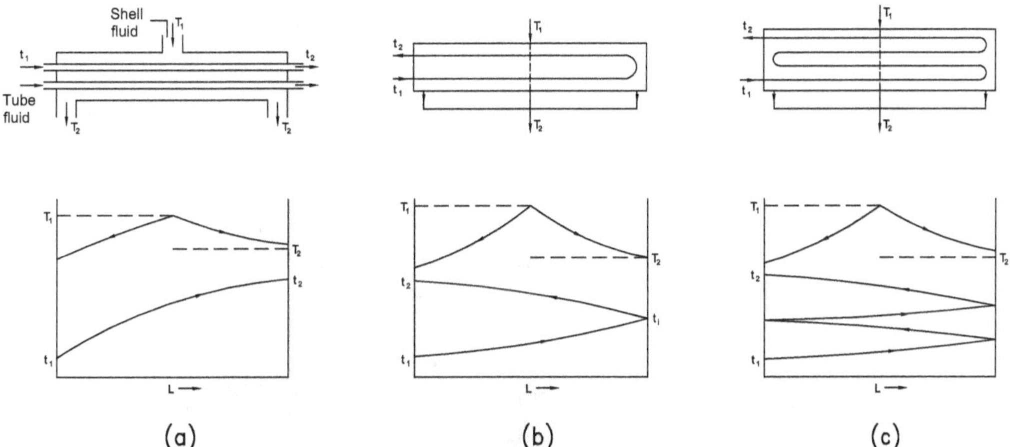

FIGURE 4.58 TEMA *J* shell flow arrangement and temperature distribution. (a) $J_{1\text{-}1}$, (b) $J_{1\text{-}2}$, and (c) $J_{1\text{-}4}$. (From Jaw, L., *Trans. ASME, J. Heat Transfer*, 56C, 408, 1964.)

FIGURE 4.59 Kettle-type reboilers. (Courtesy of Brembana Costruzioni Industriali S.p.A., Milan, Italy.)

E shell, and hence the pressure drop will be approximately one-eighth that of a comparable *E* shell. For this reason, it is used for LP-drop applications such as condensing in a vacuum [7]. For a condensing shell fluid, the *J* shell is used with two inlets for the gas phase and one central outlet for the condensates and leftover gases.

4.22.7 TEMA *K* Shell or Kettle Type Reboiler

The *K* shell is used for partially vaporizing the shell fluid. It is used as a kettle reboiler in the process industry and as a flooded chiller in the refrigeration industry. Usually, it consists of a horizontal bundle of heated U tubes or floating head placed in an oversized shell. The tube bundle is free to move and is removable. Its diameter is about 50–70% of the shell diameter. The large empty space above the tube bundle acts as a vapor-disengaging space. A kettle boiler heat exchanger of TEMA-type AKT is shown in Figure 4.4e. The liquid to be vaporized enters at the bottom, near the tubesheet, and covers the tube bundle; the vapor occupies the upper space in the shell, and the dry vapor exits from the top nozzle(s), while a weir (the vertical unperforated baffle shown in Figure 4.4e) helps to maintain the liquid level over the tube bundle. The bottom nozzle in this space is used to drain the excess liquid. A kettle boiler heat exchanger is shown in Figure 4.59.

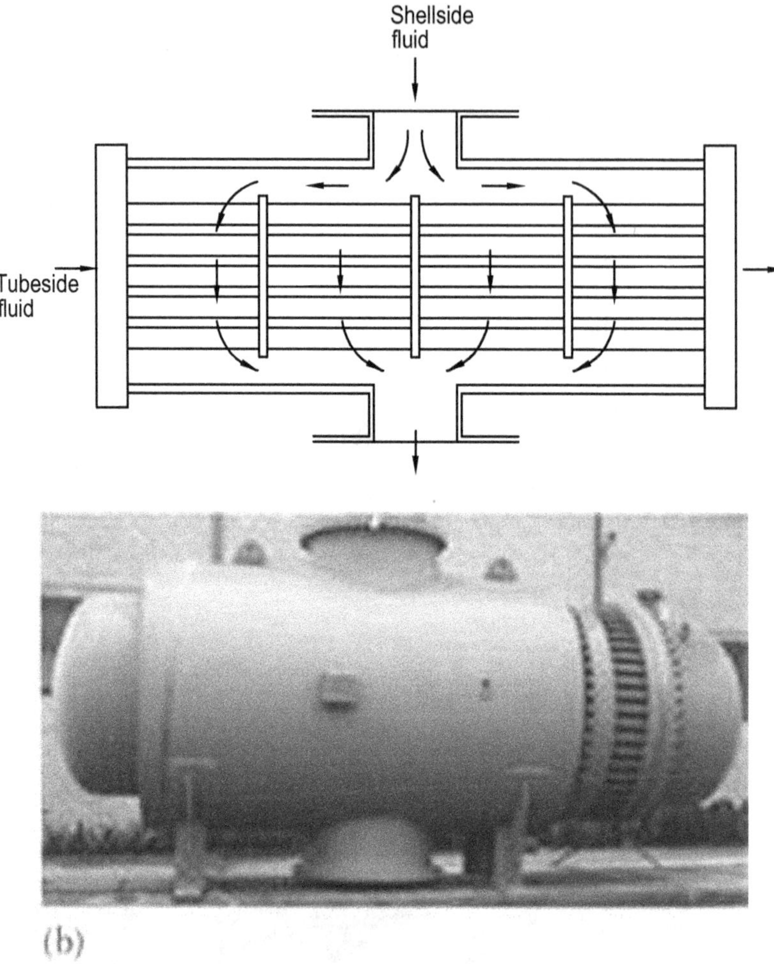

FIGURE 4.60 (a) Cross section of TEMA X shell—schematic and (b) Heat exchanger.

4.22.8 TEM *X* Shell

The *X* shell (Figure 4.60) is characterized by pure shellside crossflow. No transverse baffles are used in the *X* shell; however, support plates are used to suppress the flow-induced vibrations. It has nozzles in the middle as in the *G* shell. The shellside fluid is divided into many substreams, and each substream flows over the tube bundle and leaves through the bottom nozzle. Flow maldistributions on the shellside can be a problem unless proper provision has been made to feed the fluid uniformly at the inlet. Uniform flow distribution can be achieved by a bathtub nozzle, multiple nozzles, or keeping a clear lane along the length of the shell near the nozzle inlet [4]. The tubeside can be single pass, or two passes, either parallel crossflow or counter-crossflow. For a given set of conditions, the *X* shell has the lowest shellside pressure drop compared to all other shell types (except the *K* shell). Hence, it is used for gas heating and cooling applications and for condensing under vacuum.

4.22.9 Comparison of Various TEMA Shells

In general, *E*- and *F*-type shells are suited to single-phase fluids because of the many different baffle arrangements possible and the relatively long flow path [11]. When the shellside pressure

TABLE 4.2
Comparison of the Pressure Drop of Different TEMA Shell Types

F shell	$8\ \Delta P_{E\ shell}$
G shell	$\Delta P_{E\ shell}$
H shell	$\frac{1}{8}\ \Delta P_{E\ shell}$
J shell	$\frac{1}{8}\ \Delta P_{E\ shell}$

drop is a limiting factor, *G* and *H* shells can be used. The *G* and *H* shells are not used for shellside single-phase applications, since there is no edge over the *E* or *X* shells. They are used as horizontal thermosiphon reboilers, condensers, and in other phase-change applications. The longitudinal baffle serves to prevent flashing out of the lighter components of the shell fluid, helps flushing out noncondensables, provides increased mixing, and helps flow distribution [7]. Velocity head and pressure drop for various TEMA shells are given in Ref. [33].

4.22.9.1 Pressure Drop Comparisons among TEMA Shells

A comparison of the pressure drops of different TEMA shell types compared to TEMA *E* shell is given in Table 4.2

4.23 FRONT AND REAR HEAD DESIGNS

Head designs can vary from plain standard castings to fabricated assemblies with many special features. Two of the major considerations in the choice of heads are [6]: (1) accessibility to the tubes and (2) piping convenience. Where fouling conditions are encountered or where frequent access for inspection is desired, a head or cover plate that can be easily removed is an obvious choice. In this head, connections are located on the sides, not on the ends of removable heads. Typical open-end heads used for this purpose are called channels. They are fabricated from cylindrical shells and fitted with easily removable cover plates so that the tubes can be cleaned without disturbing piping.

4.24 BOLTED FLANGE JOINT (BFJ)

Flanged joints play a crucial role in piping systems, connecting valves with other equipment. Pipe flanges are the second-most commonly used joining mechanism after welding. Using flanges provides added flexibility, allowing easier assembly and disassembly of pipe systems. A typical flanged connection is comprised of three parts: (i) the flanges; (ii) the gaskets; and (iii) the bolting. A gasket is normally inserted between the two mating flanges to provide a tighter seal.

Proper controls must be exercised in the selection and application of all these elements to attain a joint that has acceptable leak tightness. Special techniques, such as controlled bolt tightening, are described in ASME PCC-1.

Components and materials in a BFJ include flange joints, threaded fasteners (bolts/studs, nuts, and washers), and gaskets. The goal of a BFJ is to create a tight leak-sealing load on the gasket material.

4.24.1 Heat Exchanger Gaskets

Heat exchanger gaskets come in several forms based on the type of heat exchanger or application. They can be manufactured in different sizes, shapes, with or without bars. Among the most requested styles are double-jacketed gaskets, Kammprofile, corrugated gaskets, and solid gaskets, all available in a choice of metals and filler materials. The most common heat exchanger gaskets are:

i. Double-jacketed gaskets are gaskets in which the gasket material is enclosed by an outer metal cover.
ii. Solid gaskets from metal, stamped into a gasket shape.
iii. Corrugated metal gaskets. These are metal gaskets, usually incorporating a filler material in the well of the corrugations, in which a seal is formed between the peaks of the corrugations and the mating flanges.

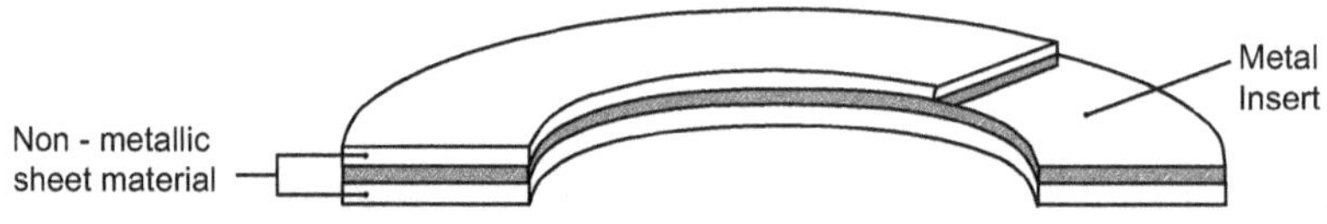

(a) Metal Reinforced non-metallic gasket (cross section).

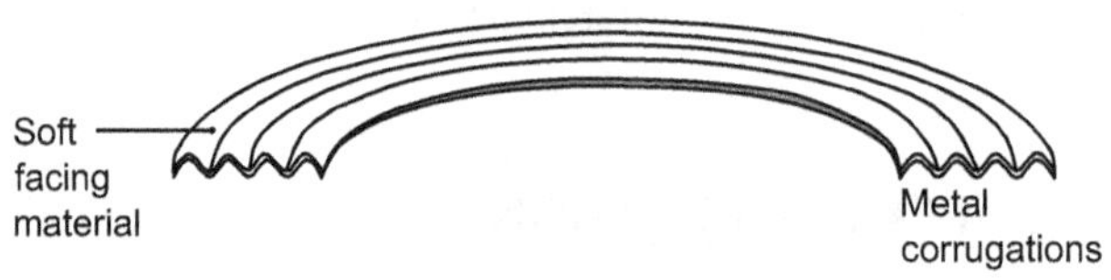

(b) Corrugated gasket.

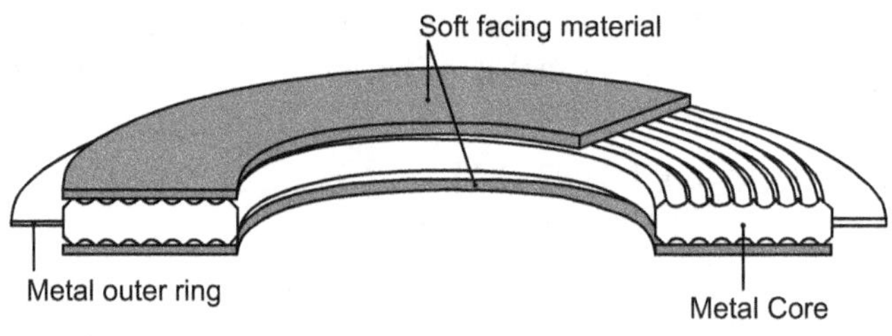

(c) Kammprofile gasket.

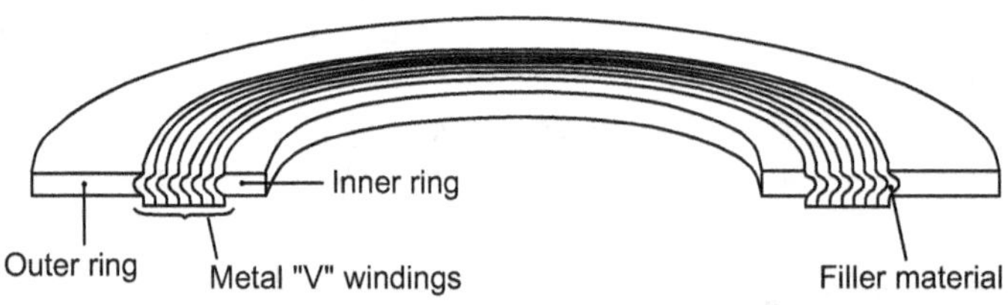

(d) Spiral wound gasket.

FIGURE 4.61 Heat exchanger gaskets profiles.

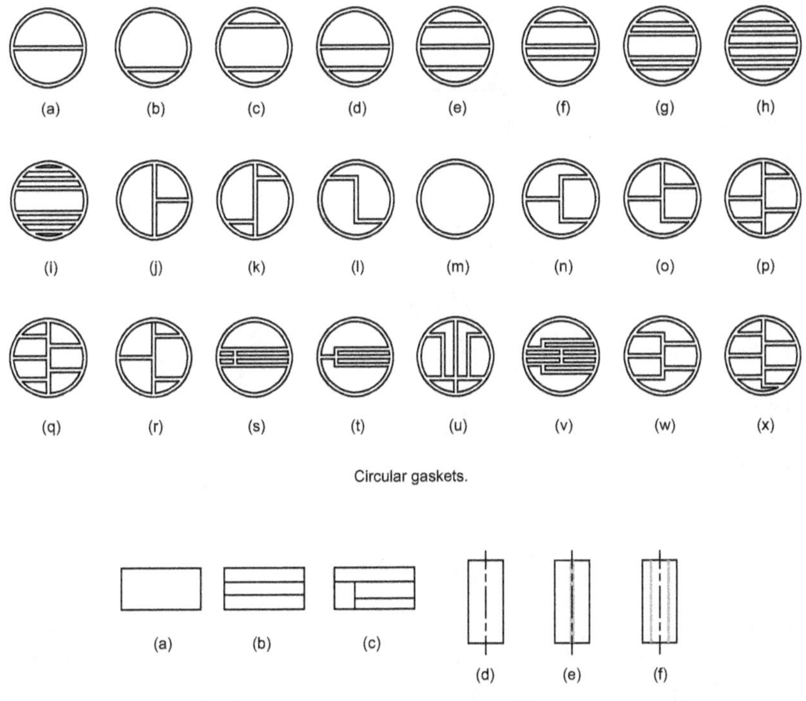

FIGURE 4.62 Profiles of heat exchanger gaskets. (Adapted from Gasket & Fastener Handbook, A Technical Guide to Gasketing & Bolted Flanges, LAMONS, Houston, Texas, pp. 1–49.)

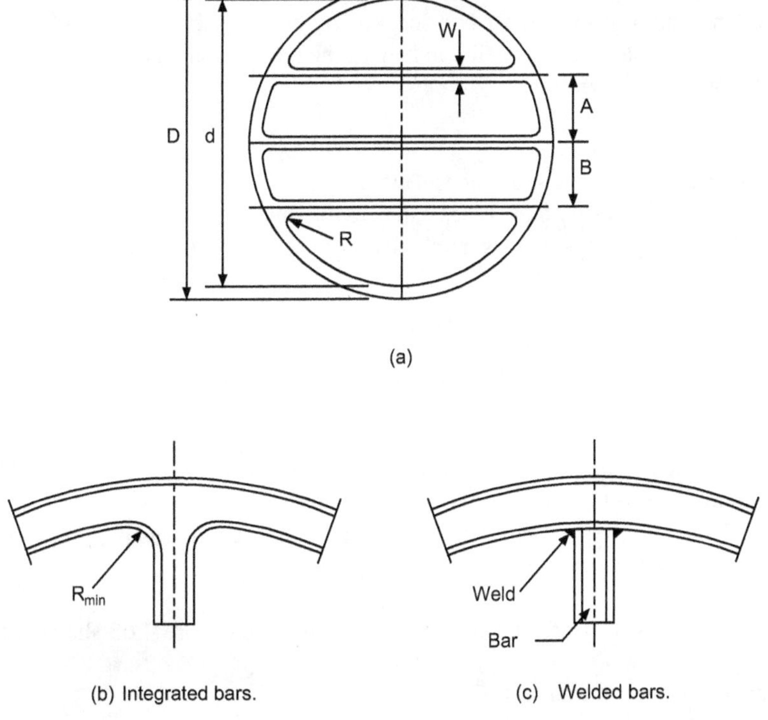

FIGURE 4.63 Metal bar heat exchanger gaskets profiles. (Adapted from https://www.tektrade.ee/assets/Uploads/Pdf/Heat-Exchanger-gaskets.pdf.)

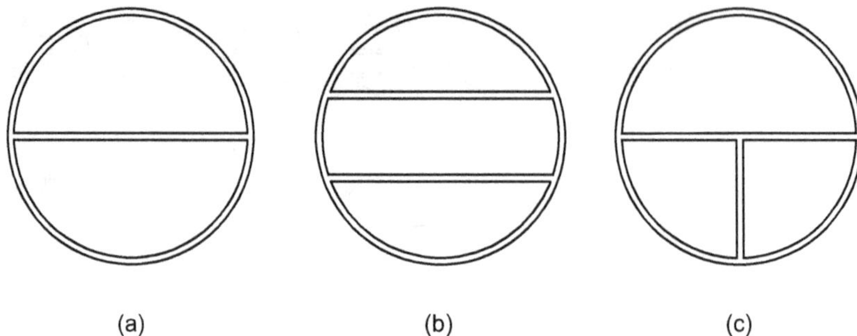

FIGURE 4.64 Welded metal bar heat exchanger gaskets profile – illustration.

iv. Kammprofile gaskets are metal gaskets with grooved/serrated faces, with or without resilient sealing layer on surfaces. These gaskets are used in areas where extreme temperatures and excessive movement due to thermal expansion exist.

The above four types of gaskets are shown in Figure 4.61 and other numerous profiles are shown in Figure 4.62.

Gaskets for heat exchangers can be manufactured with or without bars. Double-jacketed heat exchanger gaskets are created with integrated bars. A radius can be found between the bars and the internal diameter of the gaskets. Gaskets with welded bars, on the other hand, are created to eliminate the crack problems in the radius area. Figures 4.63 and 4.64 show metal bar heat exchanger gaskets.

4.24.2 Flanged Joints in Shell and Tube Heat Exchangers

Flanges are often employed to connect two sections by bolting them together so that the sections can be assembled and disassembled easily. In heat exchangers, the flange joints are used to connect together the following components:

1. Channel and channel cover
2. Heads or channels with the shell/tubesheets
3. Inlet and outlet nozzles with the pipes carrying the fluids
4. To close various openings such as manholes, peep holes, and their cover plates

4.24.3 Flanges Found in Shell and Tube Exchangers

Three types of flanges are found in shell and tube exchangers, namely,

i. Girth flanges for the shell and channel barrels
ii. Internal flanges in the floating head exchanger to allow disassembly of the internals and removal of the tube bundle
iii. Nozzle flanges where the flange and gasket standards, the size, and pressure rating will be set by the line specification

Figure 4.65 shows types of heat exchanger flanged joints and Figure 4.66 shows a few gasketed flanged joints.

A girth flange is shown in Figure 4.67.

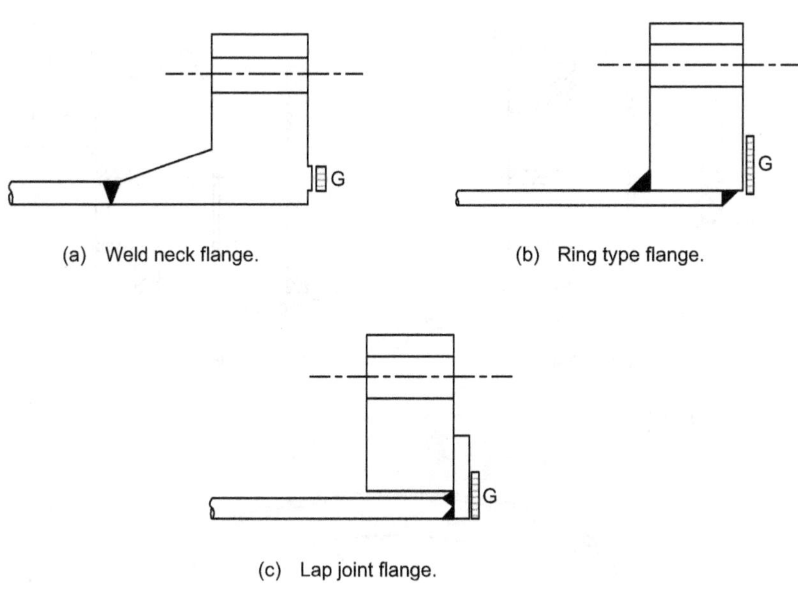

FIGURE 4.65 Heat exchanger flanged joints – illustration.

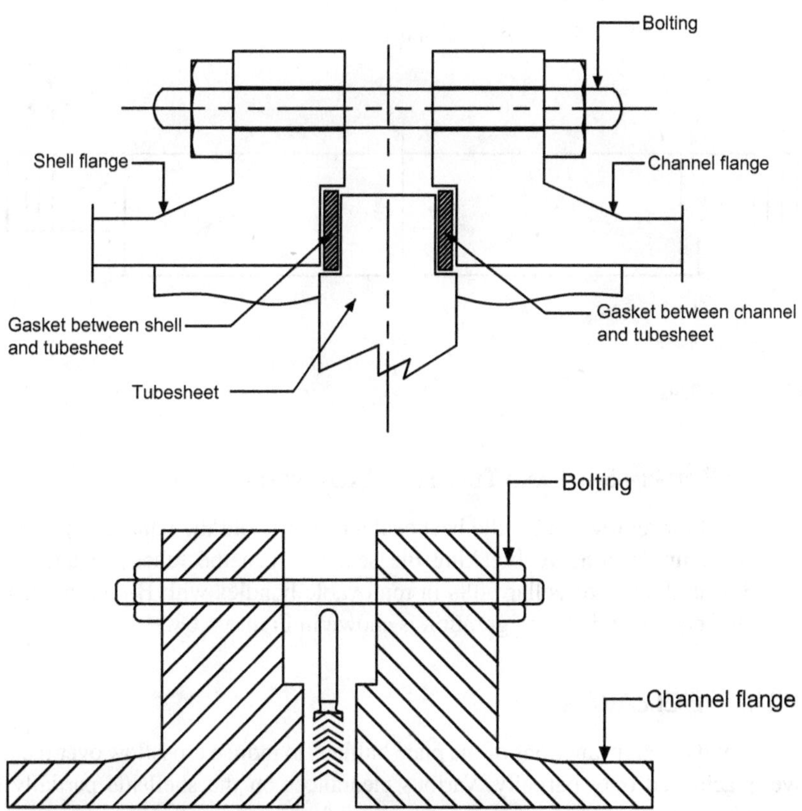

FIGURE 4.66 Gasketed flanged joints of heat exchanger – illustration.

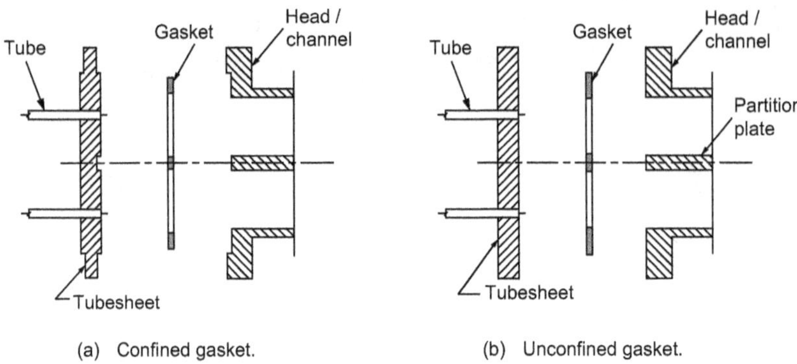

FIGURE 4.66 (Continued)

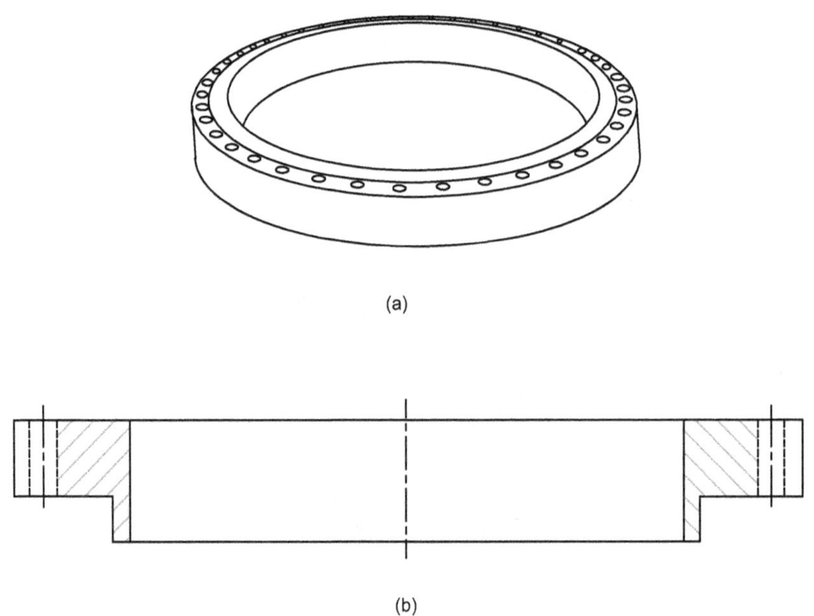

FIGURE 4.67 Grith flange.

4.24.4 Collar Bolts in Shell and Tube Heat Exchangers

Collar bolts are used for removable bundle heat exchangers to hold the bundle in place and remove the channel without interrupting or breaking the seal between the tubesheet and shell. TEMA standards recommend the use of collar bolts in removable bundles with B-type bonnet under part RCB-11.8, and the recommended configuration is shown in Figure 4.68.

4.25 SHELLSIDE CLEARANCES

Even though one of the major functions of the plate baffle is to induce crossflow over the tube bundle, this objective is achieved only partially. Various clearances on the shellside partially bypass the fluid. Bypassing is defined as a leakage flow where fluid from the crossflow stream, intended to flow through the tube bundle, avoids flowing through it by passing through an alternative, low-resistance

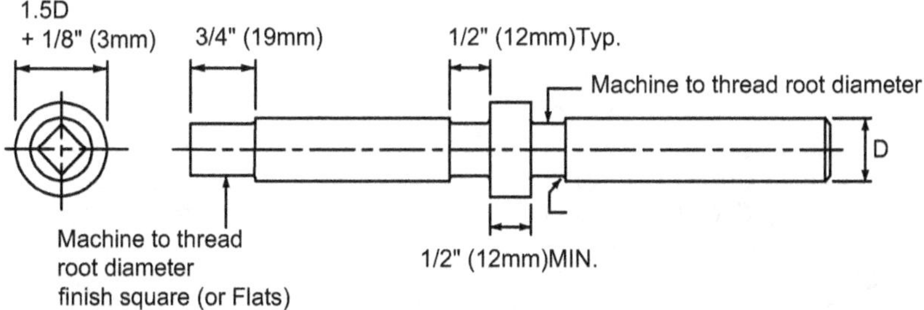

(a) Type II Collar stud (NUT type).

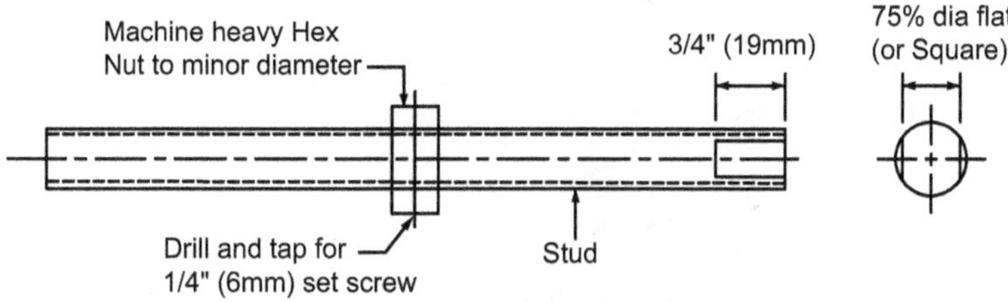

(b) Assembly and drilling details (Types I and II)

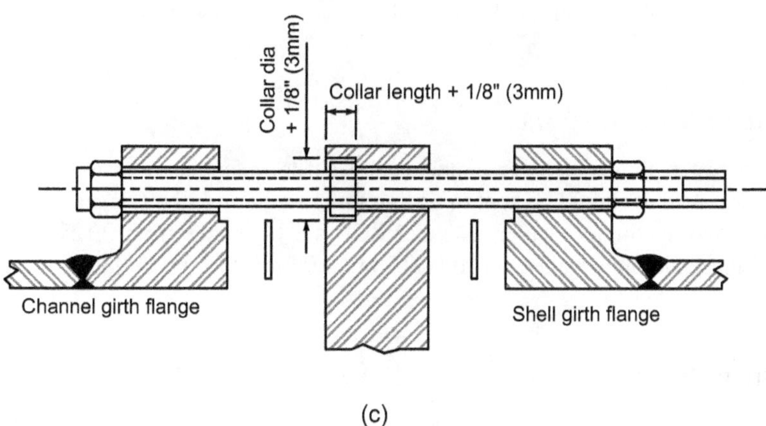

(c)

FIGURE 4.68 Collar bolts in shell and tube heat exchanger. Source: TEMA Standards [2].

flow path [34]. A major source of bypassing is the shellside clearances. Clearances are required for heat exchanger fabrication. Fabrication clearances and tolerances have been established by the TEMA, and these have become widely accepted around the world and are enforced through inspection during fabrication. Three clearances are normally associated with a plate baffle exchanger. They are (1) tube-to-baffle hole; (2) baffle to shell; and (3) bundle to shell. Additionally, in a multipass unit, the tube layout partitions may create open passages for the bypass of the crossflow stream. Since bypassing reduces heat transfer on the shellside, to achieve good shellside heat transfer, bypassing of the fluid must be reduced. These clearances and their effects on the shellside performance are discussed in detail next.

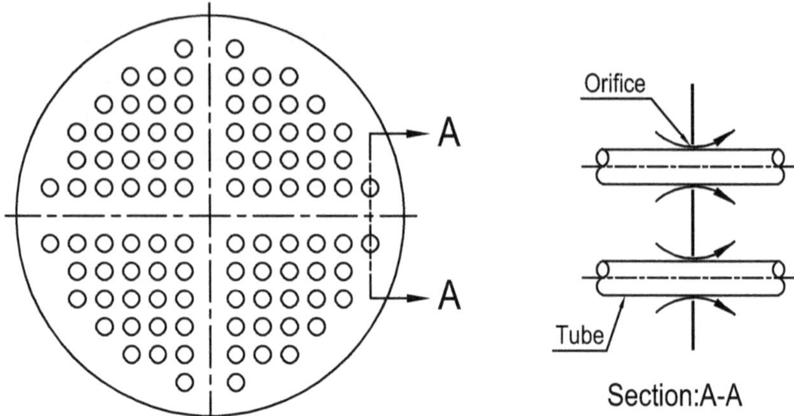

FIGURE 4.69 Tube-to-baffle leakage stream.

4.25.1 Tube-to-Baffle-Hole Clearance

The holes in the baffles must be slightly larger than the tube outer diameter to ensure easy tube insertion. The resultant clearance is referred to as the tube-to-baffle-hole clearance (Figure 4.69). It should be kept at a minimum (1) to reduce the flow-induced vibration and (2) to minimize the *A* leakage stream (various shellside bypass streams are discussed next). As per TEMA RCB-4.2, where the maximum unsupported tube length is 36 in. (914.4 mm) or less, or for tubes larger than 1.25 in. (31.8 mm) outer diameter, the clearance is 1/32 in. (0.80 mm); where the unsupported tube length exceeds 36 in. for tube outer diameter 1.25 in. or smaller, the clearance is 1/62 in. (0.40 mm).

4.25.2 Shell-to-Baffle Clearance

The clearance between the shell inner diameter and baffle outer diameter is referred to as the baffle-to-shell clearance. It should be kept to a minimum to minimize the *E* leakage stream. TEMA recommends that the maximum clearance as per RCB-4.3 [2] varies from 0.125 in. (3.175 mm) for 6 in. (152.4 mm) shell inner diameter to 0.315 in. (7.94 mm) for 60 in. (1524 mm) shell inner diameter.

4.25.3 Shell-to-Bundle Clearance

Since the tube bundle does not fill the shell, there is a shell-to-bundle clearance. This allows the so-called bypass stream *C* flowing around the bundle. Sealing strips are used to block this space and force the bypass stream to flow across the tubes. The use of sealing strips is recommended every five to seven rows of tubes in the bypass stream direction.

4.25.4 Bypass Lanes

The lanes provided for the tubeside partition ribs are a source of bypass on the shellside. The pass partition lanes are, however, placed perpendicular to the crossflow whenever possible, or the bypass lanes are usually blocked by tie-rods, which act similar to the sealing strips. Hence, the effect of the bypass lanes on thermal performance usually can be neglected.

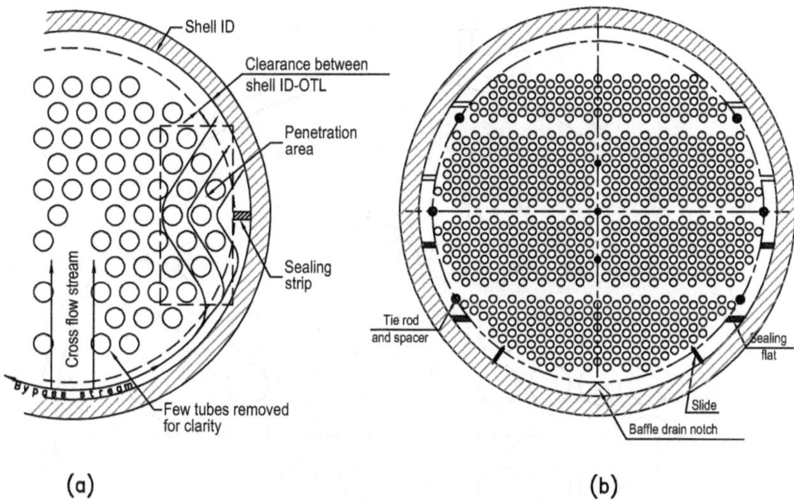

FIGURE 4.70 Sealing strips. (a) Typical flow pattern with sealing strips. (Adapted from Palen, J.W. and Taborek, J., *Chem. Eng., Symp.Ser.*, No. 92, *Heat Transfer—Philadelphia*, 65, 53, 1969.) (b) Shows attachment of sealing strip, baffle drain notch, tie-rod, and slide plate attached to the tube bundle.

4.26 SHELLSIDE FLOW PATTERN

4.26.1 Shell Fluid Bypassing and Leakage

In shell and tube exchangers with segmental plate baffles, the shellside flow is very complex due to a substantial portion of the fluid bypassing the tube bundle through various shellside constructional clearances defined earlier. Another contributing factor for bypassing is due to notches made in the bottom portion of the baffles for draining purposes. Notches are usually not required for draining because the necessary fabrication tolerances provide ample draining [7]. To achieve good shellside heat transfer, bypassing of the fluid must be reduced.

4.26.2 Bypass Prevention and Sealing Devices

Sealing devices can be employed to minimize the bypassing of fluid around the bundle or through pass partition lanes [34]. If the tube bundle-to-shell bypass clearance becomes large, such as for pull-through bundles, resulting in decreased heat transfer efficiency, the effectiveness can be restored by fitting "sealing strips." As a rule of thumb, sealing strips should be considered if the tube bundle-to-shell diameter clearance exceeds approximately 2.25 in. (30 mm). Fixed tubesheet and U-tube heat exchangers usually do not require sealing strips, but split ring and pull-through floating head designs usually require sealing strips.

Types of sealing devices: Sealing devices are strips that prevent bypass around a bundle by "sealing" or blocking the clearance area between the outermost tubes and the inside of the shell. Some common types include [9] tie-rods, sealing devices, and tie-rods.

1. Tie-rods and spacers hold the baffles in place but can be located at the periphery of the baffle to prevent bypassing.
2. Sealing strips. These are typically longitudinal strips of metal between the outside of the bundle and the shell and fastened to the baffles (Figure 4.70); they force the bypass flow back

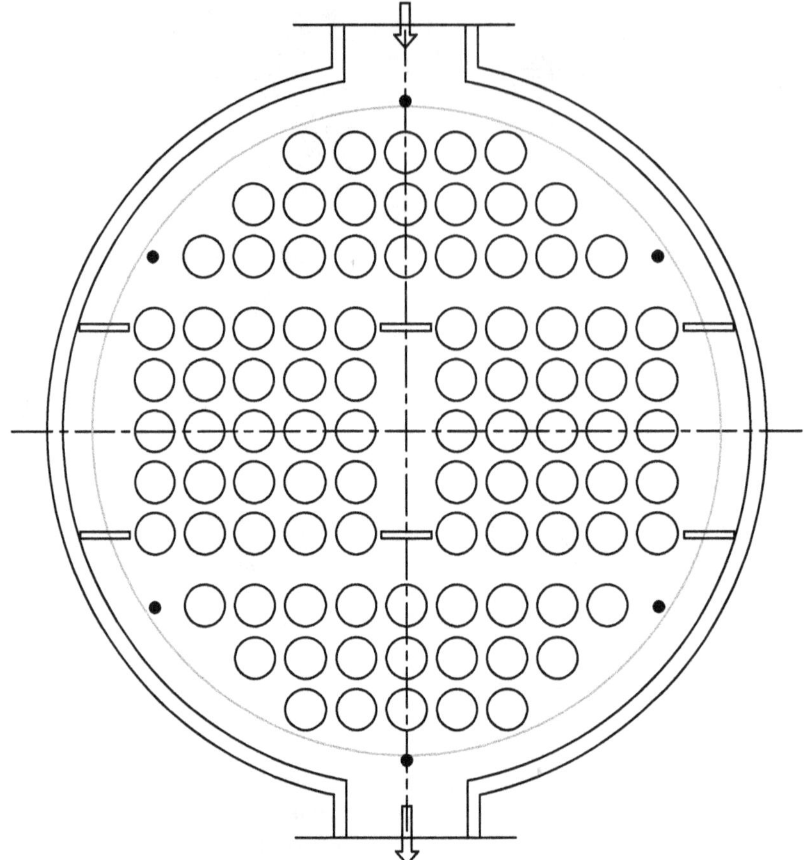

FIGURE 4.71 Shell and tube heat exchanger tube bundle assembly showing sealing devices between the shell and outer tube limit and tube bundle middle lane.

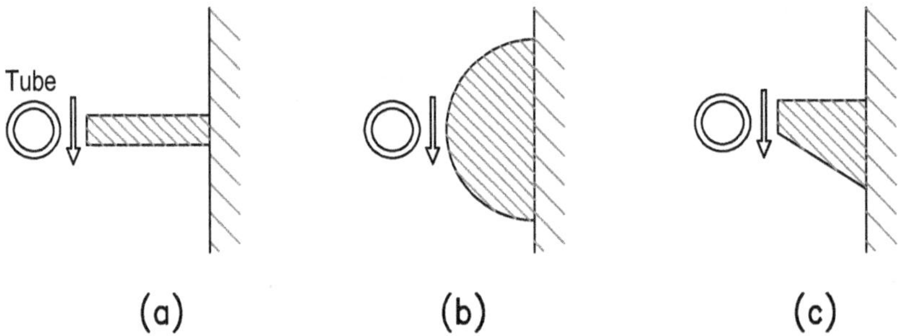

FIGURE 4.72 Forms of sealing strips. (a) Rectangular, (b) semicircular, and (c) triangular. (From Taylor, C.E. and Currie, I.G., *Trans. ASME J. Heat Transfer*, 109, 569, 1987.)

into the tube field. A typical flow pattern with sealing strips in an effective penetration area is shown in Figure 4.71.

Number of sealing strips pair in between two baffles: Bypassing of the shellside fluid can be adequately controlled by providing one sealing device for every four tube rows on the bundle periphery and by providing one sealing device for every two tube rows at bypass lanes internal

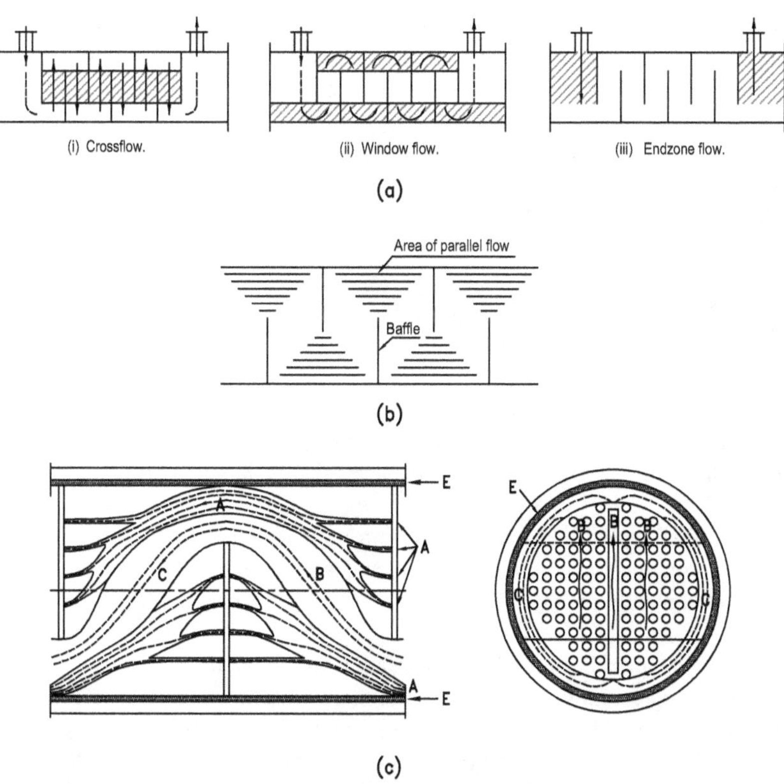

FIGURE 4.73 Shellside flow distribution. (a) Three regions of flow over the tube bundle, (b) regions of parallel flow, and (c) shows different streams flow through the tube bundle.

to the bundle such as pass partition lanes [1]. The use of sealing strips to divert flow within heat exchangers was studied by Taylor et al. [35]. They examined the variation of sealing strip shape (Figure 4.72), location, and gap size, i.e., the distance between the sealing device and the nearest tube. According to their study; (1) rectangular-shaped sealing strips are preferred; (2) sealing strips placed close together (3.6 rows apart) will provide optimum heat transfer characteristics; and (3) the most significant results were obtained when the gap size = $p - d$.

3. Tie-rods with "winged" spacers. The wings are extended longitudinal strips attached to the spacers.

Dummy tubes: Usually closed at one end, dummy tubes are used to prevent bypassing through lanes parallel to the direction of fluid flow on the shellside (Figure 4.25a). They do not pass through the tubesheet. In moderate to large exchangers, one dummy tube is as effective in promoting heat transfer as 50 process tubes [9].

4.26.3 Shellside Flow Pattern

An ideal tube bundle [the term introduced by Heat Transfer Research, Incorporated (HTRI), CA] refers to segmentally baffled circular bundles with no clearance between tubes and baffles, baffles and shell, or outer tubes and the shell, so that all fluid must flow across the tube bundle [15]. In a practical tube bundle, the total shellside flow distributes itself into a number of distinct partial streams due to varying flow resistances through the shellside clearances. This stream distribution pattern is now well established and is shown schematically in Figure 4.73. Figure 4.73a shows three regions of flow over the tube bundle, Figure 4.73b shows regions of parallelflow, and Figure 4.73c shows different streams flow through the tube bundle. This flow model was originally proposed by Tinker [36] and later modified by Palen et al. [15] for a segmentally baffled exchanger. Various streams in order of decreasing thermal effectiveness are discussed next.

A stream: This is a tube-to-baffle-hole leakage stream through the clearance between the tubes and the tube holes in the baffles (Figure 4.73c). This stream is created by the pressure difference on the sides of the baffle. As heat transfer coefficients are very high in the annular spaces, this stream is considered fully effective.

B stream: This is a crossflow stream through a tube bundle. This stream is considered fully effective for both heat transfer and pressure drop.

C stream: This is a bundle-to-shell bypass stream through the annular spaces between the tube bundle and the shell. It flows between successive baffle windows. This stream is only partially effective for heat transfer as it contacts the tubes near the tube bundle periphery.

E stream: This is a shell-to-baffle leakage stream through the clearance between the edge of a baffle and the shell. This stream is the least effective for heat transfer, particularly in laminar flow, because it may not come in contact with any tubes.

F stream: This is a tube pass partition bypass stream through open passages created by tube layout partition lanes (when placed in the direction of the main crossflow stream) in a multipass unit. This stream is less effective than the *A* stream because it comes into contact with less heat transfer area per unit volume; however, it is slightly more effective than the *C* stream.

From the earlier discussion, the fluid in streams *C*, *E*, and *F* bypasses the tubes, which reduces the effective heat transfer area. Stream *C* is the main bypass stream and will be particularly significant in pull-through bundle exchangers, where the clearance between the shell and the bundle is of necessity large. Stream *C* can be considerably reduced by using sealing strips.

4.26.4 Flow Fractions for Each Stream

Each of the streams has a certain flow fraction of the total flow such that the total pressure drop for each stream is the same. Each stream undergoes different frictional processes and has a different heat transfer effectiveness, as discussed earlier. The design of the plate baffle shell and tube exchanger should be such that most of the flow (ideally about 80%) represents the crossflow *B* stream. However, this is rarely achieved in practice. Narrow baffle spacing will result in a higher pressure drop for the *B* stream and forces more flow into the *A*, *C*, and *E* streams.

Based on extensive test data, Palen et al. [15] arrived at the flow fractions for various streams both for laminar flow and for turbulent flow. Even for a good design, the crossflow stream represents only 65% of the total flow in turbulent flow and 50% in laminar flow. Hence, the predicted performance based on the conventional LMTD method generally will not be accurate. As a result, there is no need to compute very accurate *F* factors for the various exchanger configurations [15]. If the computed values of the B stream are lower than those indicated, the baffle geometry and various clearances should be checked.

4.26.5 Shellside Performance

Many investigators have studied the shellside thermal performance. One of the earliest was Tinker [36,37]. Perhaps the most widely accepted and recognized study is known as the Bell–Delaware method [38]. The original method has been further refined by Bell [39] and presented also in Ref.

[40], and other modifications and improvements have been made by Taborek [17]. For generic details on shell and tube heat exchangers, their classification, stream analysis, and guidance for thermal design refer to Mukurjee [41] and Hewitt et al. [42].

4.27 DESIGN METHODOLOGY

4.27.1 Material Selection: ASME Code Material Requirements

All materials used for pressure-retaining parts shall be as per ASME codes.

4.27.1.1 Section II Materials

All materials used for pressure-retaining parts must meet the ASME code (ASME Boiler and Pressure Vessel Code, Section II—Material Specifications and Section VIII Div 1).

i. Part A covers Ferrous Material
ii. Part B covers Nonferrous Material
iii. Part C covers Welding Rods, Electrodes, and Filler Metals
iv. Part D covers Material Properties in both Customary and Metric units of measure

Only code-specified material should be used in vessel construction or repair. These specifications contain requirements for chemical and mechanical properties, heat treatment, manufacture, heat and product analyses, and methods of testing.

4.27.1.2 Raw Material Forms Used in the Construction of Heat Exchangers

In the construction of heat exchangers, various forms of raw materials are used. The raw material forms include plates, sheets and strips, pipes and tubes, forgings, castings, bars and rods, etc. The major use of sheets and strips is in the construction of compact heat exchangers and plate heat exchangers. Tubes are used both in compact heat exchangers and in shell and tube heat exchangers. The other forms are used extensively in the fabrication of shell and tube heat exchangers.

4.27.1.3 Material Selection for Pressure Boundary Components

A wide spectrum of materials is used in the construction of heat exchangers and pressure vessels. The materials may be metals or nonmetals such as cast iron, carbon steel, low-alloy steel, SS, martensitic, austenitic, ferritic, superferritic, duplex, superaustenitic, aluminum and aluminum alloys, copper and copper alloys, nickel and nickel alloys, titanium and titanium alloys, zirconium, tantalum, graphite, glass, Teflon, ceramics, silicon carbide, composites, etc.

4.27.1.4 Tubes

The material selection is based on the overall consideration and calculation of working pressure, temperature, flow rate, corrosion, erosion, workability, cost efficiency, viscosity, design, and other environments. A material and corrosion allowance for both the shellside and the tubeside of the exchanger are specified. Tubes are often designed with a life span significantly shorter than that of the shell or channel, since they can be retubed. The tube material must be compatible with the tubesheet material. To avoid galvanic corrosion due to an unfavorable area ratio, the tube must be cathodic to the tubesheet material.

Tubing Materials. Usually, the heat exchanger tubing can be furnished in ferrous or nonferrous metal materials. Tubes can be bare metal, finned tube or enhanced tube. Generally the most commonly selected two metals are aluminum and copper. Others include carbon steel, copper, admiralty brass, cupronickel, 304/304L SS, 316L SS, Duplex 2205, AL6XN, Duplex 2507, Hastelloy C-276,

Hastelloy C22, Nickel 200, Monel 400, Inconel, Alloy 625, titanium, zirconium and tantalum, glass, graphite, Teflon, etc. for corrosive services.

ASTM specifications for some ferrous alloy tubings for heat exchanger tubes and condenser tubes and boiler, superheater and feedwater tubes are provided in Chapter 2, Volume 2, Heat Exchangers.

4.27.1.5 Tubing Forms

The tubing forms used for shell and tube or air-cooled heat exchangers include straight tube non-finned (prime surface), straight tube finned, straight tube skip finned, rolled fins, straight tube skip finned, rolled fins, internally and externally integral finned, u-bend tubes, twisted tube, helical tubes, etc.

4.27.1.6 Tube Testing and Inspection

Standard testing and inspection of heat exchanger tubes usually include visual examination, dimensional inspection, eddy current test, hydrostatic pressure testing, pneumatic air–underwater testing, magnetic particle test, ultrasonic test, corrosion tests, mechanical tests (including tensile, flaring, flattening, and reverse flattening testing), chemical analysis (PMI), and X-ray inspection of the welds (if any).

4.27.1.7 Tubing Functional Requirements

For heat exchanger applications, the tubing must meet requirements such as (1) the tube must withstand the pressures from both the shellside and the tubeside at the operating temperature; (2) the tube must be corrosion resistant to both shellside and tubeside fluids; (3) the tube and the tubesheet materials shall be compatible to form leak-free tube-to-tubesheet joints either by rolling-in or by welding; and (4) the tube must achieve optimum economy in selection, fabrication, and service.

4.27.1.8 Standard Testing for Tubular Products

Standard tests for tubular products are visual examination; the eddy current test (ASTM E-426); hydrostatic pressure testing including pneumatic air–underwater testing (ASTM B-338/ASME SB-338); the magnetic particle test; an ultrasonic test (ASTM E-213); corrosion tests; mechanical tests, including tensile, flaring, flattening, and reverse flattening; metallographic testing; and dimensional tolerance testing. Hydrostatic pressure tests, including pneumatic air–underwater tests, corrosion tests, and dimensional tolerance tests, are discussed next.

1. Hydrostatic Pressure Testing

Hydrostatic testing is the most common test performed on tubing at the mill and is required on ASME code exchangers. According to ASTM standards, each tube shall withstand a hydrostatic pressure sufficient to subject the material to a fiber stress of 48 MPa (7000 psi) determined by the following equation for thin-walled cylinder under internal pressure:

$$P = \frac{2St}{D - 0.8t} \quad \text{or} \quad P = \frac{2SEt}{D - 0.8t} \quad \left(\text{ASME Code procedure}\right) \tag{4.7}$$

where

P is the hydrostatic pressure (MPa or psi)
S is the code-allowable stress of the tube material (MPa or psi)
E is the weld joint efficiency that is equal to 0.85 for welded tube and 1.00 for seamless tube
D is the tube outer diameter (mm or in.)
t is the thickness of the tube wall (mm or in.)

The tube should not show any evidence of leakage. The tube need not be tested at a hydrostatic pressure more than 6.9 MPa (1000 psi), unless so specified.

2. Pneumatic Test

The pneumatic air–underwater test is a superior test for detecting through-wall defects that result from tube manufacturing, compared to the hydrostatic test. Air pressure up to 1.0 MPa (150 psi) is applied to the tubing submerged in water, and the tube is observed for leaks for 5 s. This test method is dependent solely on visual observation of leaks and not on pressure drop, which is an indication of leaks during hydrostatic testing. Extremely small through-wall defects may not be detected by air passage.

4.27.1.9 Tubesheets

Most tubesheets are made from rolled plates or forged ingots. Tubesheets of small production line units may be cast. Commonly used tubesheet materials are as follows:

Carbon steel
SS, Type 304—S30400, 304L S30403
SS, Type 316—S31600, 316L S31603
Super alloy
Aluminum bronze—C61400
Muntz metal—C36500
90-10 Copper nickel—C70600
70-30 Copper nickel—C71500
Alloy 904L N08904, Alloy 825 N08825, Alloy 625 N06625, Alloy C22, Alloy C276, N10276, Alloy 400 N04400
Titanium, Grade 2, Titanium Grade 3, Titanium Grade 9, Titanium Grade 12

4.27.1.10 Baffles

Specify a material to match the nominal chemistry of the tube material and shellside fluid. If the shellside conditions favor galvanic coupling, or if shellside fluid is not conducive to baffles, corrosion rates would cause the baffles to fail before the tubes.

4.28 GUIDELINES FOR STHE DESIGN

The basic criterion that a given or designed heat exchanger should satisfy is that it should perform the given heat duty within the allowable pressure drop. The design is also to satisfy additional criteria such as [43]:

1. Withstand operating conditions, start-up, shutdown, and upset conditions that influence the thermal and mechanical design
2. Maintenance and servicing
3. Multiple shell arrangement
4. Cost
5. Size limitations

Shipping and handling may dictate restrictions on the overall size or weight of the unit, resulting in multiple shells for an application.

4.28.1 Tips for Thermal Design

1. Always specify counterflow operation for maximum performance.
2. The heat exchanger can be mounted vertically if required; however, this may require special mounting brackets.
3. If the sizing procedure suggests a nonstandard shell length, use the next larger size or consider using two shorter heat exchangers.
4. Consider using two or more heat exchangers when the heat duty is on the higher side or space for locating the heat exchanger is limited, or temperature and flow parameters are outside the recommended range for a single heat exchanger.

4.28.2 Specify the Right Heat Exchanger

When specifying an exchanger for design, various factors to be considered or questions that should be raised are listed by Gutterman [4]. A partial list includes the following:

1. Type of heat transfer, i.e., boiling, condensing, or single-phase heat transfer.
2. Since the heat exchanger has two pressure chambers, which chamber should receive the cold fluid?
3. More viscous fluid shall be routed on the shellside to obtain better heat transfer.
4. It is customary to assign the higher pressure to the tubeside to minimize shell thickness.
5. Consider various potential and possible upset conditions in assigning the design pressure and/or design temperature.
6. Pass arrangements on the shellside and tubeside to obtain maximum heat transfer?
7. Have you considered the tube size and the thickness?
8. What is the acceptable pressure drop on the tubeside and the shellside? Is the sum of the pumping cost and the initial equipment cost minimized?
9. Have you considered the maximum allowable pressure drop to obtain the maximum heat transfer?
10. Are the tubeside and shellside velocities high enough for good heat transfer and to minimize fouling but well below the limits that can cause erosion–corrosion on the tubeside, and impingement attack and flow-induced vibration on the shellside?
11. Have you considered the nozzle sizes and adequate shell escape area? Are the nozzle orientations consistent with the tube layout pattern?
12. Is the baffle arrangement designed to promote good flow distribution on the shellside and hence good heat transfer, and to minimize fouling and flow-induced vibration?
13. Does the design provide for efficient expulsion of noncondensables that may degrade the performance? (A prime example in this category is surface condensers.)
14. Is the service corrosive or dirty? If so, have you specified corrosion-resistant materials and reasonable fouling factors?
15. Does the design minimize fouling?
16. Do you want to remove the bundle? If so, are adequate space and handling facilities available for tube bundle removal?
17. Is leakage a factor to be considered? If so, did you specify (a) seal-welded tube joint; (b) rolled joint; (c) strength welded joint? Is the tube wall thickness adequate for welding? Are you specifying tube holes with grooves or without grooves?
18. What kinds of tests do you specify to prove the tube-to-tubesheet joint integrity?

All of these and numerous other factors determine the type of exchanger to be specified.

4.29 SIZING OF SHELL AND TUBE HEAT EXCHANGERS

The design of shell and tube heat exchangers involves determination of the heat transfer coefficient and pressure drop on both the tubeside and the shellside. A large number of methods [37,44–46] are available for determining the shellside performance. Since the Bell–Delaware method is considered the most suitable open-literature method for evaluating shellside performance, the method is described here. Before discussing the design procedure, tips for thermal design, and heat transfer coefficient and pressure drop are discussed, some guidelines for shellside design and points to be raised while specifying a heat exchanger are listed, followed by preliminary sizing of a shell and tube heat exchanger.

4.29.1 Guidelines for Shellside Design

Recommended guidelines for shellside design include the following [1]:

1. Accept TEMA fabrication clearances and tolerances and enforce these standards during fabrication.
2. For segmental baffles employ 20% baffle cuts.
3. Employ NTIW design to eliminate the damage from flow-induced vibration.
4. Evaluate heat transfer in the clean condition and pressure drop in the maximum fouled condition.
5. Employ sealing devices to minimize bypassing between the bundle and shell for pull-through floating heat exchanger and through pass partition lanes.
6. Ratio of baffle spacing to shell diameter may be restricted to values between 0.2 and 1.0. Baffle spacing much greater than the shell diameter must be carefully evaluated.
7. Avoid shell longitudinal baffles that are not welded to the shell; all other sealing methods are inadequate.

4.29.2 Heat Transfer Coefficient and Pressure Drop

4.29.2.1 Heat Transfer Coefficient

The tubeside heat transfer coefficient is a function of the Reynolds number, the Prandtl number, and the tube diameter. The earlier-mentioned nondimensional numbers are based on the fundamental parameters such as fluid physical properties (i.e., viscosity, thermal conductivity, and specific heat), tube diameter, and mass velocity. For turbulent flow, the tubeside heat transfer coefficient varies to the 0.8 power of tubeside mass velocity (refer to Eqns. 4.75, 4.76a, and 4.76b) . Also, the variation in liquid viscosity has a dramatic effect on the heat transfer coefficient.

The shellside heat transfer coefficient is dependent on shellside parameters like fluid velocity, shell inside diameter, baffle cut, baffle spacing, tube hole clearance in the baffle plate, various shellside clearances other than the desired fluid flow path, number of passes on shellside, shell type, etc.

4.29.2.2 Pressure Drop

Fluid pressure drop is controlled by a wide variety of design variables and the process fluid flow parameters. For the tubeside stream, the controlling variables are tube diameter, tube length, tube geometry (straight or U-tubes), number of tube passes and number of shells in series or parallel, and nozzle size. For turbulent flow, tubeside pressure drop is proportional to the square of velocity. Consequently, there will be an optimum mass velocity above which it will be wasteful to increase the velocity further. In addition to higher pumping costs, very high velocities lead to erosion. However, the pressure-drop limitation usually becomes the controlling factor long before the erosive velocity limits are attained. The minimum recommended liquid velocity inside tubes is 1.0 m/s, while the maximum is 2.5–3.0 m/s or higher for cupronickel and titanium. Since pressure drop depends on the total length of travel, as the number of tube passes increases, for a given number of tubes and a given tubeside flow rate, the pressure drop rises dramatically. Shellside fluid pressure drop is influenced by equipment design variables such as tube diameter, tube pitch, tube layout, shell diameter, baffle type, baffle cut, baffle spacing, number and size of shellside nozzles, and the exchanger shell type.

4.29.3 TEMA Specification Sheet

The TEMA specification sheet (both in metric and in FPS-English units) for shell and tube heat exchangers is shown in Figure 4.74. It is used as a basic engineering document in the thermal design, purchasing, and construction of shell and tube heat exchangers.

4.30 THERMAL DESIGN PROCEDURE

The overall design procedure of a shell and tube heat exchanger is quite lengthy, and hence it is necessary to break down this procedure into distinct steps:

1. Approximate sizing of shell and tube heat exchanger
2. Evaluation of geometric parameters, also known as auxiliary calculations
3. Correction factors for heat transfer and pressure drop
4. Shellside heat transfer coefficient and pressure drop
5. Tubeside heat transfer coefficient and pressure drop
6. Evaluation of the design, i.e., comparison of the results with the design specification

In this section, approximate sizing of the shell and tube exchanger by Bell's method is discussed first; then this is extended to size estimation, and subsequently the rating is carried out as per the Bell–Delaware method. Finally, the rated unit is evaluated.

Bell's method [43] for approximate sizing of a shell and tube heat exchanger: The approximate design involves arriving at a tentative set of heat exchanger parameters, and if the design is accepted after rating, then this becomes the final design. Various stages of approximate design include the following:

1. Compute overall heat transfer coefficient
2. Compute heat transfer rate required
3. Compute the heat transfer area required
4. Design the geometry

A flowchart for approximate sizing is given in Figure 4.75.

Estimation of heat load: The heat load is calculated in the general case from

$$q = M_h c_{p,h} \left(T_{h,i} - T_{h,o}\right) = M_c c_{p,c} \left(T_{c,o} - T_{c,i}\right) \tag{4.8}$$

Heat Exchanger Specification Sheet (FPS Units)

1				Job No.		
2	Customer			Reference No.		
3	Address			Proposal No.		
4	Plant Location			Date	Rev.	
5	Service of Unit			Item No.		
6	Size	Type	(Hor / Ver)	Connected in	Parallel	Series
7	Surf / Unit (Gross / Eff.)		Sq.ft ; Shells / Unit	Surf / Shell (Gross / Eff.)		Sq.ft
8	PERFORMANCE OF ONE UNIT					
9	Fluid Allocation		Shell Side		Tube Side	
10	Fluid Name					
11	Fluid Quantity Total	lb / hr				
12	Vapor (In / Out)					
13	Liquid					
14	Steam					
15	Water					
16	Noncondensable					
17	Temperature (In / Out)	°F				
18	Specific Gravity					
19	Viscosity, Liquid	cP				
20	Molecular Weight, Vapor					
21	Molecular Weight, Noncondensable					
22	Specific Heat	BTU / lb °F				
23	Thermal Conductivity	BTU ft / hr sq.ft °F				
24	Latent Heat	BTU / lb @ °F				
25	Inlet Pressure	psia				
26	Velocity	ft / sec				
27	Pressure Drop, Allow. / Calc.	psi	/		/	
28	Fouling Resistance (Min.)	hr sq ft °F / BTU				
29	Heat Exchanged		BTU / hr MTD (Corrected)			°F
30	Transfer Rate, Service		Clean			BTU / hr sq ft °F
31	CONSTRUCTION OF ONE SHELL				Sketch (Bundle/Nozzle Orientation)	
32			Shell Side	Tube Side		
33	Design / Test Pressure	psig	/	/		
34	Design Temp. Max/Min	°F	/	/		
35	No. Passes per Shell					
36	Corrosion Allowance	In				
37	Connections	In				
38	Size &	Out				
39	Rating	Intermediate				
40	Tube No. OD	in; Thk (Min/Avg)	in; Length	ft; Pitch	in	◄30 ▲60 ■90 ◆45
41	Tube Type			Material		
42	Shell	ID	OD in	Shell Cover	(Integ.)	(Remov.)
43	Channel or Bonnet			Channel Cover		
44	Tubesheet - Stationary			Tubesheet - Floating		
45	Floating Head Cover			Impingement Protection		
46	Baffles - Cross	Type		%Cut (Diam. / Area)	Spacing: c/c	Inlet in
47	Baffles - Long			Seal Type		
48	Support - Tube		U-Bend		Type	
49	Bypass Seal Arrangement			Tube-to-Tubesheet Joint		
50	Expansion Joint			Type		
51	ρv^2 - Inlet Nozzle		Bundle Entrance		Bundle Exit	
52	Gaskets - Shell Side			Tube Side		
53	Floating Head					
54	Code Requirements				TEMA Class	
55	Weight / Shell		Filled with Water		Bundle	
56	Remarks					
57						
58						

FIGURE 4.74 (a) TEMA specification sheet for shell and tube heat exchanger—FPS units. (b) TEMA specification sheet for shell and tube heat exchanger—metric system units. (© Standards of Tubular Exchanger Manufacturers Association, Inc., 10th edn., 2019.)

Heat Exchanger Specification Sheet (MKS Units)

1				Job No.	
2	Customer			Reference No.	
3	Address			Proposal No.	
4	Plant Location			Date	Rev.
5	Service of Unit			Item No.	
6	Size	Type	(Hor / Ver)	Connected in	Parallel Series
7	Surf / Unit (Gross / Eff.)		Sq.m ; Shells / Unit	Surf / Shell (Gross / Eff.)	Sq.m
8	PERFORMANCE OF ONE UNIT				
9	Fluid Allocation		Shell Side	Tube Side	
10	Fluid Name				
11	Fluid Quantity Total	kg / Hr			
12	Vapor (In / Out)		|	|	
13	Liquid		|	|	
14	Steam		|	|	
15	Water		|	|	
16	Noncondensable		|	|	
17	Temperature (In / Out)	°C	|	|	
18	Specific Gravity		|	|	
19	Viscosity, Liquid	cP	|	|	
20	Molecular Weight, Vapor		|	|	
21	Molecular Weight, Noncondensable		|	|	
22	Specific Heat	J / kg °C	|	|	
23	Thermal Conductivity	W / m °C	|	|	
24	Latent Heat	J / kg @ °C			
25	Inlet Pressure	kPa (abs.)			
26	Velocity	m / sec			
27	Pressure Drop, Allow, / Calc.	kPa	/	/	
28	Fouling Resistance (Min.)	Sq m °C / W			
29	Heat Exchanged		W / MTD (Corrected)		°C
30	Transfer Rate, Service		Clean		W / Sq m °C
31	CONSTRUCTION OF ONE SHELL				Sketch (Bundle/Nozzle Orientation)
32			Shell Side	Tube Side	
33	Design / Test Pressure	kPag	/	/	
34	Design Temp. Max/Min	°C	/	/	
35	No. Passes per Shell				
36	Corrosion Allowance	mm			
37	Connections	In			
38	Size &	Out			
39	Rating	Intermediate			
40	Tube No.	OD mm;Thk (Min/Avg)	mm; Length	mm; Pitch	mm ◄30 ▲60 ⊟90 ◆45
41	Tube Type		Material		
42	Shell	ID OD mm	Shell Cover	(Integ.)	(Remov.)
43	Channel or Bonnet		Channel Cover		
44	Tubesheet - Stationary		Tubesheet - Floating		
45	Floating Head Cover		Impingement Protection		
46	Baffles - Cross	Type	%Cut (Diam. / Area)	Spacing: c/c	Inlet mm
47	Baffles - Long		Seal Type		
48	Support - Tube	U-Bend		Type	
49	Bypass Seal Arrangement		Tube-to-Tubesheet Joint		
50	Expansion Joint		Type		
51	ρv^2 - Inlet Nozzle	Bundle Entrance		Bundle Exit	
52	Gaskets - Shell Side		Tube Side		
53	Floating Head				
54	Code Requirements			TEMA Class	
55	Weight / Shell	Filled with Water		Bundle	kg
56	Remarks				
57					
58					

FIGURE 4.74 (Continued)

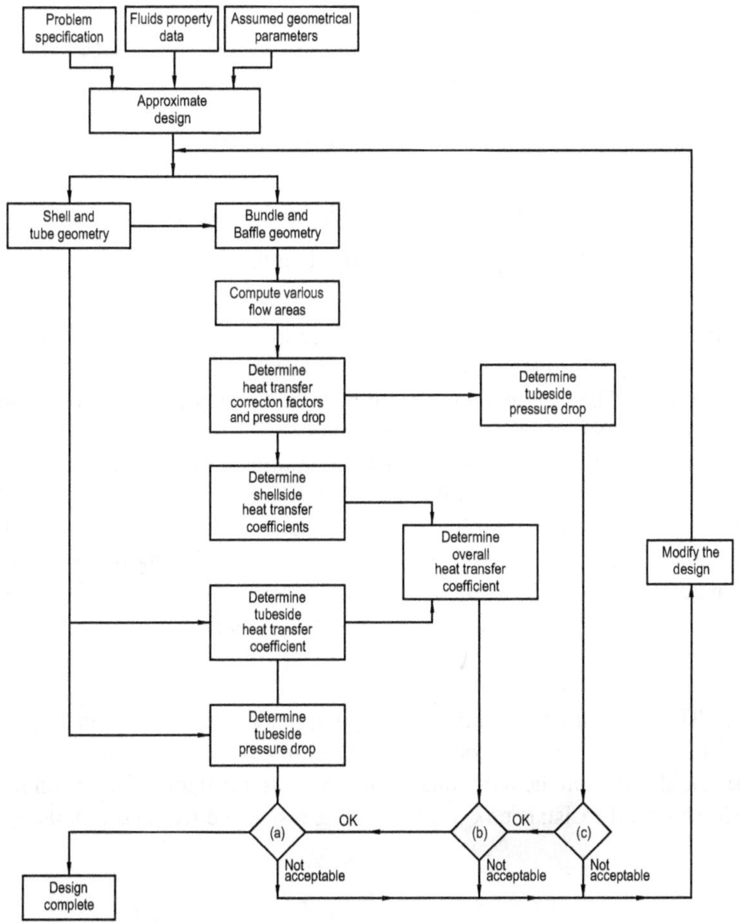

FIGURE 4.75 Flowchart for approximate sizing of STHE.

where

$c_{p,h}$ and $c_{p,c}$ are the specific heats of the hot and cold fluids
$T_{h,i}$ and $T_{h,o}$ are the inlet and outlet temperatures of the hot stream
$T_{c,i}$ and $T_{c,o}$ are the inlet and outlet temperatures of the cold stream

Estimation of log mean temperature difference: Determine the logarithmic mean temperature difference for countercurrent flow using the temperatures as defined earlier:

$$\text{LMTD} = \frac{\left(T_{h,i} - T_{c,o}\right) - \left(T_{h,o} - T_{c,i}\right)}{\ln\left[\left(T_{h,i} - T_{c,o}\right) / \left(T_{h,o} - T_{c,i}\right)\right]} \tag{4.9}$$

LMTD correction factor: Values of F can be found from the thermal relation charts given in Chapter 2 for a variety of heat exchanger flow configurations. However, for estimation purposes, a reasonable estimate may often be obtained without resorting to the charts.

1. For a single tube pass, purely countercurrent heat exchanger, $F = 1.0$.
2. For a single shell with any even number of tubeside passes, F should be between 0.8 and 1.0. (For other shell types refer to Chapter 2.)

Method to determine number of shells: Quickly check the limits:

$$2T_{h,o} \geq T_{c,i} + T_{c,o} \quad \text{hot fluid on the shell side}$$

$$2T_{c,o} \leq T_{h,i} + T_{h,o} \quad \text{cold fluid on the shell side}$$

If these limits are approached, it is necessary to use multiple 1 – 2*N* shells in series. There is a rapid graphical technique for estimating a sufficient number of 1 – 2*N* shells in series. The procedure is discussed here. The terminal temperatures of the two streams are plotted on the ordinates of an arithmetic graph paper sheet, as shown in Figure 4.76. The distance between the ordinates is arbitrary. Starting with the cold fluid outlet temperature, a horizontal line is laid off until it intercepts the hot fluid line. From that point, a vertical line is drawn to the cold fluid temperature. The process is repeated until a vertical line intercepts the cold fluid operating line at or below the cold fluid inlet temperature. The number of horizontal lines (including the one that intersects the right-hand ordinate) is equal to the number of shells in series that is clearly sufficient to perform the duty. Following this procedure will usually result in a number of shells having an overall F close to 0.8.

Estimation of U: The greatest uncertainty in preliminary calculations is estimating the overall heat transfer coefficient. Approximate film coefficients for different types of fluids are given by Bell [43]. Thus, U can be calculated from the individual values of heat transfer coefficient on the shellside (h_s) and the tubeside (h_t), wall resistance (k_w), and fouling resistance (R_{fo} and R_{fi}), using the following equation:

$$U = \frac{1}{\left[(1/h_s) + R_{fo} + (t_w/k_w)(A_o/A_m) + (R_{fi} + (1/h_t))A_o/A_i\right]} \tag{4.10}$$

where t_w is the wall thickness and A_m is the effective mean wall heat transfer area, which is approximated by the arithmetic mean, using the outside and inside radii, r_o and r_i:

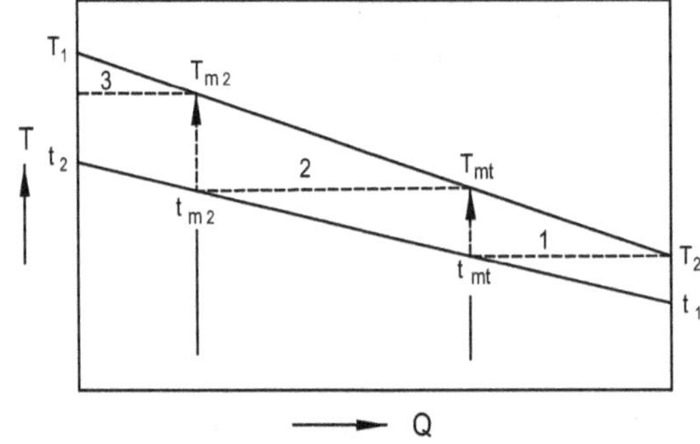

FIGURE 4.76 Procedure to determine the number of shells. *Note:* T_1 and T_2 are shellside terminal temperatures and t_1 and t_2 are tubeside terminal temperatures.

$$A_m = \pi L(r_o + r_i) \tag{4.11a}$$

$$\text{For a bare tube } \frac{A_o}{A_i} \to \frac{r_o}{r_i} \tag{4.11b}$$

$$\frac{A_o}{A_m} \to \frac{r_o}{r_o + r_i} \tag{4.11c}$$

Heat transfer coefficient for finned tubes: Bell [43] suggests that the values given for plain tubes can be usually used with caution for low-finned tubes if the controlling resistance is placed on the shellside; the values should be reduced by 10–30% if the shellside fluid is of medium or high viscosity, and by 50% if the shellside fluid is of high viscosity and is being cooled. Whitley et al. [13] suggest reducing the value for finned tubes to 90% of those of plain tubes.

Fouling resistance: Fouling resistance values may be chosen from TEMA tables in RGP-Section 10. Also refer to Chapter 2 of Volume 3 for more details on fouling resistances.

Calculation of A_o: Once q, U, LMTD, and F are known, the total outside heat transfer area (including fin area) A_o is readily found from the following equation:

$$A_o = \frac{q}{UF(\text{LMTD})} \tag{4.12}$$

Determination of shell size and tube length from heat transfer area, A_o (after Taborek [28]): The problem now arises of how to interpret the value of A in terms of tube length and shell diameter, when both values are not known. If the problem specification specifies the limitation on shell length and diameter, the problem can be simplified. In the absence of these values, A_o is given by

$$A_o = \pi d L_{ta} N_t \tag{4.13}$$

and for estimation purposes, the tube count N_t is given in terms of tube pitch, L_{tp}, by

$$N_t = \frac{0.78 D_{ctl}^2}{C_1 L_{tp}^2} \tag{4.14}$$

where C_1 is the tube layout constant given by

$$C_1 = 0.86 \quad \text{for} \quad \theta_{tp} = 30°$$

$$C_1 = 1.0 \quad \text{for} \quad \theta_{tp} = 45° \quad \text{and} \quad 90$$

Substituting Equation 4.14 into Equation 4.13, the resulting equation is given by

$$A_o = (0.78\pi)\frac{d}{C_1 L_{tp}^2}\left[L_{ta} D_{ctl}^2\right] \tag{4.15}$$

In Equation 4.15, the first term is a constant; the second term reflects the tube size and the tube layout geometry; and the third term includes the values of tube length and shell diameter (known as aspect ratio), which are the items to be determined. Heat transfer surface A_o can be obtained by various combinations of the parameters L_{ta} and D_{ctl} for any given tube layout pattern. An initially assumed aspect ratio of 8 is suggested. Some tube count tables are available in Refs. [12,14,23], and there is a tube count chart in Ref. [43] for various tube layout patterns, tube diameters, and shell diameters. This helps to calculate A_o easily. If such a source is not available, the designer must assume a rational tube length and calculate the corresponding diameter D_{ctl} and finally the shell inside diameter D_s.

4.31 DETAILED DESIGN METHOD: BELL–DELAWARE METHOD

Designing a shell and tube heat exchanger with the Bell–Delaware method is explained here. A flowchart for designing with the Bell–Delaware method is shown in Figure 4.77.

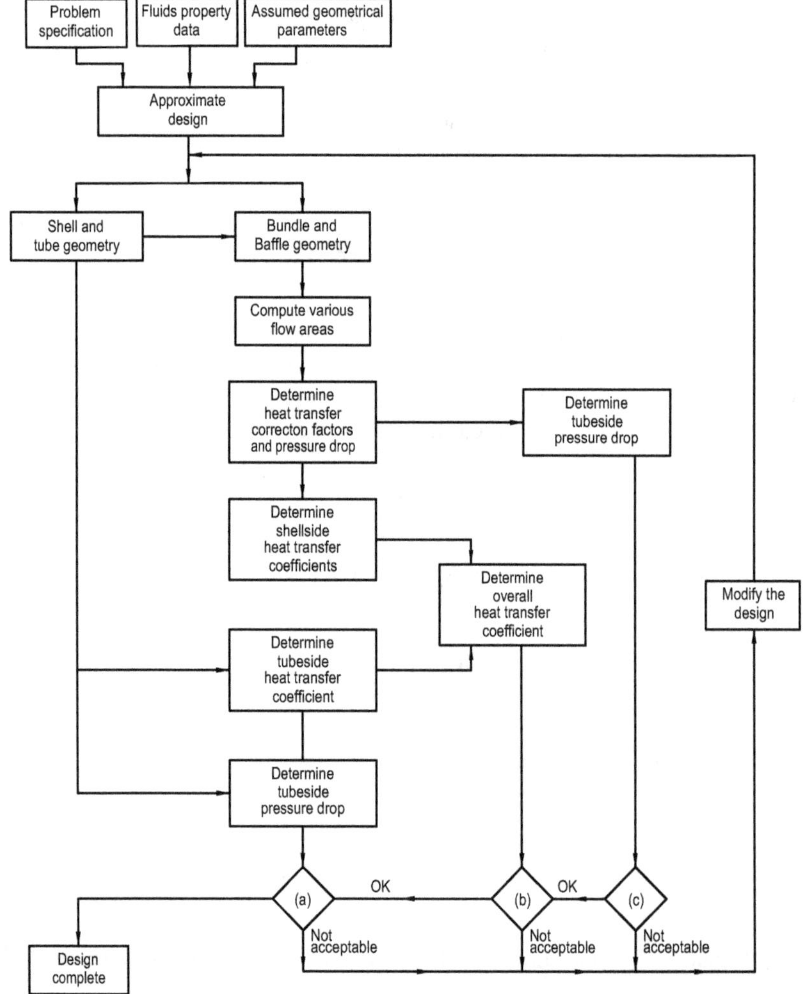

FIGURE 4.77 Flowchart for detailed design of STHE. *Note*: (a) $\Delta P_t \leq$ allowed pressure drop, (b) compare area required with area available for heat transfer, and (c) $\Delta P_s \leq$ allowable pressure drop.

4.31.1 Evaluation of Geometric Parameters

After determination of the shell inside diameter and tube length, the next step is the evaluation of geometric parameters, such as

1. Baffle and bundle geometry
2. Flow areas
3. Various flow areas for calculating various correction factors

The calculation of various geometric parameters is known as auxiliary calculations in the Bell–Delaware method [40]. These calculations are required for the determination of shellside heat transfer coefficient and pressure drop. The auxiliary calculations are defined in the following steps.

4.31.2 Input Data

The Bell–Delaware method assumes that the flow rate and the inlet and outlet temperatures (also pressures for a gas or vapor) of the shellside fluid are specified and that the density, viscosity, thermal conductivity, and specific heat of the shellside fluid are known. The method also assumes that the following minimum set of shellside geometry data is known or specified:

Tube outside diameter, d
Tube layout pattern, θ_{tp}
Shell inside diameter, D_s
Tube bank OTL diameter, D_{otl}
Effective tube length (between tubesheets), L_{ti}
Baffle cut, B_c, as a percent of D_s
Central baffle spacing, L_{bc} (also the inlet and outlet baffle spacing, L_{bi} and L_{bo}, if different from L_{bc})
Number of sealing strips per side, N_{ss}

From this geometrical information, all remaining geometrical parameters pertaining to the shellside can be calculated or estimated by methods given here, assuming that the standards of TEMA are met with respect to various shellside constructional details.

4.31.3 Shellside Parameters

Bundle-to-shell clearance, L_{bb}: A suitable tube bundle is selected based on the user's requirement, and the bundle-to-shell clearance is calculated based upon these equations:

For a fixed tubesheet heat exchanger,

$$L_{bb} = 12.0 + 0.005D_s\,(\text{mm})$$

For a U-tube exchanger,

$$L_{bb} = 12.0 + 0.005D_s\,(\text{mm})$$

Bundle diameter (D_{ctl}): This is computed from the equation

$$\begin{aligned} D_{otl} &= D_s - L_{bb} \\ &= D_{ctl} + d \end{aligned}$$

Shell length: This is taken as the overall nominal tube length, L_{to}, given by

$$L_{to} = L_{ta} + 2L_{ts}$$

where L_{ts} is tubesheet thickness. Its value may be assumed initially as 1 in. (24.4 mm) for calculation purposes.

Central baffle spacing, L_{bc}. The number of baffles N_b is required for calculation of the total number of cross passes and window turnarounds. It is expressed as

$$N_b = \frac{L_{ti}}{L_{bc}} - 1 \tag{4.16}$$

where L_{ti} and L_{bc} are the tube length and central baffle spacing, respectively. Tube length L_{ti} is defined in Figure 4.78. A uniform baffle spacing (L_{bc}) is assumed initially, equal to the shell diameter, D_s. To determine L_{ti}, we must know the tubesheet thickness. If drawings are not available, the tubesheet thickness, L_{ts}, can be roughly estimated as $L_{ts} = 0.1D_s$ with limit $L_{ts} = 25$ mm. Otherwise assume the minimum tubesheet thickness as specified in TEMA [2]. For all bundle types except U-tubes, $L_{ti} = L_{to} - L_{ts}$, whereas for U-tube bundles, L_{to} is nominal tube length.

The number of baffles is rounded off to the lower integer value, and the exact central spacing is then calculated by

$$L_{bc} = \frac{L_{ta}}{N_b + 1} \tag{4.17}$$

4.31.4 Auxiliary Calculations, Step-by-Step Procedure

Step 1: Segmental baffle window calculations. Refer to Figure 4.79, which reveals the basic segmental baffle geometry in relation to the tube field. Calculate the centriangle of baffle cut, θ_{ds}, and upper centriangle of baffle cut, θ_{ctl}.

The centriangle of baffle cut, θ_{ds}, is the angle subtended at the center by the intersection of the baffle cut and the inner shell wall as shown in Figure 4.79. It is given by

$$\theta_{ds} = 2\cos^{-1}\left(1 - \frac{2B_c}{100}\right) \tag{4.18}$$

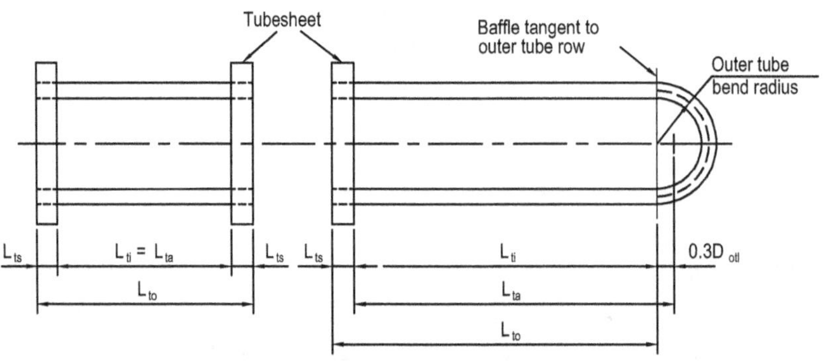

FIGURE 4.78 Shell and tube heat exchanger tube length definition.

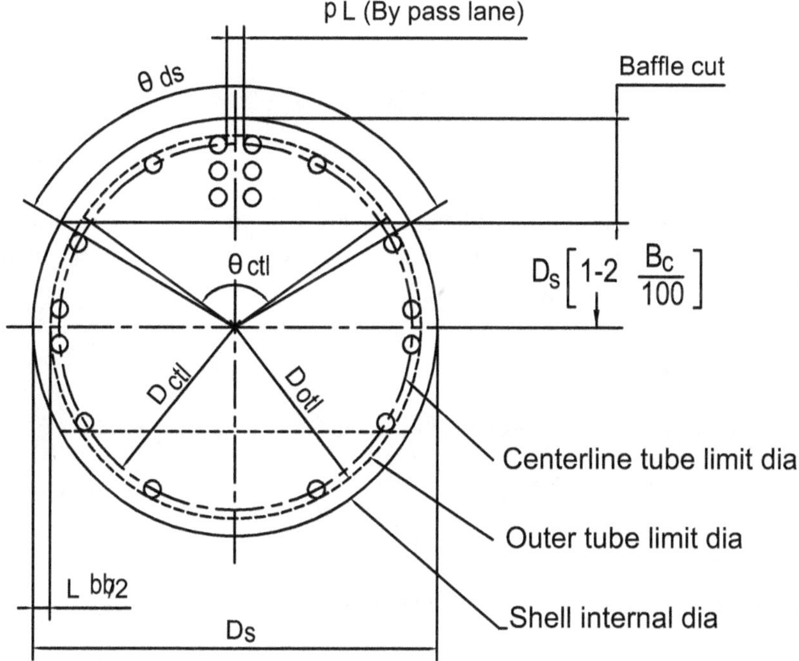

FIGURE 4.79 Basic segmental baffle geometry.

The upper centriangle of baffle cut, θ_{ctl}, is the angle subtended at the center by the intersection of the baffle cut and the tube bundle diameter, as shown in Figure 4.59. It is given by

$$\theta_{ctl} = 2\cos^{-1}\left[\frac{D_s}{D_{ctl}}\left(1 - \frac{2B_c}{100}\right)\right] \tag{4.19}$$

Step 2: Shellside crossflow area. The shellside crossflow area, S_m, is given by

$$S_m = L_{bc}\left[L_{bb} + \frac{D_{ctl}}{L_{tp,eff}}\left(L_{tp} - d\right)\right] \tag{4.20}$$

where

$$L_{bb} = D_s - D_{otl}$$

$$D_{ctl} = D_{otl} - d$$

$$L_{tp,eff} = L_{tp} \text{ for } 30° \text{ and } 90° \text{ layouts}$$

$$= 0.707L_{tp} \text{ for } 45° \text{ layouts}$$

$$L_{tp} = \text{tube pitch}$$

Basic tube layout parameters are given in Table 4.3.

Step 3: Baffle window flow areas. The gross window flow area, i.e., without tubes in the window, S_{wg}, is given by

$$S_{wg} = \frac{\pi}{4} D_s^2 \left(\frac{\theta_{ds}}{2\pi} - \frac{\sin\theta_{ds}}{2\pi} \right) \tag{4.21}$$

From the calculations of centriangle and gross window flow area, calculate the fraction of tubes in baffle window, F_w, and in pure crossflow, F_c, i.e., between the baffle cut tips as indicated in Figure 4.79 by distance $D_s[1 - 2(B_c/100)]$:

$$F_c = 1 - 2F_w \tag{4.22}$$

where F_w is the fraction of the number of tubes in the baffle window, given by

$$F_w = \frac{\theta_{ctl}}{2\pi} - \frac{\sin\theta_{ctl}}{2\pi} \tag{4.23}$$

The segmental baffle window area occupied by the tubes, S_{wt}, can be expressed as

$$S_{wt} = N_{tw} \frac{\pi}{4} d^2 \tag{4.24a}$$

$$= N_t F_w \frac{\pi}{4} d^2 \tag{4.24b}$$

The number of tubes in the window, N_{tw}, is expressed as

$$N_{tw} = N_t F_w \tag{4.25}$$

The net crossflow area through one baffle window, S_w, is the difference between the gross flow area, S_{wg}, and the area occupied by the tubes, S_{wt}. Net crossflow area through one baffle window, S_w, is given by

$$S_w = S_{wg} - S_{wt} \tag{4.26}$$

Step 4: Equivalent hydraulic diameter of a segmental baffle window, D_w. The equivalent hydraulic diameter of a segmental baffle window, D_w, is required only for pressure-drop calculations in laminar flow, i.e., if $Re_s < 100$. It is calculated by classical definition of hydraulic diameter, i.e., four times the window crossflow area S_w divided by the periphery length in contact with the flow. This is expressed in the following equation:

$$D_w = \frac{4S_w}{\pi d N_{tw} + \pi D_s \theta_{ds} / 2\pi} \tag{4.27}$$

TABLE 4.3
Tube Layout Basic Parameters

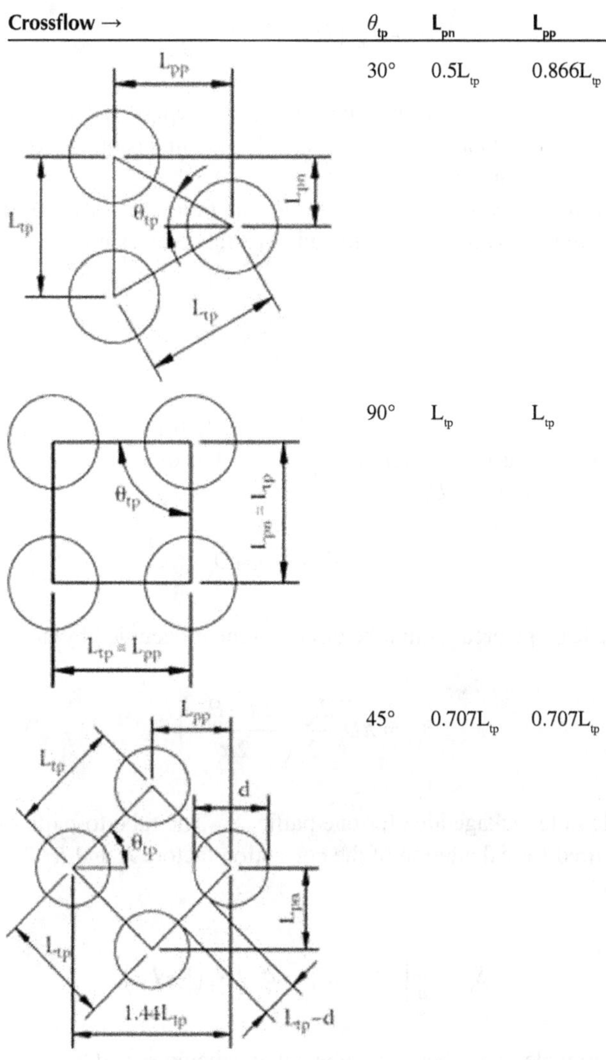

Crossflow →	θ_{tp}	L_{pn}	L_{pp}
	30°	$0.5L_{tp}$	$0.866L_{tp}$
	90°	L_{tp}	L_{tp}
	45°	$0.707L_{tp}$	$0.707L_{tp}$

Step 5: Number of effective tube rows in crossflow, N_{tcc} and baffle window, N_{tew}. The number of effective tube rows crossed in one crossflow section, i.e., between the baffle tips, is expressed as N_{tcc}:

$$N_{tcc} = -\frac{D_s}{L_{pp}}\left(1 - \frac{2B_c}{100}\right) \tag{4.28}$$

where L_{pp} is the effective tube row distance in the flow direction, which is given in Table 4.3. The effective number of tube rows crossed in the baffle window, N_{tcw}, is given by

$$N_{tcw} = \frac{0.8}{L_{pp}}\left[\frac{D_s B_c}{100} - \frac{D_s - D_{ctl}}{2}\right] \tag{4.29}$$

Step 6: Bundle-to-shell bypass area parameters, S_b and F_{sbp}. The bypass area between the shell and the tube bundle within one baffle, S_b, is given by

$$S_b = L_{bc}\left(D_s - D_{otl} + L_{pl}\right) \tag{4.30}$$

where L_{pl} expresses the effect of the tube lane partition bypass width (between tube walls) as follows: L_{pl} is 0 for all standard calculations; L_{pl} is half the dimension of the tube lane partition L_p. For estimation purposes, assume that $L_p = d$.

For calculations of the correction factors J_1 and R_1, the ratio of the bypass area, S_b, to the overall crossflow area, S_m, designated as F_{sbp}, is calculated from the expression

$$F_{sbp} = \frac{S_b}{S_m} \tag{4.31}$$

Step 7: Shell-to-baffle leakage area for one baffle, S_{sb}. The shell-to-baffle leakage area, S_{sb}, is a factor for calculating baffle leakage effect parameters J_1 and R_1. The diametral clearance between the shell diameter D_s and the baffle diameter D_b is designated as L_{sb} and given by

$$L_{sb} = 3.1 + 0.004D_s \tag{4.32}$$

The shell-to-baffle leakage area within the circle segment occupied by the baffle is calculated as

$$S_{sb} = \pi D_s \frac{L_{sb}}{2}\left(\frac{2\pi - \theta_{ds}}{2\pi}\right) \tag{4.33}$$

Step 8: Tube-to-baffle-hole leakage area for one baffle, S_{th}. The tube-to-baffle-hole leakage area for one baffle, S_{th}, is required for calculation of the correction factors J_1 and R_1. The total tube-to-baffle leakage area is given by

$$S_{tb} = \frac{\pi}{4}\left[\left(d + L_{tb}\right)^2 - d^2\right]N_t\left(1 - F_w\right) \tag{4.34}$$

where L_{tb} is the diametral clearance between tube outside diameter and baffle hole. TEMA standards specify recommended clearances as a function of tube diameter and baffle spacing. Its value is either 0.8 or 0.4.

Step 9: Calculate the shellside crossflow velocity, U_s. The shellside crossflow velocity U_s from shellside mass flow rate M_s is given by

$$U_s = \frac{M_s}{\rho_s S_m} \tag{4.35}$$

where ρ_s is the mass density of the shellside fluid. Equation 4.35 gives the shellside crossflow velocity as per the Bell–Delaware method. Since flow-induced vibration guidelines given in the TEMA standards are based on crossflow velocity as per Tinker [37], the procedure to calculate crossflow velocity is given in Appendix 4.A.

4.31.5 Shellside Heat Transfer and Pressure-Drop Correction Factors

Heat transfer correction factors: In the Bell–Delaware method, the flow fraction for each stream is found by knowing the corresponding flow areas and flow resistances. The heat transfer coefficient for ideal crossflow is then modified for the presence of each stream through correction factors. The shellside heat transfer coefficient, h_s, is given by

$$h_s = h_i J_c J_l J_b J_s J_r \tag{4.36}$$

where h_i is the heat transfer coefficient for pure crossflow of an ideal tube bank. The correction factors in Equation 4.36 are thus:

- J_c is the correction factor for baffle cut and spacing. This correction factor is used to express the effects of the baffle window flow on the shellside ideal heat transfer coefficient h_i, which is based on crossflow.
- J_l is the correction factor for baffle leakage effects, including both shell-to-baffle and tube-to-baffle leakage.
- J_b is the correction factor for the bundle bypass flow (C and F streams).
- J_s is the correction factor for variable baffle spacing in the inlet and outlet sections.
- J_r is the correction factor for adverse temperature gradient buildup in laminar flow.

The combined effect of all of these correction factors for a reasonably well-designed shell and tube heat exchanger is typically of the order of 0.6; i.e., the effective mean shellside heat transfer coefficient for the exchanger is of the order of 60% of that calculated if the flow took place across an ideal tube bank corresponding in geometry to one crossflow section. It is interesting to note that this value was suggested by McAdams [47] in 1933 and has been used as a rule of thumb [1].

Pressure-drop correction factors: The following three correction factors are applied for pressure drop:

1. Correction factor for bundle bypass effects, R_b
2. Correction factor for baffle leakage effects, R_l
3. Correction factor for unequal baffle spacing at inlet and/or outlet, R_s

4.31.6 Step-by-Step Procedure to Determine Heat Transfer and Pressure-Drop Correction Factors

Step 10: Segmental baffle window correction factor, J_c. For the baffle cut range 15–45%, J_c is expressed by

$$J_c = 0.55 + 0.72F_c \tag{4.37}$$

This value is equal to 1.0 for NTIW design, increases to a value as high as 1.15 for small baffle cut, and decreases to a value of about 0.52 for very large baffle cuts. A typical value for a well-designed heat exchanger with liquid on the shellside is about 1.0.

Step 11: Correction factors for baffle leakage effects for heat transfer, J_1, and pressure drop, R_1. The correction factor J_1 penalizes the design if the baffles are put too close together, leading to an excessive fraction of the flow being in the leakage streams compared to the crossflow stream. R_1 is the correction factor for baffle leakage effects. For computer applications, the correction factors are curve-fitted as follows:

$$J_1 = 0.44(1-r_s)+[1-0.44(1-r_s)]e^{-2.2r_{lm}} \tag{4.38}$$

$$R_1 = \exp[-1.33(1+r_s)]r_{1m}^x \tag{4.39}$$

where

$$x=[-0.15(1+r_s)+0.8] \tag{4.40}$$

The correlational parameters used are

$$r_s = \frac{S_{sb}}{S_{sb}+S_{tb}} \tag{4.41}$$

$$r_{lm} = \frac{S_{sb}+S_{tb}}{S_m} \tag{4.42}$$

where
S_{sb} is the shell-to-baffle leakage area
S_{tb} is the tube-to-baffle leakage area
S_m is the crossflow area at bundle centerline

A well-designed exchanger should have a valve of J_1 of not less than 0.6, and preferably in the range 0.7–0.9. If a low J_1 value is obtained, modify the design with wider baffle spacing, increase tube pitch, or change the tube layout to 90° or 45°. More drastic measures include change to double or triple segmental baffles, TEMA *J* shell type, or both. A typical value for R_1 is in the range of 0.4–0.5, though lower values may be found in exchangers with closely spaced baffles.

Step 12: Correction factors for bundle bypass effects for heat transfer, J_b, and pressure drop, R_b. To determine J_b and R_b, the following parameters must be known:

1. N_{ss}, the number of sealing strips (pairs) in one baffle spacing
2. N_{tcc}, the number of tube rows crossed between baffle tips in one baffle section

The expression for J_b is correlated as

$$J_b = \exp\left\{-C_{bh}F_{sbp}\left[1-(2r_{ss})^{1/3}\right]\right\} \tag{4.43}$$

where

$$C_{\mathrm{bh}} = 1.25 \text{ for laminar flow, } \mathrm{Re_s} \le 100, \text{ with the limit of } J_{\mathrm{b}} = 1, \text{ at } r_{\mathrm{ss}} \ge 0.5$$
$$= 1.35 \text{ for turbulent and trasition flow, } \mathrm{Re_s} > 100$$

The expression for R_b is given by

$$R_{\mathrm{b}} = e^{\left[-C_{\mathrm{bp}}F_{\mathrm{sbp}}\left\{1-(r_{\mathrm{ss}})^{1/3}\right\}\right]} \tag{4.44}$$

where

$$r_{\mathrm{ss}} = \frac{N_{\mathrm{ss}}}{N_{\mathrm{tcc}}} \tag{4.45}$$

with the limits of

$$\begin{aligned} R_{\mathrm{b}} &= 1 && \text{at } r_{\mathrm{ss}} \ge 0.5 \\ C_{\mathrm{bp}} &= 4.5 && \text{for laminar flow, } \mathrm{Re_s} \le 100 \\ &= 3.7 && \text{for turbulent and transition flow, } \mathrm{Re_s} > 100 \end{aligned}$$

For the relatively small clearance between the shell and the tube bundle, J_b is about 0.9; for the much larger clearance required by pull-through floating head construction, it is about 0.7. J_b can be improved using sealing strips.

A typical value for R_b ranges from 0.5 to 0.8, depending upon the construction type and number of sealing strips. The lower value would be typical of a pull-through floating head with only one or two pairs of sealing strips, and the higher value typical of a fully tubed fixed tubesheet exchanger.

Step 13: Heat transfer correction factor for adverse temperature gradient in laminar flow, J_r. J_r applies only if the shellside Reynolds number is less than 100 and is fully effective only in deep laminar flow characterized by $\mathrm{Re_s}$ less than 20. For $\mathrm{Re_s} < 20$, J_r can be expressed as

$$J_{\mathrm{r}} = \frac{1.51}{N_{\mathrm{c}}^{0.18}} \tag{4.46}$$

where N_c is the total number of tube rows crossed in the entire exchanger. N_c is given by

$$N_{\mathrm{c}} = \left(N_{\mathrm{tcc}} + N_{\mathrm{tcw}}\right)\left(N_{\mathrm{b}} + 1\right) \tag{4.47}$$

For $\mathrm{Re_s}$ between 20 and 100, a linear proportion is applied resulting in

$$J_{\mathrm{r}} = \frac{1.51}{N_{\mathrm{c}}^{0.18}} + \left(\frac{20-\mathrm{Re_s}}{80}\right)\left(\frac{1.51}{N_{\mathrm{c}}^{0.18}} - 1\right) \tag{4.48}$$

with the limit

$$J_r = 0.400 \quad \text{for Re}_s \le 100$$
$$J_r = 1 \quad \text{for Re}_s > 100$$

Step 14: Heat transfer correction for unequal baffle spacing at inlet and/or outlet, J_s. Figure 4.80 shows a schematic sketch of an exchanger where the inlet and outlet baffle spacing L_{bi} and L_{bo} are shown in comparison to the central baffle spacing, L_{bc}:

$$J_s = \frac{(N_b - 1) + (L_i^*)^{1-n} + (L_o^*)^{1-n}}{(N_b - 1) + (L_i^* - 1) + (L_o^* - 1)} \tag{4.49}$$

where

$$L_i^* = \frac{L_{bi}}{L_{bc}} \quad L_o^* = \frac{L_{bo}}{L_{bc}} \tag{4.50}$$

J_s will usually be between 0.85 and 1.0. If $L_{bi} = L_{bo} = L_{bc}$ or $L^* = L_i^* = L_0^* = 1.0$, $J_s = 1.0$. For turbulent flow, $n = 0.6$, and values of L^* larger than 2 would be considered poor design, especially if combined with a few baffles only, i.e., low N_b. In such a case, an annular distributor or other measures should be used. Typical arrangements for increasing the effectiveness of end zones are presented by Tinker [36]. For laminar flow, the correction factor is about halfway between 1 and J_s computed for turbulent conditions.

Step 15: Pressure-drop correction for unequal baffle spacing at inlet and/or outlet, R_s. R_s is given by

$$R_s = \left(\frac{1}{L_i^*}\right)^{2-n} + \left(\frac{1}{L_o^*}\right)^{2-n} \tag{4.51}$$

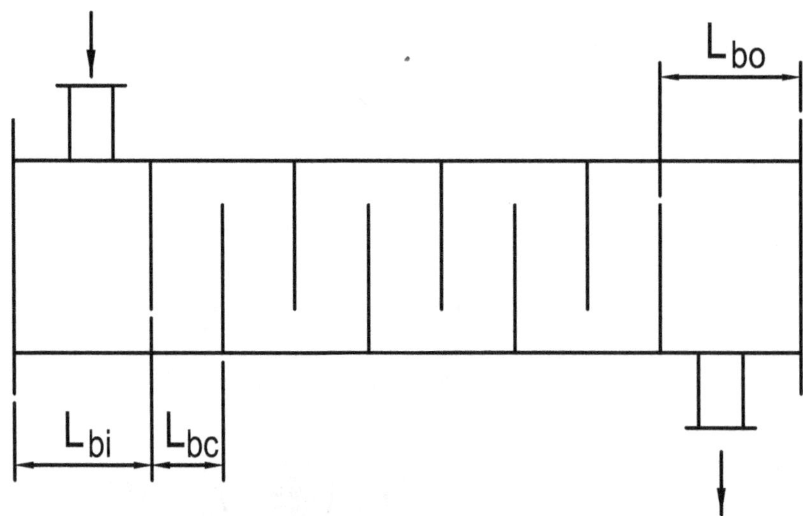

FIGURE 4.80 Typical layout of baffle spacing.

with $n = 1$ for laminar flow, $\mathrm{Re}_s \le 100$, and $n = 0.2$ for turbulent flow.

1. For $L_{bc} = L_{bo} = L_{bi}$, $R_s = 2$.
2. For the reasonable extreme case $L_{bo} = L_{bi} = 2L_{bc}$, $R_s = 1.0$ for laminar flow, and $R_s = 0.57$ for turbulent flow.
3. For a typical U-tube, $L_{bi} = L_{bc}$ and $L_{bo} = 2L_{bc}$, $R_s = 1.5$ for laminar flow, and $R_s = 3.0$ for turbulent flow.

4.31.7 Shellside Heat Transfer Coefficient and Pressure Drop

A. Shellside Heat Transfer Coefficient

1. Calculate the shellside mass velocity G_s, Reynolds number Re_s, and Prandtl number Pr_s:

$$G_s = \frac{M_s}{S_m}\ \mathrm{kg/(m^2 \cdot s)} \quad \text{or} \quad \mathrm{lb_m/(h \cdot ft^2)} \tag{4.52}$$

$$\mathrm{Re}_s = \frac{dG_s}{\mu_s} \qquad \mathrm{Pr}_s = \frac{\mu_s C_{ps}}{k_s} \tag{4.53}$$

2. Calculate the ideal heat transfer coefficient h_i given by

$$h_i = \frac{j_i C_{ps} G_s (\phi_s)^n}{\mathrm{Pr}_s^{2/3}} \tag{4.54}$$

where j_i and $(\phi_s)^n$ are defined next.

The term j_i is the ideal Colburn j factor for the shellside and can be determined from the appropriate Bell–Delaware curve for the tube layout and pitch and a typical curve. For example, $d = 0.75$ in. (19.05 mm), pitch = 1.0 in. (24.4 mm), and $\theta_{tp} = 30°$; curve fits for j_i are given by Bell [40]:

$$j_i = 1.73\ \mathrm{Re}_s^{(-0.694)} \qquad 1 \le \mathrm{Re}_s < 100 \tag{4.55a}$$

$$= 0.717\ \mathrm{Re}_s^{(-0.574)} \qquad 100 \le \mathrm{Re}_s < 1000 \tag{4.55b}$$

$$= 0.236\ \mathrm{Re}_s^{(-0.346)} \qquad 1000 \le \mathrm{Re}_s \tag{4.55c}$$

The term $(\phi_s)^n$ is the viscosity correction factor, which accounts for the viscosity gradient at the tube wall (μ_w) versus the viscosity at the bulk mean temperature (μ_s) of the fluid and is given by [48]:

$$(\phi_s)^n = \left(\frac{\mu_s}{\mu_w}\right)^{0.14} \tag{4.56}$$

For liquids, ϕ_s is greater than 1 if the shellside fluid is heated and less than 1 if the shellside fluid is cooled. In order to determine μ_w, it is essential to determine T_w, which is estimated as follows using the approximate values of h_t and h_s [48]:

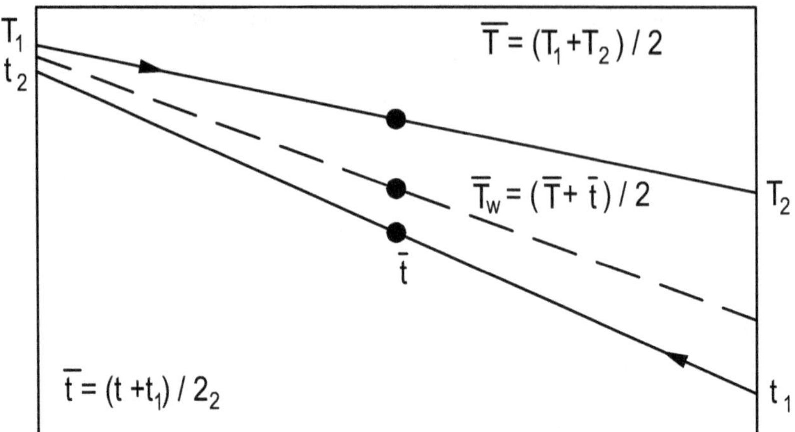

FIGURE 4.81 Mean temperature calculation method for shellside and tubeside fluids and mean metal temperature for conduction wall.

$$T_w = T_{t,av} + \frac{T_{s,av} - T_{t,av}}{1 + h_t / h_s} \tag{4.57}$$

where $T_{s,av}$ and $T_{t,av}$ denote the average mean metal temperatures of shell and tube, both of them being the arithmetic means of inlet and outlet fluid temperatures on the shellside and tubeside, respectively. The mean temperature calculation method for shellside and tubeside fluids and mean metal temperature for the conduction wall is shown graphically in Figure 4.81. An accurate equation to calculate tube mean metal temperature is given by TEMA. For gases, the viscosity is a weak function of temperature. The correction factor ϕ_s is formulated as follows:

$$\text{For gases being cooled}: \ (\phi_s)^n = 1.0 \tag{4.58}$$

$$\text{For gases being heated}: \ (\phi_s)^n = \left[\frac{(T_{s,av} + 273.15)}{(T_w + 273.15)} \right]^{0.25} \tag{4.59}$$

For a gas being heated, T_w is always higher than $T_{s,av}$ and hence the correction factor is less than 1.0. Calculate the shellside heat transfer coefficient given by

$$h_s = h_i J_c J_l J_s J_b J_r \tag{4.60}$$

B. Shellside Pressure Drop

The shellside pressure drop is calculated in the Delaware method by summing the pressure drop for the inlet and exit sections, and the internal sections after applying various correction factors. The total shellside pressure drop Δp_s consists of the pressure drop due to (1) crossflow Δp_c; (2) window regions Δp_w; and (3) entrance and exit sections Δp_e as given by Refs. [39,40]:

$$\Delta p_s = \Delta p_c + \Delta p_w + \Delta p_e \tag{4.61}$$

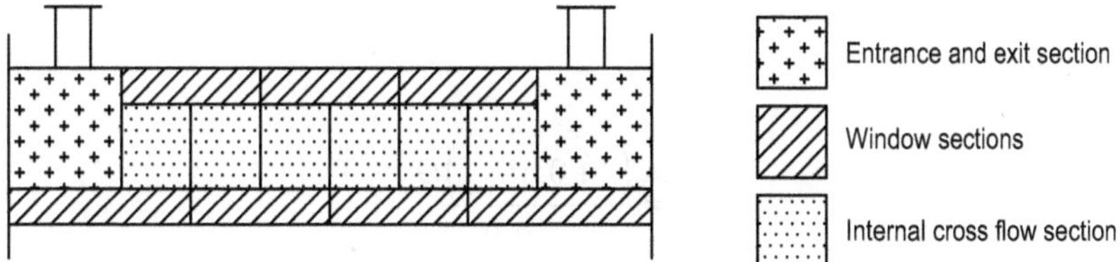

FIGURE 4.82 Elements of shellside pressure drop of a TEMA *E* shell.

The elements of shellside pressure drop are shown schematically in Figure 4.82. The crossflow pressure drop and the entrance and exit region pressure drop depend on the ideal tube bank pressure drop, given by

$$\Delta p_{\mathrm{b,i}} = 2 f_{\mathrm{s}} N_{\mathrm{tcc}} \frac{G_{\mathrm{s}}^{2}}{g_{\mathrm{c}} \rho_{\mathrm{s}}} \left(\phi_{\mathrm{s}}\right)^{-\mathrm{n}} \tag{4.62}$$

The friction factor f_{s} can be determined from the appropriate Bell–Delaware curve for the tube layout and pitch under consideration. For example, for d = 0.75 in. (19.05 mm), pitch = 1.0 in. (24.4 mm), and $\theta_{\mathrm{tp}} = 30°$, curve fits for f_{s} are given by Bell [40]:

$$f_{\mathrm{s}} = \frac{52}{\mathrm{Re}_{\mathrm{s}}} + 0.17 \quad 1 \le \mathrm{Re}_{\mathrm{s}} < 500 \tag{4.63a}$$

$$= 0.56\ \mathrm{Re}_{\mathrm{s}}^{(-0.14)} \quad 500 \le \mathrm{Re}_{\mathrm{s}} \tag{4.63b}$$

Calculate the various terms of shellside pressure drop as given next and finally calculate the overall heat transfer coefficient.

1. The pressure drop in the interior crossflow sections is affected by both bypass and leakage. Therefore, the combined pressure drop of all the interior crossflow sections is given by

$$\Delta p_{\mathrm{c}} = \left(N_{\mathrm{b}} - 1\right)\left(\Delta p_{\mathrm{b,i}} R_{\mathrm{b}} R_{\mathrm{l}}\right) \tag{4.64}$$

2. The pressure drop in the entrance and exit sections is affected by bypass but not by leakage, and by variable baffle spacing. Therefore, the combined pressure drop for the entrance and exit sections is

$$\Delta p_{\mathrm{e}} = 2\left(\Delta p_{\mathrm{b,i}}\right)\left(1 + \frac{N_{\mathrm{tcw}}}{N_{\mathrm{tcc}}}\right) R_{\mathrm{b}} R_{\mathrm{s}} \tag{4.65}$$

3. The pressure drop in the windows is affected by leakage but not by bypass. Therefore, the combined pressure drop of all the window sections is given by

$$\Delta p_{\mathrm{w}} N_{\mathrm{b}} R_{\mathrm{l}} \tag{4.66}$$

where Δp_w is given as follows:
For $Re_s \geq 100$,

$$\Delta p_w = \frac{(2+0.6N_{tcw})G_w^2}{2g_c\rho_s} \tag{4.67}$$

For $Re_s < 100$,

$$\Delta p_w = 26\frac{G_w\mu_s}{g_c\rho_s}\left[\frac{N_{tcw}}{L_{tp}-d}+\frac{L_{bc}}{D_w^2}\right]+2\frac{G_w^2}{g_c\rho_s} \tag{4.68}$$

in which the window mass velocity G_w is given by

$$G_w = \frac{M_s}{\sqrt{S_m S_w}} \tag{4.69}$$

Summing these individual effects, we obtain the equation for the total nozzle-to-nozzle shellside pressure drop:

$$\Delta p_s = \left[(N_b-1)(\Delta p_{b,i}R_b)+N_b(\Delta p_w)\right]R_l+2(\Delta p_{b,i})\left(1+\frac{N_{tcw}}{N_{tcc}}\right)R_bR_s \tag{4.70}$$

The total shellside pressure drop of a typical shell and tube exchanger is of the order of 20–30% of the pressure drop that would be calculated for flow through the corresponding heat exchanger without baffle leakage and without tube bundle bypass effects [39].

4.31.8 Tubeside Heat Transfer Coefficient and Pressure Drop

A. Tubeside Heat Transfer Coefficient

1. Calculate the tubeside mass velocity G_t, Reynolds number Re_t, and Prandtl number, Pr_t:

$$G_t = \frac{M_t}{A_t} \quad \text{for single pass} \tag{4.71a}$$

$$= \frac{M_t}{A_t/N_p} \quad \text{for } N_p \text{ pass} \tag{4.71b}$$

where

$$A_t = \frac{\pi}{4}d_i^2N_t \tag{4.71c}$$

and where A_1 is the tubeside flow area, N_p the number of tubeside passes, N_t the number of tubes, and

$$\mathrm{Re}_1 = \frac{G_t d_i}{\mu_i} \quad \mathrm{Pr}_t = \frac{\mu_t C_{pt}}{k_t} \tag{4.72}$$

2. Calculate the tubeside heat transfer coefficient, h_t: For laminar flow with Re_t of 2,100 and less, the heat transfer coefficient is determined from the Sider–Tate (1936) empirical equation for both heating and cooling of viscous liquids:

$$\frac{h_t d_i}{k_i} = 1.86\left[Re_t Pr_t \frac{d_i}{L}\right)^{0.5} \mathrm{Pr}_t^{1/3}\left(\frac{\mu_t}{\mu_w}\right)^{0.14} \tag{4.73}$$

For those cases where the Grashof number Na_r exceeds 25,000, the value of h_t obtained from Equation 4.71 must be corrected for the increase in heat transfer due to natural convection effects by multiplying by the term

$$0.8\left(1+0.015Na_r^{1/3}\right)$$

$$Na_r = \frac{\beta \Delta t d_i^2 \rho_i^2 g_c}{\mu_t^2} \tag{4.74}$$

where

β is the thermal coefficient of cubical expansion, 1/°F
Δt is the temperature difference, °F
Na_r is the Grashof number
ρ_i is the tubeside fluid density, lb/ft³
g_c is the acceleration of gravity, 4.17×10^8 ft/h²
μ_t is the viscosity at bulk mean temperature, lb/h ft

At values above 10,000, turbulent flow occurs and the heat transfer coefficient is determined from the following correlation:

Sider–Tate equation modified by McAdams:

$$\frac{h_t d_i}{k_t} = 0.027\ \mathrm{Re}_t^{0.8}\mathrm{Pr}_t^{1/3}\left(\frac{\mu_t}{\mu_w}\right)^{0.14} \tag{4.75}$$

In the intermediate region where the Reynolds number varies from 2,100 to 10,000, the relation does not follow a straight line.

Colburn equation

$$\frac{h_t d_i}{k_1} = 0.023\ \mathrm{Re}_t^{0.8}\mathrm{P}r_t^{0.4}\left(\frac{\mu_t}{\mu_w}\right)^{0.14} \tag{4.76a}$$

Dittus–Boelter equation

$$\frac{h_t d_i}{k_t} = 0.023\ \mathrm{Re}_t^{0.8} \mathrm{Pr}_t^{n} \tag{4.76b}$$

where
$n = 0.4$ for heating
$= 0.3$ for cooling

In the intermediate region where Re_t varies between 2,100 and 10,000, the relation does not follow a straight line. In this region, Kern [44] recommends the following formula:

$$\frac{h_t d_i}{k_i} = 0.116\left[\mathrm{Re}_t^{2/3} - 125\right]\left[1 + \left(\frac{d_i}{L}\right)^{2/3}\right]\mathrm{Pr}_t^{1/3}\left(\frac{\mu_t}{\mu_w}\right)^{0.14} \tag{4.77}$$

B. Tubeside Pressure Drop

Pressure drop in tubes and pipes—general: For tubes and pipes, the pressure drop in steady flow between any two points may be expressed by the following Weisbach–Darcy equation:

$$\Delta p_t = f_t\left(\frac{L}{d_i}\right)\frac{G_t^2}{2g_c\rho_t}\frac{1}{(\varphi_t)^r} \tag{4.78}$$

where

$$(\phi_t)^r = \left(\frac{\mu_t}{\mu_w}\right)^{0.14} \quad \text{for } \mathrm{Re}_t > 2100 \tag{4.79a}$$

$$= \left(\frac{\mu_t}{\mu_w}\right)^{0.25} \quad \text{for } \mathrm{Re}_t < 2100 \tag{4.79b}$$

and where L is the length of pipe between two points, d_i the inside diameter of the pipe, ρ_t the mass density of the tubeside fluid, G_t the mass velocity of the fluid inside the pipe, k_i or k_t is thermal conductivity of the tubeside fluid, and f_t the friction factor. To determine the pressure drop through the tube bundle, multiply the pressure drop by the number of tubes. For multipass arrangements, multiply by the number of tubeside passes.

The total pressure drop for a single pass consists of the following items:

1. Pressure drop in the nozzles, Δp_n, which is the sum of pressure drop in the inlet ($\Delta p_{n,i}$) and outlet nozzle ($\Delta p_{n,o}$):

$$\Delta p_n = \frac{1.5 G_n^2}{2g_c\rho_t} \tag{4.80}$$

2. Sudden contraction and expansion losses at the tube entry and exit, $\Delta p_{c,e}$, are given by

$$\Delta p_{c,e} = \frac{G_t^2}{2g_c\rho_t}\left(K_c + K_e\right)N_p \tag{4.81}$$

where K_c and K_e are the contraction and expansion loss coefficients.

3. Pressure drop through the tube bundle, Δp_t:

$$\Delta p_t = \frac{fL_{1p}G_t^2 N_p}{2g_c\rho_t d_i}\frac{1}{\phi_t^r} \tag{4.82}$$

4. Pressure drop associated with the turning losses, Δp_r, given by

$$\Delta p_r = \frac{4N_p G_t^2}{2g_c\rho_t} = 4N_p \times \text{Velocity head per pass} \tag{4.83}$$

Total tubeside pressure drop, Δp_t, is given by

$$\Delta p_t = \frac{G_t^2}{2g_e\rho_t}\left[\frac{1.5}{N_p} + \frac{fL_{1p}}{d_i}\frac{1}{\left(\phi_t\right)^r} + K_c + K_e + 4\right]N_p \tag{4.84}$$

Determination of friction factor f on the tubeside: The friction factor f has been found to depend upon Reynolds numbers only. For laminar flow in smooth pipes, the value of f can be derived from the well-known Hagen–Poiseuille equation.

Accordingly, f for laminar flow is given by

$$f = \frac{64}{\text{Re}_t} \tag{4.85}$$

For turbulent flow in smooth pipes, f can be determined from the Blasius empirical formula given by

$$f = \frac{0.3164}{\text{Re}_t^{0.25}} \tag{4.86}$$

This equation is valid for Reynolds numbers up to 1,000,000. Another formula to determine f for smooth-walled conduits in the Reynolds number range 10,000–120,000 is given by [49]

$$f = \frac{0.184}{\text{Re}_t^{0.2}} \tag{4.87}$$

Calculate the overall heat transfer coefficient, U, as per Equation 4.10 using the values of h_s and h_t from Equations 4.60 and 4.73, respectively.

Calculate the heat transfer area, A_o, required using Equation 4.12 using the U value calculated earlier.

C. Evaluation and Comparison of the Results with the Specified Values

In the evaluation stage, the calculated values of the film coefficients and the pressure drop, for both streams, are compared with the specified values. If the values match, the design is complete. If the calculated values and the specified values do not match, repeat the design with new design variables. Sometimes, even after many iterations, the design may not meet the specified performance limits in the following parameters [28]:

1. Heat transfer coefficient
2. Pressure drop
3. Temperature driving force
4. Fouling factors

Taborek [28] describes various measures to overcome limitations in these parameters. For the heat transfer-limited cases, utilize the permissible pressure drop effectively. The film coefficients can be increased by increasing the flow velocities, changing the baffle spacing, changing the number of the tubeside passes, etc. Attempts to increase the heat transfer coefficient also result in a pressure-drop increase. Designs with inherently low heat transfer coefficients, such as laminar flow or LP gases, may deserve special attention. For pressure-drop-limited cases, try alternate designs such as the following [28]:

- Double or multiple segmental baffles
- Use TEMA J or X shell
- Reduce tube length
- Increase tube pitch
- Change tube layout pattern

For the temperature–driving-force-limited cases, methods to improve the LMTD correction factor have been discussed while describing various shell types and pass arrangements. Fouling-limited cases are dealt with during the design stage by making sure that the factors promoting fouling are suppressed and/or designed for periodic cleaning.

4.31.9 Accuracy of the Bell–Delaware Method

It should be remembered that this method, though apparently generally the best in the open literature, is not extremely accurate [15]. Palen et al. [15] compare the thermohydraulic performance prediction error of Bell–Delaware methods [38,39], stream analysis methods [15], and Tinker [37].

4.31.10 Extension of the Delaware Method to Other Geometries

The Delaware method, as originally developed and as it exists in the open literature, is more or less explicitly confined to the design of fully tubed E shell configurations using plain tubes. Extension of this method to other geometries is discussed in Refs. [17,29], and these are discussed in this subsection.

4.31.10.1 Applications to Low-Finned Tubes

It is possible to apply the Delaware method to the design of such heat exchangers in a fairly straightforward way by making use of the results of Brigs and Young [50] and Briggs, Katz, and Young [51]. As shown by Briggs et al. [51], the Colburn *j* factor for low-finned tubes is slightly less than that for a plain tube in the Reynolds number range of about 2–1000; the difference is larger in the lower Reynolds number range. The fin-tube *j* factor j_f can be represented in terms of plain-tube-bank *j* factor as j_i as [17]

$$j_f = K_f j_i \tag{4.88}$$

where K_f is the correction factor whose values, taken from Ref. [50], are presented in Ref. [17]. Its approximate values at discrete points are as follows:

Re	20	70	100	200	400	600	800–1000
K_f	0.575	0.65	0.675	0.75	0.885	0.975	1.0

For the fin tube friction factor, Bell [39,40] suggests a conservative value that is 1.5 times that for the corresponding plain tube bank, whereas Taborek [17] recommends 1.4 times the friction factor of the plain tube bank. The adaptation of the Bell–Delaware method to finned tubes is explained in Refs. [17,29].

1. **Input data.** The following additional information is required for calculating the shellside thermal performance of a low-finned tube bundle:
   Diameter over the fins: D_{fo}
   Fin root diameter: D_{fr}
   Number of fins per unit of tube length: N_f
   Average fin thickness (assuming rectangular profile): L_{fs}
   Wetted surface area of finned tube per unit of tube length: A_{of}
   Tube-to-baffle hole clearance: L_{tb}
   To determine L_{tb} use D_{fo} in place of D_t. Normally, D_{fo} is equal to or slightly less than D_t
2. **Heat transfer and flow geometries.**
   The total heat transfer surface area upon which to apply the shellside heat transfer coefficient of the finned tube bundle α_{ss} is A_o, which for the finned tube is obtained from the following expression:

   $$A_o = A_{of} L_{tb} N_{tt}$$

   The equivalent projected area of an integral low-finned tube is less than that of a plain tube of the same diameter because of the openings between adjacent fins in the direction of flow. Hence, the "melt down" or equivalent projected diameter D_{req} is a function of the tube geometry and density as:

   $$D_{req} = D_{fr} + 2L_{fh} N_f L_{fs} \tag{4.89}$$

   And the fin height L_{fh} is

   $$L_{fh} = \frac{D_{fo} - D_{fr}}{2} \tag{4.90}$$

Thus, wherever the plain tube diameter D_t appears in the correlations and geometrical equations for the plain tube bank, it is replaced by D_{req} or as noted below:

S_m calculations use D_{req} in place of D_t

Re calculations use D_{req} in place of D_t

S_{wt} calculations use D_{req} in place of D_t

3. Ideal tube bank values of j_r

The method for plain tubes is applicable to low-finned tubes without modification with Re determined using D_{req} rather than D_t for Re > 1000. When Re<1000, a laminar boundary layer overlap on the fins begins to adversely affect the heat transfer. This is accounted for by the following expression, applicable only when Re<1000.

$$j_I = J_f \, j_{I\_plain} \tag{4.91}$$

4. Ideal tube bank values of f_r

The equivalent cross-flow area is larger for a finned tube bundle relative to an identical plain tube bundle because of the additional flow area between the fins. For finned tubes the friction factor is about 1.4 times larger than for a plain tube: however, the lower velocity due to the larger flow area from the open area between fins in the direction of flow is also taken into account in the Reynolds number Re. To calculate the friction factor for the low-finned tube bank, first calculate $f_{r,plain}$ using the finned tube values of Re and D_{req} from the plain tube correlation and then multiply this value by 1.4 as follows:

$$f_I = 1.4 f_{I,plain} \tag{4.92}$$

4.31.10.2 Application to other STHE construction types

Application to the no-tubes-in-window configuration: In calculations, use $N_{cw} = 0$ and the baffle configuration correction factor $J_c = 1$. Otherwise, the calculation is essentially identical to a fully tubed bundle. It is suggested that one choose a smaller baffle cut so that the free flow area through the window corresponds reasonably closely to the free crossflow area through the tube bank itself.

Application to F shells: Since the shellside is essentially split into two, it is possible to adopt the Delaware method by reducing all of the areas for flow by half compared to the case for a single shellside pass. It is assumed that the flow leakage and conduction across the longitudinal baffle are minimized.

Application to J shells: The adaptation of the method to a *J* shell is straightforward. Take one-half the length of the exchanger as a "unit" equivalent to an *E* shell and take one-half the mass flow rate.

Application to double segmental and disk and doughnut exchangers: According to Taborek [15], the Bell–Delaware method cannot be easily adapted to these baffle types, as the driving forces for bypass and leakage flow are much smaller than for segmental baffles.

4.32 SOFTWARE FOR THERMAL DESIGN OF SHELL AND TUBE HEAT EXCHANGERS

Nowadays, most exchangers are designed using commercially available software such as Xchanger Suite® of HTRI [52] and Aspen Shell & Tube Mechanical of HTFS, Oxon, UK [53], among others. Typical features of the thermal design program structure of a shell and tube heat exchanger are shown in Table 4.4, and the output program in Table 4.5. These methods are generally restricted for use by members of the relevant organizations.

4.33 SHELL AND TUBE HEAT EXCHANGERS WITH NON-SEGMENTAL BAFFLES

The baffle configuration shall be segmental or nonsegmental baffles. Based on the baffle configuration, the types of STHEs are classified as shown in Table 4.6.

4.33.1 Phillips RODbaffle Heat Exchanger

The RODbaffle exchanger is a shell and tube heat exchanger that uses an improved support system for the tubes. It consists of rods located in a predisposed manner such that they confine tube movement. RODbaffle heat exchangers are used in process industries to enhance thermohydraulic performance, eliminate flow-induced tube vibration occurrence, minimize shellside pressure losses, increase shellside flow-field uniformity, and reduce shellside fouling [54]. References [54–60] provide general and specific information on RODbaffle heat exchanger design.

TABLE 4.4
Typical Features of Thermal Design Program Structure of Shell and Tube Heat Exchangers

Specification	Description
Process parameters (input data)	Heat duty, inlet parameters, pressure drop, fouling resistances
TEMA class	TEMA *R, B, C*
TEMA exchanger types	Front head: *A, B, C, D, N*
	Shell: *E, F, G, H, J, K, X*
	Rear head: *L, M, N, P, S, T, U, W*
Tubesheet	Single tubesheet (also OTL, untubed area), double tubesheet
Tube details	Plain tube, integral low finned tube, tube material, tube diameters, and tube length. For U-tubes, U-bend details
Baffle types	Single segmental, double segmental, triple segmental, no tubes in window, disk and doughnut, RODbaffles, EMbaffle®, etc.
Plate baffle spacings	Inlet, central, and outlet baffle spacings
Impingement protection	Plate or rod type on shellside
Baffle cuts	Horizontal, vertical, rotated, % baffle cut, baffle spacing
Tubelayout patterns	Triangular, rotated triangular, square, rotated square, and pitch
Tube passes	Possible: 1–16 (odd or even). Specify the number of tube passes
Pass layout types	Quadrant, mixed, ribbon
Shell details	Diameter, length, number of shells in series/parallel
Nozzles	Number, inside diameter, orientation, etc.
Tubeside heat transfer calculation (sensible heat)	Laminar flow, turbulent flow—Colburn j factor
Single-phase frictional pressure drop	Method–Dittus–Boelter equation, Blasius equation or others
Method to calculate shellside pressure drop and film coefficients for single-phase sensible flow	Stream-analysis method, Bell–Delaware method
Special operating conditions	Start-up, transient, and shutdown

TABLE 4.5
Typical Features of Output of Thermal Design Program of Shell and Tube Heat Exchangers

Design summary	An overview of the most important variables: shell and tubeside temperatures, flow rates, heat transfer coefficients, velocities, and pressure drops. Description of key heat exchanger dimensions and geometry
TEMA specification sheet	Filling up of TEMA specification sheet (English/metric units)
Performance evaluation	Required area and distribution of resistances in clean, specified fouling, and maximum fouling conditions
Heat transfer coefficients	Film coefficients and their components
MTD and heat flux	LMTD, *F*-correction factors, heat fluxes, and heat flux limitations. Heating curves
Pressure drop	Clean and dirty conditions, velocity and pressure-drop distribution from inlet to outlet zone
Shellside flow data	Stream analysis—flow fractions and ρV^2 analysis (per TEMA)
Construction of tube bundle	Baffle and tube layout details, tube count, bundle diameter, shell-to-tube-bundle clearances, etc.
Flow-induced vibration analysis details	Critical velocity and natural frequency, fluid elastic instability and turbulent buffeting analysis, frequency matching for acoustic and vortex shedding at inlet, bundle, outlet, and user specified spans
Recap of design cases	Concise summary of alternative solutions that have been explored and their costs

TABLE 4.6
Shell and Tube Heat Exchanger with Segmental Baffles and Nonsegmental Baffles

With segmental baffles	With non-segmental baffles
1. Shell and tube heat exchanger	1. Phillips Rodbaffle heat exchanger
2. Disk and doughnut heat exchanger	2. Embaffle® heat exchanger
	3. Helixchanger® heat exchanger
	4. Twisted Tube® heat exchanger

4.33.1.1 PHILLIPS RODbaffle Exchanger Concepts

The RODbaffle exchangers (Figure 4.83) use rods, with a diameter equal to the clearance between tube rows, inserted between alternate tube rows in the bundle in both the horizontal and vertical directions. The support rods are laid out on a square pitch with no nominal clearance between tubes and support rods. Support rods are welded at each end to a fabricated circumferential baffle ring. Major components of each individual RODbaffle are support rods, baffle ring, cross-support strips, partition blockage plate, and longitudinal slide bars. The concept of a RODbaffle heat exchanger is shown in Figure 4.83a. Four different RODbaffle configurations, *W–X–Y–Z* are welded to longitudinal slide bars to form the RODbaffle cage assembly as shown in Figure 4.83b. The tube and rod layout is shown in Figure 4.83c.

4.33.1.2 Important Benefit: Elimination of Shellside Flow-Induced Vibration

Although RODbaffle heat exchangers are being increasingly used because of thermohydraulic advantages, flow-induced vibration protection of tubes remains one of the major design considerations. The RODbaffle bundle eliminates harmful flow-induced vibration by using these major design innovations:

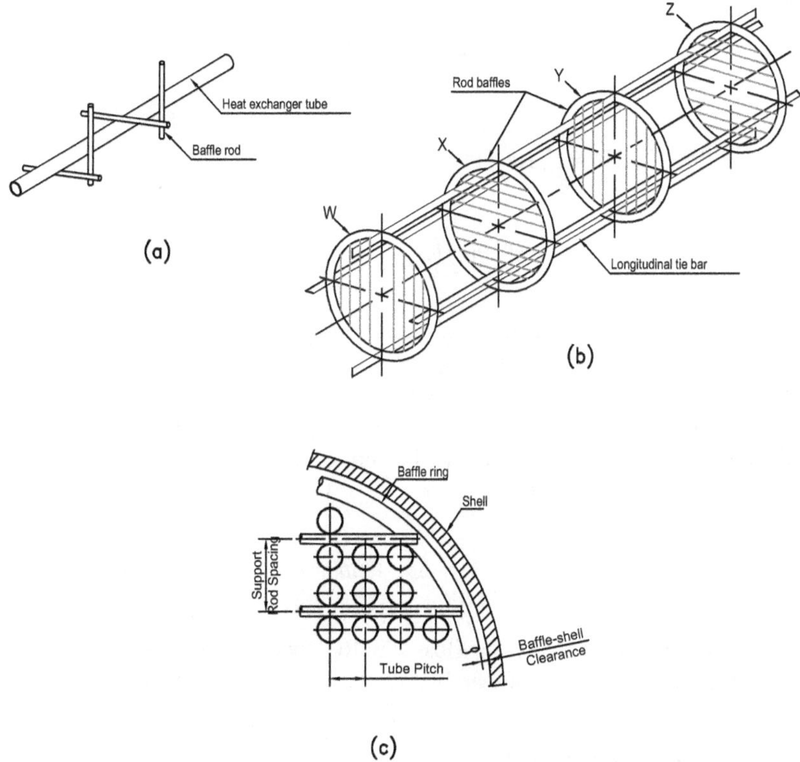

FIGURE 4.83 PHILLIPS RODbaffle heat exchanger and support. (a) Concept—schematic, (b) RODbaffle cage assembly, and (c) tube and rod layout.

1. Each tube is supported in all four directions
2. Positive four-point confinement of the tubes
3. Minimum convex point contact between rods and tubes

Under design conditions, where plate baffles are required and positive tube support is critical, combination RODbaffle–plate–baffle exchangers may be used [55].

4.33.1.3 Proven RODbaffle Applications

Typical applications for RODbaffle exchangers include gas–gas exchangers, compressor aftercoolers, gas–oil coolers, reactor feed–effluent exchangers, condensers, kettle reboilers, waste heat boilers, HF acid coolers, and carbon black air preheaters.

4.33.1.4 Operational Characteristics

The unobstructed flow path created by the support rod matrix makes the flow field in a RODbaffle heat exchanger predominantly longitudinal. As reported in Refs. [55,56], RODbaffle exchanger heat transfer rates compare favorably with double segmental plate baffle exchangers and are generally higher than in comparable triple segmental plate baffle and "NTIW" designs. Shellside pressure losses in RODbaffle exchangers are lower because neither bundle crossflow form drag nor repeated flow reversal effects are present.

4.33.1.5 Thermal Performance

For shellside, the expressions for Nusselt number, Nu, for laminar flow and turbulent flow are given by [59]

$$\mathrm{Nu} = C_L \mathrm{Re}_h^{0.6} \mathrm{Pr}^{0.4} \left(\phi_s\right)^n \quad \text{for laminar flow} \tag{4.93a}$$

$$\mathrm{Nu} = C_T \mathrm{Re}_h^{0.8} \mathrm{Pr}^{0.4} \left(\phi_s\right)^n \quad \text{for turbulent flow} \tag{4.93b}$$

where C_L and C_T are RODbaffle exchanger geometric coefficient functions for laminar flow and turbulent flow, respectively.

Pressure drop: Shellside pressure loss for flow through a RODbaffle heat exchanger bundle, excluding inlet and exit nozzles, is defined as the sum of an unbaffled frictional component, Δp_L, and a baffle flow contribution Δp_b as [54]:

$$p = \Delta p_L + \Delta p_b \tag{4.94}$$

According to Gentry [54], shellside pressure losses in RODbaffle exchangers are generally less than 25% of shellside losses produced in comparable double segmental plate baffle exchangers.

4.33.2 EMbaffle® Heat Exchanger[1]

The patented EMbaffle design uses expanded metal baffles made of plate material that has been slit and expanded. Figure 4.84a shows a section of an EMbaffle. The open structure results in low hydraulic resistance and enhanced heat transfer. With this new EMbaffle technology, the shellside fluid flows axially along the tubes, but in the vicinity of the baffles, the flow area is reduced. This creates local turbulence in the flow while breaking up the boundary layer over the tubes. The shape of the grid induces a local crossflow component on top of the longitudinal flow pattern, which together improve the heat transfer characteristics at the surface of the tubes. The breakup of the boundary layer occurs repeatedly at each expanded metal baffle along the length of the heat exchanger, resulting in lower hydraulic resistance while maintaining higher heat transfer [61]. Pressure loss is effectively converted into improved heat transfer, and compared with the segmental baffle, heat transfer at the same fluid velocity is significantly higher. Figure 4.84 shows the concept of an EMbaffle, heat exchanger tube bundle assembly, and completed tube bundle assembly.

As a direct consequence of the longitudinal flow on the shellside, tube vibration in an EMbaffle heat exchanger is effectively eliminated, significantly reducing the risk of mechanical damage. The longitudinal direction of the shellside liquid in EMbaffle heat exchangers approaches pure countercurrent flow; as a result, they can deliver higher heat duty at the same approach temperature.

4.33.2.1 Application of EMbaffle Technology

Initially developed as a potential solution for fouling services, EMbaffle technology has been proven in a wide range of applications from liquid/liquid and gas/gas to condensing and boiling. In fouling services, for instance, in crude preheat trains, the dead zones typically found with conventional segmental designs reduce the performance of the heat exchanger with an increasing pressure drop during operation.

FIGURE 4.84 EMbaffle heat exchanger. (a) (i) Sectional view of tube bundle cage assembly, (ii) Tube bundle under assembly, and (iii) Assembled tube bundle (2 passes on the tubeside). (b) (i–ii) EMbaffle heat exchanger tube bundle assembly. (Courtesy of EMbaffle B.V., LionsParc, A. van Leeuwenhoekweg 38A10, 2408 Amsterdam, the Netherlands.)

4.33.2.2 Design

The thermal design parameters for EMbaffle technology have been embedded in the industry standard HTRIX Changer™ software suite, enabling licensees and users to carry out their own thermal engineering designs. A complete range of standard and customized grid designs is available for all conventional tube sizes and most applications. Baffles are available in most sizes and materials from carbon steel, stainless, duplex and superduplex steels, high-nickel alloys, and copper alloys, including bronze, brass, and cupronickels. The EMbaffle heat exchanger utilizes TEMA tolerances and the same pressure parts as conventional segmental designs. It is compliant with all relevant international standards [61].

4.33.3 Helixchanger® Heat Exchanger[2]

The Helixchanger heat exchanger is a high-efficiency heat exchanger and a proprietary product of Lummus Technology Heat Transfer, a division of Lummus Technology. In a Helixchanger heat exchanger, the conventional segmental baffle plates are replaced by quadrant-shaped baffle plates positioned at an angle to the tube axis in a sequential arrangement to create a helical flow pattern creating a uniform velocity through the tube bundle. Figures 4.85–4.87 show the helical flow pattern on the shellside and orientation of helical baffles of a tube bundle. Shellside helical flow offers higher thermal efficiency as well as lower fouling rates as compared to the conventional segmental baffled heat exchangers [62]. Effective protection against flow-induced vibrations is achieved by both the single- and the double-helix baffle arrangement. In a double-helix arrangement, two strings of helical baffles are intertwined to reduce the unsupported tube spans, offering greater integrity against vibration without compromising the thermohydraulic performance.

4.33.3.1 Merits of Helixchanger Heat Exchanger

In a conventional shell and tube heat exchanger with segmental baffles, uneven velocity profiles, back flows, and eddies result in higher fouling and hence shorter run lengths. In a Helixchanger heat exchanger, the helical flow offers improved thermal effectiveness, enhanced heat transfer, reduced pressure drop, lower fouling, and significantly reduced flow-induced vibration hazards [62].

4.33.3.2 Applications

Helixchanger heat exchangers are in operation worldwide for varied applications in the oil and gas, refining, petrochemical, and chemical industries. The applications range from crude preheat

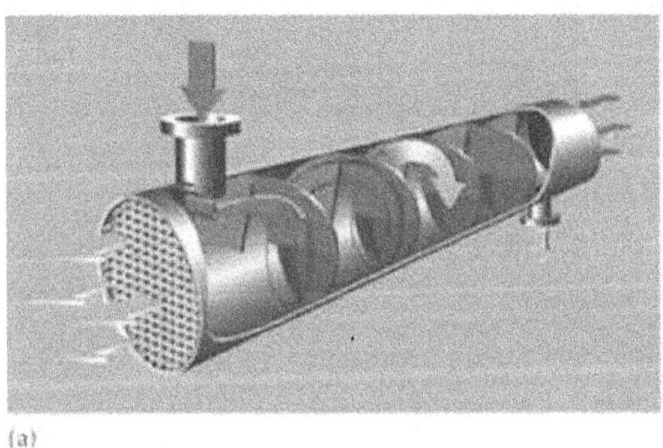
(a)

(b)

FIGURE 4.85 Helixchanger heat exchanger. (a) Shellside flow pattern (schematic) and (b) helical baffle orientation. (Courtesy of Lummus Technology Heat Transfer, A Division of Lummus Technology Inc., Bloomfield, NJ.)

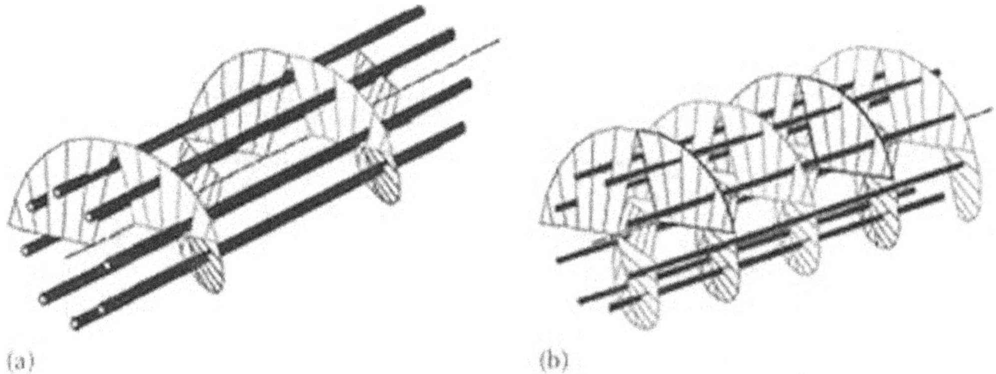

FIGURE 4.86 Helical baffles. (a) Single helical baffle and (b) double helical baffles (schematic). (Courtesy of Lummus Technology Heat Transfer, A Division of Lummus Technology Inc., Bloomfield, NJ.)

exchangers, feed preheat exchangers in delayed cokers, feed/effluent exchangers in refinery and petrochemical processes, bitumen exchangers in oil sands, major equipment in ethylene plants, to polymer solution coolers in the chemical industry.

4.33.3.3 Helixchanger Heat Exchanger: Configurations

1. Helixchanger Heat Exchanger

In a Helixchanger heat exchanger, quadrant-shaped plate baffles are used on the shellside, placed at an angle to the tube axis in a successive arrangement to create a helical flow pattern. Figure 4.88 shows the Helixchanger heat exchanger tube bundle assembly.

2. Helifin® Heat Exchanger

When a Helixchanger heat exchanger is built with low-fin tubes for the tube bundle, it is called a Helifin heat exchanger. Low-fin tubes offer an extended heat transfer surface on the outside of the tubes. For process conditions where low-fin tubes are suitable, the Helifin heat exchanger offers a more economical solution, with increased reliability against flow-induced vibration.

3 Helitower™ Heat Exchanger[3]

In certain processes where high-temperature effluent is used to preheat the feed stream, a large bank of horizontal heat exchangers is required in series arrangement. A more economical option using one or two heat exchangers with longer tubes in a vertical orientation is often better suited to achieving the desired performance. When Helixchanger heat exchangers are employed in such a vertical orientation using long tube bundles, they are called Helitower heat exchangers. Figure 4.89 shows a Helitower heat exchanger.

4.33.3.4 Performance

Frequently, the fouling mechanism responsible for the deterioration of heat exchanger performance is maldistribution of flow, wakes, and eddies caused by poor heat exchanger geometry on the shellside. In a shell and tube heat exchanger, the conventional segmental baffle geometry is largely responsible for higher fouling rates. Uneven velocity profiles, backflows, and eddies generated on the shellside of a segmental baffled heat exchanger result in higher fouling and shorter run lengths

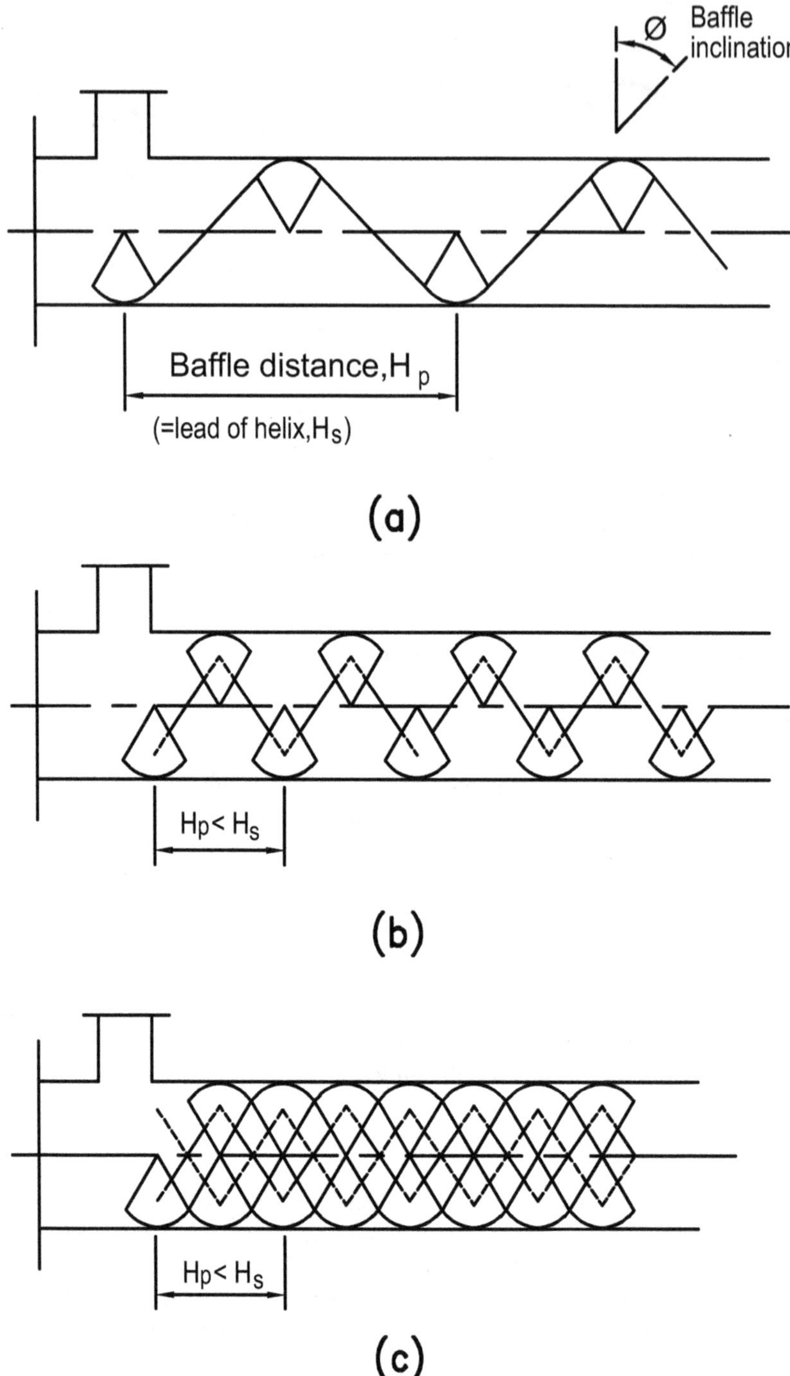

FIGURE 4.87 Orientation of single helical and double helical baffles.

FIGURE 4.88 Helixchanger heat exchanger tube bundle assembly. (a) A section of tube bundle under assembly and (b) completed tube bundle assembly. (a and b: Courtesy of Lummus Technology Heat Transfer, A Division of Lummus Technology Inc., Bloomfield, NJ.) (c) Helixchanger tube bundle under assembly and (d) Helixchanger tube bundle assembly is inserted into the shell. (c and d: Courtesy of Vermeer Eemhaven B.V., Rotterdam, the Netherlands.)

between periodic cleaning and maintenance of tube bundles. In a Helixchanger heat exchanger, the quadrant-shaped baffle plates are arranged at an angle to the tube axis, creating a helical flow pattern on the shellside. Near-plug flow conditions are achieved in a Helixchanger heat exchanger with little backflow and eddies, often responsible for fouling and corrosion. Figure 4.90 shows the flow patterns of a conventional shell and tube heat exchanger compared with a Helixchanger heat exchanger. The absence of backflows and eddies on the shellside of Helixchanger heat exchangers significantly reduces fouling while in operation. Figure 4.91 shows a comparison of typical run lengths between periodic cleaning and maintenance of the tube bundle of a conventional shell and tube heat exchanger (shorter run length) and the Helixchanger heat exchanger (longer run length).

4.33.4 Twisted Tube® Heat Exchanger[4]

Koch Heat Transfer Company's innovative Twisted Tube design avoids the need for baffles. The unique helix-shaped tubes are arranged in a triangular pattern. Each tube is firmly and frequently supported by adjacent tubes (as shown in Figure 4.92), yet fluid swirls freely along its length. This support system eliminates crossflow-induced tube vibration, which is a common problem in

FIGURE 4.89 Helitower™ heat exchanger. (Courtesy of Lummus Technology Heat Transfer, A Division of Lummus Technology Inc., Bloomfield, NJ.)

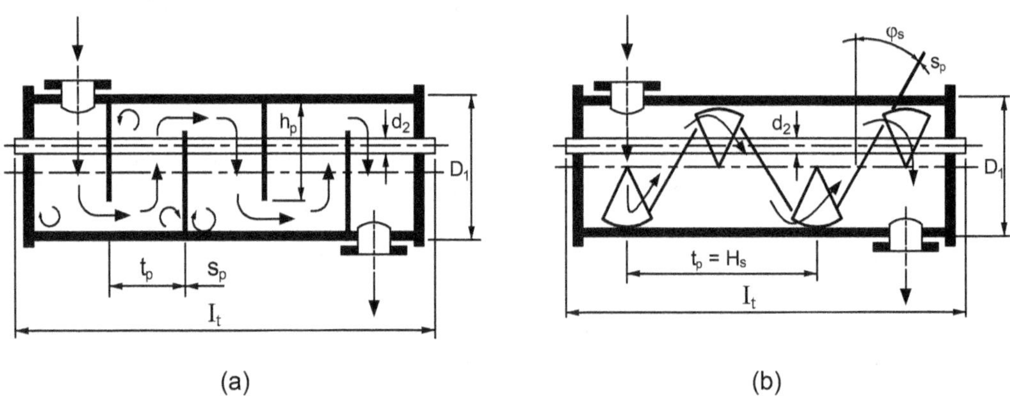

FIGURE 4.90 Shellside flow pattern. (a) Conventional shell and tube heat exchanger with segmental baffles and (b) Helixchanger heat exchanger with helical baffles. (Courtesy of Lummus Technology Heat Transfer, A Division of Lummus Technology Inc., Bloomfield, NJ.)

conventional shell and tube heat exchanger services. The twist arrangement for baffle-free support with gaps aligned between the tubes also provides for easier cleaning on the shellside. The Twisted Tube heat exchangers are round at each end, allowing for conventional tube-to-tubesheet joints to be used [63,64].

4.33.4.1 Applications

Crude preheat, feed/effluent for reformer (CCR and semi-regeneration), hydrotreater, hydrocracker, alkylation, etc., overhead condensers, reboilers (kettle and *J* shell), lean/rich amine, compressor interstage coolers, etc.

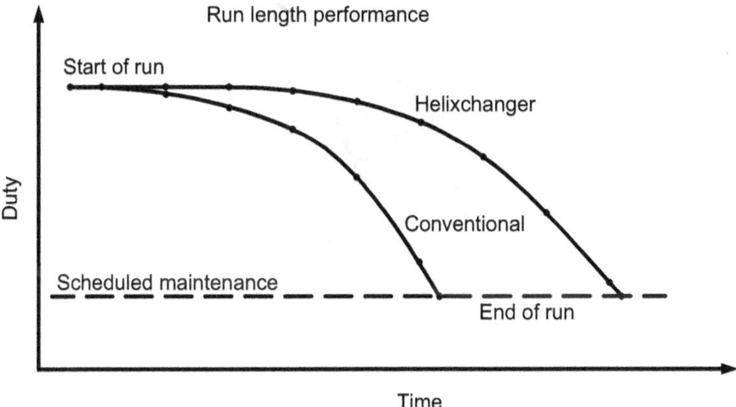

FIGURE 4.91 Comparison of run lengths between periodic cleaning and maintenance of tube bundle of conventional shell and tube heat exchanger with that of Helixchanger heat exchanger. (Courtesy of Lummus Technology Heat Transfer, A Division of Lummus Technology Inc., Bloomfield, NJ.)

4.33.4.2 Advantages

Twisted Tube tubing offers benefits such as increased heat transfer, smaller exchangers or fewer shells, elimination of flow-induced vibration, and reduced fouling. When used as retrofit bundles or exchangers, Twisted Tube tubings also offer increased capacity, lower installed costs, lower pressure drop, and extended run time between cleanings.

4.33.4.3 Merits of Twisted Tube Heat Exchanger

Twisted Tube heat exchangers provide a higher heat transfer coefficient than the conventional tubular heat exchanger for the following reasons [63]:

a. Uniform fluid distribution combined with interrupted swirl flow on the shellside induces the maximum turbulence to improve heat transfer.
b. On the tubeside, swirl flow creates turbulence resulting in higher tubeside heat transfer coefficient.
c. High localized velocities scrub the tube wall to combat fouling and offer a 40% higher tubeside heat transfer coefficient.
d. Lower pressure drop: The longitudinal swirl flow of Twisted Tube tubing reduces the high pressure drop associated with segmental baffles. Twisted Tube heat exchangers are usually shorter in length and have fewer passes for a lower pressure drop on the tubeside.
e. Baffle-free tube support and elimination of flow-induced vibration.

4.34 END CLOSURES

4.34.1 Breech-Lock™ Closure

Breech-Lock closure is used to seal the tubeside of shell and tube heat exchangers in specific applications (Figure 4.93). Breech-Lock heat exchangers are fabricated and supplied by heat exchanger manufacturers under fabrication license from ABB Lummus Heat Transfer and the Breech-Lock is a registered trademark of ABB LHT. Koch Heat Transfer Company, LP, Houston, Texas, is one of the leading manufacturers of Breech-Lock Closure heat exchangers.

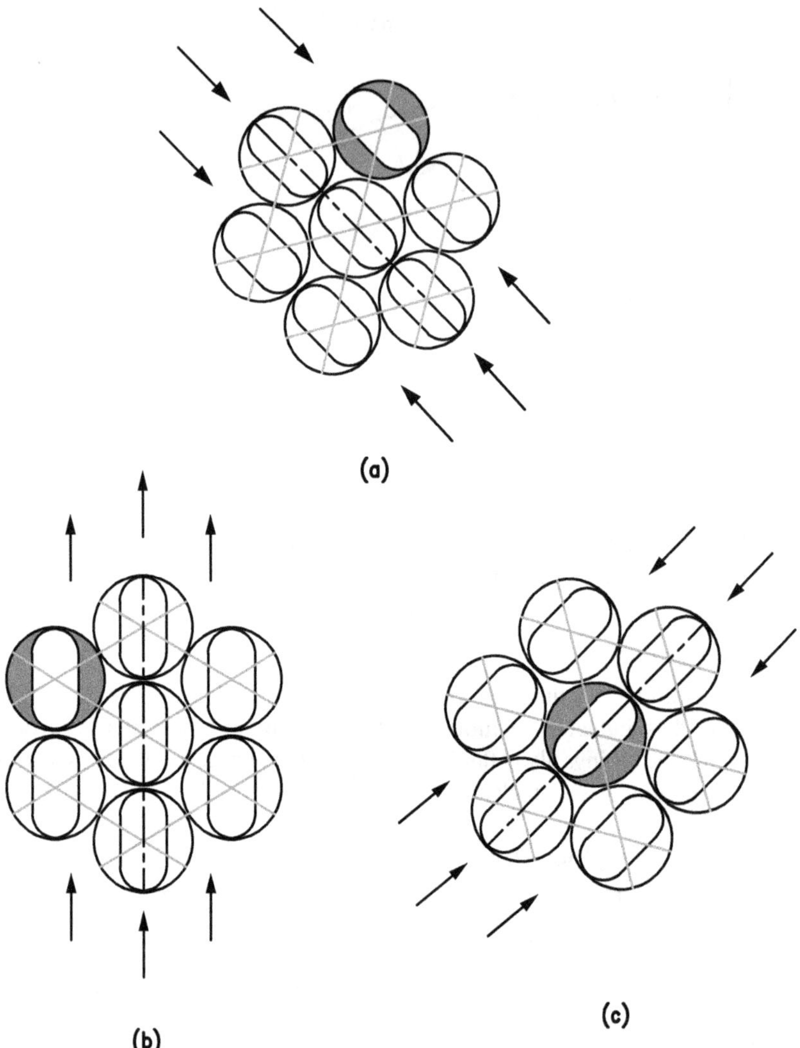

FIGURE 4.92 Twisted tube tube design which is firmly and frequently supported by adjacent tubes. Note – No baffle is used in twisted tube heat exchanger.

4.34.1.1 Easy Installation and Dismantling Jig

IMB™ has developed a unique proprietary jig assembly designed to facilitate dismantling and reassembling operations. With the IMB jig, it is possible to open the channel, remove the internals, and then reassemble and reclose the Breech-Lock exchangers in a very short time *without the need for heavy cranage.*

4.34.2 Taper-Lok® Closure

The Taper-Lok closure is typically used for high-pressure [up to 15,000 psi (1,055 kg/cm^2)] applications. The Taper-Lok closure uses separate tubeside and shellside flanges, bolting, and gasketing. The unique feature of this closure offers a reusable Taper-Lok ring that acts as a

FIGURE 4.93 Breech-Lock end closure. (Courtesy of Brembana Costruzioni Industriali S.p.A., Milan, Italy.)

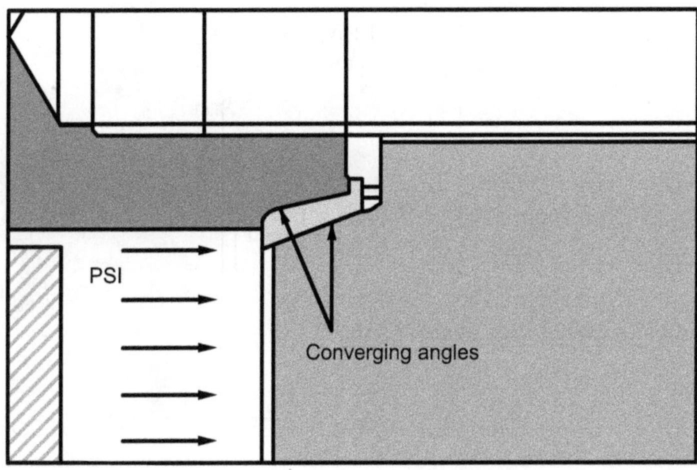

FIGURE 4.94 Taper-Lok closure metal-to-metal seal ring with dual converging tapered contact surfaces. (Courtesy Taper-Lok Corporation, Houston, TX.)

self-energizing seal. The Taper-Lok closure is shown in Figure 4.94. More details on Taper-Lok closure are given in Chapter 1 of Volume 2.

NOTES

1 EMbaffle® is a registered trademark of EMbaffle B.V., Alphen a/d Rijn, the Netherlands.
2 HELIXCHANGER and HELIFIN are registered trademarks of Lummus Technology Inc.
3 Helitower is a trademark of Lummus Technology Inc.
4 Twisted Tube is a registered trademark of Koch Heat Transfer Company, LP, and is registered in the United States and other countries worldwide.

4.A APPENDIX A

4.A.1 Reference Crossflow Velocity as per Tinker

Reference crossflow velocity as per Tinker [37] is given by

$$U_s = \frac{F_h M_s}{M A_x \rho_s} \tag{4.95}$$

where

A_x is the crossflow area within the limits of the tube bundle
F_h is the fraction of total fluid flowing through A_x of clean unit
M is the A_x multiplication factor
M_s is the mass flow rate of shellside fluid
ρ_s is the mass density of shellside fluid

$$A_x = a_x L_{bc} D_3 \tag{4.96}$$

$$F_h = \frac{1}{1 + N_h \sqrt{S}} \tag{4.97}$$

$$M = \left[\frac{1}{1 + \frac{0.7 L_{bc}}{D_1} \left[\frac{1}{M_w^{0.6}} - 1 \right]} \right]^{1.67} \tag{4.98}$$

The list of expressions required to evaluate N_h, M_w, is

$$a_1 = \frac{D_1}{D_3} \tag{4.99}$$

For units with sealing strips, a_1 is given by

$$a_1 = 1 + \left[\frac{D_1 / D_3 - 1}{4} \right] + 1.5 \left[\frac{D_1 - D_2}{D_1} \right] \tag{4.100}$$

$$a_2 = \frac{d_B - d}{d} \tag{4.101}$$

$$a_3 = \frac{D_1 - D_2}{D_1} \tag{4.102}$$

$$b_1 = \frac{(a_1 - 1)^{1.5}}{\sqrt{a_1}} \tag{4.103}$$

$$b_2 = \frac{a_2}{a_1^{1.5}} \tag{4.104}$$

$$b_3 = a_3\sqrt{a_1} \tag{4.105}$$

The terms a_x, a_4, a_5, a_6, and m are dependent on the tube layout pattern, and they are given in Table 4.7.

The expression for a_7 is

$$a_7 = a_4\left(\frac{p}{p-d}\right)^{1.5} \tag{4.106}$$

Values for a_8 are given in Table 4.8.

$$A = a_5 a_8\left(\frac{D_1}{L_{\text{bc}}}\right)\left(\frac{d}{p}\right)^2\left(\frac{p}{p-d}\right) \tag{4.107}$$

$$E = a_6\left(\frac{p}{p-d}\right)\left(\frac{D_1}{L_{\text{bc}}}\right)\left(1-\frac{h}{D_1}\right) \tag{4.108}$$

$$N_{\text{h}} = a_7 b_1 + A b_2 + b_3 E \tag{4.109}$$

$$M_{\text{w}} = m a_1^{0.5} \tag{4.110}$$

TABLE 4.7
Terms to Find Crossflow Velocity

	30°	90°	45°	60°
a_x	$\frac{0.97(p-d)}{p}$	$\frac{0.97(p-d)}{p}$	$\frac{1.372(p-d)}{p}$	$\frac{0.97(p-d)}{p}$
a_4	1.26	1.26	0.90	1.09
a_5	0.82	0.66	0.56	0.61
a_6	1.48	1.38	1.17	1.28
M	0.85	0.93	0.80	0.87

Source: Tinker, T. (1958) *Transactions of ASME* 80: 36.

Note: Values for 60° have been taken from TEMA [2].

TABLE 4.8
Baffle Cut Ratio and the Term a_8 for Crossflow Velocity

h/D_1	0.10	0.15	0.20	0.25	0.30	0.35	0.40	0.45	0.50
a_8	0.94	0.90	0.85	0.80	0.74	0.68	0.62	0.54	0.49

Source: Tinker, T. (1958) *Transactions of ASME* 80: 36.

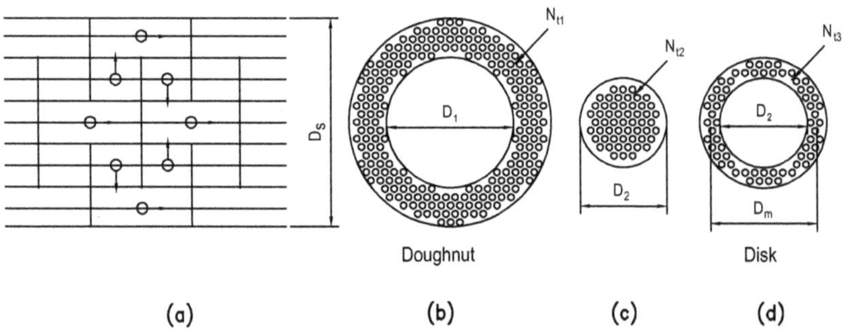

FIGURE 4.95 Disk and doughnut heat exchanger with tubes in various flow areas. *Note*: $D_m = 0.5(D_1 + D_2)$, where D_1 is disk diameter.

4.A.2 Design of Disk and Doughnut Heat Exchangers

In shell and tube heat exchangers with baffle disks and rings, known as disk and doughnut heat exchangers (Figure 4.95), the flow inside the shell is made to alternate between longitudinal and transverse (or crossflow) directions in relation to the tube bundle. This type of heat exchanger is used in some European countries. Occasionally, oil coolers at thermal power stations are of this type. It has been reported that for the same pressure drop in shellside fluid, the heat transfer coefficient obtained with disk and doughnut baffles is approximately 15% more than that obtained with segmental baffles [65]. Exchangers are fabricated for horizontal or vertical installation.

4.A.2.1 Design Method

The effect of clearance between segmental baffles and the shell on the performance of a heat exchanger has been reasonably well studied, but there is hardly any corresponding published literature available for the disk-and-doughnut-type heat exchangers. Very few open literature methods are available for the thermal design of disk and doughnut heat exchangers. Notable among them are Donohue [66], Slipcevic [67,68], and Goyal et al. [69]. Donohue's thermal design method was recently discussed by Ratnasamy [70]. In this section, the Slipcevic and Donohue's method is discussed.

4.A.2.1.1 Slipcevic Method

4.A.2.1.1.1 Design Considerations

Follow these design guidelines on the shellside [67]:

1. The spacing between the baffle disks and rings usually amounts to 20–45% of the inside shell diameter. Less than 15% is not advisable.
2. Since the heat transfer coefficient for flow parallel with the tubes is lower than that for flow normal to the tubes, space the baffle plates so that the velocity in the baffle spacing opening is higher than that of the flow normal to the tubes.

3. It is advisable to make the annular area between the disk and the shell, S_{k1}, the same as the area inside the ring, S_{k2}, such that

$$S_{k1} = \frac{\pi}{4}\left(D_s^2 - D_1^2\right) \tag{4.111}$$

$$S_{k2} = \frac{\pi}{4} D_2^2 \tag{4.112}$$

$$S_{k1} = S_{k2} \rightarrow D_s^2 = D_1^2 + D_2^2 \tag{4.113}$$

4.A.2.2 Heat Transfer

Step 1: Calculate the individual flow area for longitudinal flow:

1. Flow area in the annular region between the disk and the shell, S_1:

$$S_1 = \frac{\pi}{4}\left(D_s^2 - D_1^2 - N_{t1} d^2\right) \tag{4.114}$$

2. Flow area in the ring opening, S_2:

$$S_2 = \frac{\pi}{4}\left(D_2^2 - N_{t2} d^2\right) \tag{4.115}$$

Step 2: Calculate the effective crossflow area, S_q:

$$S_q = L_s \sum x \tag{4.116}$$

where Σx is the sum of the clear distances between the neighboring tubes closest to the mean diameter, $D_m = 0.5(D_1 + D_2)$. Determine the sum of clear distances between neighboring tubes closest to D_m, either from a drawing (as shown in Figure 4.96) or by computational methods.

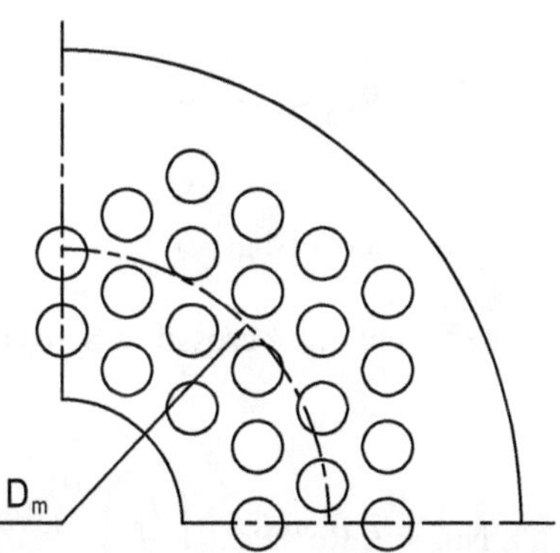

FIGURE 4.96 Mean diameter to calculate effective crossflow area of a disk and doughnut heat exchanger.

The number of tubes in the flow areas S_1, S_2, and crossflow area S_3 (to be defined later) is N_{t1}, N_{t2}, and N_{t3}, respectively. They are shown in Figure 4.95.

Step 3: Calculate the hydraulic diameter for the longitudinal flow in the annulus, D_{h1}, and in the ring opening, D_{h2}:

$$D_{h1} = \frac{4S_1}{\pi\left(N_{t1}d + D_s + D_1\right)} \tag{4.117}$$

$$D_{h2} = \frac{4S_2}{\pi\left(N_{t2}d + D_2\right)} \tag{4.118}$$

Step 4: Calculate the heat transfer coefficients h_1 and h_2 for the longitudinal flow through the cross section S_1 and S_2 from the equation of Hausen [71]:

$$\mathrm{Nu}_1 = 0.024\mathrm{Re}_1^{0.8}\mathrm{Pr}^{0.33}\left(\frac{\mu_b}{\mu_w}\right)^{0.14} \tag{4.119}$$

$$\mathrm{Nu}_2 = 0.024\mathrm{Re}_2^{0.8}\mathrm{Pr}^{0.33}\left(\frac{\mu_s}{\mu_w}\right)^{0.14} \tag{4.120}$$

where

$$\mathrm{Nu}_1 = \frac{h_1 D_{h1}}{k} \tag{4.121a}$$

$$\mathrm{Nu}_2 = \frac{h_2 D_{h2}}{k} \tag{4.121b}$$

$$\mathrm{Re}_1 = \frac{U_1 D_{h1}}{\mu_s} \tag{4.122a}$$

$$\mathrm{Re}_2 = \frac{U_2 D_{h2}}{\mu_s} \tag{4.122b}$$

and U_1 and U_2 are based on S_1 and S_2, respectively.

Step 5: Calculate the heat transfer coefficient h_3 for crossflow across the bundle from McAdams [47] for the turbulent range:

$$\mathrm{Nu}_3 = K\,\mathrm{Re}_3^{0.6}\mathrm{Pr}^{0.33}\left(\frac{T_b}{T_w}\right)^{0.14} \tag{4.123}$$

where $K = 0.33$ for the staggered tube arrangement and $K = 0.26$ for the in-line tube arrangement.

To calculate Re_3, use U_3 calculated using the crossflow area S_q:

$$Re_3 = \frac{U_3 d}{\mu_s} \tag{4.124}$$

Alternately, Slipcevic recommends calculating the heat transfer coefficient of the individual tube row with the corresponding flow cross section. From these heat transfer coefficients, calculate the weighted mean average of heat transfer coefficients.

Step 6: Calculate the heat transfer area in the longitudinal flow section A_1, A_2, crossflow section A_3, and total heat transfer area, A:

$$A_1 = \pi d L N_{t1} \tag{4.125a}$$

$$A_2 = \pi d L N_{t2} \tag{4.125b}$$

$$A_3 = \pi d L N_{t3} \tag{4.125c}$$

where

$$N_{t3} = N_{t2} - \left(N_{t1} + N_{t2}\right) \tag{4.126}$$

$$A = A_1 + A_2 + A_3 \tag{4.127a}$$

$$= \pi d N_t L \tag{4.127b}$$

Step 7: Calculate the heat transfer coefficient h referred to the entire heat transfer surface A:

$$h = \frac{h_1 A_1 + h_2 A_2 + h_3 A_3}{A} \tag{4.128}$$

4.A.2.3 Shellside Pressure Drop

Refer to Slipcevic [68] for pressure-drop calculation.

4.A.2.4 Shortcomings of Disk and Doughnut Heat Exchanger

The disadvantages of this design include the following [7]: (1) All the tie-rods to hold baffles are within the tube bundle, and (2) the central tubes are supported by the disk baffles, which in turn are supported only by tubes in the overlap of the larger diameter disk over the doughnut hole.

4.A.3 NORAM RF™ Radial Flow Gas Heat Exchanger

The NORAM RF radial flow gas heat exchanger (by NORAM Engineering and Constructors Limited, Vancouver, British Columbia, Canada) offers a fundamental improvement in gas heat exchanger performance over designs traditionally used in the sulfuric acid industry. The RF design

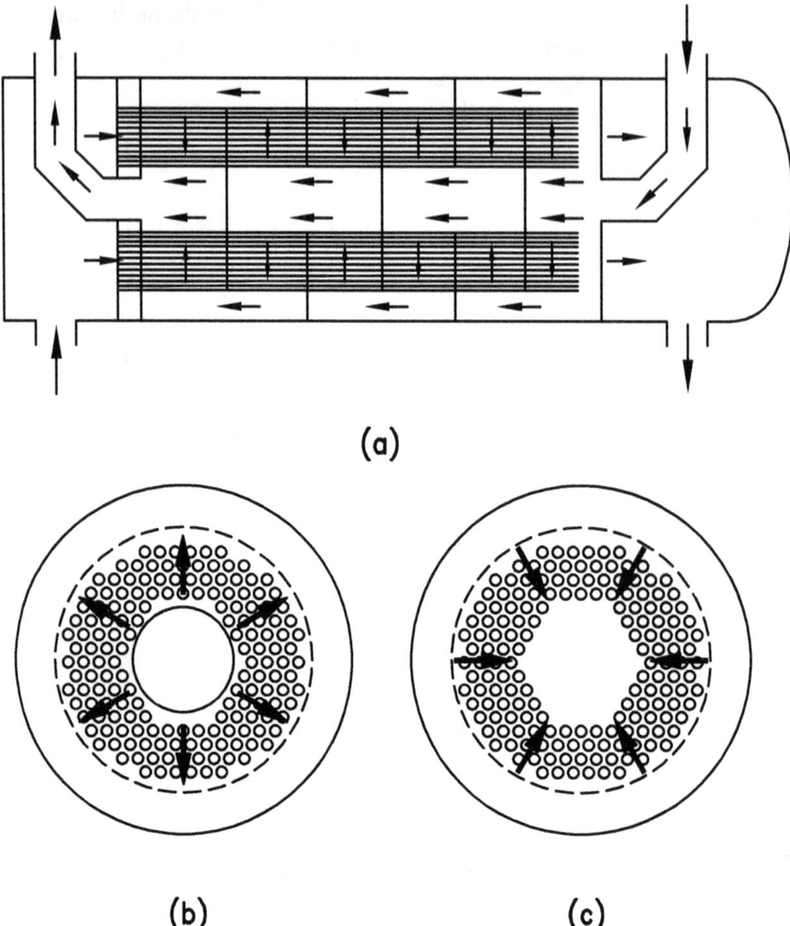

FIGURE 4.97 NORAM RF radial flow gas heat exchanger. (Courtesy of NORAM Engineering and Constructors Limited, Vancouver, British Columbia, Canada.)

far surpasses existing single and double segmental heat exchanger design in heat transfer rates, at a significantly lower pressure loss.

In the RF gas exchanger, tubes are arranged in an annular pattern using disk and doughnut baffles (Figure 4.97). Flow is directed from an inner core outward across the tube bundle and from an outer annulus inward across the tube bundle. The gas turns in the inner core and outer annulus are not restricted by tubes, so that pressure losses are minimized. Heat transfer rates in the RF gas exchanger are maximized because all the heat transfer surface is fully utilized. Shell gas flow is always perpendicular to the tubes providing the maximum heat transfer coefficients.

4.A.3.1 Tube Layout

The RF design makes effective use of the heat transfer surface by placing tubes only in areas of crossflow. The gas flows radially across the tube bundle between an open core and an open outer annulus, directed by disk and doughnut baffles. All tubes are located in an annular domain between the core and the outer annulus.

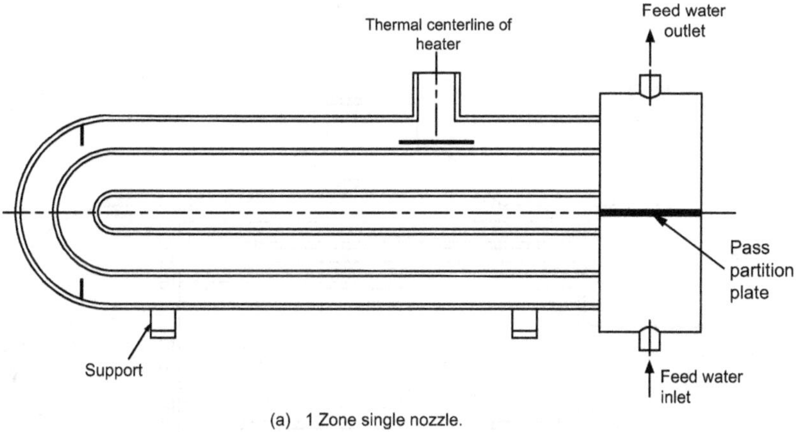

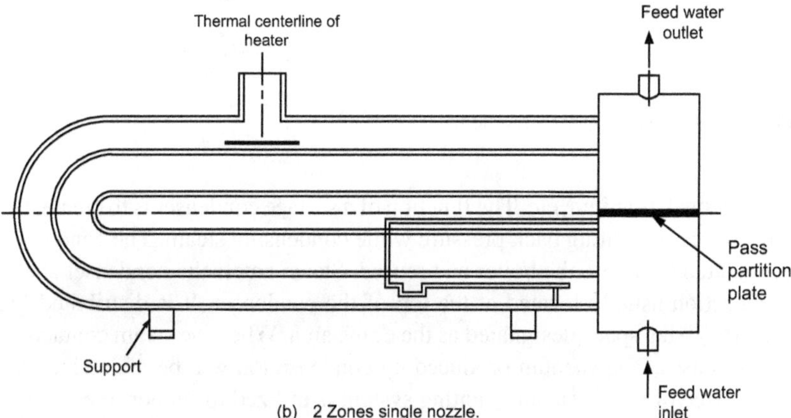

FIGURE 4.98 Single zone and 2 zones feedwater heater(schematic).

4.A.4 Closed Feedwater Heaters

A feedwater heater is a heat exchanger designed to preheat feedwater by means of condensing steam extracted or bled from a steam turbine. Feedwater heaters are used in a regenerative water-system cycle to improve the thermodynamics efficiency, resulting in reductions of fuel consumption and thermal pollution. The boiler feedwater is heated up by steam extracted from suitable turbine stages. The heater is classified as closed, since the tubeside fluid remains in a closed circuit and does not mix with the shellside condensate, as is the case with open feedwater heaters. They are unfired since the heat transfer within the vessel does not occur by means of combustion, but by convection and condensation. Figure 4.98 shows single-zone and two-zone feedwater heaters (schematic). They are designed as per the HEI standard for closed feedwater heaters [72] or other national standards. For design of power plant heat exchangers, refer to Ref. [73].

4.A.5 Steam Surface Condenser

The condenser is simply a large heat exchanger with tubes usually horizontally mounted. The condenser has thousands of small tubes that are made out of admiralty brass, copper, ferritic/super

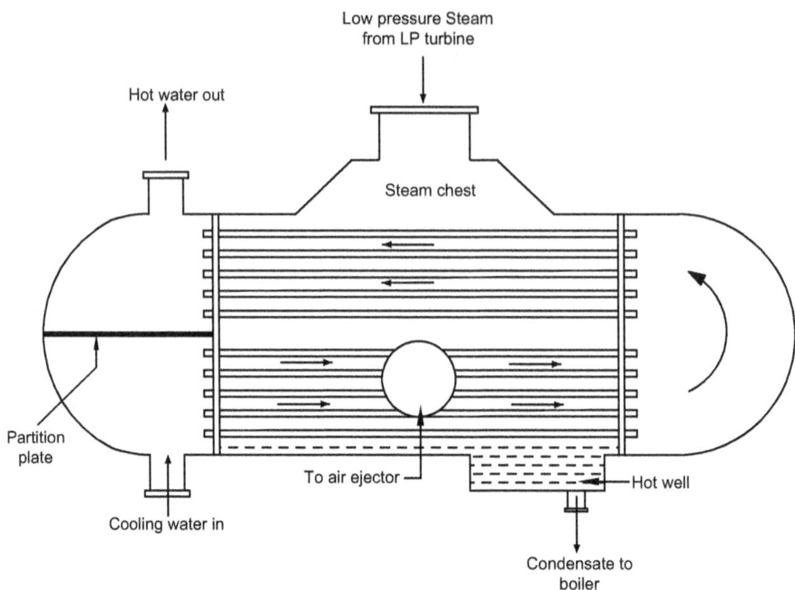

FIGURE 4.99 Steam surface condenser(schematic).

austenitic stainless steel, titanium, etc. The function of a surface condenser is to create the lowest possible turbine or process operating back pressure while condensing steam. The condensate generated is usually recirculated back into the boiler and reused. Steam enters the condenser shell through the steam inlet connection usually located at the top of the condenser. It is distributed longitudinally over the tubes through the space designated as the dome area. When the steam contacts the relatively cold tubes, it condenses. The vacuum produced by condensation will be maintained as long as the condenser is kept free of air. A vacuum venting system is utilized to support the condenser vacuum by continually removing any air entering the system. The condensate is continually removed from the hotwell by condensate pump(s) and is discharged into the condensate system. The air in the system, generally due to leakage in piping around shaft seals, valves, etc., enters the condenser and mixes with the steam. The saturated air is removed from the condenser by the vacuum venting equipment such as steam jet air ejectors, liquid ring vacuum pumps, or a combination of both. It is necessary to continuously remove air from the system in order to maintain the desired vacuum. An increasing amount of air in the condenser would reduce its capacity and cause the pressure to rise. For standards on steam surface condensers refer to Ref. [74]. A typical steam surface condenser is shown in Figure 4.99 (schematic).

Nomenclature

A	total heat transfer area on the shellside
A_n	crossflow area within the limits of tube bundle
A_w	area of one baffle window
A_1	heat transfer area for longitudinal flow in the annulus between the shell and the disk
A_2	heat transfer area for longitudinal flow through the ring opening
A_3	heat transfer area for crossflow
a_1	bypass clearance constant
a_2	baffle hole clearance constant
a_3	baffle edge clearance constant

a_1–a_8	factors required to determine F_h
b_1,b_2,b_3	factors required to determine F_h
C_L	laminar heat transfer bundle geometry function
C_T	turbulent heat transfer bundle geometry function
D_1	disk diameter
D_1	shell inside diameter (D_s)
D_2	diameter of ring opening
D_2	baffle diameter
D_3	diameter of OTL (D_{otl})
d_B	baffle hole diameter
D_m	mean diameter for calculating the crossflow area = 0.5 (D_1 + D_2)
D_s	shell inside diameter
d	tube outer diameter (for finned tubes, equal to fin root diameter)
F_h	fraction of total fluid flowing within the limits of tube bundle
g_c	acceleration due to gravity
h	baffle cut
h_1	heat transfer coefficient for the longitudinal flow corresponding to Nu_1
h_2	heat transfer coefficient for the longitudinal flow corresponding to Nu_2
h_3	heat transfer coefficient for the crossflow corresponding to Nu_3
K	heat transfer correlation coefficient
L	tube length
L_{bc}	baffle spacing
M	multiplication factor in Equation 4.86
M_s	mass flow rate of shellside fluid
M_wA_x	multiplier to obtain geometric mean of A_x and A_w
N_1	total number of tubes
N_{t1}	number of tubes in annulus between the shell and the disk
N_{t2}	number of tubes in the ring opening
N_{t3}	number of tubes in crossflow
Nu	Nusselt number
Nu_1	Nusselt number to calculate h_1
Nu_2	Nusselt number to calculate h_2
Nu_3	Nusselt number to calculate h_3
p	tube pitch
Pr	Prandtl number
Re_h	Reynolds number
Re_1	Reynolds number in the longitudinal flow area S_1
Re_2	Reynolds number in the longitudinal flow area S_2
Re_3	Reynolds number in the longitudinal flow area S_3
S	exchanger size ratio = D_1/p
S_1	cross-sectional area for longitudinal flow through the annulus between the shell and the disk
S_2	cross-sectional area for longitudinal flow through the ring opening
S_{k1}	annular area between the disk and the shell
S_{k2}	area inside the ring
S_q	crossflow area through tube bundle
S_s	mean longitudinal flow area = 0.5(S_1 + S_2)
T_b	bulk mean temperature of the shellside fluid
T_w	tube wall temperature
U_1	shellside velocity through the cross-sectional area S_1
U_2	shellside velocity through the cross-sectional area S_2
U_3	shellside velocity through the cross-sectional area S_3
Δp_L	pressure drop due to unbaffled frictional component

Δp_b	pressure drop due to baffle flow contribution
μ_s	viscosity at bulk mean temperature of the shellside fluid (kg/m s)
μ_w	viscosity at tube wall temperature of the shellside fluid (kg/m s)
$(\phi_s)^n$	bulk-to-wall viscosity correction $(\mu_s/\mu_w)^{0.14}$
ρ_s	density of shellside fluid

REFERENCES

1. Minton, P.E. (1990) Process heat transfer. *Proceedings of the 9th International Heat Transfer Conference, Heat Transfer 1990*, Jerusalem, Israel, Paper No. KN-22, Vol. 1, pp. 355–362.
2. TEMA (2019) *Standards of Tubular Exchanger Manufacturers Association*, 10th edn. Tubular Exchanger Manufacturers Association, Tarrytown, NY.
3. American Society of Mechanical Engineers (2021) *ASME Boiler and Pressure Vessel Code, Section VIII, Division 1—Pressure Vessels*. American Society of Mechanical Engineers, New York.
4. Gutterman, G., (1980). Specify the right heat exchanger, *Hydrocarbon Process*, April, 161–163.
5. www.thermopedia.com/content/1121/
6. Drake, C.E., Carp, J.R. (1960) Shell and tube heat exchangers. *Chemical Engineering* 36: 165–170.
7. Shah, R.K., Sekulic, D.P. (1998) Heat exchangers. In: *Handbook of Heat Transfer Applications*, Rohsenow, W. M., Hartnett, J. P., Ganic, E. N. (Eds.), 3rd edn., pp. 17.1–17.169, Chapter 17. McGraw-Hill, New York.
8. Bell, K.J., Mueller, A.C. (1984) *Wolverine Heat Transfer Data Book II*. Wolverine Division of UOP Inc., Decatur, AL.
9. Lord, R.C., Minton, P.E., Sulusser, R.P. (1970) Design of exchangers. *Chemical Engineering* January: 96–118.
10. Saunders, E.A.D. (1983) Features relating to thermal design. In: *Heat Exchanger Design Handbook*, Vol. 4 (E. U. Schlunder, editor-in-chief), Section 4.2.4. Hemisphere, Washington, DC.
11. (a) Larowski, A., Taylor, M.A. (1982) *Systematic Procedures for Selection of Heat Exchangers*, C58/82, pp. 32–56. Institution of Mechanical Engineers, London, UK; (b) Larowski, A., Taylor, M.A. (19830 Systematic procedure for selection of heat exchangers. IMechE, *Proceedings of the Institution of Mechanical Engineers* 197A: 51–69.
12. Phadke, P.S. (1984) Determining tube counts for shell and tube exchangers. *Chemical Engineering* 91: 65–68.
13. Whitly, D.L., Ludwig, E.E. (1961) Rate exchangers this computer way. *Petroleum Refinery* 1961: 147–156.
14. Fraas, A.P., Ozisik, M.N. (1964) *Heat Exchanger Design*. John Wiley & Sons, New York.
15. Palen, J.W., Taborek, J. (1969) Solution of shellside flow, pressure drop and heat transfer by stream analysis method. *Chemical Engineering, Symposium Series*, No. 92, *Heat Transfer—Philadelphia* 65: 53–63.
16. Fanaritis, J.P., Bevevino, J.W. (1979) How to select the optimum shell and heat exchanger. In: *Process Heat Exchange, Chemical Engineering Magazine* (V. Cavaseno, ed.), pp. 5–13. McGraw-Hill, New York.
17. Taborek, J. (1983) Shell and tube heat exchanger design—Extension of the method to other shell, baffle, and tube bundle geometries. In: *Heat Exchanger Design Handbook*, Vol. 3 (E.U. Schlunder, editor-in-chief), , Section 3.3.11. Hemisphere, Washington, DC.
18. Gupta, J.P. (1986) *Fundamentals of Heat Exchanger and Pressure Vessel Technology*. Hemisphere, Washington, DC.
19. Singh, K.P. (1988) Mechanical design of tubular heat exchangers—An appraisal of the state-of-the-art. In: *Heat Transfer Equipment Design* (R.K. Shah, C. Subbarao, R.M. Mashelekar, eds.), pp. 71–87. Hemisphere, Washington, DC.
20. Boyer, R.C., Pase, G.K. (1980) The energy saving NESTS concept. *Heat Transfer Engineering* 2: 19–27.
21. Butterworth, D. (1992) New twists in heat exchange. *Chemical Engineering* September: 23–28.

22. IAEA (1999) *Assessment and management of aging of major nuclear power plant components important for the Safety Steam Generator*, IAEA-TEC DOC 981, pp. 1–181. International Atomic Energy Agency, Vienna, Austria.
23. Ebert, H.W. (2000) Improving the reliability of tube-to-tubesheet joints. *Welding Journal of the American Welding Society* 79(9): 47–49.
24a. Yokell, S. (1990) *A Working Guide to Shell and Tube Heat Exchangers*, McGraw-Hill, New York.
24b. Yokell, S. (1973) Double tubesheet heat exchanger design stops shell-tube leakage. *Chemical Engineering* May: 133–136.
25. Woodward, A.R., Howard, D.L., Andrews, E.F.C. (1991) Condensers, pumps and cooling water plant. In *Modern Power Station Practice, Vol. C, Turbines, Generators, and Associated Plant* (D.J. Littler, E.J. Davies, H.E. Johnson, F. Kirkby, P.B. Myerscouh, W. Wright, eds.), 3rd edn., Chapter 4. Pergamon Press, New York.
26. Spencer, T.C. (1987) Mechanical design and fabrication of exchangers in the United States. *Heat Transfer Engineering* 8: 58–61.
27. Gollin, M. (1989) Heat exchanger design and rating. In: *Handbook of Applied Thermal Design* (E.C. Guyer, ed.), pp. 7-24–7-36. McGraw-Hill, New York.
28. Taborek, J. (1983) Design procedures for segmentally baffled heat exchangers. In: *Heat Exchanger Design Handbook*, Vol. 3 (E.U. Schlunder, editor-in-chief), pp. 3.3.10-1–3.3.10–8. Hemisphere, Washington, DC.
29. Mueller, A.C. (1985) Process heat exchangers. In: *Handbook of Heat Transfer Applications*, 2nd edn. (W.M. Rohsenow, J.P. Hartnett, E.N. Ganic, eds.), pp. 4-78–4-173. McGraw-Hill, New York.
30. www.globalspec.com/learnmore/processing_equipment/heat_transfer_equipment/heat_exchangers
31. Singh, K.P., Soler, A.I. (1984) *Mechanical Design of Heat Exchangers and Pressure Vessel Components*. Arcturus Publishers, Cherry Hill, NJ.
32. Jaw, L. (1964) Temperature relations in shell and tube exchangers having one pass splitflow shells. *Transactions of ASME, Journal of Heat Transfer* 86C: 408–416.
33. Grant, I.D.R. (1980) Shell and tube exchangers for single phase applications. In: *Developments in Heat Exchanger Technology—1* (D. Chisholm, ed.), pp. 11–40. Applied Science Publishers, London, UK.
34. Martin, D.J., Haseler, L.E., Hollingsworth, M.A., Mathew, Y.R. (1990) The use of sealing strips to block bypassing flow in a rectangular tube bundle. *Heat Transfer*, Jerusalem, Israel, Paper No. 14-HX-23, pp. 33–138.
35. Taylor, C.E., Currie, I.G. (1987) Sealing strips in tubular heat exchangers. *Transactions of ASME, Journal of Heat Transfer* 109: 569–573.
36. Tinker, T. (1951) Shellside characteristics of shell and tube heat exchangers—Proceedings on general discussion on heat transfer, Parts I, II and III. *Institution of Mechanical Engineers London* 97–116.
37. Tinker, T. (1958) Shellside characteristics of shell and tube heat exchangers—A simplified rating system for commercial heat exchangers. *Transactions of ASME* 80: 36–52.
38. Bell, K.J. (1963) *Final report of the cooperative research program on shell and tube heat exchangers, Bull. No. 5*, University of Delaware Engineering Experiment Station, New York, June 1963.
39. Bell, K.J. (1981) Delaware method for shellside design. In: *Heat Exchangers—Thermal Hydraulic Fundamentals and Design* (S. Kakac, A.E. Bergles, F. Mayinger, eds.), pp. 581–618. Hemisphere/McGraw-Hill, Washington, DC.
40. Bell, K.J. (1988) Delaware method for shellside design. In: *Heat Transfer Equipment Design* (R.K. Shah, E.C. Subbarao, R. A. Mashelkar, eds.), pp. 145–166. Hemisphere, Washington, DC.
41a. Mukherjee, R. (1998) Effectively design shell and tube heat exchangers. *Chemical Engineering Progress, American Institute of Chemical Engineers* 94(2): 21–37.
41b. Mukherjee, R. (2004) *Practical Thermal Design of Shell and Tube Heat Exchangers*. Begell House, Inc., Madison Avenue, New York, NY.
42. Hewitt, G.F., Shires, G.L., Bott, T.R. (1994) *Process Heat Transfer*. CRC Press, Boca Raton, FL.
43. Bell, K. J., Approximate sizing of shell and tube heat exchangers, in *Heat Exchanger Design Handbook*, Vol. 3 (E. U. Schlunder, editor-in-chief), Hemisphere, Washington, DC, 1983, Section 3.1.4.
44. Kern, D.Q. (1950) *Process Heat Transfer*. McGraw-Hill, New York.

45. Donohue, D.A. (1955/1956) Heat exchanger design. In: *Petroleum Refiner: Part 1, Types and Arrangements* 34: 94–100; *Part 2, Heat Transfer* 34: 9–8; *Part 3, Fluid Flow and Thermal Design* 34: 175–184; *Part 4, Mechanical Design* 35: 155–160.
46. Devore, A. (1961) Try this simplified method for rating baffled heat exchangers. *Hydrocarbon Process: Petroleum Refining* 40: 221–233.
47. McAdams, W.H. (1954) *Heat Transmission*, 3rd edn., p. 229. McGraw-Hill, New York.
48. Taborek, J. (1983) Ideal tube bank correlations for heat transfer and pressure drop. In: *Heat Exchanger Design Handbook*, Vol. 3, Section 3.3.7 (E.U. Schlunder, editor-in-chief). Hemisphere, Washington, DC.
49. Kays, W.M., London, A.L. (19840 *Compact Heat Exchangers*, 3rd edn. McGraw-Hill, New York.
50. Briggs, D.E., Young, E.H. (1963) Convection heat transfer and pressure drop of air flowing across triangular pitch bank of finned tubes. *Chemical Engineering Progress Symposium Series*, No. 41, *Heat Transfer—Houston,* 59; 1–10.
51. Briggs, D.E., Katz, D.L., Young, E.H. (1963) How to design finned tube heat exchangers. *Chemical Engineering Progress* 59: 49–59.
52. Heat Transfer Research, Inc. (HTRI) *Xchanger Suite6® software*. Heat Transfer Research, Inc. (HTRI), Alhambra, CA.
53. *Aspen Shell & Tube Exchanger software*, Heat transfer and fluid flow service (HTFS), Oxon, UK.
54. Gentry, C.C. (1990) ROD baffle heat exchanger technology. *Chemical Engineering Progress* July: 48–57.
55. Gentry, C.C. (1993) ROD baffle heat exchanger design methods. *Sixth International Symposium on Transport Phenomena in Thermal Engineering*, Seoul, Korea, May 9–13, 1993, pp. 301–306.
56. Gentry, C.C., Young, R.K., Small, W.M. (1982) RODbaffle heat exchanger thermal and hydraulic predictive methods. *Proceedings of the 7th International Heat Transfer Conference*, Munich, Germany, Vol. 6, pp. 197–202. Hemisphere, Washington, DC,
57. Gentry, C.C., Young, R.K., Small, W.M. (1984) RODbaffle heat exchanger thermal and hydraulic predictive methods for bare and low finned tubes. *Proceedings of the 22nd National Heat Transfer Conference, Heat Transfer*, Niagara Falls, NY, 1984, *AIChE Symposium Series* 80: 104–109.
58. Gentry, C.C., Young, R.K., Small, W.M. (1982) A conceptual RODbaffle nuclear steam generator design. *1982 ASME Joint Power Generation Conference, Nuclear Heat Exchanger Sessions*, Denver, CO, October 17–21, 1982.
59. Hesselgraeaves, J.E. (1988) *Proceedings of the 2nd UK National Conference on Heat Transfer*, Glasgow, UK, Vol. 1, pp. 787–800. Institution of Mechanical Engineers, London, UK.
60. Taborek, J. (1989) Longitudinal flow in tube bundles with grid baffles. In: *Heat Transfer*, pp. 72–78. Philadelphia, PA.
61. (a) *EMbaffle® Technology, a major advance in heat exchanger technology* (www.shellglobalsolutions.com/EMbaffle); (b) *EMbaffle® Technology, Shell Global Solutions* (www.shellglobalsolutions.com/EMbaffle).
62. (a) Chunangad, K.S., Master, B.I., Thome, J.R., Tolba, M.B. (1999) Helixchanger heat exchanger: Single-phase and two-phase enhancement. *Proceedings of the International Conference on Compact Heat Exchangers and Enhancement Technology for the Process Industries*, , pp. 471–477. Begell House, New York; (b) Zhang, Z., Fang, X. (2006) Comparison of heat transfer and pressure drop for the helically baffled heat exchanger combined with three-dimensional and two-dimensional finned tubes. *Heat Transfer Engineering* 27(7): 17–22.
63. (a) *Design and Application of TWISTED TUBE® Heat Exchangers*, Koch Heat Transfer Company, Houston, TX (www.kochheattransfer.com); (b) *Heat Transfer Innovators, Byron Black, Brown fintube innovators*, pp. 1–9.
64. Butterworth, D., Guy, A.R., Welkey, J.J. (1996) Design and application of twisted tube heat exchangers, advances in industrial heat transfer. *IChemE* 87–95.
65. Short, B.E. (1943) *Heat Transfer and Pressure Drop in Heat Exchangers*, Publication No. 4324, pp. 1–54. University of Texas, Austin, TX.
66. Donohue, D.A. (1969) Heat transfer and pressure drop in heat exchangers. *Industrial Engineering Chemistry* 41: 2499–2511.
67. Slipcevic, B. (1976) Designing heat exchangers with disk and ring baffles. *Sulzer Technical Reviews* 3: 114–120.

68. Slipcevic, B. (1978) Shellside pressure drop in shell and tube heat exchangers with disk and ring baffles. *Sulzer Technical Reviews* 60: 28–30.
69. Goyal, K.P., Gupta, B.K. (1984) An experimental performance evaluation of a disk and doughnut type heat exchanger. *Transactions of ASME, Journal of Heat Transfer* 106: 759–765.
70. Ratnasamy, K. (1987) Exchanger design using disc and donut baffles. *Hydrocarbon Processes* April: 63–65.
71. Hausen, H. (1950) *Warmeubertrangung im Gegenstrom und Kreuzstrom*. Springer-Verlag, Berlin, Germany.
72. Heat Exchange Institute (2015) *Standards for Closed Feedwater Heaters*, 9th Edition. Heat Exchange Institute, Cleveland, Ohio.
73. Heat Exchange Institute (2013) *Standards for Shell and Tube Exchangers*, 5th edn. Heat Exchange Institute, Cleveland, OH.
74. Heat Exchange Institute (2017) *Standards for Steam Surface Condensers*, 12th Edition. Heat Exchange Institute, Cleveland, Ohio.

Suggested Reading

Bell, K.J. (1988) Overall design methodology for shell and tube exchangers. In: *Heat Transfer Equipment Design* (R.K. Shah, E.C. Subbarao, R.A. Mashelkar, eds.), pp. 131–144. Hemisphere, Washington, DC.

Escoe, K.A. (1994) *Mechanical Design of Process Systems, Vol. 2, Shell and Tube Heat Exchangers, Rotating Equipment, Bins, Silos, Stacks*. Gulf Publishing Company, Houston, TX.

Gentry, G.C., Gentry, M.C., Scanlon, G.E. (1999) *Heating the Process and Fluid Flow, Shell and Tube Heat Exchanger Applications*, pp. 95–101. PTQ Autumn.

Gulley, D. (1996) Troubleshooting shell and tube heat exchangers. *Hydrocarbon Process* September: 91–98.

McNaughton, K.J. (Ed.) (1986) *The Chemical Engineering Guide to Heat Transfer, Vol. I. Plant Principles*. Hemisphere/McGraw-Hill, New York.

Mehra, D.K. (1986) Shell and tube exchangers. In: *The Chemical Engineering Guide to Heat Transfer, Vol. I, Plant Principles* (K.J. McNaughton and the staff of Chemical Engineering, eds.), pp. 43–52. Hemisphere/McGraw-Hill, New York.

www.linkedin.com/pulse/collar-bolts-shell-tube-heat-exchangers-baher-elsheikh/

Saunders, E.A.D. (1989) *Heat Exchangers: Selection, Design and Construction*. Addison Wesley Longman, Reading, MA.

Soler, A.I. (1985) Expert system for design integration—Application to the total design of shell and tube heat exchangers. *Proceedings of the ASME, Thermal/Mechanical Heat Exchanger Design, Karl Gardner Memorial Session*, Vol. 118 (K.P. Singh, S.M. Shenkman, eds.), pp. 135–138. ASME PVP, Washington, DC.

Taborek, J. (1978) Heat exchanger design. *Heat Transfer* 4: 269–285.

Yokell, S. (1986) Heat exchanger tube-to-tubesheet connections. *The Chemical Engineering Guide to Heat Transfer, Vol. I, Plant Principles* (K.J. McNaughton, ed.), pp. 76–92. Hemisphere, New York.

5 Boiling, Condensation, and Steam Generation

5.1 INTRODUCTION

Boiling, condensation, and evaporation heat transfer occur in many engineering applications, such as in power plant condensers, boilers, and steam generators, which are all important components in conventional and nuclear power stations. In the design of these condensers and evaporators, proper correlations must be selected to calculate the condensing and boiling heat transfer coefficients. The most common type of condensation involved in heat exchangers is surface condensation, where a cold wall, at a temperature lower than the local saturation temperature of the vapor, is placed in contact with the vapor. Boiling is a complex phenomenon, and boiling heat-transfer coefficients are difficult to predict with any certainty. Whenever possible, experimental values obtained for the system being considered should be used, or values for a closely related system. The formation of steam bubbles along a heat transfer surface has a significant effect on the overall heat transfer rate. Some of the boiling heat transfer configurations are shown in Figure 5.1. An important reference source on phase change heat transfer is Wolverine Engineering Data Book III [1] and Ref. (2).

5.2 BOILING HEAT TRANSFER

The boiling process commencing from nucleate boiling to film boiling is graphically represented in Figure 5.2 [3]. This figure illustrates the effect of boiling on the relationship between the heat flux and the temperature difference between the heat transfer surface and the fluid passing it. The following discussion is based on DoE Fundamentals Handbook: Thermodynamics, Heat Transfer, and Fluid Flow [3].

Four regions, namely, natural convection boiling, nucleate boiling, partial film boiling, and film boiling are represented in Figure 5.2 [3]. The first and second regions show that as heat flux increases, the temperature difference (surface to fluid) does not change very much. Better heat transfer occurs during nucleate boiling than during natural convection. As the heat flux increases, the bubbles become numerous enough that partial film boiling (part of the surface being blanketed with bubbles) occurs. This region is characterized by an increase in temperature difference and a decrease in heat flux. The increase in temperature difference thus causes total film boiling, in which steam completely blankets the heat transfer surface.

5.2.1 Nucleate Boiling

In nucleate boiling, steam bubbles form at the heat transfer surface and then break away and are carried into the main stream of the fluid. Such movement enhances heat transfer because the heat generated at the surface is carried directly into the fluid stream. Once in the main fluid

DOI: 10.1201/9781003352044-5

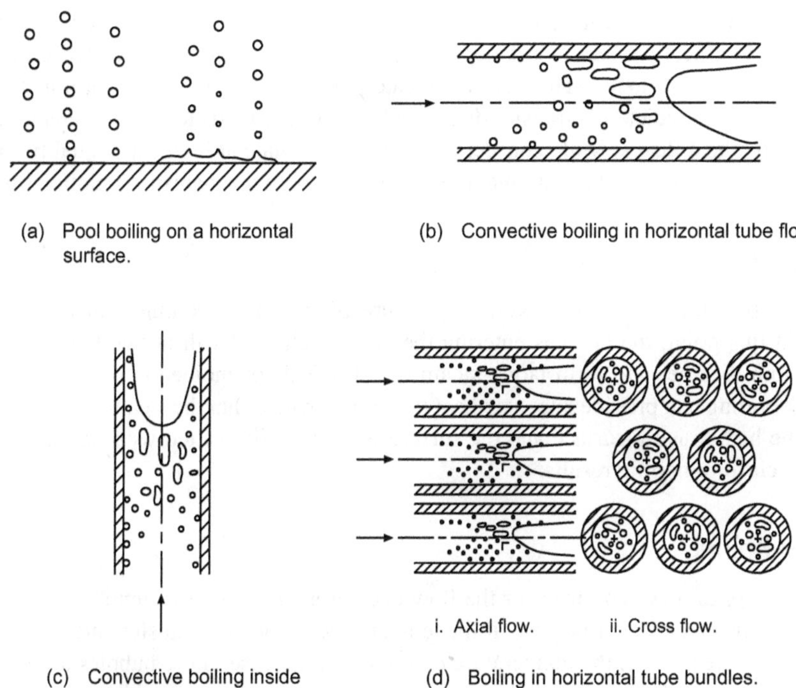

(a) Pool boiling on a horizontal surface.
(b) Convective boiling in horizontal tube flow.
(c) Convective boiling inside vertical tubes.
i. Axial flow.
ii. Cross flow.
(d) Boiling in horizontal tube bundles.

FIGURE 5.1 Some of the boiling heat transfer configurations.

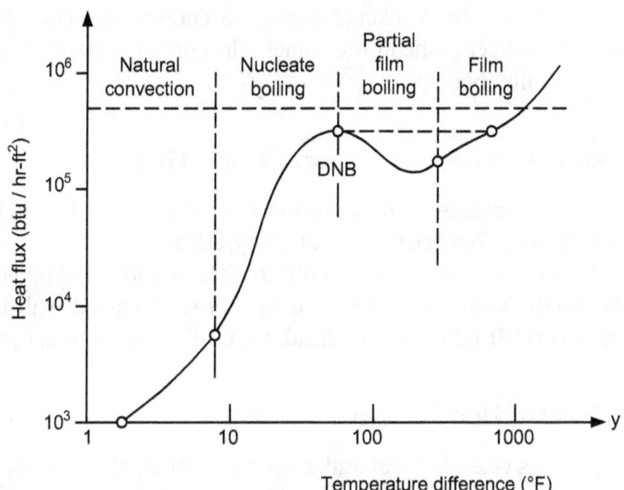

FIGURE 5.2 Boiling Curve. *Note*: DNB, departure from nucleate boiling. (Adapted from DOE Fundamentals Handbook: Thermodynamics, Heat Transfer, and Fluid Flow, Volume 2 of 3, DOE-HDBK-1012/2-92 JUNE 1992 [3]. https://www.steamtablesonline.com/pdf/Thermodynamics-Volume2.pdf.)

stream, the bubbles collapse because the bulk temperature of the fluid is not as high as the heat transfer surface temperature where the bubbles were created. This heat transfer process is sometimes desirable because the energy created at the heat transfer surface is quickly and efficiently "carried" away.

In nucleate pool boiling, heat transfer is a strong function of heat flux, instead, in forced convective evaporation, heat transfer is less dependent on heat flux while its dependence on the local vapor quality and mass velocity. Thus, both nucleate boiling and convective heat transfer must be taken into account to predict heat transfer data. Nucleate boiling tends to be dominant at low vapor qualities and high heat fluxes, while convection tends to dominate at high vapor qualities and mass velocities and low heat fluxes. For intermediate conditions, both mechanisms are often important.

5.2.2 Bulk Boiling

As system temperature increases or system pressure drops, the bulk fluid can reach saturation conditions. At this point, the bubbles entering the coolant channel will not collapse. The bubbles will tend to join together and form larger steam bubbles. This phenomenon is referred to as bulk boiling. Bulk boiling can provide adequate heat transfer provided that the steam bubbles are carried away from the heat transfer surface and the surface is continually wetted with liquid water. When this cannot occur film boiling results.

5.2.3 Film Boiling

When the pressure of a system drops or the flow decreases, the bubbles cannot escape as quickly from the heat transfer surface. Likewise, if the temperature of the heat transfer surface is increased, more bubbles are created. As the temperature continues to increase, more bubbles are formed than can be efficiently carried away. The bubbles grow and group together, covering small areas of the heat transfer surface with a film of steam. This is known as partial film boiling. As the area of the heat transfer surface covered with steam increases, the temperature of the surface increases dramatically, while the heat flux from the surface decreases. This unstable situation continues until the affected surface is covered by a stable blanket of steam, preventing contact between the heat transfer surface and the liquid in the center of the flow channel. The condition after the stable steam blanket has formed is referred to as film boiling.

5.2.4 Departure from Nucleate Boiling and Critical Heat Flux

In practice, if the heat flux is increased, the transition from nucleate boiling to film boiling occurs suddenly, and the temperature difference increases rapidly, as shown by the dashed line in Figure 5.2. The point of transition from nucleate boiling to film boiling is called the point of departure from nucleate boiling, commonly written as DNB. The heat flux associated with DNB is commonly called the critical heat flux (CHF). In many applications, CHF is an important parameter.

5.2.5 Subcooled Boiling Heat Transfer

Subcooled flow boiling occurs when the local wall temperature during the heating of a subcooled liquid is above the saturation temperature of the fluid and sufficiently high for nucleation to occur. Subcooled boiling is characterized by vapor formation at the heated wall as isolated bubbles or as a bubbly layer along the wall. The bubbles are swept into the subcooled core by the liquid and then condense.

5.2.6 Boiling in a Vertical Tube

There are various hydrodynamic conditions encountered when a liquid is evaporated in a confined channel, in this case round tubes and methods to predict their heat transfer coefficients. First of all, consider a vertical tube heated uniformly along it at a relatively low heat flux with subcooled liquid entering the tube from the bottom and then completely evaporated over the length of the tube, as shown in Figure 5.3 [4]. While the liquid is being heated up to its saturation temperature at the local

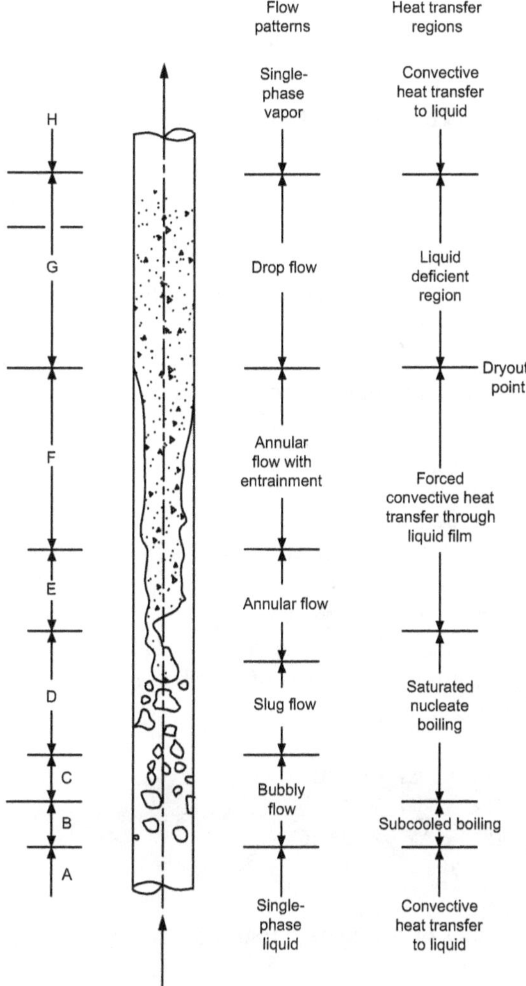

FIGURE 5.3 Heat transfer regions in convective boiling in a vertical tube from Collier and Thome (5).

pressure at that height in the tube, the wall temperature initially is below that necessary for nucleation (zone A).

Then, the wall temperature rises above the saturation temperature and boiling nucleation takes place in the superheated thermal boundary layer on the tube wall, such that subcooled flow boiling occurs in zone B with the vapor bubbles condensing as they drift into the subcooled core. The liquid then reaches its saturation temperature and saturated boiling in the form of bubbly flow begins in zone C. Saturated boiling continues through the slug flow regime (zone D), the annular flow regime (zone E), and then the annular flow regime with liquid entrainment in the vapor core (zone F). At the end of zone F, the annular film is either dried out or sheared from the wall by the vapor, a point that is referred to as the onset of dryout or simply dryout. Above this point, mist flow in the form of entrained droplets occurs with a large increase in wall temperature for this instance of an imposed wall heat flux (zone G).

The temperature of the continuous vapor phase in zone G tends to rise above the saturation temperature and heat transfer is via four mechanisms: single-phase convection to the vapor, heat transfer to the droplets within the vapor, heat transfer to droplets impinging on the wall, and thermal radiation from the wall to the droplets. Because of this non-equilibrium effect, droplets continue to exist in the vapor phase beyond the point of $x = 1$, all the way to the beginning of zone H where all the liquid has been evaporated and heat transfer is by single-phase convection to the dry vapor at transfer regions in convective boiling in a vertical tube from Collier and Thome [5].

5.2.7 Flow Boiling inside Horizontal Plain Tubes

Flow patterns formed during the generation of vapor in horizontal evaporator tubes are shown in Figure 5.4 (adapted from Collier and Thome [5]). The schematic representation of a horizontal tubular channel heated by a uniform low heat flux and fed with liquid just below the saturation temperature for a relatively low inlet velocity illustrates the sequence of flow patterns that might be observed. Asymmetric distributions of the vapor and liquid phases due to the effects of gravity introduce new complications compared to vertical upflow. Important points to note from a heat transfer standpoint are the possibility of complete drying or intermittent drying of the tube wall around part of the tube perimeter, particularly in slug and wavy flow and for annular flow with partial dryout.

5.2.8 Boiling Heat Transfer

Pioneering work on boiling was carried out in 1934 by S. Nukiyama [6], who used electrically heated nichrome and platinum wires immersed in liquids in his experiments. Nukiyama was the first to identify different regimes of pool boiling using his apparatus. He noticed that boiling takes different forms depending on the value of the wall superheat temperature ΔT_{sat} (also known as the excess temperature), defined as the difference between the wall temperature, T_{wall}, and the saturation temperature, T_{sat}. Four different boiling regimes of pool boiling, as shown in Figure 5.1 (based on the excess temperature), are observed [6]:

i. Natural convection boiling $\Delta T_{sat} < 5°C$
ii. Nucleate boiling $5°C < \Delta T_{sat} < 30°C$
iii. Transition boiling $30°C < \Delta T_{sat} < 200°C$
iv. Film boiling $200°C < \Delta T_{sat}$

Refer to Ref. [7] for correlations of boiling heat transfer.

5.2.8.1 Nucleate Boiling Correlations

The Rohsenow correlation for nucleate pool boiling is as follows.

The most widely used correlation for the rate of heat transfer in the nucleate pool boiling was proposed in 1952 by Rohsenow [8]:

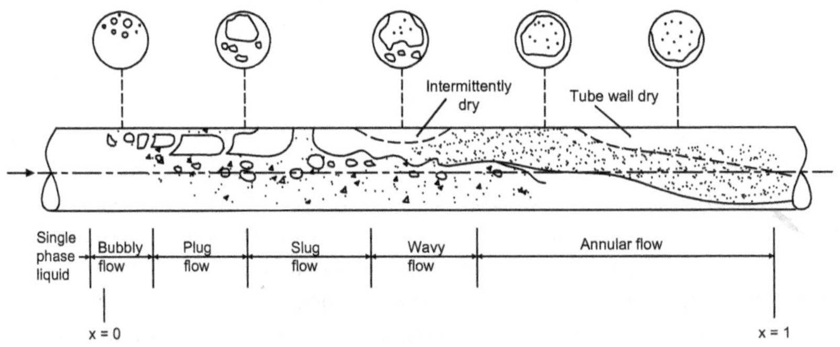

FIGURE 5.4 Flow patterns during evaporation in a horizontal tube from Collier and Thome (5).

$$\frac{Q}{A} = q = \left[\frac{C_1 \Delta T}{h_{fg} Pr^n C_{sf}}\right]^3 \mu_1 h_{fg} \left[\frac{g(\rho_1 - \rho_\upsilon)}{g_0 \sigma}\right]^{0.5} \tag{5.1}$$

where

q is nucleate pool boiling heat flux (W/m^2)
c_1 is specific heat of liquid (J/kg K)
ΔT is excess temperature (°C or K)
h_{fg} is enthalpy of vaporization (J/kg)
Pr is Prandtl number of the liquid
n is experimental constant equal to 1 for water and 1.7 for other fluids
C_{sf} is surface fluid factor, for example, water and nickel have a C_{sf} of 0.006
μ_1 is dynamic viscosity of the liquid (kg/m.s)
g is gravitational acceleration (m/s^2)
g_0 is force conversion factor (kgm/Ns^2)
ρ_1 is density of the liquid (kg/m^3)
ρ_v is density of vapor (kg/m^3)
σ is surface tension–liquid–vapor interface (N/m)

As can be seen from Eq. 5.1, $\Delta T \propto (q)^{1/3}$. This very important proportionality shows the increasing ability of the interface to transfer heat.

5.2.8.2 Flow Boiling or Forced Convection Boiling

In flow boiling (or forced convection boiling), fluid flow is forced over a surface by external means such as a pump and buoyancy effects. Therefore, flow boiling is always accompanied by other convection effects. Conditions depend strongly on geometry, which may involve external flow over heated plates and cylinders or internal (duct) flow. Flow boiling is also classified as either external or internal flow boiling, depending on whether the fluid is forced to flow over a heated surface or inside a heated channel. The two-phase flow in a tube exhibits different flow boiling regimes, depending on the relative amounts of the liquid and vapor phases. Therefore, internal forced convection boiling is commonly referred to as two-phase flow.

5.2.8.3 Chen's Correlation for Flow Boiling

In 1963, Chen [9] proposed the first flow boiling correlation for evaporation in vertical tubes to attain widespread use. Chen's correlation includes both the heat transfer coefficients due to nucleate boiling and forced convective mechanisms. It must be noted that the nucleate pool boiling correlation of Forster and Zuber [10] is used to calculate the nucleate boiling heat transfer coefficient, h_{FZ}, and the turbulent flow correlation of Dittus-Boelter is used to calculate the liquid-phase convective heat transfer coefficient, h_1.

The nucleate boiling suppression factor, S, is the ratio of the effective superheats to wall superheat. It accounts for decreased boiling heat transfer because the effective superheat across the boundary layer is less than the superheat based on wall temperature. The two-phase multiplier, F, is a function of the Martinelli parameter χ_t

$$h_{tp} = h_{fz} S + h_l F \tag{5.2}$$

$$h_{tp} = S\frac{kl^{0.79}cl^{0.45}\rho l^{0.49}gc^{0.25}\Delta T^{0.24}\Delta P^{0.74}}{\sigma^{0.5}\mu l^{0.29}Hfg^{0.24}\rho v^{0.24}} + F\,Re\,l^{0.8}Prl^{0.4}k_l/D_H \tag{5.3}$$

............1................ ....2............

Note:

Part 1 is nucleate pool boiling due to Forster–Zuber [10]
Part 2 is forced convection due to Dittus–Boelter

where

h_{tp} is the two-phase heat transfer coefficient
S is the nucleate boiling suppression factor (0.00122 for *F–Z* correlation)
F is the two–phase multiplier, correction factor, which is defined as the ratio of the true two-phase Reynolds number to the single-phase
Re is Reynolds number to the single-phase
k_l is thermal conductivity of liquid
C_l is specific heat of liquid
g_c is gravitational conversion factor
ΔT is wall super heat
ΔP is difference in saturation and wall superheat pressures
σ is *s* urface tension
μ_l is viscosity
H_{fg} is enthalpy of vaporization
D_H is hydraulic diameter

$$S = \frac{1}{1 + 0.00000253Re_{tp}^{1.17}}$$

$$F = \left(\frac{1}{\chi_t} + 0.213\right)^{0.736}$$

5.2.8.4 McAdams Correlation for Nucleate Boiling

In fully developed nucleate boiling with saturated coolant, the wall temperature is determined by local heat flux and pressure and is only slightly dependent on the Reynolds number. For subcooled water at absolute pressures between 0.1–0.6 MPa, the McAdams correlation gives [11]:

$$q = 0.074\left[T_w - T_{sat}\right]^{3.86} \tag{5.4}$$

where q is heat transfer, $\left[\frac{Btu}{hr}.ft^2\right]$ and $T_w - T_{sat}$ is in °F

5.2.8.5 Thom Correlation for Flow Boiling

The Thom correlation is for flow boiling (subcooled or saturated at pressures up to about 20 MPa) under conditions where the nucleate boiling contribution predominates over forced convection. This correlation is useful for a rough estimation of expected temperature difference given the heat flux [12]:

$$T_w - T_{sat} = 22.5\,q^{0.5} e^{-P/8.7} \tag{5.5}$$

where
  q is the heat flux $(\mathrm{MW/m^2})$
  P is the pressure (MPa)

5.2.9 Boiling Heat Transfer Enhancement Methods

Boiling heat transfer enhancement is caused by an increase in active nucleation site density. It is expected that a heat transfer surface with the microporous structure will decrease vapor bubble departure diameters. Small vapor bubbles will be generated from the boiling heat transfer enhanced surface. The simplest way is to increase the surface roughness. Cavities, especially re-entrant cavities, are considered to be effective in keeping nucleation sites active. Generally, such heat transfer enhancement of the surface is very effective for bubble nucleation at low heat flux; however, the heat transfer enhancement becomes weak at high heat flux due to a partial dry-out in cavities.

The requirements for boiling heat transfer are higher heat transfer coefficient, smaller wall superheat for the onset of nucleate boiling, and higher CHF. The wall superheat greatly depends on the heat transfer surface characteristics, such as surface roughness, cavities, microporous structure, wettability, etc.

With the advances in manufacturing techniques, mechanical microstructures such as micro pin-fins have been applied for boiling heat transfer enhancement. Examples are the GEWA tubes produced by Wieland Werke AG and Thermoexel tubes by Hitachi Cable, Ltd, with re-entrant cavities manufactured by a mechanical process, and High Flux tubes produced by Union Carbide with a porous structure made by sintering and thermal spraying. Another method of boiling heat transfer enhancement that is different from the manufacture of the surface configuration is to control the wettability. For example, it was shown that in a TiO_2-coated super-hydrophilic surface with dipping or sputtering processes that the TiO_2-sputtered surface produced excellent heat-transfer characteristics in the nucleate boiling region and higher CHF than the non-coated surface [13,13.1].

The enhancement techniques adopted for internal/external evaporation include [14]:

i. Microfin tubes
ii. Twisted tape inserts
iii. Corrugated tubes
iv. Tubes with porous coatings

The discussion on enhanced boiling surfaces is based on Ref. [14].

5.2.9.1 Microfin Tubes

Most microfins have approximately a trapezoidal cross-sectional shape with a rounded top and rounded corners at the root. Microfin tubes are available primarily in copper. For instance, Wolverine Tube Inc. is one of the major manufacturers of copper microfin tubes for the air-conditioning and refrigeration industries. In addition, microfin tubes are also becoming available in other materials, such as aluminum, carbon steel, steel alloys, etc. For horizontal applications, heat transfer enhancement ratios are as high as three to four times those at low mass velocities while falling off towards their internal area ratio at high mass velocities. Pressure drop ratios most often range from about 1.0 at low mass velocities up to about a maximum of about 1.5 at high mass velocities. Hence, microfins are very attractive from a heat transfer augmentation to pressure drop penalty point-of-view. Figure 8.8 in Chapter 8 shows the microfin tube geometry (schematic).

5.2.9.2 Twisted Tape Inserts

Twisted tape inserts are made by twisting a metal strip and fit loosely in the tubes to allow for standard tube wall dimensional tolerances. Heat transfer augmentation ratios are typically in the range from 1.2 to 1.5, while two-phase pressure drop ratios are often as high as 2.0 as the tape divides the flow channel into two smaller cross-sectional areas with smaller hydraulic diameters. Twisted tapes have seen some applications in both horizontal and vertical units. Figure 8.17 in Chapter 8 shows twisted tape inserts.

5.2.9.3 Corrugated Tubes

Figure 8.19 in Chapter 8 shows corrugated tube geometry (schematic) with its characteristic dimensions. Corrugated tubes are manufactured in many metals: copper, copper alloys, carbon steels, stainless steels, and titanium. Heat transfer ratios are usually between 1.2–1.8 with performances matching microfin tubes at high mass velocities but with much larger two-phase pressure drop ratios, which are on the order of twice those of a plain tube.

5.2.9.4 Porous Coated Tubes

These have heat transfer performances similar to those for nucleate pool boiling and are on the order of 5–10 times plain tube performance in vertical tubes. For evaporation in horizontal tubes, the porous coating is only effective for annular flows but not for stratified flows where part of the tube perimeter is dry. Figure 8.27 in Chapter 8 shows a porous coated tube surface (schematic).

5.2.9.5 Flow Boiling in Vertical Microfin Tubes

The performance in the vertical orientation is slightly less than that of the microfin tube compared to its horizontal orientation over the range of vapor qualities from 0.23 to 0.5. Heat transfer augmentation is on the order of 2 or more compared to a plain tube.

5.2.9.6 Applications of Boiling Heat Transfer inside Enhanced Tubes

In vertical tubes, the most important applications and potential benefits for use of enhancements for evaporation are in the petrochemical industry and hence for vertical thermosyphon reboilers. These units typically evaporate from 10% to 35% of the flow, while the inlet is slightly subcooled from the static head with respect to the liquid level in the distillation tower. In horizontal tubes, the most important applications of enhancements are to direct-expansion evaporators in refrigeration, air-conditioning, and heat pump units. The standard practice for enhanced boiling inside tubes is to define the internal heat transfer coefficient based on the nominal area at the maximum internal diameter, i.e., at the root of the fins for microfin tubes or that of the plain tube surface for a twisted tape insert [14].

5.2.9.7 Heat Transfer Enhancement Methods for Pool Boiling

Numerous types of surface treatment and structuring have been utilized to enhance pool boiling [15]. Low fin tubes have been the standard enhancement technique for pool boiling of refrigerants and organics. Heat transfer coefficients (based on total area) are typically greater than those for the reference plain tube. Examples of commercial structured boiling surfaces are provided in Figure 5.5

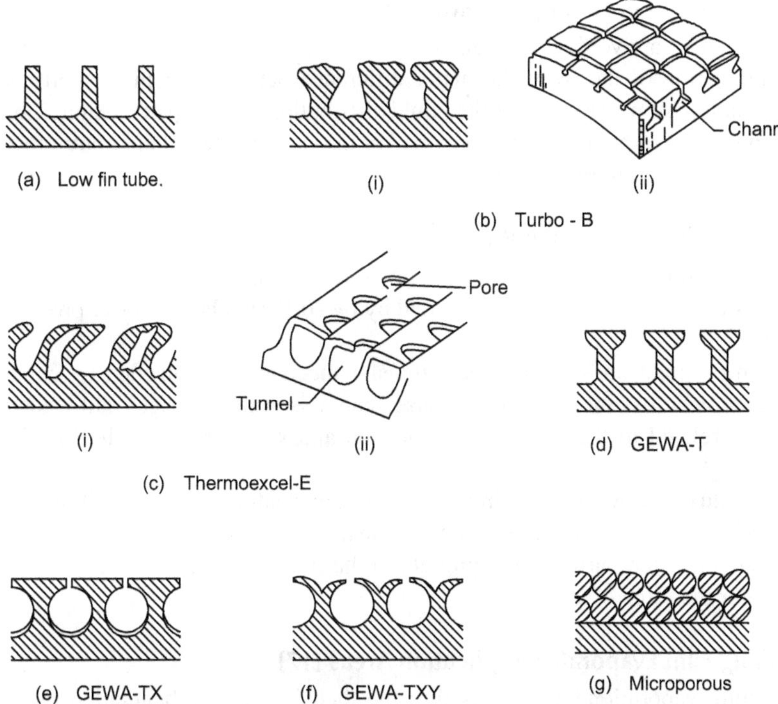

FIGURE 5.5 Examples of commercial structured pool boiling enhancement surfaces. (Partially adapted from Boiling Heat Transfer on External Surfaces, Chapter 9, (Ed.) John R Thome, *Engineering Data Book III*, Wolverine Tube, Inc., pp. 9.1–9.38).

5.2.10 Falling Film Evaporator

A falling film evaporator is an industrial device to concentrate solutions, especially with heat-sensitive components. The evaporator is a special type of heat exchanger. In general, evaporation takes place inside vertical tubes, but there are also applications where the process fluid evaporates on the outside of horizontal or vertical tubes. In all cases, the process fluid to be evaporated flows downwards by gravity as a continuous film. The fluid creates a film along the tube walls, progressing downwards (falling)—hence the name. The fluid distributor has to be designed carefully in order to maintain an even liquid distribution for all tubes along which the solution falls.

Evaporation occurs inside vertical tubes through a film of liquid established using one of several liquid distribution devices, the selection of which depends on the wetting rate on the tubes. This is one of the most commonly used evaporator types due to the simplicity of operation, dependability of performance, low temperature difference required, and high heat transfer coefficients developed in the film.

5.2.10.1 Working of a Falling Film Evaporators

In falling film evaporators, the liquid product usually enters the evaporator at the head of the evaporator. In the head, the product is evenly distributed into the heating tubes. The liquid enters the heating tube and forms a thin film on the tube wall where it flows downwards at boiling temperature and is partially evaporated. In most cases, steam is used for heating the evaporator. The product and the vapor both flow downwards in a parallel flow. This gravity-induced downward movement is increasingly augmented by the co-current vapor flow. The separation of the concentrated product from its vapor takes place in the lower part of the heat exchanger and the vapor/liquid separator.

5.2.10.2 Construction of Falling Film Evaporators

Falling film evaporators are vertical shell and tube heat exchangers. Typical TEMA types are BEM, NEN, or a combination of the two. The major difference between a typical shell and tube heat exchanger and a falling film evaporator is the liquid distribution at the top of the unit. Liquid entering the top of the unit passes either through a spray ball or through an internal distributor to ensure equal distribution of the liquid to all of the tubes [16].

5.2.10.3 Falling Film Evaporation Process

The discussion on falling film evaporators is based on Ref. [17].

Falling film evaporation is a process controlled by two different heat transfer processes [17]:

(1) Thin film evaporation—this is a heat transfer mechanism controlled by conduction and/or convection across the film where phase change is at the interface and whose magnitude is directly related to the thickness of the film and whether or not the film is laminar or turbulent.
(2) If the heat flux is above that required for onset of nucleation, nucleate boiling is also present, where bubbles grow in the thin film at the heated wall (or in the re-entrant channels of an enhanced structured surface) and migrate to the interface. The film normally flows downward under the force of gravity.

5.2.10.4 Falling Film Evaporator Application Areas [17]

i. Falling film evaporation has been used on the shellside of large heat pump systems.
ii. The desalination industry also exploits falling film evaporators, typically utilizing plain, horizontal tube bundles.
iii. In large tonnage air-separation plants, massive vertical coiled-tube-in-shell units with shellside falling film evaporation are used to take advantage of the close temperature approaches that can be attained to save on energy consumption. Here the tubes are nearly horizontal in their spiral within the coil.
iv. Falling film evaporators have also been tested in ocean thermal energy conversion (OTEC) pilot plants to achieve a closer temperature approach between the evaporating fluid and the heating fluid, and hence attain higher cycle thermal efficiency.
v. Falling film evaporation has also been exploited in absorbers and vapor generators of absorption heat pump systems.
vi. One of the significant advantages of falling film operation is the large reduction in liquid charge in the evaporator. Furthermore, higher heat transfer performance can also be attained.

5.2.10.5 Falling Film Evaporation on a Single Horizontal Tube

Figure 5.6 shows a schematic illustration of falling film evaporation on a single horizontal tube, with nucleate boiling occurring in the falling film [17]. Hence, both thin falling film evaporation and nucleate boiling play a role in the heat transfer process. In a falling film evaporator, an array of horizontal tubes arranged in a matrix is used with the liquid falling from tube to tube. Vertical falling film evaporators have been used for many years in the petrochemical industry, in the chemical industry to evaporate fluids under vacuum conditions where the liquid static head from the distillation column would otherwise create too much subcooling for efficient operation as a vertical or horizontal thermosyphon reboiler. They are also used to evaporate temperature-sensitive fluids and to remove volatiles from mixtures.

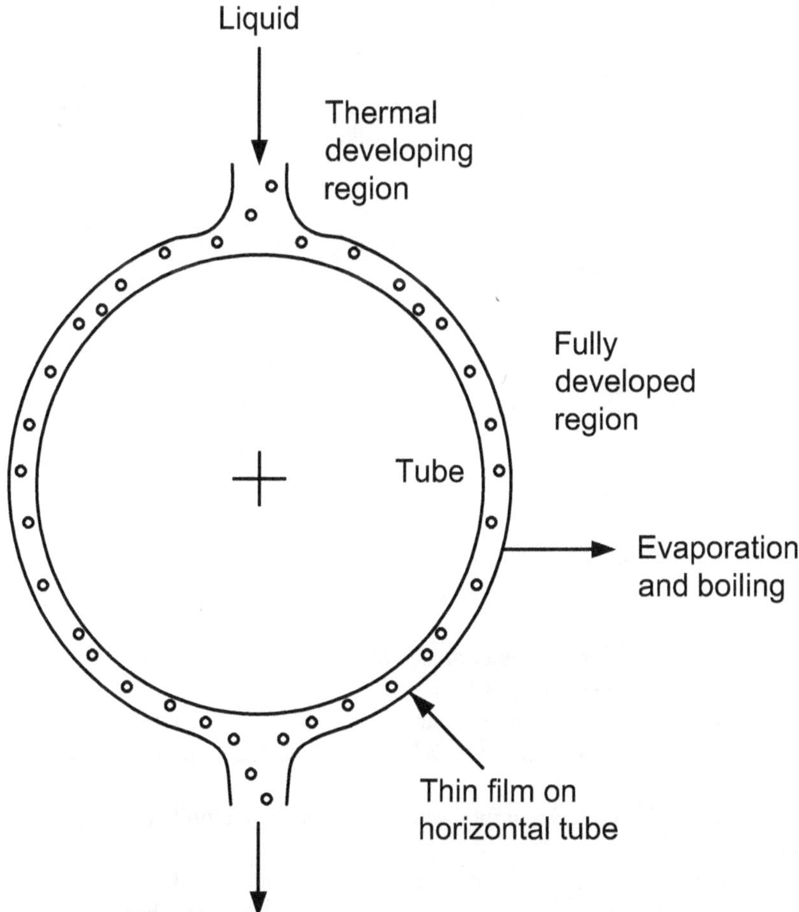

FIGURE 5.6 Falling film evaporation on a heated horizontal tube with nucleate boiling. (Adapted from Thome, J.R. (Ed.) (2004) Falling Film Evaporation, Chapter 14. In: *Engineering Data Book III*, pp. 14.1–14.14. Wolverine Tube, Inc., Decatur, AL.)

5.2.10.6 Horizontal, Shellside Falling Film Evaporator

Horizontal falling film evaporators are to some extent similar to kettle-type steam generators in that the liquid is fed to the bundle overhead, using sprinklers or trays. The unevaporated liquid can be removed from the bottom of the bundle by placing a nozzle at the bottom of the shell as shown in Figure 5.7 [17].

For petrochemical applications, shellside falling film evaporation on horizontal tube bundles is advantageous because enhanced boiling tubes, such as the Turbo-Bii or Turbo-Biii, can be utilized and hence a much more compact design is obtained compared to vertical plain tube units. Additionally, such a unit can also take advantage of multiple tube passes for the heating fluid, further improving heat transfer performance and compactness. For petrochemical applications, the potential advantages of using horizontal falling film units as opposed to vertical units can be summarized as follows [17]:

i. Heat transfer coefficients on plain horizontal tubes are higher than those for vertical tubes since the heated flow length is much shorter.
ii. A horizontal bundle can have multiple tube passes of the heating fluid to significantly increase its heat transfer coefficient compared to a single shell pass in vertical units.

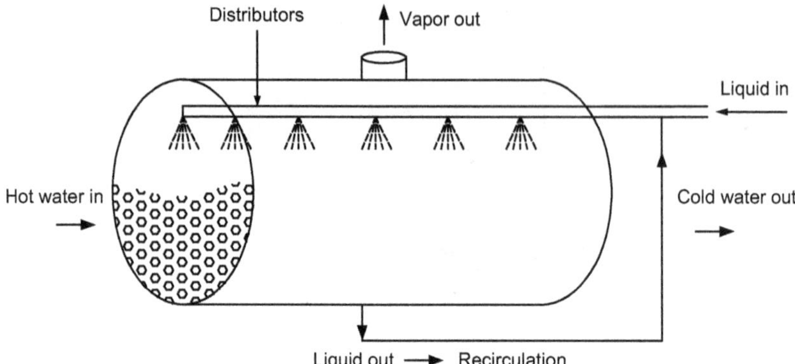

FIGURE 5.7 A horizontal shell and tube falling film evaporator. (Adapted from Thome, J.R. (Ed.) (2004) Falling Film Evaporation, Chapter 14. In: *Engineering Data Book III*, pp. 14.1–14.14. Wolverine Tube, Inc., Decatur, AL.)

5.2.10.7 Intertube Falling Film Modes

The intertube flow modes are classified from observations as follows [17]:

a. Droplet mode. The flow is in droplet mode when there is only a flow of liquid in the form of distinct droplets between the tubes.
b. Droplet-columns mode. This intermediate mode is present when at least one stable column exists between the tubes in addition to falling droplets. A column is a continuous liquid link between tubes.
c. Column mode. This mode is simply when there is only liquid flow in columns between the tubes.
d. Column-sheet mode. In this intermediate mode, both columns and a liquid sheet are simultaneously flowing between the tubes at different locations along the tubes.
e. Sheet mode. This mode is when the fluid flows uniformly between the tubes as a continuous film or sheet.

Intertube flow modes are shown in Figure 5.8 [17].

5.2.10.8 Thermal Design Considerations of Falling Film Evaporators

Special aspects must be taken into account when designing a horizontal falling film evaporator such as [17]:

i. Choice of the most appropriate enhanced tube for the fluid to be handled. Note that conventional low-finned tubes should not be used since they tend to inhibit longitudinal spreading of the liquid film along the tube.
ii. Choice of the optimum tube bundle layout such as the number of tubes and their length, bundle width and height, tube pitch and layout, and number of tube passes.

5.2.10.9 Evaporator of the Refrigeration System

The evaporator is one of the five essential components in the refrigeration system, together with the condenser, compressor, expansion valve, and refrigerant. In the evaporator, the refrigerant boils off by absorbing energy from the warmer secondary fluid, thus reducing the temperature. The secondary fluid may be a gas or a liquid, depending on the system.

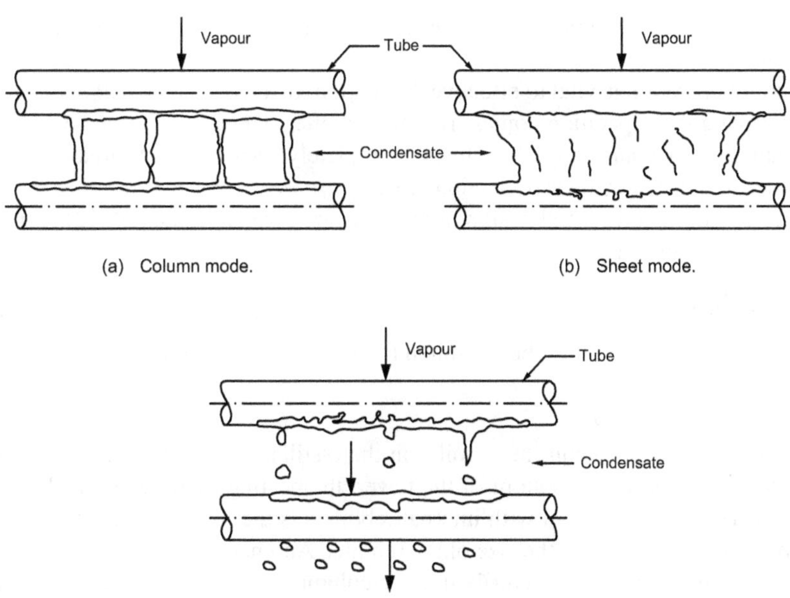

FIGURE 5.8 Intertube falling film flow modes in a tube bundle. (Adapted from Thome, J.R. (Ed.) (2004) Falling Film Evaporation, Chapter 14. In: *Engineering Data Book III*, pp. 14.1–14.14. Wolverine Tube, Inc., Decatur, AL.)

a. Boiling Regimes

The boiling process depends on factors such as mass flow, vapor content, and the temperature difference between the refrigerant and the heating surface.

b. Pool Boiling

The heat transfer coefficient for pool boiling has a characteristic curve that displays heat flux versus the temperature difference between the evaporating and the secondary media which is similar to the boiling curve of steam.

Preferably, a BPHE evaporator should be designed to operate below the critical heat flux to avoid the unstable performance of partial film boiling and the low heat transfer coefficient of complete film boiling.

5.2.11 Reboilers

Reboilers are heat exchangers typically used in distilling processes to increase component separation. They are often provided by steam, to evaporate the volatile components of a product that has been drawn from a fractionating tower. There are several types of reboilers, including jacketed kettle, kettle, internal reboiler, and thermosyphon reboiler. These designs differ from one another in the way the product is heated with steam.

5.2.11.1 How Does a Reboiler Work?

In distillation processes, a liquid that is often composed of different components (e.g., light and heavier hydrocarbons) is drawn from the distillation column into a reboiler, where partial or full boiling occurs. As a result, a vapor of lighter components is produced and returned to the column to drive the distillation separation. The reboiler is positioned at the base of the distillation column to supply the heat.

Proper reboiler operation is vital to effective distillation. In a typical classical distillation column, all the vapor driving the separation comes from the reboiler. The reboiler receives a liquid stream from the column bottom and may partially or completely vaporize that stream. Steam usually provides the heat required for the vaporization [18].

General reviews of the forms and design of reboilers are given in Refs. [19–22]. For general discussion of reboilers refer to Refs. [23–25].

5.2.11.2 Types of Reboiler

There is a wide variety of types of reboiler suited to different applications.

5.2.11.3 Internal Reboilers

The simplest approach is to mount the reboiler in the distillation tower itself. Here, boiling takes place in the pool of liquid at the bottom of the tower, the heating fluid being inside the bundle of tubes as shown. The major problem with internal reboilers is the limitation imposed by the size of the distillation column. This limits the size of the reboiler. Another problem sometimes encountered is that of mounting the bundle satisfactorily into the column. The problem of size restriction can be overcome if compact heat exchangers such as a plate-fin or printed circuit heat exchanger are used.

5.2.11.4 Kettle Reboilers

The layout of a kettle reboiler is illustrated schematically in Figure 5.9. Kettle reboilers allow heating fluid to pass through the tubes that are surrounded by the liquid, which needs to be distilled. Steam (often used as a heating medium) or any other heating fluid (such as thermal oil), flows through the tube bundle and exits as condensate, while the liquid from the base of the tower flows through the shellside. The tower processes and collects the vapor; the liquid by-product drops to the bottom of the kettle and is withdrawn as per the process design. Kettle reboilers are widely used in the petroleum and chemical industries.

5.2.11.5 Thermosyphon Reboilers

There are two primary types of thermosyphon reboiler: vertical and horizontal. These reboilers work on a simple principle based on the difference of densities of liquid and vapor for recirculation between the outlet and inlet line. An external pump is not needed to inject the vapor into the column.

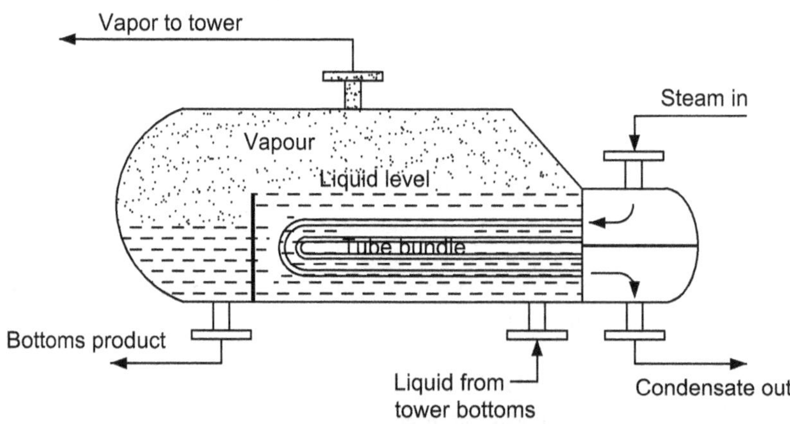

FIGURE 5.9 Kettle reboiler.

5.2.11.6 Reboiler Design Considerations

When selecting a reboiler heater, some critical choices are made. These include whether to vaporize on the shellside or tubeside, rely on gravity flow or forced flow, orient the unit vertically or horizontally, opt for a flow-through or kettle design, or install baffles inside the tower.

5.2.11.7 Thermal Performance of Kettle Type Reboilers

Kettle reboilers, and other submerged bundle equipment, are essentially pool boiling devices, and their design is based on data for nucleate boiling. The tube arrangement, triangular or square pitch, will not have a significant effect on the heat-transfer coefficient. A tube pitch of between 1.5–2.0 times the tube outside diameter should be used to avoid vapor blanketing. Long thin bundles will be more efficient than short thick bundles.

5.2.11.8 Disengagement of Vapor and Liquid

The shell should be sized to give adequate space for disengagement of the vapor and liquid. The shell diameter required will depend on the heat flux. The freeboard between the liquid level and shell should be at least 0.25 m.

5.2.11.9 Check for Maximum Vapor Velocity

To avoid excessive entrainment, the maximum vapor velocity u_v at the liquid surface should be less than that given by the expression [19,20,23]:

$$u_v < 0.2\left[\frac{\rho_L - \rho_v}{\rho_v}\right]^{1/2} \tag{5.6}$$

5.2.11.10 Boiling Heat Transfer Coefficients

In the design of vaporizers and reboilers, the designer will be concerned with two types of boiling: pool boiling and convective boiling. Pool boiling is the name given to nucleate boiling in a pool of liquid, such as in a kettle-type reboiler or a jacketed vessel. Convective boiling occurs where the vaporizing fluid is flowing over the heated surface, and heat transfer takes place both by forced convection and nucleate boiling, such as in forced circulation or thermosyphon reboilers.

5.2.11.11 Boiling Heat Transfer Coefficient Estimation

The correlation given by Forster and Zuber [26] can be used to estimate pool boiling coefficients:

$$h_{pb} = 0.00122\left[\frac{k_L^{0.79} C_{pL}^{0.45} \rho_L^{0.49}}{\sigma^{0.5} \mu_L^{0.29} \lambda^{0.24} \rho_v^{0.24}}\right]\left(T_w - T_N\right)^{0.24}\left(P_w - P_s\right)^{0.75} \tag{5.7}$$

where

h_{pb} is nucleate, pool, boiling coefficient, W/ m^2 °C

k_L is liquid thermal condcutivity, W/ m^2C

ρ_L is liquid density, kg/ m^3

μ_L is liquid viscosity, Ns/ m^2

λ is latent heat, J/kg

ρ_v is vapor density, kg/ m^3

T_w is wall surface temperature,
T_N is saturation termperature of boiling liquid,
ρ_w is saturation pressure correspoinding to the wall temperature, T_w, N/ m^2
ρ_s is saturation pressure correspoinding, T_S, N/ m^2
σ is surface tension, N/m

5.2.11.12 Critical Heat Flux

It is important to check that the design, and operating, heat flux is well below the critical flux. The maximum heat flux achievable with nucleate boiling is known as the critical heat flux. In a system where the surface temperature is not self-limiting, such as a nuclear reactor fuel element, operation above the critical flux will result in a rapid increase in the surface temperature, and in an extreme situation the surface will melt. This phenomenon is known as " burn-out." The heating media used for process plants are normally self-limiting; for example, with steam the surface temperature can never exceed the saturation temperature. Care must be taken in the design of vaporizers to ensure that the critical flux can never be exceeded. The critical flux is reached at low temperature differences: around 20–30°C for water and 20–50°C for light organics.

5.2.11.13 Check for Critical Heat Flux

In SI units, Zuber's equation can be written as [27]:

$$q_c = 0.131\lambda\left[g\sigma(\rho_L - \rho_v)\rho_v^2\right]^{1/4} \tag{5.8}$$

where q_c is maximum, critical heat flux, W/ m^2
g is gravitational acceleration, 9.81 m/ s^2

Mostinski [28] also gives a reduced pressure equation for predicting the maximum critical heat flux:

$$q_c = 3.67 \times 10^4 P_c \left(\frac{P}{P_c}\right)^{0.35} + \left[1 - \left(\frac{P}{P_c}\right)\right]^{0.9} \tag{5.9}$$

The reduced pressure correlation given by Mostinski [28] is simple to use and gives values that are as reliable as those given by more complex equations.

$$h_{nb} = 0.104\left(P_c\right)^{0.69}(q)^{0.7}\left[1.8\left(\frac{P}{P_c}\right)^{0.17} + 4\left(\frac{P}{P_c}\right)^{1.2} + 10\left(\frac{P}{P_c}\right)^{10}\right] \tag{5.10}$$

where

P is operating pressure, bar
P_c is liquid critical pressure, bar
q is heat flux, W/ m^2
Note: $q = h_{nb}\left(T_w - T_s\right)$

Mostinski's equation is convenient to use when data on the fluid physical properties are not available.

The modified Zuber equation can be written as

$$q_{cb} = K_b\left(\frac{P_t}{d_o}\right)\left(\frac{\lambda}{\sqrt{N_t}}\right)\left[\sigma g\left(\rho_L - \rho_v\right)\rho_v^2\right]^{0.25} \tag{5.11}$$

where

q_{cb} is maximum (critical) heat flux for the tube bundle, W/m^2
K_b is 0.44 for square pitch arrangements and is 0.41 for equilateral triangular pitch arrangements
P_t is tube pitch
d_o is tube outside diameter
N_t is total number of tubes in the bundle

Palen and Small [29] suggest that a factor of safety of 0.7 be applied to the maximum flux estimated from the equation.

5.3 CONDENSATION

Condensation is defined as the phase change from a vapor state to a liquid state. When vapor comes in contact with the surface of an object kept at a temperature lower than the saturation temperature, it condenses into a liquid with the release of latent heat. From a thermodynamics standpoint, condensation occurs when the enthalpy of the vapor is reduced to the state of saturated liquid. The heat transfer accompanied by condensation is called condensation heat transfer. Condensation can take place on the outside of surfaces, such as on plates, horizontal or vertical tubes, and tube bundles. The components that utilize condensation heat transfer include the steam turbine condenser in use for coal-based thermal power plants, nuclear power plants, and the refrigeration and air conditioning units condensers, etc. Energy in the latent heat form must be removed from the condensation area during the phase change process. Condensation can be classified as bulk condensation and surface condensation.

Condensation may be of two types, film condensation and dropwise condensation. If the liquid (condensate) wets the surface, a smooth film is formed and the process is called film type condensation. In this process, the surface is blocked by the film, which grows in thickness as it moves down the plate. However, if the liquid does not wet the system, drops are formed on the surface in a random fashion. This process is called dropwise condensation.

Consider a vertical flat plate which is exposed to a condensable vapor as shown in Figure 5.10 [30]. If the temperature of the plate is below the saturation temperature of the vapor, condensate will form on the surface and flow down the plate due to gravity. It should be noted that a liquid at its boiling point is a saturated liquid and the vapor in equilibrium with the saturated liquid is saturated vapor. A liquid or vapor above the saturation temperature is described as superheated.

In film-wise condensation, the surface over which the steam condenses is wettable and hence, as the steam condenses, a film of condensate is formed. Generally, film-wise condensation results in low heat transfer rates as the film of condensate impedes the heat transfer. The thickness of the film formed depends on many parameters including orientation of the surface, viscosity, and rate of condensation.

5.3.1 Film Flow Regimes

Filmwise condensation heat transfer is associated mainly with the thermal resistance of the liquid film, which depends on its thermal conductivity, thickness, and velocity distribution. Thickness and velocity distribution are determined by the mode of film flow on the wetted surface, which is

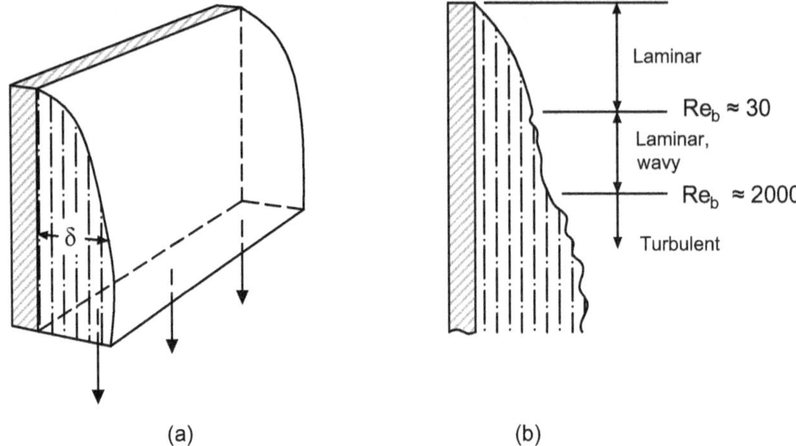

FIGURE 5.10 Filmwise condensation over a vertical surface.

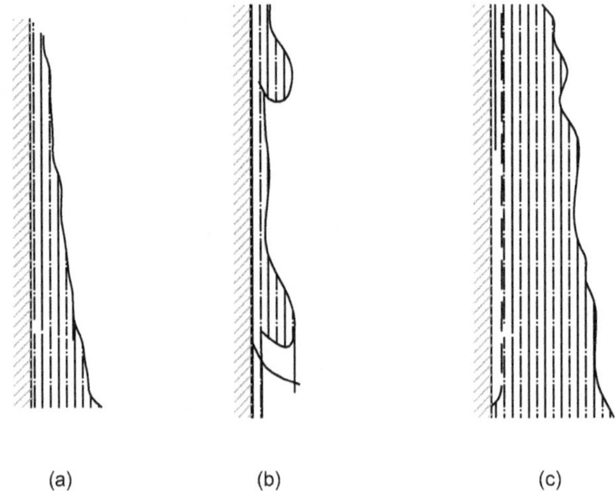

FIGURE 5.11 Three modes of film condensation (a) laminar, (b) wavy and (c) turbulent along a vertical plate surface (Adapted from www.sciencedirect.com/topics/engineer ing/condensation-heat-transfer.)

determined by a balance of the gravity force, friction, inertia forces, and the surface tension. Assume condensation on a vertical wall surface. The thickness of the condensate film and velocity are small on the upper part of the surface, where a laminar mode would prevail. As the flow rate in the film increases in the lower parts, there are wavy motions, and the turbulent regime prevails. Accordingly, there are three modes of film flow along the vertical or inclined surface (Figure 5.11) [31]:

i. Laminar ($Re < Re_b$)
ii. Wavy ($Re_b < Re < Re_{cr}$)
iii. Turbulent (turbulent wavy) ($Re > Re_{cr}$)

Filmwise condensation heat transfer was first modeled by Nusselt in 1916 who obtained the analytical solution to the problem of condensation of a saturated vapor on a vertical plain surface.

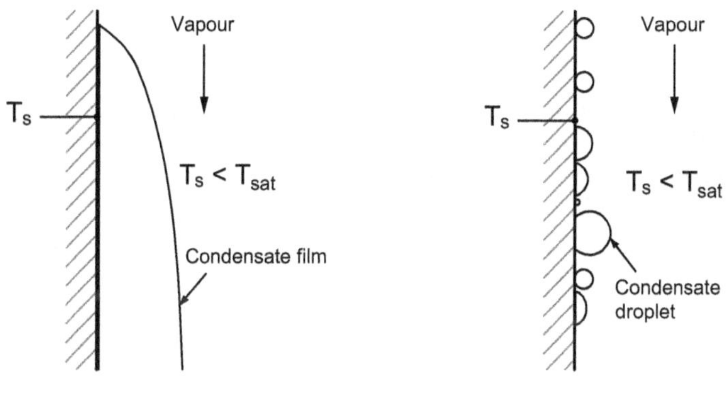

FIGURE 5.12 Condensation over a vertical surface (a)Filmwise and (b) Dropwise.

5.3.2 Dropwise Condensation

Dropwise condensation takes place when the surface over which condensation takes place is non-wettable. In this mode, when steam condenses, droplets are formed. When the drops become larger, they simply fall under gravity. Dropwise condensation is shown in Figure 5.12 along filmwise condensation. While dropwise condensation is known for the very high heat transfer coefficients, it is not considered to be suitable for deliberate employment in process equipment. In dropwise condensation, high heat transfer rates are achieved and hence, often, special materials must be employed on the heat transfer surface to attain dropwise condensation. Various surface coatings, such as gold, silicones, and Teflon, have been used in the industry to maintain dropwise condensation, however none of these methods has reached any considerable success. Because the effectiveness of such coatings gradually decreases due to oxidation and fouling, film condensation occurs after a period of time. Another reason for losing the effectiveness of dropwise condensation is the accumulation of droplets on the condenser surface. The heat transfer rate sharply decreases because of the accumulated droplets. Therefore, most condensers are designed on the assumption of being film condensation. Practically, these techniques for dropwise condensation are not easy for sustained dropwise condensation. For these reasons, in many instances, film condensation is assumed because the film condensation is sustained on the surface and it is comparatively easy to quantify and analyze [32].

5.3.3 Bulk Condensation

Vapor condenses as droplets suspended in a gas phase in the bulk condensation. When condensation takes place randomly within the bulk of the vapor, it is called homogeneous condensation. If condensation occurs with foreign particles in the vapor, this type of bulk condensation is defined as heterogeneous condensation. Fog is a typical example of this type of condensation.

5.3.4 Surface Condensation

Surface condensation occurs when the vapor contacts a surface whose temperature is below the saturation temperature of the vapor. Surface condensation has a wide application area in the industry. It is classified as filmwise or dropwise condensation.

5.3.5 Heat Transfer Due to Condensation on a Plane Vertical Surface

Condensation heat transfer plays an important role in many engineering applications, notably thermal power generation, process industries, refrigeration, and air-conditioning. The problem of calculating the heat transfer rate for a plane vertical surface and for a horizontal cylinder with uniform surface temperatures, and where the condensate flow is laminar and governed only by gravity and viscous forces, was solved by Nusselt in 1916. For the well-known *Nusselt equations* for the vertical plane surface as shown in Figure 5.10, the mean value of Nusselt number is [31–35]:

$$\mathrm{Nu} = 0.943\left\{\frac{\rho\Delta\rho\, g\, h_{1g} L^3}{\eta\lambda\Delta\mathrm{T}}\right\}^{1/4} \tag{5.12}$$

and for the horizontal tube

$$\mathrm{Nu} = 0.728\left\{\frac{\rho\Delta\rho\, g\, h_{1g} d^3}{\eta\lambda\,\Delta\mathrm{T}}\right\}^{1/4} \tag{5.13}$$

where
ρ is the liquid density
g is acceleration due to gravity
h_{1g} is the latent heat of evaporation
η is the liquid viscosity
λ is the liquid thermal conductivity
$\Delta\rho$ is the density difference between vapor and condensate
$\Delta\mathrm{T}$ is the vapor-to-surface temperature difference
L is the plate height
d is the tube diameter

5.3.6 Modes of Condensation

In in-tube condensation, the process takes place inside enclosed channels, typically with forced flow conditions. Four basic mechanisms of condensation are generally recognized [36]:

i. Dropwise
ii. Filmwise
iii. Direct contact
iv. Homogeneous

Figure 5.13 illustrates the above condensation modes.

For clarity some of the above terms are discussed again.

In *dropwise condensation*, the drops of liquid form from the vapor at particular nucleation sites on a solid surface, and the drops remain separate during growth until carried away by gravity or vapor shear.

In *filmwise condensation*, the drops initially formed quickly coalesce to produce a continuous film of liquid on the surface through which heat must be transferred to condense more liquid.

In *direct contact condensation*, the vapor condenses directly on the (liquid) coolant surface, which is sprayed into the vapor space.

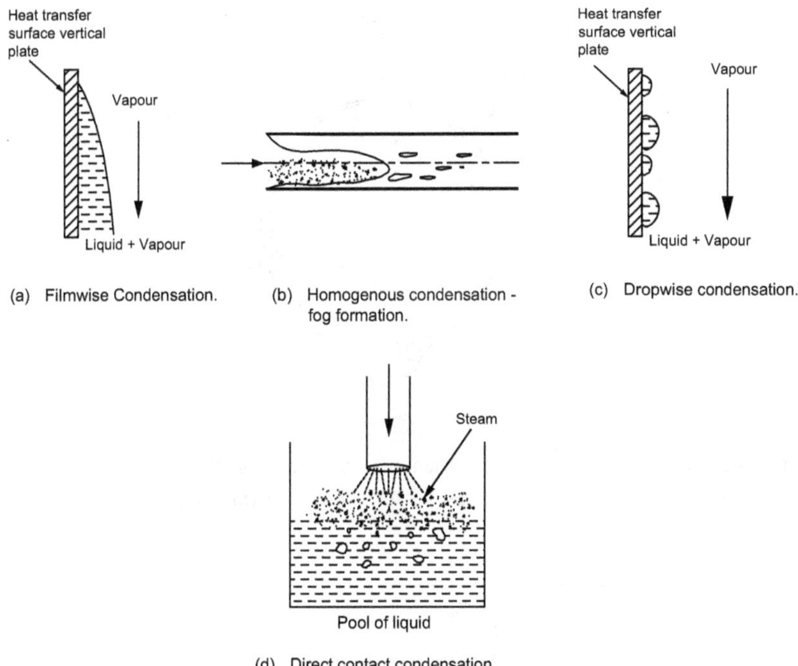

FIGURE 5.13 Various modes of condensation. (a) Filmwise condensation. (b) Homogenous condensation - fog formation. (c) Dropwise condensation. (d) Direct contact condensation (Adapted from Thome, J.R. (Ed.) (2004) Condensation on External Surfaces, Chapter 7. In: *Engineering Data Book III* pp. 7.1–7.39. Wolverine Tube, Inc.)

In *homogeneous condensation*, the liquid phase forms directly from supersaturated vapor, away from any macroscopic surface; it is however generally assumed, in practice, that there are particles of dust or mist particles present in the vapor to serve as nucleation sites.

Direct contact condensation is a very efficient process, however it results in mixing the condensate with the coolant. Therefore, it is useful only in those cases where the condensate is easily separated, where there is no desire to reuse the condensate, or where the coolant and condensate are in fact the same substance. Direct contact condensation is also shown in Figure 1.33 in Chapter 1.

Condensate forming as suspended droplets or mist in a subcooled vapor is called *homogeneous* condensation, of which the most common example in nature is fog. Homogeneous condensation is primarily of concern in fog formation in equipment, i.e. it is to be avoided, and is not a design mode.

Filmwise condensation is the only one of the above processes of particular industrial interest.

5.3.6.1 Laminar Film Condensation on a Vertical Plate

Figure 5.10 depicts the process of laminar film condensation on a vertical plate from quiescent vapor [36]. The film of condensate begins at the top and flows downward under the force of gravity, adding additional new condensate as it flows. The flow is laminar and the thermal profile in the liquid film is assumed to be fully developed from the leading edge. Thus, the temperature profile across the film is linear and heat transfer is by one-dimensional heat conduction across the film to the wall.

5.3.6.2 Laminar Film Condensation on a Horizontal Tube

Applying the Nusselt integral approach to laminar film condensation on a vertical isothermal plate, a similar process on the outside of a single, horizontal isothermal tube can be analyzed. Condensation on the outside of horizontal tube bundles is often used for shell-and-tube heat exchanger applications

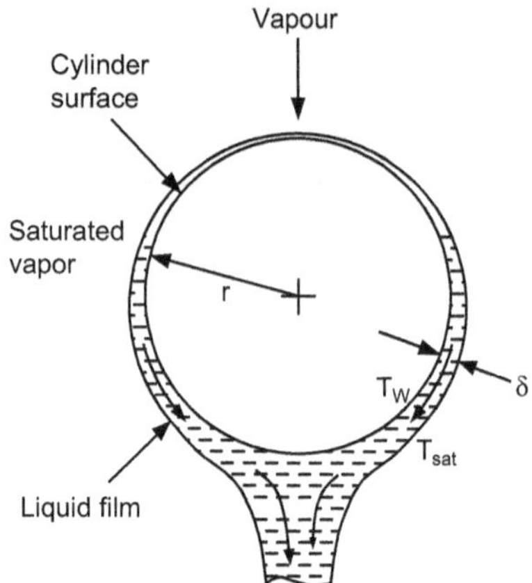

FIGURE 5.14 Condensation model for film condensation on a horizontal tube. (Adapted from Thome, J.R. (Ed.) (2004) Condensation on External Surfaces, Chapter 7. In: *Engineering Data Book III* pp. 7.1–7.39. Wolverine Tube, Inc.)

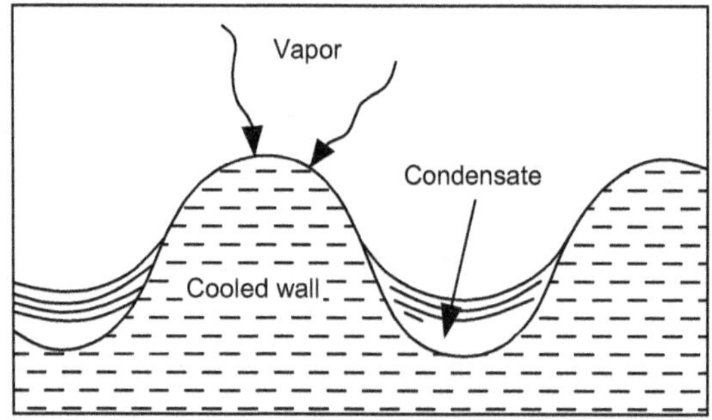

FIGURE 5.15 Gregorig effect on a liquid film on a fluted surface. (Adapted from Thome, J.R. (Ed.) (2004) Condensation on External Surfaces, Chapter 7. In: *Engineering Data Book III* pp. 7.1–7.39. Wolverine Tube, Inc.)

and the first step is the analysis of a single tube. The flow is nearly always laminar on a single tube because of the short cooling length around the perimeter and is illustrated in Figure 5.14 [36].

5.3.6.3 Condensation on Horizontal Tube Bundles

During condensation on a tube bundle, the condensate from the above tubes drains onto the tubes below, increasing the amount of condensate flowing on each tube in addition to the new condensate formed on that particular tube. The flow regimes formed by the condensate as it flows from one tube to that directly below it in an array of horizontal tubes are similar to Figure 5.8 [36]. The regimes encountered are described as follows for increasing film mass flow rate [36]:

a. Droplet mode. The liquid flows from tube to tube as individual droplets, often in rapid succession at uniform intervals along the bottom of the upper tube.
b. Column mode. At higher flow rates, the jets of droplets coalesce to form individual liquid columns that extend from the bottom of the upper tube to the top of the lower tube.
c. Sheet mode. At even higher flow rates, the columns become unstable and form short patches of liquid sheets that flow from one tube to the next.
d. Spray mode. Under vapor shear conditions with its complex flow field around tubes, the above flow modes may be interrupted and the liquid carried away by the vapor. In this case, there is significant entrainment of liquid droplets into the vapor flowing between the tubes and hence a spray flow is formed.

5.3.6.4 Condensation on Low-Finned Tubes and Tube Bundles

Integral low-finned tubes have been utilized for enhancing condensation for more than half a century. The geometry of a low-finned tube is illustrated in Figure 8.9 in Chapter 8. The fins are helical around the tube with a small axial pitch. For condensation, the optimum fin density depends on the particular fluid, primarily the surface tension, and varies from 19 fins/in. (19 fpi or 748 fins/m) up to 42 fpi (1653 fins/m). Fin heights depend on the fin density and the particular tube metal, ranging from about 0.66 to 1.50 mm (0.026 to 0.059 in.). The most typical fin thickness is 0.305 mm (0.012 in.).

5.3.6.5 Role of Surface Tension on Film Condensation on Low-Finned Tubes

For condensate formed on a horizontal low-finned tube, the two forces acting on the liquid film in quiescent vapor conditions are gravity and surface tension. Here, surface tension draws the liquid from the fin tip towards its root and dominates the gravity force on the liquid with its downward influence. Instead, in the root area between the fins, the radius of the film around the circumference of the tube is relatively large and uniform; thus, gravity dominates in this direction and governs the drainage of condensate from the tube, while surface tension tends to promote retention of the condensate between the fins. The first to exploit this surface tension induced flow in film condensation was Gregorig [36], and this has become to be known as the *Gregorig effect*, as illustrated in Figure 5.15 [36,37].

5.3.7 Condensation in Horizontal Tubes

Condensation in horizontal tubes may involve partial or total condensation of the vapor. Depending on the application, the inlet vapor may be superheated, equal to 1.0 or below 1.0. Hence, the condensation process path may first begin with a dry wall desuperheating zone, followed by a wet wall desuperheating zone, then a saturated condensing zone, and finally a liquid subcooling zone. The condensing heat transfer coefficient is a strong function of local vapor quality, increasing as the vapor quality increases, and also a strong function of mass velocity, increasing as the mass velocity increases. As opposed to external condensation, in-tube condensation is independent of the wall temperature difference ($T_{sat} - T_w$) for most operating conditions, except at low mass flow rates [38]. Condensation in horizontal tubes is shown in Figure 5.16

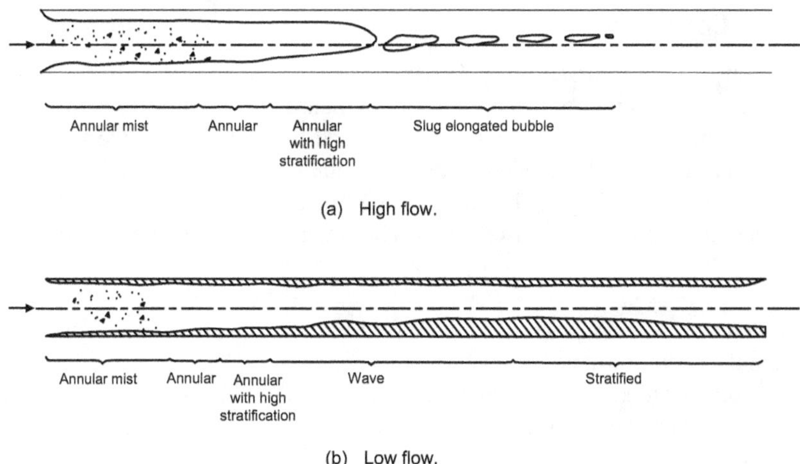

FIGURE 5.16 Typical flow patterns encountered for condensation inside horizontal tubes. (Adapted from Thome, J.R. (Ed.) (2004) Condensation Inside Tubes, Chapter 8. In: *Engineering Data Book III*, pp. 8.1–8.27. Wolverine Tube, Inc.)

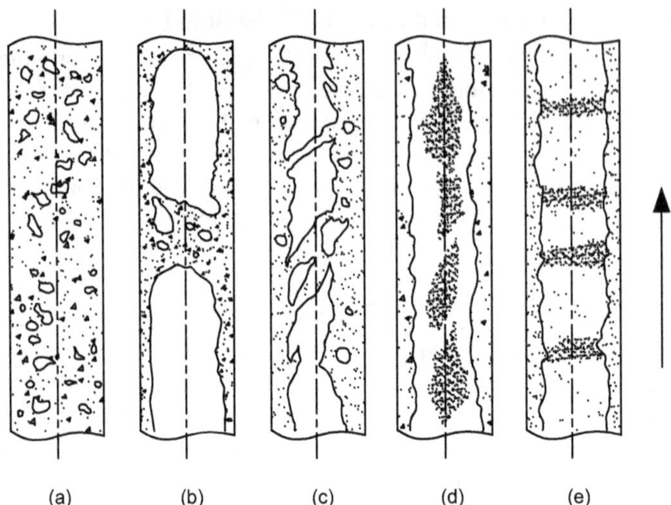

FIGURE 5.17 Two-phase flow patterns in vertical upflow. (Adapted from Palen, J.W., Breber, G., Taborek, K. (1979) Prediction of flow regimes in horizontal tube-side condensation. *Heat Transfer Engineering* 1(2): 47–57.)

5.3.8 Flow Regimes for Condensation in Tubes

Figure 5.17 (adapted from Palen, Breber, and Taborek [39]) illustrates the two-phase flow patterns typical of condensation in vertical tubes. In the top diagram at high mass flow rates, the flow takes on the annular flow regime, where the liquid film is on the perimeter of the wall, the vapor is in the central core and some liquid is entrained in the vapor from the tips of waves on the interface of the film. At low flow rates, as depicted in the lower diagram, at the entrance region annular flow is formed but this quickly transforms to intermittent flow with its characteristic large-amplitude waves washing the top of the tube or to stratified-wavy flow with smaller amplitude waves. If liquid does not span the cross-section of the tube, vapor may reach the end of the tube without condensing.

5.3.9 Surface Condenser

Surface condensation of a saturated or superheated pure vapor in contact with a cooled surface takes place when the temperature of the surface t_w is below the vapor saturation temperature t_{sat}. The exhausted steam from the LP turbines is condensed by passing over the tube bank through which cooling water passes. The condenser must maintain a sufficiently low vacuum in order to increase the power plant efficiency. The vacuum pumps maintain a sufficient vacuum in the condenser by extracting air and uncondensed gases. The steam condensers are broadly classified into two types [40]:

i. Surface condensers. In surface condensers, there is no direct contact between the exhaust steam and the cooling water.
ii. Jet condensers (or mixing type condensers). In jet condensers there is direct contact between the exhaust steam and cooling water.

A surface condenser is shown in Figure 4.99 in Chapter 4.

5.4 TWO-PHASE FLOW PATTERNS

5.4.1 Boiling in Vertical Tubes

For co-current upflow of two-phase gas and liquid in a vertical tube, the liquid and gas phases distribute themselves into several recognizable flow structures. These are referred to as flow patterns and are depicted in Figure 5.18 and can be described as follows [41]:

a. ***Bubbly flow.*** Numerous bubbles are observable as the gas is dispersed in the form of discrete bubbles in the continuous liquid phase.
b. ***Slug flow.*** With increasing gas void fraction, the proximity of the bubbles is very close, such that bubbles collide and coalesce to form larger bubbles, which are similar in dimension to the tube diameter. They are commonly referred to as Taylor bubbles after the instability attributed to that name. Taylor bubbles are separated from one another by slugs of liquid, which may include small bubbles.
c. ***Churn flow.*** Increasing the velocity of the flow, the structure of the flow becomes unstable with the fluid traveling up and down in an oscillatory fashion but with a net upward flow. The instability is the result of the relative parity of the gravity and shear forces acting in opposing directions on the thin film of liquid of Taylor bubbles.
d. ***Annular flow.*** Once the interfacial shear of the high-velocity gas on the liquid film becomes dominant over gravity, the liquid is expelled from the center of the tube and flows as a thin film on the wall (forming an annular ring of liquid), while the gas flows as a continuous phase up the center of the tube. This flow regime is particularly stable and is the desired flow pattern for two-phase pipe flows.
e. ***Wispy annular flow.*** When the flow rate is further increased, the entrained droplets may form transient coherent structures as clouds or wisps of liquid in the central vapor core.
f. ***Annular or Mist flow.*** At very high gas flow rates, the annular film is thinned by the shear of the gas core on the interface until it becomes unstable and is destroyed, such that all the liquid in entrained as droplets in the continuous gas phase, analogous to the inverse of the bubbly flow regime. Impinging liquid droplets intermittently wet the tube wall locally.

5.4.2 Vertical Upflow in an Evaporator Tube

For vertical upflow, the flow pattern typically begins in the bubbly flow regime at the inlet at the onset of nucleate boiling in the tube. This onset of nucleate boiling may begin in the subcooled zone of the tube, where bubbles nucleate in the superheated thermal boundary layer on the heated tube wall but tend to condense in the subcooled core. The onset of nucleate boiling may also be delayed to local vapor qualities greater than zero in the case of a subcooled inlet and a low heat flux. After bubbly flow the slug flow regime is entered and then the annular flow regime with its characteristic annular film of liquid. This film eventually dries out or the film is entrained by the interfacial vapor shear, taking the flow into the mist flow regime. The entrained liquid droplets may persist in the flow past the point of the vapor quality equal to 1.0.

5.4.3 Flow Patterns in Boiling in Horizontal Tubes

Flow patterns for flow of gas and liquid in a horizontal tube are shown in Figure 5.4 and are categorized as follows [41]:

a. ***Bubbly flow.*** The gas bubbles are dispersed in the liquid with a high concentration of bubbles in the upper half of the tube due to their buoyancy. When shear forces are dominant, the bubbles tend to disperse uniformly in the tube.

b. ***Stratified flow.*** At low liquid and gas velocities, complete separation of the two phases occurs. The gas goes to the top and the liquid to the bottom of the tube, separated by an undisturbed horizontal interface. Hence the liquid and gas are fully stratified in this regime.
c ***Stratified-wavy flow.*** Increasing the gas velocity in a stratified flow, waves are formed on the interface and travel in the direction of flow.
d. ***Intermittent flow.*** Further increasing the gas velocity, these interfacial waves become large enough to wash the top of the tube. This regime is characterized by large-amplitude waves intermittently washing the top of the tube with smaller amplitude waves in between. Intermittent flow is also a composite of the plug and slug flow regimes. These subcategories are characterized as follows:
   - ***Plug flow.*** This flow regime has liquid plugs that are separated by elongated gas bubbles. The diameters of the elongated bubbles are smaller than the tube such that the liquid phase is continuous along the bottom of the tube below the elongated bubbles. Plug flow is also sometimes referred to as *elongated bubble flow*.
   - ***Slug flow.*** At higher gas velocities, the diameters of elongated bubbles become similar in size to the channel height. The liquid slugs separating such elongated bubbles can also be described as large-amplitude waves.
e. ***Annular flow.*** At even larger gas flow rates, the liquid forms a continuous annular film around the perimeter of the tube, similar to that in vertical flow but the liquid film is thicker at the bottom than the top.
f. ***Mist flow.*** Similar to vertical flow, at very high gas velocities, all the liquid may be stripped from the wall and entrained as small droplets in the now continuous gas phase.

5.4.4 Parameters of Two-Phase Fluid Flow Phenomena

For details on parameters of two-phase fluid flow phenomena refer to Refs. [2,3,42,43].

5.4.4.1 Void Fraction

A parameter of particular importance when evaluating the pressure loss in steam-water flows is void fraction. The void fraction can be defined by time-averaged flow area ratios or local-volume ratios of steam to the total flow. The void fraction α is one of the most important parameters used to characterize two-phase flows. It is the key physical value for determining numerous other important parameters, such as the two-phase density and the two-phase viscosity, for obtaining the relative average velocity of the two phases, and is of fundamental importance in models for predicting flow pattern transitions, heat transfer, pressure drop, etc. Various geometric definitions are used for specifying the void fraction: local, chordal, cross sectional or area-based, and volumetric [1]. The area-based void fraction, α, can be defined as the ratio of the time-averaged steam flow cross-sectional area (A_s) to the total flow area ($A_s + A_w$):

$$\alpha = \frac{A_s}{A_s + A_w} \tag{5.14}$$

Using the simple continuity equation, the relationship between quality, x, and void fraction, α is:

$$\alpha = \frac{x}{x + (1 - x)\frac{\rho_g}{\rho_f} S} \tag{5.15}$$

where

S is ratio of the average cross-sectional velocities of steam and water (referred to as slip)
ρ_g is saturated steam density, lb/ft³ (kg/m³)
ρ_f is saturated water density, lb/ft³ (kg/m³)

5.4.4.2 Homogeneous Flow Model Applied to In-tube Flow

A homogeneous fluid is a convenient concept for modeling of two-phase pressure drops; it is a pseudo fluid that obeys the conventional design equations for single-phase fluids and is characterized by suitably averaged properties of the liquid and vapor phase. The homogeneous design approach is presented below. The total pressure drop of a fluid is due to the variation of kinetic and potential energy of the fluid and that due to friction on the walls of the flow channel. Thus, the total pressure drop Δp_t is the sum of the static pressure drop (elevation head) Δp_s, the momentum pressure drop (acceleration) Δp_m, and the frictional pressure drop Δp_f [3,42]:

$$\Delta p_t = \Delta p_s + \Delta p_m + \Delta p_f \tag{5.16}$$

The static pressure drop for a homogeneous two-phase fluid is:

$$\Delta p_s = \rho_H \, g \, H \sin\theta \tag{5.17}$$

where H is the vertical height, θ is the angle with respect to the horizontal, and the homogeneous density ρ_H is

$$\rho_H = \rho_L(1-\varepsilon_H) + \rho_G \varepsilon_H \tag{5.18}$$

and ρ_L and ρ_G are the liquid and gas (or vapor) densities, respectively. The homogeneous void fraction ε_H is determined from the quality x as

$$\varepsilon_H = \frac{1}{1 + \left[\dfrac{u_G (1-x) \rho_G}{u_L \, x \, \rho_L}\right]} \tag{5.19}$$

where u_G / u_L is the velocity ratio, or slip ratio (S), and is equal to 1.0 for a homogeneous flow. The momentum pressure gradient per unit length of the tube is:

$$\left(\frac{dp}{dz}\right)_m = \frac{d(m_t / \rho_H)}{d_Z} \tag{5.20}$$

The most problematic term is the frictional pressure drop, which can be expressed as a function of the *two-phase friction factor* f_{tp}, and for a steady flow in a channel with a constant cross-sectional area is:

$$\Delta p_f = \frac{2 f_{tp} L m^2_{total}}{d_i \rho_{tp}} \tag{5.21}$$

The friction factor may be expressed in terms of the Reynolds number by the Blasius equation:

$$f_{tp} = \frac{0.079}{Re^{0.25}} \quad (5.22)$$

where the Reynolds number, Re is

$$Re = \frac{\dot{m}_m d_i}{\mu_{tp}} \quad (5.23)$$

The viscosity for calculating the Reynolds number can be chosen as the viscosity of the liquid phase or as a quality-averaged viscosity μ_{tp} :

$$\mu_{tp} = x\mu_G + (1 - x)\mu_L \quad (5.24)$$

This correlation is suitable for mass velocities greater than 2000 kg/m^{2}s (1,471,584 lb/h ft^2) in the case of the frictional pressure drop calculations and for mass velocities less than 2000 kg/m^{2}s (1,471,584 lb/h ft^2) and $(\rho_L/\rho_G) < 10$ for gravitational pressure drop calculations. Generally speaking, this correlation should be used at highly reduced pressures and very high mass velocities.

5.4.4.3 Head Loss Due to Fluid Friction

Water at saturation conditions may exist as both a fluid and a vapor. This mixture of steam and water can cause unusual flow characteristics within fluid systems. There are several techniques used to predict the head loss due to fluid friction for two-phase flow. Two-phase flow friction is greater than single-phase friction for the same conduit dimensions and mass flow rate. The difference appears to be a function of the type of flow and results from increased flow speeds. Two-phase friction losses are experimentally determined by measuring pressure drops across different piping elements. The two-phase losses are generally related to single-phase losses through the same elements. One accepted technique for two-phase friction loss based on the single-phase loss involves the two-phase friction multiplier, *R*, which is defined as the ratio of the two head loss divided by the head loss evaluated using saturated liquid properties [3]:

$$R = \frac{H_{f2p}}{H_{f1p}} \quad (5.25)$$

where

R is two-phase friction multiplier (no units)
H_{f2p} is two-phase head loss due to friction (m/ft)
H_{f1p} is single-phase head loss due to friction (m/ft)

The friction multiplier (*R*) has been found to be much higher at lower pressures than at higher pressures. The two-phase head loss can be many times greater than the single-phase head loss.

5.4.4.4 Flow Instability

Unstable flow can occur in the form of flow oscillations or flow reversals. Flow oscillations are variations in flow due to void formations or mechanical obstructions from design and manufacturing. Flow oscillations are undesirable for several reasons [3].

First, sustained flow oscillations can cause undesirable forced mechanical vibration of components. This can lead to failure of those components due to fatigue.

Second, flow oscillations can cause system control problems of particular importance in liquid-cooled nuclear reactors because the coolant is also used as the moderator.

Third, flow oscillations affect the local heat transfer characteristics and boiling.

5.4.4.5 Instabilities in Steam-Generating Systems

Instability in two-phase flow refers to the set of operating conditions under which sudden changes in flow direction, reduction in flow rate, and oscillating flow rates can occur in a single flow passage. Such unstable conditions in steam-generating systems can result in [2]:

1. Unit control problems, including unacceptable variations in steam drum water level
2. CHF/DNB/dryout
3. Tube metal temperature oscillation and thermal fatigue failure
4. Accelerated corrosion attack

Two of the most important types of instabilities in steam generator design are excursive instability, including Ledinegg and flow reversal, and density wave/pressure drop oscillations. The first is a static instability evaluated using steady-state equations, while the last is dynamic in nature, requiring the inclusion of time-dependent factors [2].

5.4.4.6 Pipe Whip

If a pipe were to rupture, the reaction force created by the high-velocity fluid jet could cause the piping to displace and cause extensive damage to components, instrumentation, and equipment in the area of the rupture. This characteristic is similar to an unattended garden hose or fire hose "whipping" about unpredictably. This type of failure is analyzed to minimize damage if pipe whip were to occur in the vicinity of safety-related equipment [3].

5.4.4.7 Water Hammer

Water hammer is a liquid shock wave resulting from the sudden starting or stopping of flow. It is affected by the initial system pressure, the density of the fluid, the speed of sound in the fluid, the elasticity of the fluid and pipe, the change in velocity of the fluid, the diameter and thickness of the pipe, and the valve operating time [3].

The initial shock of suddenly stopped flow can induce transient pressure changes that exceed the static pressure. If the valve is closed slowly, the loss of kinetic energy is gradual. If it is closed quickly, the loss of kinetic energy is very rapid. A shock wave results because of this rapid loss of kinetic energy. The shock wave caused by water hammer can be of sufficient magnitude to cause physical damage to piping, equipment, and personnel. Water hammer in pipes has been known to pull pipe supports from their mounts, rupture piping, and cause pipe whip [3].

5.4.4.8 Pressure Spike

A pressure spike is the resulting rapid rise in pressure above static pressure caused by water hammer. The highest pressure spike attained will be at the instant the flow is changed and is governed by the following equation [3]:

$$\Delta P = \frac{\rho c \Delta v}{g_c} \tag{5.26}$$

where

ΔP is pressure spike (lbf/ft^2)

ρ is density of the fluid (lbm/ft^3)

c is velocity of the pressure wave (speed of sound in the fluid) (ft/s)

Δv is change in velocity of the fluid (ft/s)

g_c is gravitational constant 32.17 (lbm-ft / lbf–s^2)

5.4.4.9 Steam Hammer

Steam hammer is similar to water hammer except it is for a steam system. Steam hammer is a gaseous shock wave resulting from the sudden starting or stopping of flow. Steam hammer is not as severe as water hammer for three reasons [3]:

1. The compressibility of the steam dampens the shock wave.
2. The speed of sound in steam is approximately one-third the speed of sound in water.
3. The density of steam is approximately 1600 times less than that of water.

The items of concern that deal with steam piping are thermal shock and water slugs (i.e., condensation in the steam system) as a result of improper warm up.

5.4.4.10 Operational Considerations

Water and steam hammer are not uncommon occurrences in industrial plants. Flow changes in piping systems should be done slowly as part of good operator practice. To prevent water and steam hammer, operators should ensure liquid systems are properly vented and ensure gaseous or steam systems are properly drained during start-up. When possible, initiate pump starts against a closed discharge valve, and open the discharge valve slowly to initiate system flow. If possible, start-up smaller capacity pumps before larger capacity pumps. Use warm-up valves around main stream stop valves whenever possible. If possible, close pump discharge valves before stopping pumps. Periodically verify proper function of moisture traps and air traps during operation [3].

5.5 STEAM GENERATION

5.5.1 Boiling Process and Steam Generation

The process of boiling water to make steam is a familiar phenomenon. As heat is added to water, the temperature of the water increases. When the water temperature reaches the boiling point, or saturation temperature, some of the water begins to vaporize to steam.

5.5.1.1 Saturated Water and Saturated Steam

When water just begins to boil, it is called saturated water. As more heat is added (at constant pressure), the fluid temperature will remain at the saturation temperature until all of the water is converted to steam. The heat input or enthalpy necessary to convert saturated water to saturated steam is called the heat of vaporization. The conversion of water to steam requires much more energy beyond that required to reach the boiling point.

5.5.1.2 Boiling Point

The term boiling point is most frequently used to identify conditions at atmospheric pressure (29.92 inches of mercury.) For instance, the boiling point of water at atmospheric pressure is 100°C (212°F). The boiling point is actually a function of pressure and increases as pressure increases. At higher pressures, more heat energy is required to raise the fluid temperature to the boiling point.

5.5.1.3 Enthalpy

The amount of heat energy contained in the fluid is termed enthalpy and is measured in BTUs/lb.

5.5.1.4 Heat of Vaporization

The points at which all of the water has been converted to steam are indicated by the saturated steam line. The heat input or enthalpy necessary to convert saturated water to saturated steam is termed the heat of vaporization and is indicated for a given temperature by the horizontal constant pressure lines.

5.5.1.5 Steam Quality

The measure of how far the conversion from saturated water to saturated steam has progressed is called quality. Quality is the percent by weight of vapor in a steam/water mixture. As more water is converted to steam, quality increases. Water on the saturated water line has a quality of 0%. Superheated and saturated steams have a quality of 100%. Water that has been heated to saturation and has sufficient additional heat added to convert half of it to steam has a quality of 50%.

5.5.1.6 Superheated Steam

If still more heat is added to saturated steam, the temperature will again begin to rise. The fluid in this area is said to be superheated steam. The superheaters derive their name from their function of heating steam above the saturation curve. Steam is sometimes referred to as having a number of degrees of superheat. The number of degrees of superheat describes how far the steam has been heated above the saturation curve.

5.5.2 Boiling Process

In general terms, boiling is the heat transfer process where heat addition to a liquid no longer raises its temperature under constant pressure conditions; the heat is absorbed as the liquid becomes a gas. This process, which takes place at constant pressure and constant temperature, is known as boiling. The heat transfer rates are high, making this an ideal cooling method for surfaces exposed to the high heat input rates found in fossil fuel boilers, concentrated solar energy collectors, and nuclear reactor fuel bundles. An important reference source on steam generator boiling process is B&W Company's *Steam Its Generation and Use* [2].

5.5.2.1 Boiling Process and Fundamentals

The boiling point, or saturation temperature, of a liquid can be defined as the temperature at which its vapor pressure is equal to the total local pressure. The saturation temperature for water at atmospheric pressure is 212°F (100°C). This is the point at which net vapor generation occurs and free steam bubbles are formed from a liquid undergoing continuous heating. This saturation temperature (T_{sat}) is a unique function of pressure. The thermophysical characteristics of water include the enthalpy (or heat content) of water, the enthalpy of evaporation (also referred to as the latent heat of vaporization), and the enthalpy of steam. As the pressure is increased to the critical pressure [3200 psi (22.1 MPa)], the latent heat of vaporization declines to zero and the bubble formation associated with boiling no longer occurs. Instead, a smooth transition from liquid to gaseous behavior occurs with a continuous increase in temperature as energy is applied. Two other definitions are also helpful in discussing boiling heat transfer [2]:

1. Subcooling. For water below the local saturation temperature, this is the difference between the saturation temperature and the local water temperature ($T_{sat} - T$).

2. Quality, x. This is the flowing mass fraction of steam (frequently stated as percent steam by weight or %SBW after multiplying by 100%):

$$x = \frac{\dot{m}_{steam}}{\dot{m}_{water} + \dot{m}_{steam}} \tag{5.27}$$

where

m_{steam} is steam flow rate, lb/h (kg/s)
m_{water} is water flow rate, lb/h (kg/s)
Thermodynamically, this can also be defined as:

$$x = \frac{H - H_s}{H_{fs}} \, or \, \frac{H - H_f}{H_g - H_f} \tag{5.28}$$

where

H is local average fluid enthalpy, Btu/lb (J/kg)
H_f is enthalpy of water at saturation, Btu/lb (J/kg)
H_g is enthalpy of steam at saturation, Btu/lb (J/kg)
H_{fg} is latent heat of vaporization, Btu/lb (J/kg)

When boiling is occurring at saturated, thermal equilibrium conditions, Equation 5.27 provides the fractional steam flow rate by mass. For subcooled conditions where $H < H_f$, quality (x) can be negative and is an indication of liquid subcooling. For conditions where $H > H_g$, this value can be greater than 100% and represents the amount of average superheat of the steam. However, the boiling phenomenon poses special challenges such as [2]:

1. The sudden breakdown of the boiling behavior at very high heat input rates
2. The potential flow rate fluctuations which may occur in steam–water flows
3. The efficient separation of steam from water

5.5.2.2 Steam Generation

The process of boiling water to make steam is a familiar phenomenon. Thermodynamically speaking, the heat energy used results in a change of phase from the liquid to vapor state. A steam-generating system has to provide a continuous and uninterrupted heat source for this conversion. For a given pressure, steam heated above the saturation temperature is called superheated steam, whereas water cooled below the saturation temperature is called subcooled water. If the boiling water is a closed system, then after converting all the water to steam, new heat energy coming into the system is used for increasing steam temperature, i.e., superheating the saturated steam.

5.5.2.3 High-Pressure Steam Generation

Technical and economic considerations indicate that the most efficient and economical way of producing high-pressure steam is to supply heat to relatively small-diameter tubes containing a continuous flow of water. Two inherently different boiling systems are used in order to accomplish this task [44]:

i. A steam drum
ii. A once-through steam generator (OTSG)

5.5.2.4 Steam Drum

In this system, the drum serves as the point of separation of steam from water. Subcooled water enters the tube, to which heat is applied. As the water flows through the tube, it is heated to the boiling point. Consequently, bubbles are formed and wet steam is generated. In most boilers, a steam–water mixture leaves the tube and enters the steam drum. The remaining water is then mixed with the makeup (replacement) water, returned back to the heated tube, and the process is repeated.

5.5.2.5 Once-Through Steam Generator (OTSG)

In an OTSG, subcooled water enters the heated tube and gets converted to steam along the flow path. The point where water turns into steam depends on the water flow rate (boiler load) and heat input rate. With close control of both flow rate and heat input rate, all of the water is evaporated withing the tube and only steam leaves the tube. Therefore, there is no need for having a steam drum. At very high pressures, a point is reached where water no longer exhibits boiling behavior. Above this pressure (approximately 221 bar or 3200 psi), the water temperature keeps increasing with added heat. Boilers designed to operate above this critical pressure are referred to as supercritical boilers. Figure 5.18 presents a schematic view of steam generators with steam drum and in tube steam generation, namely, OTSG [44].

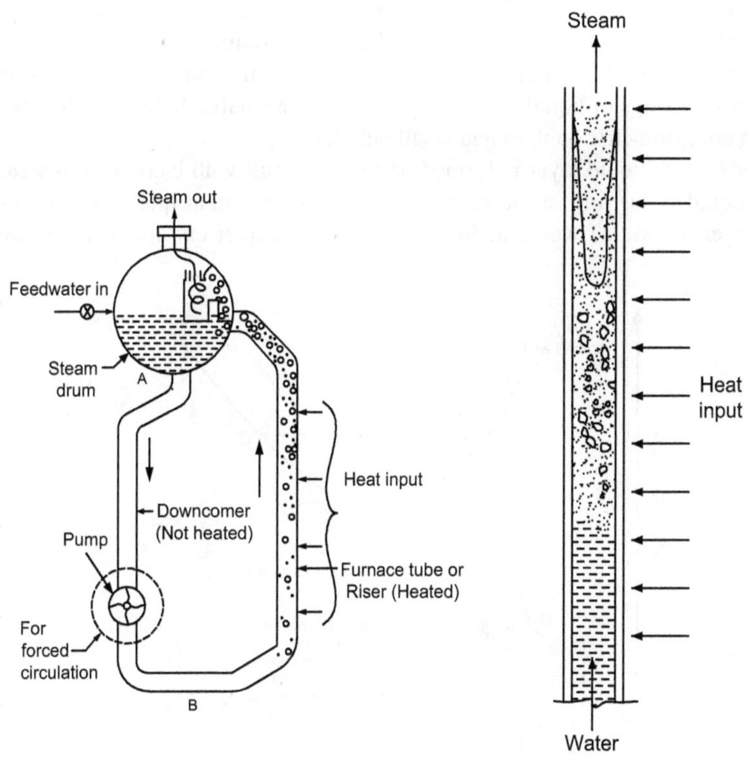

FIGURE 5.18 A schematic view of steam generators with steam drum and OTSG.

5.5.2.6 Boiling Curve of Steam Generation

The boiling curve shown in Figure 5.19 for a steam generator [2] provides the results of a heated wire in a pool, although the characteristics are similar for most situations. The heat transfer rate per unit area, or heat flux, is plotted versus the temperature differential between the metal surface and the bulk fluid.

5.5.2.7 Boiling Heat Transfer

At low-temperature differences, when the liquid is below its boiling point, heat is transferred by natural convection and follows nucleate boiling. Nucleate boiling takes place when the surface temperature is hotter than the saturated fluid temperature by a certain amount but where the heat flux is below the critical heat flux followed by transition boiling.

5.5.2.8 Flow Pattern–: Upward, Co-current Steam–Water Flow in a Heated Vertical Tube

A steam–water mixture progresses through a series of flow structures or patterns: bubbly, intermediate, and annular. Flow pattern–upward, co-current steam–water flow in a heated vertical tube is described hereunder (adopted from B&W's *Steam Its Generation and Use* [2]) and it is illustrated in Figure 5.20:

1. *Bubbly flow.* Relatively discrete steam bubbles are dispersed in a continuous liquid water phase. Bubble size, shape, and distribution are dependent upon the flow rate, local enthalpy, heat input rate, and pressure.
2. *Intermediate flow.* This is a range of patterns between bubbly and annular flows; the patterns are also referred to as slug or churn flow. They range from: (a) large bubbles, approaching the tube size in diameter, separated from the tube wall by thin annular films and separated from each other by slugs of liquid which may also contain smaller bubbles to (b) chaotic mixtures of large nonsymmetric bubbles and small bubbles.
3. *Annular flow.* A liquid layer is formed on the tube wall with a continuous steam core; most of the liquid is flowing in the annular film. At lower steam qualities, the liquid film may have larger amplitude waves adding to the liquid droplet entrainment and transport in the

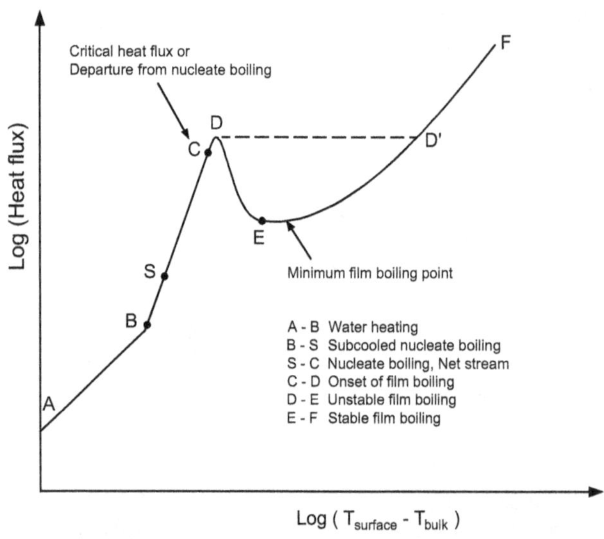

FIGURE 5.19 Boiling curve – heat flux versus applied temperature difference for a steam generator. (Adapted from Dipak Sarkar, Thermal Power Plant: Design and Operation, CHAPTER 2 – Steam Generators, Elsevier. 2015.)

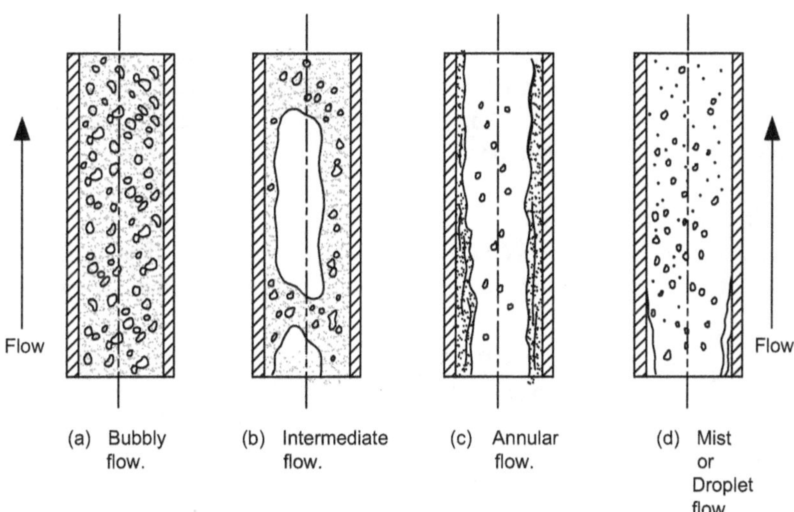

FIGURE 5.20 Flow pattern – upward, co-current steam-water flow in a heated vertical tube. (Adapted from Kitto, J.B., Stultz, S.C. (Eds.) (2005) *Steam Its Generation and Use*, Forty-first edition, First printing. The Babcock & Wilcox McDermott Company pp. 5.1–5.21.)

continuous steam core. At high qualities, the annular film becomes very thin, bubble generation is suppressed, and the large-amplitude waves disappear.

4. *Mist flow.* A continuous steam core transports entrained water droplets which slowly evaporate until a single-phase steam flow occurs.

5.5.2.9 Flow Boiling

Flow or forced convective boiling is found in virtually all steam-generating systems. Figure 5.21 shows boiling water in a long, uniformly heated, circular tube.

5.5.2.10 Boiling Heat Transfer Evaluation

Engineering design of steam generators requires the evaluation of water and steam heat transfer rates under boiling and nonboiling conditions. In addition, the identification of the location of critical heat flux (CHF) is important where a dramatic reduction in the heat transfer rate could lead to [2]:

1. Excessive metal temperatures potentially resulting in tube failures
2. An unacceptable loss of thermal performance
3. Unacceptable temperature fluctuations leading to thermal fatigue

5.5.2.11 Nucleate and Convective Boiling

Heat transfer in the saturated boiling region occurs by a complex combination of bubble nucleation at the tube surface (nucleate boiling) and direct evaporation at the steam–water interface in annular flow (convective boiling). At low steam qualities, nucleate boiling dominates, while at higher qualities, convective boiling dominates. While separate correlations are available for each range, the most useful relationships cover the entire saturated boiling regime.

5.5.2.12 Critical Heat Flux Phenomena

Critical heat flux (CHF) is one of the most important parameters in steam generator design. CHF denotes the set of operating conditions (mass flux, pressure, heat flux, and steam quality) covering the transition from the relatively high heat transfer rates associated with nucleate or forced convective boiling to the lower rates resulting from transition or film. These operating conditions have been found to be geometry specific. CHF encompasses the phenomena of departure from nucleate boiling (DNB), burnout, dryout, and boiling crisis. One objective in recirculating boiler and nuclear reactor designs is to avoid CHF conditions.

5.5.2.13 CHF Criteria

A number of criteria are used to assess the CHF margins in a particular tube or tube bundle geometry. These include the CHF ratio, flow ratio, and quality margin, defined as follows [2]:

1. CHF ratio = minimum value of $$\frac{CHF\ heatflux}{upsetheat\ flux} \tag{5.29}$$
2. Flow ratio = minimum value of $$\frac{min.design\ mass\ flux}{massflux\ at\ CHF} \tag{5.30}$$
3. Quality margin

5.5.2.14 Factors Affecting CHF

Critical heat flux phenomena under flowing conditions found in fossil fuel and nuclear steam generators are affected by a variety of parameters. The primary parameters are the operating conditions and the design geometries. The operating conditions affecting CHF are pressure, mass flux, and steam quality, and design geometry factors include flow passage dimensions and shape, flow path obstructions, heat flux profile, inclination, and wall surface configuration.

5.5.2.15 Use of Ribbed Tubes

B&W has developed two general types of rib configurations as given below [2]:

1. Single-lead ribbed tubes for small internal diameters used in once-through subcritical pressure boilers
2. Multi-lead ribbed tubes for larger internal diameters used in natural circulation boilers. Both of these ribbed tubes have shown a remarkable ability to delay the breakdown of boiling

The ribbed bore tubes provide a balance of improved CHF performance at an acceptable increase in pressure drop without other detrimental effects. The ribs generate a swirl flow resulting in a centrifugal action which forces the water to the tube wall and retards entrainment of the liquid.

5.5.3 Steam Generators

5.5.3.1 Nucleate Boiling

Nucleate boiling takes place when the surface temperature is hotter than the saturated fluid temperature by a certain amount but where the heat flux is below the critical heat flux. For water, as shown in the graph in Figure 5.20, nucleate boiling occurs when the surface temperature is higher than the saturation temperature (T_S) by between 10°C (18°F) to 30°C (54°F). The critical heat flux is the peak on the curve between nucleate boiling and transition boiling. The heat transfer from surface to liquid is greater than that in film boiling [44].

5.5.3.2 Pool Boiling

When a large volume of liquid is heated by a submerged heating surface and the motion is caused by free convection currents stimulated by agitation of the rising vapor bubbles, it is called pool boiling, i.e., in pool boiling, the fluid is not forced to flow by a pump or any other fluid-moving device. Any motion of the fluid is due to natural convection currents and the motion of the bubbles under the influence of buoyancy.

5.5.3.3 Departure from Nucleate Boiling (DNB)

Heat flux is extremely high, as, for example, with water at atmospheric pressure heat flux is about 1500 kW/m^2. The temperature difference, ΔT, between the boiling liquid and the heating surface at which the critical heat flux occurs, is known as the critical temperature difference. During operation of the boiler it is difficult to theoretically predict the heat transfer coefficients of nucleate boiling that would ensure ΔT remains below the critical temperature difference.

5.5.3.4 Forced Convection Boiling

Besides pool boiling, vapor is also generated by passing liquid through a tube heated either by firing fuel, as in a once-through boiler, or by condensing steam, as is commonly used in process industries. This method of generating vapor is known as forced-convection boiling. Both heat transfer and pressure drop are affected by the behavior of the two-phase flow, which keeps on changing along the tube due to the gradual evaporation of liquid.

5.5.3.5 Once-Through Boiler

In a once-through tube steam generation, liquid below or at the saturation point enters at the bottom and gradually evaporates until a dry or superheated vapor leaves at the top, as shown in Figure 5.21 [44].

The water enters the tube as a subcooled liquid and convection heat transfer cools the tube. The point of incipient boiling is reached. This results in the beginning of subcooled boiling and bubbly flow. The fluid temperature continues to rise until the entire bulk fluid reaches the saturation temperature and nucleate boiling occurs (point 2).

While boiling heat transfer continues throughout, a point is reached in the annular flow regime where the liquid film on the wall becomes so thin that nucleation in the film is suppressed (point 3).

Eventually, an axial location (point 4) is reached where the tube surface is no longer wetted and CHF or dryout occurs. This is typically associated with a temperature rise. The exact tube location and magnitude of this temperature, however, depend upon a variety of parameters, such as the heat flux, mass flux, geometry, and steam quality.

The following formula, from Roshenow and Griffith [45], gives the heat transfer value of q_{m} in SI units, i.e., W/m^2:

$$q_m/(\rho_V\lambda) = 0.0121\{(\rho_L - \rho_V)/\rho_V\}^{0.6} \tag{5.31}$$

where

ρ_L is density of liquid, kg/m^3
ρ_V is density of vapor, kg/m^3
λ is latent heat of the fluid, kJ/kg

Once the critical heat flux data are determined they are compared with the required heat flux data. For the design to be acceptable, the critical heat flux, which is dependent on the following parameters, must always be greater than the heat flux generated in the course of boiling: operating pressure, steam quality, i.e., surface tension, subcooling, etc., type of tube (rifled or plain bore), tube diameter, flux profile around the tube, steam/water flow in the tube, and angle of inclination of the tube.

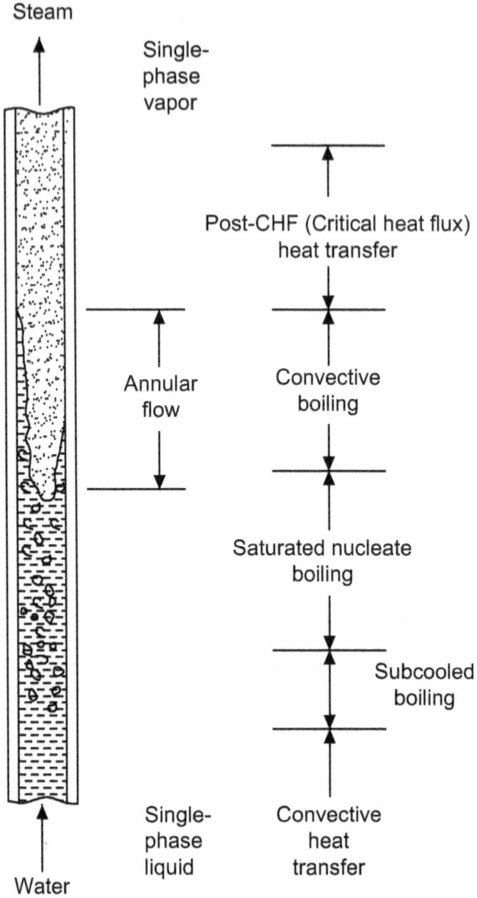

FIGURE 5.21 Once-through steam generation (Adapted from Dipak Sarkar, Thermal Power Plant: Design and Operation, CHAPTER 2 – Steam Generators, Elsevier, 2015.)

In Figure 5.21, the first and the fourth regions are extremes consisting of liquid-only and vapor-only regions, respectively:

a. In these regions in turbulent flow the phenomena of heat transfer to water in pipes is described by the Dittus-Boelter equation as follows:

$$Nu = 0.023\ Re^{0.8}\ Pr^{0.4} \tag{5.32}$$

b. For superheated steam through pipes, the equation is

$$Nu = 0.0133\ Re^{0.84} Pr^{0.33} \tag{5.33}$$

where

Nu is Nusselt number, hd/k
Re is Reynolds number, $\rho Vd/\mu$
Pr is Prandtl number, $C_p\mu/k$
h is convective heat transfer coefficient, W/m²k

ρ is density of fluid, kg/m^3
V is velocity of fluid, m/s
d is internal diameter of pipe, m
μ is dynamic viscosity of fluid, kg/m.s
C_p is specific heat of fluid, kJ/kg.K
k is thermal conductivity of fluid, W/m.K

REFERENCES

1. Thome, J.R. (Ed.) (2004) *Engineering Data Book III*. Wolverine Tube, Inc., Decatur, AL.
2. Kitto, J.B., Stultz, S.C. (Eds.) (2005) *Steam Its Generation and Use*, Forty-first edition, First printing. The Babcock & Wilcox McDermott Company pp. 5.1–5.21.
2.1 Thome, J.R. (Ed.) (2004) *Boiling Heat Transfer, Two-Phase Flow and Circulation, Chapter 5, Engineering Data Book III*. Wolverine Tube, Inc., Decatur, AL.
3. U.S. Department of Energy (1992) *DOE Fundamentals Handbook: Thermodynamics, Heat Transfer, and Fluid Flow*, DOE-HDBK-1012/2–92. U.S. Department of Energy, June 1992.
3.1 DOE Fundamentals Handbook: Thermodynamics, Heat Transfer, and Fluid Flow.
4. Thome, J.R. (Ed.) (2004) *Boiling Heat Transfer Inside Plain Tubes, Chapter 10, Engineering Data Book III*, pp. 10.1–10.29. Wolverine Tube, Inc., Decatur, AL, 2004.
5. Collier, J.G., Thome, J.R. (1994) *Convective Boiling and Condensation*, 3rd Edition. Oxford University Press, Oxford.
6. Nukiyama, S. (1934) The maximum and minimum values of the heat Q transmitted from metal to boiling water. *Journal of the Japan Society of Mechanical Engineering* 37(206): 367–374.
7. www.nuclear-power.com/nuclear-engineering/heat-transfer/boiling-and-condensation/boiling-heat-transfer/ © Nuclear Power plant
8. Rohsenow, W.M. (1952) A method of correlating heat-transfer data for surface boiling of liquids. *Transactions of ASME* 74(6): 969–975.
9. Chen, J.C. (1963) *A Correlation for Boiling Heat Transfer of Saturated Fluids in Convective Flow*, ASME Paper 63-HT-34. 6th National Heat Transfer Conference, Boston, August 11–14, 1963.
10. Forster, H.K., Zuber, N. (1955) Dynamics of vapor bubbles and boiling heat transfer. *AIChE Journal* 1(4): 531–535.
11. McAdams, W.M. (1954) *Heat Transmission*, 3rd Edition. McGraw-Hill, New York.
12. Rohsenow, W., Hartnet, J., Cho, Y. (1998) *Handbook of Heat Transfer*, 3rd edn. McGraw-Hill Book Company.
13. Koizumi, Y., Shoji, M., Monde, M., Takata, Y., Nagai, N. (Eds.) Chapter 6 Boiling. In: *Topics on Boiling: From Fundamentals to Applications*, , pp. 443–777. Elsevier.
13.1 www.sciencedirect.com/topics/engineering/boiling-heat-transfer
14. Thome, J.R. (Ed.) (2004) Boiling Heat Transfer Inside Plain Tubes, Chapter 11. In: *Engineering Data Book III*, pp. 11.1–11.22. (Wolverine Tube, Inc., Decatur, AL.
15. www.thermopedia.com/content/575/
16. https://thermalkinetics.net/evaporation-equipment/falling-film-tubular-evaporator
17. Thome, J.R. (Ed.) (2004) Falling Film Evaporation, Chapter 14. In: *Engineering Data Book III*, pp. 14.1–14.14. Wolverine Tube, Inc., Decatur, AL.
18. https://rccostello.com/wordpress/boilers/distillation-part-2-reboilers/
19. Sinnott, R.K. (2008) *Heat Transfer Equipment, Chapter 12, Chemical Engineering Design: Principles, Practice and Economics of Plant and Process Design*. Elsevier Inc.
20. www.globalspec.com/reference/22825/203279/chapter-12-heat-transfer-equipment
21. McKee, H.R. (1970) Thermosyphon Reboilers—A Review. *Industrial Engineering Chemistry* 62(12): 76–82.
22. Palen, J.W. (1983) Shell and tube reboilers. Section 3.6. In: *Heat Exchanger Design Handbook*. Hemisphere Publishing Corporation, New York.
23. https://msubbu.in/ln/design/II/DesignII-Lecture-02-ReboilerDesign.pdf
24. www.thermopedia.com/content/1078/
25. www.sterlingtt.com/products-services/shell-and-tube-heat-exchangers/reboilers/

26. Forster, H.K., Zuber, N. (1955) Dynamics of vapor bubble growth and boiling heat transfer. *AIChE Journal* 1: 531–535.
27. Zuber, N. (1961) The dynamics of vapor bubbles in nonuniform temperature fields. *International Journal of Heat and Mass Transfer* 2(1–2): 83–98.
28. Mostinski, L. (1963) Application of the Rule of Corresponding States for Calculation of Heat Transfer and Critical Heat Flux to Boiling Liquids. *British Chemical Engineering Abstracts*: Folio no. 150, 580.
29. Palen, J.W., Small, W.M. (1964). A new way to design kettle and inernal reboilers. *Hydrocarbon Processing* 43(11): 199–208.
30. www.sciencedirect.com/topics/engineering/condensation-heat-transfer
31. Kirillov, P.L., Ninokata, H. (2017) Heat transfer in nuclear thermal hydraulics. In: *Thermal-Hydraulics of Water Cooled Nuclear Reactors*, pp. 357–492.
32. http://nptel.ac.in/courses/103103032/module6/lec25/4.html
33. Rose, J.W. (2011) *Condensation, overview*. DOI: 10.1615/AtoZ.c.condensation_overview. Article added: 2 February 2011, article last modified: 9 February 2011.
34. Nusselt, W. (1916) Die Oberflächenkondensation des Wasserdampfes. *Zeitschrift Ver. Deutch. Ing.* 60: 541–546, 569–575.
35. www.thermopedia.com/content/653
36. Thome, J.R. (Ed.) (2004) Condensation on External Surfaces, Chapter 7. , In: *Engineering Data Book III* pp. 7.1–7.39Wolverine Tube, Inc.
37. Gregorig, R. (1954) Film condensation on finely rippled surface with consideration of surface tension. *Z. Agnew. Math. Phys.* 5: 36–49.
38. Thome, J.R. (Ed.) (2004) Condensation Inside Tubes, Chapter 8. In: *Engineering Data Book III*, pp. 8.1–8.27. Wolverine Tube, Inc.
39. Palen, J.W., Breber, G., Taborek, K. (1979) Prediction of flow regimes in horizontal tube-side condensation. *Heat Transfer Engineering* 1(2): 47–57.
40. www.nuclear-power.com/nuclear-engineering/heat-transfer/boiling-and-condensation/condensation/
41. Thome, J.R. (Ed.) (2004) Two-Phase Flow Patterns, Chapter 12. In: *Engineering Data Book III*, pp. 12.1–12.21. Wolverine Tube, Inc., Decatur, AL.
42. Thome, J.R. (Ed.) (2004) Two-Phase Pressure Drops, Chapter 13. In: *Engineering Data Book III*, pp. 13.1–13.34. Wolverine Tube, Inc., Decatur, AL.
43. Thome, J.R. (Ed.) (2004) Void Fractions in Two-Phase Flows, Chapter 17. In: *Engineering Data Book III*, pp. 17.1–17.33. Wolverine Tube, Inc., Decatur, AL, 2004.
44. Dipak Sarkar, Thermal Power Plant: Design and Operation, CHAPTER 2 – Steam Generators, Elsevier. 2015. https://staff.emu.edu.tr/devrimaydin/Documents/MENG446/Chapter/2
45. Rohsenow, W.M., Griffith, P. (1956) Correlation of maximum heat flux data for boiling of saturated liquids. *Chemical Engineering Progress Symposium Series* 52(18): 47–49.

BIBLIOGRAPHY

Chen, J.C. (1962) *A correlation for boiling heat transfer to saturated fluids in convective flow*. United States: N. p., 1962. Web. doi:10.2172/4636495.

Collier, J.G., Thome, J.R. (1994) *Convective Boiling & Condensation,* Third Ed. Oxford University Press, Oxford, United Kingdom.

Thome, J.R. (Ed.) (2004) Enhanced single phase laminar tube side flows and heat transfer, Chapter 4. In: *Engineering Data Book III*, pp. 4.1–4.27. Wolverine Tube Inc., Decatur, AL.

https://en.wikipedia.org/wiki/Nucleate_boiling#:~:text=Nucleate%20boiling%20is%20a%20type,below%20the%20critical%20heat%20flux.

Nukiyama, S. (1934) The maximum and minimum values of heat Q transmitted from metal to boiling water under atmospheric pressure. *Journal of the Japan Society of Mechanical Engineering* 37: 367–374 (in Japanese) (trans. in *International Journal of Heat and Mass Transfer* [1966] 9: 1419–1433).

Rohsenow, W.M. (1951) A method of correlating heat transfer data for surface boiling of liquids. *Journal of Heat Transfer* 74: 969–976.

Rohsenow, W.M. (1952) A method of correlating heat-transfer data for surface boiling of liquids. *Transactions of ASME* 74(6): 969–975.

6 Regenerators and Waste Heat Recovery Devices

6.1 INTRODUCTION

Waste heat is generated from a variety of industrial systems distributed throughout a manufacturing plant. The largest sources of waste heat for most industries are exhaust and flue gases and heated air from heating systems such as high-temperature gases from burners in process heating; lower temperature gases from heat treating furnaces, dryers, and heaters; and heat from heat exchangers, cooling liquids, and gases. Waste heat recovery is an old technology dating back to the first open hearths and blast furnace stoves as shown in Figure 6.1 [1]. Manufacturing and process industries such as glass, cement, primary and secondary metals, etc., account for a significant fraction of all energy consumed. Much of this energy is discarded in the form of high-temperature flue gas exhaust streams. Recovery of waste heat from flue gas by means of heat exchangers known as regenerators can improve the overall plant efficiency and serves to reduce the national energy needs and conserve fossil fuels.

Currently, interest in storage-type regenerators has been renewed due to their applications in heat recovery, heat storage, and general energy-related problems [2]. The objective of this chapter is to acquaint readers with various types of regenerators, construction details, and thermal and mechanical design. Additionally, some industrial heat recovery devices for waste heat recovery are discussed.

6.1.1 Regeneration Principle

For many years, the regeneration principle has been applied to waste heat recovery by preheating the air for blast furnaces and boiler. In the former, the regeneration is achieved with periodic and alternate blowing of the hot and cold streams through a fixed matrix of checkered brick. During the hot flow period, the matrix receives thermal energy from the hot gas and transfers it to the cold stream during the cold stream flow. In the latter, heat exchange takes place between the flue gases and the air through metallic or ceramic walls. The two gas streams may flow either in parallel directions or opposite directions. However, counterflow is mostly preferred because of its high thermal effectiveness. The inclusion of a recuperator such as an air preheater in a steam power plant improves the boiler efficiency and overall performance of the power plant. The layout can be either staggered or inline.

6.1.2 Regenerators in Thermodynamic Systems and Others

The addition of regeneration as a thermodynamic principle improves the overall performance of gas turbine power plants, steam power plants, and heat exchangers embodying the thermodynamic

DOI: 10.1201/9781003352044-6

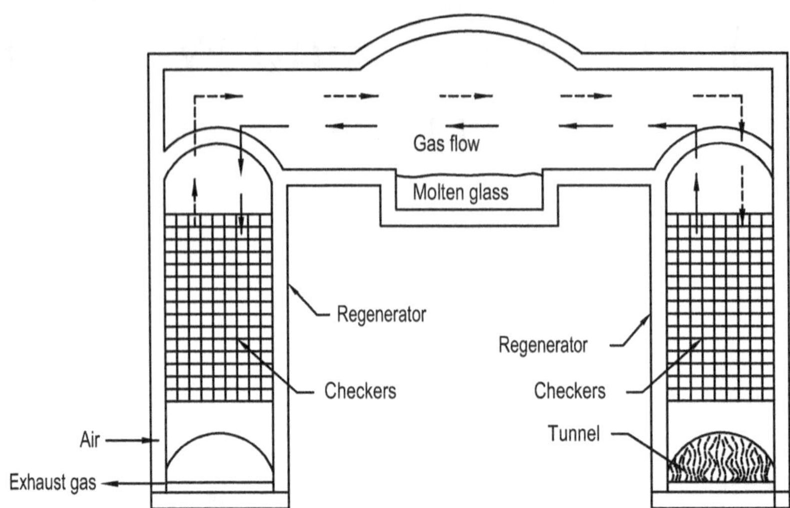

FIGURE 6.1 Regenerative furnace. *Note*: Dotted line shows air flow.

cycle of Stirling, Ericssion, Gifford, McMohan, and Vuilleumier [3]. Regenerators are also used as dehumidifiers for air-conditioning applications, cryogenic separation processes, and in noncatalytic chemical reactors such as the Wisconsin process for the fixation of nitrogen and the Wulff process for the pyrolysis of hydrocarbon feedstocks to produce acetylene and ethylene. The principle of regeneration applied to a gas turbine plant is described next.

6.1.3 Gas Turbine Cycle with Regeneration

The simple gas turbine plant consisting only of a compressor, combustion chamber, and turbine has the advantages of light weight and compactness. The single improvement that gives the greatest increase in thermal efficiency is the addition of a regenerator for transferring thermal energy from the hot turbine exhaust gas to the air leaving the compressor, especially when it is employed in conjunction with intercooling during compression. The addition of a regenerator results in a flat fuel economy versus load characteristic, which is highly desirable for the transportation-type prime movers such as gas turbine locomotives, marine gas turbine plants, and aircraft turboprops. A gas turbine regenerative cycle is shown in Figure 6.2.

6.1.4 Waste Heat Recovery Application

Substantial gains in fuel efficiency can be achieved by recovering the heat contained in the flue gas by the following four means [4]:

1. Reheating process feedstock
2. Waste heat boiler and feed water heating for generating steam (low-temperature recovery system)
3. Preheating the combustion air (high-temperature recovery system)
4. Space heating—A rotary heat exchanger (wheel) is mainly used in building ventilation or in the air supply/discharge system of conditioning equipment. The wheel transfers the energy (cold or heat) contained in exhaust air to the fresh air supply to indoor. It is one important piece of equipment and key technology in the field of construction energy saving

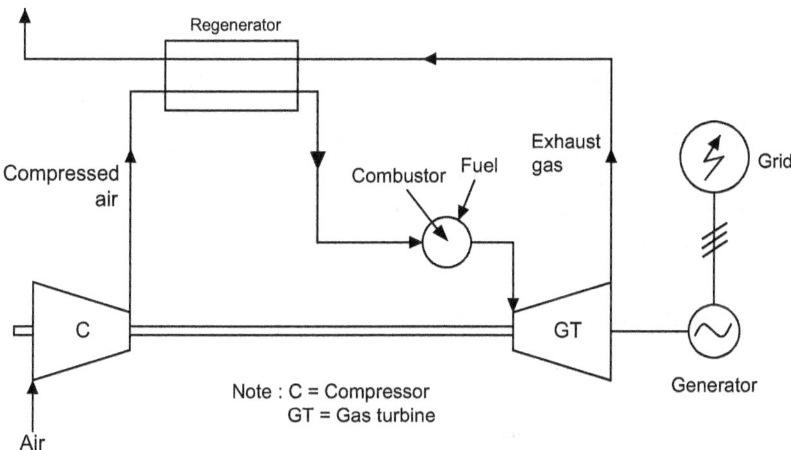

FIGURE 6.2 Gas turbine regenerative cycle.

Each of these methods of heat recovery has its own merits. According to Liang et al. [4], feedstock preheating is often best suited for continuous counterflow furnaces. Applications of this method, however, are often limited by high capital costs and large space requirements. Steam generation is an effective means for heat recovery when the demand for steam corresponds well with the availability of flue gas. Typical methods of steam generation include simple forced recirculation cycle, exhaust heat recovery with economizer, and exhaust heat recovery with a superheater and economizer. Combustion air preheating is the most adaptable of heat recovery systems because it requires minimum modification to the existing system. This improves the system efficiency, and preheating the cold combustion air reduces the fuel requirement.

6.1.5 Benefits of Waste Heat Recovery

Benefits of waste heat recovery can be broadly classified into two categories as follows.

6.1.5.1 Direct Benefits

Recovery of waste heat has a direct effect on the efficiency of the process. This is reflected in fuel savings.

6.1.5.2 Indirect Benefits

1. Reduction in pollution: A number of toxic combustible wastes such as carbon monoxide gas, sour gas, oil sludge, plastic chemicals, etc., released to the atmosphere if burnt in the incinerators are avoided.
2. Reduction in sizes of flue gas handling equipment such as fans, stacks, ducts, and burners since waste heat recovery reduces fuel consumption, which leads to a reduction in the flue gas produced.
3. Reduction in energy consumption by auxiliary equipment due to a reduction in equipment sizes.

6.1.5.3 Fuel Savings due to Preheating Combustion Air

Recuperators of high-temperature furnaces can provide and increase the process efficiency. Fuel savings depends on parameters such as [5]:

1. Exhaust gas temperature
2. Preheat combustion air temperature
3. Effectiveness of the recuperator

The higher the percentage of exhaust gas temperature, the higher the percentage of fuel that can be saved. For a flue gas temperature of 1350°C and an air preheated temperature of 1100°C, the theoretical fuel savings achievable are approximately 65% of fuel consumption of an unrecuperated furnace operating at the same temperature levels.

6.2 HEAT EXCHANGERS USED FOR REGENERATION

The heat exchanger used to preheat combustion air is called either a recuperator or a regenerator. Thermodynamically, the exhaust gas thermal energy is in part recuperated or regenerated, and this same thermodynamic function is served regardless of the type of heat exchanger employed. The thermodynamic principle of regeneration is discussed in the following.

6.2.1 Recuperator

A recuperator is a convective heat transfer-type heat exchanger where the two fluids are separated by a conduction wall through which heat transfer takes place due to convection or a combination of radiation and convective design. The fluids flow simultaneously and remain unmixed. There are no moving parts in the recuperator. Some examples of recuperators are tubular, plate-fin, and extended surface heat exchangers. Recuperators are used when the flue gas is clean and uncontaminated. A convective recuperator, namely an air preheater of a boiler, is shown in Figure 6.3.

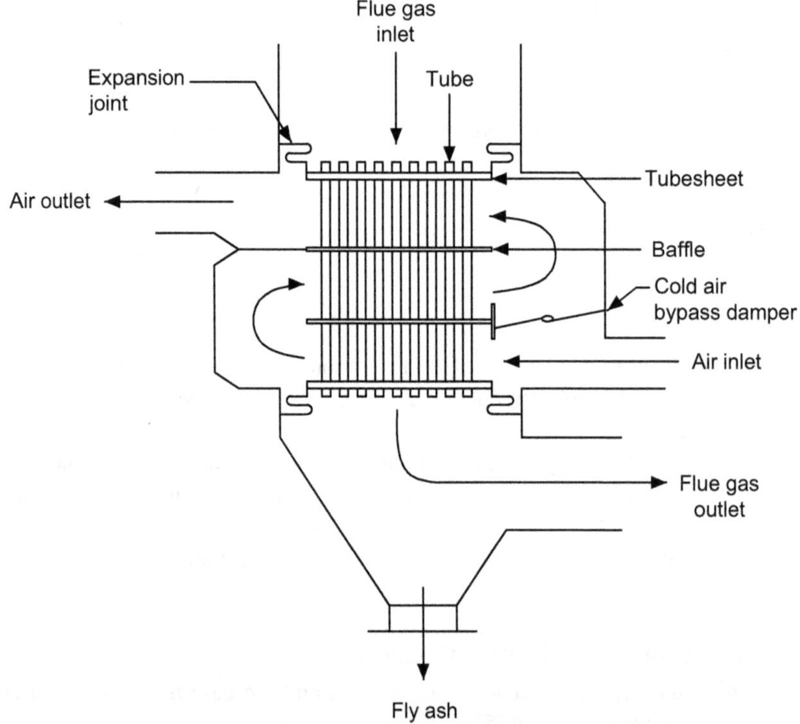

FIGURE 6.3 A convective recuperator Air preheater.

6.2.1.1 Merits of Recuperators

Recuperators have the advantages of (1) ease of manufacture; (2) stationary nature; (3) uniform temperature distribution and hence less thermal shock; and (4) absence of a sealing problem. However, their use is limited due to the requirement of temperature-resistant material to withstand the high-temperature flue gas and to retain its shape under operating temperature and pressure. Also, recuperators are subject to degradation in heat recovery performance by fouling from exhaust gas-borne volatiles and dust.

6.2.2 Regenerator

As mentioned above, a regenerator consists of a matrix through which the hot stream and cold stream flow periodically and alternately. First, the hot fluid gives up its heat to the regenerator. Then the cold fluid flows through the same passage, picking up the stored heat. The passing of a hot fluid stream through a matrix is called hot blow and a cold flow is called a cold blow. Thus, by regular reversals, the matrix is alternatively exposed to hot and cold gas streams, and the temperature of the packing, and the gas, at each position fluctuates with time.

Figure 6.4 shows the temperature field in a regenerator, for both the fluid and matrix at the instant of flow reversal. The upper curve represents the temperature of the fluid and matrix at the end of the

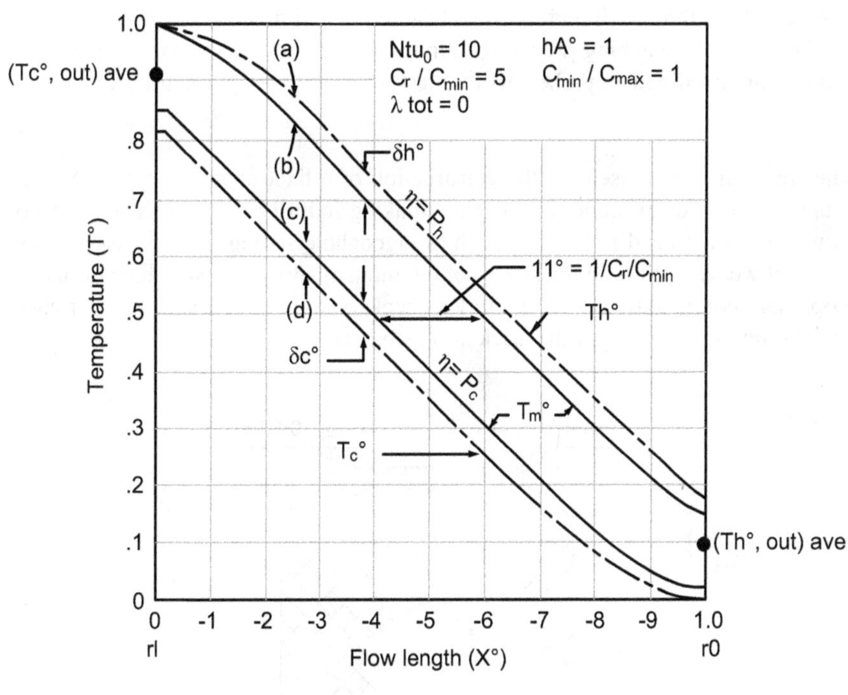

FIGURE 6.4 Balanced regenerator temperature distribution at switching instants. (From Mondt, J.R., *Trans. ASME J. Eng. Power*, 86, 121, 1964.)

hot blow and the start of the cold blow. The lower curves represent temperature conditions at the end of the cold blow and the start of hot blow. At any particular station along the length of the matrix, the temperatures may fluctuate between the upper and lower curves in a time-dependent relationship.

Regenerator types. Since the matrix is alternately heated by the hot fluid and cooled by the cold fluid, either the matrix must remain stationary and the gas streams must be passed alternately or the matrix must be rotated between the passages of the hot and cold gases. Hence, the regenerator may be classified according to its position with respect to time as (1) fixed-matrix or fixed-bed regenerators in which hot and cold gases are alternately routed through the same unit and (2) rotary regenerators, i.e., a rotating wheel with a core matrix in which the hot and cold gases flow on opposite sides of the wheel in a counterflow manner.

6.2.3 Fixed-Matrix or Fixed-Bed-Type Regenerator

The fixed-matrix or fixed-bed or storage-type regenerator is a periodic-flow heat transfer device with a high thermal capacity matrix through which the hot fluid stream and cold fluid stream pass alternately. To achieve continuous flow, at least two matrices are necessary, as shown in Figure 6.1. The flow through the matrix is controlled by valves. According to the number of beds employed, the fixed-matrix regenerators are classified into two categories: (1) single-bed and (2) dual-bed valved. In a dual-bed valved type, shown in Figure 6.5, initially matrix A is heated by the hot fluid and matrix B is cooled by the cold fluid. After a certain interval of time, the valves are operated so that the hot fluid flows through B and transfers heat to it. Cold fluid, however, flows through A and the fluid is heated. The switching process continues periodically. Some examples of storage-type heat exchangers are fixed-bed air preheaters for blast furnace stoves, glass furnaces, and open hearth furnaces.

Shape: The shape of matrix used for the construction of a fixed-matrix regenerator is dependent upon the application. For example, in the glass-making industry, the matrix is often constructed from ceramic bricks arranged in patterns such as pigeonhole setting, closed basket weave setting, and open basket weave setting. In the steel-making industry, the gases are much cleaner and a variety of proprietary configurations such as Andco checkers, McKee checkers, Kopper checkers, and Mohr checkers are used in place of the brick arrangements.

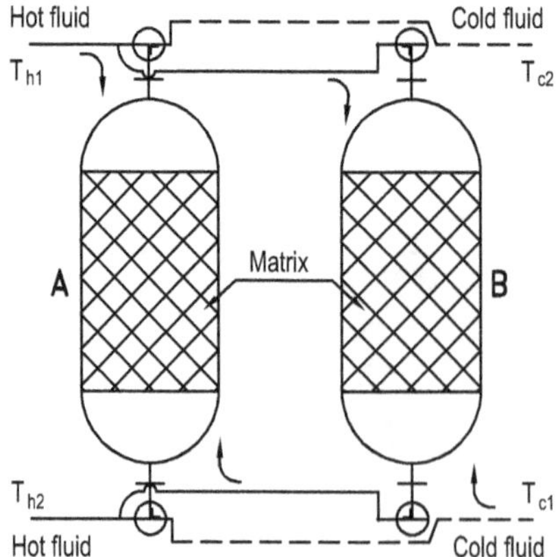

FIGURE 6.5 Dual-bed valved-type regenerator. (Adapted from Mondt, J.R., Regenerative heat exchangers: The elements of their design, selection and use, Research Publication GMR-3396, General Motors Research Laboratories, Warren, MI, 1980.)

Construction features: A fixed regenerator is generally constructed of either a porous matrix or checkers. The porous matrix forms a long, tortuous path for the flowing fluid in order to provide the largest possible contact surface between the regenerator matrix and the fluid. Checkers are blocks with holes pierced through them. Additional details on checkers and settings are provided next.

6.2.3.1 Fixed-Matrix Surface Geometries

Noncompact surface geometry: The commonly used checker shapes have a surface area density range of 25–42 m^2/m^3 (8–13 ft^2/ft^3). The checker flow passage (referred to as flue) size is relatively large primarily due to the fouling problem. A typical heat transfer coefficient in such a passage is about 5 W/m^2 K (1 Btu/h ft^2°F) [3].

Compact surface geometry: The surface geometries used for the compact fixed-matrix regenerator are similar to those used for rotary regenerators, but in addition quartz pebbles, steel, copper, or lead shots, copper wool, packed fibers, powders, randomly packed woven screens, and crossed rods are used.

6.2.3.2 Size

The fixed-bed regenerators are usually very large heat exchangers, some having spatial dimensions up to 50 m height and having unidirectional flow periods of many hours.

6.2.3.3 Merits of Fixed-Bed Regenerators

Fixed-bed regenerators consisting of a packed bed of refractory offer the following inherent advantages:

1. If loosely packed, the bed material is free to expand thermally; hence, thermal stresses are low.
2. Regenerators of this type are easily equipped so that the bed materials can be removed, cleaned, and replaced.
3. Unlike a recuperator, accumulated fouling does not reduce the heat exchanger capability of a regenerator; it merely increases the resistance to flow [6].

The major disadvantages of fixed-bed regenerators are the size, additional complexity and cost associated with the flow-switching mechanisms.

6.2.4 Rotary Regenerators

A rotary regenerator [Figures 6.6(a) and (b)], also known as a heat wheel, consists of a rotating matrix through which the hot and cold fluid streams flow continuously. The rotary regenerator is also called a periodic-flow heat exchanger since each part of the matrix, because of its continuous rotation, is exposed to a regular periodic flow of hot and cold gas streams. The rotary regeneration principle is achieved by two means: (1) the flow through the matrix is periodically reversed by rotating the matrix, and (2) the matrix is held stationary, whereas the headers are rotated continuously. Both approaches are rotary because for either design approach, heat transfer performance, pressure drop, and leakage considerations are the same, and rotary components must be designed for either system. The examples for rotary regenerators are (1) the Rothemuhle type and (2) the Ljungstrom type. In the Ljungstrom type, the matrix rotates and the connecting hood is stationary, whereas in the Rothemuhle type the connecting hood rotates while the heat transfer matrix is stationary—these two types are shown in Figures 6.6(c) and 6.6(d). The rotary regenerators are also used in vehicular gas turbine engines and as dehumidifiers in HVAC applications.

6.2.4.1 Salient Features of Rotary Regenerators

The periodic flow rotary regenerator is characterized by features such as the following:

1. A more compact size (β = 8800 m^2/m^3 for rotating type and 1600 m^2/m^3 for fixed-matrix type), shape, desired density, porosity, and low hydraulic diameter can be achieved by pressing the metal strips, wire mesh, or sintered ceramic and hence less expensive surface per unit transfer area is achieved.
2. The porous matrix provides a long, tortuous path and hence a large area of contact for the flowing fluids.
3. The absence of a separate flow path such as tubes or plate walls but the presence of seals to separate the gas stream in order to avoid mixing due to pressure differential.
4. The presence of moving parts such as the rotary core in rotary regenerators and alternate closing and opening of valves in a fixed regenerator.
5. With rotary regenerators, high effectiveness can be obtained since the matrix can be heated up to nearly the full exhaust gas temperature.
6. To achieve very high thermal effectiveness approaching unity, the thermal capacity of the matrix should be very large compared to the working fluids. This requirement restricts the use of regenerators exclusively to gaseous applications.
7. Bypass leakage from the high-pressure cold stream to the low-pressure hot stream and carry-over loss from one stream to the other with flow reversal or rotary nature of the matrix. However, the bypass leakage problem is less in a dehumidifier for an air-conditioning application.
8. Application to both high temperatures (800–1100°C) for metal matrix, and 2000°C for ceramic regenerators for services like gas turbine applications, melting furnaces or steam power plant heat recovery, and cryogenic applications (–20°).
9. Operating pressure of 5–7 bar for gas turbine applications and low pressure of 1–1.5 bar for air dehumidifier and waste heat recovery applications.
10. Regenerators have self-cleaning characteristics because the hot and cold gases flow in the opposite directions periodically through the same passage. As a result, compact regenerators have minimal fouling. If heavy fouling is anticipated, regenerators are not used.
11. Normally, a laminar flow condition prevails, due to the small hydraulic diameter.

6.2.4.2 Types of Rotary Regenerators

Though there are two types of rotary regenerators—(1) disk type (shown in Figures 6.6a)—the most commonly used configuration is of the disk type. The disk-type matrix consists of alternate layers of corrugated, flat, thin metal strips wrapped around a central hub or ceramic pressing in a disk shape. Gases flow normal to the disk. In an ideal circumstance without maldistribution, the single-disk design is favored due to less seal length and lower seal leakage. Depending upon the applications, disk-type regenerators are variously referred to as heat wheel, thermal wheel, Munter wheel, or Ljungstrom wheel. A schematic of a disk-type rotary regenerator with ducts is shown in Figures 6.6(b) and (c). The drum-type matrix consists of heat exchanger material in a hollow drum shape. Gases flow radially through the drum. The cost of fabricating a drum-type regenerator is much higher than that for a disk-type regenerator, and hence the drum is not used in any applications.

6.2.4.3 Drive to Rotary Regenerators

The matrix in the regenerator is rotated by a hub shaft or a peripheral ring gear drive. Most ceramic regenerators have been driven at the periphery, probably because of their brittleness.

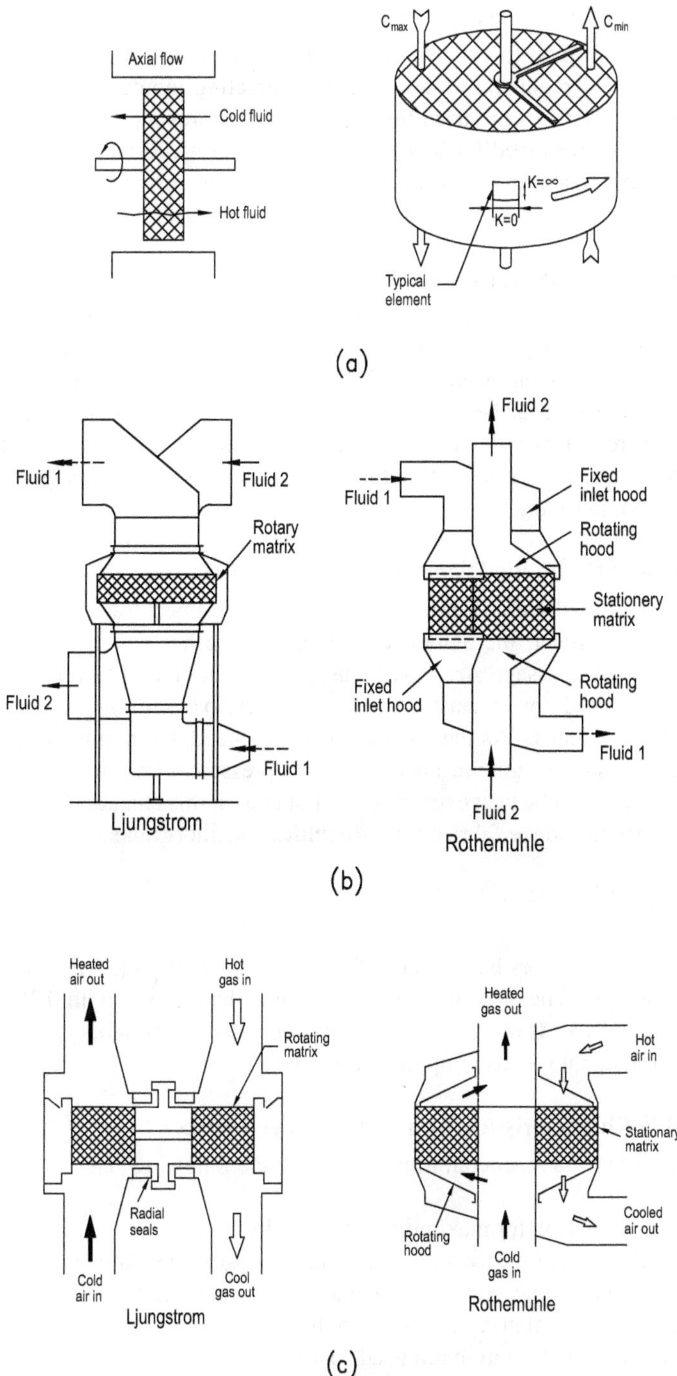

FIGURE 6.6 Rotary regenerator- (a) Disk type (From Coppage, J.E. and London, A.L., *Trans. ASME*, 75, 779, 1953) and (From Bahnke, G.D. and Howard, C.P., *Trans. ASME J. Eng. Power*, 86, 105, 1964), (b) and (c) Ljungstrom type and Rothemuhle type.

6.2.4.4 Operating Temperature and Pressure

The regenerators are designed to cover an operating temperature range from cryogenic to very high temperatures. Metal regenerators are used for operating temperatures up to about 870°C (1600°F). Ceramic regenerators are used for higher temperatures, up to about 2000°C (3600°F). Regenerators are usually designed for low-pressure applications. Rotary regenerators are limited to operating pressures of about 615 kPa or 90 psi and even lower pressures for fixed-matrix regenerators.

6.2.4.5 Surface Geometries for Rotary Regenerators

The rotary regenerator surfaces consist of many uninterrupted passages in parallel. The most common are triangular, rectangular, circular, or hexagonal smooth continuous passages. Details on the foregoing surface geometries are provided in Ref. [2]. Interrupted passage surfaces (such as strip fins, louver fins) are not used because a transverse flow leakage will be present if the two fluids are at different pressures. Hence, the matrix generally has continuous, uninterrupted flow passages and the fluid is unmixed at any cross section for these surfaces. Some surface geometries for rotary regenerators are shown in Figure 6.7.

6.2.4.6 Influence of Hydraulic Diameter on Performance

Packings having a small hydraulic diameter provide the highest heat transfer coefficients; for laminar flow, $h \propto 1/D_h$. However, the associated fouling by the hot gas may limit the size of the flow passages. Very small passage size may be used in air ventilation heat recovery regenerators compared to in those used for exhaust gas heat recovery, particularly from coal-fired exhaust gases or glass furnace exhausts. As the passage size is reduced, the number of passages required increases, while the passage length must decrease if the pressure drop is to be maintained constant. The conventional shell and tube heat exchanger cannot exploit this characteristic very far as a large number of small tubes introduce fabrication difficulties and short tube lengths prevent the use of counterflow.

6.2.4.7 Size

Typical power-plant regenerators have a rotor diameter of up to 10 m (33 ft) and rotational speeds in the range of 0.5–3 rpm. The air-ventilating regenerators have rotors with 0.25–3 m (0.8–9.8 ft) diameters and rotational speeds of up to 10 rpm. The vehicular regenerators have diameters of up to 0.6 m (24 in.) and rotational speeds of up to 18 rpm [3].

6.2.4.8 Desirable Characteristics for a Regenerative Matrix

Desirable characteristics of a regenerator matrix include the following [7]:

1. A large and solid matrix, for maximum heat capacity.
2. A porous matrix without obstruction, to minimize possible blocking and contamination.
3. A large, finely divided matrix, to achieve maximum heat transfer rate.
4. A small, highly porous matrix, for minimum flow losses.
5. A small, dense matrix, for minimum dead space.

Other desirable matrix properties include negligible thermal conduction in the direction of fluid flow, to minimize longitudinal conduction, and maximum specific heat, for high thermal capacity.

6.2.4.9 Total Heat Regenerators

Up to this point, consideration has only been given to the transfer of sensible heat between two fluid streams and the intermittent storage of thermal energy, as sensible heat, in a solid matrix. A number of

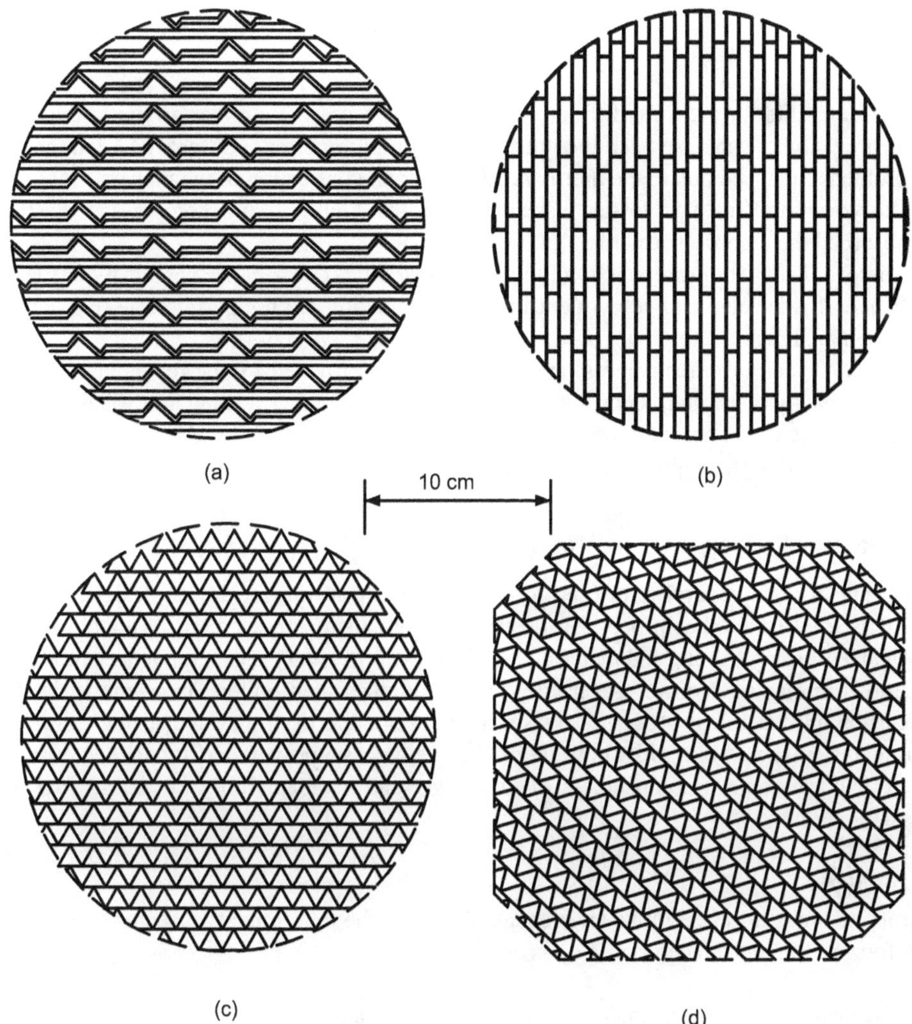

FIGURE 6.7 Surface geometries for rotary regenerator matrix. (a) Nonuniform complex profile, (b) rectangular, and (c and d) triangular. (From Mondt, J.R., Regenerative heat exchangers: The elements of their design, selection and use, Research Publication GMR-3396, General Motors Research Laboratories, Warren, MI, 1980.)

variations on these conditions are possible; for example, rotary regenerators are designed to transfer both sensible heat and latent heat of vapors mixed with the gas stream, known as total regenerators. They are intended mainly for air-conditioning applications and can employ both absorbent fibrous materials and nonabsorbent materials like plastics.

6.2.4.10 Merits of Regenerators

Among their advantages, regenerators

1. Can use compact heat transfer materials
2. Can use less expensive heat transfer surfaces
3. Are self-cleaning because of periodic flow reversals
4. Can use simpler header designs

In contrast, there are several major disadvantages of the periodic flow regenerators. These are as follows:

1. Seals suitable for pressure differentials of 4–7 bar represent a major problem; the necessity of provision of seals between the hot fluids due to high-pressure differential, and the leakage problem enhanced due to the thermal expansion and contraction of the matrix.
2. Carryover and leakage losses, especially for the high-pressure compressed air that has absorbed the compressor work.
3. There is always some amount of mixing of the two fluids due to carryover and bypass leakages. Where this leakage and subsequent fluid contamination are not permissible (e.g., cryogenic systems), a regenerator is not used.
4. The high thermal effectiveness, approaching unity, provided by the regenerator demands a heat capacity of the matrix considerably larger than that of the working fluid. This requirement restricts the use of regenerators to gases only.
5. The rotary designs require a drive and support system.

6.2.5 Rotary Regenerative Air Preheater

An air preheater is a general term used to describe any heat transfer device designed to heat air before it is used in another process, for example, combustion in a boiler. Available in a broad range of sizes, arrangements, and materials, the Ljungström® rotary regenerative air preheater finds applications in electrical power-generating plants, fluidized bed boilers, large industrial boilers, hydrocarbon and chemical processes, waste incinerators and drying systems, flue gas and other reheating systems, etc., due to its high thermal effectiveness, proven performance and reliability, effective leakage control, compactness of its design, and its adaptability to various fuel burning process.

6.2.5.1 Types

Construction types of regenerative air preheaters include the traditional two-sector type, and three-sector, four-sector, and concentric types as described below:

i. Bi-Sector Type—with single gas and single air stream
ii. Tri-Sector Type—with single gas stream but two air streams (primary air stream and secondary air stream)
iii. Quad-Sector Type—with single gas stream and one primary air stream sandwiched between two secondary air streams for leakage reduction.

An image of a Howden air preheater is provided in Figure 6.8. Leading air preheater manufacturers include Howden, UK, ALSTOM Power Air Preheater Company and Balcke-Dürr GmbH, Germany. Salient features of air preheaters are discussed later.

6.2.5.2 Design Features

Metallic heat transfer surfaces are contained in the rotor that turns at 1–3 rpm, depending on the size of the unit. A typical rotor under refurbishment is shown in Figure 6.9. The rotor housing and rotor have sealing members to form separate gas and air passages through the heat exchanger. The rotor drive unit, cleaning device mechanism, rotor bearing assemblies, and sealing surface adjuster are all located externally and are readily accessible while the unit is in operation. The overall design of the rotary regenerative heat exchanger lends itself to modularization. This helps reduce the time and effort required for on-site erection and maintenance.

FIGURE 6.8 Picture of rotary air preheater. (© Howden Group Limited.)

FIGURE 6.9 Refurbishment of air preheater rotor matrix. (© Howden Group Limited.)

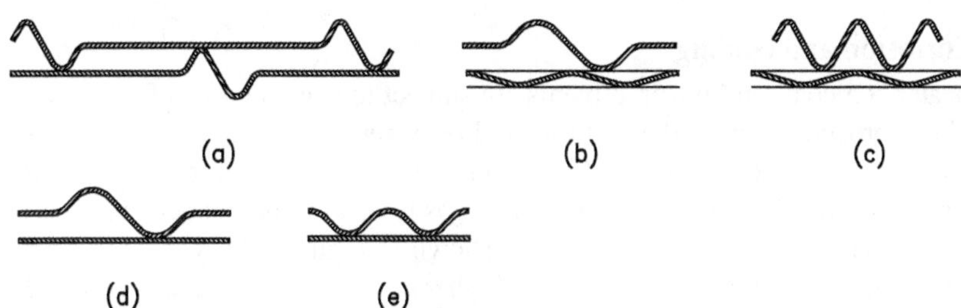

FIGURE 6.10a Regenerator heating elements profile.

6.2.5.3 Heating Elements

With a range of individual profiles, simple "open" profiles have a lower tendency for fouling but correspondingly lower heat transfer properties, while more complex profiles are more compact and induce turbulence in the gas flow to improve heat transfer. The choice of heating element is vital to ensure that the optimum combination of heat transfer and pressure drop is achieved. Some profiles of heating elements are provided in Figure 6.10a. In applications where heat transfer surfaces are

FIGURE 6.10b Enameled elements of rotary regenerator. (© Howden Group Limited.)

exposed to highly corrosive atmospheres and low exit gas temperatures, alloy steels or porcelain enamel coatings can be utilized. Selective heat exchanger enameled elements of the Howden make are shown in Figure 6.10b.

6.2.5.4 Corrosion and Fouling

All air preheaters on coal- and oil-fired plants are subject to some degree of corrosion and fouling caused by the approach to either the sulfuric acid or water dew point. Cold end corrosion can be caused by the high sulfur content in the fuel oil in combination with very low combined cold end temperatures, or due to insufficient ash in the flue gas to absorb the condensing acid formed from the sulfur in the fuel. The cold end corrosion is shown in Figure 6.11. Cold end corrosion can be minimized through the use of a cold end layer of higher grade steel or enamel coating. Hot end elements are prone to fouling as a result of large particles of fused ash becoming lodged in the element passages and smaller particles compacting behind them.

6.2.5.5 Heat Exchanger Element Baskets

Baskets are designed and manufactured to ensure that the elements remain tightly packed to avoid damage caused by vibration while the air preheater or gas–gas heater is operating. A typical heat exchanger basket of the Howden air preheater element is shown in Figure 6.12.

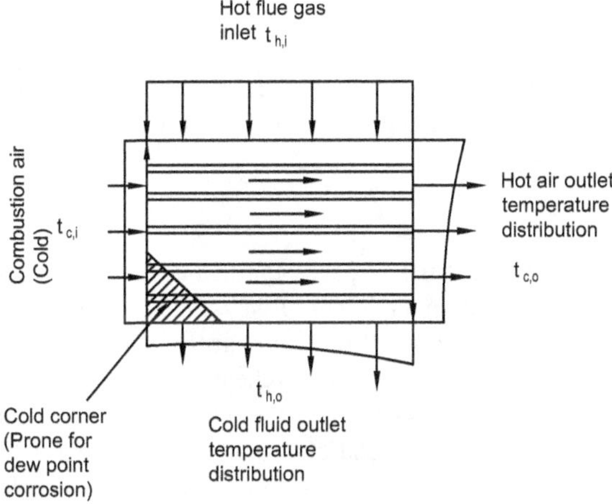

FIGURE 6.11 Illustration of cold end corrosion in an air preheater.

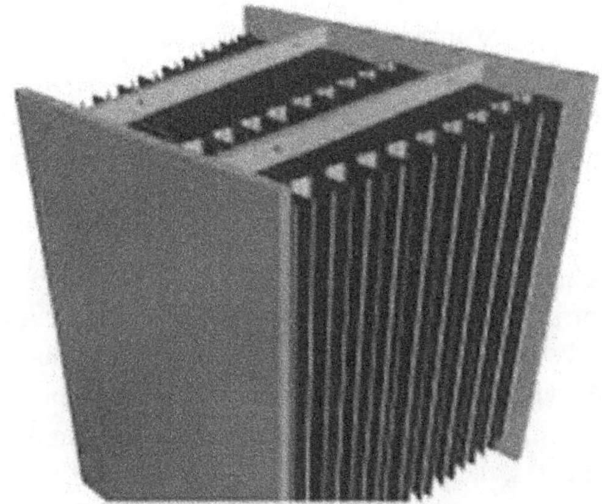

FIGURE 6.12 Rotary regenerator heat transfer elements basket. (© Howden Group Limited.)

6.2.6 Seals and Sealing System Components

6.2.6.1 Radial Seals and Sector Plates

Radial seals and sector plates are located at the hot and cold ends of the air preheater. The radial seals that separate the individual rotor compartments, and as the rotor turns the seals pass in close proximity to the sector plates located between the air and gas sides of the air preheater. The purpose of these seals is to reduce the area available for leakage from the air to the gas side between the rotor and the air preheater housing.

6.2.6.2 Axial Seals and Sealing Plates

Axial seals, used in conjunction with bypass seals, minimize leakage by passing radially around the rotor shell. The axial seals are mounted on the outside of the rotor shell and seal against the axial seal plates mounted on the air preheater housing.

6.2.6.3 Circumferential Seals and Circumferential Sealing Ring

Circumferential seals are mounted on the rotor and the sealing ring connected to the housing. The circumferential seals and sealing ring prevent air and gas from bypassing the heating surface through the space between the rotor and the housing shell. They also prevent air and gas from flowing axially around the rotor.

6.2.6.4 Leakage

Air preheaters suffer from leakage drift, i.e., the significant increase in leakage over a period of time. This can affect boiler operation in a number of ways, such as increasing fan power, increasing velocities in the precipitators, reducing the flow of hot air to the mills, or shrinking the draught fan margins. Hence, reducing and maintaining low air preheater leakage is vital for the overall performance of the thermal system. Seals can wear due to soot blowing, corrosion, erosion, and contact with the static sealing surfaces on start-up and/or shutdown. Seal wear and seal settings should be checked as per schedule so that seals can be reset to proper clearances or replaced should they exhibit excessive wear. Sealing plate surfaces may also wear due to contact with the seals and erosion, and they may also become out of level and out of plane. Seal plate wear should also be repaired, and plate alignments should be verified as soon as the need is detected. By fitting a sealing such as that of the Howden VN sealing system, these problems can be reduced or eliminated. Figure 6.13 shows a computer model of a Howden VN air preheater. The system improves the seal design on both the rotor and the casing, and dispenses with the need for actuated sector plates. In addition to better heat recovery and improved thermal performance, a peripheral benefit of using the Howden VN sealing system is a reduction in maintenance requirements on the air preheater.

6.2.6.5 Alstom Power Trisector Ljungström® Air Preheater

Designed for coal-fired applications, the *Alstom Power* trisector Ljungström® air preheater permits a single heat exchanger matrix to perform two functions: coal drying and combustion air heating. Because only one gas duct is required, the need for ductwork, expansion joints, and insulation is greatly reduced as compared with a separate air-heating system. Equipment layout is simplified, less structural steel is needed to install the system, and less cleaning equipment is required. The design has three sectors: one for flue gas, the second sector for primary air that dries the coal in the pulverizer, and the third sector is for preheating the secondary air that goes to the boiler for combustion.

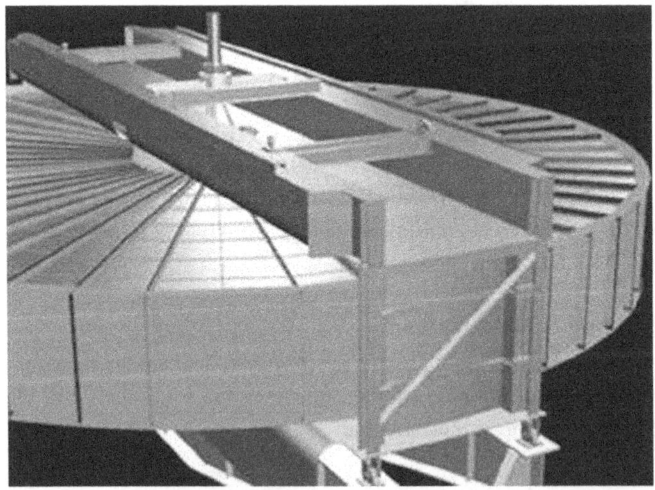

FIGURE 6.13 Computer model of Howden VN rotary air preheater-ducting removed. (© Howden Group Limited.)

6.3 COMPARISON OF RECUPERATORS AND REGENERATORS

1. A recuperator is easier to build and rugged in design. For a given size, the recuperator is less effective than the regenerator due to lower overall mean temperature difference. The regenerator, on the other hand, is smaller and, for its size, a much more efficient heat exchanger. Much higher surface area, higher heat transfer coefficients because of the random nature of flow through the packing, and a higher overall temperature difference as the top of the packing is heated to the exhaust gas temperature [7] are found.
2. Because a recuperator is not rotating and is at a constant temperature, there is less thermal shock and many normal materials can be used. Because of periodic flow, regenerators are subjected to thermal shock.
3. A recuperator does not have the problem of sealing between the hot and cold gas streams.

6.4 CONSIDERATIONS IN ESTABLISHING A HEAT RECOVERY SYSTEM

Although the addition of a regenerator is highly attractive from a thermodynamic point of view, its bulk, shape, mass, or cost may be such as to nullify the thermodynamic advantages. Optimum design therefore will call for a careful consideration of the following factors [4]:

1. Heat recovery—Quality and quantity. Quality: Usually, the higher the temperature, the higher the quality and more cost effective is the heat recovery. Quantity: In any heat recovery situation, it is essential to know the amount of heat recoverable and also how it can be used.
2. Compatibility with the existing process [8].
3. Initial capital cost.
4. Economic benefits.
5. Life of the equipment.
6. Maintenance and cleaning requirement.
7. Controllability and production scheduling.

Other limitations include manufacturing limitations, and weight, space, and shape limitations.

6.4.1 Compatibility with the Existing Process System

In order for a heat exchanger to be integrated into a process, the system design must be compatible with the process design parameters. One key area of compatibility is pressure drop. Many combustion processes are designed to use the natural draft of a stack to remove the products of combustion from the process. The heat recovery equipment designed for such a process either must be designed within the pressure drop limitations imposed by natural draft or must incorporate an educator to overcome these pressure-drop restrictions [8].

6.4.2 Economic Benefits

It is necessary to evaluate the selected waste heat recovery system on the basis of financial analysis such as investment, depreciation, payback period, fuel savings, rate of return, etc.

6.4.3 Capital Costs

The capital costs include the heat exchanger costs and the ancillary equipment required for the functioning of the system. Such ancillary equipment includes the following:

Blowers: Both the flue gas and the preheat air streams would require blowers to overcome pressure losses in the system.

Ducting: Ducting is between the furnace and the heat exchanger to conduit the flue gas and the preheat air.

Controls: The proposed heat recovery system would be operated in conjunction with the demand of the furnace. A control mechanism would be established for this purpose.

Energy savings: The benefits of energy savings due to the heat recovery system should offset the capital costs on heat exchanger, blowers, ducting and control mechanisms, and maintenance cost. A break-even analysis (capital costs vs. fuel savings) will help in this regard. The desired economic benefits can be accomplished by designing a heat exchanger to [8]

1. Maximize the equipment life
2. Maximize the energy cost savings
3. Minimize the equipment capital costs

6.4.4 Life of the Exchanger

The success of waste heat recovery depends on the maximum equipment life against the hostile environment of exhaust gases. The factors that affect the heat exchanger life are [8]

1. Excessive thermal stress
2. Creep
3. Thermal fatigue and thermal shock
4. High-temperature gaseous corrosion

Corrosion and stress cause accelerated failure in both metal and ceramic recuperators, resulting in leakage, which degrades the performance.

6.4.5 Maintainability

Ideally, a heat exchanger should provide a long service life and the equipment installation should facilitate easy maintenance. Where necessary, the design should incorporate provisions to isolate the heat exchanger from the system so that inspection, maintenance, repairs, and replacement can be made without interrupting the process, and the ability to clean the fouling deposits from the heat transfer surfaces should be provided for those processes in which fouling will take place [8].

6.5 REGENERATOR CONSTRUCTION MATERIAL

An important requirement of a regenerator for waste heat recovery is life and extended durability. Regenerators are required to work under hostile environments including (1) elevated temperatures; (2) fouling particulates; (3) corrosive gases and particulates; and (4) thermal cycling [9]. To achieve this requirement, proper material selection is vital. Much of the waste heat is in the form of high temperature (1900–3000°F), and the exhaust gases are corrosive in nature. Therefore, two important considerations for selecting material for heat exchangers of a heat recovery system are (1) strength and stability at the operating temperature, and (2) corrosion resistance. Other parameters include low cost, formability, and availability.

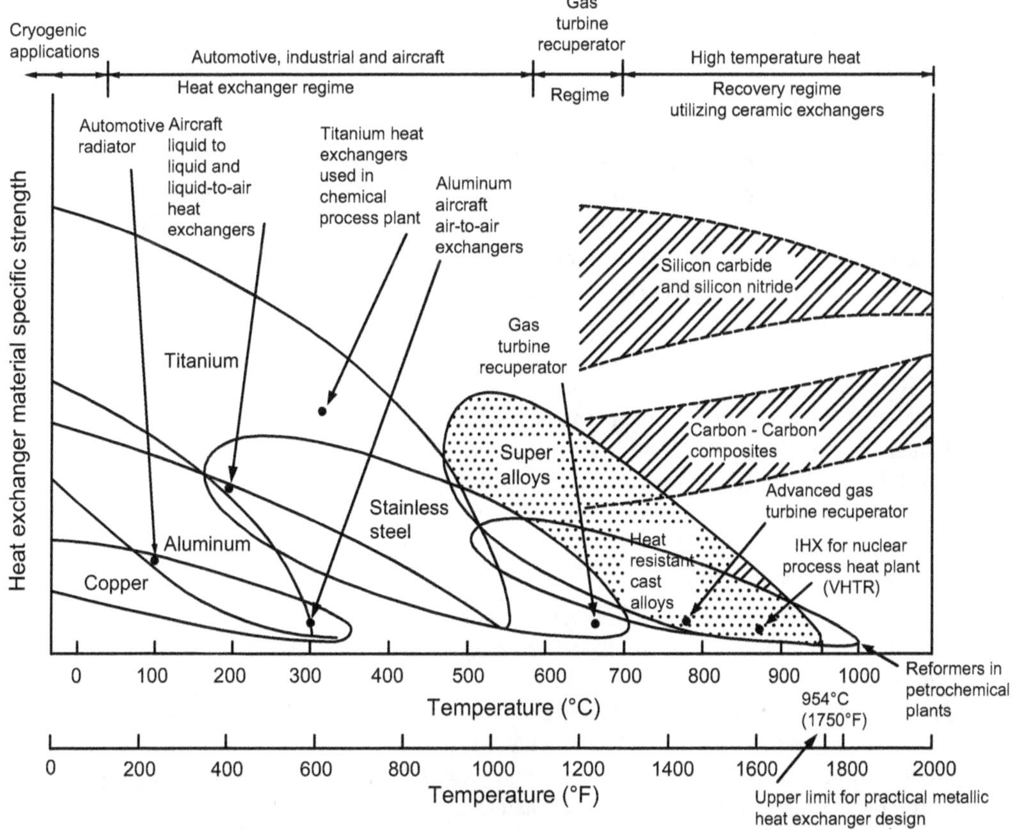

FIGURE 6.14 Approximate temperature range for various heat exchanger materials. (From McDonald, C.F., *Trans. ASME J. Eng. Power*, 102, 303, 1980.)

6.5.1 Strength and Stability at the Operating Temperature

Metals are ideal for recuperator and regenerator construction because of their ductility and ease of fabrication. Generally, carbon steel at temperatures above 425°C (800°F) and stainless steel at temperatures above 650°C (1200°F) begin to rapidly oxidize, and if the exhaust stream contains corrosive constituents, there will be a severe corrosion attack [10]. Approximate temperature ranges for various heat exchanger materials are shown in Figure 6.14 [11]. In general, above 1600°F, metal heat exchangers have limitations due to excessive costs from the expensive alloys required, requirements of temperature control devices or loss of effectiveness due to air dilution, and maintenance problems due to high thermal stresses [12]. In some cases, the high-temperature problem is overcome by diluting the flue gas with a portion of cold air to be preheated. However, dilution lowers the heat transfer from the flue gas stream.

6.5.2 Corrosion Resistance

Materials for high-temperature heat exchangers should exhibit high-temperature gaseous and liquid corrosion resistance to the carburizing, sulfidizing, oxidizing, and other effects of combustion products, and to the coal ash slag and fuel impurities. Regenerators used for heat recovery from the gases generated by boilers, furnaces, heaters, etc., should be protected against "cold end corrosion" caused by condensation of sulfuric acid and water on the heat transfer surface and the volatiles and corrosive fluxes in the exhaust gas. This problem is overcome either by ceramic units or with the use of the porcelain enameled exchanger.

6.5.3 Ceramic Heat Exchangers

Ceramic units can be very cost effective above 1600°F since they have the potential for greater resistance to creep and oxidation at very high temperatures, and do not require temperature controls or air dilution because of their high melting points [12]. At present, a ceramic heat exchanger is probably the most economical solution for high temperatures in excess of 800°C (1475°F) applications where the specific strength of metallic materials decreases very rapidly. The ceramic recuperators are available as plate-fin and bayonet tube exchangers. Material properties that favor ceramic as regenerator materials are:

1. Moderate strength, generally inexpensive, high thermal shock resistance and lower high temperature creep, good corrosion resistance, and excellent erosion resistance properties.
2. Low thermal expansion characteristics; this simplifies the sealing problem. About one-fifth the density of steel, high specific heats, and higher thermal conductivity.

6.5.3.1 Low Gas Permeability

The inability to join ceramic shapes reliably, once the parts have been densified, is a serious impediment to the design of large assemblies such as heat exchangers. Among other drawbacks, the application of ceramic materials in recuperators is limited by the inherent brittleness, the higher fabrication costs associated with producing ceramic heat exchangers, and the low thermal conductivity of many oxide ceramics [13]. According to McNallan et al. [13], because of their higher thermal conductivities and corrosion resistance compared with other ceramics, silicon carbide-based ceramics have received the most attention as heat exchanger materials.

6.5.4 Ceramic–Metallic Hybrid Recuperator

To overcome the high-temperature heat recovery problem, a ceramic–metallic hybrid recuperator is used. This consists of a ceramic recuperator operating in series with a conventional plate-fin metallic recuperator. An air dilution system is employed between the two recuperators to overcome the high-temperature oxidation problem of the metallic recuperator. This system is capable of handling flue gas temperatures up to 1370°C (2520°F).

6.5.5 Regenerator Materials for other than Waste Heat Recovery

The regenerator construction materials used for applications other than waste heat recovery are discussed by Shah [3]. They include the following:

1. In the air-conditioning and industrial process heat recovery applications, rotary regenerators are made from knitted aluminum or stainless steel wire matrix.
2. For cryogenic applications, ordinary carbon steels become brittle, so materials such as austenitic stainless steels, copper alloys, certain aluminum alloys, nickel, titanium, and a few other metals that retain ductility at cryogenic temperatures must be used. Cryogenic materials are covered in the chapter on material selection and fabrication (Chapter 2 of Volume 2).
3. Plastics, paper, and wool are used for regenerators operating below 65°C (150°F).

6.6 THERMAL DESIGN: THERMAL-HYDRAULIC FUNDAMENTALS

6.6.1 Surface Geometrical Properties

A consistent set of geometry-related symbols and factors facilitates comparative studies. The following factors can be used for all geometries, either bluff or laminar [2].

1. Porosity, σ = void volume/total volume
2. Area density, β = surface area/total volume
3. Hydraulic radius, r_h = flow area/wetted perimeter = σ/β
4. Flow area, A_x = porosity × frontal area = σA_{fr}
5. Cell density, N_c = number of holes or cells per unit frontal area

Reference [2] presents these surface geometrical factors for several surface geometries—square, hexagonal, circular, rectangular, and triangular—and the same is provided in Table 6.1. The surface geometrical properties, D_h ($4r_h$), L (flow length), σ, and β are the same on both fluid sides. They are related to each other as follows:

$$D_h = \frac{4A_o L}{A} = \frac{4\sigma}{\beta} \quad \sigma = \frac{A_o}{A_{fr}}, \quad \beta = \frac{4\sigma}{D_h} \tag{6.1}$$

TABLE 6.1
Surface Geometrical Properties for Rotary Regenerators Matrix

	Cell density N_c (cells/In²)	Porosity p_{or}	Area density β (1/m)	Hydraulic radius r_h (m)
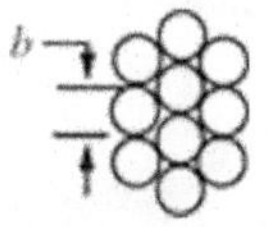	—	0.37–0.39	$\frac{6(1-p_{or})}{b}$	$\frac{b(p_{or})}{6(1-p_{or})}$
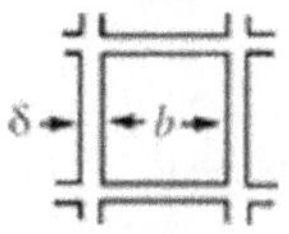	$\frac{1}{(b+\delta)^2}$	$\frac{b^2}{(b+\delta)^2}$	$\frac{4b}{(b+\delta)^2}$	$\frac{b}{4}$
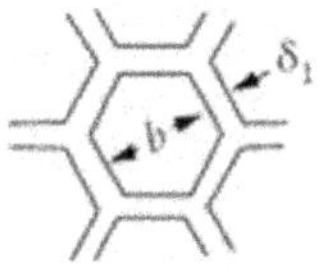	$\frac{2}{\sqrt{3}(b+\delta)^2}$	$\frac{b^2}{(b+\delta)^2}$	$\frac{4b}{(b+\delta)^2}$	$\frac{b}{4}$
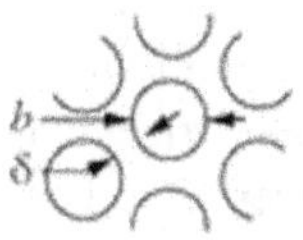	$\frac{2}{\sqrt{3}(b+\delta)^2}$	$\frac{\pi b^2}{2\sqrt{3}(b+\delta)^2}$	$\frac{2\pi b}{\sqrt{3}(b+\delta)^2}$	$\frac{b}{4}$
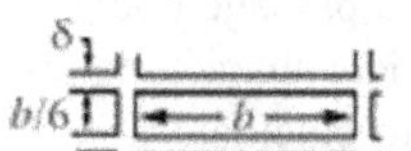	$\frac{1}{\left(\frac{b}{6}+\delta\right)(b+\delta)}$	$\frac{b^2/6}{\left(\frac{b}{6}+\delta\right)(b+\delta)}$	$\frac{7b/3}{\left(\frac{b}{6}+\delta\right)(b+\delta)}$	$\frac{b}{14}$
	$\frac{4\sqrt{3}}{(2b+3\delta)^2}$	$\frac{4b^2}{(2b+3\delta)^2}$	$\frac{24b}{(2b+3\delta)^2}$	$\frac{b}{6}$

If the frontal area is not 50% for each fluid, the disk frontal area, minimum free flow area, and heat transfer area on each side will not be the same. The method to calculate frontal area and heat transfer surface area on the hot and cold sides, area for longitudinal conduction, and mass is described by Shah [14], and various parameters are explained as per Shah's method. If the flow area ratio split for both fluids is in the ratio $x{:}y$, where $x + y = 1$, and if the face seal and hub coverage are specified as $a\%$ of the total face area $A_{fr,t}$, then the frontal areas on each side, namely, $A_{fr,h}$

$$A_{fr,t} = \frac{\pi}{4}\left(D^2 - d^2\right) \quad A_{fr} = A_{fr,t}\left(1 - \frac{a}{100}\right) \tag{6.2}$$

$$A_{fr,h} = \frac{x}{x+y}A_{fr} \quad A_{fr,c} = \frac{y}{x+y}A_{fr} \tag{6.3}$$

where

$A_{fr} = A_{fr,h} + A_{fr,c}$

D is the regenerator disk outer diameter at the end of the heat transfer surface

d is the hub diameter

The effective total volume of the regenerator is given by

$$V = \left(A_{fr,h} + A_{fr,c}\right)L \tag{6.4}$$

With the known porosity σ, the minimum free flow area on each side is calculated from $A_o = \sigma A_{fr}$. The heat transfer surface areas on the hot and cold sides are given by

$$A_h = \frac{x}{x+y}\beta V \quad A_c = \frac{y}{x+y}\beta V \tag{6.5}$$

The matrix mass M_r is calculated from the following equation for known matrix material density, ρ_r:

$$M_r = \rho_r V\left(1-\sigma\right) = \rho_r\left(1-\sigma\right)A_{fr,1}L \tag{6.6}$$

6.6.2 Correlation for j and f

For j and f values of commonly used surface geometries, refer to Mondt [2] and Kays and London [15]. Basic heat transfer and flow friction design data are presented for three straight, triangular passage, glass ceramic heat exchanger surfaces in London et al. [16]. These surfaces have heat transfer area density ratios ranging from 1300 to 2400 fr^2/ft^3 corresponding to a passage count of 526 to 2215 passages/in.^{2} Table 6.2 provides a comparison of the surface geometrical characteristics. Figure 6.15 provides a description of the idealized triangular geometry used to calculate D_h and α. Glass ceramic heat exchangers are of importance to vehicular gas turbine technology as they give promise of allowing low mass production costs for the high-effectiveness rotary regenerator required for most vehicular turbine engine concepts now under development.

If manufacturing tolerances are controlled as specified, the recommended j factor for Re < 1000 is given by

TABLE 6.2
Surface Geometrical Properties for Triangular Geometry Glass Ceramic Heat Exchangers

Core description/Case no.	505A	503A	504A
Passage count, N_c	526	1008	2215
Number/in.^{2}			
Porosity, σ	0.794	0.708	0.644
Hyd. diam., D_h, 10^{-3} ft	2.47	1.675	1.074
Area density, α ft^2/ft^3	1285	1692	2397
Cell height/width, d^*	0.731	0.709	0.708
L/D_h	101	149.5	233

Source: London, A.L. et al. (1970) *Trans. ASME J. Eng. Power* 92A: 381.

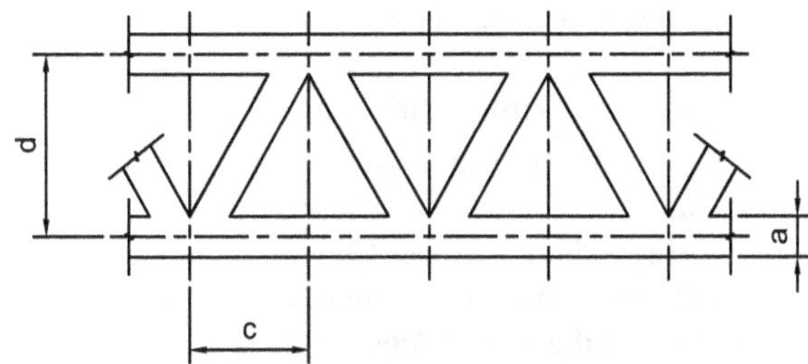

FIGURE 6.15 Idealized triangular geometry for glass ceramic heat exchanger. (From London, A.L. et al., *Trans. ASME J. Eng. Power*, 92A, 381, 1970.)

$$j = \frac{3.0}{\text{Re}} \tag{6.7}$$

This is approximately 10% lower than in theory to allow for some passage nonuniformity (±20% of flow area).

The recommended j factor for Re < 1000 is given by

$$j = \frac{14.0}{\text{Re}} \tag{6.8}$$

This is 5% higher than in theory to allow a small margin for the walls and the variation of passage cross section along flow length.

6.7 THERMAL DESIGN THEORY

The thermal design theory of recuperators is simple and quite straightforward. The effectiveness number of transfer units (ε-NTU) method is used to analyze heat transfer in recuperators. In contrast, the theory of the periodic-flow-type regenerator is much more difficult. Despite the simplicity of the differential equations under classical assumptions, their solution has proved to be challenging, and performances of counterflow regenerators have been widely investigated numerically as well as analytically [17]. Various solution techniques have been tried for solving the thermal design problem, and they are discussed here.

6.7.1 Regenerator Solution Techniques

For the steady-state behavior of a regenerator, Nusselt [18] has given an exact solution, which consists of an infinite series of integrals. However, this solution is complicated for practical purposes, and hence several solution techniques have been developed. These are as follows:

1. Approximate solution methods by simplifying the parameters defining the regenerator behavior; examples include Hausen's [19] heat pole method, which approximates the integral equation for the initial temperature distribution of the matrix. Hausen expressed the performance result in terms of two dimensionless parameters called the *reduced length* and the *reduced period*.
2. Empirical effectiveness relation correlated by treating the regenerator by an equivalent recuperator, e.g., that of Coppage and London [20].

6.7.1.1 Open Methods: Numerical Finite-Difference Method

Prior to the availability of general-purpose digital computers, regenerator thermal behavior was analyzed by restricting the range of parameters. With the availability of digital computers, Lamberston [21], Bahnke and Howard [22], Mondt [23], and Chung-Hsiung Li [24], among others, used a numerical finite-difference method for calculation of the regenerator thermal effectiveness. In this approach, the regenerator and the gas streams are represented by two separate but dependent heat exchangers with the matrix stream in crossflow with each gas stream as shown in Figure 6.16.

6.7.1.2 Closed Methods

In the closed methods, the reversal condition is implicitly incorporated in the mathematical model for solving the differential equations both for the hot period and the cold period. The closed methods used to solve the Nusselt integral equation are

1. Collocation method of Nahavandi and Weinstin [25] and Willmott and Duggan [26]
2. Galerkin method of Baclic [17] and Baclic et al. [27]
3. Successive integral method (SIM) of Romie and Baclic [28]

6.7.2 Basic Thermal Design Methods

For design purposes, a regenerator is usually considered to have attained regular periodic flow conditions. The solution to the governing differential equations is presented in terms of the regenerator effectiveness as a function of the pertinent dimensionless groups. The specific form of these dimensionless groups is to some extent optional, and the two most common forms are (1) the effectiveness and modified number of transfer units (ε-NTU$_0$) method (generally used for rotary regenerators), whereby

$$\varepsilon = \phi\left[\mathrm{NTU}_0, C^*, C_r^*, (hA)^*\right] \tag{6.9}$$

and (2) the reduced length-reduced period (Λ–Π) method (generally used for fixed-matrix regenerators) in which

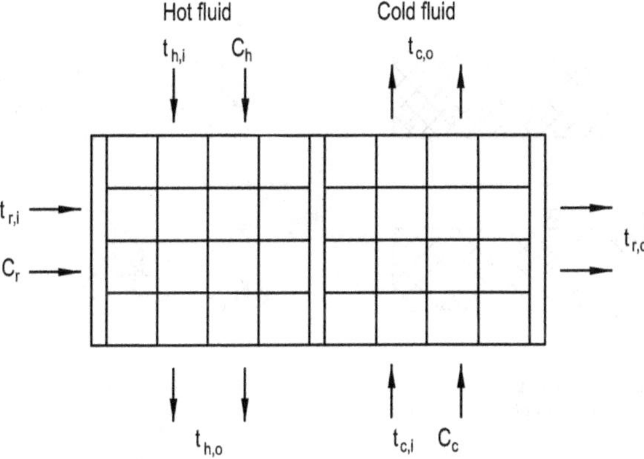

FIGURE 6.16 Regenerator matrix as the third stream. (From Lambertson, T.J., *Trans. ASME*, 80, 586, 1958.)

TABLE 6.3
Equivalence between the ε-NTU$_o$ and Λ–Π Design Methods

ε-NTU$_o$ method	Λ–Π method
$\mathrm{NTU_o} = \dfrac{\Lambda_c/\Pi_c}{1/\Pi_c + 1/\Pi_h}$	$\Lambda_h = C^*\left[1+\dfrac{1}{(hA)^*}\right]\mathrm{NTU_o}$
$C^* = \dfrac{\Pi_c/\Lambda_c}{\Pi_c/\Lambda_h}$	$\Lambda_c = \left[1+(hA)^*\right]\mathrm{NTU_o}$
$C_r^* = \dfrac{\Lambda_c}{\Pi_c}$	$\Pi_h = \dfrac{1}{C_r^*}\left[1+\dfrac{1}{(hA)^*}\right]\mathrm{NTU_o}$
$(hA)^* = \dfrac{\Pi_c}{\Pi_h}$	$\Pi_h = \dfrac{1}{C_r^*}\left[1+(hA)^*\right]\mathrm{NTU_o}$

$$\varepsilon = \phi\left[\Lambda_h, \Lambda_c, \Pi_h, \Pi_c\right] \tag{6.10}$$

These two methods are equivalent, as shown in Ref. [20] and Shah [29]; that is, the thermal effectiveness of a single fixed regenerator is equal to that of the rotary regenerator with the same values of the four parameters. This one-to-one correspondence between the two dimensionless methodologies, as shown in Table 6.3, allows the results obtained in the form of either Equation 6.9 or 6.10 to be used for both types of regenerators.

The ε-NTU$_0$ method is mainly used for rotary regenerator design and analysis, and was first developed by Coppage and London [20]. Their model is explained next.

6.7.2.1 Coppage and London Model for a Rotary Regenerator

The ε-NTU$_0$ method was used for the rotary regenerator design and analysis by Coppage and London [20] with the following idealizations.

1. The thermal conductivity of the matrix is zero in the gas and air flow directions, and infinite in the direction normal to the flow.

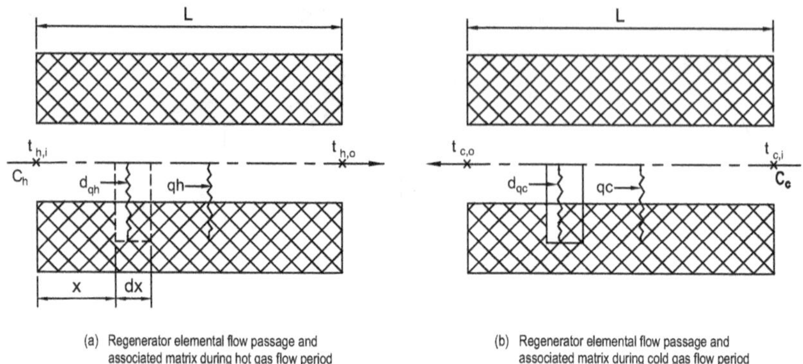

FIGURE 6.17 Regenerator basic heat transfer model. Regenerator elemental flow passage and associated matrix during (a) hot flow period and (b) cold gas flow period. (From Coppage, J.E. and London, A.L., *Trans. ASME*, 75, 779, 1953.)

2. The specific heats of the fluids and the matrix material are constant with temperature.
3. No mixing of the fluids occurs during the switch over from hot to cold flow.
4. The convective conductances between the fluids and the matrix are constant with flow length.
5. The fluids flow in counterflow directions.
6. Entering fluid temperatures are uniform over the flow cross section and constant with time.
7. Regular periodic conditions are established for all matrix elements, and heat losses to the surroundings are negligible.
8. Fluid carryover during the switching operation and the pressure leakage effects are negligible.

On the basis of these idealizations, the following differential equations and boundary conditions may be expressed. For the hot gas flow, energy balance on the element dx, as shown in Figure 6.17, is given by

$$\frac{dq_{\mathrm{h}}}{dx} = \frac{M_{\mathrm{r}} C_{\mathrm{r}}}{L} \frac{\partial t_{\mathrm{r}}}{\partial \tau} \tag{6.11}$$

$$\frac{dq_{\mathrm{h}}}{dx} = -\left(M_{\mathrm{h}} c_{\mathrm{p,h}} \frac{\partial t_{\mathrm{h}}}{\partial x} + \frac{m'_{\mathrm{h}} c_{\mathrm{p,h}}}{L} \frac{\partial t_{\mathrm{h}}}{\partial \tau} \right) \tag{6.12}$$

Equation 6.11 represents the energy received by the matrix in terms of its thermal capacitance. The right-hand side of Equation 6.12 represents the energy given up by the hot gas, and m'_h/L represents the mass of the hot gas retained in a flow length dx. The convective heat transfer rate equation is

$$dq_{\mathrm{h}} = \frac{h_{\mathrm{h}} A_{\mathrm{h}}}{L} \left(t_{\mathrm{h}} - t_{\mathrm{r}} \right) dx \tag{6.13}$$

Elimination of dq_{h} yields the equation

$$M_{\mathrm{h}} c_{\mathrm{p,h}} \frac{\partial t_{\mathrm{h}}}{\partial x} + \frac{m'_{\mathrm{h}} c_{\mathrm{p,h}}}{L} \frac{\partial t_{\mathrm{h}}}{\partial \tau} = \frac{M_{\mathrm{r}} c_{\mathrm{r}}}{L} \frac{\partial t_{\mathrm{r}}}{\partial \tau} = \frac{-h_{\mathrm{h}} A_{\mathrm{h}}}{L} \left(t_{\mathrm{h}} - t_{\mathrm{r}} \right) \tag{6.14}$$

For the cold gas flow, a similar equation results as follows:

$$M_h c_{p,c} \frac{\partial t_c}{\partial x} + \frac{m'_c c_{p,c}}{L} \frac{\partial t_c}{\partial \tau} = -\frac{M_r c_r}{L} \frac{\partial t_r}{\partial \tau} = \frac{-h_c A_c}{L}\left(t_c - t_r\right) \tag{6.15}$$

The boundary conditions are as follows:

$$\begin{aligned} &\textit{For interval of hot flow}: \quad t_{h,i} = \textit{constant at } x = 0 \\ &\textit{For interval of cold flow}: \quad t_{c,i} = \textit{constant at } x = L \end{aligned}$$

6.7.2.2 Thermal Effectiveness

The overall heat transfer performance of the regenerator is most conveniently expressed as the heat transfer "effectiveness" ε, which compares the actual heat transfer rate to the thermodynamically limited maximum possible heat transfer rate. With this definition, the regenerator thermal effectiveness is simply the dimensionless time-average cold fluid outlet temperature. This is true under the condition that the cold stream is weaker, that is, it has the lower heat capacity rate. This definition (for $C_c \le C_h$ as is the case for the recuperator) results in

$$\varepsilon = \frac{t_{c,o} - t_{c,i}}{t_{h,i} - t_{c,i}} \tag{6.16}$$

where $t_{c,o}$ is the bulk average temperature of the cold air stream after passage through the regenerator. The effectiveness is expressed as a function of four nondimensional parameters given by

$$\varepsilon = \phi\left[C^*, C_r^*, (hA)^*, \text{NTU}_o\right] \tag{6.17}$$

Definition of $\text{NTU}_o, C^*, C_r^*, (hA)^*$

NTU_o is the modified number of transfer units, defined as

$$\text{NTU}_o = \text{NTU}_c \left[\frac{1}{1 + (hA)^*}\right] \tag{6.18}$$

The heat capacity rate ratio C^* is simply the ratio of the smaller to larger heat capacity rate of the fluid steams so that C^* should be less than or equal to 1.

$$C^* = \frac{C_{min}}{C_{max}} = \frac{(Mc_p)_{min}}{(Mc_p)_{max}} = \frac{C_c}{C_h} \tag{6.19}$$

C_r^* is the matrix heat capacity rate C_r normalized with respect to C_{min}

$$C_r^* = \frac{C_r}{C_{min}} = \frac{M_r c_r N}{\left(Mc_p\right)_{min}} \quad (6.20)$$

where

M_r is the mass of the matrix
c_r is the specific heat of the matrix material
N is the speed of the matrix in revolution per unit time
$(hA)^*$ is the ratio of the convective conductance on the C_{min} side to that on the C_{max} side:

$$(hA)^* = \frac{(hA)\text{on the } C_{min}\text{ side}}{(hA)\text{on the } C_{max}\text{ side}} = \frac{(hA)_c}{(hA)_h} \quad (6.21)$$

For the special case of $C_r/C_c = \infty$, the behavior becomes identical in form to that of a counterflow direct-type exchanger and its effectiveness is given by

$$\varepsilon_{cf} = \frac{1 - e^{\left[-\text{NTU}_o(1-C^*)\right]}}{1 - C^* e^{\left[-\text{NTU}_o(1-C^*)\right]}} \quad \text{for } C^* < 1 \quad (6.22a)$$

$$= \frac{\text{NTU}_o}{\left(1 + \text{NTU}_o\right)} \quad \text{for } C^* = 1 \quad (6.22b)$$

The following formula for thermal effectiveness was first empirically correlated by Lambertson [21] and later modified by Kays and London [15]:

$$\varepsilon = \varepsilon_{cf}\left[1 - \frac{1}{9C_r^{*1.93}}\right] \quad (6.23)$$

Equation 6.23 agrees with accurate numerical results within ±1% for the following ranges [14]: (1) $3 \le \text{NTU}_o \le 9$, $0.90 \le C^* \le 1$, $1.25 \le C_r^* \le 5$; (2) $2 \le \text{NTU}_o \le 14$, $C^* = 1$, $C_r^* \ge 1.5$; (3) $\text{NTU}_o \le 20$, $C^* = 1$, C_r^* 2; and (4) the complete range of NTU_o, $C^* = 1$, $C_r^* \ge 5$. In all cases, $0.25 \le (hA)^* \le 4$. Note that the value of C^* is limited to between 0.9 and 1.0.

The regenerator effectiveness increases with NTU_o asymptotically and approaches unity for specified values of C^* and C_r^*. Thus, in this case, the periodic-flow-type regenerator would have the same performance as a counterflow direct-type unit possessing the same hot-side and cold-side transfer areas and the same convection heat transfer coefficients, providing only negligible thermal resistance offered by the wall structure. To provide a sense of magnitude for these parameters, the following tabulations were presented by these authors to indicate extreme ranges of values to be expected in gas turbine design work:

$$\varepsilon = 50\% - 90\%$$

$$\frac{C_c}{C_h} = 0.90 - 1.00$$

$$\frac{C_r}{C_c} = 1 - 10$$

$$(hA)^* = \frac{(hA)_c}{(hA)_h} = 0.2 - 1$$

$$\mathrm{NTU_c} = 2 - 20(\approx \text{twice } \mathrm{NTU_o})$$

$$\mathrm{NTU_o} = 1 - 10$$

6.7.2.3 Heat Transfer

For rotary regenerators, the magnitude of the heat transfer rate between the gas and the matrix during a flow period is necessarily the same for both flow periods, and therefore

$$q = \varepsilon\left(Mc_p\right)_c\left(t_{h,i} - t_{c,i}\right) \tag{6.24a}$$

in which $t_{h,i}$ and $t_{c,i}$ are constant inlet temperatures of the hot and cold gases. For stationary regenerators, the heat transfer rate is

$$q = \varepsilon\left(Mc_p\tau\right)_c\left(t_{h,i} - t_{c,i}\right) \tag{6.24b}$$

6.7.2.4 Parameter Definitions

Various parameter definitions and their equivalents for the stationary and rotary regenerators are provided in Table 6.4.

$$\frac{C_a}{C_b} \le 1 \quad \text{or} \quad \frac{\Lambda_b \Pi_a}{\Lambda_a \Pi_b} \le 1 \tag{6.25}$$

6.7.2.5 Classification of Regenerators

Based on the values of C^* and $(hA)^*$, or Λ–Π, regenerators are classified into eight types [29,30]: (1) balanced regenerator; (2) unbalanced regenerator; (3) symmetric regenerator; (4) unsymmetric regenerator; (5) symmetric and balanced regenerator; (6) unsymmetric and balanced regenerator; (7) symmetric and unbalanced regenerator; and (8) unsymmetric and unbalanced regenerator. For a balanced regenerator, the heat capacity of the fluids blown through the regenerator is equal and a regenerator is termed symmetric if the reduced length of each period is equal. The unsymmetric and unbalanced regenerator operation is the most general one, and the others are

TABLE 6.4
Parameter Definitions and Equivalents for Stationary and Rotary Regenerators

	Stationary	Rotary
NTU_o	$\dfrac{\Lambda_c}{1+(hA)^*}$	$\text{NTU}_c\left[\dfrac{1}{1+(hA)^*}\right]$
C^*	$\dfrac{(Mc_p\tau)_c}{(Mc_p\tau)_h}$	$\dfrac{(Mc_p)_c}{(Mc_p)_h}$
C_r^*	$\dfrac{M_r c_r}{(Mc_p\tau)_c}$	$\dfrac{M_r c_r N}{(Mc_p)_c}$
$(hA)^*$	$\dfrac{(hA\tau)_c}{(hA\tau)_h}$	$\dfrac{(hA)_c}{(hA)_h}$

Note: While defining various parameters, throughout this chapter, it is assumed that the cold fluid is the C_{min} fluid and accordingly, subscripts c for cold fluid and h for hot fluid are allotted.

merely subsets [30]. These designations are shown in Table 6.5. The utilization factors U_h on the hot side and U_c on the cold side are given by

$$U_c = \left(\frac{\Lambda}{\Pi}\right)_c = \frac{C_r}{C_c} \tag{6.26a}$$

$$U_h = \left(\frac{\Lambda}{\Pi}\right)_h = \frac{C_r}{C_h} \tag{6.26b}$$

6.7.3 Additional Formulas for Regenerator Effectiveness

Tables of thermal effectiveness are given by many sources: Kays and London [15], Baclic [17], Lamberston [21], Bhanke and Howard [22], and Romie [31], among others. Methods of computing the thermal effectiveness are described by, for example, Baclic [17], Willmott and Duggan [26], Baclic et al. [27], Romie and Baclic [28], Refs. [32a,32b], and many others. Additional formulas and closed form solutions from a few of these references are given next.

6.7.3.1 Balanced and Symmetric Counterflow Regenerator

Baclic [17] obtained a highly accurate closed form expression for the counterflow regenerator effectiveness for the balanced and symmetric regenerator; that is, for $C^* = 1$ and $(hA)^* = 1$ by the Galerkin method:

$$\varepsilon = \frac{\Lambda}{\Pi}\left[\left(2\sum_{j=o}^{N} B_j - 1\right)\right] \tag{6.27}$$

TABLE 6.5
Designation of Regenerators

Terminology	Λ–Π method	NTU$_o$ method
1. Balanced regenerator (defined in terms of utilization factor U or heat capacity rate ratio C^*)	$U_h = U_c$ $U_h/U_c = 1$ $\Lambda_h/\Pi_h = \Lambda_c/\Pi_c$	$C_h = C_c$ $C_h/C_c = 1$ $C^* = 1$
2. Unbalanced regenerator	$U_h \neq U_c$ $U_h/U_c \neq 1$ $\Lambda_h/\Pi_h \neq \Lambda_c/\Pi_c$	$C_h \neq C_c$ $C_h/C_c \neq 1$ $C^* \neq 1$
3. Symmetric regenerator (defined in terms of (i) reduced length Λ or reduced period Π or (ii) thermal conductance ratio $(hA)^*$)	$\Lambda_h = \Lambda_c$ $\Pi_h = \Pi_c$	$(hA)_h = (hA)_c$ $(hA)^* = 1$
4. Unsymmetric regenerator	$\Lambda_h \neq \Lambda_c$ $\Pi_h \neq \Pi_c$	$(hA)_h \neq (hA)_c$ $(hA)^* \neq 1$
5. Symmetric and balanced regenerator	$\Lambda_h = \Lambda_c$, $\Pi_h = \Pi_c$ $U_h = U_c$ i.e., $\Lambda_h/\Pi_h = \Lambda_c/\Pi_c$	$(hA)^* = 1$ $C^* = 1$

where B_j is determined by solving the set of equations of the form

$$\sum_{j=0}^{N} a_{jk} B_j = 1 \tag{6.28}$$

where

$$a_{jk} = \frac{(k+1)!(j+1)!}{(k+j+1)!}\left[1+(k+j+1)!\sum_{m=0}^{k}\frac{(-1)^m}{(k-m)!}V_{j+m+2}\left(\Pi,\frac{\Lambda}{\Lambda^{j+m+l}}\right)\right] \tag{6.29}$$

$$V_m(x,y) = e^{[-(x+y)]}\sum_{n=m-1}^{\infty}\binom{n}{m-1}\left(\frac{y}{x}\right)^{n/2} I_n\left(2\sqrt{xy}\right) \quad m \geq 1 \tag{6.30a}$$

$$\binom{n}{m-1} = \frac{n!}{(m-1)!(n-m+1)!} \tag{6.30b}$$

For $N = 2$, the equation for ε simplifies to the following equation:

$$\varepsilon = C_r^* \frac{1+7\beta_2 - 24\{B-2[R_1 - A_1 - 90(N_1+2E)]\}}{1+9\beta_2 - 24\{B-6[R-A-20(N-3E)]\}} \tag{6.31}$$

where

$$B = 3\beta_3 - 13\beta_4 + 30(\beta_5 - \beta_6) \tag{6.32}$$

and

$$R = \beta_2\left[3\beta_4 - 5(3\beta_5 - 4\beta_6)\right] \tag{6.33}$$

$$A = \beta_3\left[3\beta_3 - 5(3\beta_4 + 4\beta_5 - 12\beta_6)\right] \tag{6.34}$$

$$N = \beta_4\left[2\beta_4 - 3(\beta_5 + \beta_6 + 3\beta_5^2)\right] \tag{6.35}$$

$$E = \beta_2\beta_4\beta_6 - \beta_2\beta_5^2 - \beta_5^2\beta_6 + 2\beta_3\beta_4\beta_5 - \beta_4^3 \tag{6.36}$$

$$N_1 = \beta_4\left[\beta_4 - 2(\beta_5 + \beta_6)\right] + 2\beta_5^2 \tag{6.37}$$

$$A_1 = \beta_3\left[\beta_3 - 15(\beta_4 + 4\beta_5 - 12\beta_6)\right] \tag{6.38}$$

$$R_1 = \beta_2\left[\beta_4 - 15(\beta_5 - 2\beta_6)\right] \tag{6.39}$$

$$\begin{aligned} \beta_i &= V_i \frac{\left(2\text{NTU}_0, 2\dfrac{\text{NTU}_0}{C_r^*}\right)}{(2\text{NTU}_0)^{i-1}} \\ &= \frac{V_i(\Pi, \Lambda)}{\Lambda^{i-1}} \quad \text{for } i = 2,3,\ldots,6 \end{aligned} \tag{6.40}$$

$$V_i(x,y) = e^{[-(x,y)]} \sum_{n=i-1}^{\infty} \binom{n}{i-1} \left(\frac{y}{x}\right)^{n/2} I_n\left(2\sqrt{xy}\right) \quad i \geq 1 \tag{6.41}$$

Regenerator effectiveness ε, computed from Equation 6.31, as a function of NTU_0 and C_r^* is presented in Table 6.6. Note that the results are valid for $C^* = 1$ and $(hA)^* = 1$. It is emphasized again that the regenerator effectiveness ε from Equation 6.23 is valid for $0.9 \leq C^* \leq 1$ and $C_r^* > 1.25$, while that of Equation 6.31 is valid not only for $C^* = 1$ but for all values of C_r^*. Exact asymptotic values of ε for $\Lambda \to \infty$ and $\Pi/\Lambda \to 0$, when $\varepsilon = \Lambda/(2 + \Lambda)$ holds, are also included in Table 6.6.

TABLE 6.6
Symmetric and Balanced Counterflow Regenerator Effectiveness $\varepsilon = \varphi(\Lambda, \Pi)$ (as Calculated from Equation 6.31)

Π/Λ Λ	0.0	0.1	0.2	0.3	0.4	0.5	0.6	0.7	0.8	0.9	1.0
1.0	1/3	0.3332	0.3329	0.3323	0.3315	0.3304	0.3292	0.3277	0.3260	0.3241	0.3221
1.5	3/7	0.4283	0.4276	0.4264	0.4248	0.4227	0.4202	0.4173	0.4139	0.4102	0.4061
2.0	1/2	0.4996	0.4986	0.4968	0.4943	0.4912	0.4874	0.4830	0.4780	0.4725	0.4665
2.5	5/9	0.5551	0.5537	0.5513	0.5481	0.5440	0.5391	0.5333	0.5269	0.5197	0.5120
3.0	3/5	0.5994	0.5977	0.5949	0.5910	0.5861	0.5802	0.5733	0.5655	0.5570	0.5477
3.5	7/11	0.6357	0.6338	0.6305	0.6261	0.6204	0.6137	0.6058	0.5970	0.5872	0.5766
4.0	2/3	0.6659	0.6638	0.6602	0.6553	0.6490	0.6416	0.6329	0.6232	0.6124	0.6006
4.5	9/13	0.6915	0.6892	0.6853	0.6800	0.6732	0.6652	0.6559	0.6454	0.6337	0.6210
5.0	5/7	0.7134	0.7109	0.7068	0.7011	0.6940	0.6855	0.6756	0.6645	0.6521	0.6385
5.5	11/15	0.7324	0.7293	0.7255	0.7195	0.7121	0.7032	0.6928	0.6811	0.6681	0.6537
6.0	3/4	0.7491	0.7463	0.7418	0.7356	0.7279	0.7187	0.7080	0.6958	0.6822	0.6672
6.5	13/17	0.7637	0.7609	0.7562	0.7498	0.7419	0.7324	0.7215	0.7089	0.6948	0.6792
7.0	7/9	0.7768	0.7738	0.7690	0.7625	0.7544	0.7447	0.7335	0.7206	0.7061	0.6900
7.5	15/19	0.7884	0.7854	0.7804	0.7738	0.7656	0.7558	0.7444	0.7313	0.7164	0.6997
8.0	4/5	0.7989	0.7958	0.7908	0.7840	0.7757	0.7658	0.7543	0.7409	0.7257	0.7036
8.5	17/21	0.8084	0.8052	0.8001	0.7933	0.7849	0.7749	0.7633	0.7497	0.7342	0.7167
9.0	9/11	0.8171	0.8138	0.8086	0.8017	0.7933	0.7833	0.7715	0.7578	0.7421	0.7242
9.5	19/23	0.8250	0.8216	0.8164	0.8094	0.8010	0.7909	0.7791	0.7653	0.7493	0.7311
10.0	5/6	0.8322	0.8288	0.8235	0.8165	0.8080	0.7980	0.7862	0.7722	0.7560	0.7375
10.5	21/25	0.8388	0.8354	0.8300	0.8230	0.8146	0.8046	0.7927	0.7787	0.7623	0.7435
11.0	11/13	0.8450	0.8415	0.8361	0.8290	0.8206	0.8106	0.7988	0.7847	0.7681	0.7491
11.5	23/27	0.8506	0.8471	0.8417	0.8346	0.8262	0.8163	0.8044	0.7903	0.7736	0.7543
12.0	6/7	0.8559	0.8524	0.8469	0.8398	0.8315	0.8216	0.8098	0.7956	0.7788	0.7592
12.5	25/29	0.8603	0.8573	0.8517	0.8447	0.8364	0.8266	0.8147	0.8005	0.7836	0.7638
13.0	13/15	0.8654	0.8618	0.8562	0.8492	0.8410	0.8312	0.8194	0.8052	0.7882	0.7682
13.5	27/31	0.8697	0.8661	0.8605	0.8535	0.8453	0.8356	0.8239	0.8096	0.7925	0.7723
14.0	7/8	0.8737	0.8701	0.8644	0.8575	0.8494	0.8397	0.8280	0.8138	0.7966	0.7762
14.5	29/33	0.8775	0.8738	0.8682	0.8612	0.8532	0.8436	0.8320	0.8177	0.8004	0.7799
15.0	15/17	0.8811	0.8773	0.8717	0.8648	0.8568	0.8473	0.8358	0.8215	0.8041	0.7834
15.5	31/35	0.8844	0.8807	0.8750	0.8681	0.8602	0.8508	0.8393	0.8251	0.8076	0.7868
16.0	8/9	0.8876	0.8838	0.8781	0.8713	0.8635	0.8542	0.8427	0.8285	0.8110	0.7900
17.0	17/19	0.8934	0.8896	0.8839	0.8771	0.8695	0.8604	0.8490	0.8348	0.8172	0.7959
18.0	9/10	0.8967	0.8948	0.8891	0.8824	0.8749	0.8660	0.8548	0.8406	0.8229	0.8014
19.0	19/21	0.9034	0.8995	0.8938	0.8872	0.8799	0.8711	0.8601	0.8459	0.8282	0.8065
20.0	10/11	0.9077	0.9038	0.8981	0.8916	0.8844	0.8758	0.8649	0.8509	0.8331	0.8111
25	25/27	0.9245	0.9205	0.9148	0.9087	0.9024	0.8947	0.8844	0.8707	0.8528	0.8302
30	15/16	0.9360	0.9319	0.9263	0.9207	0.9151	0.9081	0.8984	0.8851	0.8673	0.8442
35	35/37	0.9445	0.9403	0.9347	0.9295	0.9246	0.9182	0.9090	0.8961	0.8785	0.8552
40	20/21	0.9509	0.9467	0.9411	0.9363	0.9320	0.9261	0.9174	0.9048	0.8875	0.8640
45	45/47	0.9559	0.9517	0.9462	0.9417	0.9379	0.9325	0.9241	0.9119	0.8948	0.8713
50	25/26	0.9600	0.9557	0.9503	0.9461	0.9427	0.9378	0.9297	0.9177	0.9009	0.8775
60	30/31	0.9662	0.9619	0.9565	0.9529	0.9502	0.9459	0.9383	0.9269	0.9106	0.8874
70	35/36	0.9707	0.9663	0.9611	0.9578	0.9558	0.9519	0.9447	0.9337	0.9180	0.8951
80	40/41	0.9740	0.9697	0.9645	0.9616	0.9600	0.9565	0.9497	0.9390	0.9239	0.9014
90	45/46	0.9767	0.9723	0.9672	0.9646	0.9634	0.9602	0.9536	0.9432	0.9236	0.9065

(continued)

TABLE 6.6 (Continued)
Symmetric and Balanced Counterflow Regenerator Effectiveness ε = φ(Λ, Π) (as Calculated from Equation 6.31)

Π/Λ Λ	0.0	0.1	0.2	0.3	0.4	0.5	0.6	0.7	0.8	0.9	1.0
100	50/51	0.9788	0.9744	0.9693	0.9670	0.9662	0.9632	0.9568	0.9467	0.9325	0.9109
150	75/76	0.9852	0.9808	0.9759	0.9744	0.9747	0.9726	0.9667	0.9574	0.9452	0.9258
200	100/101	0.9835	0.9840	0.9792	0.9782	0.9791	0.9774	0.9719	0.9629	0.9520	0.9348
300	150/151	0.9917	0.9873	0.9826	0.9821	0.9837	0.9824	0.9772	0.9686	0.9592	0.9456
400	200/201	0.9934	0.9889	0.9843	0.9841	0.9860	0.9850	0.9799	0.9714	0.9629	0.9522
500	250/251	0.9944	0.9899	0.9853	0.9853	0.9874	0.9865	0.9815	0.9731	0.9651	0.9568
600	300/301	0.9950	0.9905	0.9860	0.9861	0.9884	0.9876	0.9826	0.9743	0.9665	0.9602
800	400/401	0.9959	0.9914	0.9869	0.9871	0.9896	0.9889	0.9840	0.9758	0.9682	0.9650
1000	500/501	0.9964	0.9919	0.9874	0.9877	0.9903	0.9897	0.9848	0.9766	0.9692	0.9634
2000	1000/1001	0.9974	0.9928	0.9884	0.9890	0.9917	0.9912	0.9865	0.9784	0.9711	0.9770
∞	1.0000	1.0000	1.0000	1.0000	1.0000	1.0000	1.0000	1.0000	1.0000	1.0000	1.0000

Source: Baclic, B.S. (1985) *Trans. ASME J. Heat Transfer* 107: 214.

Note: For the ε-NTU$_o$ method, $\frac{\Pi}{\Lambda}=\frac{1}{C_r^*}$ and $\Lambda = 2\text{NTU}_o$.

6.7.3.2 Reduced Length–Reduced Period (Λ–Π) Method

6.7.3.2.1 *Counterflow Regenerator*

The asymmetric-unbalanced counterflow regenerator problem is solved by Baclic et al. [27] by adopting Galerkin model. Based on these assumptions, an energy balance provides two equations applicable during the hot gas flow period and two equations during the cold gas flow period:

$$\frac{1}{\Lambda_h}\frac{\partial t_h}{\partial \xi} = -\frac{1}{\Pi_h}\frac{\partial t_r}{\partial \eta} = t_r - t_h \tag{6.42}$$

$$\pm\frac{1}{\Lambda_c}\frac{\partial t_c}{\partial \xi} = -\frac{1}{\Pi_c}\frac{\partial t_r}{\partial \eta} = t_r - t_c \tag{6.43}$$

The regenerator can operate with either parallelflow (plus sign in Equation 6.43) or counterflow arrangement (minus sign in Equation 6.43). In Equations 6.42 and 6.43, ξ is the fractional distance along the flow path in the regenerator matrix of length L and η is the fractional completion of a respective gas flow period. Temperatures of the hotter gas, colder gas, and the solid matrix are denoted by t_h, t_c, and t_r, respectively. The expressions for four parameters Λ_h and Λ_c (reduced lengths), Π_h and Π_c (reduced periods) are given in Table 6.7.

The differential equations (6.42) and (6.43) describe the regenerator operation when the appropriate boundary conditions are specified. These are constant gas inlet temperatures, $t_{h,I}$ and $t_{c,i}$, at the opposite ends ($\xi = 0$ and 1) of the regenerator matrix, and the condition stating that the matrix temperature field at the end of one gas flow period is the initial matrix temperature distribution for the subsequent gas flow period.

TABLE 6.7
Expressions for Λ_h, Λ_c, Π_c, and Π_h

$\Pi_c = \frac{(hA\tau)_c}{M_r C_r}$	$\Pi_h = \frac{(hA\tau)_h}{M_r c_r}$	$\Lambda_c = \frac{(hA)_c}{(Mc_p)_c}$	$\Lambda_h = \frac{(hA)_h}{(Mc_p)_h}$

The subscript 1 is assigned to the weaker gas flow period such that the respective $U_1 = (\Lambda/\Pi)_1$ ratio is the smaller of the two ratios $U_h = (\Lambda/\Pi)_h$ and $U_c = (\Lambda/\Pi)_C$. The ratio (Λ/Π) is termed the *utilization factor*. The main advantage of the utilization factors is the fact that these parameters do not contain the fluid to matrix heat transfer coefficients of the respective flow periods. Thus, U_1 is defined as

$$U_1 = \min\{U_C, U_h\} \tag{6.44}$$

The effectiveness is given by

$$\varepsilon = \frac{1}{U_1}\sum_{m=0}^{M}\frac{x_{2m} - x_{1m}}{(m+1)!} \quad \text{for } m = 0,1,2,3,\ldots,M \tag{6.45}$$

The unknown coefficients x_{1m} and x_{2m} are determined from the following set of algebraic equations:

$$\sum_{m=0}^{M}\left[-A_{mk}(\Pi_1,\Lambda_1)x_{1m} + B_{mk}x_{2m}\right] = C_k \quad k = 0,1,2,3,\ldots,M \tag{6.46}$$

$$\sum_{m=0}^{M}\left[B_{mk}x_{1m} - A_{mk}(\Pi_2,\Lambda_2)x_{2m}\right] = 0 \quad k = 0,1,2,3,\ldots,M \tag{6.47}$$

where

$$A_{mk}(\Pi_j,\Lambda_j) = \sum_{j=0}^{M}\frac{(-1)^i}{(k-i)!}\frac{V_{i+m+2}(\Pi_j,\Lambda_j)}{\Lambda_j^{i+m+1}} \quad j = 1,2 \tag{6.48}$$

$$B_{mk} = \frac{1}{(m+k+1)!} \tag{6.49}$$

$$C_k = \frac{1}{(k+1)!} - \sum_{j=0}^{k}\frac{(-1)^i}{(k-i)!}\frac{V_{i+2}(\Pi_1,\Lambda_1)}{\Lambda_1^{i+1}} \tag{6.50}$$

For any fixed M, one needs $2(M + 1)$ equations for $M + 1$ unknowns x_{1m} and for $M + 1$ unknowns x_{2m}. For any practical purpose $M = 2$ will be adequate.

6.7.3.3 Razelos Method for Asymmetric-Unbalanced Counterflow Regenerator

The forms produced by the Galerkin method of Baclic et al. [27] are not amenable for hand calculation methods. They can be solved by either a computer or a programmable calculator. A hand calculation method was presented by Razelos [33]. In this approach, the unsymmetric regenerator is approximated using an "equivalent" symmetric regenerator. The Razelos approximate method can be used to calculate the counterflow regenerator effectiveness for the complete range of C^* and NTU_o, and $C_r^* \geq 1$, $0.25 \leq (hA)^* \leq 4$. The method follows the following steps:

1. For the ε-NTU_o method, compute the range of $\text{NTU}_{o,m}$ and C_r^* for an equivalent balanced regenerator as follows from the specified values of NTU_o, C_r^*, and C^*.

$$\text{NTU}_{o,m} = \frac{2\text{NTU}_o\left(C^*\right)}{1+C^*} \tag{6.51}$$

$$C^*_{r,m} = \frac{2C_r^*C^*}{1+C^*} \tag{6.52}$$

2. For the Λ–Π method, calculate Λ_m and Π_m as

$$\Pi_m = \frac{2}{\left(\Pi_h^{-1}+\Pi_c^{-1}\right)} \quad \Lambda_m = \frac{2\Pi_m}{\left[(\Pi/\Lambda)_h + (\Pi/\Lambda)_c\right]} \tag{6.53}$$

3. The equivalent balanced regenerator effectiveness ε_r is evaluated from Equation 6.23 or 6.31 for $\text{NTU}_{o,m}$ and $C^*_{r,m}$ or from Table 6.6 for Λ_m and Π_m. For example, the equivalent form of Equation 6.23 is given by

$$\varepsilon_r = \frac{\text{NTU}_{o,m}}{1+\text{NTU}_{o,m}}\left[1-\frac{1}{9\left(C^*_{r,m}\right)^{1.93}}\right] \tag{6.54}$$

4. The actual regenerator effectiveness ε is then given by

$$\varepsilon = \frac{1-e^{\left\{\varepsilon_r\left(C^{*2}-1\right)/\left[2C^*\left(1-\varepsilon_r\right)\right]\right\}}}{1-C^*e^{\left\{\varepsilon_r\left(C^{*2}-1\right)/\left[2C^*\left(1-\varepsilon_r\right)\right]\right\}}} \tag{6.55}$$

or

$$\varepsilon = \frac{1-e^{\left[\varphi\varepsilon_r/\left(1-\varepsilon_r\right)\right]}}{1-\left[(\Lambda/\Pi)_h/(\Lambda/\Pi)_c\right]e^{\varphi\varepsilon_r/\left(1-\varepsilon_r\right)}} \tag{6.56}$$

where

$$\phi = \frac{\left[(\Lambda/\Pi)_h^2 - (\Lambda/\Pi)_c^2\right]}{2(\Lambda/\Pi)_h (\Lambda/\Pi)_c} \tag{6.57}$$

where C^* is the original specified value, which is less than unity.

A sample calculation [17] for $\Lambda_h = 18$, $\Lambda_c = 15.5$, $\Pi_h = 16$, and $(\Lambda/\Pi)_c = 1.2$ yields $\Pi_m = 4960/347$ and $\Lambda_m = 5760/347$. From Table 6.6, by linear interpolation, one obtains $\varepsilon_r = 0.8215$, since $(\Lambda/\Pi)_m = 31/36$ for this case, while $\phi = -0.0645833$ from the equation. With these values, $\varepsilon = 0.847047$ from the equation. The exact value of the effectiveness for this example is 0.847311, which is just 0.03% higher than the result obtained from this approximate procedure.

6.7.4 Influence of Longitudinal Heat Conduction in the Wall

In the foregoing analysis, the influence of longitudinal heat conduction was neglected. For a high-thermal-effectiveness regenerator with a large axial temperature gradient in the wall, longitudinal heat conduction reduces the regenerator effectiveness and the overall heat transfer rate. Bahnke and Howard [22] showed that an additional parameter λ, the longitudinal conduction parameter, would account for the influence of longitudinal heat conduction in the wall. Here, λ is defined as

$$\lambda = \frac{k_w A_{k,t}}{LC_{min}} \tag{6.58}$$

where

k_w is the thermal conductivity of the matrix wall
$A_{k,t}$ is the total solid area available for longitudinal conduction

$A_{k,t}$ is given by [14]

$$A_{k,t} = A_{k,h} + A_{k,c} = A_{fr} - A_o = A_{fr}(1-\sigma) \tag{6.59}$$

Introducing the longitudinal conduction parameter, a generalized expression for thermal effectiveness in the functional form is given by

$$\varepsilon = \left\{\text{NTU}_o, C^*, C_r^*, (hA)^*, \lambda\right\} \tag{6.60}$$

Lowered effectiveness because of longitudinal conduction, $\Delta\varepsilon$, can be calculated in several ways. Parameter definitions for λ and its equivalents for the stationary and rotary regenerators are given in Table 6.8.

6.7.4.1 Bahnke and Howard Method

Bahnke and Howard [22] illustrated the influence of thermal conductivity on thermal effectiveness by a factor called the *conduction effect*. The conduction effect is defined as the ratio of the difference

TABLE 6.8
Parameter Definitions for λ

	Stationary	Rotary
λ	$\dfrac{k_w A_{k,t}}{L(MC_p)_c}\left(\dfrac{\tau_c+\tau_h}{\tau_c}\right)$	$\dfrac{k_w A_{k,t}}{L(MC_p)_c}$

between effectiveness with no conduction effect and with conduction effect to the no conduction effectiveness, as given by

$$\frac{\Delta\varepsilon}{\varepsilon}=\frac{\varepsilon_{\lambda=0}-\varepsilon_{\lambda}}{\varepsilon_{\lambda=0}} \tag{6.61}$$

Bahnke and Howard tabulated results for ε for a wide range of NTU_o, C_r^*, C_r, and λ. Their results are valid over the following range of dimensionless parameters:

$$1.0 \ge \frac{C_{min}}{C_{max}} \ge 0.90$$

$$1.0 \le \frac{C_r}{C_{min}} \le \infty$$

$$1 \le NTU_o \le 100$$

$$1.0 \ge (hA)^* \ge 0.25$$

$$1.0 \ge A_s^* \ge 0.25$$

$$0.01 \le \lambda \le 0.32$$

Bahnke and Howard's results for $C^* = 1$ can be accurately expressed by [29]:

$$\varepsilon = C_\lambda \varepsilon_{\lambda=0} \tag{6.62}$$

where $\varepsilon_{\lambda=0}$ is given by Equation 6.23 or 6.31, and C_λ is given by

$$C_\lambda=\frac{1+NTU_o}{NTU_o}\left[1-\frac{1}{1+NTU_o(1+\lambda\psi)/(1+\lambda NTU_o)}\right] \tag{6.63}$$

in which

$$\psi=\left(\frac{\lambda NTU_o}{1+\lambda NTU_o}\right)^{0.5}\tanh\left[\frac{NTU_o}{\left[\lambda NTU_o/(1+\lambda NTU_o)\right]^{0.5}}\right] \tag{6.64a}$$

$$\approx \left(\frac{\lambda \mathrm{NTU_o}}{1+\lambda \mathrm{NTU_o}} \right)^{0.5} \quad \text{for NTU}_o \geq 3 \tag{6.64b}$$

6.7.4.2 Romie's Solution

Romie [34] expressed the effects of longitudinal heat conduction on thermal effectiveness in terms of a parameter G_L. The factor G_L is derived from the analysis of axial conduction in counterflow recuperators by Hahnemann [35]. G_L is given by

$$G_L = 1 - \left[1 - \frac{\tanh(b)}{b} \right] \left(\frac{\mathrm{NTU}}{b} \right)^2 \quad \text{for } C^* = 1, (hA)^* = 1 \tag{6.65}$$

where

$$b = \mathrm{NTU_o} \left[1 + \frac{1}{(\lambda \mathrm{NTU_o})} \right]^{0.5} \tag{6.66}$$

The expression for $\varepsilon_{\lambda=0}$ is given by

$$\varepsilon_{\lambda \neq 0} = \frac{1 - e^{-G_L \mathrm{NTU_o}(1-C^*)}}{1 - C^* e^{-G_L \mathrm{NTU_o}(1-C^*)}} \quad C^* < 1 \tag{6.67a}$$

$$= \frac{G_L \mathrm{NTU_o}}{(1 + G_L, \mathrm{NTU_o})} \quad C^* = 1 \tag{6.67b}$$

Figure 6.18 shows G_L as a function of the product $\mathrm{NTU_o}$ for several values of $\mathrm{NTU_o}$. The curves are for $(hA)^* = 1$ and $C^* = 0.9$ and 1. The same curves were obtained using $(hA)^* = 0.5$, thus indicating that G_L is a very weak function of $(hA)^*$ [34].

6.7.4.3 Shah's Solution to Account for the Longitudinal Conduction Effect

For $0.9 \leq C^* \leq 1$, Shah [36] has provided a closed-form formula to take into account the longitudinal conduction effect. However, the following more general method is recommended for $C^* < 1$ by Shah [14]:

1. Use the Razelos method to compute $\varepsilon_{r\lambda=0}$ for an equivalent balanced regenerator using the procedure from Equations 6.51, 6.52, and 6.54.
2. Compute C_λ from Equation 6.63 using $\mathrm{NTU_{o,m}}$ and λ.
3. Calculate $\varepsilon_{r\lambda\neq0} = C_\lambda \varepsilon_{r\lambda=0}$.
4. Finally, ε is determined from Equation 6.55 or 6.56 with ε_λ replaced by $\varepsilon_{r\lambda\neq0}$.

This procedure yields a value of ε accurate within 1% for $1 \leq \mathrm{NTU_o} \leq 20$ and $C_r^* \geq 1$ when compared to Bahnke and Howard's results.

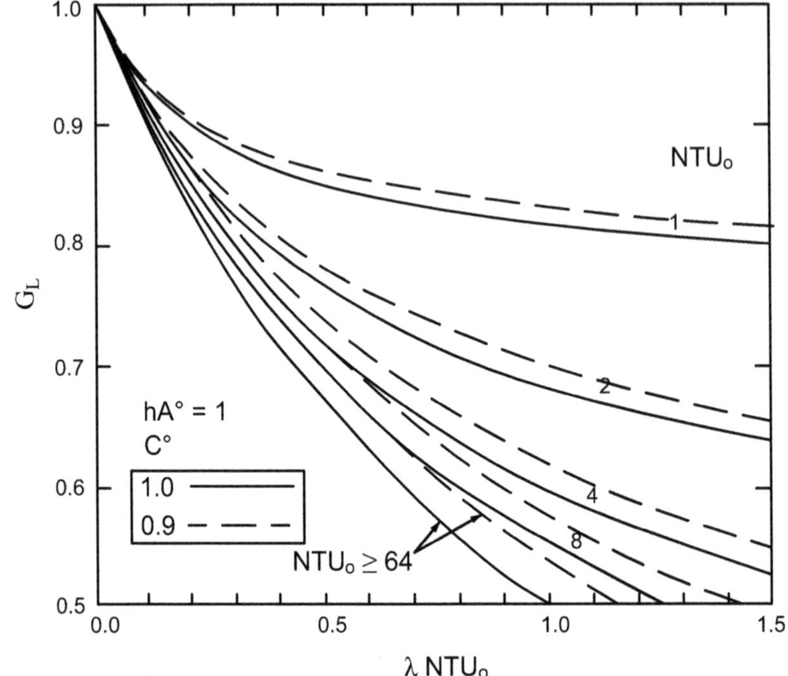

FIGURE 6.18 Parameter G_L for longitudinal conduction effect. (From Romie, F.E., *Trans. ASME J. Heat Transfer*, 113, 247, 1991.)

6.7.5 Fluid Bypass and Carryover on Thermal Effectiveness

Fluid bypass and carryover on thermal effectiveness are discussed in Refs. [14,37].

6.7.6 Regenerator Design Methodology

To design a successful regenerator, the engineer must consider many design factors including the following [2]:

1. Describe the different types of regenerative heat exchangers.
2. Discuss the basic heat transfer process and the correlation of cyclic temperature distributions resulting from storage of thermal energy.
3. Relate the regenerator to classical counterflow arrangements to enable trade-offs of heat transfer, pressure drop, size, and alternative surfaces.
4. Discuss the effects of leakage on both performance and design.
5. Consider the effects of pressure loads, leakage control, and rotation on mechanical design.

6.7.6.1 Primary Considerations Influencing Design

Primary considerations influencing the design of a regenerator especially for aircraft/vehicular gas turbine applications may be summarized as follows [38]:

1. Manufacturing limitations
2. Cost limitations
3. Maintenance and cleaning requirements
4. Mechanical design problems

5. Weight, shape, and size limitations
6. Limitations on flow friction-power expenditure
7. Desired exchanger heat transfer effectiveness

6.7.6.2 Rating of Rotary Regenerators

The basic steps involved in the analysis of a rating problem are the determination of surface geometrical properties, matrix wall properties, fluid physical properties, Reynolds numbers, j (or Nu) and f factors corrected for property variations, heat transfer coefficients, NTU_0, C^*, C_r^*, and pressure drops. The procedure presented for regenerators parallels that for compact heat exchangers. Therefore, the rating procedure is not repeated here. However, for details on a rating procedure from first principles consult Shah [14].

6.7.6.3 Sizing of Rotary Regenerators

The sizing problem for a rotary regenerator is more difficult. Similar to recuperator design, the sizing of a regenerator involves decisions on material, surface geometry selection, and mechanical design considerations. Once these are decided, the problem reduces to determination of the disk diameter, division of flow area, disk depth, and disk rotational speed to meet the specified heat transfer and pressure drops [14]. Sometimes a limitation on the size is also imposed. One method for a new design would be to assume a regenerator size, consider it as a rating problem, and evaluate the performance as outlined for compact recuperators. Iterate on the size until the computed performance matches the specified performance. A detailed sizing procedure is outline in Ref. [14].

6.8 MECHANICAL DESIGN

After the heat transfer surface has been chosen, sizing studies completed, and the regenerator selected to be compatible with the overall system requirements, the next step involves mechanical design. Mechanical designs must consider items such as [2] (1) thermal distortions; (2) leakages; (3) pressure loadings, and in the case of the rotary types; (4) a drive system; and (5) sealing. An important source book on regenerator mechanical design is Mondt [2].

6.8.1 Single-Bed and Dual-Bed Fixed Regenerators

Single-bed and dual-bed regenerators can be relatively simple to design. If the bed is composed of loose pellets, grids or screens are necessary to retain the pellets within a container. The grid mesh must be fine enough to retain the smallest pellet in the bed, yet the mesh must be porous enough to keep pressure drops low. If the retaining grids are not strong enough at the operating temperature, the bed will "sag" and loosen the packing.

Leakages: Both the single-bed and dual-bed valved regenerators have displacement leakage, that is, displacement of the fluid entrained in the matrix voids at the switching instant. This fluid is trapped and "displaced" into the other fluid stream and hence is known as displacement leakage. In addition to displacement leakage, fixed-bed regenerators may have small primary leakage of one fluid into the other through leaks in closed switching valves.

6.8.2 Rotary Regenerators

The mechanical design of a rotary regenerator requires (additionally to the recuperator) the following items [39]:

1. A subdivision of the casing into a rotor and a stator
2. Leakage and an effective seal between the two gases

3. Means for supporting the rotor in the stator
4. A drive for the rotor

Items (2) and (4) are discussed next.

6.8.2.1 Leakages of Gases

Primary leakages in a rotary regenerator can be classified as (1) labyrinth leakage; (2) carryover or displacement leakage; and (3) structural leakage.

Labyrinth or seal leakage: This is leakage through the seal stem from the fact that the cold gases are at much higher pressures than the hot gases. This is especially true in the case of a gas turbine with a high-pressure ratio.

Carryover or displacement leakage: In addition to direct leakage through seals, there is inevitably a "carryover" or displacement leakage due to air trapped in the matrix as it passes through the gas side.

Structural primary leakage: Structural leakages take place through any gaps existing in the matrix and the housing.

Any leakage is a serious loss, because it is the air that has absorbed compressor work but escapes to the atmosphere without passing through the turbine. It is more important than pressure loss, as it affects the net work severely [40]. Leakage is dependent mainly on pressure ratio and carryover on matrix rpm. Leakage due to seal leakage and displacement leakage can amount to 3–4% of the airflow out of the compressor [41]. The effect of a given fractional leakage depends on the values of exchanger effectiveness and cycle pressure ratio. As an example, for a cycle with typical temperatures, component efficiencies, and regenerator effectiveness of 90%, a leakage of 5% will reduce the net output by about 11%, while a leakage of 10% will reduce it by twice that amount [40].

6.8.2.2 Seal Design

In a rotary regenerator, the stationary seal locations control the desired frontal areas for each fluid and also serve to minimize the primary leakage from the high-pressure side to the low-pressure side. The sealing arrangement must adequately prevent leakage flow, and at the same time not introduce high frictional resistance for rotation. Because the matrix is alternately exposed to hot and cold gases (Figure 6.19), thermal expansion and contraction add to the sealing problem. Typical seals used in a rotary regenerator include radial seals, axial seals, and circumferential seals. While designing seals, the pressure forces under the seal faces, which must be balanced, and allowances to be made for the thermal distortion of the rotor [42]. Sealing arrangements are illustrated diagrammatically in Figure 6.20. Certain considerations for seal design are as follows [43]:

1. The seals must be efficient at very high temperatures and pressures but still allow freedom of movement.
2. The seals must accommodate the expansion and contraction accompanying temperature variations and still prevent leakage.

6.8.2.3 Drive for the Rotor

A design problem encountered with the regenerator is that of supporting and driving the matrix at low speed. The regenerator usually rotates at a speed of approximately 20–30 rpm [41]. An electric drive will be ideal, because of its low rpm and small power requirement.

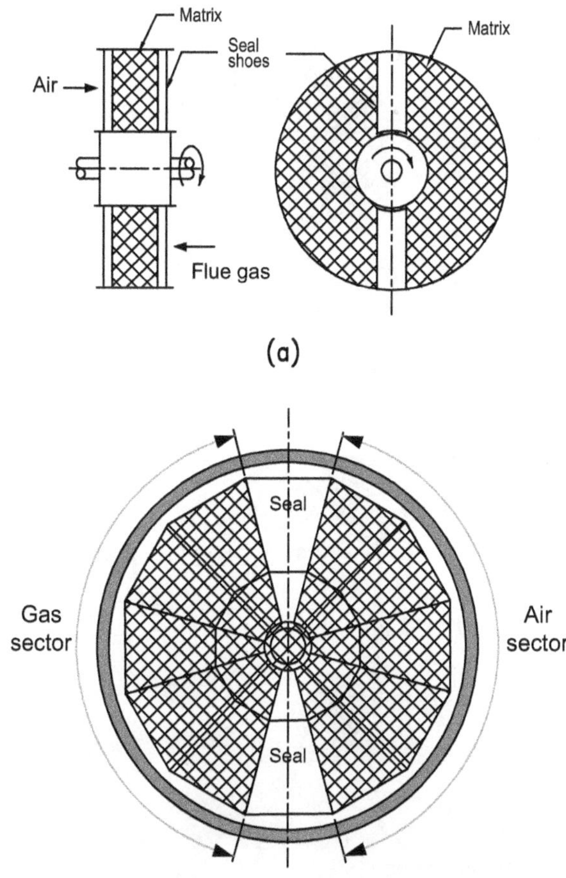

FIGURE 6.19 Rotary regenerator seal for separating hot and cold streams sections.

6.8.3 Thermal Distortion and Transients

A regenerator is subjected to thermal shock and thermal distortion due to unsteady thermal conditions. However, this can be minimized through the use of ceramic material. A regenerator cannot be designed assuming that all operations will be at steady-state conditions. In a vehicular gas turbine system, transients are common. Even industrial regenerators are occasionally exposed to starting and stopping transients. Gentle or long-duration transients are desirable to minimize thermal distortions and potential thermal stresses [2]. More problems are usually faced during the warm-up of a regenerator than during stopping because, on starting, the matrix and rotor respond much faster than the surrounding housing. If the housing does not accommodate the matrix and the rotor to expand during warm-up, the housing may constrain the matrix growth and cause thermal stresses. The concept of thermal "turn down" or thermal distortion is shown in Figure 6.21

6.8.3.1 Thermal Turndown

The typical air preheater may be as much as 60 feet in diameter. When the APH rotor is heated from a cold condition (blue), thermal expansion (yellow) can cause the rotor to droop or "turn down" up to 3 inches on the periphery. According to Storm Technologies Inc., UK. (www.storm-technolog ies.com/) knowing the amount of turndown is important when presetting the seal position before

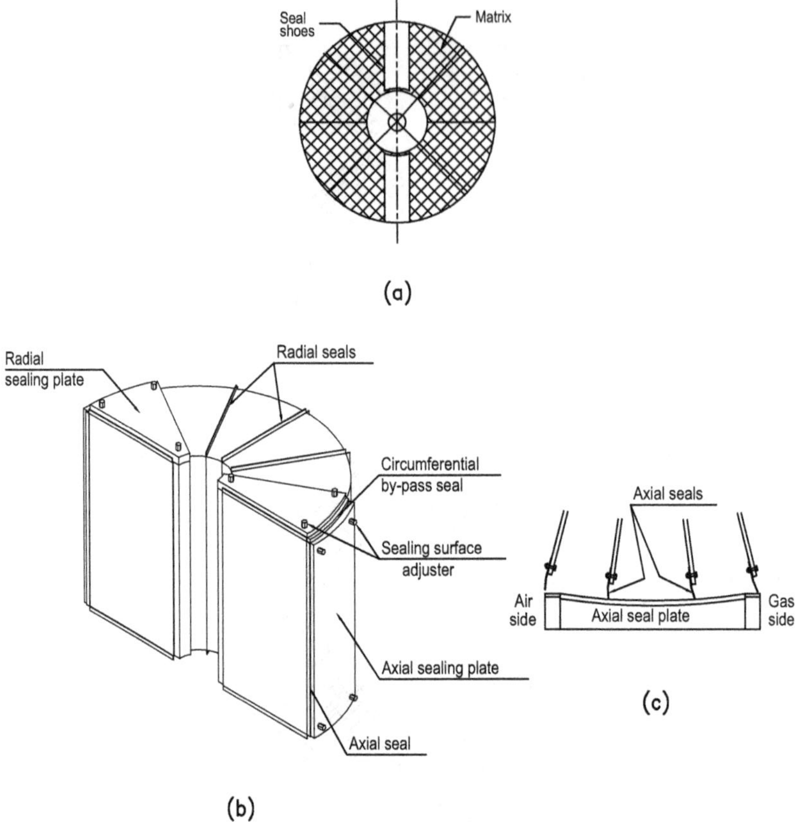

FIGURE 6.20 Seals for rotary regenerator.

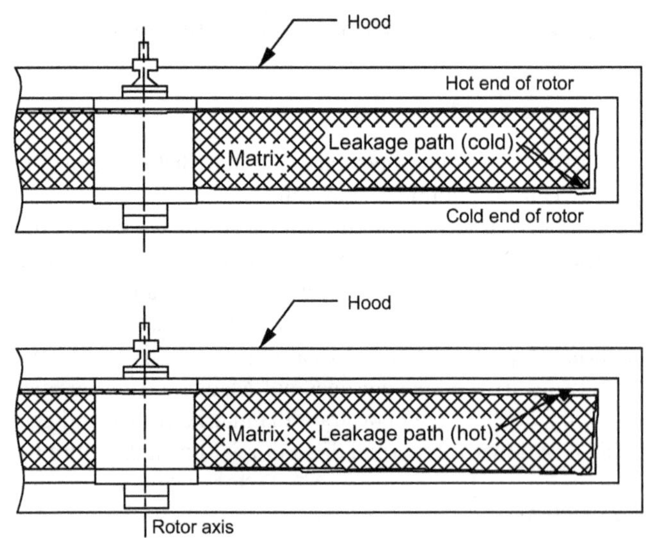

FIGURE 6.21 Concept of thermal 'turn down' or thermal distortion of rotary marix and seals. (Adapted from Storm Technologies Inc., UK, www.storm-techn ologies.com/.)

operation, because seal positions will change as the rotor warms to its operating temperature. This thermal distortion droop opens gaps in the sealing surfaces that separate the cold incoming air from the outlet gases as well as in the sealing surfaces around the circumference of the APH. Turndown changes the gaps between both the radial and circumferential seals and their respective sealing surfaces where gas and air bypass the rotor and heat exchange elements. Two penalties to boiler performance occur with excessive radial seal leakage. The first is the thermal losses associated with the leakage air cooling the APH. The second is the additional auxiliary horsepower consumed by the fans for pushing more flow.

Other Causes of Seal Leakages

Erosion caused by fly ash has resulted in the rapid loss of a heat exchange element as well as damage to perimeter seals, radial seals, and rotor diaphragms. The high ash content associated with many of the coals used at plants experiencing these problems is obviously a contributing factor. However, two other factors with regard to erosion are actually more important than ash content: abrasiveness of ash and ash velocity [44].

The abrasiveness of fly ash increases as the amount of silica and alumina increases. One way to defeat high ash velocity is to increase the fineness of the coal particles leaving the pulverizers and balancing the coal and air flows to each of the burners. Also, pay careful attention to excessive turbulence of the flue gas entering the APH by adding flow straighteners and turning vanes as determined by computerized flow modeling studies.

6.8.3.2 Pressure Forces

Pressures in the regenerator system exert forces on the matrix or structure. There are two sources of pressure forces in a rotary regenerator [2]: One source of pressure forces is due to pressure difference between the high-pressure cold air and the low-pressure hot gases. This pressure force depends on the seal arrangement. The seals for a disk can be arranged so that the net force on the support is essentially zero. The second source results from pressure drop of the fluids flowing through the matrix. This pressure force is the product of the total pressure drop and the frontal area.

6.9 GAS-SIDE FOULING IN REGENERATORS AND INDUSTRIAL HEAT TRANSFER EQUIPMENT

If fouling occurs in the presence of a flue gas or combustion gas stream, the process is known as gas-side fouling. Combustion gases typically contain ash, unburned hydrocarbons, and such elements as sodium, sulfur, chlorine, potassium, calcium, and magnesium, which are all potential foulants. As the gases pass through the heat exchanger, these substances can be deposited on the heat transfer surface. The types of heat exchangers in which gas-side fouling can be a problem are grouped as (a) recuperators; (b) regenerators; (c) direct contract exchangers; (d) fluidized bed exchangers; and (e) heat pipe exchangers [45]. Of the six categories of fouling,namely, precipatation fouling, particular fouling, chemical reaction fouling, corrosion fouling, biological fouling, and solidification fouling, all but crystalllization fouling can occur in gas-side fouling. Particulate fouling is most likely to be associated with gas-side fouling but chemical reaction fouling, corrosion fouling, and solidification can also be improtant.

6.9.1 Gas-Side Fouling Characteristics and Mechanisms

The basic phenomena of gas-side fouling involve transport to the surface, attachment to the surface, and (in some cases) removal from the surface. Dewpoint condensation is a very important deposition mechanism in gas-side fouling. In low-temperature dewpoint condensation, which takes place at temperatures below 500°F, the problem of low-temperature condensation and corrosion is avoided by making certain that the surface temperature always remains above the acid dewpoint temperature of the gas stream.

Materials considerations are very important in several gas-side fouling regimes. In those cases where corrosion is a problem, stainless steels, super alloys, glass, and a variety of coatings have been used with varying degrees of success. There is also considerable interest in the use of ceramic materials to withstand the hostile environments of dirty gases at elevated temperatures, i.e., greater than 1500°F.

6.9.2 Economic Impact of Gas-Side Fouling

An important consideration in the operation of industrial heat-transfer equipment in dirty-gas environments is the economic impact of gas-side fouling [45]:

i. All fossil fuels are sulfur-bearing fuels. As such, if the exit flue-gas temperature from the air heater falls below the acid dew point of the as-fired fuel, sulfurous or sulfuric acid will form, causing back-end corrosion in the air heater. As a consequence, both maintenance cost and plant outages will increase
ii. Capital costs incurred for installing additional heat recovery surfaces may offset the savings obtained by reducing flue-gas temperature
iii. Increased maintenance costs
iv. Loss of production
v. Energy losses

6.10 WASTE HEAT RECOVERY EQUIPMENT AND SYSTEMS

6.10.1 A Heat Recovery System

A heat recovery system should be considered if at least one of these conditions is fulfilled; it may then be economical. Usually, recovery of 65% of exhaust heat can be accomplished with a reasonable payback period. Industrial waste heat generally occurs in the following forms [46]:

i. Sensible heat of solids, liquids, and gases
ii. Latent heat contained in water vapor or other type of vapors and gases
iii. Radiation and convection from hot surfaces

6.10.2 Considerations When Adopting Waste Heat Recovery

Waste heat is rejected heat released from a process at a temperature that is higher than the temperature of the plant air. In contemplating waste heat recovery, take into account the following considerations [46]:

i. Compare the supply and demand for heat.
ii. Determine how easily the waste heat source can be accessed.
iii. Assess the distance between the source and demand.
iv. Evaluate the form, quality, and condition of the waste heat source.
v. Determine the temperature gradient and the degree of heat upgrade required.
vi. Determine any regulatory limitations regarding the potential for product contamination, health and safety.
vii. Perform suitability and economic comparisons (using both the payback period and annuity method evaluations) on the short-listed heat recovery options.

6.10.3 Methods of Waste heat Recovery

Recovering industrial waste heat can be achieved via numerous methods. The heat can either be "reused" within the same process or transferred to another process. Ways of reusing heat locally include using combustion exhaust gases to preheat combustion air or feedwater in industrial boilers. By preheating the feedwater before it enters the boiler, the amount of energy required to heat the water to its final temperature is reduced. Typical technologies used for air preheating include recuperators, furnace regenerators, burner regenerators, rotary regenerators, and passive air preheaters. Heat exchangers are most commonly used to transfer heat from combustion exhaust gases to combustion air entering the furnace. For effective or viable waste heat recovery (WHR), the following requirements are to be met [47]:

1) An accessible source of waste heat
2) An economical and cost-effective recovery technology
3) Use for the recovered energy

The above requirements are illustrated in Figure 6.22

6.10.3.1 Limitations

A heat recovery system should be considered if at least one of the above conditions is fulfilled; it may then be economical. Usually, recovery of 65% of exhaust heat can be accomplished with a reasonable payback period. However, recent developments now allow heat recovery from even small temperature gradient streams, and a suitable application should be investigated.

6.10.3.2 Factors Affecting Waste Heat Recovery Feasibility

Evaluating the feasibility of waste heat recovery requires characterizing the waste heat source and the stream to which the heat will be transferred. Important waste stream parameters that must be determined include [47]:

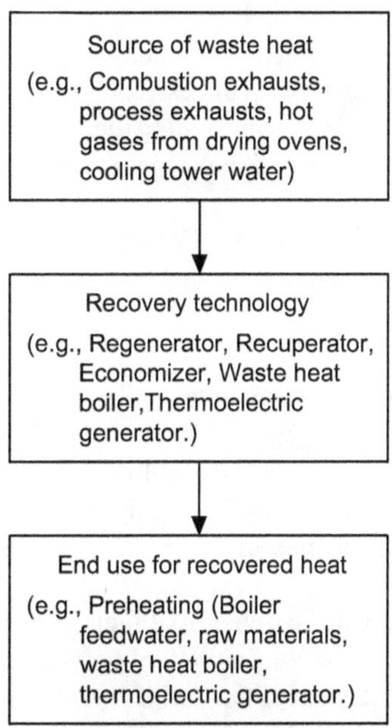

FIGURE 6.22 WHR concept and requirements for its viability.

i. Heat quantity
ii. Heat/temperature quality
iii. Composition of the combustion products/waste heat source
iv. Minimum allowed temperature
v. Cleanliness of flue gas/exhaust gas
vi. Operating schedules, availability, and other logistics

These parameters allow for analysis of the quality and quantity of the stream and also provide insight into possible materials/design limitations.

6.10.3.3 Heat Quantity

The quantity, or heat content, is a measure of how much energy is contained in a waste heat stream, while quality is a measure of the usefulness of the waste heat. The quantity of waste heat contained in a waste stream is a function of both the temperature and the mass flow rate of the stream [47]:

$$Q = mC_p dT \tag{6.68}$$

where Q is the waste heat loss, m is the waste gas stream mass flow rate, and dT is the expected temperature drop of the waste stream.

6.10.3.4 Waste Heat Temperature/Quality

The waste heat temperature is a key factor determining waste heat recovery feasibility. Waste heat temperatures can vary significantly, with cooling water returns having low temperatures around 100–200°F (40–90°C) and glass melting furnaces having flue temperatures above 2,400°F (1,320°C). In order to enable heat transfer and recovery, it is necessary that the waste heat source temperature is higher than the heat sink temperature. Moreover, the magnitude of the temperature difference between the heat source and sink is an important determinant of waste heat's utility or "quality" [47].

6.10.3.5 Cleanliness of Flue gas/Exhaust Gas

The quality of exhaust gases from industrial heating processes depends on many factors related to the operation and design of heating equipment. For example, the presence of highly corrosive fluxing agents (e.g., chlorides, fluorides, etc.), particulates (e.g., metal oxides, carbon or soot particles, fluxing materials, slag, aluminum oxide, magnesium oxide, manganese), and combustibles (e.g., H_2, hydrocarbons, etc.) affect the cleanliness of exhaust gases.

6.10.3.6 Waste Heat Recovery Operational Considerations

Some issues which affect the performance, operation, and maintenance of heat recovery devices, including heat exchangers, are as follows:

i. Gasside fouling. The characterization of the combustion gases including the gaseous species, temperature, velocity, particle loading, particle composition, and particle size distribution is a very important consideration in gas-side fouling.
ii. Certain gas-side fouling deposits can be highly corrosive even at low temperatures, and combustion gases with abrasive particles can cause severe erosion if proper precautions are not taken.
iii. Acid and water dew point condensation. Fossil fuels such as hydrocarbon oil, coal, and many other fuels contain sulfur. The sulfur in the fuel combines with the oxygen in the air during combustion to form sulfur dioxide, SO_2, and if there is sufficient excess air sulfur trioxide, SO_3, will also be formed. The SO_3 can combine with water to form sulfuric acid, H_2SO_4, which condenses onto heat transfer surfaces if they are below the acid dewpoint temperature.

6.10.3.7 Quality of Waste Heat

In Refs. [48, 49], the potential for waste heat recovery technology based on the following five temperature ranges is dealt with in detail:

1. Ultra-low temperature: below 250°F. An example of the lower temperature for this range is the temperature of a cooling medium such as cooling tower water or other water used for cooling systems.
2. Low temperature: 250–450°F.
3. Medium temperature: 450–1,200°F.
4. High temperature: 1,200–1,600°F. The temperature of the exhaust gases discharged into the atmosphere from heating equipment depends on the process temperature and whether a WHR system is used to reduce the exhaust gas temperature. The temperature of discharged gases varies approximately from as low as 150°F to as high as 3,000°F.

6.11 COMMERCIALLY AVAILABLE WASTE HEAT RECOVERY DEVICES

Apart from recuperators and regenerators, various types of heat recovery devices have been available for many years. Commercially available heat recovery devices for combustion air preheaters include the following [47,48,50,51]:

1. Fluid-bed regenerative heat exchangers
2. Fluidized-bed waste heat recovery systems
3. Vortex-flow direct-contact heat exchangers
4. Bayonet tube heat exchanger systems
5. Regenerative burner systems
6. Porcelain-enameled flat-plate heat exchangers
7. Radiation recuperators
8. Heat-pipe exchangers
9. Air preheaters
10. Economizers
11. Waste heat recovery boilers
12. Direct contact heat exchangers
13. Thermocompressors
14. Combined heat and power
15. Combined heat and power—topping and bottoming cycle
16. Thermoelectric generators
17. Plate heat exchanger energy banks
18. Rotary heat exchangers for space heating

Because of the multitude of different industrial heating processes and exhaust system designs, there is no one universal type of waste heat recovery device best suited for all applications. The above types of industrial waste heat recovery devices are discussed hereunder and readers may refer for more details to Refs. [47,48,50,51].

6.11.1 Fluid-Bed Regenerative Heat Exchangers

In fluid-bed regenerative heat exchangers, fluidized beds of pellets recover heat from hot exhaust gas and transfer this heat to cold combustion air [52]. The basic fluid-bed heat exchanger consists of an insulated cylindrical tower incorporating an upper and a lower chamber, as shown in Figure 6.23. Each chamber is fitted with several horizontal perforated trays. Hot exhaust gases enter the upper

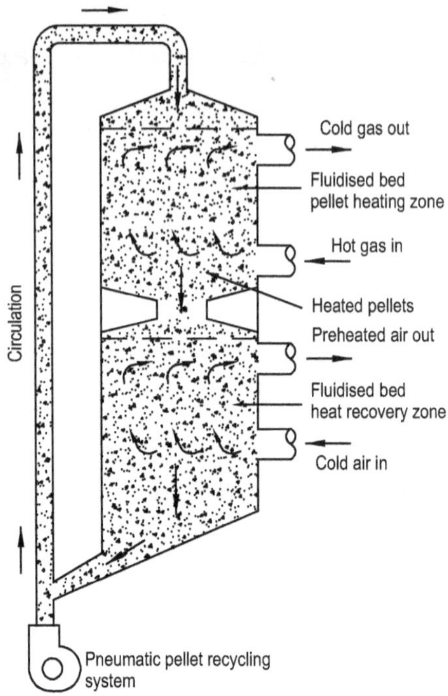

FIGURE 6.23 Fluid-bed regenerator.

chamber near its base and rise to the outlet near the top, heating counterflow alumina pellets as they fall. The heated pellets pass through an aperture into the lower chamber, where they give up their heat to the cold air. Air enters the lower chamber near the base and exits near the top. Direct contact of gas and air streams with the heat exchange media, namely, the alumina pellets, coupled with controlled dwell time at the trays, results in high heat transfer rates.

6.11.2 Fluidized-Bed Waste Heat Recovery (FBWHR) Systems

FBWHR systems employ a heat transfer particulate medium, which is heated as it falls through the upward-flowing flue gases in a raining bed heat exchanger [9,53]. This type of heat exchanger is known as a fluidized-bed heat exchanger (FBHE). FBHEs offer the potential for economic recovery of high-temperature (2000–3000°F) heat from the flue gases of industrial processes such as steel soaking pits, aluminum remelt furnaces, and glass melting furnaces.

An FBHE consists of horizontal finned heat exchanger tubes with a shallow bed of fine inert particles, which move upward with gas flow and give up the heat to the finned tubes, which in turn transfer heat to cold air passing through them. A thin steel plate with small perforations supports the bed and distributes the hot gas evenly. Among the advantages of an FBHE are enduring hostile environments such as elevated temperature, fouling particulates, corrosive gases, corrosive particulates, and thermal cycling [9]. Most surfaces are kept clean by fluidizing action. Figure 6.24 shows fluidized-bed waste heat recovery systems.

6.11.3 Vortex-Flow Direct-Contact Heat Exchangers

In the vortex-flow direct-contact heat exchanger, the heat conduction material wall of conventional metal or high-temperature ceramic material is replaced by a vortex-induced fluid dynamic

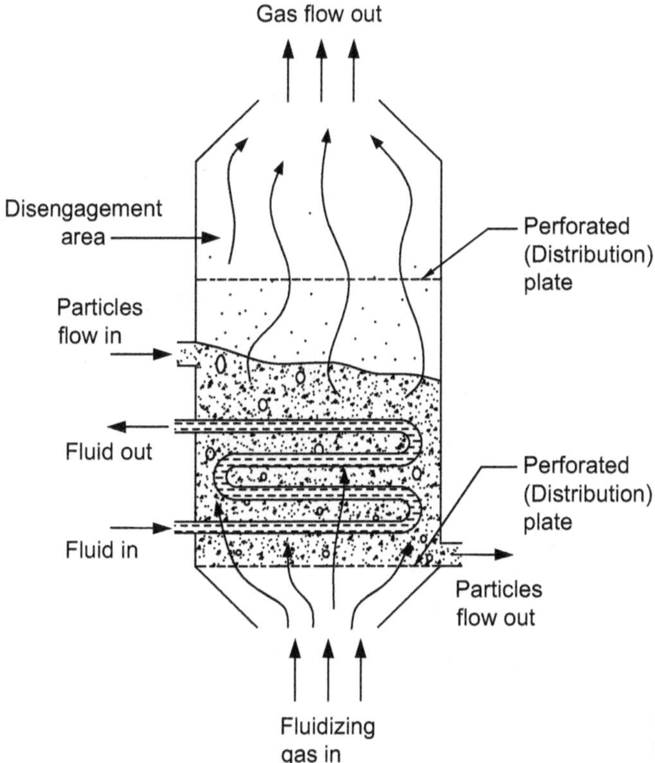

FIGURE 6.24 Fluidized-bed waste heat recovery systems.

gas boundary, which separates the tangentially rotating hot and cold gas flows between which heat exchange is desired [54].

The VFHE is a cylindrical cavity into which separate hot flue gas and cold preheat air gas flows are injected. Injection of the gases is such that a vortex-flow pattern is established in the cavity, providing confinement of hot fluid to the central region of the cavity surrounded by cold air stream. Solid particles of 20–200 μm diameters as a medium to exchange heat between hot and cold streams are injected into the hot spiraling gas stream. Due to centrifugal action, these particles pass into the surrounding cold gas rotating flow, to which the particles transfer heat. Upon reaching the outside cavity surface, the particles are entrained in the outermost region of the vortex flow and removed with a portion of the cold flow, which exits at the periphery (exhaust end wall). The VFHE requires a particle injection and separation subsystem to collect and recycle the particle to the vortex chamber for continued heating and cooling cycles.

6.11.4 Bayonet Tube Heat Exchangers for Gases

The bayonet tube heat exchanger is a tubular form consisting of two concentric tubes, the inner tube open at both ends positioned inside the outer tube open only at one end, as shown in Figure 6.25. The fluid can either flow by entering the inner tube and exiting the annulus termed as flow A or flow B, through the annulus and exiting the inner tube, the fluid flow is driven by the pressure difference between the inlet and outlet of the bayonet tube, and it is suitable when the fluid to be heated or cooled is accessible from one side only and is free from bending and axial compressive stresses.

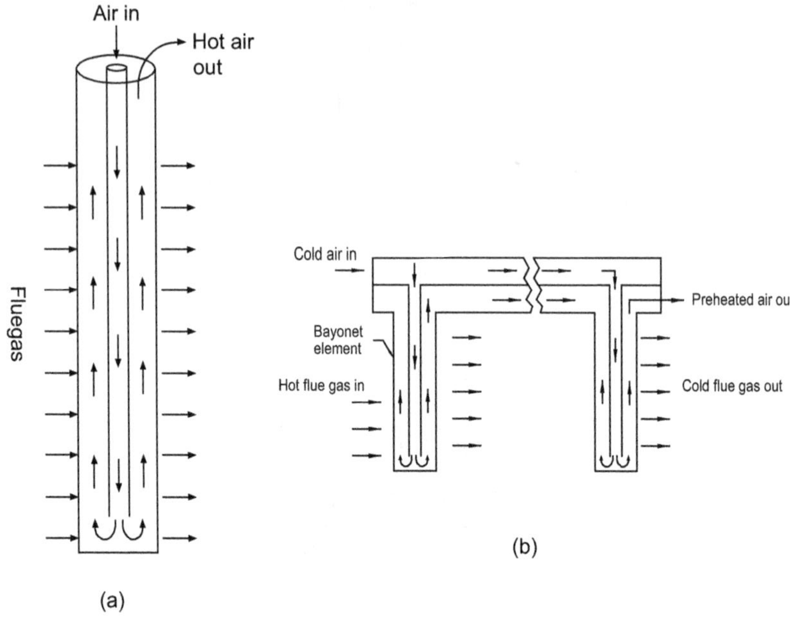

FIGURE 6.25 Bayonet tube heat exchanger for gases.

6.11.5 Ceramic Bayonet Tube Heat Exchangers

Babcock and Wilcox (B&W) and the Department of Energy (DOE) developed high-temperature burner duct recuperator (HTBDR) systems capable of recovering waste heat from high-temperature flue gas using ceramic bayonet elements [55,56]. The HTBDRs consist of silicon carbide ceramic bayonet tubes suspended into the exhaust gas flow stream, where they recover the heat from the exhaust and preheat the combustion air up to a temperature of 1090°C (2000°F). Combustion air is directed through the elements at the inner tube or the annulus from heavily insulated plena located above the tubesheets. The tubes of each element are free to expand or contract under the influence of temperature variations.

6.11.6 Regenerative Burners

The "reGen" regenerative burner, an all-ceramic high-temperature burner, close coupled to a compact, fast-cycle, ceramic regenerator, provides air preheats in excess of 85% of the process temperature in fuel-fired applications up to 1399°C (2550°F) [7]. Figure 6.26 shows a complete regenerative unit comprised of two burners, two regenerators, a reversing valve, and the related pipings. While one of the burners fires, drawing the preheated air fed through its regenerator, exhaust gas is drawn through the other burner and down into its associated regenerator, then discharged to the atmosphere. After a sufficient interval, the reversing operation takes place.

6.11.7 Porcelain-Enameled Flat-Plate Heat Exchangers

Porcelain-enameled heat exchangers offer cost-effective solutions to problems of heat recovery from extremely corrosive gases that preclude the use of carbon steel or common types of stainless steel [57]. Porcelain enamel is a glass coating material that is applied to a metal substrate to protect the base metal. The base metal gives strength and rigidity; the metal can be steel, cast iron, aluminum, or copper, but mostly steel is used on a larger scale. The glass coating offers significant corrosion

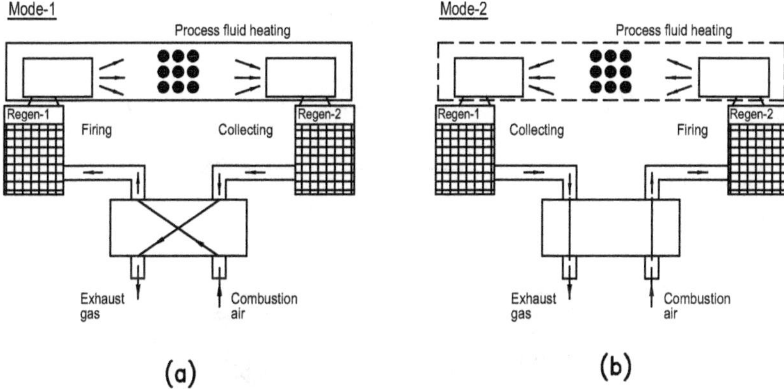

FIGURE 6.26 Regenerative burner – flow reversal.

resistance to the corrosive flue gas. The porcelain-enameled coating finds application in flat-plate heat exchangers of the open channel air preheater (OCAP) type for advanced heat recovery. The OCAP is a plate-type exchanger characterized by a nonweld construction of the heat transfer core. The enameled plates are assembled into a floating construction. Elastic springs are used between all adjacent plates to ensure uniform distribution of the dead weights and to allow for thermal expansion and contraction in operation. This floating core is enclosed in a rigid frame, again without any welding or damage to the porcelain enamel layer.

6.11.8 Metallic Radiation Recuperator

Metallic recuperators for industrial furnaces are convective units, operating at flue gas temperatures up to a maximum of 1100°C (2000°F) [58]. They are radiation-type units, that is, the heat transfer mode is by radiation, operating at temperature ranges of nearly 900–1500°C (2700°F). In radiation-type recuperators, parallelflow is essential to control the operating temperatures of the heating surface. A combination of parallelflow and counterflow is required to reach the highest recuperator efficiencies of air temperature above 800°C (1500°F). Typical radiation recuperators include a double-shell parallelflow radiation recuperator consisting of two concentric cylinders, as shown in Figure 6.27 [48]—through the inner shell the flue gas flows and through the annulus the cold air flows in a parallel direction—as well as cage-type radiation recuperators for very large high-temperature furnaces, with a combination of parallelflow and counterflow units. The reason for the use of parallel flow is that recuperators frequently serve the additional function of cooling the duct carrying away the exhaust gases and consequently extending its service life.

6.11.9 Convective Recuperator

In this unit, hot gases are carried through tubeside, while the incoming air to be heated passes through shellside surrounding the tubes and passes over the hot tubes one or more times in a direction normal to their axes, i.e., crossflow. Shell and tube type recuperators are generally more compact and have greater effectiveness than radiation recuperators because of the larger heat transfer area made possible through the use of multiple tubes and multiple passes of the gases.

6.11.10 Combined Radiation and Convection Recuperator

For maximum effectiveness of heat transfer, hybrid recuperators are used. These are combinations of radiation and convective designs, with a high-temperature radiation section followed by a convective

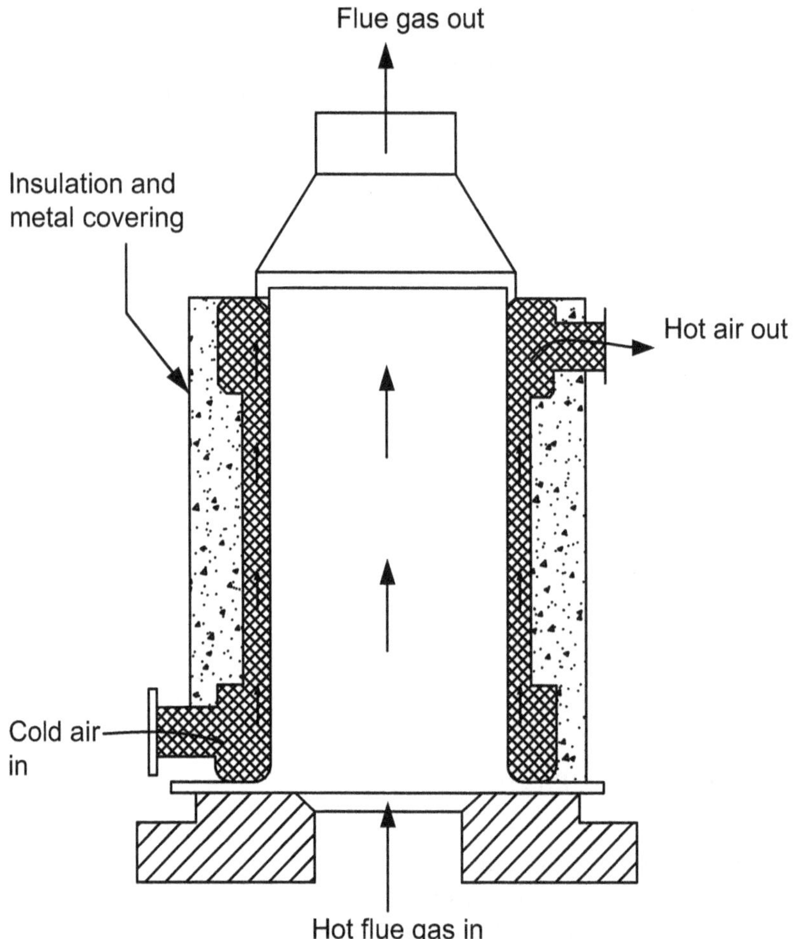

FIGURE 6.27 Metallic Radiation recuperator.

section. A combined radiation/convection recuperator is shown in Figure 6.28 [50]. The system includes a radiation section followed by a convection section in order to maximize heat transfer effectiveness. In this technology, hot exhaust gas is fed into a larger shelf and then split into smaller diameter tubes.

6.11.11 Heat-Pipe Heat Exchangers

The heat-pipe heat exchanger used for gas–gas heat recovery is essentially a bundle of finned heat pipes assembled like a conventional air-cooled heat exchanger. The heat pipe consists of three elements: (1) a working fluid inside the tubes; (2) a wick lining inside the wall; and (3) a vacuum-sealed finned tube. Because the pipe is sealed under a vacuum, the working fluid is in equilibrium with its own vapor. The purpose of the wick is to transport the working fluid contained within the heat pipe, from one end to the other by capillary action.

The heat-pipe heat exchanger consists of an evaporative section through which the hot exhaust gas flows and a condensation section through which the cold air flows. These two sections are separated by a separating wall as shown in Figure 6.29 [59]. The working of a heat-pipe heat exchanger is as follows. Heat transfer by the hot flue gas at the evaporative section causes the working fluid

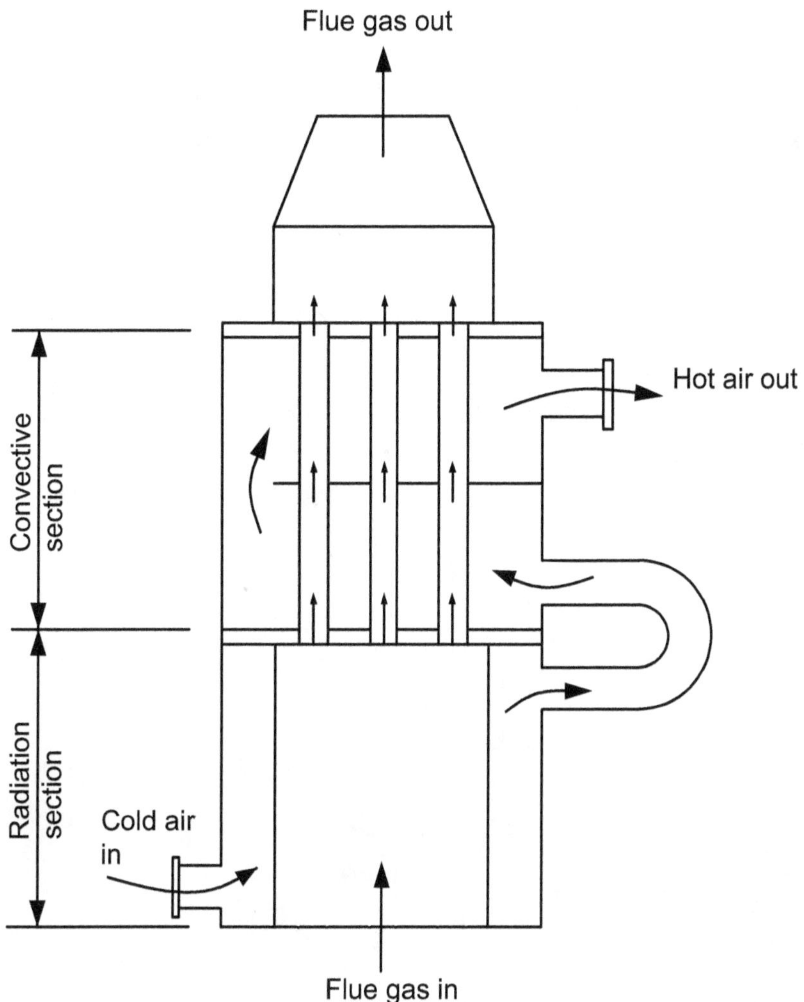

FIGURE 6.28 Combined radiation/convection recuperator.

contained within the wick to evaporate and the increase in pressure causes the vapor to flow along the central vapor region to the condensation section, where it condenses, and gives up the latent heat to the cold air. The condensate then drains back to the evaporative section by capillary action in the wick. A boiling type heat pipe is shown in Figure 6.30 [48], and Figure 6.31 shows a finned U-tube heat pipe.

6.11.11.1 Merits of Heat-Pipe Heat Exchangers

The heat-pipe exchanger is a lightweight compact heat recovery system. It does not need mechanical maintenance as there are no moving parts. Also, it does not need input power for its operation.

6.11.11.2 Application

The heat pipes are used in following industrial applications: (1) process fluid to preheating of air; (2) preheating of boiler combustion air; (3) heating, ventilating, and air-conditioning (HVAC) systems; (4) recovery of waste heat from furnaces; (5) reheating of fresh air for hot air driers;

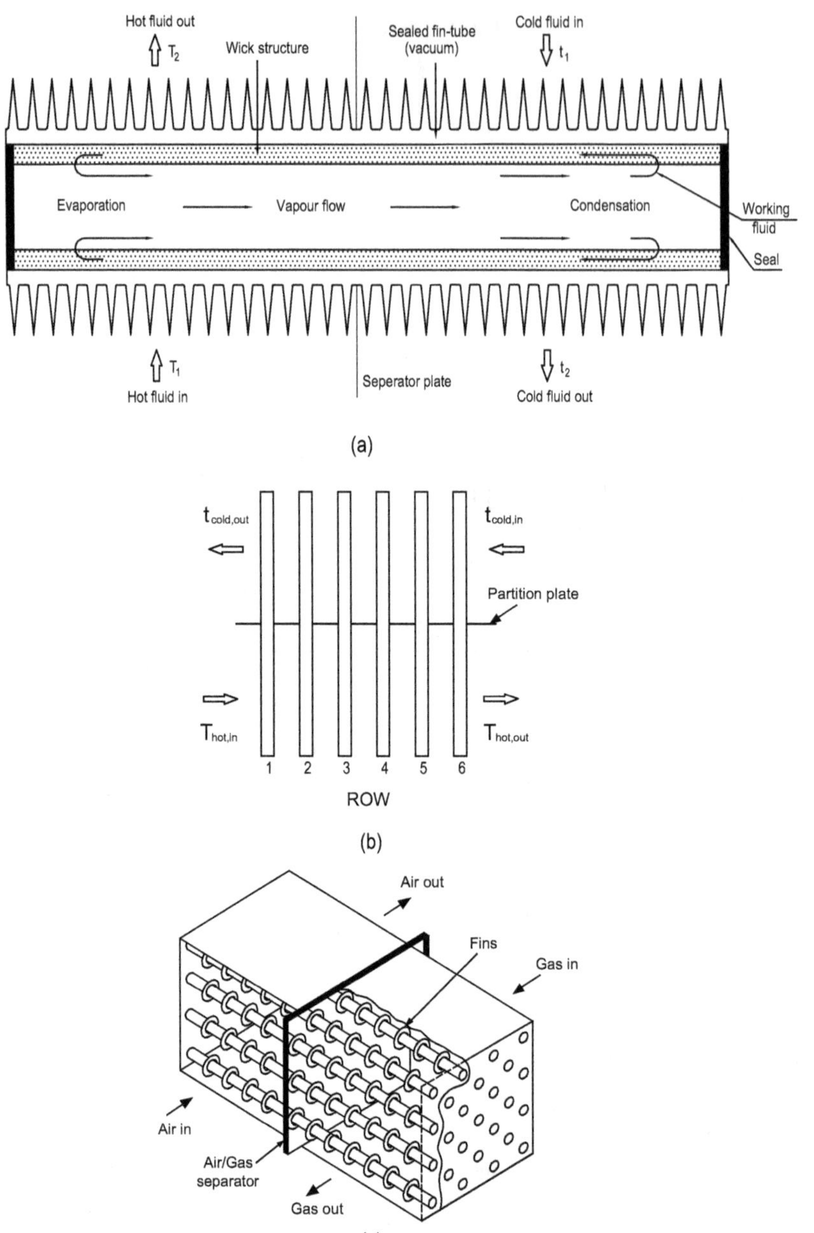

FIGURE 6.29 Heat-pipe heat exchanger. (a) Principle of working, (b) heat-pipe tube arrangement, and (c) heat-pipe heat exchanger (schematic).

(6) preheating of boiler feed water with waste heat recovery from flue gases in the heat-pipe economizers; and (7) reverberatory furnaces (secondary recovery).

6.11.12 Economizers

In the case of a boiler system, in an economizer the incoming feedwater absorbs heat from the exiting flue gases, raising the temperature of the feedwater as close as possible to but lower than its saturation temperature prior to entering into the evaporating circuit. The remaining heat in the flue gas

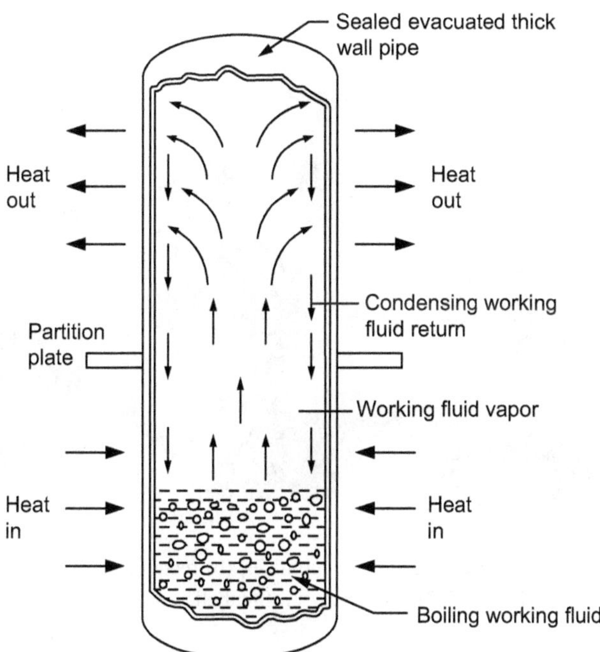

FIGURE 6.30 Boiling type heat pipe.

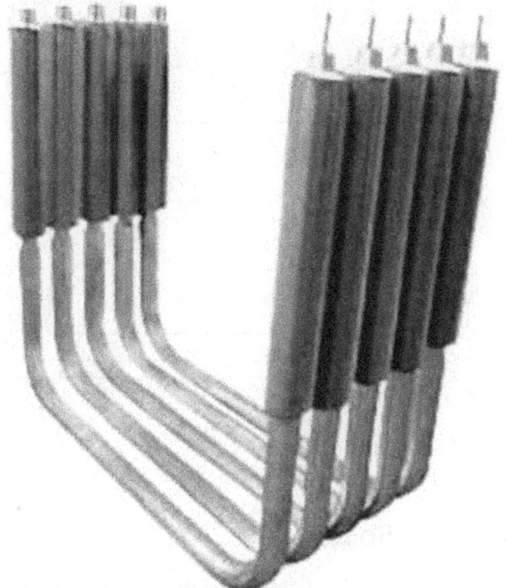

FIGURE 6.31 Heat-pipe heat exchanger finned U-tube. (From Thermofin, www.thermofin.net).

downstream of the economizer is recovered in an air heater to raise the temperature of combustion air before it enters the furnace. In both these cases, there is a corresponding reduction in the fuel requirements of the boiler. For every 22°C reduction in flue gas temperature by passing through an economizer or a preheater, there is a 1% saving of fuel in the boiler. In other words, for every 6°C rise in feed water temperature through an economizer or 20°C rise in combustion air temperature through an air preheater, there is a 1% saving of fuel in the boiler [59]. Some types of industrial economizers are shown in Figures 6.32 and 6.33. Figure 6.34 shows an economizer for a utility boiler.

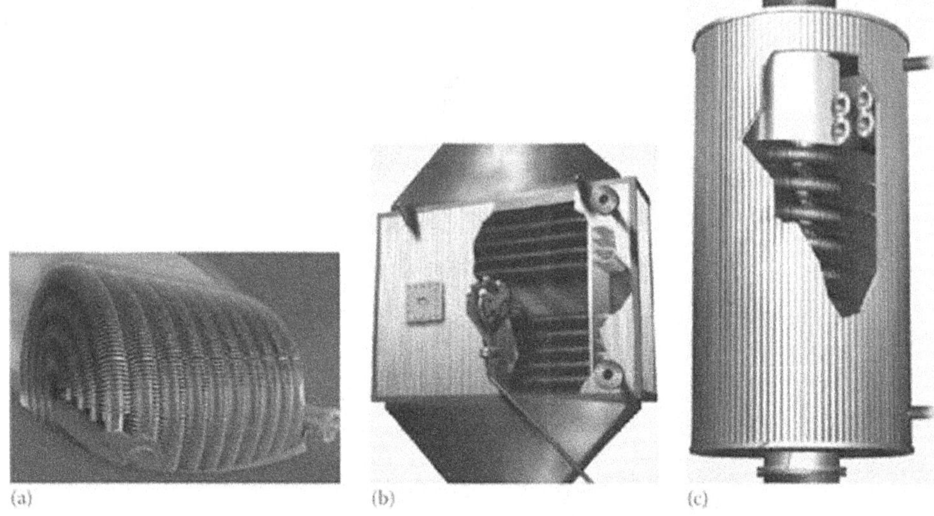

FIGURE 6.32(a) Economizer. (a) Coiled-tube economizer, (b) rectangular economizer, and (c) cylindrical economizer. (Courtesy of Fintube, LLC, Tulsa, OK; www.fintubellc.com)

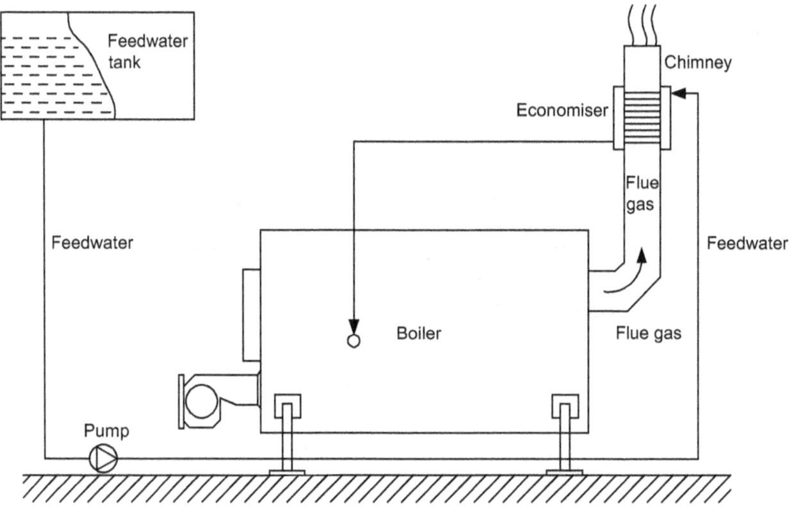

FIGURE 6.32(b) Economizer of an industrial boiler.

6.11.13 Waste Heat Recovery Boilers

Waste heat recovery boilers are ordinarily water tube boilers in which the hot exhaust gases from gas turbines, incinerators, etc., pass over a number of parallel tubes containing water. The water is vaporized in the tubes and collected in a steam drum from which it is drawn off for use as heating or processing steam. Because the exhaust gases are usually in the medium temperature range, and in order to conserve space, a more compact boiler can be produced if the water tubes are finned in order to increase the effective heat transfer area on the gas side. A waste heat recovery boiler contributes in terms of improvement of thermal efficiency, energy saving, environmental protection, etc. Figure 6.35 shows a mud drum, a set of tubes over which the hot gases make a double pass, and a steam drum which collects the steam generated above the water surface.

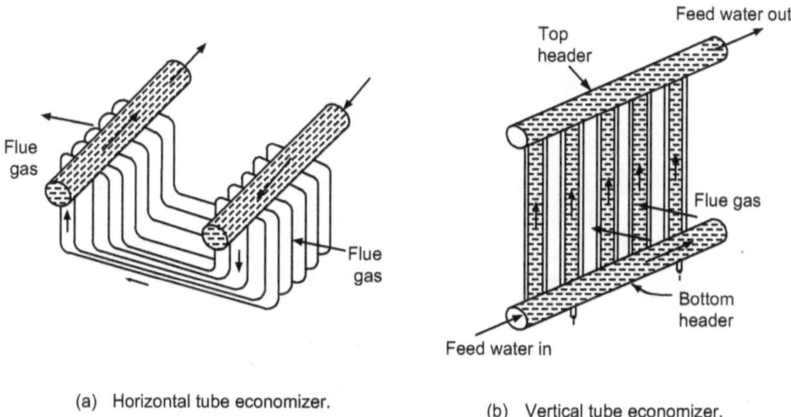

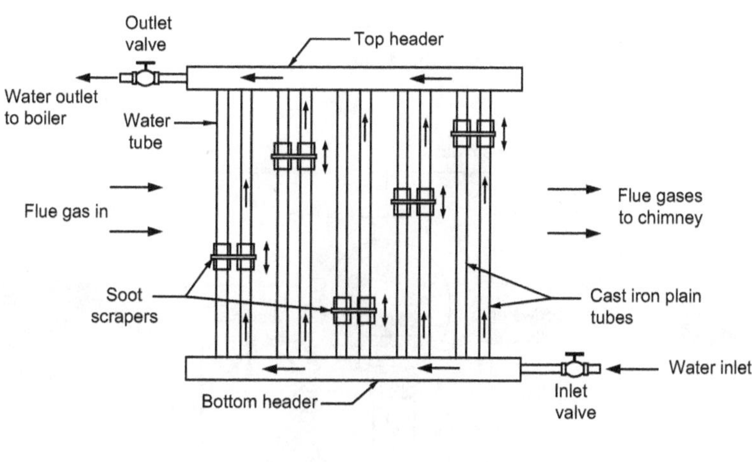

FIGURE 6.33 (a) Horizontal and (b and c) Vertical tubes economizer.

6.11.14 Direct Contact Heat Exchangers

Direct contact heat exchangers allow the two heat transfer media to come directly together, i.e., in the absence of a solid surface separating the two media. This principle finds wide use in a deaerator of a steam generation station. It essentially consists of a number of trays mounted one over the other. Steam is supplied below the packing while the cold water is sprayed at the top. The steam is completely condensed in the incoming water, thereby heating it. A direct contact heat exchanger is shown in Figure 6.36. Some other direct contact heat exchangers are illustrated in Chapter 1.

6.11.15 Thermally Activated Technologies (TATs)

TATs consist of equipment using thermal energy for heating, cooling, humidity control, and power (mechanical and electric). These technologies include absorption chillers or refrigerating equipment, desiccant systems for humidity control, and organic Rankine type power generation systems.

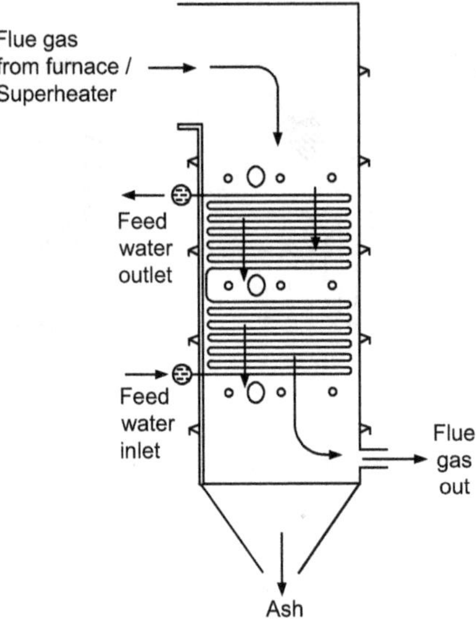

FIGURE 6.34 Location of Economizer of utility boiler.

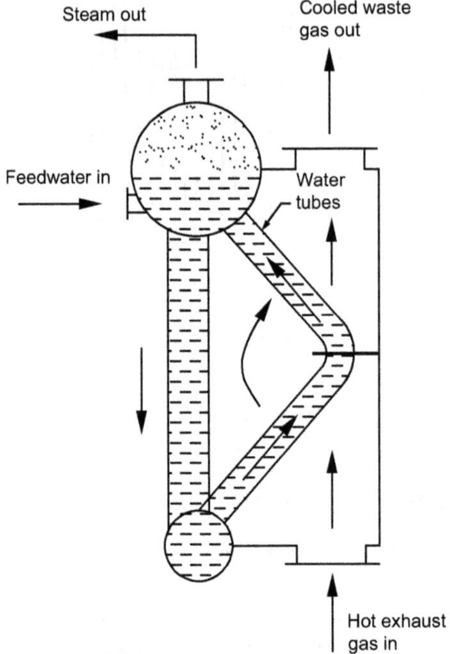

FIGURE 6.35 Waste Heat Recovery Boiler.

6.11.16 Combined Heat and Power System (CHP)

CHP generally consists of a prime mover, a generator, a heat recovery system, and electrical interconnection equipment configured into an integrated system. CHP is a form of distributed generation, which, unlike central station generation, is located at or near the energy-consuming facility. CHP's inherent higher efficiency and its ability to avoid transmission losses in the delivery of electricity

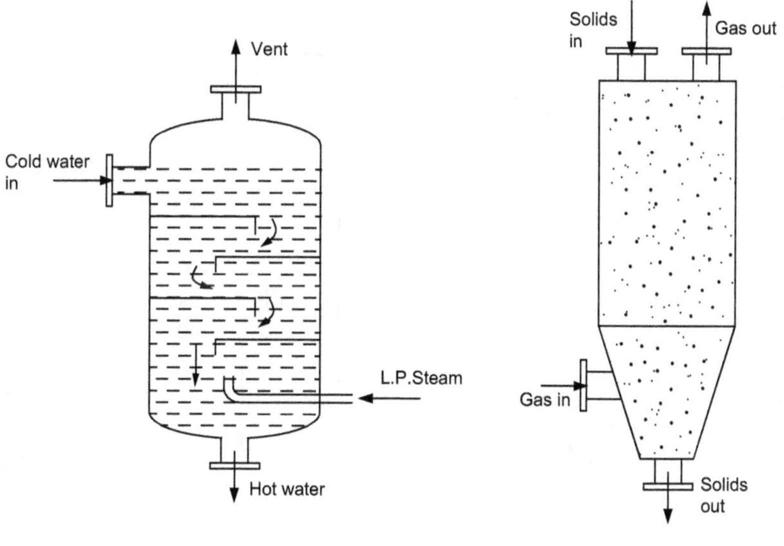

FIGURE 6.36 Direct contact heat exchanger.

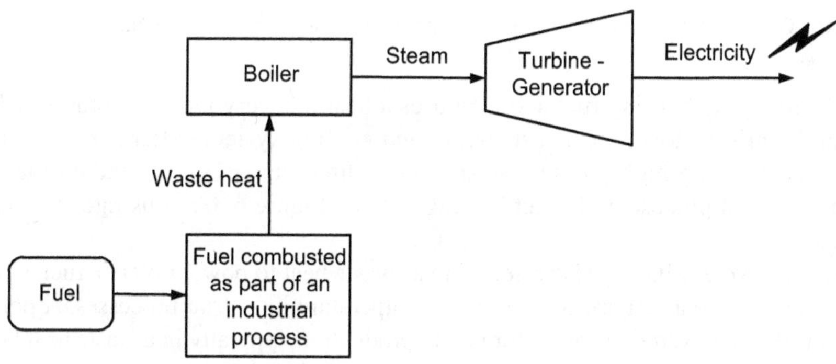

FIGURE 6.37 Combined Heat and Power (CHP) Cycle.

from the central station power plant to the user result in reduced primary energy use and lower greenhouse gas (GHG) emissions.

6.11.16.1 Common CHP Configurations

The two most common CHP system configurations are:

i. Combustion turbine, or reciprocating engine, with heat recovery unit
ii. Steam boiler with steam turbine

A combined heat and power system is shown in Figure 6.37 [60].

6.11.16.2 Topping and Bottoming Cycle

The most common CHP configuration is known as a topping cycle, where fuel is first used in a heat engine to generate power, and the waste heat from the power generation equipment is then recovered to provide useful thermal energy. As an example, a gas turbine or reciprocating engine

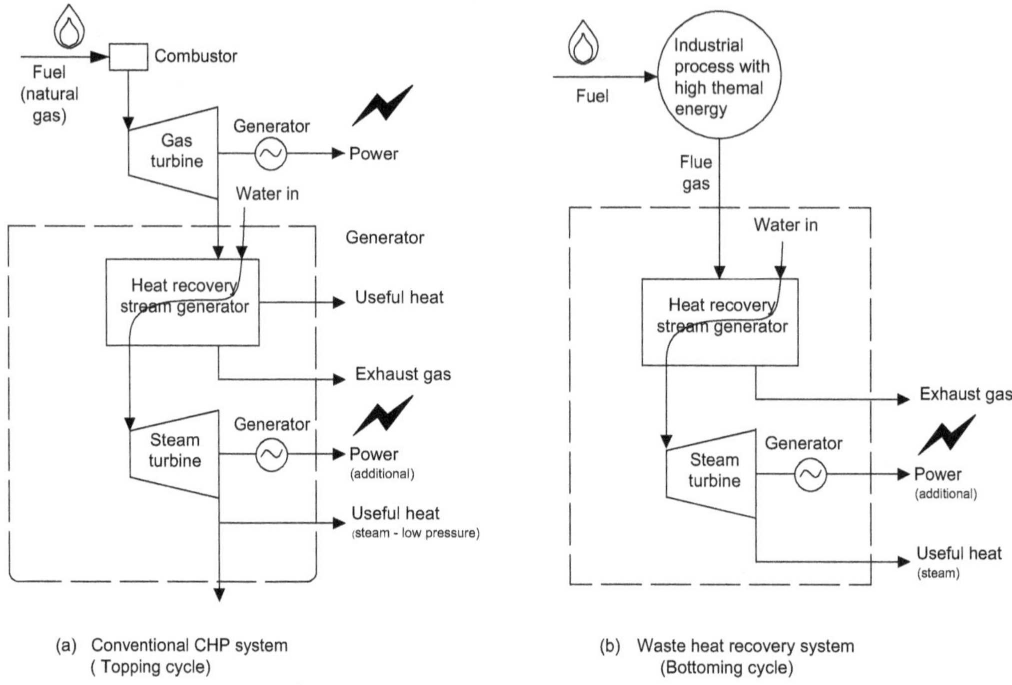

FIGURE 6.38 Combined Heat and Power (CHP) topping cycle and bottoming cycle.

generates electricity by burning fuel and then uses a heat recovery unit to capture useful thermal energy from the prime mover's exhaust stream and cooling system. Alternately, steam turbines generate electricity using high-pressure steam from a fired boiler before sending lower pressure steam to an industrial process or district heating system. Figure 6.38a illustrates the typical CHP topping cycle [61].

In a bottoming cycle, which is also referred to as waste heat to power (WHP), fuel is first used to provide thermal input to a furnace or other high-temperature industrial processes. A portion of the rejected heat is then recovered and used for power production, typically in a waste heat boiler/steam turbine system. The energy associated with waste heat would otherwise be wasted. The generated electricity may be used on-site for the building or facility or transferred off-site to the power grid. Figure 6.38b illustrates the CHP bottoming cycle.

6.11.16.3 Heat Recovery Steam Generator (HRSG)

A heat recovery steam generator (HRSG) is a system used to recover the waste heat from the exhaust of a power generation plant, specifically gas turbine exhaust. It consists of several heat recovery sections such as an evaporator, super heater, economizer, and steam drum, which are very large. By looking at the configuration of a HRSG in Figure 6.39, it can be pointed out that the superheater is placed in the path of the hottest gas upstream of the evaporator and the economizer is placed downstream of the evaporator in the coolest region.

6.11.17 Waste Heat to Power Technologies

Waste heat to power (WHP) is the process of capturing heat discarded by an existing process and using that heat to generate electricity. WHP technologies fall under the WHR category. In general, the least expensive option for utilizing waste heat is to re-use this energy in an on-site thermal process. If it is not feasible to recover energy from a waste heat stream for another thermal process, then a WHP system may be an economically attractive option. Waste heat to power is the process of

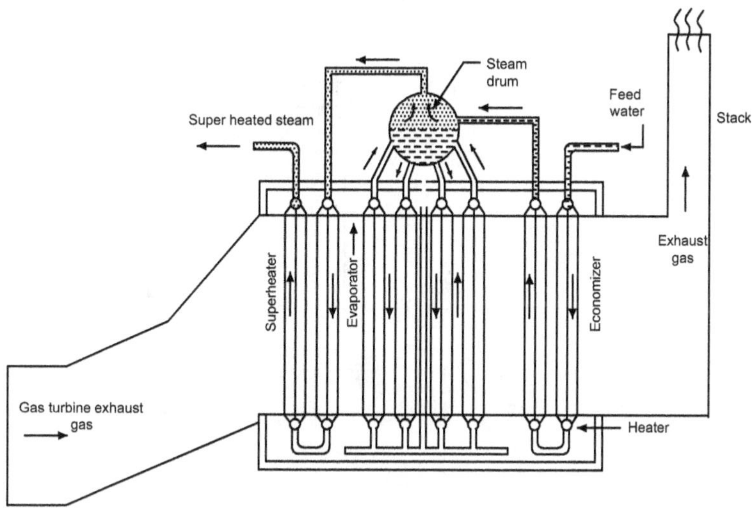

FIGURE 6.39 Heat recovery steam generator (HRSG).

capturing heat discarded by an existing industrial process and using that heat to generate power (see Figure 6.37). Commonly used WHP technologies are [61,61.1]:

a. Rankine cycle (RC)
b. Organic Rankine cycle (ORC)
c. Kalina cycle (KC)
d. Supercritical CO_2 cycle

6.11.18 Thermoelectric Generation

Thermoelectric (TE) materials are semiconductor solids that allow direct generation of electricity when subject to a temperature differential. These systems are based on the Seebeck effect—when two different semiconductor materials are subject to a heat source and heat sink, a voltage is created between the two semiconductors. Conversely, TE materials can also be used for cooling or heating by applying electricity to dissimilar semiconductors. A thermoelectric power generator is shown in Figure 6.40.

6.11.19 Thermocompressor

In many cases, very low-pressure steam is reused as water after condensation for lack of any better option of reuse. Also, it becomes feasible to compress low-pressure steam by very high-pressure steam and reuse it as medium-pressure steam. The major energy in steam is its latent heat value and thus thermocompressing would result in a large improvement in waste heat recovery. The thermocompressor is a simple piece of equipment with a nozzle where high-pressure steam is accelerated into a high-velocity fluid (Figure 6.41). This entrains the low-pressure steam by momentum transfer and then recompresses in a divergent venturi. It is typically used in evaporators where the boiling steam is recompressed and used as heating steam.

6.11.20 Mueller Temp-Plate® Energy Recovery Banks

A Mueller Temp-Plate energy recovery bank (Figure 6.42) transfers the heat content of waste gas to the media used in the process or to an intermediate heat transfer solution. High-grade heat can be efficiently recovered using a Mueller Temp-Plate energy recovery bank.

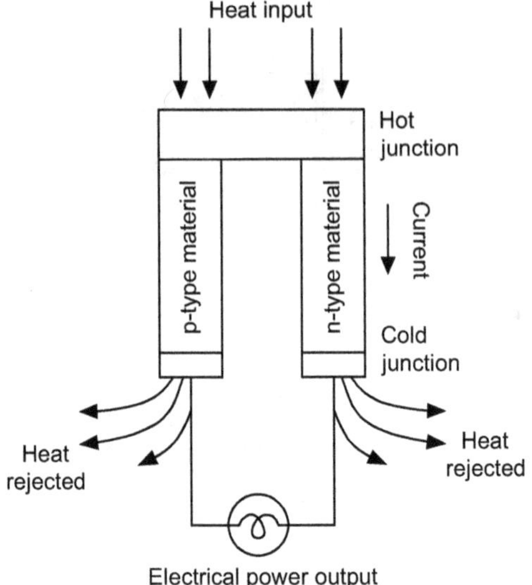

FIGURE 6.40 Thermoelectric generation.

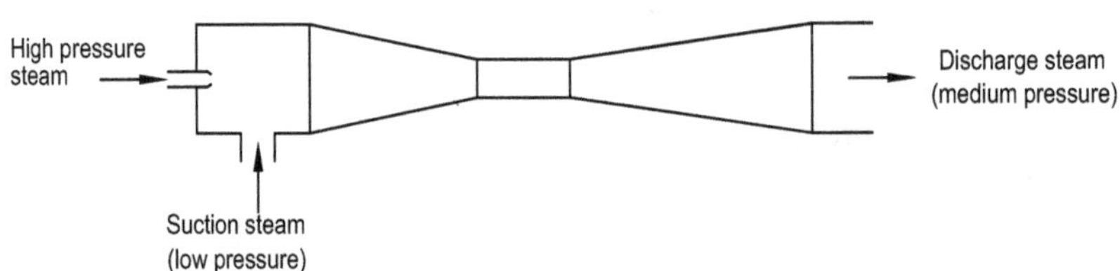

FIGURE 6.41 Thermocompressor.

6.11.21 Rotary Heat Exchangers for Space Heating

Rotary heat exchangers are well-proven means of high-efficiency energy recovery, typically recovering up to 80% in typical HVAC (space heating and ventilation) applications. The combination of heat and humidity recovery by rotary heat exchangers is a highly efficient energy-saving method. A typical summer application is dehumidification of warm and humid supply air to reduce the energy consumption of the downstream cooling equipment. During winter, operation of this feature recovers moisture from the exhaust air to reduce the humidification load.

Leading rotary heat exchanger manufacturers for HVAC applications include Hovalwerk AG, 9490 Vaduz, Liechtenstein, Reznor UK Ltd, Klingenburg GmbH, Germany, Munters International Inc., FL, and Fläkt Woods AB, SE-551 84, Jönköping.

6.11.21.1 Working Principle

The rotor with its axial, smooth air channels serves as a storage mass, half of which is heated by the warm air stream and half of which is cooled by the cold air stream, in a counterflow arrangement. Consequently, the temperature of the storage mass varies depending on the rotor depth in the axial direction and the angle of rotation. In a counterflow arrangement, the rotating, air-permeable storage mass is heated and cooled alternately by the heat-releasing and heat-absorbing air streams. Depending on the air conditions and the surface of the storage mass, moisture also may be transferred in this process. The supply and extract air streams must be adjacent and pass through

FIGURE 6.42 Mueller Temp-Plate energy recovery banks. (Courtesy of Mueller, Heat Transfer Products, Springfield, MO.)

the heat exchanger simultaneously. A typical rotary wheel for HVAC applications is shown in Figure 6.43.

6.11.22 Cleaning Devices

For the regenerative heat exchangers, there are several types of cleaning devices. They are selected according to the degree of fouling of the rotor. Under normal conditions, the rotor has a high self-cleaning effect, on account of the continuously changing directions of air flow. The cleaning methods include: (1) compressed air cleaning; (2) compressed air and water cleaning; (3) steam cleaning; and (4) compressed air and warm water cleaning.

6.11.23 Limitations of Currently Available WHR Technologies

While considering/adopting a waste heat recovery technology, device-specific opportunities and limitations are to be considered and taken into account for its effectiveness. Limitations of currently available WHR technologies are described in Table 6.9 [49].

6.11.24 Key Barriers to Waste Heat Recovery

Key restrictions are presented by cost, heat stream composition, temperature, process and application specific constraints, and inaccessibility/transportability of certain heat sources. In order to promote heat recovery practices, several efforts could be made to reduce system costs, optimize heat exchange materials, heat transfer rates, low temperature recovery, and available end uses for waste heat.

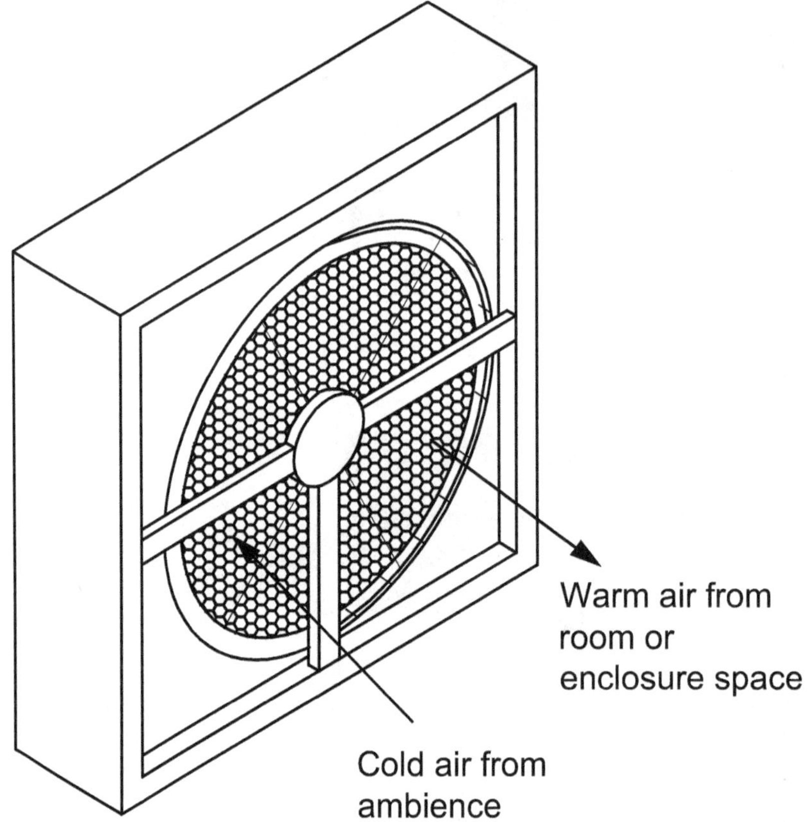

FIGURE 6.43 Rotary wheel for HVAC application.

6.11.25 Materials Selection Considerations for WHR Equipment

A number of factors must be considered to select appropriate materials to improve the energy efficiency of the equipment while extending its life at the minimum cost. These factors include mechanical properties for strength considerations, thermal stability, oxidation or hot corrosion resistance, use of cast or fabricated components, the metallurgical stability of the material, and material availability. Technical data describing the properties of heat-resistant alloys are necessary guides for selection. Service experience with high-temperature equipment is needed to judge the relative significance of the many variables involved [62,62.1].

6.11.25.1 Carburization

Chromium, nickel, and silicon are the three major elements that confer resistance to carbon absorption. Nickel and silicon lower the maximum solubility of carbon and nitrogen. Carburization is usually of concern, because highly carburized alloys become brittle.

6.11.25.2 Sulfidation

Low or moderate nickel with high chromium content minimizes sulfidation attack at high temperatures. With the exception of alloy HR-160, less than 20% nickel content is preferred.

TABLE 6.9
Limitations of Currently Available WHR Technologies, High- and Ultra-High-Temperature Ranges

Equipment	Limitations and barriers
Metallic recuperators	Upper temperature limit of 1,600°F
	Economically justifiable heat recovery efficiency limit of 40–60%
	High maintenance costs for use with gases containing particulates, condensable vapors, or combustible material
	Reduced life expectancy in applications where the mass flow and temperature of the fluids vary or are cyclic
	Fouling and corrosion of heat transfer surfaces
	Difficulty in maintaining or cleaning the heat transfer surfaces
Ceramic recuperators	Life expectancy due to thermal cycling and possibility of leaks from high-pressure side
	Initial cost
	Relatively high maintenance
	Size limitations—difficult to build large size units
Recuperative burners	Lower heat recovery efficiency (usually less than 30%)
	Temperature limitation—exhaust gas temperature less than 1,600°F
	Limited size availability (usually for burners with less than 1 MM Btu/h)
	Cannot be applied to processes where exhaust gases contain particles and condensable vapors
Stationary regenerators	Large system footprint
	Declining performance over time
	Plugging of exhaust gas passages when the gases contain particulates
	Chemical reaction of certain exhaust gas constituents with the heat transfer surfaces
	Possibility of leakage through dampers and moving parts
	Cost can be justified only for high-temperature (>2,000°F) exhaust gases and larger size (>50 MM Btu/h firing rate)
Rotary regenerators	Seal failure between the high-pressure and low-pressure gases (air)
	Plugging of exhaust gas passages when the gases contain particulates
	High pressure drop compared to recuperators
	Maintenance and operation reliability issues for rotary mechanism
Regenerative burners	Large footprint for many applications
	Complicated controls with dampers that cannot be completely sealed
	Difficult pressure control for the furnace
	Cost competitiveness
	Plugging of the bed when the gases contain particulates Require frequent cleaning of the media and the bed
Heat recovery steam generators—boilers	Limited to use for large-size systems (usually higher than 25 MM Btu/h)
	Limited to use with only clean and particulate-free exhaust gases
	Only viable for plants with need for steam use
	Initial cost is very high compared to other options such as recuperators

Source: [49].

Nomenclature

A_C	matrix transfer area on cold side, m^2 (ft^2)
A_{fr}	frontal area, m^2 (ft^2) $= A_{fr,h} + A_{fr,c}$
$A_{fr,c}$	frontal area on cold side, m^2 (ft^2)
$A_{fr,h}$	frontal area on hot side, m^2 (ft^2)
$A_{fr,t}$	total face area, m^2 (ft^2)
A_h	matrix transfer area on hot side, m^2 (ft^2)
$A_{k,l}$	total solid area for longitudinal conduction, m^2 (ft^2)

A_{mk}	constants defined by Equation 6.48
a_{jk}	constants defined by Equation 6.29
A_o	minimum free flow area, m² (ft²) = σA_{fr}
$\%a$	face seal and hub coverage
B_j	constants defined by Equation 6.28
B_{mk}	constants defined by Equation 6.49
b	cell geometry dimension, m (ft); or function defined by Equation 6. 66
C	heat capacity rate (Mc_p), J/°C (Btu/h · °F)
C^*	heat capacity rate ratio (it should be less than or equal to 1) = C_{min}/C_{max}
C_c	fluid stream capacity rate (Mc_p) for cold gas, W/°C (Btu/h · °F)
C_h	fluid stream capacity rate (Mc_p) for hot gas, W/°C (Btu/h · °F)
C_k	constant defined by Equation 6.50
C_{max}	maximum of C_c or C_h, W/°C (Btu/h · °F)
C_{min}	minimum of C_c or C_h, W/°C (Btu/h · °F)
$c_{p,c}$	specific heat of cold gas, J/kg · °C (Btu/lbm · °F)
$c_{p,h}$	specific heat of hot gas, J/kg · °C (Btu/lbm · °F)
C_r	capacity rate of the rotor matrix ($M_r c_r$), W/°C (Btu/h · °F)
$= M_r c_r N$	for rotary regenerator, W/°C (Btu/h · °F)
c_r	specific heat of the matrix material, J/kg · °C (Btu/lbm · °F)
C^*	matrix heat capacity rate ratio
$C^*_{r,m}$	matrix heat capacity rate ratio = $(M_r c_r)/C_{min}$
$C^*_{r,m} = C^*_r$	for equivalent balanced regenerator
C_λ	longitudinal conduction correction factor
D	regenerator disk outer diameter at the end of the heat transfer surface, m (ft)
D_h	hydraulic diameter, ($4r_h$), m (ft)
d	hub diameter, m (ft)
d^*	cell height/width
f	Fanning friction factor
G_L	parameter to account the effects of longitudinal heat conduction on thermal effectiveness as defined in Equation 6.65
hA	thermal conductance W/°C (Btu/h · °F)
$(hA)^*$	ratio of the convective conductance on the C_{min} side to C_{max} side = $(hA)_c/(hA)_h$
h_c	convective conductance for cold gas, W/m² °C (Btu/h ft² °F)
h_h	convective conductance for hot gas, W/m² · °C (Btu/h · ft² · °F)
$I_n(\cdot)$	modified Bessel function of the first kind and *n*th order
j	Colburn factor for heat transfer
i, j, k, n	integers
k_w	thermal conductivity of the matrix wall
L	matrix flow length, m (ft)
M	order of the trial solution, Equation 6.45
M_c	mass flow rate of cold gas, kg/s (lbm/h)
m_c	mass of the cold gas retained in the matrix flow length *dx*, kg (lbm)
$(Mc_p)_c$	capacity rate of cold gas, W/°C (Btu/h · °F)
$(Mc_p)_h$	capacity rate of hot gas, W/°C (Btu/h · °F)
M_h	mass flow rate of hot gas, kg/s (lbm/h)
m'_h	mass of the hot gas retained in the matrix flow length *dx*, kg (lbm)
M_r	matrix mass, kg (lbm)
$M_r c_r$	thermal capacitance of matrix, W/°C (Btu/h · °F)
N	rpm of the rotary matrix
N_c	cell density
Nu	Nusselt number, hD_h/k
NTU	number of transfer units, hA/C
NTU_c	number of transfer units on the cold side (≈twice NTU_o)
NTU_h	number of transfer units on the hot side (≈twice NTU_o)

NTU_o	modified number of transfer units
$NTU_{o,m}$	NTU_o for an equivalent balanced regenerator
Q	total heat duty of the exchanger, W s (Btu)
Q	heat transfer rate, W (Btu/h)
Re	Reynolds number
r_h	hydraulic radius, m (ft) = $D_h/4$
t_c	temperature of the cold gas, °C (°F)
t_h	temperature of the hot gas, °C (°F)
t_r	temperature of the matrix, °C (°F)
$t_{c,i}$, $t_{c,o}$	cold fluid terminal temperatures, °C (°F)
$t_{h,i}$, $t_{h,o}$	hot fluid terminal temperatures, °C (°F)
U_c	utilization factor on the cold side $(\Lambda/\Pi)_c$
U_h	utilization factor on the cold side $(\Lambda/\Pi)_h$
U_1	$(\Lambda/\Pi)_1$, the smaller of the two ratios U_C and U_h
x_{1m}, x_{2m}	unknown coefficients defined by Equation 6.45
x, y	variables as defined in Equation 6.2
V	$(A_{fr,h} + A_{fr,c})L$, m^3 (ft^3), volume of regenerator
$V_i(x, y)$	special functions defined by Equation 6.41
ß	area density, m^2/m^3(ft^2/ft^3)
δ	matrix stock thickness, m (ft)
ε	thermal effectiveness
ε_{cf}	counterflow regenerator thermal effectiveness
ε_r	thermal effectiveness of reference regenerator
$\Delta\varepsilon$	lowered effectiveness because of longitudinal conduction
$\varepsilon_{r,\lambda=0}$	equivalent balanced regenerator neglecting longitudinal heat conduction
$\varepsilon_{r,\lambda\neq 0}$	$C_\lambda \varepsilon_{r,\lambda=0}$
ξ	fractional distance along the flow path in the regenerator matrix of length L
η	fractional completion of a respective gas flow period
Λ	reduced length
Λ_m	mean reduced length
λ	longitudinal conduction parameter
λ_c	longitudinal conduction on the cold side
λ_h	longitudinal conduction on the hot side
Π	reduced period
Π_m	mean reduced period
ρ_r	matrix material density
φ	function defined by Equation 6.57
ψ	function defined by Equation 6.64
σ	porosity = A_o/A_{fr}
τ	flow period, s
τ_c	duration of cold flow period, *s*
τ_h	duration of hot flow period, *s*
τ_L	time for one revolution, $\tau_c + \tau_{h,} s$

Subscripts

c	cold
h	hot
i	in
m	mean
o	out
r	matrix

REFERENCES

1. Denniston, D.W. (1985) Waste recovery in the glass industry. In: *Industrial Heat Exchangers Conference Proceedings* (A.J. Hayes, W.W. Liang, S.L. Richlen, E.S. Tabb, eds.), pp. 13–20. American Society for Metals, Metals Park, OH.
2. Mondt, J.R. (1980) *Regenerative heat exchangers: The elements of their design, selection and use, Research Publication GMR-3396*. General Motors Research Laboratories, Warren, MI.
3. Shah, R.K., Sekulic, D.P. (1998) Heat exchangers. In: *Handbook of Heat Transfer Applications* (Rohsenow, W.M., Hartnett, J.P., Ganic, E.N., eds.), 3rd edn., pp. 17.1–17.169, Chapter 17. McGraw-Hill, New York.
4. Liang, W.W., Tabb, E.S. (1985) CRI's advanced heat transfer systems program. In: *Industrial Heat Exchangers Conference Proceedings* (A.J. Hayes, W.W. Liang, S.L. Richlen, E.S. Tabb, eds.), pp. 29–36. American Society for Metals, Metals Park, OH.
5. Coombs, M.G., Strumpf, H.J., Kotchick, D.M. (1985) A ceramic finned plate recuperator for industrial applications. In: *Industrial Heat Exchangers Conference Proceedings* (A.J. Hayes, W.W. Liang, S.L. Richlen, E.S. Tabb, eds.), pp. 63–68. American Society for Metals, Metals Park, OH.
6. Newby, J.N. (1985) The Regen regenerative burner—Principles, properties and practice. In: *Industrial Heat Exchangers Conference Proceedings* (A.J. Hayes, W.W. Liang, S.L. Richlen, E.S. Tabb, eds.), pp. 143–152. American Society for Metals, Metals Park, OH.
7. Walker, G. (1983) *Cryocoolers, Part 2: Applications*, p. 44. Plenum Press, New York.
8. Bugyis, E.J., Taylor, H.L. (1985) Heat exchange in the steel industry. In: *Industrial Heat Exchangers Conference Proceedings* (A.J. Hayes, W.W. Liang, S.L. Richlen, E.S. Tabb, eds.), pp. 3–12. American Society for Metals, Metals Park, OH.
9. Hoffman, L.C., Williams, H.W. (1985) Fluid bed waste recovery performance in a hostile environment. In: *Industrial Heat Exchangers Conference Proceedings* (A.J. Hayes, W.W. Liang, S.L. Richlen, E.S. Tabb, eds.), pp. 87–94. American Society for Metals, Metals Park, OH.
10. Hayes, A.J., Richlen, S.L. (1985) The Department of Energy's advanced heat exchangers program. In: *Industrial Heat Exchangers Conference Proceedings* (A.J. Hayes, W.W. Liang, S.L. Richlen, E.S. Tabb, eds.), pp. 21–28. American Society for Metals, Metals Park, OH.
11. McDonald, C.F. (1980) The role of the ceramic heat exchanger in energy and resource conservation. *Transactions of ASME, Journal of Engineering for Gas Turbines and Power* 102: 303–315.
12. Kleiner, R.N., Coubrough, L.E. (1985) Advanced high performance ceramic heat exchanger designs for industrial heat recovery applications. In: *Industrial Heat Exchangers Conference Proceedings* (A.J. Hayes, W.W. Liang, S.L. Richlen, E.S. Tabb, eds.), pp. 51–56. American Society for Metals, Metals Park, OH.
13. McNallan, M.J., Liang, W.W., Rothman, M.F. (1985) An investigation of the hot-corrosion of metallic and ceramic heat exchanger materials in chlorine contaminated environments. In: *Industrial Heat Exchangers Conference Proceedings* (A.J. Hayes, W.W. Liang, S.L. Richlen, E. S. Tabb, eds.), pp. 293–298. American Society for Metals, Metals Park, OH.
14. Shah, R.K. (1988) Counterflow rotary regenerator thermal design procedures. In: *Heat Transfer Equipment Design* (R.K. Shah, E.C. Subbarao, R.M. Mashelikar, eds.), pp. 267–296. Hemisphere, Washington, DC.
15. Kays, W.M., London, A.L. (1984) *Compact Heat Exchangers*, 3rd edn. McGraw-Hill, New York.
16. London, A.L., Young, M.B.O., Stang, J.H. (1970) Glass-ceramic surfaces, straight triangular passages—Heat transfer and flow friction characteristics. *Transactions of ASME, Journal of Engineering for Gas Turbines and Power* 92A: 381–389.
17. Baclic, B.S. (1985) The application of the Galerkin method to the solution of the symmetric and balanced counterflow regenerator problem, *Transactions of ASME, Journal of Heat Transfer* 107: 214–221.
18. Nusselt, W. (1927) Die Theorie des Winderhitzers [Theory of the air heater]. *Z. Ver. Dt. lng.*, 71: 85–91.
19. Hausen, H. (1929) Uber die Theorie des Warmeaustausches in Regeneratoren. *Z. Angew. Math. Mech.* 9: 173–200. Also, in VervollstandigteBerechnung des Warmeaustausches in Regeneratoren, *Z. Ver. Dtsch. lng.* 2 (1942).
20. Coppage, J.E., London, A.L. (1953) The periodic-flow regenerator—A summary of design theory. *Transactions of ASME* 75: 779–787.
21. Lambertson, T.J. (1958) Performance factors of a periodic flow heat exchanger. *Transactions of ASME* 80: 586–592.

22. Bahnke, G.D., Howard, C.P. (1964) The effect of longitudinal heat conduction on periodic flow heat exchanger performance. *Transactions of ASME, Journal of Engineering for Gas Turbines and Power 86*, 105–120 (1964).
23. Mondt, J. R., Vehicular gas turbine periodic flow heat exchanger solid and fluid temperature distributions, *Transactions of ASME, Journal of Engineering for Gas Turbines and Power 86*, 121–126 (1964).
24. Li, C.-H., A numerical finite difference method for performance evaluation of a periodic flow heat exchanger, *Transactions of ASME, Journal of Engineering for Gas Turbines and Power 105*, 611–617 (1983).
25. Nahavandi, A.H., Weinstein, A.S. (1961) A solution to the periodic flow regenerative heat exchanger problem. *Applied Scientific Research* 10: 335–348.
26. Willmott, A.J., Duggan, R.C. (1980) Refined closed methods for the contra-flow thermal regenerator problem. *International Journal of Heat and Mass Transfer* 23: 655–662.
27. Baclic, B.S., Dragutinovic, G.D. (1991) Asymmetric-unbalanced counterflow thermal regenerator problem: Solution by the Galerkin method and meaning of dimensionless parameters. *International Journal of Heat and Mass Transfer* 34: 483–498.
28. Romie, F.E., Baclic, B.S. (1988) Methods for rapid calculation of the operation of asymmetric counterflow regenerator. *Transactions of ASME, Journal of Heat Transfer* 110: 785–788.
29. Shah, R.K. (1981) Thermal design theory for regenerators. In: *Heat Exchangers: Thermal-Hydraulic fundamentals and Design* (S. Kakac, A.E. Bergles, F. Mayinger, eds.), pp. 721–763. Hemisphere/McGraw-Hill, New York.
30. Heggs, P.J. (1986) Calculation of high temperature regenerative heat exchangers. In: *High Temperature Equipment* (A.E. Sheindlin, ed.), pp. 115–149. Hemisphere, Washington, DC.
31. Romie, F.E. (1990) A table of regenerator effectiveness. *Transactions of ASME, Journal of Heat Transfer* 112: 497–499.

32a. Hill, A., Willmott, A.J. (1987) A robust method for regenerative heat exchanger calculations. *International Journal of Heat and Mass Transfer* 30: 241–249.

32b. Hill, A., Willmott, A.J. (1989) Accurate and rapid thermal regenerator calculations. *International Journal of Heat and Mass Transfer* 32: 465–476.

33. Razelos, P. (1979) An analytical solution to the electric analog simulation of the regenerative heat exchanger with time varying fluid inlet temperatures. *WarmeStoffubertrag* 12: 59–71.
34. Romie, F.E. (1991) Treatment of transverse and longitudinal heat conduction in regenerators. *Transactions of ASME, Journal of Heat Transfer* 113: 247–249.
35. Hahnemann, H.N. (1948) *Approximate calculations of thermal ratios in heat exchangers including conduction in the direction of flow, National Gas Turbine Establishment Memorandum No. M36, TPA3/TIB*. Ministry of Supply, Millbank, London, UK.
36. Shah, R.K. (1975) A correlation for longitudinal heat conduction effects in periodic-flow heat exchangers. *Transactions of ASME, Journal of Engineering for Gas Turbines and Power* 97A: 453–454.
37. Shah, R.K. (1985) Compact heat exchangers. In: *Handbook of Heat Transfer Applications*, 2nd edn. (W.M. Rohsenow, J.P. Hartnett, E. N. Ganic, eds.), pp. 4-174–4-311. McGraw-Hill, New York.
38. London, A.L., Ferguson, C.K. (1949) Test results of high performance heat exchanger surfaces used in aircraft intercoolers and their significance for gas turbine regenerator design. *Transactions of ASME* 71: 17–26.
39. Bowden, A.T., Hryniszak, W. (1953) The rotary regenerative air preheater for gas turbines. *Transactions of ASME* 769–777.
40. Shepherd, D.G. (1960) *Introduction to the Gas Turbines*, 2nd edn. Constable and Company, London, UK.
41. Bathe, W.W. (1984) *Fundamentals of Gas Turbines*. John Wiley & Sons, New York.
42. Harper, D.B. (1957) Seal leakage in the rotary regenerator and its effect on rotary regenerator design for gas turbines. *Transactions of ASME* 79: 233–245.
43. Vincent, E.J. (1956) *The Theory and Design of Gas Turbines and Jet Engines*. McGraw-Hill, New York.
44. www.powermag.com/air-preheater-seal-upgrades-renew-plant-efficiency/
45. Marner, W.J., Suitor, J.W. *A Survey of Gas-Side Fouling in Industrial Heat-Transfer Equipment*. Final Report. Prepared for US Department of Energy Pacific Northwest Laboratory Battelle Memorial Institute Through an Agreement with National Aeronautics and Space Administration by Jet Propulsion Laboratory California Institute of Technology Pasadena, California.

46. Her Majesty the Queen in Right of Canada (2002) *Energy Efficiency Planning and Management Guide (Aussidisponible en français sous le titre): PEEIC Guide*. Cat. No. M92–239/2001E. Her Majesty the Queen in Right of Canada, pp. 1–185.
47. BCS Incorporated (2008) *Waste Heat Recovery: Technology and Opportunities in U.S. Industry*, March 2008, pp. 1–71. Prepared by BCS Incorporated.
48. US Department of Energy (2014) *Industrial Waste Heat Recovery: Potential Applications, Available Technologies and Crosscutting R&D Opportunities*. Crosscutting R&D Opportunities, ORNL/TM-2014/622, pp. 1–72, January 2014, for the US Department of Energy, Oak Ridge National Laboratory, Oak Ridge, Tennessee.
49. US Department of Energy (2015) *Chapter 6: Innovating Clean Energy Technologies in Advanced Manufacturing, Quadrennial Technology Review 2015*, pp. 1–39. Waste Heat Recovery Systems, US Department of Energy.
50. UNEP. *Thermal Energy Equipment: Waste Heat Recovery Energy Efficiency Guide for Industry in Asia* –www.energyefficiencyasia.org, pp. 1–18. UNEP.
51. Hayes, A.J., Liang, W.W., Richlen, S.L., Tabb, E.S. (Eds.) (1985) *Industrial Heat Exchangers Conference Proceedings*. American Society of Metals, Metals Park, OH.
52. Schadt, H.F. (1985) High-efficiency fluid-bed heat exchanger recovers heat from combustion gases. In: *Industrial Heat Exchangers Conference Proceedings* (A.J. Hayes, W.W. Liang, S.L. Richlen, E.S. Tabb, eds.), pp. 71–80. American Society for Metals, Metals Park, OH.
53. Williams, H.W. (1985) Fluid bed waste heat recovery operating experience on aluminum melting furnace. In: *Industrial Heat Exchangers Conference Proceedings* (A.J. Hayes, W.W. Liang, S.L. Richlen, E.S. Tabb, eds.), pp. 241–248. American Society for Metals, Metals Park, OH.
54. Rodgers, R.J. (1985) Vortex flow direct contact heat exchanger conceptual design and performance. In: *Industrial Heat Exchangers Conference Proceedings* (A.J. Hayes, W.W. Liang, S.L. Richlen, E. S. Tabb, eds.), pp. 101–108. American Society for Metals, Metals Park, OH.
55. Godbole, S.S., Gilbert, M.L., Snyder, J.E. (1985) Design verification of the DOE/B&W HTBDR control system using interactive simulation. In: *Industrial Heat Exchangers Conference Proceedings* (A.J. Hayes, W.W. Liang, S.L. Richlen, E.S. Tabb, eds.), pp. 45–50. American Society for Metals, Metals Park, OH.
56. Luu, M., Grant, K.W. (1985) Thermal and fluid design of a ceramic bayonet tube heat exchanger for high temperature waste heat recovery. In: *Industrial Heat Exchangers Conference Proceedings* (A.J. Hayes, W.W. Liang, S.L. Richlen, E.S. Tabb, eds.), pp. 159–174. American Society for Metals, Metals Park, OH.
57. Hackler, C.L., Dinulescu, M. (1985) Porcelain enamelled flat plate heat exchangers—Engineering and application. In: *Industrial Heat Exchangers Conference Proceedings* (A.J. Hayes, W.W. Liang, S.L. Richlen, E.S. Tabb, eds.), pp. 381–384. American Society for Metals, Metals Park, OH.
58. Seehausen, J.W. (1985) Thermal-hydraulic performance of radiation recuperators. In: *Industrial Heat Exchangers Conference Proceedings* (A.J. Hayes, W.W. Liang, S.L. Richlen, E.S. Tabb, eds.), pp. 175–180. American Society for Metals, Metals Park, OH.
59. Reay, D.A. (1980) Heat exchangers for waste heat recovery, International Research and Development Co., Ltd., Newcastle upon Tyne, U.K. In: *Developments in Heat Exchanger Technology*—1 (D. Chisholm, ed.), pp. 233–256. Applied Science Publishers, London, UK.
60. US EPA (2012) *Waste heat to power systems*, pp. 1–9. May 30, 2012. US EPA.
61. CHP Industrial Bottoming and Topping Cycle with Energy Information Administration Survey Data Paul Otis, August 14, 2015 Independent Statistics & Analysis www.eia.gov U.S. Energy Information Administration Washington, DC 20585. August 2015, pp. 1–18.

61.1 www.eia.gov/workingpapers/pdf/chp-Industrial_81415.pdf

62. U.S. Department of Energy. Energy Efficiency and Renewable Energy (2004) *Materials Selection Considerations for Thermal Process Equipment, A Best Practices Process Heating Technical Brief*, pp. 1–6. U.S. Department of Energy. Energy Efficiency and Renewable Energy, DOE/GO-102004-1974 November 2004.

62.1 www.energy.gov/sites/default/files/2014/05/f15/proc_heat_tech_brief.pdf

BIBLIOGRAPHY

Atthey, D.R., Chew, P.E. (1986) Calculation of high temperature regenerative heat exchangers. In: *High Temperature Equipment* (A.E. Sheindlin, ed.), pp. 73–113. Hemisphere, Washington, DC.

Chapter 2 – Steam Generators, https://staff.emu.edu.tr/devrimaydin/Documents/MENG446/Chapter/2

Kilkovský, B., Jegla, Z. (2016) Preliminary design and analysis of regenerative heat exchanger. *Chemical Engineering Transactions* 52: 655–660.

https://global.kawasaki.com/en/energy/solutions/energy_plants/waste_heat_boiler.html#:~:text=Waste%20Heat%20Recovery%20Boiler%20is,recovered%20heat%20into%20useful%20and

Hausen, H. (1983) *Heat Transfer in Counterflow, Parallel Plow and Cross Flow*, 2nd edn. McGraw-Hill, New York.

Iliffe, C.E. (1948) Thermal analysis of the contra-flow regenerative heat exchanger. *Proceedings of the Institution of Mechanical Engineers* 159: 363–371.

Jakob, M. (1957) *Heat Transfer*, Vol. II. John Wiley & Sons, New York.

Johnson, J.E. (1948) *Regenerator heat exchangers for gas turbines*. R. A. E. Report Aero. 2266 S.D. 27: 1–72.

Kroger, P.G. (1967) Performance deterioration in high effectiveness heat exchangers due to axial heat conduction effects. *Advances in Cryogenic Engineering* 12: 363–372.

London, A.L., Kays, K.M. (1951) The liquid-coupled indirect-transfer regenerator for gas-turbine plants. *Transactions of ASME* 73 : 529–542.

Romie, F.E. (1979) Periodic thermal storage: The regenerator. *Transactions of ASME, Journal of Heat Transfer* 101: 726–731.

Romie, F.E. (1988) Transient response of rotary regenerators. *Transactions of ASME, Journal of Heat Transfer* 110: 836–840.

Schmidt, F.W., Willmott, A.J. (1981) *Thermal Energy Storage and Regeneration*. McGraw-Hill, New York.

Shah, R.K. (1981) Compact heat exchanger design procedures. In: *Heat Exchangers: Thermal-Hydraulic Fundamentals and Design* (S. Kakac, A.E. Bergles, F. Mayinger, eds.), pp. 495–536. Hemisphere, Washington, DC.

Shah, R.K. (1992) Compact heat exchanger technology and applications, Key Note Lecture. *Eleventh Heat and Mass Transfer Conference*, pp. 1–25. Indian Society for Heat and Mass Transfer, Madras, India.

Willmott, A.J. (1969) The regenerative heat exchanger computer representation. *International Journal of Heat and Mass Transfer* 12: 997–1014.

www.aidic.it/cet/16/52/110.pdf

www.opexworks.com/KBase/Energy_Management/Termal_Energy_Management/Waste_Heat_Recovery/Commercial_Waste_Heat_Recovery_Devices.h

7 Plate Heat Exchangers and Spiral Plate Heat Exchangers

Dr. Richard Seligman, the founder of APV International, introduced the first commercially successful gasketed plate heat exchanger design in 1923. A plate heat exchanger (PHE) essentially consists of a number of corrugated metal plates provided with gaskets and corner ports to achieve the desired flow arrangement. Each fluid passes through alternate channels. The plates are clamped together in a frame that includes connections for the fluids. Since each plate is generally provided with peripheral gaskets to provide sealing arrangements, the plate heat exchangers are called gasketed plate heat exchangers. Extensive information on the plate heat exchanger is provided by Focke [1], Cooper et al. [2], Raju et al. [3,4], and Shah et al. [5]. Leading PHE manufacturers include Alpha Laval, Pune, India and Sweden, APV Co., Tonawanda, NY, Paul-Muller Company, Springfield, MO, GEA PHE Systems, Sarstedt, Germany, Tranter, Inc., Wichita Falls, TX, APV/SPX, HRS Heat Exchangers Ltd, Herts, United Kingdom, ITT STANDARD, Cheektowaga, NY, Polaris Plate Heat Exchangers, Tinton, Falls, NJ, Kelvion Holding GmbH, Exchanger Industries Limited, SWEP, SONDEX® etc., SEC Heat Exchangers, Belfast PEI, Canada, etc., among others.

7.1 PLATE HEAT EXCHANGER CONSTRUCTION—GENERAL

A plate heat exchanger (PHE) is usually comprised of a stack of corrugated or embossed metal plates in mutual contact, each plate having four apertures serving as inlet and outlet ports, and seals designed so as to direct the fluids in alternate flow passages. The heat transfer between the two fluids takes place through the plates. The plate pack is assembled between a frame plate and a pressure plate and compressed by tightening bolts. The plates are fitted with a gasket which seals the channel and directs the fluids into alternate channels. The plate corrugation promotes fluid turbulence and supports the plates against differential pressure.

A PHE is shown in Figure 7.1. In a PHE, the stack of plates is held together in a frame by a pressure arrangement as shown in Figure 7.2, and its components and assembly details are shown in Figure 7.3 (a PHE assembly is also shown in Figure 1.7 in Chapter 1). The flow passages are formed by adjacent plates so that the two streams exchange heat while passing through alternate channels. When assembled, the spacing between adjacent plates ranges from 1.3 to 6.4 mm. The number and size of the plates are determined by the flow rate, physical properties of the fluids, pressure drop, and temperature program. The plate corrugations promote fluid turbulence and support the plates against differential pressure. The standard PHE design is "single pass" with all four connections on one frame plate, as shown in Figure 7.4

The periphery of each plate is grooved to house a molded gasket, each open to the atmosphere in the event of leakage, as shown in Figure 7.5. Gaskets are generally cemented in, but snap-on gaskets are available that do not require cement. Gasket failure cannot result in fluid intermixing but merely

DOI: 10.1201/9781003352044-7

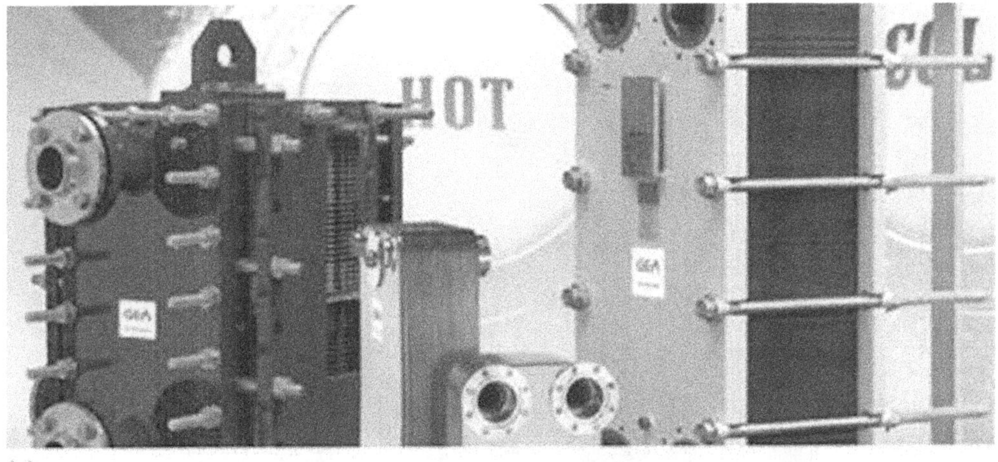

(a)

(b)

FIGURE 7.1 PHE. (a) Collection of different types of PHE. (b) NT series plate heat exchanger. (Courtesy of GEA PHE Systems, Sarstedt, Germany).

in leakage to the atmosphere. Proper selection of gasket material and operating conditions will eliminate the leakage risk. Frame assembly tightly holds the plates pack. It ensures optimum compression and leak tightness. If the plates are correctly assembled, the edges form a "honeycomb" pattern, as shown in Figure 7.6 and a wrong plate assembly is shown in Figure 7.7.

The elements of the frame are fixed plate, compression plate, pressing equipment, and connecting ports. As shown in Figure 7.5, every plate is notched at the center of its top and bottom edges so

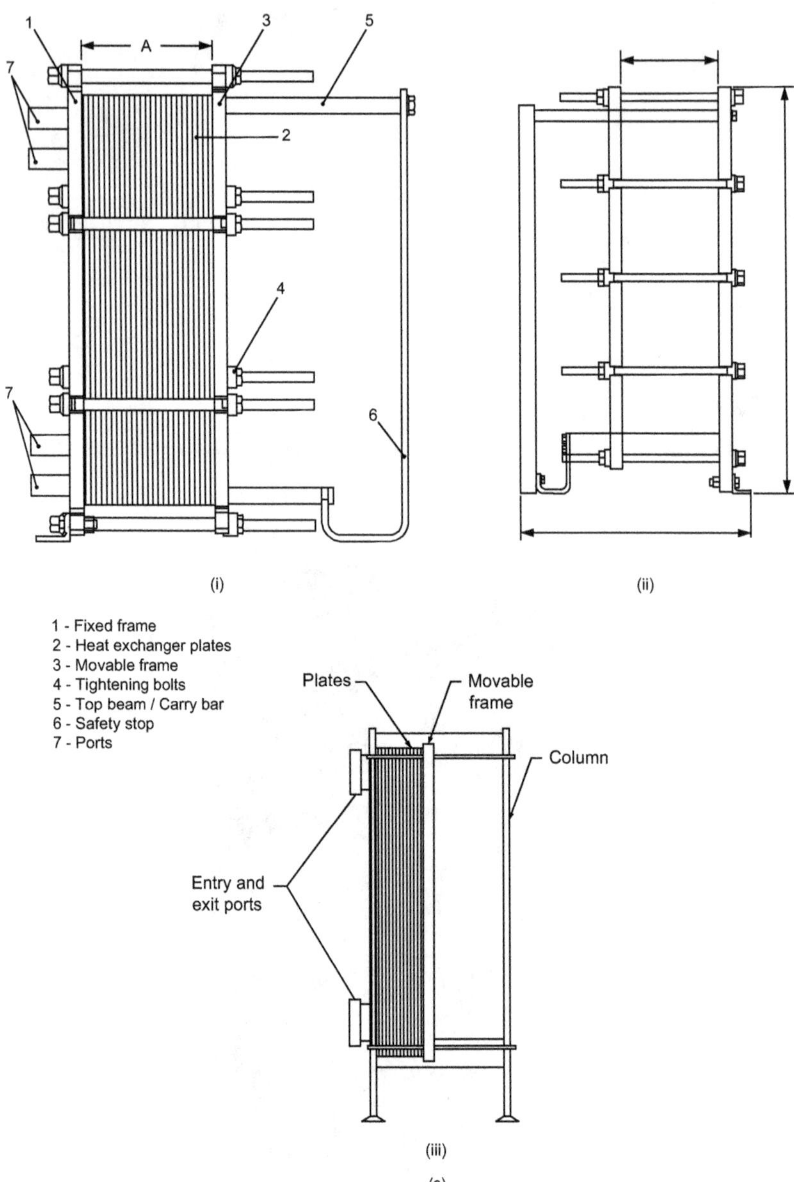

FIGURE 7.2 Plate heat exchanger assembly.

that it may be suspended from the top carrier bar and guided by the bottom guide bar, and the plates are free to slide along the bars. The movable head plate is similarly notched and free to slide along both bars. Both the upper carrying bar and the lower guiding bar are fixed to the support columns as shown in Figure 7.3. The plate pack is tightened by means of either a mechanical or hydraulic tightening device. Connections are located in the frame cover or, if either or both fluids make more than a single pass within the unit, the frame and pressure plates. By including intermediate separating/connecting plates, three or more separate fluid streams can be handled. Frames are usually free standing; for smaller units, they are attached to structural steel work. Salient constructional features of PHE and the resulting advantages and benefits are given in Table 7.1.

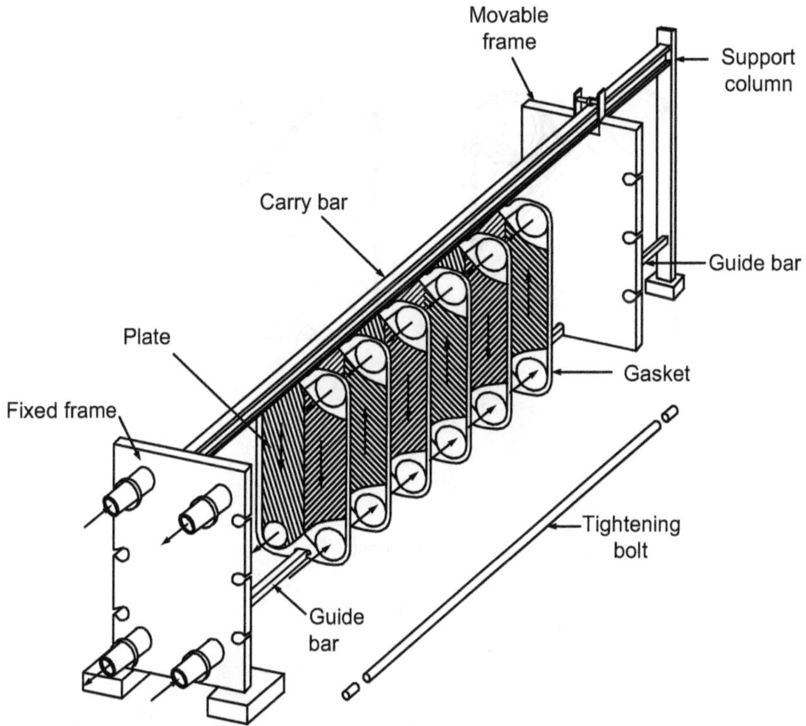

FIGURE 7.3 Plate heat exchanger assembly details.

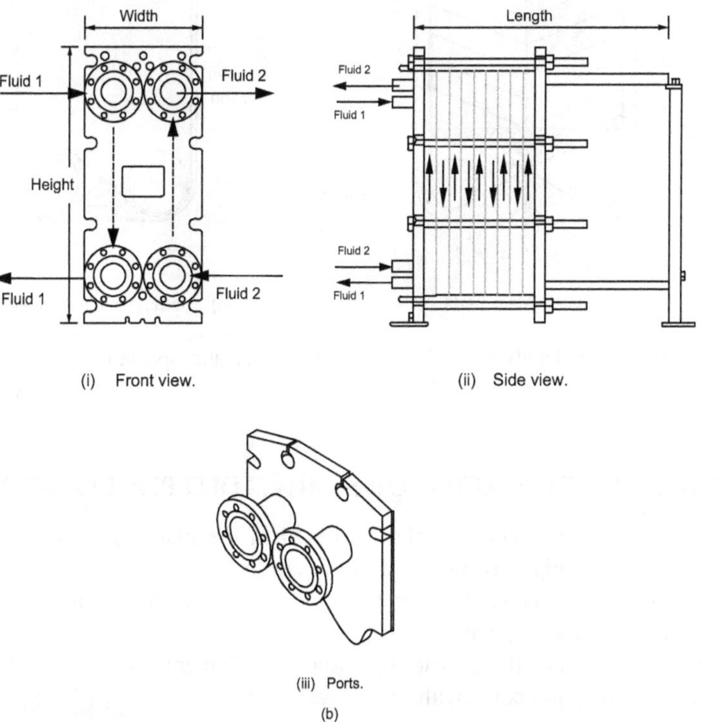

FIGURE 7.4 Standard single pass PHE design with all four ports connections on one frame plate.

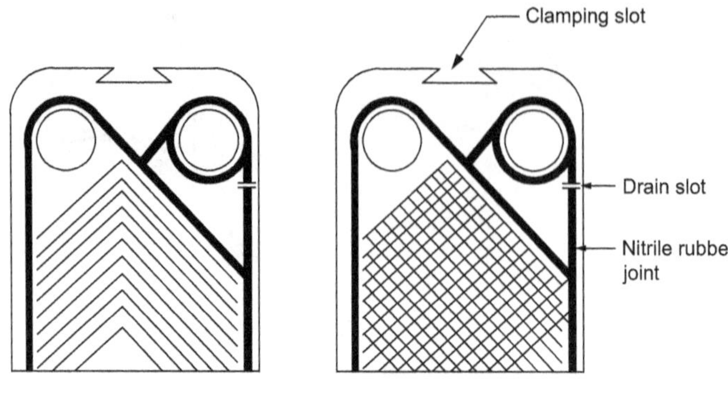

(a) Plate corrugations.

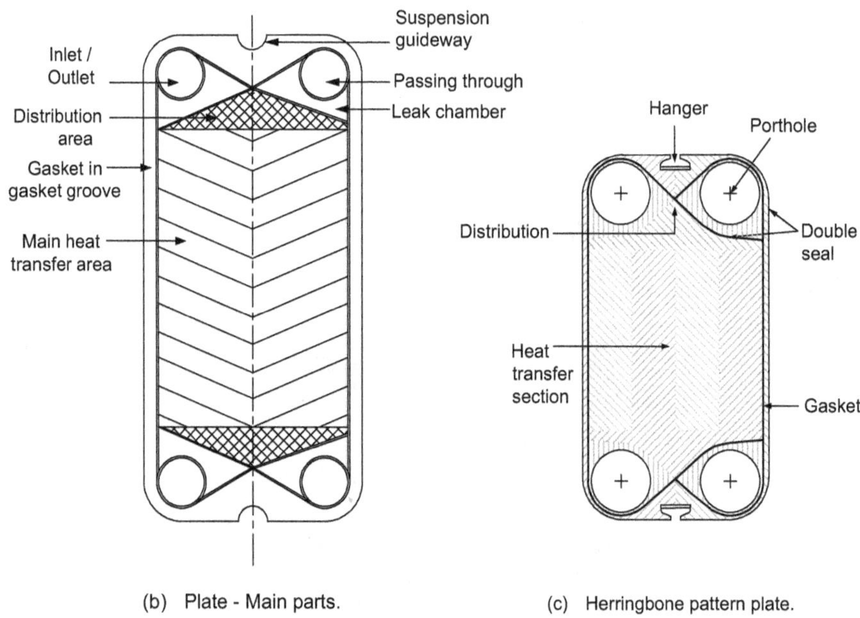

(b) Plate - Main parts.

(c) Herringbone pattern plate.

FIGURE 7.5 The PHE plate details and gasket design open to the atmosphere in the event of fluids leakage.

7.2 DEFINITIONS OF GASKETED OR SEMIWELDED PHE COMPONENTS

Connection plate: an intermediary "endplate" located in the plate pack that permits additional nozzles, additional fluids, and redirection of flow patterns.

Divider plate: a plate that changes the direction of the flow of the fluid in a two-pass or larger heat exchanger. Also called a turning plate.

Fixed endplate: a fixed plate that provides pressure containment and locations for the nozzles or connections; it may or may not come with feet.

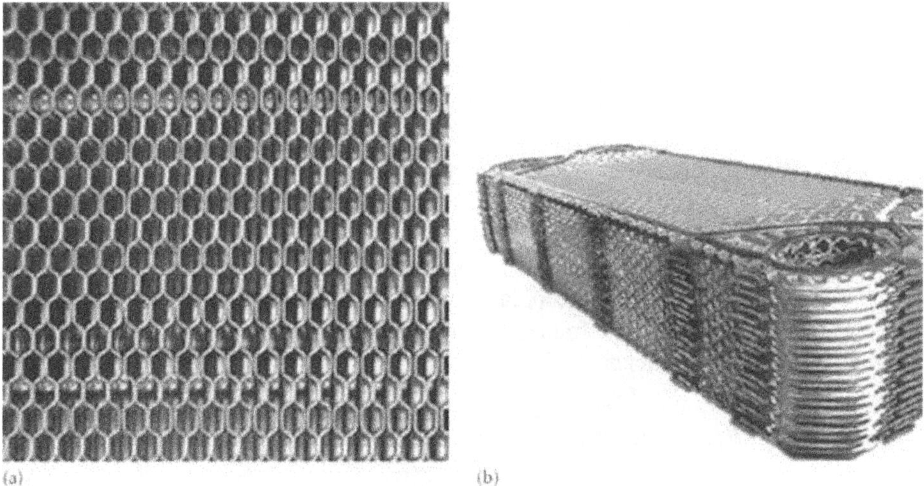

FIGURE 7.6 PHE plate pack shows a continuous honeycomb pattern. (Courtesy of HRS Heat Exchangers Ltd, Herts, U.K.)

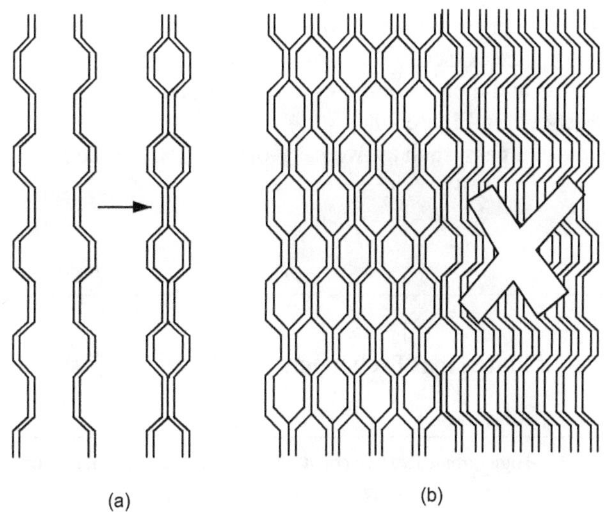

FIGURE 7.7 PHE plate pack (a) shows a correct plate pack and (b) wrong plates assembly.

Frame: A structural support and pressure-containment components. The components may consist of a fixed endplate, a movable endplate, upper carrying and lower guide bars, a support column, and frame compression bolts.

Frame compression bolt: a bolt assembly used to compress the fixed endplate, movable endplate, and heat transfer plates to affect a pressure seal.

Gasket: a sealing element between two plates or semiwelded plate pairs.

Heat transfer plate: a thin corrugated plate that makes up the plate pack and is in contact with the process fluids.

Movable endplate: a movable plate that provides pressure containment and locations for the nozzles or connections.

Plate pack: a collection of all gasketed or semiwelded heat transfer plates in the frame.

Semiwelded plate pair: two adjacent heat transfer plates welded together. The weld replaces the gasket between the two adjacent plates. A gasket is required between each plate pair.

Support column: the structural component that supports the upper carrying and lower guide bars of the frame.

Lower guide bar: a structural component that aligns the heat transfer plates and movable endplate.

Upper carrying bar: a structural component that supports the heat transfer plates, movable endplate, and internal fluids.

Name plate: the type of unit, manufacturing number, and manufacturing year can be found on the name plate. Pressure vessel details in accordance with the applicable pressure vessel code are also given. The name plate is fixed to the frame plate, most commonly, or the pressure plate. The name plate can be a steel plate or a sticker label.

7.3 FLOW PATTERNS AND PASS ARRANGEMENT

In a PHE, several types of flow patterns can be achieved:

1. Series flow arrangement, in which a stream is continuous and changes direction after each vertical path, that is, *n* pass–*n* pass with individual passes per channel, as shown in Figure 7.8a.
2. Single-pass looped arrangement: U-arrangement, Z-arrangement. Both fluids flow countercurrently through parallel passages that make up a single pass, as shown in Figure 7.8b.
3. Multipass with equal passes (series flow pattern), wherein the stream divides into a number of parallelflow channels and then recombines to flow through the exit in a single stream, that is, *n* pass–*n* pass, as shown in Figure 7.8c.
4. Multipass with unequal pass, such as two pass–one pass, and three pass–one pass, as shown in Figure 7.8d.

TABLE 7.1
Construction Features of Plate Heat Exchangers and Their Benefits

Feature	Advantages
PHE concept	High turbulence, true counterflow, and hence efficient heat transfer and very close approach temperature; educed fouling. Low weight and smaller foundations
Modular construction	Flexibility for future expansion and easy to modify for altered duties
Fixed frame and movable pressure plates	Easy accessibility for maintenance
Metallic contact between plates	Rigidity of plate pack is enhanced and hence minimized vibrations and less noise
Thin channels	Quick process control and low hold-up volume. Light weight
Each plate is individually gasketed	Cross-contamination is eliminated. Easy to inspect
Glue-free gaskets	Simplified regasketing while still in frame, maintenance friendly feature, reduced gasket changing time

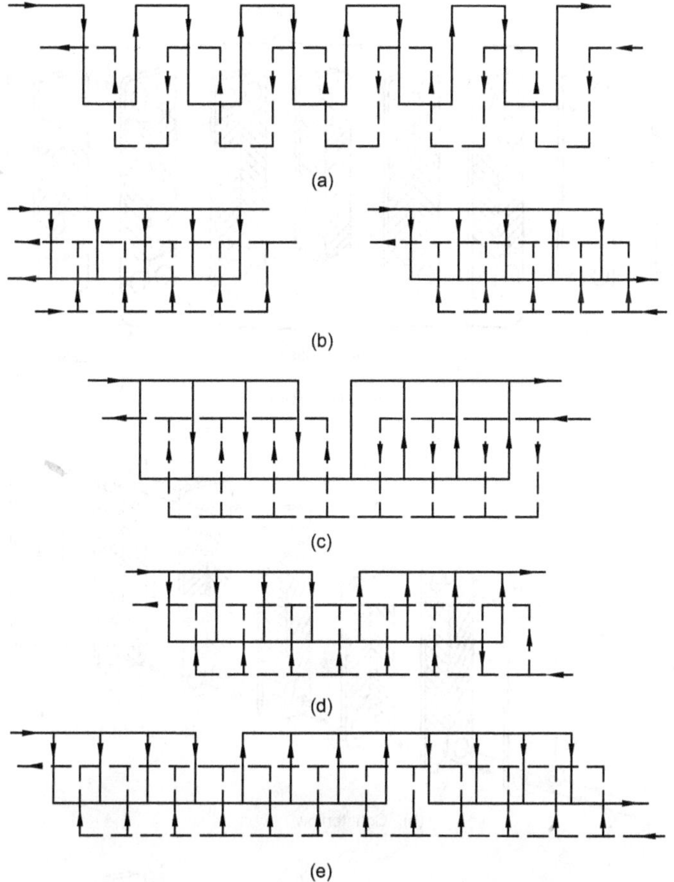

FIGURE 7.8 PHE flow arrangements. (a) Series flow *n* pass–*n* pass; (b) single-pass looped: U-arrangement, Z-arrangement; (c) multipass with equal pass (*n* pass–*n* pass); (d) multipass with unequal pass, 2 pass–1 pass; and (e) multipass with unequal pass, 3 pass–1 pass.

In an assembled unit, the flow patterns, i.e., parallel flow and counter flow, are shown in Figure 7.9, and Figure 7.10 shows flow over a plate surfaces.

7.4 USEFUL DATA ON PHE

Useful data on plate exchangers are provided in Table 7.2.

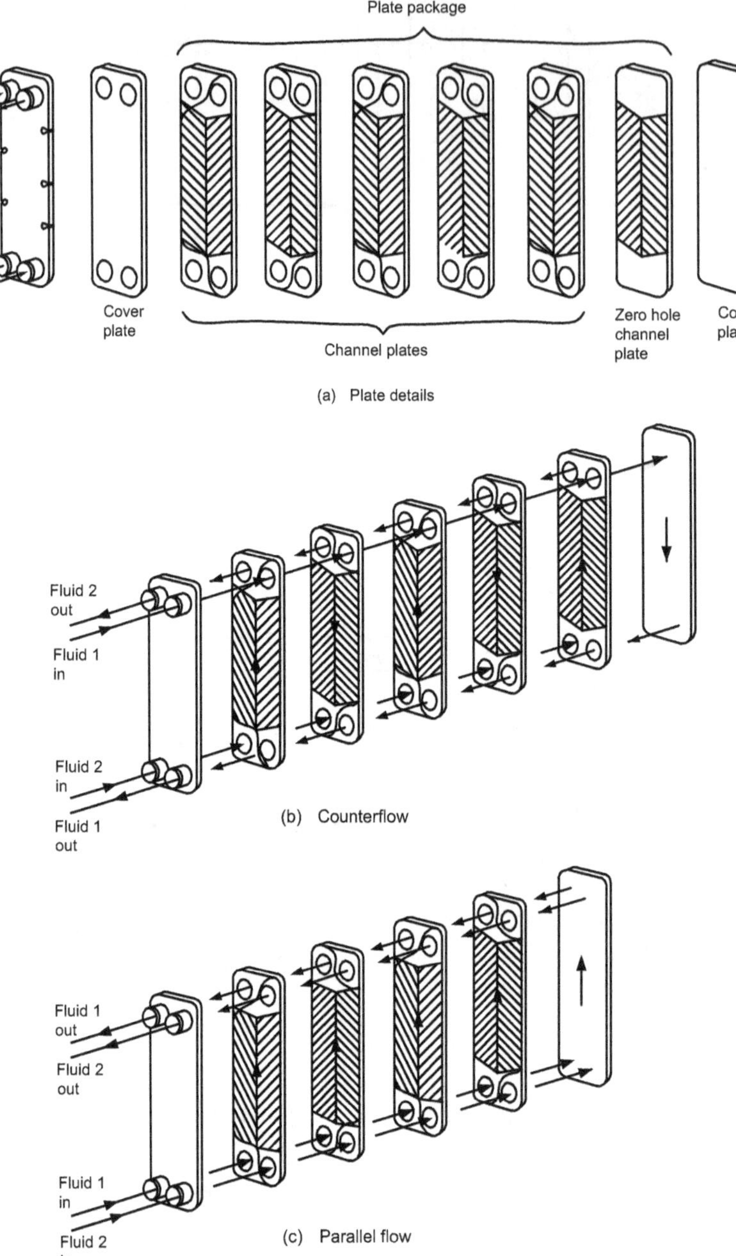

FIGURE 7.9 PHE flow pattern –parallel flow.

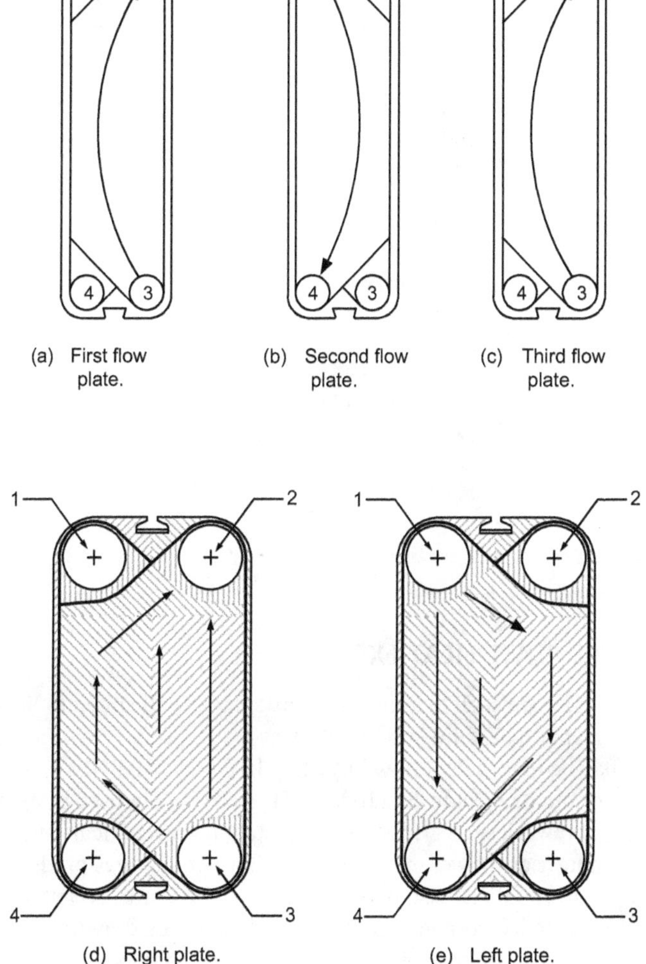

FIGURE 7.10 Flow over plate surface.

TABLE 7.2
Useful Data on PHE

Unit		Plates	
Size	1540–2500 m^2	Thickness	0.5–1.2 mm
Number of plates	Up to 700	Size	0.03–2.2 m
Port size	Up to 39 cm	Spacing	1.5–5.0 mm
		Corrugation depth	3–5 mm

7.5 STANDARD PERFORMANCE LIMITS

Standard performance limits for Alpha-Laval 1 PHE are [6]:

Maximum operating pressure	25 bar (360 psi)
With special construction	30 bar (435 psi)
Maximum temperature	160°C (320°F)
With special gaskets	200°C (390°F)
Maximum flow rate	3,600 m^3/h (950,000 USG/min)
Heat transfer coefficient	3500–7500 W/m^2·°C (600–1,300) BTU/ft^2·h·°F
Heat transfer area	0.1–2,200 m^2 (1–24,000 ft^2)
Maximum connection size	450 mm (18 in.)
Thermohydraulic data	
Temperature approach	As low as 1°C
Heat recovery	As high as 93%
Heat transfer coefficient	3,000–7,000 W/m^2·°C (water–water duties with normal fouling resistance)
NTU	0.3–4.0
Pressure drop	30 kPa per NTU

Note: Refer to manufacturer's catalogue for the latest performance lists.

7.6 WHERE TO USE PLATE HEAT EXCHANGERS

The characteristics of PHEs are such that they are particularly well suited to liquid–liquid duties in turbulent flow [7]. Incidentally, a fluid sufficiently viscous to produce laminar flow in a shell and tube heat exchanger may be in turbulent flow in a PHE. In the 1930s, PHEs were introduced to meet the hygienic demands of the dairy industry. Today, PHEs are widely used in many fields, including heating and ventilating, breweries, dairy, food processing, pharmaceuticals and fine chemicals, petroleum and chemical industries, power generation, offshore oil and gas production, onboard ships, pulp and paper production, etc. PHEs also find applications in water-to-water closed-circuit cooling-water systems using potentially corrosive primary cooling water drawn from sea, river, lake, or cooling tower to cool a clean, noncorrosive secondary liquid flowing in a closed circuit [7]. In this application, titanium (99.8%)–palladium (0.2%)-stabilized titanium is used because of its outstanding corrosion resistance.

7.6.1 Plate Evaporator

Compact and economically efficient, the plate evaporator/condenser replaces conventional large and expensive falling film units. Its deep channels, large ports, and laser welding allow vacuum and low-pressure evaporation and condensing for both aqueous and organic systems. Plate evaporators are used for condensing in the chemical, sugar, fats, and oil industries.

7.6.2 Plate Condenser

A compact design plate condenser replaces traditional large units. Customized connections for large volumes of vapor, specific plate pattern, and an asymmetric plate gap to optimize heat transfer and minimize pressure drop make it suitable for condensation. A PHE condenser flow over a plate surface is shown in Figure 7.11

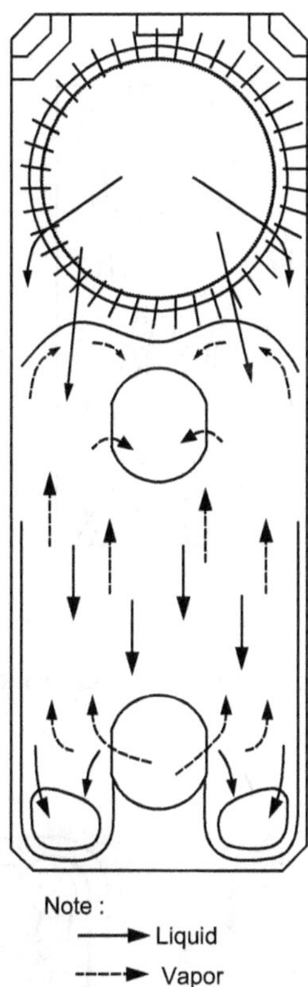

FIGURE 7.11 Flow over PHE condenser plate surface.

7.7 APPLICATIONS FOR WHICH PLATE HEAT EXCHANGERS ARE NOT RECOMMENDED

PHEs are not recommended for the following services [3]:

1. Gas-to-gas applications.
2. Fluids with very high viscosity may pose flow distribution problems, particularly when cooling is taking place; flow velocities less than 0.1 m/s are not used because they give low heat transfer coefficients and low heat exchanger efficiency.
3. Less suitable for vapors condensing under vacuum.

7.7.1 Benefits Offered by Plate Heat Exchangers

High turbulence and high heat transfer performance: The embossed plate pattern promotes high turbulence at low fluid velocities (Figure 7.12). On plates with the washboard pattern, turbulence is promoted by the continuously changing flow direction and velocity [8]. Plates with the herringbone pattern are assembled with the corrugations pointing in alternate directions, which imparts a swirling motion to the fluids. This high turbulence results in very high heat transfer coefficients.

Reduced fouling: High turbulence and its scrubbing action of the fluids, absence of stagnant areas, uniform fluid flow, and the smooth plate surface reduce fouling and the need for frequent cleaning.

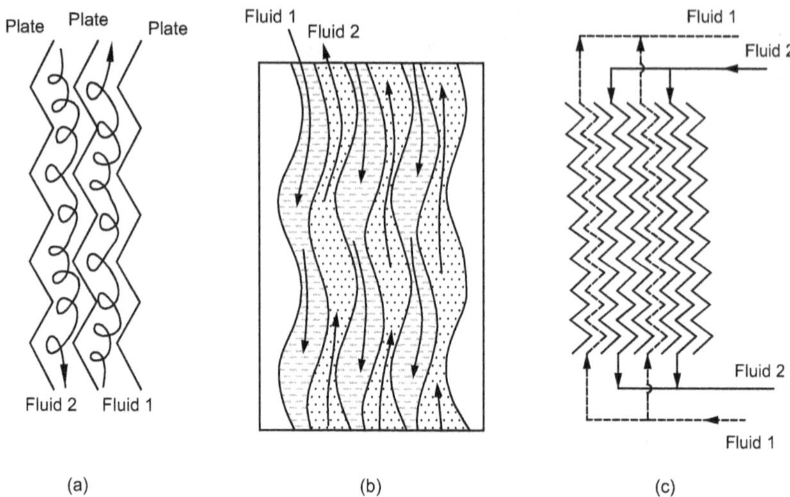

FIGURE 7.12 PHE—turbulence promoted by embossed plate pattern.

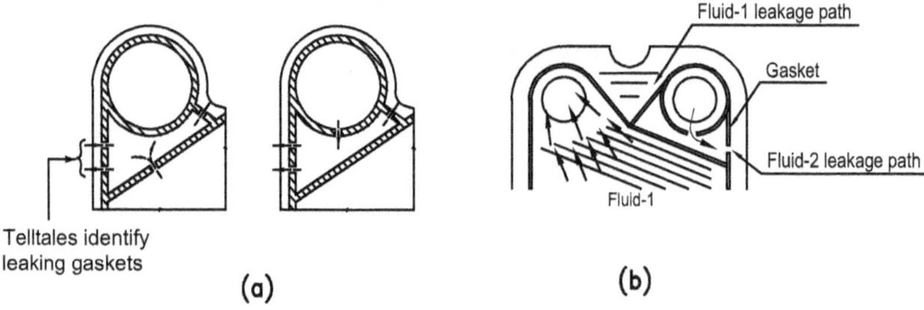

FIGURE 7.13 PHE gasket design and configuration eliminates cross-contamination of fluids by venting of gasket space.

Cross-contamination eliminated: In PHE, each medium is individually gasketed. The space between gaskets is vented to the atmosphere, eliminating the possibility of any cross-contamination of fluids (Figure 7.13).

True counterflow: In PHE, fluids can be made to flow in opposite directions, resulting in a greater and more effective temperature difference (Figure 7.14).

Close approach temperature: In PHE, very close approach temperatures of 1–2°F are possible because of the true counterflow, advantageous flow rate characteristics, and high heat transfer efficiency of the plates.

Multiple duties with a single unit: It is possible to heat or cool two or more fluids within the same unit by simply installing intermediate divider sections between the heat transfer plates (Figure 7.15).

Multi-pass design

The standard PHE design is "single pass," with all four connections on one frame plate. However, for certain applications it is preferable to have a "multi-pass" design, effectively creating two or more plate packs within one frame. This is achieved by adding special "turning" plates within the

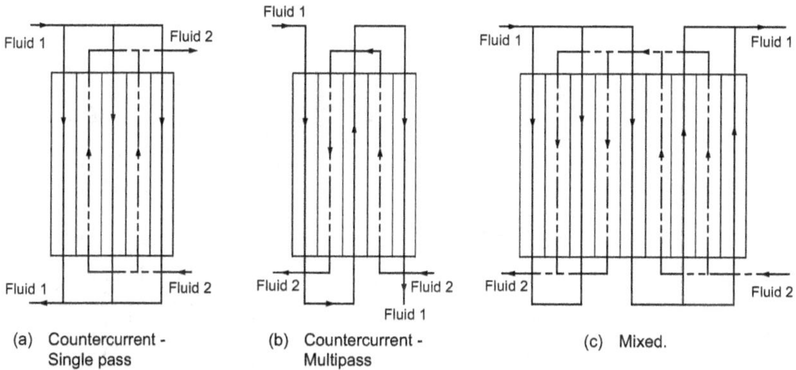

FIGURE 7.14 Two fluids flow in true counterflow.

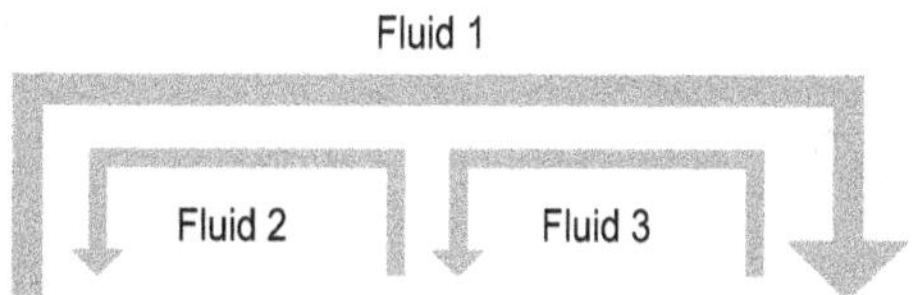

FIGURE 7.15 PHE—capability of handling of multiple duties/fluids in one unit.

pack to create more than one pass of the process fluids, i.e., multi-pass sections can be created by using turning plates with one, two, or three unholed ports. The main purpose is to change the flow direction of one or both fluids.

Expandable: Due to modular construction, true flexibility is unique to the PHE both in initial design and after installation (Figure 7.16).

Easy to inspect and clean, and less maintenance: A PHE can be easily opened for inspection, cleaning, and gasket replacement. By simply removing the compression bolts and sliding away the movable end frame, one can visually inspect the entire heat transfer surface. Easy cleanability and a thin layer of product ensure a good bacteriological effect during pasteurization, which are advantages in the dairy industry, where PHEs are cleaned daily.

Temperature cross: A true countercurrent flow allows a PHE to operate with crossing temperatures, meaning the cold medium can have a higher exit temperature than the exiting hot medium, as shown in Figure 7.17

Lightweight: The PHE unit is lighter in total weight than other types of heat exchangers because of reduced liquid volume space and less surface area for a given application.

High-viscosity applications: Because the PHE induces turbulence at low fluid velocities, it has practical application for high-viscosity fluids. Note that a fluid sufficiently viscous to produce laminar flow in a shell and tube heat exchanger may well be in turbulent flow in PHE.

Saves space and servicing time: The PHE fits into an area one-fifth to one-half of that required for a comparable shell and tube heat exchanger. The PHE can be opened for inspection, maintenance, or

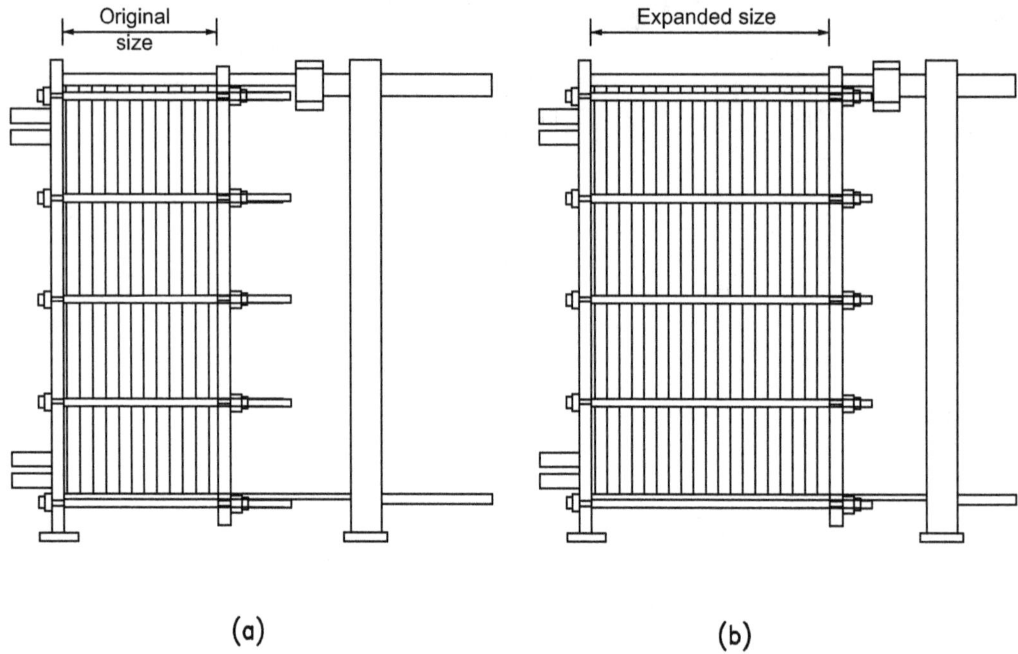

FIGURE 7.16 PHE—Flexibility of plates pack addition (expansion of PHE) to meet higher thermal duty.

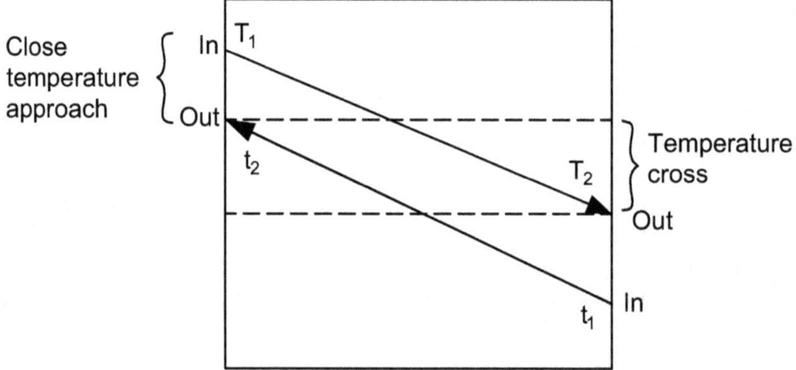

FIGURE 7.17 Temperature cross in a PHE. *Note*: The figure shows hot and cold fluids temperature distribution.

rebuilding all within the length of the frame, while the shell and tube heat exchanger with removable tube bundle requires double its length to remove the tube bundle.

Lower liquid volume: Since the gap between the heat transfer plates is relatively small, a PHE contains only low quantities of process fluids. The benefit is reduced cost due to lower volume requirements for often costly refrigerants, coolants, or process fluids. Since the product remains in the heat exchanger for a short period of time, the process can be easily stopped or the temperature can be changed quickly with minimum impact on the product.

Less operational problems: In PHEs, flow-induced vibration, noise, and erosion–corrosion due to impingement attack are not present as in shell and tube heat exchangers.

Lower cost: PHEs are generally more economical than other types of equivalent duty heat exchangers due to the higher thermal efficiency and lower manufacturing costs.

Quick process control: Owing to the thin channels created between the two adjacent plates, the volume of fluid contained in PHE is small; it quickly reacts to the new process condition and is thereby easier to control.

With the various types of gasketed plate heat exchanger available, there are advantages and disadvantages in choosing a gasketed plate heat exchanger.

Advantages of plate heat exchangers:

1. Heat transfer precision—improved temperature approach, true countercurrent flow, 80–90% less hold-up volume.
2. Low cost—low capital investment, installation costs, limited maintenance and operating costs.
3. Greatest reliability—less fouling, stress, wear, and corrosion.
4. Least energy consumption for most process effect, reduced cleaning.
5. Easy to expand capacity—adjustable plates on existing frames.

Disadvantages of plate heat exchangers:

1. Poor sealing would cause leakage occurrence, which will cause product contamination and replacement hassle.
2. Limited pressure use, generally not more than 1.5 MPa.
3. Limited operating temperature due to temperature resistance of the gasket material.
4. Small flow path, and not suited for gas-to-gas heat exchange or steam condensation.
5. High blockage occurrence, especially with suspended solids in fluids.
6. The flow resistance is larger than for the shell and tube.

7.8 LIMITATIONS OF PLATE HEAT EXCHANGERS

While plate exchangers have pressure drops comparable to those found in shell and tube units, they are generally confined to operation at lower pressures and temperatures due to the use of elastomer gaskets for sealing. Commonly stated limits have been 300°F (149°C) and 300 psi [9]. The maximum allowable working pressure is also limited by the frame strength and plate deformation resistance. Manufacturers produce a low-cost frame for low-pressure duties, say up to 85 psig (6 atmg), and a more substantial frame for higher pressures, for a given plate size. All PHEs used in the chemical and allied industries are capable of operating at 85 psig, most at 142 psig, many at 230 psig, and some at pressures as high as 300 psig [7]. The limitations of temperature and pressure are overcome in heat exchangers with the lamella heat exchanger or by a brazed plate exchanger or welded heat exchangers. Other limitations of PHEs include the following:

1. Large differences in fluid flow rates of two streams cannot be handled.
2. The gaskets cannot handle corrosive or aggressive media.
3. The standard PHEs cannot handle particulates that are larger than 0.5 mm. This restricts the use of heat exchangers with multiphase flow. If free-flowing-style plates are used, a slightly larger particulate size of 12 mm can be managed.
4. Gaskets always increase the leakage risk.

7.9 COMPARISON BETWEEN A PLATE HEAT EXCHANGER AND A SHELL AND TUBE HEAT EXCHANGER

In many applications, the PHE has replaced the shell and tube heat exchanger (STHE) for services within the former's operational limits. For identical duties (liquid–liquid service) and in those cases when the working limits of the gaskets are not exceeded, the merits of the PHE over the STHE are as follows: (a) can handle multiple duties and future expansion is possible (20–30%); (b) high heat transfer ratio; (c) low operating weight ratio; (d) low hold-up volume; (e) no welds; (e) no flow-induced vibrations; (f) access for inspection is easy and less time consuming; (g) true countercurrent flow is achieved; (h) very close approach temperature can be achieved; (i) due to high turbulence/swirl motion, less fouling is encountered; and (j) leakage is easily detected and it is easy to replace the plate and/or gasket. (Note: For a typical railroad application, a shell and tube oil cooler with an operating pressure of 8 bar and temperature of 135°C [approx.] was replaced by a gasketed PHE which solved the problem of frequent tube failure and hence the mixing of cooling water with lube oil.)

7.10 PLATE HEAT EXCHANGER: DETAILED CONSTRUCTION FEATURES

With a brief discussion of PHE construction, additional construction features and materials of construction for components such as plates, gaskets, frame, and connectors are discussed next.

7.10.1 Plate

7.10.1.1 Plate Pattern

Plates are available in a variety of corrugated or embossed patterns. The basic objective of providing corrugation to the plates is to impart high turbulence to the fluids, which results in a very high heat transfer coefficient compared to those obtainable in a shell and tube heat exchanger for similar duties. These embossing patterns also result in increased effective surface areas and provide additional strength to the plates by means of many contact points over the plates to withstand the differential pressure that exists between the adjacent plates. Plate thickness as low as 0.6 mm (0.024 in.) can therefore be used for working pressures as high as 230 psig, particularly when using the cross-corrugated, herringbone pattern [7].

7.10.1.2 Types of Plate Corrugation

Over 60 different plate patterns have been developed worldwide. The pattern and geometry are proprietary. The most widely used corrugation types are the *intermating troughs* or *wash board* and the *chevron* or *herringbone* pattern. Figure 7.18 shows the chevron pattern that is most widely used. The construction features of these two patterns are discussed next.

7.10.1.3 Intermating Troughs Pattern

In the intermating troughs or washboard pattern, the corrugations are usually pressed deeper than the plate spacing. The plates nestle into one another when the plate pack is assembled. The plate spacing is maintained by dimples, which are pressed into the crests and troughs and contact one another on adjacent plates. Since the washboard type has fewer contact points, and due to its greater corrugation depth than the herringbone type, it operates at lower pressures. The maximum channel gap varies from 3 to 5 mm and the minimum channel gap varies from 1.5 to 3 mm. Typical liquid velocity range in turbulent flow is from 0.2 to 3 m/s, depending upon the required pressure drop [5].

7.10.1.4 Chevron or Herringbone Trough Pattern

In the herringbone pattern (Figure 7.18a), the corrugations are pressed to the same depth as the plate spacing. This is the most common type in use currently. The corrugation depth generally varies from

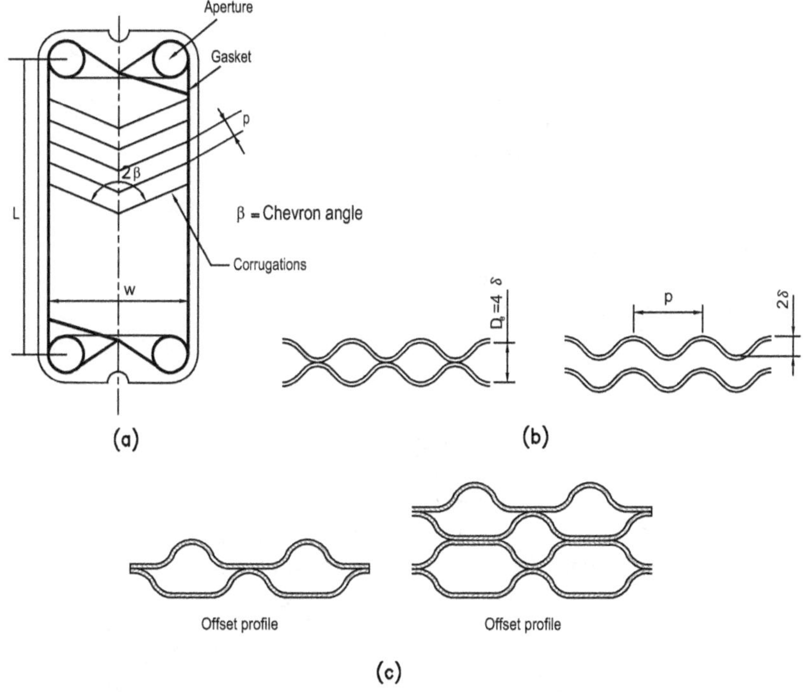

FIGURE 7.18 (a) Chevron plate—schematic showing Chevron angle, (b) gap between two plates-chevron troughs, and (c) two adjoining plates with different corrugation profile in contact.

TABLE 7.3
Common PHE Plate Materials

Stainless steel AISI 304	Incoloy 825
Stainless steel AISI 316	Monel 400
Avesta SMO 254	Hastelloy B
Titanium, titanium—0.2%	Hastelloy C-276
Palladium stabilized	Aluminum brass 76/22/2
Tantalum	Cupronickel (70/30)
Inconel 600	Cupronickel (90/10)
Inconel 625	Diabon F 100

3 to 5 mm. Typical liquid velocities (in turbulent flow) range from 0.1 to 1 m/s [5]. The chevron angle is reversed on adjacent plates so that when the plates are clamped together the corrugations cross one another to provide numerous contact points. Possible profiles of two adjoining plates in contact are shown in Figure 7.18c. The herringbone type therefore has greater strength than the washboard type, which enables it to withstand higher pressures with smaller plate thickness [10].

7.10.1.5 Plate Materials

Materials that are suitable for cold pressing and corrosion resistant are the standard materials of construction. Table 7.3 gives a list of the most common materials used for fabrication of plates. Carbon steel is rarely used due to its poor corrosion resistance.

TABLE 7.4
Gasket Materials and Maximum Operating Temperature

Styrene butadiene rubber (80°C)
Nitrile rubber (140°C)
Ethylene propylene rubber, EPDM (150°C)
Resin-cured butyl rubber (140°C)
Fluorocarbon rubbers (180°C)
Fluoroelastomer (Viton) (100°C)
Compressed asbestos fiber (CAF) (260°C)
Silicon elastomers (low-temperature applications)

7.10.2 Gasket Selection

When selecting the gasket material, the important requirements to be met are chemical and temperature resistance coupled with good sealing properties and shape over an acceptable period of life [3]. Much work has been done to develop elastomer formulations that increase the temperature range and chemical resistance of gaskets. Typical gasket materials and their maximum operating temperatures are given in Table 7.4.

7.10.3 Bleed Port Design

With gasket/bleed port design, fluids will not intermix (other than a through-plate failure) when the plates are properly gasketed and the unit is assembled in accordance with prescribed instructions and design specifications. Liquid flowing on the surface of each plate flows on the inside of the boundary gasket. Should one of the liquids leak beyond a boundary gasket, it will flow to the outside of the unit through the bleed ports, preventing intermixing.

7.10.4 Frames

Frames are classified [8] as: (1) B frame, suspension-type frame used with larger PHEs, (2) C frame, compact cantilever-type frame used with smaller PHEs, and (3) F frame, an intermediate size suspension type frame. The frame is usually constructed in carbon steel and painted for corrosion resistance. Where stringent cleanliness requirements apply, such as in pharmaceuticals and in dairy, food, and soft drinks industries, the frame may be supplied in stainless steel. Stainless steel clad frames are available for highly corrosive environments. The units are normally floor mounted, but small units may be wall-mounted.

7.10.5 Nozzles

Nozzles are located in one or both end plates. In a single-pass arrangement, both the inlet and outlet ports for both fluids are located in the fixed head end, and hence the unit may be opened up without disturbing the external piping. However, with multipass arrangements, the ports must always be located on both heads. This means that the unit cannot be opened up without disturbing the external piping at the movable head end. To be corrosion resistant, nozzles are usually made from the same material as the plate. Typical nozzle materials include stainless steel type 316, rubber clad, titanium, Incoloy 825, and Hastelloy C-276.

7.10.6 Tie Bolts

The tie bolts are usually made of 1% Cr-0.5% Mo low-alloy steel. The packing of large units may be compressed by hydraulic, pneumatic, or electric tightening devices.

7.10.7 Connector Plates

It is possible to process three or more fluids in a single gasketed PHE. This is achieved by means of connector plates. This practice is widely used in food processing and permits heating, cooling, and heat recovery of the fluids in a single unit [10].

7.10.8 Connections

Connections are usually made of the same material as the plates to avoid galvanic corrosion damage. Rubber-lined connections can be used for some tasks.

7.10.9 Installation

A PHE is mounted vertically on the floor with a level foundation that is strong enough that no settling occurs which could cause a loading strain on the connections. The heat exchanger must be installed with clearance on both sides for maintenance (Figure 7.19).

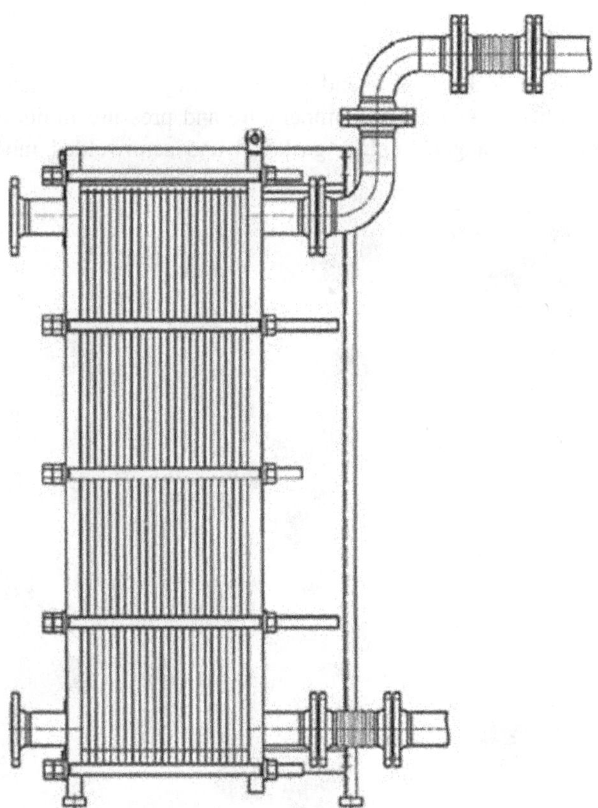

FIGURE 7.19 The opening space of the PHE, which is between the pressure plate and support, should not be obstructed by fixed piping. (Courtesy of HRS Heat Exchangers Ltd, Herts, U.K.)

7.11 BRAZED PLATE HEAT EXCHANGER

The brazed PHE evolved from the conventional PHE in answer to the need for a compact PHE for high-pressure and high-temperature tasks. Like the gasketed PHE, the brazed PHE is constructed of a series of corrugated metal plates but without the gaskets, tightening bolts, frame, or carrying and guide bars. It consists of stainless steel plates and two end plates. The plates are brazed together in a vacuum oven to form a complete pressure-resistant unit. The two fluids flow in separate channels. This compact design can easily be mounted directly on piping without brackets and foundations. Brazed PHEs accommodate a wide range of temperatures, from cryogenic to 200°C. Because of the brazed construction, the units are not expandable but get their reputation from their relatively compact size. Available with copper or nickel brazing, these units have a number of critical advantages: a sealed, compact system, high temperature and pressure capability, gasket-free construction, high thermal efficiency, and ideal for refrigeration and process applications. Figure 7.20 shows brazed PHEs, and the overall dimensions of a brazed PHE are shown in Figure 7.21.

7.12 OTHER TYPES OF PLATE HEAT EXCHANGERS

Continued developments and product innovations are taking place among various manufacturers to overcome limitations of pressure and temperature, and to handle products of high viscosity, high fibrous contents, and high corrosivity. As a result, various forms of PHEs have come into the market. Some other types of PHEs are shown in Table 7.5 and their details are discussed next.

7.12.1 All-Welded Plate Exchangers

This design entirely eliminates the gaskets and by developing a fully welded plate exchanger further enhances the reliability as well as the temperature and pressure limits of the gasketed plate exchanger. Welded plate exchangers, unlike gasketed and semiwelded models, can neither be

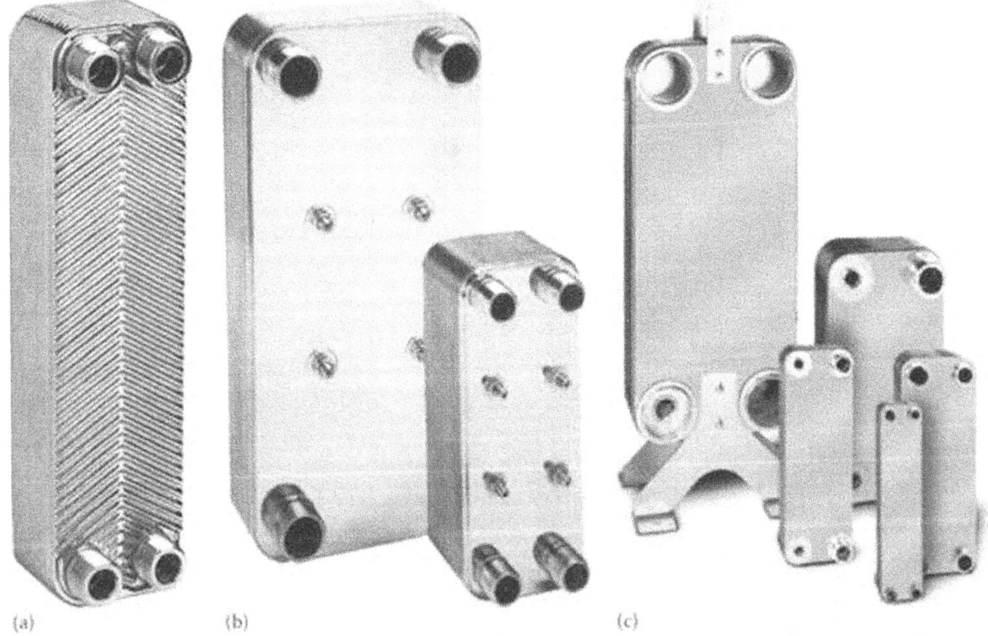

FIGURE 7.20 A collection of brazed plate heat exchangers. (Courtesy of GEA PHE Systems, Sarstedt, Germany.)

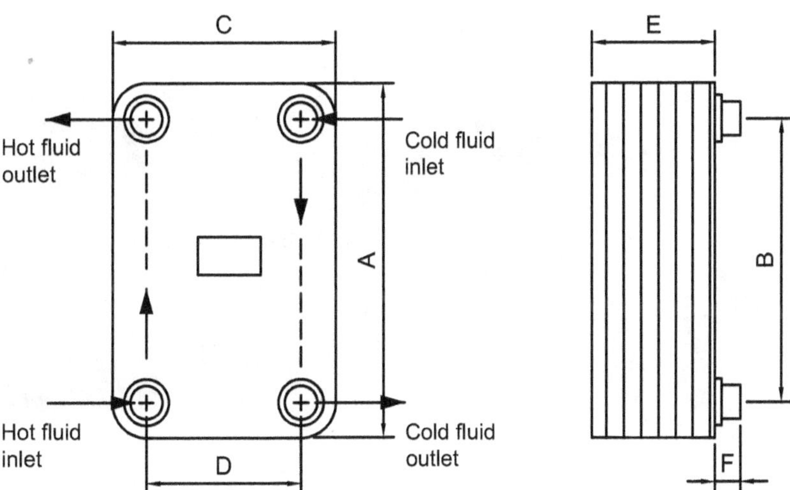

FIGURE 7.21 Overall dimensions of brazed plate heat exchanger.

expanded in surface area nor be cleaned readily by mechanical methods. Only chemical cleaning is possible.

7.12.2 Supermax® and Maxchanger® Plate Heat Exchangers

Tranter's Supermax shell and plate heat exchangers and Maxchanger welded plate heat exchangers require less space than the equivalent shell and tube exchangers. Turbulent flow induced by the corrugated and dimpled plate patterns produces high heat transfer rates. This high efficiency allows compact exchangers with a 1°C (2°F) temperature approach. Another benefit is the small hold-up volume that offers fast start-up times and close following of process changes.

Supermax (shown in Figure 7.22) and Maxchanger (shown in Figure 7.23) exchangers can be applied to applications involving liquids, gases, steam, and two-phase mixtures. This includes aggressive media, which are beyond the capability of traditional gasketed plate and frame heat exchangers. In addition to efficiency, the units offer cost effectiveness and minimal maintenance. The Supermax shell and plate heat exchanger is designed for pressures up to 100 barg (1450 psig) and at temperatures up to 538°C (1000°F) for standard range units. In refrigeration and cryogenic service, the exchangers require a low refrigerant charge. They are also resistant to freezing because of high fluid turbulence induced by the corrugated plate pattern. Supermax wide temperature/pressure ratings offer good performance with natural refrigerants such as ammonia and carbon dioxide. Fluids can undergo phase change on either the plate or shellside.

Pairs of chevron-type plates are placed back-to-back and fabricated into a cassette by full automatic perimeter welding of adjacent port holes. Cassettes are then placed together and perimeter welded to each other, producing an accordion-like core that is highly tolerant to thermal expansion. The plate pack is then inserted in a cylindrical shell. Fluid diverters positioned between the shell and the plate pack ensure proper flow through the shell side channels. End plates, nozzles, and top and bottom covers are welded to the shell to form a pressure vessel of high integrity. Extra-large nozzle sizes can be accommodated on the shell side of the exchanger. Plates can also be arranged to form multiple passes. Figure 7.24 shows a collection of Tranter plate heat exchangers—Maxchanger (foreground) and Supermax (right), welded plate heat exchangers spiral (left), Superchanger® gasketed P&F (back), and Platecoil® prime surface.

FIGURE 7.22 Supermax shell and plate heat exchanger. (a) Collection of units (b) comparison with a shell and tube heat exchanger, and (c) Assembly details. (Courtesy of Tranter, Inc., Wichita Falls, TX.)

FIGURE 7.23 Maxchanger welded plate heat exchanger. (Courtesy of Tranter, Inc., Wichita Falls, TX.)

FIGURE 7.24 Collection of Tranter plate heat exchangers: Maxchanger (foreground) and Supermax (right), welded plate heat exchangers spiral (left), Superchangergasketed PHE (back), and Platecoil prime surface bank. (Courtesy of Tranter, Inc., Wichita Falls, TX.)

7.12.3 Wide-Gap Plate Heat Exchanger

Wide-gap PHE provides a free-flow channel for liquids and products containing fibers, coarse particles, or high-viscosity liquids that normally clog or cannot be satisfactorily treated in shell and tube heat exchangers. These plates have a draw depth two to five times greater than conventional plates, permitting unrestricted passage of coarse particles and fibers. Wide-gap plates economically recover heat from hard-to-handle waste streams in a variety of industries, including pulp and paper, sugar processing, alcohol production, grain processing, chemicals, latex, polymer slurries, textiles, and ethanol distilling.

7.12.4 GEABloc Fully Welded Plate Heat Exchanger

GEABloc combines two different plate corrugations in an innovative way (Figure 7.25). The plates are rotated through 90° to one another and then welded together to produce two different cross-flow channels. The frame is made of four pillars, a bottom plate and a top plate together with four side pressure plates, and connections mounted in the pressure plates. All frame components are bolted together for easy dismantling to clean and maintain the plate pack. GEABloc is available in various corrugation designs and sizes for a wide range of applications.

7.12.5 Free-Flow Plate Heat Exchanger

Free-flow PHEs feature a special plate geometry that offers an uninterrupted wide flow path for fouling fluids, viscous fluids, or those containing fibrous materials (fruit juices containing fibers and pulps, waste water in the paper and pulp, textile, and sugar industries, as well as highly viscous products). Their special features include the constant width gap cross section between the individual plates and the rough-wave profiling of the plates. The clearance between the plates is up to 12/

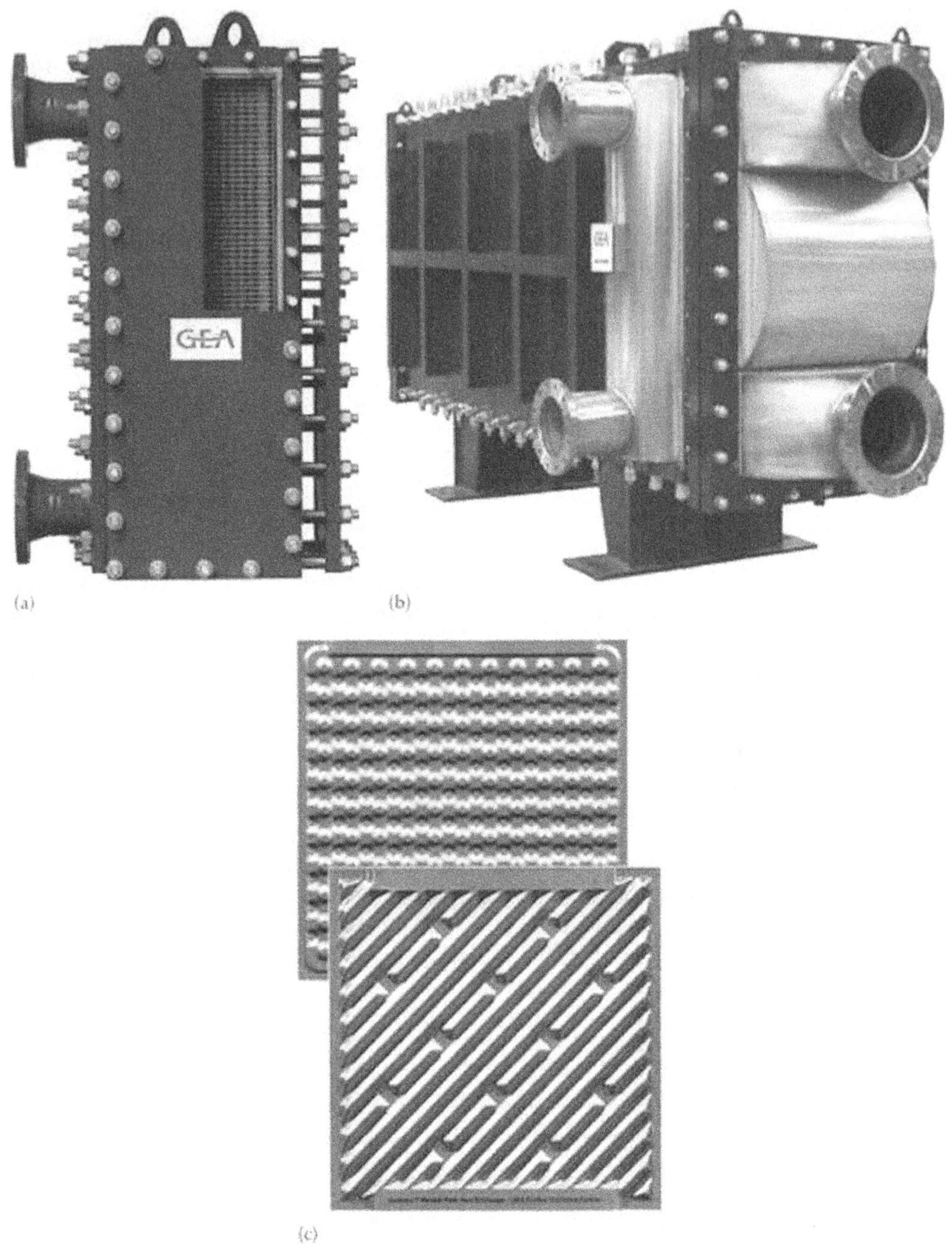

FIGURE 7.25 (a) GEA Bloc welded plate heat exchanger, (b) GEABox fully welded PHE, and (c) Chevron and free flow plate corrugations pattern. (Courtesy of GEA PHE Systems, Sarstedt, Germany.)

15 mm. The free-flow design is an improved design compared to the conventional wide-gap plate exchanger, since there are no contact points in the flow path of free-flow plates that restrict the flow. Various patterns of free-flow plates are shown in Figure 7.26. A comparison of plate patterns and the flow path of conventional and Mueller freeflow plates is shown in Figure 7.27.

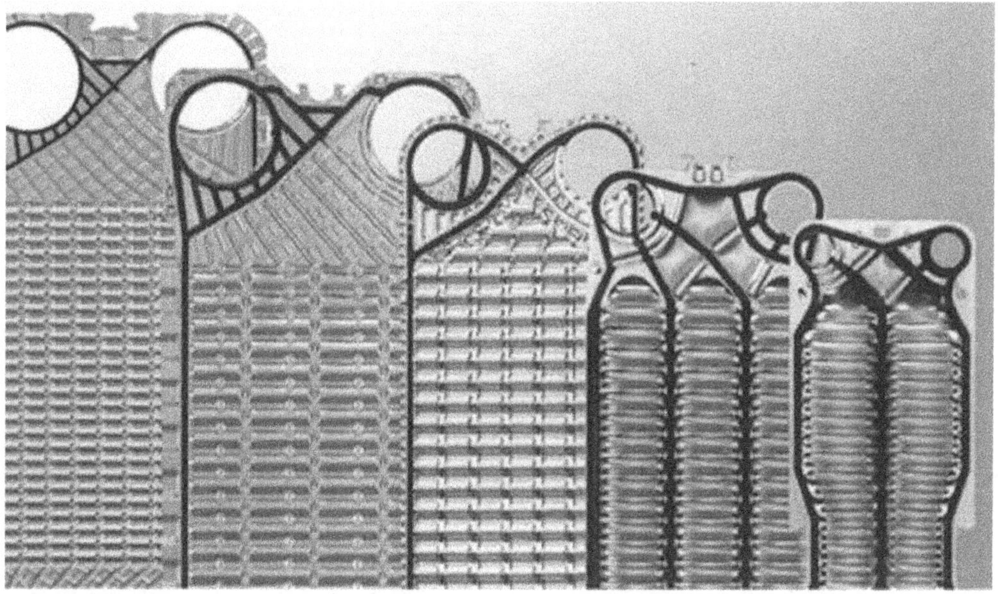

FIGURE 7.26 A collection of free flow plates. (Courtesy of GEA PHE Systems, Sarstedt, Germany.)

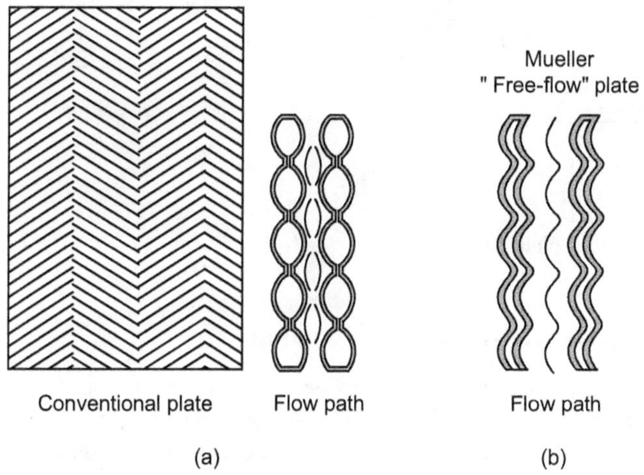

FIGURE 7.27 Comparison of plate patterns and flow path of conventional and Mueller Accu-Therm™ free flow plate. (Courtesy of Mueller, Heat Transfer Products, Springfield, MO.)

7.12.6 Flow-Flex Tubular Plate Heat Exchanger

In the Alfa Laval Flow-Flex, the tubular PHE, the plate pattern builds the tube channels with a free cross section on one side and conventional plate channels on the other. This allows Flow-Flex to handle asymmetrical duties, that is, dissimilar flow rates at a ratio of at least two to one (for the same fluid properties and pressure drop). This vibration-free construction is also suitable for low-pressure condensing and vaporizing duties and for fibrous and particle-laden fluids.

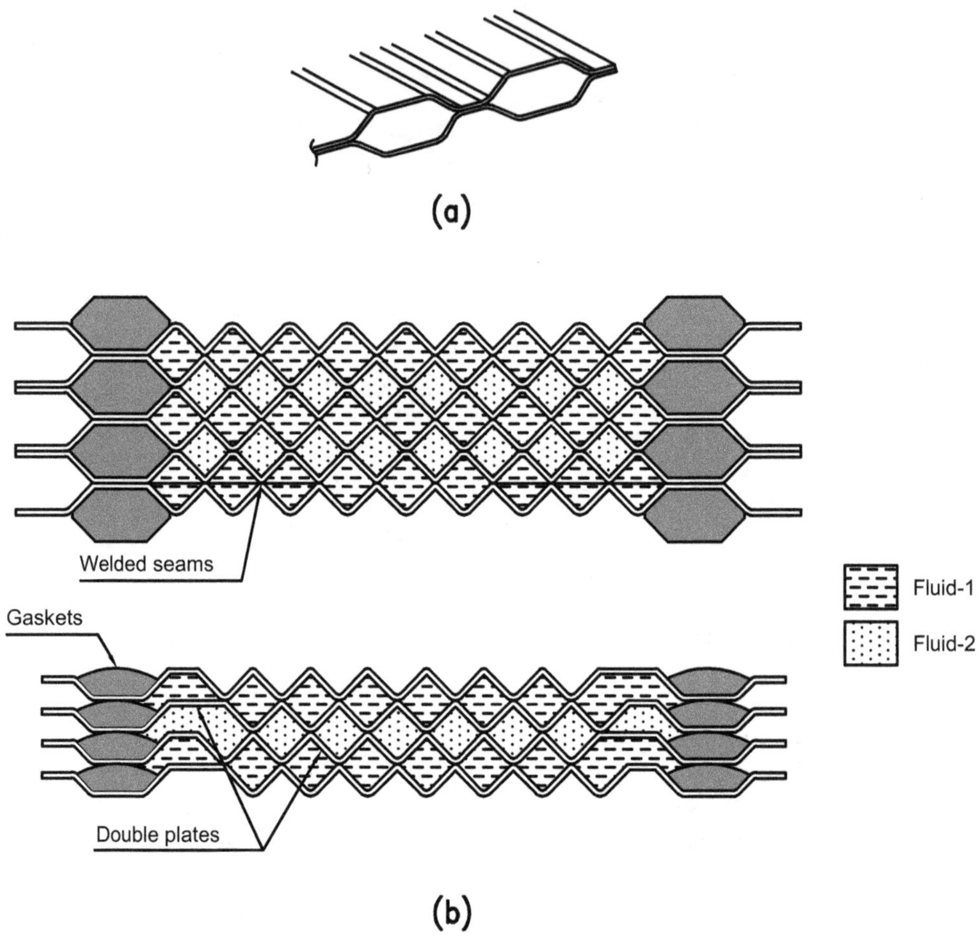

FIGURE 7.28 Semiwelded plate flow design.

7.12.7 Semiwelded or Twin-Plate Heat Exchanger

Semiwelded gasketed plate heat exchangers are used when gaskets are not compatible as one of the process media. Semiwelded GPHEs can also take a higher design pressure compared to fully gasketed plate-and-frame heat exchangers. The semiwelded or twin-plate heat exchanger plates offer the same advantages as standard PHE units, yet overcome pressure limitations and avoid chemical resistance to gaskets. Semiwelded plates are formed from two cassettes that are laser welded together. The cassettes are separated by standard gaskets, such as EPDM, NBR, or Viton, as shown in Figures 7.28 and 7.29. The welded pairs allow for aggressive fluids or refrigerants between plates and the only gaskets in contact with the aggressive medium are two circular porthole gaskets between the welded plate pair. The other medium flows across gasketed channels. Semilaser welded PHEs offer superior sealing capability, easy cleaning, and comply with sanitary standards on the welded side. The use of semiwelded units is increasing in the chemical, petroleum refining, and refrigeration industries [11].

7.12.7.1 Applications

Semi-welded GPHEs are used in refrigeration applications as evaporators, condensers, desuperheaters, cascade duties, economizers/subcoolers, etc.

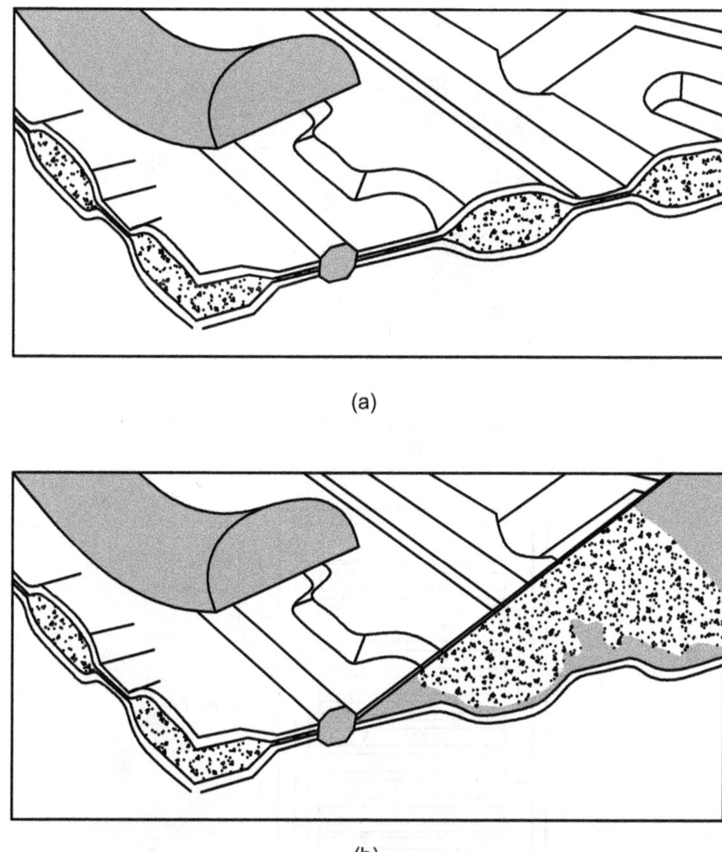

FIGURE 7.29 Semi-welded plateflow design: (a) light duty and (b) heavy duty. (Courtesy of ITT Standard, Cheektowaga, NY.)

7.12.8 Double-Wall Plate Heat Exchanger

The double-wall PHE is intended for use where the two fluids on each side of the plates should not mix due to possible contamination, an undesirable reaction, environmental effects, or it would be hazardous. The double-wall PHE consists of two plates with a small air gap between them (Figure 7.30). This gap is open to the atmosphere and is not in contact with either of the media, helping guard against intermixing of the media. The double-wall heat exchanger plates are designed for use in cooling/heating duties in dairy, pharmaceutical, brewery, and beverage installations where it is crucial to prevent any cross-contamination between the two media.

7.12.9 Compabloc Fully Welded Heat Exchanger

The Alfa Laval or other OEM's Compabloc is a fully welded compact heat exchanger designed for a wide range of process and utility duties. The heart of the Compabloc is a stack of corrugated heat transfer plates in 316L stainless steel, or other high-grade material. The plates are laser welded and form a compact core. This core is then enclosed and supported by four corner girders, top and bottom heads, and four side panels. These components are bolted together and can be quickly taken apart for inspection, service, and cleaning. The Alfa Laval Compabloc is a fully welded compact heat exchanger and is shown in Figure 7.31 [12,13].

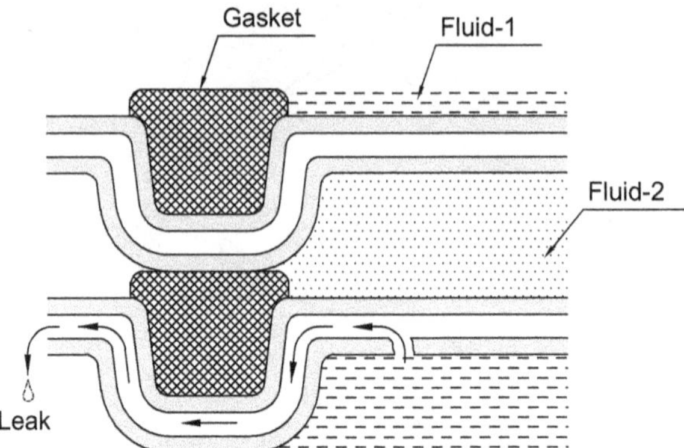

FIGURE 7.30 Double-wall units prevent intermixing of fluids due to a plate failure.

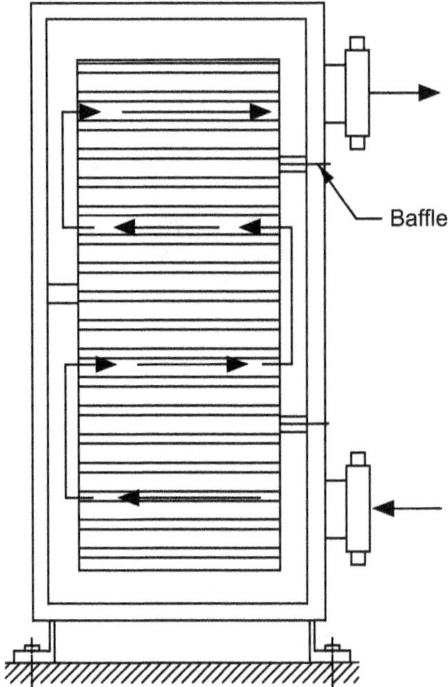

FIGURE 7.31 Compabloc fully welded compact plate heat exchanger.

7.12.9.1 Working Principle

The two media in the Compabloc heat exchanger flow in alternately welded channels between the corrugated plates. These corrugated plates promote high turbulence, which provides high heat transfer efficiency and helps minimize fouling. The media flows in a cross-flow arrangement within each pass, while the overall flow arrangement is countercurrent for a multi-pass unit (if required the unit can also be designed with overall co-current operation). Each pass is separated from the adjacent passes by a pressed baffle which forces the fluid to turn between the plate pack and the panel.

7.12.10 Diabon Plate Heat Exchanger

Alfa Laval Diabon® plate-and-frame heat exchangers are used for highly corrosive media, combining high-efficiency heat transfer benefits with the exceptional corrosion resistance of graphite material. Diabon plate heat exchangers are the ideal solution for processes in which metallic plates with low corrosion resistance cannot live up to service life requirements, and where the heat transfer efficiency of heat exchangers that use materials such as glass and Teflon® is unacceptably low. Compared to other solutions, such as graphite blocks, Diabon provides the additional benefits of reduced fouling and full access to the heat transfer surface. Diabon is a dense, synthetic resin-impregnated high-quality graphite with a fine and evenly distributed pore structure, and can be used with corrosive media up to 390°F [9,14].

7.12.10.1 PTFE Rope Gasket

Diabon plate heat exchangers are all supplied with a PTFE rope gasket. When the heat exchanger is assembled, this gasket rope is flattened to a very thin film of approximately 0.008 in. The thinness of the gasket means that the area in contact with the chemical is very limited, resulting in an extremely long service life. To date, the gasket has proved resistant to attack by all known chemicals, as well as resisting very high temperatures (>356°F). In addition, the gasket can be kept in storage for virtually unlimited periods.

7.12.10.2 Applications

Heaters, coolers, interchangers, condensers, and evaporators for corrosive media, especially in the treatment of hydrochloric acid (HCl), sulfuric acid (H_2SO_4), hydrofluoric acid (HF), mixed acids (HNO_3/HF), phosphoric acid (H_3PO_4/P_2O_5), and other organic and inorganic media.

7.12.11 Glue-Free Gaskets (Clip-On Snap-On Gaskets)

The clip-on snap-on gaskets are glue-free gaskets designed to perform in heavy industrial applications in the same manner as traditional glued gaskets and recommended wherever regular cleaning is necessary or aggressive fluids shorten gasket life. The glue-free gaskets (Figure 7.32) are attached to the plate by fasteners situated at regular intervals around the plate periphery. The gaskets simply slip into place during fitting and slip out for regasketing procedures, eliminating gluing procedures. All these contribute to reduced service costs and downtime. Typical ratings are 338°F (170°C) at low pressures and 357 psig (25 barg) at moderate temperatures, depending on application conditions.

FIGURE 7.32 Glueless gasket. (Courtesy of ITT Standard, Cheektowaga, NY.)

7.12.12 AlfaNova 100% Stainless Steel Plate Heat Exchanger

AlfaNova is a 100% stainless steel PHE. It comprises a number of corrugated stainless steel plates, a frame plate, and a pressure plate. The plate pack is brazed by AlfaFusion, a new technology patented by Alfa Laval. The fusion-brazed PHE, a new class of PHE, offers extremely high mechanical strength and greater thermal fatigue resistance than conventional brazed units. It is also hygienic, corrosion resistant, and fully recyclable. Its 100% stainless steel construction enables AlfaNova to withstand temperatures up to 550°C (1020°F).

7.12.13 Plate Heat Exchanger with Electrode Plate

A recent innovation is a standard construction of PHE that includes at least one electrode plate for cooling applications by a refrigerant. The electrode plate includes outer electrode surfaces on each side thereof to produce an electric field. The effect of the electric field is an increase in the heat transfer rate between the refrigerant and the heat transfer fluid.

7.12.14 Plate Heat Exchanger with Flow Rings

A design similar to a standard PHE includes flow rings. The design consists of a plurality of stamped plates but on each plate a flow ring and an extended surface fin structure are present. Each of the flow rings includes radial flow openings for evenly distributing the fluid between the plates. The plates are stacked with additional fin structure and ring pairs between each pair of plates to achieve the thermal requirements of the particular application. The advantages of this design are compactness, low cost, and thermal efficiency.

7.12.15 AlfaRex™ Gasket-Free Plate Heat Exchanger

The AlfaRex is a gasket-free welded PHE for high-pressure and high-temperature applications where gaskets are unacceptable. Its working temperature can be from −58°F to 650°F and pressure up to 600 psig. The AlfaRex is ideal for batch reactor temperature control systems, aggressive media, and cyclical applications. Though it overcomes the gasket problem, it cannot be opened up for cleaning.

7.12.16 Sanitary Heat Exchangers

For food, dairy, and pharmaceutical processing, sanitary heat exchangers combine low maintenance, high efficiency, and reliable separation of fluids. Plate gaps in these exchangers are sized to reduce fouling; the main pattern creates the necessary turbulence for effective heat transmission.

7.12.17 EKasic® Silicon Carbide Plate Heat Exchangers

EKasic silicon carbide PHEs are used in the chemical industry or similar sectors, especially where highly corrosive media must be heated and cooled, or evaporated and condensed. EKasic silicon carbide PHEs are also used in applications that must withstand the wear of particle-laden liquids. The very compact PHEs offer very high transfer performance in a small space. EKasic PHEs permit long service lives, high reliability, and improved product quality. The heat exchangers are further characterized by outstanding thermal shock resistance and resistance to cavitation.

7.12.18 Deep-Set Gasket Grooves

Polaris gasket grooves reduce the risk of gasket failure. In traditional designs, grooves are shallow, which exposes more of the gasket to pressure exerted by the product. The deep-set

Polaris groove exposes less gasket area to product pressure, which dramatically increases the gasket's reliability.

7.13 THERMOHYDRAULIC FUNDAMENTALS OF PLATE HEAT EXCHANGERS

The design of PHEs is highly specialized in nature, considering the variety of plate corrugations (there are more than 60 different plates) available to suit varied duties. Typical plate patterns are shown in Figure 7.33. The plate design varies from manufacturer to manufacturer, and hence the thermohydraulic performance. Unlike tubular heat exchangers, for which design data and methods are easily available, PHE design continues to be proprietary in nature. Manufacturers have developed their own empirical correlations for the prediction of thermal performance applicable to the exchangers marketed by them. Therefore, specific and accurate characteristics of specific plate patterns are not available in the open literature.

7.13.1 High- and Low-Theta Plates

A plate having a high chevron angle (about 60°–65°) provides high heat transfer combined with high pressure drop, whereas a plate having a low chevron angle, β (about 25°–30°), provides a low heat transfer combined with low pressure drop [7]. Figure 7.34 shows a schematic of high- and low-theta plates. Manufacturers specify the plates having low values of chevron angle as low-theta plates and plates having high values of chevron angle as high-theta plates. Theta is used by manufacturers to denote the number of heat transfer units (NTUs). The expression for theta is given by

$$\theta_c = \mathrm{NTU}_c = \frac{(UA)_c}{\left(mc_p\right)_c} \tag{7.1a}$$

$$\theta_h = \mathrm{NTU}_h = \frac{(UA)_h}{\left(mc_p\right)_h} \tag{7.1b}$$

The values of θ achieved per pass for various plate types vary enormously (1.15 to about 4), and the NTU for high-theta plates is of the order of 3.0 and for low-theta plates 0.5.

Salient features of high- and low-theta plates are hereunder:

a. *High-theta plates*: Corrugated with obtuse chevron angles, generate very high turbulence, extremely high heat transfer rates, high pressure loss, etc.

FIGURE 7.33 Collection of plate heat exchanger plates. (Courtesy of GEA PHE Systems, Sarstedt, Germany.)

TABLE 7.5
Salient Features of Some Newer Types of PHE

Description	Features
Brazed plate heat exchanger (BHE)—It is constructed of a series of corrugated metal plates brazed together in a vacuum oven to form a complete pressure-resistant unit. Absence of gaskets, tightening bolts, frame, or carrying and guide bars	A sealed, compact system, high temperature and pressure capability, gasket-free construction, high thermal efficiency and ideal for refrigeration and process applications. The units are not expandable
Shell and plate heat exchanger—Pairs of chevron-type plates are placed back-to-back and fabricated into a cassette. Cassettes are then placed together and perimeter welded to each other. The plate pack is then inserted into the cylindrical shell	Can handle liquids, gases, steam, and two-phase mixtures. In addition to efficiency, the units offer cost-effectiveness and minimal maintenance. Designed for pressures up to 100 barg (1450 psig) and temperatures up to 538°C (1000°F)
Welded plate heat exchanger—This design entirely eliminates the gaskets by developing a fully welded plate exchanger	It enhances the reliability as well as the temperature and pressure limits of the gasketed plate exchanger. Like gasketed PHE, it can neither be expanded in surface area nor be cleaned readily by mechanical methods. Only chemical cleaning is possible
Wide-gap plate heat exchanger—It provides a free-flow channel for liquids and products containing fibers or coarse particles or high-viscosity liquids that normally clog or cannot be satisfactorily treated in shell and tube heat exchangers	Wide-gap plates economically recover heat from hard-to-handle waste streams in a variety of industries, including pulp and paper, sugar processing, alcohol production, grain processing, chemicals, latex, polymer slurries, textiles, and ethanol distilling
Free-flow plate heat exchanger—Compared to the conventional wide-gap plate exchanger, there are no contact points in the flow path of free flow plates which restrict the flow otherwise. Hence, it offers an uninterrupted wide flow path for fouling or viscous fluids or those containing fibrous materials	The corrugation pattern provides the right balance of high efficiency in heat transfer, clogging resistance, and low pressure drop, providing a cost advantage over larger, more expensive shell and tube heat exchanger
Semiwelded or twin-plate heat exchanger—Semiwelded plates are formed from two cassettes that are laser welded together. The cassettes are separated by standard gaskets. The welded pairs allow for aggressive fluids or refrigerants between plates. The other medium flows across gasketed channels	It offers superior sealing capability, easy cleaning, and complies with sanitary standards on the welded side. Use of semiwelded units is increasing in the chemical, petroleum refining, and refrigeration industries
Double-wall plate heat exchanger—It consists of two plates with a small air gap between them. This gap is open to the atmosphere and not in contact with either of the media helping guard against intermixing of the media	The double-wall PHEs are designed for use in cooling/heating duties in dairy, pharmaceutical, brewery, and beverage installations where it is crucial to prevent any cross-contamination between the two media
Diabon F graphite plate heat exchanger—The Alfa Laval Diabon F graphite PHE has graphite plates developed for use with media normally too corrosive for exotic metals and alloys	The graphite plates offer excellent corrosion resistance and good heat transfer properties in combination with low thermal expansion

b. *Low-theta plates*: Corrugated with acute chevron angles, generate lower turbulence, lower heat transfer rates, and less pressure loss.

There are two plate corrugations (L and H). The two plate corrugations form three different channels (L, M, and H). the optimal channel type is selected on the basis of the temperature program to be satisfied and the maximum permissible pressure drop.

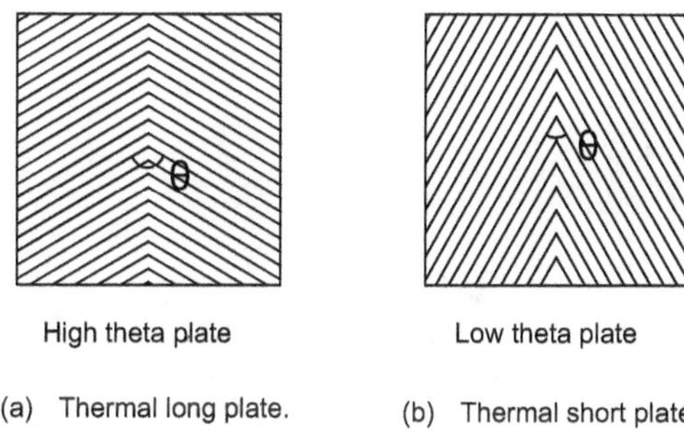

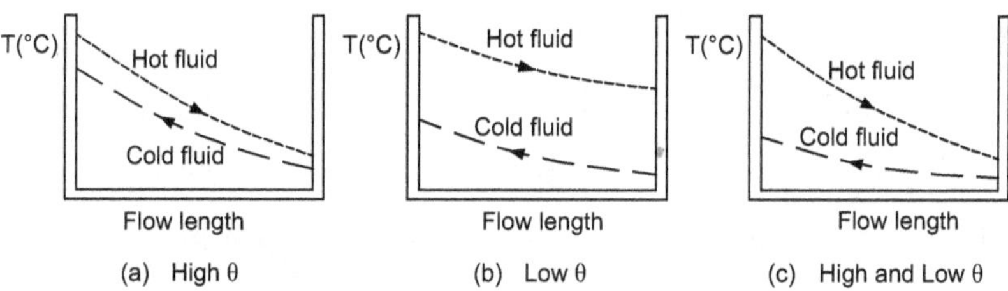

FIGURE 7.34 High-theta and low-theta plates (schematic) and temperature distribution of both the fluids.

7.13.2 Thermal Mixing

One of the thermal design problems associated with PHE is the exact matching of the thermal duties; it is very difficult to achieve the required thermal duty and at the same time utilize the available pressure drop fully [3]. This problem is overcome through a procedure known as thermal mixing. Thermal mixing provides designers with a better opportunity to utilize the available pressure drop without excessive surface, and with fewer standard plate patterns [10]. Thermal mixing is achieved by two methods:

1. Using high- and low-theta plates
2. Using horizontal and vertical plates

7.13.2.1 Thermal Mixing Using High- and Low-Theta Plates

In this method, the pack of plates may be composed of all high-theta plates (say, $\beta = 30°$) or all low-theta plates (say, $\beta = 60°$), or a combination of high- and low-theta plates arranged alternatively in the pack to provide an intermediate level of performance. Thus, two-plate configurations provide three levels of performance plates, as shown in Figure 7.35.

7.13.2.2 Thermal Mixing Using Horizontal and Vertical Plates

In this method, two combinations of geometric patterns are selected to provide three levels of performance plates, as shown in Figure 7.36 [8]:

1. Horizontal-style plates: Accu-Therm™ "H" style plates have a horizontal herringbone embossing. These plates have higher heat transfer coefficients and slightly larger pressure drop.

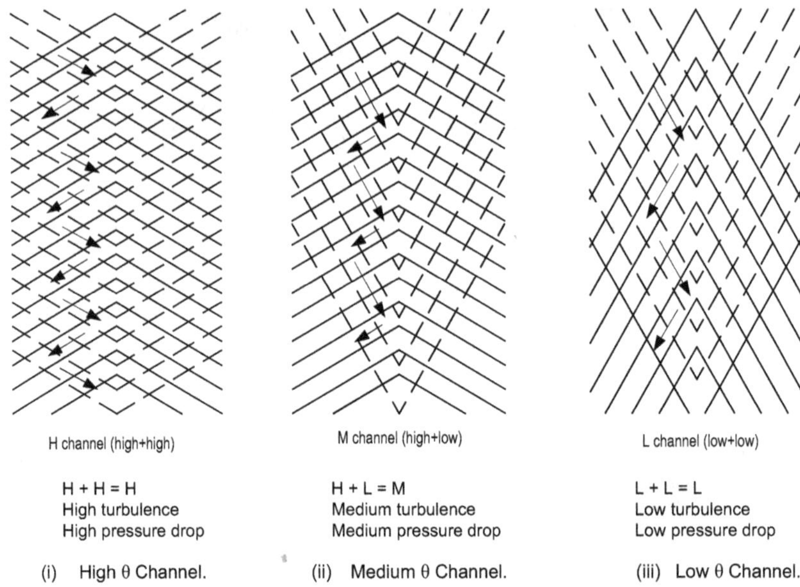

FIGURE 7.35 Thermal mixing combination of high-theta and low-theta plate arrangements and flow pattern.

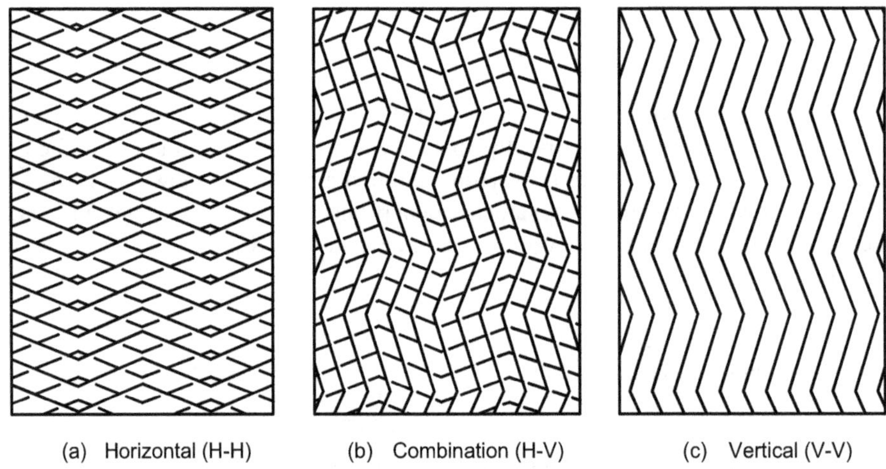

FIGURE 7.36 Thermal mixing using combination of plate geometries. (a) Horizontal–horizontal, (b) horizontal–vertical, and (c) vertical–vertical.(Courtesy of Mueller, Heat Transfer Products, Springfield, MO.)

2. Vertical-style plates: Accu-Therm™ "V" style plates have a vertical herringbone embossing. These plates have lower pressure drop and slightly lower heat transfer coefficients.
3. Combination-style plates: Accu-Therm™ "H" and "V" style plates have been combined to obtain an intermediate thermal and pressure-drop performance.

7.13.3 Flow Area

Close spacing of the plates with nominal gaps ranging from 2 to 5 mm (0.08 to 0.02 in.) gives hydraulic mean diameters in the range of 4–10 mm (0.15–0.4 in.) [10]. The plates are embossed

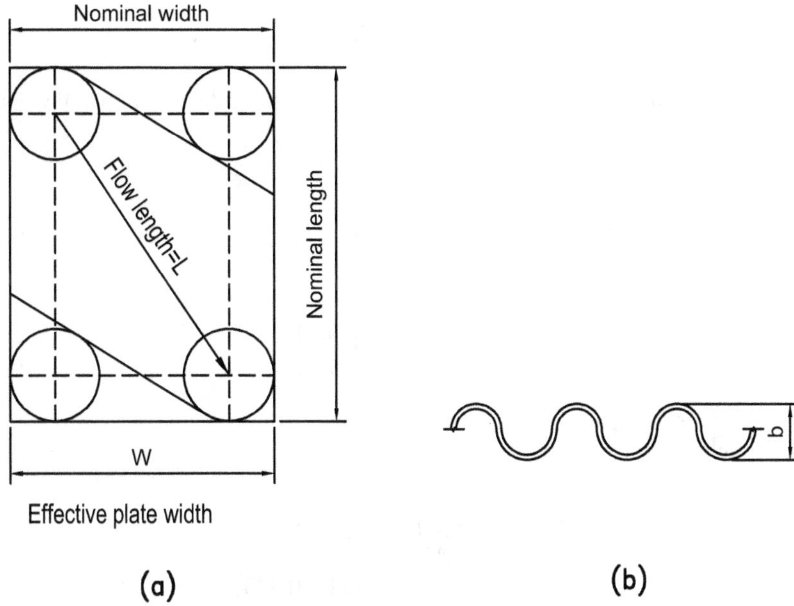

FIGURE 7.37 PHE, major plate dimensions: W-Plate width and L-Effective flow length.

so that very high degrees of turbulence are achieved. Critical Reynolds numbers are in the range of 10–400, depending on the geometry. These factors contribute to produce very high heat transfer coefficients. Nominal velocities for "waterlike" liquids in turbulent flow are usually in the range 0.3–1.0 m/s (1–3.1 ft/s), but true velocities may be higher by a factor of three or four due to the effect of the corrugations [7]. All heat transfer and pressure-drop relationships are, normally, based on channel velocity. The channel velocity is calculated by dividing the flow per channel by the channel cross-sectional area [15]. The channel cross-sectional area, A_s, is given by

$$A_s = Wb \tag{7.2}$$

where

b is the mean plate gap

W is the effective plate width (gasket to gasket)

Dimension W is shown schematically in Figure 7.37.

7.13.4 Heat Transfer and Pressure-Drop Correlations

Plate heat exchanger calculation method

To solve a thermal problem, we need to know several parameters. Further data can then be determined. The six most important parameters include:

i. The amount of heat to be transferred (heat load)
ii. The inlet and outlet temperatures on the primary and secondary sides
iii. The maximum allowable pressure drop on the primary and secondary sides
iv. The maximum operating temperature

v. The maximum operating pressure
vi. The flowrate on the primary and secondary sides

If the flow rate, specific heat, and temperature difference on one side are known, the heat load can be calculated.

7.13.4.1 Heat Transfer Correlations

A general correlation for turbulent flow is given by [7]

$$\mathrm{Nu} = C\,\mathrm{Re}^{\mathrm{n}}\mathrm{Pr}^{\mathrm{m}}\left(\frac{\mu_{\mathrm{b}}}{\mu_{\mathrm{w}}}\right)^{x} \tag{7.3}$$

Typical reported values are

$$\begin{aligned} &C = 0.15 - 0.40 \\ &n = 0.65 - 0.85 \\ &m = 0.30 - 0.45 \quad (\text{usually } 0.333) \\ &x = 0.05 - 0.20 \end{aligned}$$

One of the most widely used plates has the following relationship [7]:

$$\mathrm{Nu} = 0.374\ \mathrm{Re}^{0.668}\mathrm{Pr}^{0.33}\left(\frac{\mu_{\mathrm{b}}}{\mu_{\mathrm{w}}}\right)^{0.15} \tag{7.4}$$

where the Reynolds number is based on equivalent diameter, D_{e}, defined by

$$D_{\mathrm{e}} = \frac{4Wb}{2(W+b)} \tag{7.5a}$$

$$= 2b \text{ since } b \text{ is very small compared to } W \tag{7.5b}$$

The expression for hydraulic diameter D_{h} for PHE is given by

$$D_{\mathrm{h}} = \frac{4 \times \text{minimum free flow area}}{\text{wetted perimeter}} \tag{7.6a}$$

$$\approx \frac{2b}{\phi^{*}} \tag{7.6b}$$

where ϕ^* is the ratio of actual (developed) surface area to projected surface area.

However, to calculate the Reynolds number, the hydraulic diameter expressed by Equation 7.6 is not used because it is difficult to measure the parameter ϕ^* [5].

Buonopane and Troupe [16] present generalized relationships for a number of geometries for laminar flow. It seems that the following Sieder–Tate type relationship applies [7]:

$$\mathrm{Nu} = C_1\left(\mathrm{Re\ Pr}\frac{D_\mathrm{h}}{L}\right)^{0.333}\left(\frac{\mu_\mathrm{h}}{\mu_\mathrm{w}}\right)^{0.14} \tag{7.7}$$

where

C_1 = 1.86–4.50 depending on geometry, replace D_h by D_e

L is the effective plate length (as shown in Figure 7.37)

More correlations are summarized in Refs. [4,5,7,15].

7.13.4.2 Pressure Drop

Friction factor, f: The fanning friction factor, f, is of the form [17]

$$f = \frac{C_2}{\mathrm{Re}^y} \tag{7.8}$$

where

C_2 is a constant characterizing each type of plate

y covers different ranges

For a typical plate, the f factor is given by [17]

$$f = \frac{2.5}{\mathrm{Re}^{0.3}} \tag{7.9}$$

Usher [18] reported that the friction factor for plate exchangers is 10–60 times that developed inside tubular exchangers at similar Reynolds numbers for turbulent flow. Marriott [7] observed that this value can be up to 400 times. Though the friction factors are high compared to those of the tubular units, the pressure drop will be less for the following reasons [7]:

1. The nominal velocities are low, and nominal plate lengths do not exceed about 6 ft (1829 mm), so that the term $(G^2/2g_c)L$ in the pressure-drop equation is very much smaller than would be the case in a tubular unit.
2. Only a few passes will achieve the required NTU value, so that the pressure drop is effectively utilized for heat transfer, and losses due to unfruitful flow reversals are minimized.

7.13.4.3 Expression for Pressure Drop

Generalized equation for pressure drop: The pressure drop in a PHE consists of the following components [5]:

1. Empirically, the pressure drop associated with the inlet and outlet manifolds and ports, Δp_m, is approximately 1.5 times the inlet velocity head per pass,

$$\Delta p_\mathrm{m} = 1.5\left(\frac{\rho U_\mathrm{m}^2}{2g_\mathrm{c}}\right)_i N_\mathrm{p} \tag{7.10}$$

where
N_p is the number of fluid ports
U_m is the velocity through the ports

2. The pressure drop associated within the plate passages is given by

$$\Delta p_c = \underbrace{\frac{4fLG^2}{2g_c D_c}\left(\frac{1}{\rho}\right)_m}_{\text{friction effect}} + \underbrace{\left(\frac{1}{\rho_o} - \frac{1}{\rho_i}\right)\frac{G^2}{g_c}}_{\text{momentum effect}} \tag{7.11}$$

where

$$\left(\frac{1}{\rho}\right)_m = \frac{1}{2}\left(\frac{1}{\rho_o} + \frac{1}{\rho_i}\right) \tag{7.12}$$

and where D_e equivalent diameter of flow passages, G is the mass velocity and L is the flow passage length. The flow passage length is equal to the distance (diagonal) between the centers of inlet and outlet ports [15]. For liquids, the momentum effect is negligible, and $(1/\rho)_m \approx 1/\rho_m$.

3. The pressure drop due to an elevation change is given by

$$\Delta P_n = \pm \frac{\rho_m g L}{g_c} \tag{7.13}$$

where
the + stands for vertical upward flow
the – stands for vertical downward flow
g is the gravitational acceleration
g_c is the proportionality constant in Newton's second law of motion, $g_c = 1$ and dimensionless in SI units, or $g_c = 32.174$ lbmft/lbf s^2

The total pressure drop on one side of the plate exchanger is the sum of Δp's of Equations 7.10, 7.11, and 7.13.

Approximate equation for pressure drop: The general correlation for calculating pressure drop in a plate exchanger after accounting for the Sieder–Tate viscosity correction factor has the general form

$$\Delta p = \frac{4fLG^2}{2g_c \rho D_e}\left(\frac{\mu_w}{\mu_b}\right)^{0.14} \tag{7.14}$$

7.13.5 Specific Pressure Drop or Jensen Number

In assessing the performance of any heat exchanger, the *specific pressure drop* (J) can be used [19]. This is defined as the pressure drop per NTU, that is,

$$J = \frac{\Delta p}{NTU} = \frac{\Delta p}{\theta} \tag{7.15}$$

Jensen [19] reports that optimum values for J for water–water duties of commercially available plates are close to 4.5 psi/NTU.

7.14 PHE THERMAL DESIGN METHODS

There are two different aspects of PHE design [2]:

1. The design of individual plate types so that they conform to specific performance and operational characteristics
2. The calculation of the number and arrangement of such plates in order to satisfy the thermal and pressure-drop requirements

The thermal design method can be based on either the LMTD or ε-NTU method. Buonopane et al. [20] described the LMTD design method, and Jackson and Troupe [21] described the ε-NTU method, and both methods are described in Ref. [4] with the ε-NTU method in Ref. [5].

7.14.1 LMTD Method due to Buonopane et al. [20]

This method gives the heat transfer area needed in a PHE for both looped and series flow arrangements, given the temperature program, flow rates, and physical characteristics of the plates. The steps involved in the method are given next.

For series flow (Figure 7.8a):

1. Determine the inlet and outlet temperatures for both fluids.
2. Estimate LMTD for counterflow arrangement.
3. Estimate the Reynolds number for each stream, assuming an exchanger containing one thermal plate, one pass for each stream, as given:

$$Re = \frac{(G/n_s)D_e}{\mu} \tag{7.16}$$

where n_s is the number of substreams. For series flow, $n_s = 1$.

4. Estimate the heat transfer coefficient on both sides.
5. Estimate the overall heat transfer coefficient taking into account the wall thermal resistance.
6. Estimate the total heat transfer area from $A = q/U$ LMTD F.
7. Estimate the number of plates from $N = A/A_p$, where A_p is the area of a plate and $2A_p$ is the area per channel.

For parallelflow or looped flow (Figure 7.8b):
Repeat steps 1–7 already given.

8. From the number of thermal plates calculated in step 7, n_s is determined for both fluids. For odd N, values of n_s will be equal for both fluids, whereas for even N, n_s will be different for both fluids, and one fluid will have an additional substream compared to the other.

9. The values of n_s determined in step 8 are compared with the corresponding values assumed in step 3. If the calculated values do not agree with the assumed values, steps 3–9 are to be repeated, replacing the assumed values with the values calculated from step 8 until there is agreement between the two.

7.14.2 ε-NTU Approach

Jackson and Troupe [21] presented an ε-NTU method for the design of PHEs. The following steps illustrate the method:

1. Calculate the heat load, q, and from it determine the inlet and outlet temperatures for both fluids.
2. Calculate the bulk mean temperature and determine the thermo physical fluid properties. Estimate the heat capacity rate ratio, C^*.
3. Estimate the heat transfer effectiveness, ε, using the relation

$$\varepsilon = \frac{C_h\left(t_{h,i} - t_{h,o}\right)}{C_{min}\left(t_{h,i} - t_{c,i}\right)} = \frac{C_c\left(t_{c,o} - t_{c,i}\right)}{C_{min}\left(t_{h,i} - t_{c,i}\right)} \tag{7.17}$$

4. Assume an exchanger containing an infinite number of channels and find the required NTU using the appropriate ε-NTU relation.
5. Estimate the Reynolds number for each stream, assuming an exchanger containing one thermal plate, and one pass for each stream.
6. Calculate the heat transfer coefficient on both sides. Estimate the overall heat transfer coefficient, taking into account the wall thermal resistance.
   For series flow (Figure 7.8a):
7. Estimate the approximate number of thermal plates using the equation

$$N = \frac{\text{NTU}\left(mc_p\right)_{min}}{UA_p\Delta t_m} \tag{7.18}$$

   where Δt_m is the mean temperature difference.
8. Assuming an exchanger of $N + 1$ channels, determine NTU from the appropriate ε-NTU relationship.
9. Recalculate N from Equation 7.18.
10. Repeat the calculations in steps 8 and 9 until the calculated value of N in step 9 matches the assumed value in step 8.

For parallelflow or looped flow (Figure 7.8b): In a design involving looped flow patterns, the overall coefficient requires recalculation during each iteration because the channel flow rates become less with the addition of channels in parallel. The calculation procedure for looped flow is as follows:

1. Assuming an exchanger of N thermal plates, calculate the overall heat transfer coefficient as in step 6.
2. Estimate the approximate number of thermal plates using Equation 7.18.

TABLE 7.6
Specification Sheet for PHE

Description	Hot side	Cold side
Fluid circulated		
Flow rate		
Temperature in		
Temperature out		
Operating pressure		
Maximum pressure drop		
Specific heat		
Specific gravity or density		
Thermal conductivity		
Viscosity in		
Viscosity out		
Required gasket material		
Required plate material		
ASME code requirement		

3. Assuming an exchanger of $N + 1$ channels, determine NTU from the appropriate ε-NTU relationship for looped flow.
4. Recalculate the overall heat transfer coefficient as per step 6 and recalculate N with Equation 7.18.
5. Repeat the calculations in steps 9 and 10 until the calculated value of N matches with the assumed value.

7.14.3 Specification Sheet for PHE

The specification sheet for PHE is given in Table 7.6. The thermal design parameters and construction details of PHE adapted from Mueller, Heat Transfer Products, Springfield, USA, product literature are discussed next.

7.15 MECHANICAL DESIGN

7.15.1 ASME Code Dec VIII Div 1 Rules for Construction, Mandatory Appendix 45

These rules cover the minimum requirements for design, fabrication, assembly, inspection, testing, and documentation of gasketed, semiwelded, welded, and brazed plate heat exchangers (PHEs). These rules cover the common types of PHEs and their elements but are not intended to limit the configurations or details to those illustrated or otherwise described herein.

7.15.2 Materials of Construction

All pressure-containing parts shall be constructed using ASME code approved materials. Metallic and nonmetallic materials not permitted by this Division may be used specifically for heat transfer plates within the PHE, provided there is an applicable Code Case published for the limited use of this material as heat transfer plates within a plate pack.

7.15.3 Pressure Test Requirements

A PHE shall be hydrostatically tested in accordance with UG-99, or pneumatically tested in accordance with UG-100. The heat transfer plates shall not be included when determining the lowest stress ratio.

7.15.4 Manufacturer's Data Reports

A Manufacturer's Data Report shall be completed by the manufacturer for each PHE, or same-day production of identical vessels in accordance with Mandatory Appendix 35.

7.15.5 Design Considerations

1. Plate Packs Using Gaskets. Gasketed plate packs shall be designed to contain pressurized fluid without leaking to a pressure of at least 1.3 times the MAWP. The MAWP of gasketed plate packs may be determined without performing proof testing or design calculations for the gasketed plate pack, provided the following requirements are met:

a. The MAWP for the plate heat exchanger shall be determined considering all other pressure-retaining parts, including the endplates, bolting, and nozzles.
b. The nominal thickness of a single-wall heat transfer plate or the combined thickness of a double-wall heat transfer plate shall not be less than 0.014 in. (0.35 mm).
c. The heat exchanger shall not be used in lethal service.

2. Fully Welded PHEs. The MAWP of fully welded PHEs may be determined using methods found in UG-101.

7.15.6 Design Pressure

The PHE shall be designed, fabricated, and tested in accordance with the requirements of ASME Code Section VIII, Division 1. The exchanger shall be code stamped for MPa (psi) design pressure at the maximum—°C (°F) and minimum—°C(°F) fluid temperatures specified. The test pressure will be 130% of the design pressure. The exchanger shall be designed to withstand full design pressure in one circuit with zero pressure in the opposite circuit.

Design requirements: The heat exchanger performance, in clean condition, shall be in accordance with ARI 400-2001.

Plates: Plates shall be fabricated of _____material (Type 304 or 316 stainless steel, titanium, Incoloy, Hastalloy, or other material) with a minimum plate thickness of 0.5 mm (nominal).

Gaskets: Gaskets shall be _____ (nitrile butadiene rubber, hypalon, viton, neoprene, or other material) as specified by the user.

Connections: All inlet and outlet connections shall be designed to accept either ANSI flanged or IPS threaded connections. Studded port connections shall be lined with metal or the same elastomer as provided on the plates.

Frame: Fixed and movable end frames shall each be constructed so as to eliminate any need for adding stiffeners to provide reinforcement for less frame thickness. Frames shall be made from SA-516 or SA-515-70. The frame assembly shall be coated with corrosion-resistant paint.

Compression bolts: Compression bolts, nuts, and washers shall be galvanized, low-alloy steel (SA-193-B7/SA-194-2H).

Future expansion: The exchanger shall have frame capacity for future expansion of a minimum of 20%.

7.15.7 Plate Hanger

Plates shall be of a one-piece design, without the need for loose removable plate hanger components. The plates shall be supported by a Type 304 stainless steel plate hanger. The heat exchanger shall have a plate-positioning system that will prevent shifting during tightening of the plate pack and during unit operation.

Shroud: The top and sides of the plate pack shall be entirely enclosed within a protective shroud.

Overall size: The exchanger shall not exceed—mm (in.) height,—mm (in.) width, and—mm (in.) length.

7.16 MAINTENANCE PRACTICES

Damage to equipment can be caused by external forces, corrosion, chemical action, erosion, material exhaustion, water hammer, thermal and/or mechanical shock, freezing, wrong transport, lifting, etc. [22].

i. The heat exchanger may only be used with the fluids specified on the datasheet.
ii. The hot medium may not flow through the exchanger without the cold medium flowing through. This is to prevent damage to the exchanger.
iii. Sudden pressure and temperature changes should be avoided.
iv. When a heat exchanger (filled with water or a water mixture) which is not in operation is exposed to temperatures below zero, the plates can become deformed.
v. If a danger of frost occurs, the heat exchanger should be drained completely.
vi. "Cold leakage" is caused by a sudden change in temperature. The sealing properties of certain elastomers are temporarily reduced when the temperature changes suddenly. No action is required as the gaskets should re-seal after the temperature has stabilized.

7.17 GASKET FAILURES

Gasket failures are generally a result of

i. Outlived its life
ii. Excessive exposure to ozone
iii. High operating temperature—above the temperature limit of the material
iv. Exposure to pressure surges
v. Chemical attack
vi. Physical damage, resulting from incorrect assembly work.

Gaskets can pose several problems such as leakage and corrosion. Conditions that can cause gasket leakage include [23]:

i. Improper tightening
ii. Incorrect plate installation
iii. Pressure surges from the pump or other components of the system
iv. Use of the wrong gasket material
v. Long periods of use in a high-pressure or high-temperature application

7.18 PERFORMANCE DETERIORATION

A decrease in performance is generally a result of [22]:

i. Plate surfaces require cleaning or de-scaling
ii. Pumps or associated controls have failed
iii. Plate channels blocked
iv. Liquid flow rate not as per design specification
v. Associated chiller/cooling tower/boiler undersized or dirty
vi. Cooling water temperature to the exchanger is higher than the design
vii. Heating media temperature to the exchanger is lower than the design
viii. Steam flow not sufficient—control valve malfunction
ix. Steam trap broken or jammed—unit becomes filled with condensate
x. Plate package has been assembled incorrectly
xi. Unit is running in co-current flow, instead of counter current—check with contract drawing and alter pipework if necessary. Check direction of pump flows
xii. Air lock has developed in the plate package or pipework

7.19 CORROSION OF PLATE HEAT EXCHANGERS

It is general practice for plate exchanger manufacturers to use only corrosion-resistant materials for a PHE, dictated by factors such as product purity and to minimize corrosion fouling. However pitting of plates take place in the fluid's flow area and corrosion around port holes, as shown in Figure 7.38. Erosion is not a governing factor, since pressure-drop limitations will generally determine the maximum permissible fluid velocities. As the plates are very thin (0.6–1.2 mm) compared to the tube thickness (maximum permitted up to 4.2 mm in the case of 19.5 mm diameter tube dimensions, conforming to TEMA Standards [24]), corrosion allowances normally recommended for tubular units are not relevant for plate units. In the design of a tubular unit, a corrosion allowance of 0.125 mm/year is used; for plate units the corrosion rate should not exceed 0.05 mm/year [3].

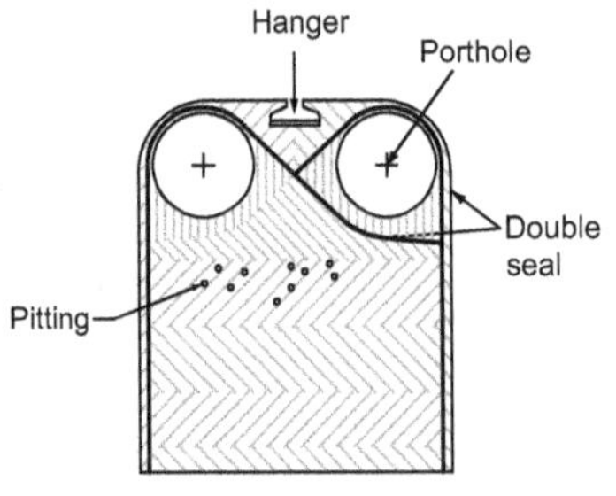

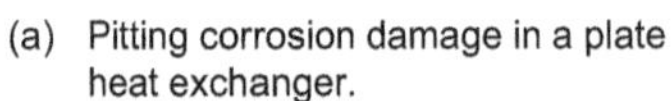
(a) Pitting corrosion damage in a plate heat exchanger.

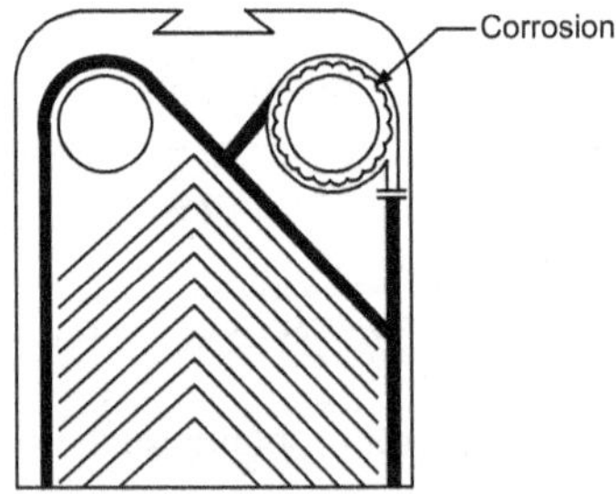

(b) Crevice corrosion under the gasket of a plate heat exchanger.

FIGURE 7.38 Illustration of corrosion of fluid port and pitting of plate surface.

7.20 FOULING

The types of fouling generally that occur in the plate fin heat exchanger include sedimentation (solids in suspension), biological (marine or organic), scaling or deposition (calcium carbonate), and corrosion. Strainers for the inlet port or the inspection port in the movable cover are available to protect the plate pack from large particulates clogging the channels. Hence, the fouling factors required in PHEs are small compared with those commonly used in shell and tube designs [7]. They should be normally 20–25% of those used in shell and tube exchangers. For design purposes, fouling values not greater than one-fifth of the published tubular figures are recommended by Cooper et al. [2]. Fouling factors for PHEs are tabulated by Marriott [7]. Marriott states that under no circumstances, even under low pressure-drop conditions, is a total fouling resistance exceeding 0.00012 $m^2 \cdot °C/W$ or 0.0006 $ft^2/°F \cdot h/Btu$ recommended.

7.21 CONDITIONS CAUSING WATER HAMMER

Four conditions have been identified as causes of violent reactions from water hammer [25]: (1) hydraulic shock; (2) thermal shock; (3) flow shock; and (4) differential shock.

Water hammer occurs when the installation pipelines carry incompressible fluids such as water, ethylene glycol, etc., and the fluid flow suddenly changes its velocity. Abruptly stopping the fluid flow produces a substantial pressure rise. In a BPHE, the water hammer will cause a bulge in the front or back plate, resulting in internal/external leakage, as illustrated in Figure 7.39. To avoid or eliminate these problems, the designer can install an air chamber or a water hammer arrester. Another way to control water hammer is to use valves with controlled closing times or controlled closing characteristics, i.e., to avoid water hammer, do not use fast-closing valves.

The pressure in the system can be seen to the right, for a "standard" quick closing valve, and to the left, for a slow closing time-controlled valve as shown in Figure 7.40 [25, 26].

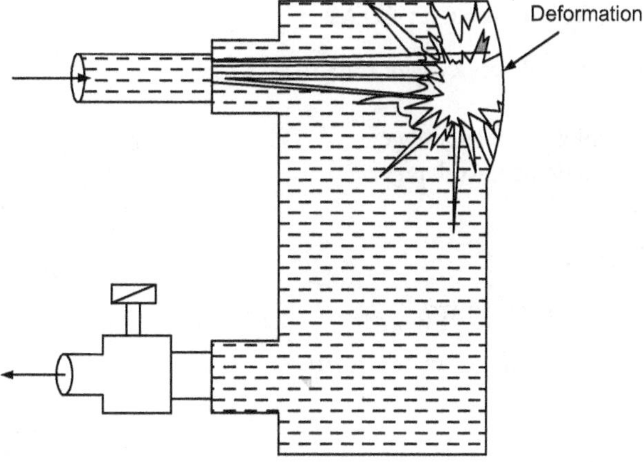

FIGURE 7.39 Water hammer causing deformation in a BPHE. (Adapted from www.swep.net/refrigerant-handbook/8.-practical-advice/qw/.)

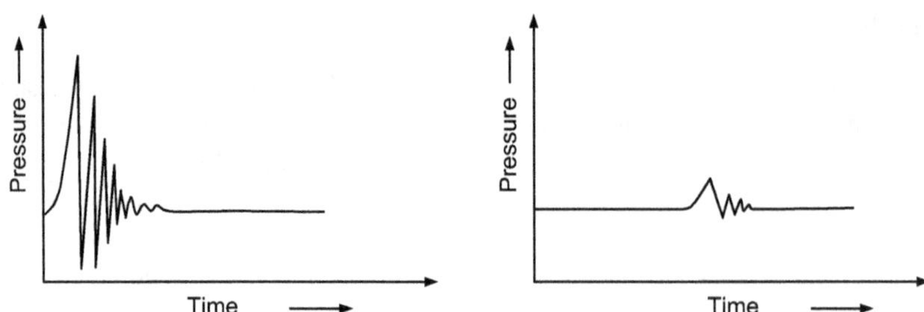

FIGURE 7.40 The pressure in the system can be seen to the right, for a "standard" quick closing valve, and to the left, for a slow closing time-controlled valve. (Adapted from www.swep.net/refrigerant-handbook/8.-practical-advice/qw/.)

7.22 PLATE HEAT EXCHANGER SERVICES

Proper service of a gasketed plate heat exchanger, performed at the right time, prevents unplanned interruptions. Whatever your equipment's type or age, there are options to safeguard or enhance its operation [27]:

i. Performance audits—qualified engineers will assess current performance levels and recommend improvements
ii. Cleaning—on- and off-site mechanical cleaning, CIP systems, and cleaning agents
iii. Reconditioning and repair—from tune-up to complete reconditioning
iv. Service equipment—Cleaning-in-place units, backflush valves, filters, and special tools
v. Spare parts from OEM—quality spare parts and service kits
vi. Upgrading—replacing or modifying existing equipment can significantly increase performance
vii. Troubleshooting—if a user experience issues with heat exchangers, find out why and prevent them from happening again
viii. Performance agreements—allow to tailor service solutions based on specific needs

7.22.1 Performance Monitoring by OEMs

The two most common types of services are [28]:

1. Troubleshooting a plate heat exchanger
2. Reconditioning of plate heat exchanger

7.23 CLEANING OF PHE

7.23.1 Manual Cleaning of Plates

The procedure for manual cleaning of plates includes:

1. Dismantle the exchanger
2. Open the unit
3. Clean each plate separately
4. Wipe off the mating surface
5. Inspect and install each plate, then the unit may be closed

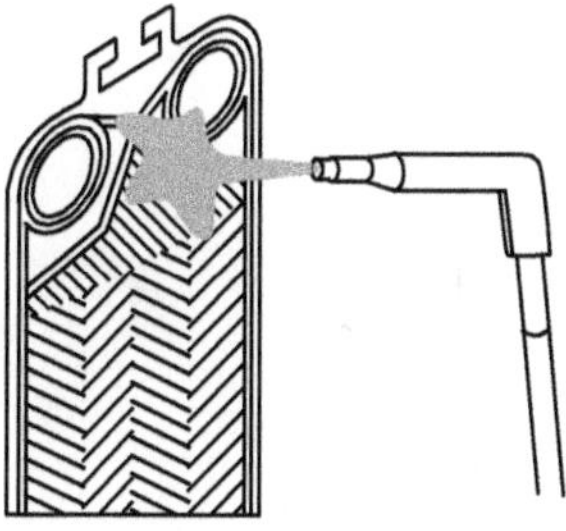

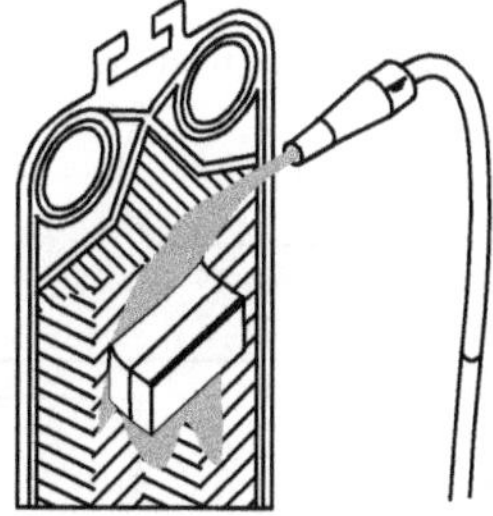

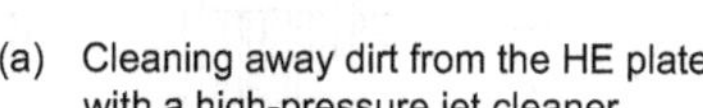

(a) Cleaning away dirt from the HE plate with a high-pressure jet cleaner.

(b) Cleaning of a HE plate with a soft brush under running water.

FIGURE 7.41 Manual cleaning of PHE plates.

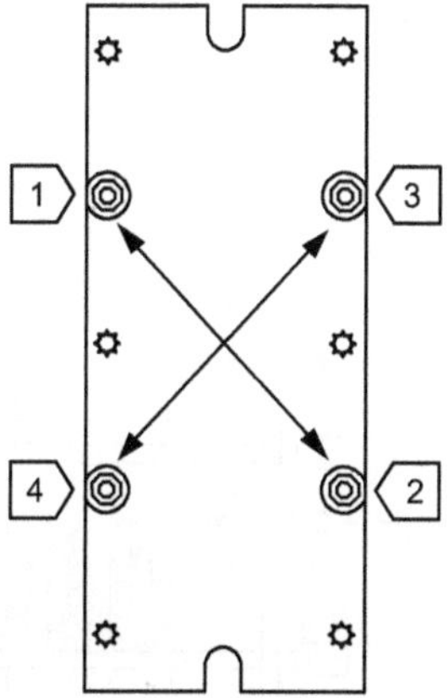

FIGURE 7.42 Schematic of flange bolts tightening scheme.

Figure 7.41 shows manual cleaning.

For tightening bolts in flanged connections, which is shown in Figure 7.42, make sure that a torque wrench is used for applying the correct torque.

7.23.2 Chemical Cleaning: Cleaning-in-Place (CIP)

A CIP system allows cleaning of the unit without disassembling, which is quick, cost-effective, and easy cleaning. Chemicals used for cleaning must be compatible with the construction materials. The cleaning is accomplished by circulating a suitable cleaning solution through the plate heat exchanger instead of opening it. The cleaning solution must be able to dissolve the fouling on the plates. Back-flushing in reverse flow with a diverter valve is also an effective method for cleaning the port area without opening the heat exchanger unit. Fibers that can pose a problem for standard PHEs can be dealt with by wide-gap or free-flow plate designs. CIP is shown schematically in Figure 7.43

Caution: Do not use chlorine or chlorinated water to clean stainless steel. Do not use phosphoric or sulfamic acid for cleaning titanium plates.

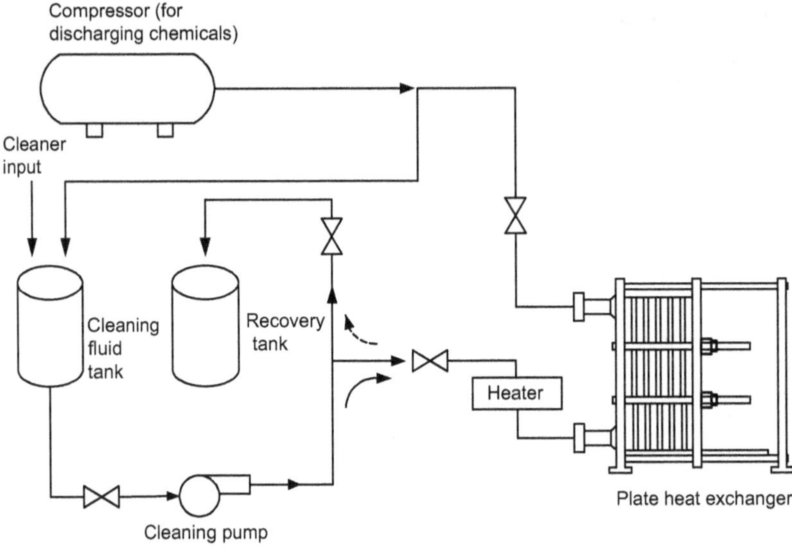

FIGURE 7.43 Cleaning-in-place.

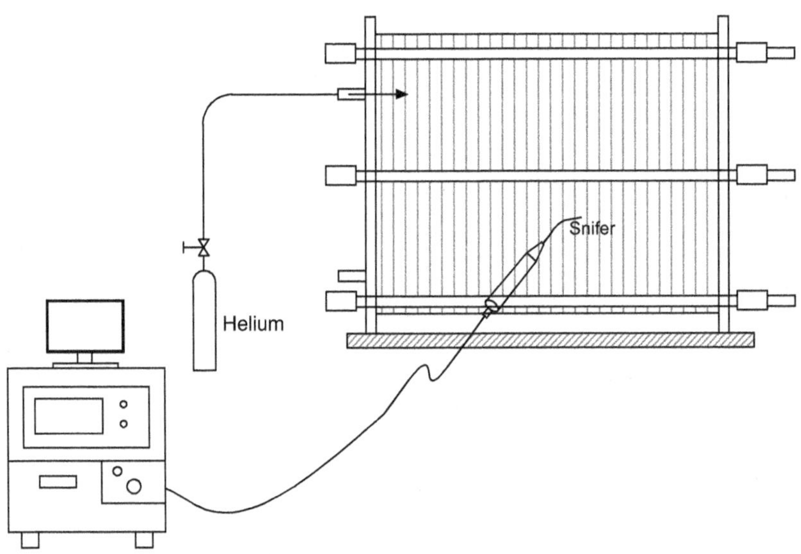

FIGURE 7.44 Leak testing of PHE.

7.24 LEAK TESTING OF PHE

The gasket inspection, also known as the leak detection process, is conducted to find external points of leakage due to defective or brittle gaskets. For this, the plate heat exchanger is pressurized up to the maximum test gas pressure, then a handheld sensor is passed externally down the length of the gaskets. This enables leaks to be detected with pinpoint precision and then remedied. Since these external leaks are usually the result of embrittled gaskets, the unit can, if necessary, be clamped even more tightly. Leak testing of PHE is shown in Figure 7.44 [29].

7.25 SPIRAL PLATE HEAT EXCHANGERS

A spiral plate heat exchanger (SPHE or SHE) is fabricated by rolling a pair of relatively long strips of plate to form a pair of spiral passages (Figures 7.45 and 7.46). Channel spacing is maintained uniformly along the length of the spiral passages by means of spacer studs welded to the plate strips prior to rolling. It can be made with channels 5–25 mm wide, with or without studs. The spiral channels are welded shut on their ends in order to contain the respective fluids. An overall gasket is applied to the cover. The covers are attached to the spiral element by means of forged hook bolts and

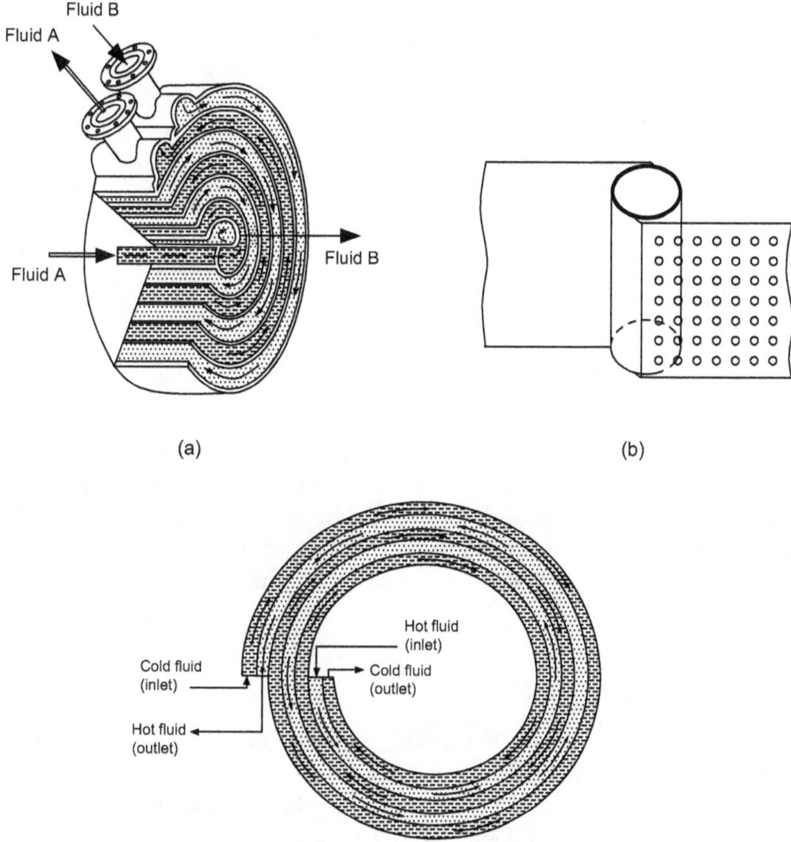

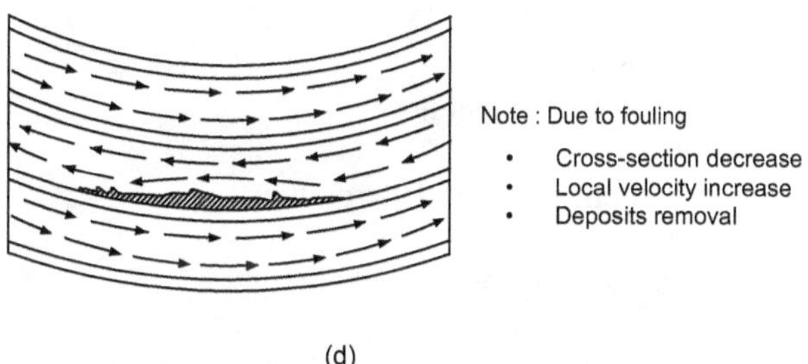

FIGURE 7.45 Spiral plate heat exchanger. (a) Unit (b) two plates design (with studded or plain design), (c) SPHE flow pattern and (d) scrubbing action of fluid flow.

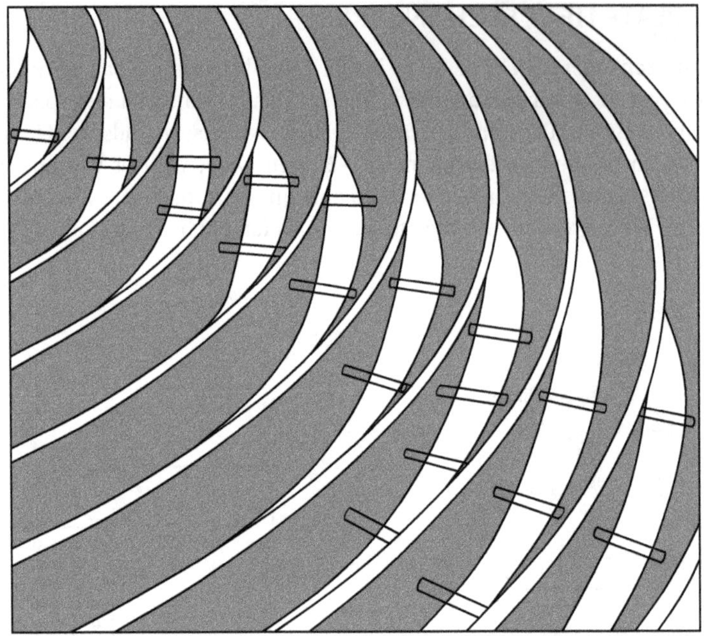

FIGURE 7.46 Stud welded Spiral plate heat exchanger.

adapters. The hook bolt engages the bevel at the back of the flange ring and the adapter engages the rim at the edge of the cover. A header is welded on the outer end of each passage to accommodate the respective peripheral nozzle. For most services, both fluid flow channels are closed by alternate channels welded at both sides of the spiral plate. In some applications, one of the channels is left completely open, and the other closed at both sides of the plate. These two types of construction prevent the fluids from mixing.

7.25.1 Features of Flows in an SPHE

i. The continuously curving, single-channel geometry creates high turbulence and hence minimizes fouling from the start.
ii. The fluid is fully turbulent at a much lower velocity than in straight tube heat exchangers, and each fluid travels at constant velocity throughout the whole unit. This eliminates the risk of dead spots and stagnation for more efficient heat transfer.

7.25.2 Flow Arrangements and Applications

a. Possible flow configurations include
   i. Type A: counterflow or cocurrent flow
   ii. Type B: crossflow
   iii. Type C: counter-crossflow or parallel-crossflow

These flow arrangements are shown schematically in Figure 7.47.

b. Flow patterns as per spiral plate construction arrangements
   1. Both fluids in spiral flows; this arrangement can accommodate the media in full counterflow. General uses are for liquid to liquid, condensers, and gas coolers.

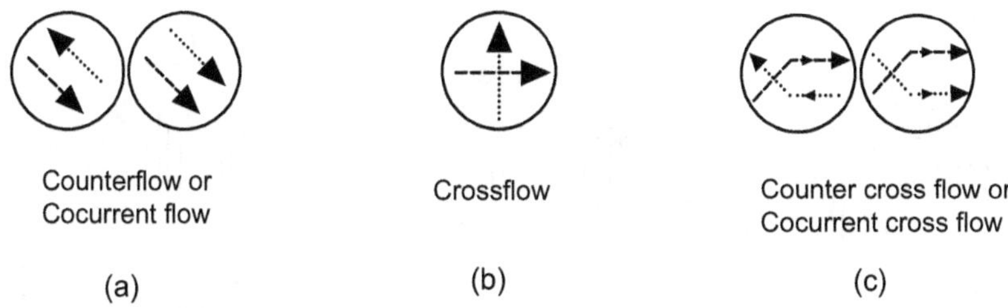

FIGURE 7.47 Representation of fluids flow arrangements in a SPHE.

2. One fluid in spiral flow and the other in axial flow across the spiral; general uses are as condensers, reboilers, and gas coolers and heaters.
3. One fluid in spiral flow and the other in a combination of axial and spiral flow; general uses are as condensers (with built-in aftercoolers) and vaporizers.

7.25.3 Types of SPHE

Based on flow patterns in the plates channel, there are types of SPHE, namely, Types 1, 2, 3 and 4, which are shown in Figure 7.48 [30.1,30.2]. The details of construction, flow arrangements, and areas of applications are discussed below.

7.25.3.1 Type 1 Spiral Heat Exchanger (Countercurrent Flow)

The Type 1 is preferred when one or both fluids are fouling. Since the fluid flows continuously each in a single channel, the force of the fluid acts against any deposits, "pushing" them through the channel and out the other end. In a few cases, Type 1 can be used as a condenser, for example when the cooling media are heavily fouling, or a very close temperature approach is required. Type 1 is almost always a liquid/liquid heat exchanger. In most cases it is mounted horizontally and can be opened at each end. Each channel circuit is seal welded on one side and open on the other. The open channel is sealed by the gasket face of the end cover. The fluid in the two channels always follows the spiral direction, from the outside towards the center, or the center to the outside. The flow arrangement is countercurrent. The unit is preferably mounted horizontally. In the vertical position there is a risk of solid material settling by gravity at the bottom of the channel.

Applications

i. Fouling liquids containing solids, fibers, liquors, slurries, and sludges.
ii. Liquid/liquid: Preheating, heating, cooling, interchanging, heat recovery.

7.25.3.2 Type 2 Spiral (Crossflow)

The Type 2 spiral is used for two-phase duties, both condensing and evaporation (re-boiling). In this configuration the spiral is always in the vertical position. The channel with the fluid being condensed or evaporated is usually open at each end so, unlike in Type 1, the fluid flows across the spiral rather than following the spiral. The liquid on the other circuit, whether heating or cooling, flows along the spiral similar to Type 1 and this channel is seal welded at each end. The Type 2 spiral is always crossflow rather than countercurrent flow. This means that it is thermally less efficient than Type 1. Generally, there is a very low pressure drop on the condensing circuit so it is very well suited to vacuum condensation duties.

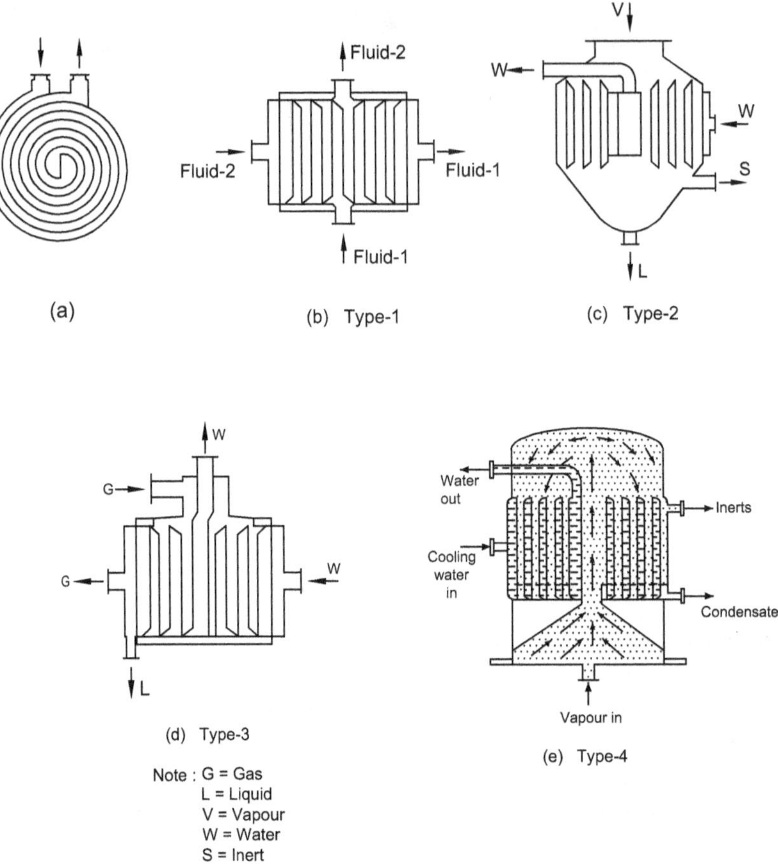

FIGURE 7.48 Alfa Laval SPHE types and their flow arrangements (L, condensate; G, gas; W, liquid; V, vapor; and S, noncondensables).

7.25.3.3 Type 3 Spiral Heat Exchanger

The Type 3 SHE is often called a "steam heater" and the construction features like a hybrid of Type 1 and Type 2. The main feature is that the liquid being heated by the steam is generally fouling, therefore the circuit is designed to be accessible for cleaning, which is not the case with the normal Type 2. In addition, the countercurrent nature of Type 3 allows tighter temperature approaches than are achievable with the cross-flow nature of Type 2.

7.25.3.4 Type 4 Spiral Heat Exchanger

Type 4 is an SHE integrated into a column, often as part of a distillation process. In most cases the spiral operates similarly to the Type 2 version, where the operation is in crossflow and the open channel allows an extremely low pressure drop in vacuum situations. Another extremely important feature and benefit is that the two fluids are never in physical contact so there is no cross-contamination, which often happens in traditional columns. The SHE is always mounted vertically and the diameter of the spiral is the same as the diameter of the column in cases where several spirals are stacked in series. The efficiency of the SHE allows the column to be both shorter in height and smaller in diameter than traditional solutions, thus offering major savings in infrastructure and installation costs.

7.25.4 Construction Material

Spiral plate exchangers are fabricated from any material that can be cold worked and welded. Typical construction materials include carbon steel, stainless steels, Hastelloy B and C, nickel and nickel alloys, aluminum alloys, titanium, and copper alloys. To protect against corrosion from cooling water, the surface is given a baked phenolic resin coatings, among others. Anodic elements may also be wound into the assembly to anodically protect surfaces against corrosion [31].

7.25.5 Thermal Design of Spiral Plate Heat Exchangers

The thermal design procedures of SPHEs are covered by Minton [31]. The system of ordinary differential equations for the temperature distribution in SPHEs has been solved numerically by Chaudhury et al. [32] to obtain the efficiency. The LMTD correction factor F is a function of number of transfer units, number of turns n, and the heat capacity rate ratio. It was found that the LMTD correction factors, when plotted versus NTU per turn, fall approximately on a single curve and that curve can be represented by the simple formula [32]:

$$F = \frac{n}{\mathrm{NTU}} \tanh\left(\frac{NTU}{n}\right) \tag{7.19}$$

7.25.6 Mechanical Design of Spiral Plate Heat Exchangers

Spiral exchangers can be furnished in accordance with most pressure vessel codes. Sizes range from 0.5 to 350 m^2 of heat transfer surface in a single spiral body. The maximum design pressure is normally limited to 150 psi for the following reasons, although for smaller diameters the pressure may sometimes be higher [31]:

1. They are normally designed for the full pressure of each passage.
2. Because the turns of the spiral are of relatively large diameter, each turn must contain its design pressure.
3. The plate thickness is somewhat restricted.
4. Limitations of construction material govern design temperature.

7.25.7 Applications for Spiral Plate Heat Exchangers

The SPHE is particularly suited for such liquid-to-liquid duties as:

1. Clogging slurries such as PVC slurries
2. Clogging particle-laden and fibrous media such as TMP condensate in pulp and paper production
3. Clogging and erosive media such as alumina, hydrate slurries, etc.
4. Fouling media such as waste water or sewage sludge
5. High-viscosity media such as heavy oil
6. Non-Newtonian fluids such as fermenting froth in pharmaceuticals processing

7.25.8 Advantages of Spiral Plate Exchangers

SPHEs have a number of advantages over conventional shell and tube exchangers:

1. The spiral heat exchanger approaches the ideal in heat exchanger design. Media can be arranged in full counterflow.
2. The exchanger is well suited for heating or cooling viscous fluids because its *L/D* ratio is lower than that of tubular exchangers [31].
3. They are easily maintainable. By removing the covers of the spiral exchanger, the entire lengths of the passages are easily accessible for inspection or mechanical cleaning, if necessary.
4. Spiral plate exchangers avoid problems associated with differential thermal expansion in non-cyclic service [31].
5. In axial flow, a large flow area affords a low pressure drop, which becomes especially important when condensing under vacuum [31].

7.25.9 Limitations

Besides the pressure limitation noted earlier, the spiral plate exchanger also has the following disadvantages [31]:

1. Repairing an SPHE in the field is difficult; however, the possibility of leakage in a spiral is less because it is generally fabricated from much thicker plate than tube walls.
2. Spiral plate exchangers are not recommended for service in which thermal cycling is frequent. When used for such services, the unit sometimes must be designed for higher stresses.
3. The SPHE usually should not be used when a hard deposit forms during operation, because the spacer studs prevent such deposits from being easily removed by drilling.

Nomenclature

A	total heat transfer area, m^2 (ft^2)
A_p	heat transfer area of a plate, m^2 (ft^2)
b	mean plate gap, m (ft)
C	constant in Equation 7.3
C_1	constant in Equation 7.7
C_2	constant in Equation 7.8
C_{min}	minimum of $(mc_p)_h$, and $(mc_p)c$, W/°C (Btu/h·°F)
C_c	specific heat of cold fluid, J/kg °C (Btu/lbm.°F)
C_h	specific heat of hold fluid, J/kg °C (Btu/lbm.°F)
C_p	specific heat, J/kg·°C (Btu/lbm.°F)
C^*	heat capacity rate ratio = $(mc_p)_{min}/(mc_p)_{max}$
D_e	equivalent diameter of flow passages, m (ft)
D_h	hydraulic diameter, m (ft)
F	LMTD correction factor
f	Fanning friction factor
G	mass velocity based on minimum flow area, kg/m^2·s (lbm/h·ft^2)
g_c	acceleration due to gravity or proportionality constant 9.81 m/s^2 (32.17 ft/s^2) = 1 and dimensionless in SI units
h	convective heat transfer coefficient, W/m^2·°C (Btu/ft^2 h·°F)
J	Jenson number (of specific pressure drop), Pa/NTU (lbf/ft^2/NTU)
j	Colburn heat transfer factor
k	thermal conductivity, W/m°C (Btu/h·ft·°F)
L	effective flow passage length, m (ft)
LMTD	log mean temperature difference, °C (°F)
m	mass flow rate of fluid, kg/s (Lb/s)
m, n	exponents in Equation 7.3
N	number of thermal plates
n	number of turns of spiral plate heat exchanger, Equation 7.19

Nu	Nusselt number
N_p	number of fluid ports, Equation 7.10
n_s	number of substreams
Pr	Prandtl number
p	pressure, Pa (lbf/ft^2)
Q	total heat duty of the exchanger, W·s (Btu)
p	pitch, center distance between two troughs or chevron pitch
q	heat transfer rate, W (Btu/h)
Re	Reynolds number
R_r	fouling resistance, m^2·°C/W (ft^2·°F·h/Btu)
t	fluid static temperature, °C (°F)
U	overall heat transfer coefficient, W/m^2·°C (Btu/h·fr^2·°F)
U_m	fluid velocity through port or manifold, m/s (ft/h)
W	width of the plate (gasket to gasket), m (ft)
x	exponent in Equation 7.3
y	exponent in Equation 7.8
β	chevron angle
Δp	pressure drop, Pa (lbf/ft^2)
Δp_c	pressure drop within the plate passages, Pa (lbf/ft^2)
Δp_h	pressure drop due to elevation, Pa (lbf/ft^2)
Δp_m	pressure drop associated with the inlet and outlet manifolds and ports, Pa (lbf/ft^2)
Δt_{lm}	log mean temperature difference, °C (°F)
Δt_m	mean temperature difference, °C (°F)
ε	thermal effectiveness
μ_b	viscosity at bulk mean temperature, Pa·s (lbm/h·ft)
μ_w	viscosity at wall temperature, Pa·s (lbm/h·ft)
ϕ^*	ratio of actual (developed) surface area to projected surface area.
ρ	fluid density, kg/m^3 (lbm/ft^3)
θ	number of transfer units (NTU)

Subscripts

c	cold
h	hot
i	inlet
m	mean
o	outlet
min	minimum
max	maximum

REFERENCES

1. Focke, W.W. (1983) *Plate heat exchangers: Review of transport phenomena and design procedures*, CSIR Report CENG 445. CSIR, Pretoria, South Africa.
2. Cooper, A., Usher, J.D. (1983) Plate heat exchangers. In: *Heat Exchanger Design Handbook*, Vol. 3, (E.U. Schlunder, editor-in-chief), Section 3.7. Hemisphere, Washington, DC.
3. Raju, K.S.N., Bansal, J.C. (1983) Plate heat exchangers and their performance. In: *Low Reynolds Number Flow Heat Exchangers* (S. Kakac, R.K. Shah, A.E. Bergles, eds.), pp. 899–912. Hemisphere, Washington, DC.
4. Raju, K.S.N., Bansal, J.C. (1983) Design of plate heat exchangers. In: *Low Reynolds Number Flow Heat Exchangers* (S. Kakac, R.K. Shah, A.E. Bergles, eds.), pp. 913–932. Hemisphere, Washington, DC.

5. Shah, R.K., Focke, W.W. (1988) Plate heat exchangers and their design theory. In: *Heat Transfer Equipment Design* (R.K. Shah, D.C. Subbarao, R.M. Mashelekar, eds.), pp. 227–255. Hemisphere, Washington, DC, 1988,.
6. *Alpha-Laval product literature.* M/s, Alpha-Laval (India) Limited, Pune, India.
7. Marriott, J. (1979) Where and how to use plate heat exchangers. In: *Process Heat Exchange, Chemical Engineering Magazine* (V. Cavaseno, ed.), pp. 156–162. McGraw-Hill, New York.
8. *Mueller Accu-Therm Plate heat exchangers.* AT-1601–4, Paul Muller Company, Springfield, MO.
9. assets.alfalaval.com/documents/p507ac7c5/alfa-laval-alfa-laval-diabon-product-leaflet-en.pdf
10. Saunders, E.A.D. (1989) Gasketed plate heat exchangers. In: *Heat Exchangers: Selection, Design and Construction*, Chapter 16. Addison Wesley Longman, Reading, MA.
11. Trom, L. (1996) Heat exchangers: Is it time for a change? *Chemical Engineering* February: 70–73.
12. www.google.com/search?q=compabloc+fully+welded+heat+exchanger
13. www.alfalaval.in/products/heat-transfer/plate-heat-exchangers/welded-plate-and-block-heat-exchangers/compabloc/
14. www.alfalaval.com/microsites/gphe/types/diabon/
15. Caeiula, L., Rudy, T.M. (1983) Prediction of plate heat exchanger performance. In: *AiChE Symposium Series Heat Transfer*, pp. 76–89. Seattle, WA.
16. Buonopane, R.A., Troupe, R.A. (1969) A study of the effects of internal rib and channel geometry in rectangular channels, Part I: Pressure drop & Part II: Heat transfer. *AiChE Journal* July: 585–596.
17. APV (2008) *Heat Transfer Handbook, A History of Excellence*. APV Co., Getzville, NY.
18. Usher, J.D. (1979) Evaluating plate heat exchangers. In: *Process Heat Exchange, Chemical Engineering Magazine* (V. Cavaseno, ed.), pp. 145–149. McGraw-Hill, New York.
19. Jenson, S. (1960) Assessment of heat transfer data. *Chemical Engineering Progress Symposium Series, Heat Transfer—1960* 56: 195–201.
20. Buonopane, R.A., Troupe, R.A., Morgan, J.C. (1963) Heat transfer design method for plate heat exchangers. *Chemical Engineering Progress* 59: 57–61.
21. Jackson, B.W., Troupe, R.A. (1966) Plate exchanger design by ε-ntu method. *Heat Transfer Los Angeles, Chemical Engineering Progress Symposium Series* 62: 185–190.
22. *Installation & operation manual: hrs tubular heat exchangers*, pp. 1–18. HRS Heat Exchangers Ltd, Watford, UK.
23. www.alfalaval.sg/service-and-support/ten-top-tips/ten-top-tips-gasketed-plate-heat-exchangers/
24. *Standards of the Tubular Exchanger Manufacturers Association*, 10th edn. (2019) Tubular Exchanger Manufacturers Association, Tarrytown, NY.
25. www.swep.net/refrigerant-handbook/8.-practical-advice/qw/
26. www.swep.net/refrigerant-handbook/refrigerant-handbook/
27. www.alfalaval.com/microsites/gphe/services/
28. www.alfalaval.sg/products/heat-transfer/plate-heat-exchangers/gasketed-plate-and-frame-heat-exchangers/heat-exchanger/how-plate-heat-exchanger-work/
29. www.resom.com/fileadmin/user_upload/Hydrogen_Leak_Detection.pdf
30.1 *Alfa Laval spiral heat exchangers*, pp. 1–11.
30.2 https://cesehsa.com.mx/cesehsa/wp-content/uploads/2021/09/PPI00424ENSHE.pdf
31. Minton, P.E. (1970) Designing spiral plate heat exchangers. *Chemical Engineering* 77: 103–112.
32. Chowdhury, K., Linkmeyer, H., Bassiouny, M.K., Martin, H. (1985) Analytical studies on the temperature distribution in spiral plate heat exchangers. *Chemical Engineering Progress* 19: 183–190.

Bibliography

Bond, M.P. (1981) Plate heat exchangers for effective heat transfer. *Chemical Engineering* 367: 162–166.

Cowan, C.T. (1979) Choosing materials of construction for plate heat exchangers—I. In: *Process Heat Exchange, Chemical Engineering Magazine* (V. Cavaseno, ed.), pp. 165–167. McGraw-Hill, New York.

Cowan, C.T. (1979) Choosing materials of construction for plate heat exchangers—II. In: *Process Heat Exchange, Chemical Engineering Magazine* (V. Cavaseno, ed.), pp. 168–169. McGraw-Hill, New York.

Cross, P.H. (1979) Preventing fouling in plate heat exchangers. In: *Process Heat Exchange, Chemical Engineering Magazine* (V. Cavaseno, ed.), pp. 211–214. McGraw-Hill, New York.

Edwards, M.F. (1983) Heat transfer in plate heat exchangers at low Reynolds numbers. In: *Low Reynolds Number Flow Heat Exchangers* (S. Kakac, R.K. Shah, A.E. Bergles, eds.), pp. 933–947. Hemisphere, Washington, DC.

Marriott, J. (1971) Where and how to use plate heat exchangers. *Chemical Engineering* 78: 127–133.

Saunders, E.A.D. (1989) Thermal appraisal: Gasketed plate heat exchangers. In: *Heat Exchangers: Selection, Design and Construction*, Chapter 16. Addison Wesley Longman, Reading, MA.

Shah, R.K., Kandlikar, S.G. (1986) The influence of the number of thermal plates on plates heat exchanger performance. In: *Current Researches in Heat and Mass Transfer, A Compendium and a Festschrift for Professor Arcot Ramachandran*, pp. 267–288. Hemisphere, Washington, DC.

www.alfalaval.com/globalassets/documents/products/heat-transfer/plate-heat-exchangers/gasketed-plate-and-frame-heat-exchangers/industrial/instruction-manual-gphe-small-en.pdf

www.alfalaval.my/products/heat-transfer/plate-heat-exchangers/gasketed-plate-and-frame-heat-exchangers/heat-exchanger/how-plate-heat-exchanger-work/

https://core.ac.uk/download/pdf/196519648.pdf

www.kelvion.com/fileadmin/user_upload/Kelvion/Downloads/Products/Plate_Heat_Exchangers/Gasketed/Kelvion_Operation_Manual_Gasketed_PHE_EN.pdf

8 Heat Transfer Augmentation

8.1 INTRODUCTION

The study of improved heat transfer performance is referred to as heat transfer enhancement or augmentation. Enhancement of heat transfer plays a significant role in maximizing the efficiency of many applications such as heat exchangers, power generation, and microelectronics. Augmentation of the convective heat transfer coefficient will minimize the size and cost of the heat exchanger. The enhancement of heat transfer has concerned researchers and practitioners since the earliest documented studies of heat transfer [1]. In recent years, increasing energy and material costs have provided significant incentives for the development of energy-efficient heat exchangers. For enhancing the heat transfer in a heat exchanger, and minimizing the surface area with materials cost, different methods can be implemented such as the passive method, active method, and hybrid method. As a result, considerable emphasis has been placed on the development of various augmented heat transfer surfaces and devices [2,3]. This chapter describes some of the practical considerations and advantages regarding the use of enhanced tubes and tube inserts in tubular heat exchangers and provides some guidelines for identifying their applications. Also, heat transfer augmentation devices are discussed in detail without specific correlations for j and f factors. Important reference sources on heat transfer augmentation devices are Bergles [4] and Engineering Data Book III, Wolverine Tube Inc. [5].

8.1.1 Benefits of Heat Transfer Augmentation

An enhanced surface is more efficient in transferring heat than what might be called the standard surface. Enhanced tubes are used extensively in the refrigeration and air-conditioning, power, petroleum refinery, process industries, etc. Designing enhanced tubular heat exchangers results in a much more compact design than conventional plain tube units, obtaining not only thermal, mechanical, and economical advantages for the heat exchanger, but also for the associated support structure, piping, and/or skid package unit, and also notably reduced cost for shipping and installation of all these components (which often bring the installed cost to a factor of two to three times that of the exchanger itself in petrochemical applications) [5]. The compact enhanced designs also greatly reduce the quantities of the two fluids resident within the exchanger, which can be an important safety consideration. The application of an enhancement device can result in benefits such as:

1. A decrease in heat transfer surface area, size, and hence weight of a heat exchanger for a given heat duty and pressure drop
2. An increase in heat transfer for a given size, flow rate, and pressure drop

DOI: 10.1201/9781003352044-8

3. A reduction in pumping power for a given size and heat duty
4. A reduction in the approach temperature difference
5. An appropriate combination of these points

While considering these benefits, the associated flow friction changes are also to be taken into account [2]. The principal advantage of introducing an augmentation is the possibility of substantially increasing thermal duty to meet the needs of new process conditions or production goals. This can be achieved either by:

1. Installing removable inserts inside the tubes—Use of an internal augmentation in turn reduces the length and weight of the unit
2. Replacing a removable tube bundle with a new enhanced tube bundle
3. Replacing the heat exchanger with a new enhanced tube heat exchanger of the same size or smaller

Heat transfer enhancement techniques generally reduce the thermal resistance either by increasing the effective heat transfer surface area or by generating turbulence. Sometimes these changes are accompanied by an increase in the required pumping power, which results in higher cost. The effectiveness of a heat transfer enhancement technique is evaluated by the thermal performance factor, which is a ratio of the change in the heat transfer rate to change in friction factor.

8.2 HEAT TRANSFER MODE

The heat transfer augmentation techniques can be applied for various methods of convective heat transfer mode such as

1. Single-phase convection with flow phenomena such as laminar and turbulent flow
2. Phase change phenomena such as boiling and condensation

Enhancement techniques can be applied to both internal and external heat exchanger tube surfaces. The flow media can be gas, single-phase liquid, phase change, or two-phase flow.

8.2.1 Augmented Surfaces

For a two-fluid exchanger, augmentation should be considered for the fluid stream that has the controlling thermal resistance. If both resistances are approximately equal, augmentation may be considered for both sides of the exchanger. The majority of two-fluid exchanger applications are served by four basic exchanger types [6]: (1) shell and tube; (2) tube-fin; (3) brazed plate-fin; and (4) plate heat exchangers (PHEs). If enhancement is employed, it must be applied to the following three basic flow geometries [6]:

1. Internal flow in circular tubes
2. External flow normal to tubes and tube banks
3. Channel flow in closely spaced parallel plate channels, e.g., brazed plate-fin heat exchangers and PHEs

8.3 PRINCIPLE OF SINGLE-PHASE HEAT TRANSFER ENHANCEMENT

Single-phase flow can be classified as internal or external flow. It can be further classified as turbulent or laminar flow. The principal methods of enhancement in laminar or turbulent flow are the use of a secondary heat transfer surface and the disruption of the velocity gradient and temperature profile [7]. The three possible mechanisms for single-phase heat transfer enhancement are decreasing the thermal boundary layer, increasing flow interruptions, and increasing the velocity gradient. For example, the rate of heat transfer between a fluid and finned surface can be expressed in a general form as [8]

$$q = h\eta_o A\Delta T \tag{8.1}$$

where
- h is the heat transfer coefficient
- η_0 is the overall surface efficiency
- A is the total heat transfer area
- ΔT is the surface–fluid temperature difference

In Equation 8.1, $h\eta_0 A$ represents the surface conductance. Obviously, high performance may be achieved by using a heat transfer surface with a high value of $\eta_0 A$, h, or both. Special surface geometries provide enhancement by establishing a higher $h\eta_0 A$ per unit surface area. Three basic principles are employed to increase $h\eta_0 A$ or hA [6,8,9]:

1. Increase of A without appreciably changing h, e.g., finned tubes.
2. Increase in h without an appreciable area A increase, e.g., surface roughness or turbulence promoters. This principle is further explained in the next section.
3. Increase of both h and A, e.g., extended surfaces in compact heat exchangers have increased heat transfer coefficient by means of surface promoters such as perforations, louvers, and corrugations or displaced promoters such as canted tubes.

8.3.1 Increase in Convection Coefficient Without an Appreciable Area Increase

A surface that results in performance enhancement due to an increase in the heat transfer coefficient h without an appreciable transfer area is known as an enhanced heat transfer (EHT) surface. EHT surfaces are vital to the development of heat transfer augmentation devices. The underlying principle of the EHT surface is as follows. As a reasonable approximation, the heat transfer coefficient is given by [8]

$$h = \frac{k}{\Delta} \tag{8.2}$$

where
- k is the thermal conductivity of the fluid
- Δ is the thermal boundary layer thickness

From Equation 8.2, it can be seen that h can be increased by altering the flow pattern near the surface in order to reduce the thermal boundary layer thickness.

8.3.2 Laminar Flow inside a Tube

In fluid dynamics, laminar flow is characterized by smooth or regular paths of fluid particles. The fluid flows in parallel layers (with minimal lateral mixing), with no disruption between the layers [10]. A schematic of laminar flow is shown in Figure 8.1a. Laminar flow generally results in low heat transfer coefficients compared to turbulent flow. The fluid velocity and temperature vary across the entire flow channel width so that the thermal resistance is not just in the region near the wall as in turbulent flow. The most commonly used methods for predicting fully developed laminar flow heat transfer coefficients inside smooth, round tubes can be derived from first principles. For a uniform heat flux wall boundary condition (H) of fully developed laminar flow (both thermally and hydrodynamically), the Nusselt number is given by [5]

$$\mathrm{Nu_H} = \frac{hd_\mathrm{i}}{k} = 4.364 \tag{8.3}$$

The Nusselt number $\mathrm{Nu_H}$ is based on the hydraulic diameter, d_H (where d_i is tube inside diameter, h is convective heat transfer coefficient for a plain tube, and k is thermal conductivity of the fluid). As can be seen, the heat transfer coefficient is not a function of the Reynolds number or the Prandtl number.

Similarly, for a uniform wall temperature wall boundary condition (T) of fully developed laminar flow, the Nusselt number, $\mathrm{Nu_T}$, is given by [5]

$$\mathrm{Nu_T} = \frac{hd_\mathrm{i}}{k} = 3.657 \tag{8.4}$$

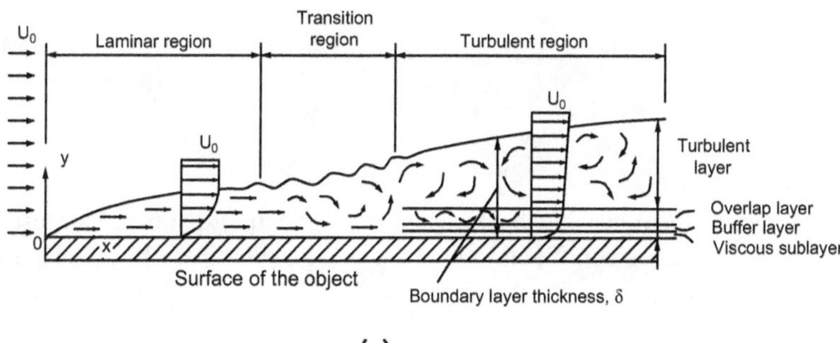

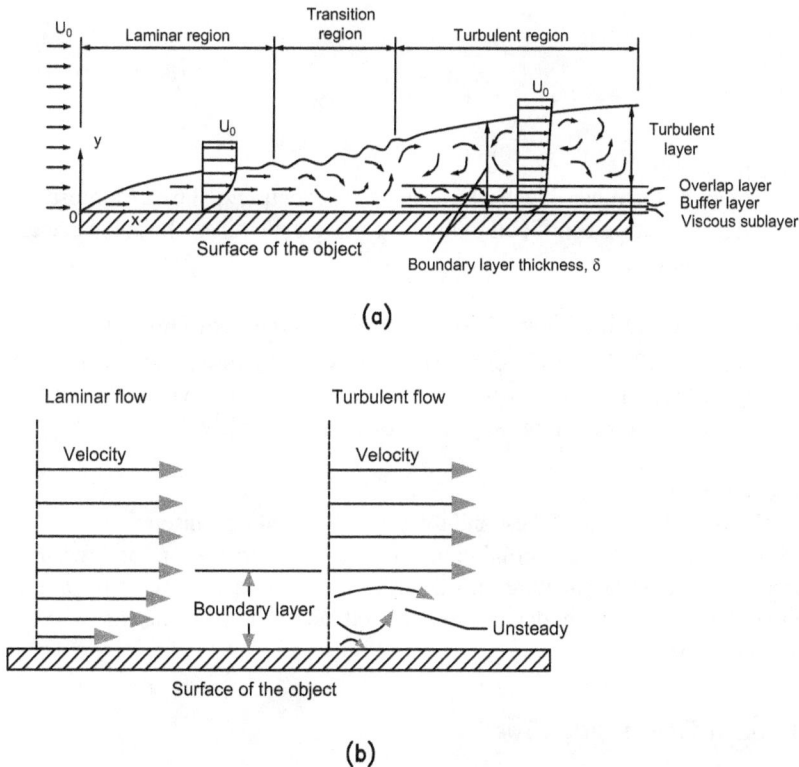

FIGURE 8.1 (a) Laminar flow over a surface and (b). Turbulent flow over a surface.

It can be seen from Eq. 8.6 that the heat transfer coefficient is not a function of the Reynolds number or the Prandtl number. The tubular Reynolds number Re is defined as:

$$\mathrm{Re} = \frac{\dot{m} d_i}{\mu} \tag{8.5}$$

The mass velocity $\dot{m}$ is in kg/m^{2}s and is obtained by dividing the mass flow rate in kg/s by the cross-sectional area of the tube in m^2. The Prandtl number Pr is obtained from the physical properties of the fluid and is defined as:

$$\mathrm{Pr} = \frac{c_p \mu}{k} \tag{8.6}$$

The hydrodynamic entrance length z_{eh} to arrive at fully developed hydrodynamic flow can be estimated by the following expression:

$$\frac{Z_{eh}}{d_i} = 0.03\,\mathrm{Re} \tag{8.7}$$

For the thermal developing length when the velocity profile is already fully developed before the heat transfer zone begins, the thermal entrance length Z_{et} starting from that location is given by the expression:

$$\frac{Z_{et}}{d_i} = 0.034\,\mathrm{Re}\,\mathrm{Pr} \tag{8.8}$$

To overcome the fact that laminar flow heat transfer coefficients (Eq. 8.6) are almost independent of fluid velocity and performance is difficult to improve using plain tubes, the enhancement techniques as given below are required.

8.3.2.1 Single-Phase Laminar flow Heat Transfer Enhancement Devices

Single-phase laminar flow heat transfer enhancement devices are discussed in Chapter 4 of Ref. [5]. Some of the single-phase laminar flow heat transfer enhancement devices and methods are given below and underlying enhancement principles are discussed later [5]:

i. Extended surfaces
ii. Internally ribbed or finned tubes, and tubes with twisted tape inserts
iii. Swirl flow devices—large internal helical fins or ribs, corrugations and twisted tapes which impart a swirl effect on the fluid, fins and ribs, corrugations, twisted tape inserts
iv. Laminar flow displacement devices like twisted tape, wire mesh inserts, wire matrix
v. Corrugated tube

8.3.3 Turbulent Flow Inside a Tube

In fluid dynamics, turbulent flow is characterized by the fluid's irregular movement of particles. In contrast to laminar flow, the fluid does not flow in parallel layers, the lateral mixing is very high, and there is disruption between the layers. Turbulent flow is characterized by recirculation, eddies, and apparent randomness. In turbulent flow, the speed of the fluid at a point continuously changes in both magnitude and direction. Turbulent flow exhibits a low-velocity flow region known as the laminar sublayer immediately adjacent to the wall, with velocity approaching zero at the wall. Most of the thermal resistance occurs in this low-velocity region. Any roughness or enhancement technique (e.g., swirl flow devices such as wire coil inserts, spiral inserts, and protrusions) that disturbs the laminar sublayer will enhance the heat transfer. A schematic of turbulent flow is shown in Figure 8.1(b) [10,11]. For fully developed flow in a smooth tube, the dimensionless laminar sublayer thickness, y^*, is given by

$$y^* = \frac{y\sqrt{G_c\tau_o/\rho}}{\nu} = 5 \tag{8.9}$$

Using the shear stress τ_o and ν for a smooth tube, the ratio of laminar sublayer thickness y^* to tube diameter d_i is given by

$$\frac{y}{d_i} = 25\,\mathrm{Re}^{-0.875} \tag{8.10}$$

For example, if Re = 30,000 and d_i = 1.0 in. (25.4 mm), the laminar sublayer thickness is 0.003 in. (0.0762 mm). Any roughness or enhancement technique that disturbs the laminar sublayer will enhance the heat transfer.

A turbulent flow generally results in higher heat transfer coefficients compared to laminar flow, as shown in Figure 8.2.

8.3.3.1 Single-Phase Turbulent Flow Heat Transfer Enhancement Devices

Single-phase turbulent flow heat transfer enhancement devices are discussed in Chapter 5 of Ref. [5]. Some of the single-phase turbulent flow heat transfer enhancement devices and methods are given below and underlying enhancement principles are discussed later [5]:

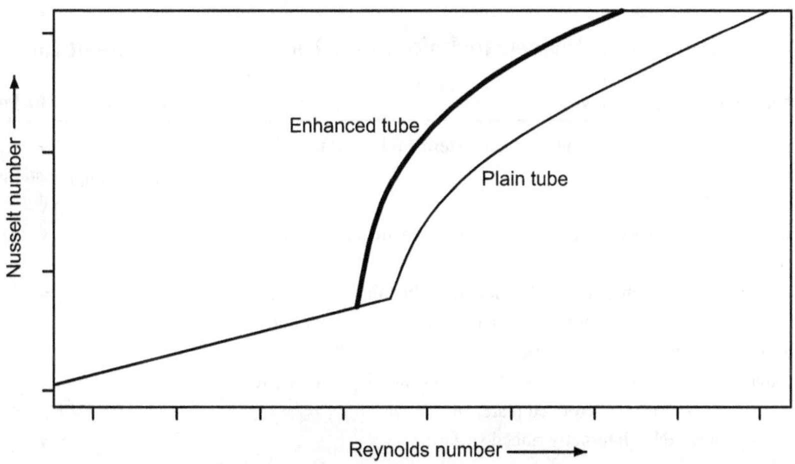

FIGURE 8.2 Comparision of heat transfer between laminar flow and turbulent flow.

i. Corrugated tubes
ii. Internally ribbed or finned tubes (extended surfaces), tubes with twisted tape inserts
iii. Surface roughness elements
iv. Swirl flow devices include internal helical fins or ribs, corrugations, twisted tapes, artificial roughness, and coil inserts
v. Turbulators such as spherical ball, twisted tape, helical wire, wire mesh, perforated plate insert, etc.

8.3.4 Comparison of Heat Transfer Enhancement Devices for Laminar and Turbulent Flows

Table 8.1 shows the heat transfer enhancement devices for laminar flow and turbulent flows.

8.4 APPROACHES AND TECHNIQUES FOR HEAT TRANSFER ENHANCEMENT

Heat transfer enhancement mechanisms can improve the heat exchanger effectiveness of internal and external flows. Typically, they increase fluid mixing by increasing flow vorticity, unsteadiness, or turbulence, or by limiting the growth of the thermal boundary layer close to the heat transfer surface. The approaches followed to enhance heat transfer include:

1. Heat transfer surface modifications/placement of displaced flow devices, etc., such as extended surfaces, microfin tube, enhanced tubes, microgrooved copper tube, surface roughness, surface protuberances, displaced enhancement devices, swirl flow devices, turbulators, and treated surfaces.
2. Interruption such as slits or offset fins that interrupt the boundary layer, restarting it, creating secondary flows, etc.
3. Surface roughness elements which accelerate the transition from laminar flow to turbulent flow.
4. Surface as ridges or 3D shapes (profiles such as circular or semicircular, cube, pyramid, etc.) which generate secondary or unsteady flows. Inserts placed inside flow channel so as to improve the fluid flow near the heat transfer surface or devices which create rotating and/or secondary flow inside tubes.

TABLE 8.1
Heat Transfer Enhancement Devices/Techniques for Laminar and Turbulent Tube Flow

Sl. no.	Enhancement method/device	Laminar flow	Turbulent flow
1.	Extended/finned surfaces on internal and external tube surfaces	√	√
1.1	Microfin tubes	Phase change applications	
2.	Surface roughness elements		√
3.	Boundary layer displacement devices/laminar flow displacement devices		√
4.	Swirl flow devices such as internal helical or ribbed fin, corrugation, twisted tape, surface roughness elements, inlet vortex generator, axial core inserts with screw type winding, etc.	√	√
5.	Turbulators like spherical ball, twisted tape, helical wire, wire mesh/ wire matrix, extruded perforated plate, etc.		√
6.	Internally finned tube, internally ribbed surface	√	√
7.	Twisted tape insert	√	
8.	Corrugated tube	√	√

TABLE 8.2
Passive Enhancement Techniques, Underlying Mechanism of Heat Transfer Enhancement and Enhancement Devices

Technique	Principle of heat transfer enhancement	Devices
Extended surfaces	Interrupts the boundary layer, restarting it, creating secondary flows, add extra heat transfer surface, etc.	Tube-fin exchangers with external or internal short fins, plate-fin exchangers such as wavy fin, offset strip fin, perforated fin and louvered fin, microfin tube, integral low-fin tubes (external)
Treated surfaces	Promotes turbulence	Fine-scale alternation of the surface finish or provision of either continuous or discontinuous coating
Surface roughness	Accelerates transition from laminar flow to turbulent flow	Random sand-grain type roughness to discrete protuberances
Surface protuberances	Generates secondary or unsteady flows; the ribs cause flow separation and reattachment	Protuberances such as ridges or 3D shapes (profiles such as circular or semicircular, cube, pyramid, etc.)
Displaced enhancement devices	Interrupting the development of boundary layer of the fluid flow and increasing the degree of turbulence; causes periodic mixing of the gross flow	Tube inserts such as wire turbulator, swirl strips, and twisted tapes. Displaced flow enhancement devices, such as streamline shape, disks, static mixer, and meshes or brushes, and wire coil inserts
Swirl flow devices	Creates rotating and/or secondary flow inside tube	Twisted tape inserts, internal fin or rib, corrugated surfaces, corrugated tubes with circumferential indentations, surface roughness elements, inlet vortex generator, axial core inserts with screw type winding, etc.
Turbulators	Alters fluid flow, maximizing contact with the tube wall, and creating turbulent flow	Spherical ball, twisted tape, helical wire, wire mesh/wire matrix, extruded perforated plate, etc.

8.4.1 Passive and Active Methods

Enhancement techniques based on these approaches may be classified as (1) passive methods, which require no direct application of external power, and (2) active methods, which require external power. The passive methods include extended surfaces, treated surfaces, rough surface or surface protuberances, displaced enhancement devices, swirl flow devices, surface tension devices, additives for fluids, additives for gases, etc. Table 8.2 provides the passive enhancement techniques, underlying mechanism of heat transfer enhancement, and enhancement devices. Active methods are discussed later.

The merits of passive techniques for heat exchanger applications include that passively enhanced tubes are relatively easy to manufacture, cost-effective for many applications, and can be used for retrofitting existing units, whereas active methods, such as vibrating tubes, are costly and complex [12].

8.5 SINGLE-PHASE LAMINAR TUBESIDE FLOW HEAT TRANSFER ENHANCEMENT METHODS

Based on Chapter 4 of Ref. [5], a brief review of single-phase tubeside laminar flow heat transfer enhancement devices/techniques is presented below. Laminar flow enhancement is quite different than turbulent flow in that the fluid in a laminar flow does not "mix" unless it is forced to flow to the center of the channel or vice versa. Methods for heat transfer enhancement mechanisms for laminar flows inside tube flow are briefly discussed below [5].

8.5.1 Area Increase Due to Extended Surface

One way to increase laminar heat transfer coefficients is to increase the surface area in contact with the fluid to be heated or cooled.

8.5.2 Swirl Flow Devices

Swirl of the flow is known to augment heat transfer. Large internal helical fins or ribs, corrugations, and twisted tapes impart a swirl effect on the fluid. This tends to increase the effective flow length of the fluid through the tube, which increases heat transfer and pressure drop. For twisted tape inserts, the effect of swirl on augmentation plays an important role. With regard to small fins or corrugations, they are often not sufficient to create a swirl effect and hence may not offer any or much in the way of enhancement.

8.5.3 Laminar Flow Displacement Devices

The displacement of the laminar flow from the heat transfer wall is a particularly important heat transfer mechanism for augmenting heat transfer. This is, for instance, done by placing a twisted tape in the path of the flow that forces the flow away from the wall and allows fresh bulk fluid to come into contact with the wall. For even higher thermal performance than a twisted-tape insert, one should consider the use of the wire mesh inserts/matrix such as those of Cal Gavin Ltd. that have been very successfully applied to enhance many laminar flows.

8.5.4 Wire Mesh Inserts

The wire mesh inserts developed by Cal Gavin Ltd. have been successfully applied to laminar flows now for several decades in a wide range of applications and are an excellent option for enhancing heat transfer in laminar flows. These inserts tend to yield much higher heat transfer and pressure drop augmentation ratios than twisted tape inserts. The detailed heat transfer effect is discussed later.

8.5.5 Twisted Tape Inserts

A twisted tape insert tends to sit loosely within a tube without making "thermal" contact with the tube wall, and hence the surface area of the twisted tape itself is not taken into consideration for heat transfer purposes, instead only its effect on the heat transfer coefficient of the tube's inner surface. Twisted tape inserts fit rather loosely inside a tube (thus they have poor thermal contact with the inner tube wall but can be easily installed and removed) and hence the surface area of a twisted tape is not considered to be a heat transfer surface area, only that of the plain tube perimeter in which it is installed.

8.5.6 Corrugated Tube

A corrugated tube is defined geometrically by the corrugation pitch, p, and the corrugation depth, e. The axial corrugation pitch is related to the internal diameter d_i, helix angle β relative to the axis of the tube, and the number of starts n_s. The internal area ratio of a corrugated tube relative to a plain tube of the same diameter d_i is slightly larger than one. The internal fins can be of various cross-sectional shapes. Most industrial tubes have fins (or ribs or ridges) with a trapezoidal cross-sectional profile (wider at the base than at the tip of the fin and with rounded corners).

8.6 SINGLE-PHASE TURBULENT FLOW HEAT TRANSFER AUGMENTATION METHODS FOR INSIDE TUBE FLOWS

Based on Chapter 5 of Ref. [5], methods for heat transfer enhancement devices/techniques for turbulent flows inside tube flow are discussed below.

8.6.1 Extended Surface Area

One way to increase turbulent heat transfer coefficients is to increase the surface area in contact with the fluid to be heated or cooled. Most industrial tubes have fins (or ribs or ridges) with a trapezoidal cross-sectional profile (wider at the base than at the tip of the fin and with rounded corners).

8.6.2 Surface Roughness

The internal roughness of the tube surface is well known to increase the turbulent heat transfer coefficient but usually has little or no effect on laminar heat transfer, except in extreme cases such as very small channel sizes. The artificial roughness can be developed by employing a corrugated surface which improves the heat transfer characteristics by breaking and destabilizing the thermal boundary layer. This can be done by keeping the height of the roughness elements small in comparison with the duct dimensions.

8.6.3 Swirl Flow Devices

Swirl flow devices include internal helical fins or ribs, corrugations, twisted tapes, artificial roughness, and coil insert which impart a swirl effect on the fluid. For internal helical fins, ribs, and corrugations however, the effect of swirl tends to decrease or disappear altogether at higher helix angles since the fluid flow then simply passes axially over the fins or ribs. For twisted tape inserts, the effect of swirl on augmentation plays an important role.

8.6.4 Boundary Layer Displacement Devices

The displacement of the turbulent boundary layer is a particularly important heat transfer mechanism for augmenting heat transfer. For essentially a two-dimensional flow, it shows the separation of the flow as it passes over a transverse rib (creating a small recirculation zone in front of the rib), the formation of a recirculation zone behind the rib, flow reattachment on the base wall, and then flow up and over the next rib. Recirculation eddies are formed above these flow regions.

8.6.5 Twisted Tape Inserts

The enhancement is defined geometrically by the thickness of the tape δ and its twist ratio, y. The twist ratio is defined as the axial length for a 180° turn of the tape divided by the internal diameter of the tube. This is the most common definition used in research literature and that used here. It is also common to use the axial length for a complete 360° turn. Twisted tape inserts normally fit loosely inside a tube (thus they have poor thermal contact with the inner tube wall) and hence the surface area of a twisted tape is not considered to be the heat transfer surface area, only that of the plain tube perimeter in which it is installed [5].

8.7 DISCUSSION OF VARIOUS ENHANCEMENT TECHNIQUES/DEVICES

8.7.1 Extended Surfaces

Extended surfaces are routinely employed in compact heat exchangers and shell and tube exchangers either for liquids or for gases. Gas-side heat transfer coefficients are relatively low in comparison to those for liquids and phase change applications. Thus, extended surfaces are widely use to enhance tubular exchangers for gas-side service by increasing the surface area of the tubes. The fins are helically wrapped and welded and are typically solid or segmented. Round and elliptical studs are also used but not as extensively as fins. Finned tube banks are used in crossflow heat exchangers with the dirty gases flowing over the outside of the tubes. In addition to the tube geometry, the layout geometry is also an important parameter. The layout can be either staggered or inline. The application of extended surfaces for liquid and gas service is discussed next.

8.7.2 Extended Surfaces for Gases

In forced convection heat transfer between a gas and a liquid, the heat transfer coefficient of the gas may be of the order of 1/50–1/10 that of the liquid [9]. The use of extended surfaces will reduce the gas side thermal resistance. However, the resulting gas side resistance may still exceed that of the liquid.

Enhanced surface geometries on tube-fin exchangers: Enhanced surface geometries that have been used on circular tube-fin heat exchangers include [9]: (1) plain circular fin; (2) slotted fin; (3) punched and bent triangular projections; (4) segmented fin; and (5) wire-loop extended surface. Some of the enhanced fin geometries of individual fins are shown in Figure 8.3 and a serrated fin tube is shown in Figure 8.4. All of these geometries provide enhancement by the periodic development of thin boundary layers on small-diameter wires or flat strips, followed by their dissipation in the wake region between elements.

Enhanced surface geometries of plate-fin exchangers: Many enhanced surface geometries such as wavy fin, offset strip fin, perforated fin, and louvered fin are used on plate-fin exchangers. To be effective, the enhancement technique must be applicable to low Reynolds number flows. Two basic concepts have been extensively employed to provide enhancement [5,9]:

i. Special channel shapes, such as the wavy channel, which provide mixing due to secondary flows or boundary layer separation within the channel
ii. Repeated growth and wake destruction of boundary layers, e.g., offset strip fin, louvered fin, pin fin, and to some extent the perforated fin

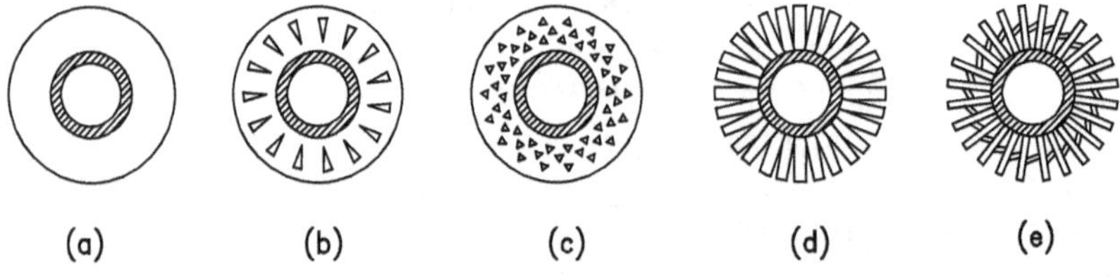

FIGURE 8.3 Externally enhanced finned tube geometries: (a) plain circular fin, (b) slotted fin, (c) punched and triangular projections, (d) serrated or segmented fin, and (e) wire-loop extended surface.

FIGURE 8.4 External enhancement-serrated fin. (Courtesy of Fintube, LLC, Tulsa, OK; www.fintubellc.com)

8.7.3 Extended Surfaces for Liquids

Extended surfaces used with liquids may be on the inner or outer surface of the tubes. Because liquids have higher heat transfer coefficients than gases, fin efficiency considerations require shorter fins with liquids than with gases.

Internally extended surfaces: Widely varied integral internally finned tubes (Figures 8.5 and 8.6) have been developed for tubeside heat transfer augmentation, and are commercially available in copper and aluminum. Most are for single-phase flows. In recent years, tubes with special internal geometries for augmentation of boiling in refrigeration systems have been developed. Typical commercial names include Thermofin tubes A, B, EX, and HEX. Other possible methods for tubeside augmentation include internal roughness geometries and inserts such as twisted tape inserts and wire coil inserts. Some forms of internal enhancement tubes are shown in Figure 8.7

Microfin tube: The current trend for enhancement of in tube evaporation of refrigerants is toward the use of *internal microfin tubes*. The reasons for their popularity include that (1) they increase heat transfer significantly while only slightly increasing pressure drop and (2) the amount of extra material required for microfin tubing is much less than that required for other types of internally finned tubes [13]. Consider microfin tubes when shellside heat transfer is controlling, expensive materials of construction are required, debottlenecking an existing exchanger, retrofitting or upgrading with new tube material, and meeting a stringent space or weight requirement.

Microfin tubes are available primarily in copper and also are becoming available in other materials, such as aluminum, carbon steel, steel alloys, etc. For instance, Wolverine Tube Inc. and HPT (high-performance tube) (Fine-Fin® types include 30FPISmoothBore, 28FPISmoothBore, 26FPISmoothBore, 36FPISmoothBore, 43FPISmoothBore) are among the major manufacturers of microfin tubes for the air-conditioning and refrigeration industries/chemical process industries. Possible tube materials are carbon steels, titanium, zirconium, nickel alloys, stainless steel, monel alloy400, copper nickel, etc. Wolverine seamless microfin tubes are produced by drawing a plain copper tube over a mandrel to form helical fins. This production method allows microfins to be produced from about 0.1 to 0.4 mm (0.004–0.016 in.) in height. The most favorable helix angles for heat transfer and pressure drop range from about 7° to 23°, but 18° seems to be most popular [1,14]. Instead, welded microfin tubes are

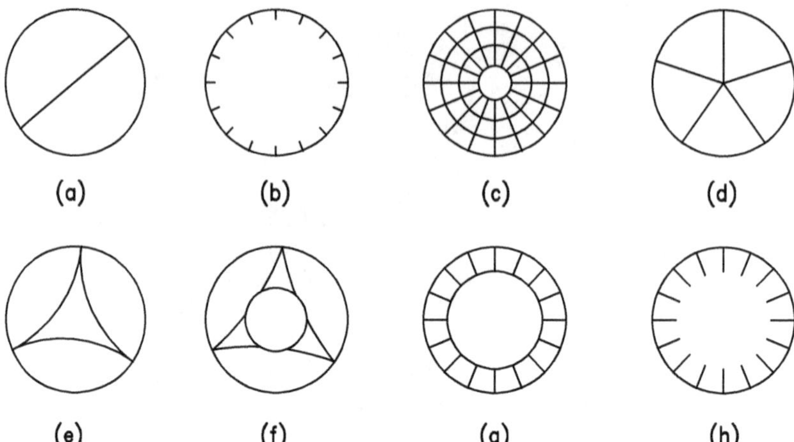

FIGURE 8.5 Representative commercially available tubes with internal enhancement. (a) Twisted tape insert in plain tube, (b) spiral internally finned tube, (c) quintuplex finned tube, (d) star-shaped insert in plain tube, (e) amatron internally finned tube, (f) corrugated tube, (g) duplex internally finned tube, and (h) straight internally finned tube. (Adapted from Marner, W.J. et al., *Trans. ASME J. Heat Transfer*, 105, 358, 1983.)

FIGURE 8.6 Internal enhancement tubes—X-ID tubes (shows microfin tube, repeated rib roughness, and spirally fluted tube). (Courtesy of Fintube, LLC, Tulsa, OK; www.fintubellc.com)

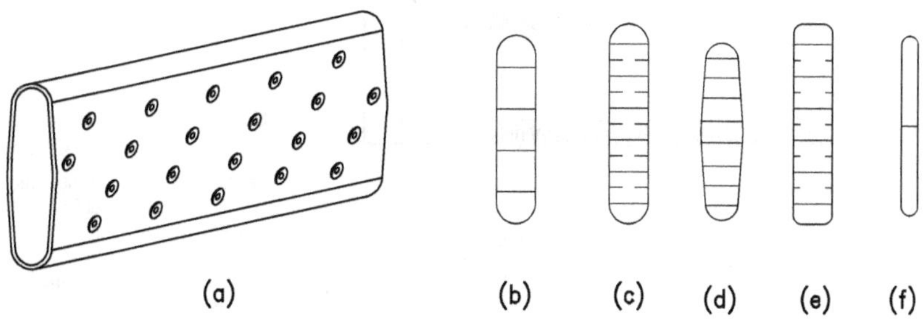

FIGURE 8.7 Representative commercially available aluminum tubes with internal enhancement.

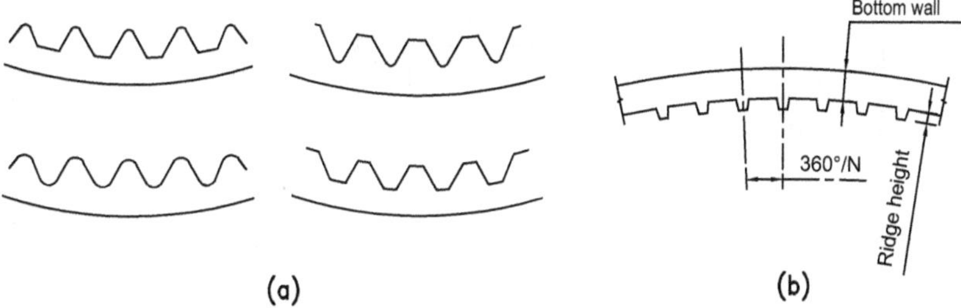

FIGURE 8.8 Fin profiles of microfin tubes used in refrigerant evaporators and condensers. (a) Fin profiles—triangular, wavy, and trapezoidal and (b) fin details. *Note*: N is integral fins density.

formed from a copper strip, whose microfin geometry has been embossed by a rolling operation. This manufacturing method allows for a much wider range of microfin geometries to be produced, including 3D fin geometries. Figure 8.8 shows the fin profile of a microfin tube. The parameters which define its geometry are outside diameter, maximum internal diameter, helix angle, fin height and thickness, and fin cross-sectional shape. Most microfins have approximately a trapezoidal cross section with a rounded top and rounded corners at the root. Other shapes are triangular, rectangular, and screw type, with a fin height of about 0.2–0.3 mm (0.008–0.012 in.).

Integral low-fin tubes: Low-fin tubes are generally 0.75–1 in. (19.2–25.4 mm) OD with a typical maximum fin height of 0.059 in. (1.5 mm), the fin thickness is 0.3 mm (0.012 in.), and fin densities are 11, 16, 19, 26, 28, 30, 32, 36 fpi, etc. Up to 19 fpi, the fin profiles are deeper, made of softer materials, and from 26 to 36 fpi shallow fin profiles made in materials such as stainless steels, titanium, nickel, zirconium, etc. Tube materials include carbon steel, austenitic stainless steels such as 304, 304L, 316, 316L, 321, etc.; duplex stainless steel; 6 Mo super austenitic stainless steel such as AL-6XN, copper alloys; nickel alloys such as Monel and cupronickel; titanium; and zirconium. An integral low-fin tube is ideally suited for a shell and tube exchanger. The outside surface of the low-fin tube is typically three times that of a plain tube. Tube diameters are the same as plain tubes; this permits low-fin and plain tubes to be interchangeable. A typical integral low-fin tube is shown in Figure 8.9.

The integral low-fin tube is also effective with compressed gases, namely, in intercoolers and aftercoolers, and they are employed in condensing and evaporating duties, particularly with refrigerants and hydrocarbons [8]. Low-fin tubes are used for applications such as [15]:

1. Sensible heat applications like cooling of gases or viscous fluids.
2. Condensing fluids with relatively low surface tension; low-fin tubes can enhance condensing heat transfer typically by about 30%.

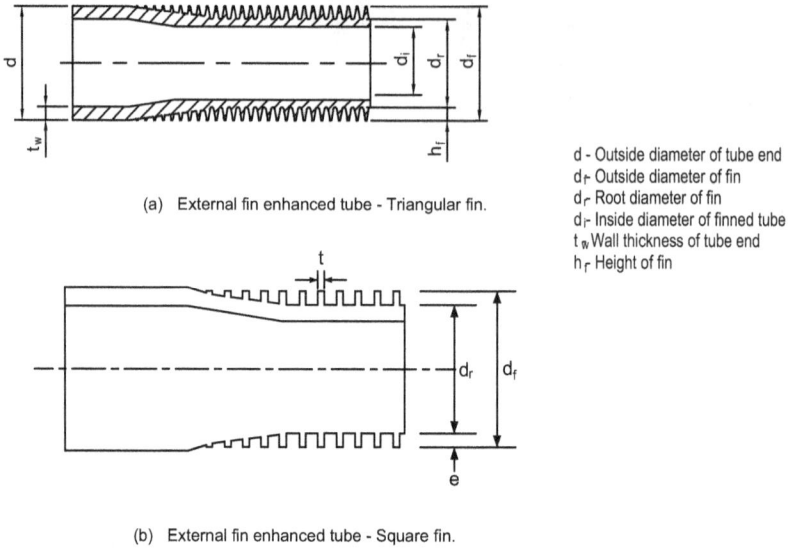

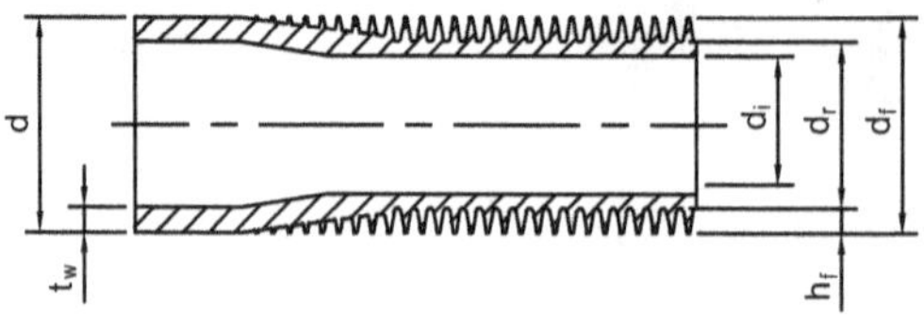

(a) External fin enhanced tube - Triangular fin.

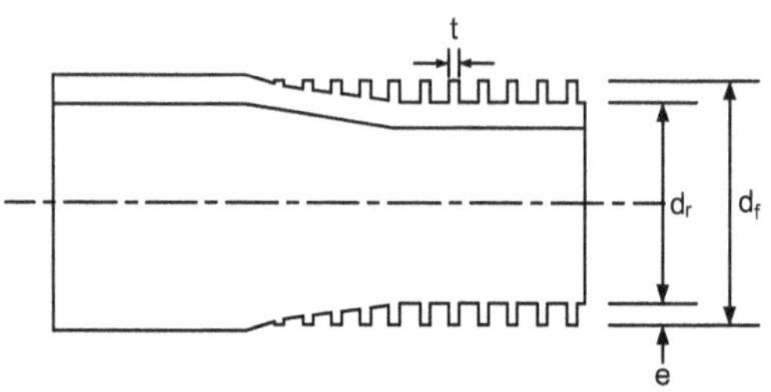

(b) External fin enhanced tube - Square fin.

FIGURE 8.9 Integral low-fin tube.

3. Boiling—low-fin tubes permit nucleate boiling at lower temperature differences than can be achieved with plain tubes.
4. Fouling fluids—fouling is not accelerated with low-fin tubes; it is often reduced. Low-fin tubes are more easily cleaned by chemical or mechanical means than are plain tubes.

8.7.4 Rough Surfaces

Rough surfaces refer to tubes and channels having roughness elements in the form of regularly repeated ridge-like protrusions perpendicular to the stream direction. The configuration is generally chosen to promote turbulence rather than to increase the heat transfer surface area. The surface roughness may be applied to any of the usual prime or extended heat exchange surfaces [1,9], flat plates, circular tubes, and annuli having roughness on the outer surface of the inner tube and fins.

Rough surfaces are produced in many configurations, ranging from random sand-grain type roughness to discrete protuberances. Helical repeated rib roughness as shown in Figure 8.10 is readily manufactured and results in good heat transfer performance in single-phase turbulent flow without severe pressure drop [16]. The principal parameters are ridge height e, pitch p, and helix angle β (Figure 8.10a). The internal area ratio relative to a plain tube of the same diameter ranges from about 1.3 to 2.0. The internal fins can be of various cross-sectional shapes such as square, rectangle, triangular, semicircular, arc, sine wave, etc. [17]; Figure 8.10(d) shows square, semicircle, and round profile repeated rib ridges. Among the leading helical internal fin tube manufacturers are Fintube LLC and Wolverine Tube Inc. (Helical internal fins or ribs are referred to as ridges in Wolverine Tube literature which includes low-finned tubes such as Turbo-Chil® and S/T Trufin® and to enhanced boiling and condensing tubes, such as the various versions of Turbo-B® and Turbo-C®.)

Repeated rib roughness tubes are shown in Figure 8.11. The main effect on heat transfer is that the ribs cause flow separation and reattachment (as shown in Figure 8.11a) and flow pattern over helical ribs (as shown in Figure 8.12). Both 2D and 3D integral roughness elements are also possible. Within the category of surface roughness, a 3D roughness of the sand-grain type appears to offer the greatest performance [17]. Roughness elements have been produced by the traditional processes of machining, forming, casting, or welding [8]. Various inserts or wrap-around structures such as wire-wound fin tube (Figure 8.13) can also provide surface protuberances [1].

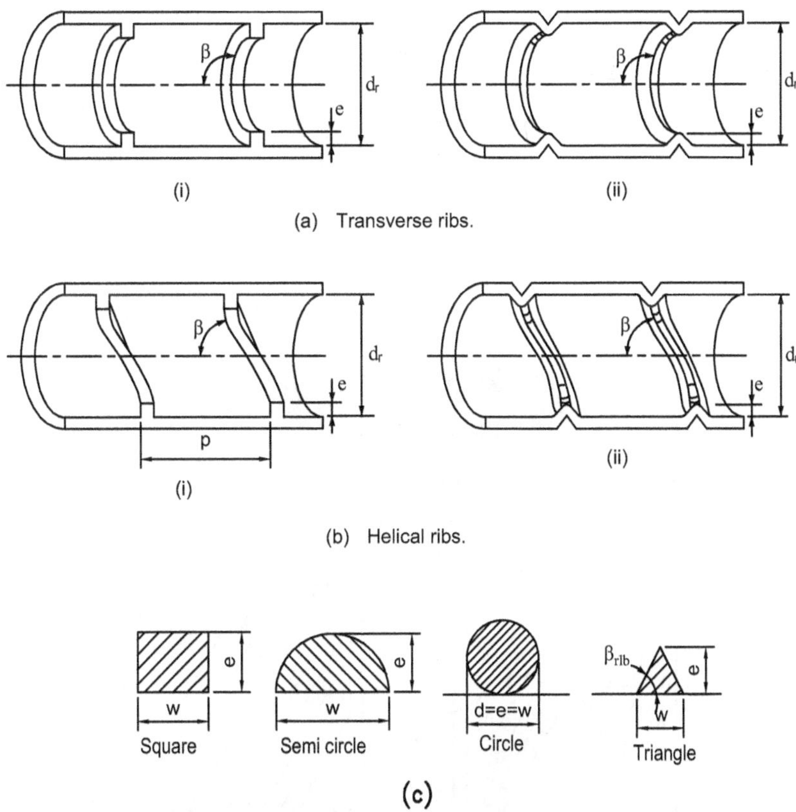

FIGURE 8.10 Internal enhancement by repeated rib roughness. (a) Square ridge, (b) semi circle, (c) circular, and (d) details of repeated rib roughness. (*Note*: *d*-tube outer diamter or coil diameter, d_i-tube inner diameter-*d* of item *c* is to be read as d_i, *e*-protuberance thickness, h_r-rib depth, *L*-unfinned tube length, *p*-rib pitch, *t*-rib thickness, t_w-tube wall thickness, *W*-protuberance width and β helix angle of the rib.). After Ravigururajan, T. S. and Bergles, A. E. [17]

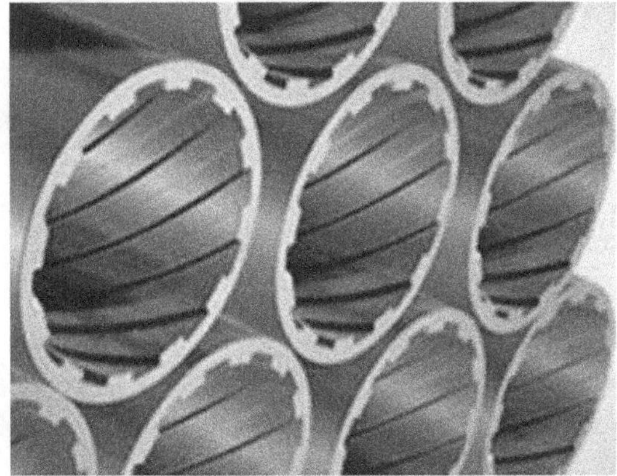

FIGURE 8.11 Internal enhancement by repeated rib roughness—X-ID tubes with repeated rib roughness. (Courtesy of Fintube, LLC, Tulsa, OK; www.fintubellc.com)

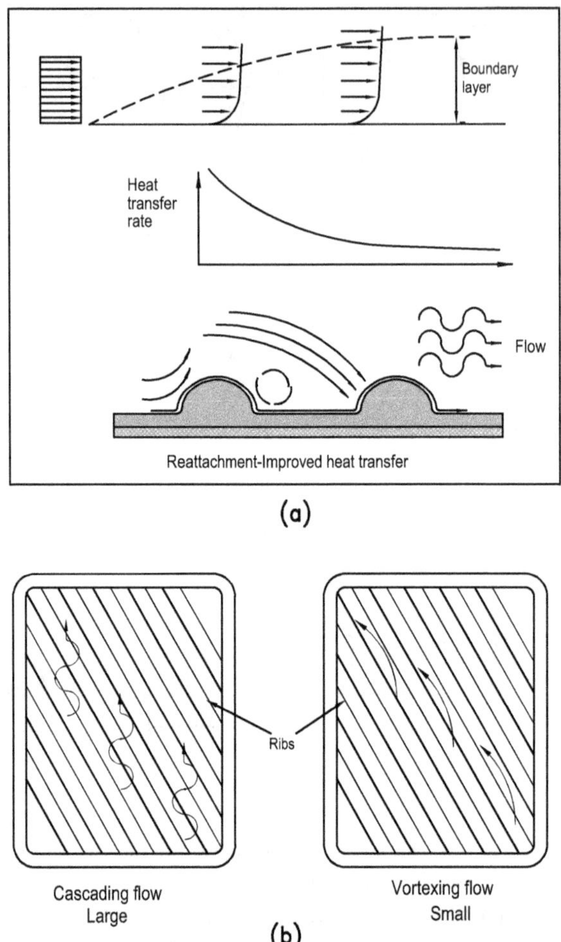

FIGURE 8.12 Principle of heat transfer enhancement of an X-ID tube: (a) developing boundary layer over smooth surface and disruption of boundary layer over ribs and (b) flow pattern over ribs. (Courtesy of Fintube, LLC, Tulsa, OK; www.fintubellc.com)

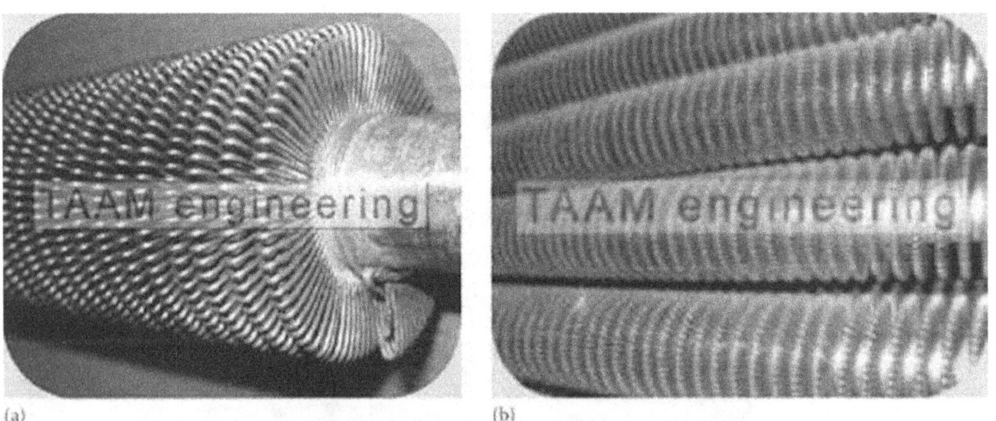

FIGURE 8.13 (a and b) Wire–loop wound fin tube. (Courtesy of TAAM Engineering, Sinnar, Maharastra, India.)

Applications: Surface roughness is attractive for turbulent-flow applications with high Prandtl number fluids [17]. In laminar-flow applications, small-scale roughness is not effective, but twisted tape inserts and internally finned tubes appear to be the favored augmentation techniques [5].

8.7.5 Tube Inserts and Displaced Flow Enhancement Devices

This class refers to devices that are inserted inside a smooth tube. The tube inserts are relatively low in cost, and relatively easy to insert and to take out of the tubes for cleaning operations.

8.7.5.1 Enhancement Mechanism

Tube inserts can create one or combinations of the following conditions, which are favorable for enhancing the heat transfer with a consequent increase in the flow friction [18]:

1. Interrupting the development of the boundary layer of the fluid flow and increasing the degree of flow turbulence
2. Increasing the effective heat transfer area if the contact between the tube inserts and the tube wall is excellent
3. Generating rotating and/or secondary flow

8.7.5.2 Forms of Insert Device

Various forms of insert devices are [9] as follows:

1. Devices that cause the flow to swirl along the flow length, e.g., swirl strips and twisted tapes
2. Extended surface insert devices that provide thermal contact with the tube wall (this may also swirl the flow), e.g., star inserts
3. Wall-attached insert devices that mix the fluid at the tube wall
4. Displaced insert devices that are displaced from the tube wall and cause periodic mixing of the gross flow

8.7.5.3 Displaced Flow Enhancement Devices

Displaced flow enhancement devices are inserted into the flow channel so as to improve the fluid flow near the heat transfer surface [19]. These devices include [9] streamline shape, disks, static mixer, and meshes or brushes, and wire coil inserts.

Coil inserts: Coil inserts provide an efficient and inexpensive way to enhance the heat transfer inside heat exchanger tubes [20]. Figure 8.14 shows a schematic of a coil insert and Figure 8.15 shows a wire coil matrix as a turbulator. In applications that require frequent cleaning, their easy-to-retrofit characteristic offers a significant advantage over integrally enhanced tube types; however, the coil inserts can have an inherent disadvantage if imperfect contact exists between the coil and the tube wall. For inserts, the parameters that affect the enhancement are the disruption height, the disruption spacing or pitch, the helix angle, and the disruption shape [12]. Wire coil enhances the heat transfer in turbulent flow efficiently. It performs better in turbulent flow than in laminar flow. Tube inserts normally have pull rings or attachments to install them, fix them in place, and to remove them for cleaning.

8.7.6 hiTRAN Thermal Systems

Poor tubeside performance can usually be avoided by considering the use of heat transfer enhancement devices. Engineered devices such as hiTRAN Matrix Elements, as shown in Figure 8.16a,

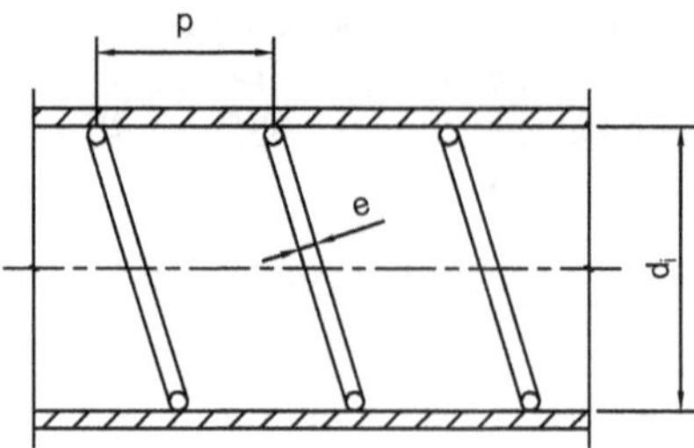

FIGURE 8.14 Wire coil insert details.

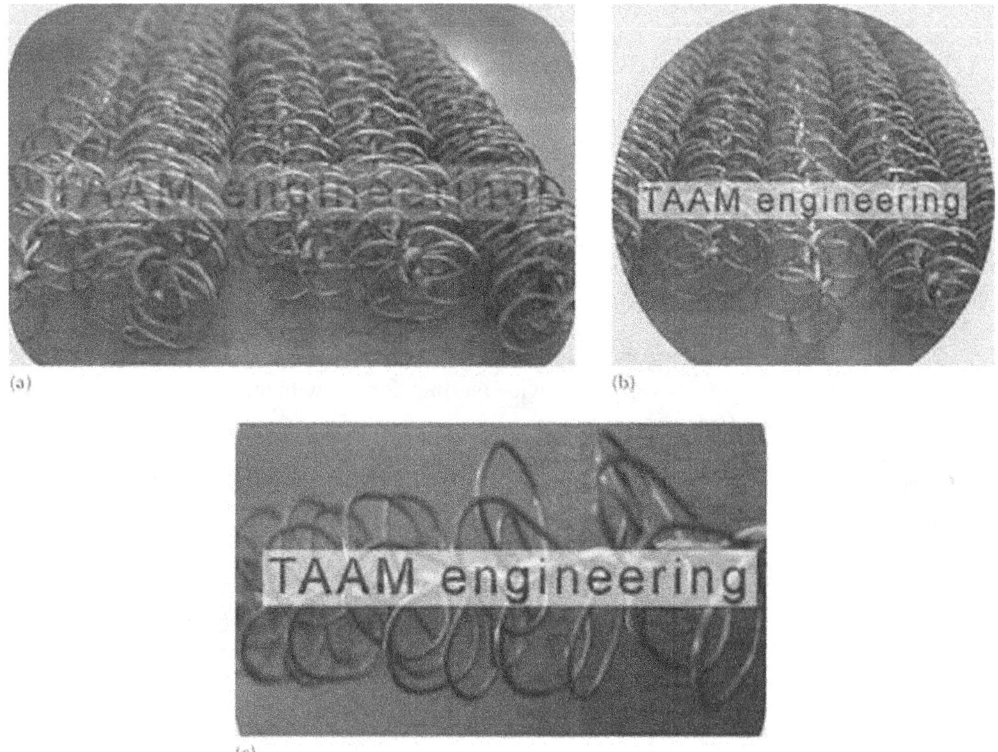

FIGURE 8.15 Wire matrix turbulators. (Courtesy of TAAM Engineering, Sinnar, Maharastra, India.)

invariably provide increased heat transfer coefficients relative to plain tubes. A graphical representation of plain tube and hiTRAN Matrix Elements enhanced performance range is provided in Figure 8.16b. When fluid flows through a plain tube, the fluid nearest to the wall is subjected to frictional drag, which has the effect of slowing down the fluid at the wall. This laminar boundary layer can significantly reduce the tubeside heat transfer coefficient and consequently, the performance of the heat exchanger. Inserting correctly profiled hiTRAN Matrix Elements which consists of a unique wire turbulator into the tube will disrupt the laminar boundary layer, creating additional fluid shear and mixing, thereby minimizing the effects of frictional drag. The working principle of hiTRAN Matrix Elements is shown in Figure 8.16c; hiTRAN Matrix Elements are particularly effective at enhancing heat transfer efficiency in tubes operating at low Reynolds numbers (laminar to transitional flow); although the heat transfer increase is greatest in the laminar flow region (up to 20

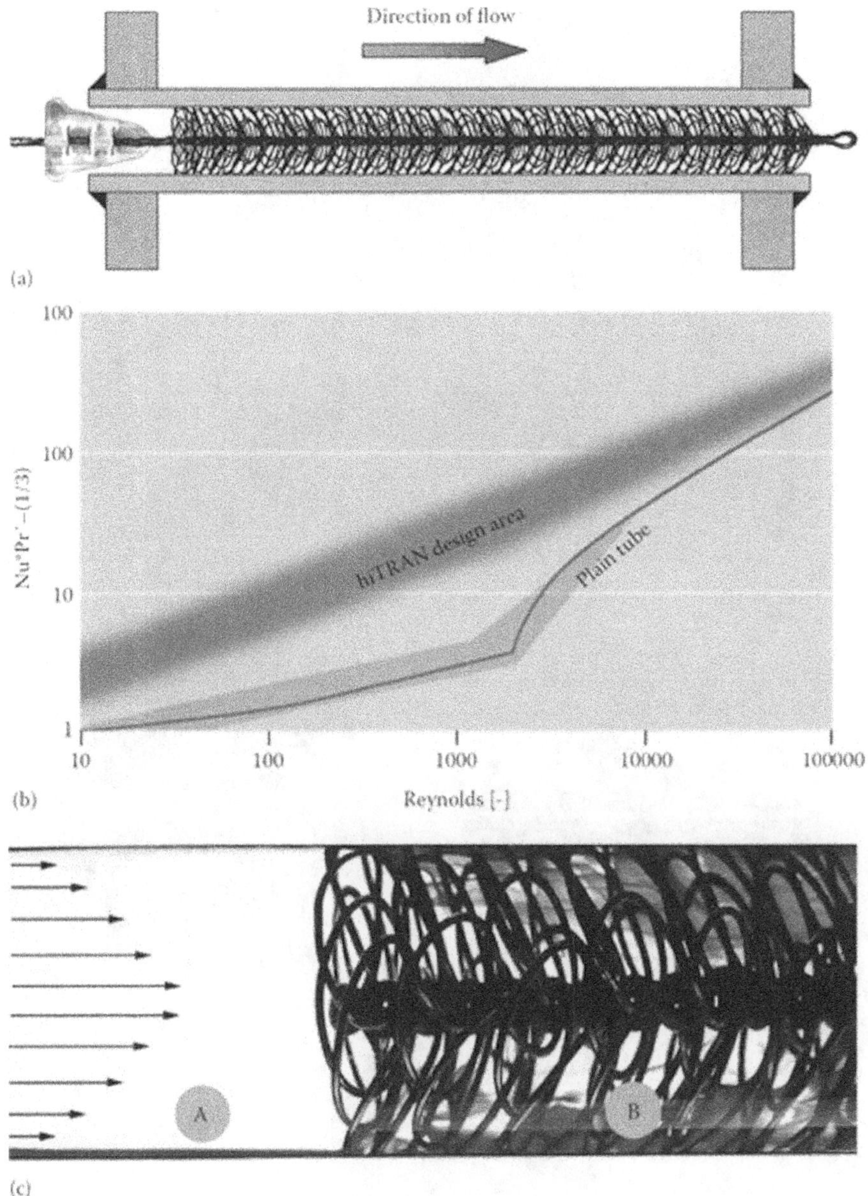

FIGURE 8.16 hiTRAN Matrix Elements—wire matrix turbulator. (a) Placement of hiTRAN matrix element in a tube; (b) hiTRAN system performance graph; (c) flow disruption caused by hiTRAN Matrix Elements; (A) laminar flow conditions; (B) turbulence caused by the use of hiTRAN matrix element tube inserts. (d) Placement of hiTRAN Matrix Elements in a shell and tube heat exchanger; (e) hiTRANMatrix Elements being held in place with key wire, core anchor, or fixing rods. (f) Shell and tube heat exchanger comparison with and without hiTRAN systems; (g) air-cooled lube oil cooler comparison with and without hiTRAN System. (Courtesy of Cal Gavin Ltd, Alcester, Warwickshire, U.K; www.calgavin.com)

(d)

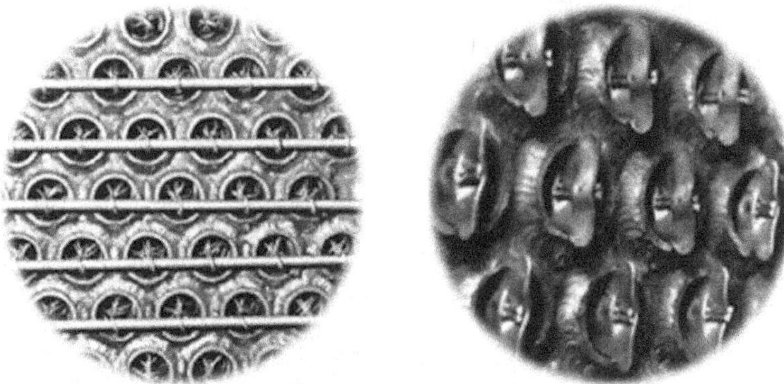

hiTRAN Matrix Elements held in place with key wire hiTRAN Matrix Elements held in place with core anchors

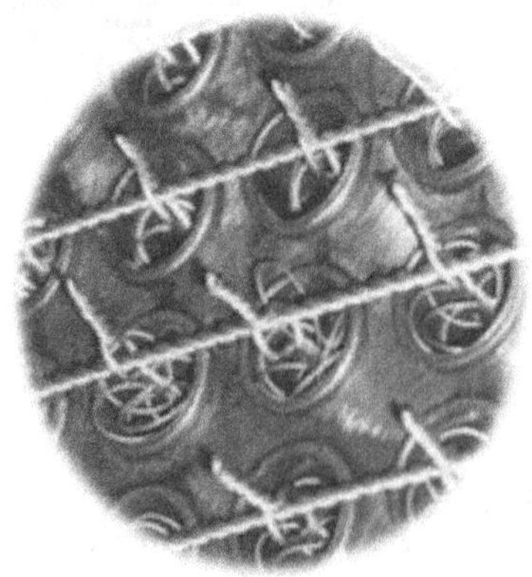

(e) hiTRAN Matrix Elements held in place with fixing rods

FIGURE 8.16 (Continued)

Improved Design for the Shell and Tube Heat Exchanger Cooling Heavy Cycle Gas Oil Fluid Catalytic Cracking Unit of an Oil Refinery

Design Comparison	Plain Tube	hiTRAN Enhanced
TEMA designation	BEM	BEM
Shell diameter (mm)	1524	689
Number of tubes	1828	371
Number of tube passes	8	1
Tube length (mm)	6096	6096
Length of flow path (m)	48.8	6.1
Tube diameter (mm)	25.4	25.4
Effective surface area (m^2)	874	178.5

Performance Details	Plain Tube	hiTRAN Enhanced
Prandtl number (in/out)	170/3800	170/3800
Reynolds number (in/out)	306/14	190/8
Overall service co-efficient (W/m^2)	40	182
Tubeside co-efficient (W/m^2)	51	295
Tubeside pressure drop (kPa)	70	70

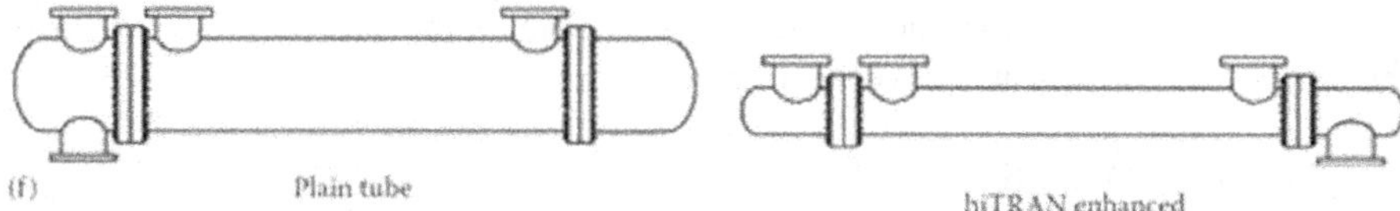

(f) Plain tube hiTRAN enhanced

Improved Design for an Air-Cooled Lube Oil Cooler Installed on a Gas Turbine-Driven Compressor

Design Comparison	Plain Bore-Finned Tube	hiTRAN System
Number of tubes/row	46	30
Number of rows of tubes	6	3
Tube length (mm)	7925	3350
Number of tube passes	6	1
Flow length (m)	47.55	3.35
Heat transfer rate (W/m^2 K) finned surface	3.29	20.95
Number and size off fans (mm)	2 × 2250	2 × 1250
Approximate total fan power (kW)	11.8	5.0
Plot area (m^2)	23.40	8.12
Finned surface (m^2)	3058.3	563.2
Oil pressure drop (kPa)	71	71
Weight (kg)	8500	2200

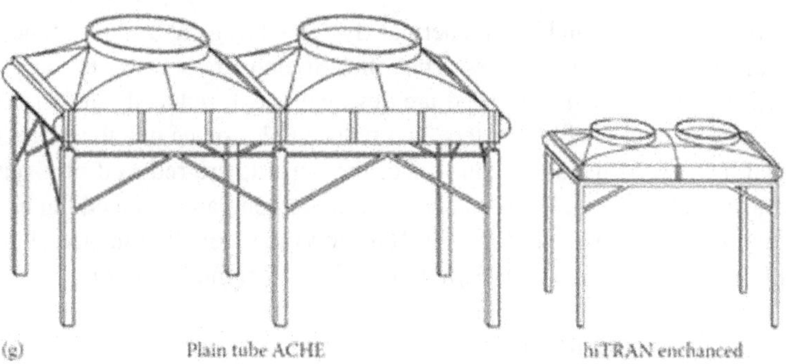

(g) Plain tube ACHE hiTRAN enchanced

FIGURE 8.16 (Continued)

times), significant benefits can be obtained in the transitional flow regime (up to 15 times), and turbulent flow regime (up to three times). Cal Gavin has installed hiTRAN Systems in heat exchangers operating with Reynolds numbers from 1 to over 100,000. While there is an increase in frictional resistance associated with hiTRAN systems, the amount of enhancement is such that solutions can be found which offer increased heat transfer at equivalent or lower pressure drop than a plain tube.

8.7.6.1 New Heat Exchanger Design

Many tubular heat exchanger designs can be improved by the application of hiTRAN Thermal Systems. The best results are obtained where the tubeside heat transfer is limiting, and substantial reductions in required surface area can be achieved. hiTRAN Thermal Systems can provide a practical and economical solution to problems such as fouling mitigation, maldistribution, turndown, performance, energy savings, etc.

8.7.6.2 Maldistribution

Where tubeside maldistribution is suspected to be a problem, hiTRAN Thermal Systems have proved beneficial, both in new designs and in retrofit. In retrofits, the increased flow resistance of the hiTRAN Thermal System can be used to correct an existing maldistribution problem. In addition to correcting the maldistribution, the retrofit will also increase the heat transfer rate leading to significantly improved performance at very modest cost.

8.7.6.3 Turndown Operation

For fully turbulent flow, the performance of a heat exchanger will be proportional to the flowrate. If, however, turndown conditions are such that the tubeside flow regime drops into the transitional or laminar region, the performance can become more difficult to predict and control. Small changes in throughput can lead to large changes in heat exchanger outlet temperatures.

8.7.6.4 Installation of hiTRAN Matrix Elements

hiTRAN Matrix Elements are easily fitted and removed. Elements are specially designed to be flexible for ease and speed of installation, even where access is very confined. In respect to U-tubes, hiTRAN Elements can be fitted into the straight leg by pushing instead of being pulled into place. The elements will be manufactured with a looser fit and one leg of each tube will have the element fitted in the opposite direction. hiTRAN Matrix Elements can be easily fitted (Figure 8.16d) in the field provided the tubeside has been effectively cleaned and held in place with fixing rods, as shown in Figure 8.16e.

8.7.6.5 Benefits

The enhanced performance of hiTRAN Thermal Systems brings a range of benefits in the design and operation of tubular heat exchangers. This, in turn, delivers benefits to the process plant as a whole. A selection of the benefits provided are described next: reduced number of shells in series, lower weight due to smaller shell diameters, and reduced plot space due to shorter tube lengths. As far as air-cooled heat exchangers are concerned, the benefits are reduced plot space requirement, fewer tube rows and tube passes, lower fan power requirement, and reduced fan noise. Figure 8.16f shows the reduction in shell size and Figure 8.16g shows the reduction in size of an air-cooled heat exchanger for typical cases due to installation of hiTRAN Thermal Systems.

8.7.6.6 Materials of Construction

hiTRAN Matrix Elements can be manufactured from a wide range of materials, including most grades of stainless steel, low-carbon steel, hastelloy, titanium, tantalum, monel, inconel, and copper-based alloys for general corrosion resistance; the most common material used is stainless steel Grades 304 and 316.

8.7.7 Inserts for Heating/Cooling Mode Operation

In the heating mode operation, the rotating flow is noted to have a favorable centrifugal convection effect, which can increase the heat transfer coefficient, whereas in the cooling mode operation, the rotating flow may have an adverse centrifugal convection effect, which even may decrease the convection heat transfer coefficient [18]. Thus, tube inserts that can generate rotating flow, such as swirl strips or twisted tapes, are generally not used in oil coolers; instead, coil or spiral springs are used, since spiral springs usually do not generate rotating flow.

8.7.8 Swirl Flow Devices

Swirl flow devices include a number of geometric arrangements or tube inserts for forced flow that create rotating and/or secondary flow inside tube. Such devices include inlet vortex generators, twisted tape inserts, and axial core inserts with screw-type windings [1,18]. The augmentation is attributable to several effects [1]: increased path length of flow, secondary flow effects, and, in the case of tapes, fin effects.

8.7.8.1 Twisted Tape Insert

In the design of a laminar flow heat exchanger, twisted tape (Figure 8.17) can be used effectively to enhance the heat transfer. Twisted tapes are simple enhancement devices which can be installed into new or existing tubular exchangers. Twisted tape enhancement is achieved by inducing swirl flow in the tubeside fluid and offers a modest increase in heat transfer at low additional pressure loss. The width of the metal strip of twisted tape is equal to or less than the internal diameter of the duct, depending upon whether the tape is loosely, snugly, or tightly fitted inside the tube. The twist ratio is defined as the axial length of tape necessary to make a 180° turn divided by the tube internal diameter. A twisted tape insert mixes the bulk flow well and therefore performs better in a laminar flow than any other insert. Compared with wire coil, twisted tape is not effective in turbulent flow because it blocks the flow and therefore the pressure drop is large. They are fully removable for cleaning. They are secured at the tube ends by rods or thick wires, as shown in Figure 8.18.

8.7.8.2 Corrugated Surfaces

Corrugated tubes (e.g., Wolverine Korodense®) are associated with circumferential indentation (as shown in Figures 8.19 and 8.20) or spiral indentation (spirally fluted) tubes as shown in Figure 8.21. Figure 8.22 shows a collection of fluted enhanced fin tubes. Corrugated tubes are manufactured from

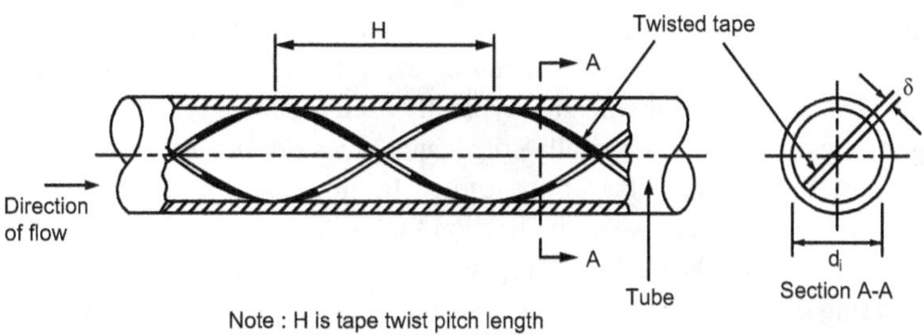

FIGURE 8.17 Twisted tape details.

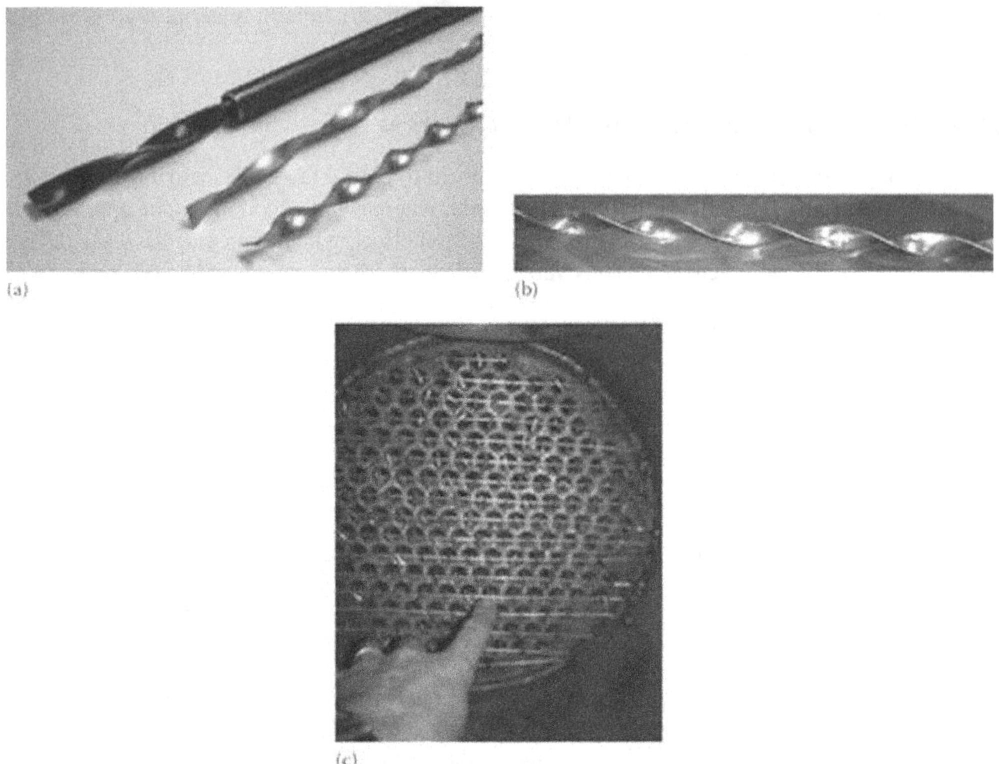

FIGURE 8.18 Twisted tapes installation. (a and b). Twisted tapes and (c) twisted tapes installed on tubeside held by fixing rods. (Courtesy of Peerless Mfg. Co. of Dallas, TX, makers of Alco and Bos-Hatten brands of heat exchangers.)

copper, copper alloys, carbon steels, stainless steels, and titanium. The corrugations are defined by the corrugation pitch, corrugation depth, and the number of corrugations. Multistart corrugated tubes with large corrugation depths are typically referred to as fluted tubes. Most corrugated tubes have a single start of corrugation that is defined by its depth e, axial pitch p, and helix angle β, as shown in Figure 8.18. The axial corrugation pitch is related to the internal diameter d_i, helix angle β relative to the axis of the tube, and the number of starts n_s, by the following equation

$$p = \pi d_i / n_s \tan\beta \tag{8.11}$$

The external diameter over the corrugations on the outside of the tube is equal to that of the plain ends of the tube. The internal diameter is taken as the external diameter less twice the tube wall thickness. The maximum internal diameter is typically used to define the internal heat transfer coefficient. Among the corrugated tubes, spirally corrugated tubes provide several advantages over other rough surfaces. They have easier fabrication, limited fouling, and higher enhancement in the heat transfer rate compared to the increase in the friction factor [21].

Flows in corrugated channels are associated with secondary flows, suppression of the secondary flow by counteracting centrifugal forces, and destruction of the secondary flow by the onset of turbulence [8]. In condensing services, corrugated tubes increase the heat transfer coefficient because only a thin film of the liquid condensate is left on the crests, while most of it drains into the troughs

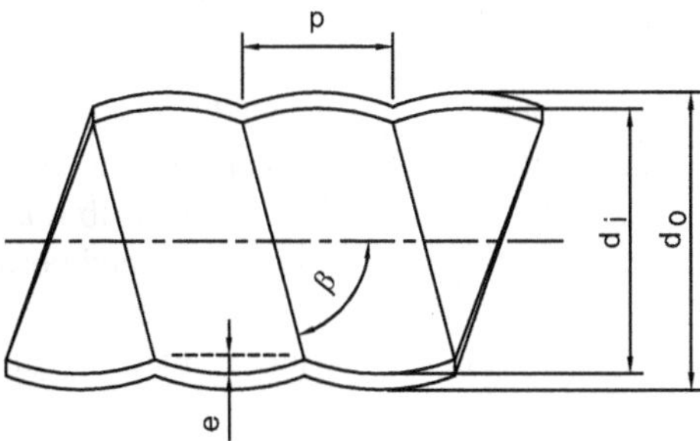

FIGURE 8.19 Corrugated tube details.

FIGURE 8.20 Corrugated tube—TEK tube—SA-214 heat exchanger tubes. (Courtesy of Fintube, LLC, Tulsa, OK; www.fintubellc.com)

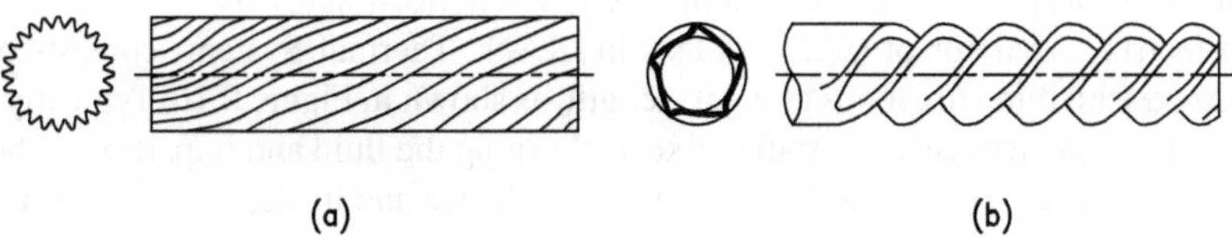

FIGURE 8.21 Spirally fluted tubes. (a) External helical rib and (b) both internally and externally corrugated tube (schematic).

FIGURE 8.22 Spirall0c Products, Inc., Windsor, CT.)

due to the surface tension and gravity effects. The thin film offers very little resistance, and hence the condensing coefficient is very high.

8.7.8.3 Doubly Enhanced Surfaces

If enhancement is applied to both the inner and outer tube surfaces, a doubly enhanced tube results. Doubly enhanced tubes are beneficial in cases where maximum reduction of heat exchanger size is of critical importance. The ID fins with helical rib contribute added surface area (30–40% approximately) but more significantly the change in the fluid flow pattern, namely, they promote mixing and turbulent flow. Applications for such doubly enhanced tubes include heat exchangers using sea water as a cooling medium, chemical industries, and condenser and evaporator tubes. Typical forms of doubly enhanced tubes include the following [9]: (1) helical rib roughness on inner surface and integral fins on outer surface; (2) internal fins on inner surface and porous boiling coating on outer surface; (3) twisted tape insert on inner surface and integral fins on outer surface; (4) corrugated roughness on inner and outer surfaces; and (5) corrugated tubes. Figure 8.23 shows some doubly enhanced tubes.

8.7.9 Turbulators

One means of achieving this desirable turbulent flow is through devices called turbulators, which alter a fluid's flow, maximizing contact with the tube wall. Turbulators are designed to create and maintain turbulent flow and help to increase the tube-side heat transfer efficiency. In tubular heat exchanger systems heat transfer occurs at the tube wall. A common problem in tubular heat exchangers is the development of a thermal boundary layer due to flow stagnation of the heating medium (gas or liquid) near the tube wall. This thermal boundary layer restricts the convective heat transfer around the tube wall considerably. By increasing the turbulence intensity, turbulators reduce the development of this thermal boundary layer and create greater contact of the heating medium with the tube wall. There are four common types of turbulators seen in heat exchangers: ball, twisted tape, spring, and matrix [22,23]; these turbulators are shown in Figure 8.24. Wire matrix turbulator is shown in Figure 8.15. Helically shaped strips are inserted into heat exchanger tubes, where they create turbulence. Flow over a helically shaped strip is shown in Figure 8.25. An example of a wire matrix turbulator is hiTRAN Thermal Systems discussed in Section 8.7.6.

The wire matrix turbulator of CALGAVIN's hiTRAN® Thermal Systems consists of a twisted wire rod adorned with thin wire loops along its length, as shown in Figure 8.16. Typically made from stainless steel, these devices serve as static mixers, churning the fluid and imparting turbulence. The wire matrix functions as a displaced flow enhancement device and its details were discussed earlier.

8.7.9.1 Turbulator for Fire Tube Boiler

Turbulators are made from a narrow, thin metal strip bent and twisted in zig-zag fashion to allow periodic contact with the tube wall. A popular technique for enhancement of heat transfer in fire tube boilers is to introduce turbulators into the boiler tubes to increase the gas side heat transfer coefficients, thereby increasing heat recovery from flue gases [24]. This will reduce fuel consumption and improve boiler efficiency.

8.7.10 Surface Tension Devices

Surface tension devices consist of relatively thick wicking material or grooved surfaces on heat transfer devices to direct the flow of liquid in boiling and condensing.

8.7.11 Additives for Liquids

Additives for liquids are liquid droplets for boiling, solid particles in single-phase flow, and gas bubbles for boiling.

8.7.12 Additives for Gases

Additives for gases are liquid droplets or solid particles, either dilute phase (gas–solid suspensions) or dense phase (fluidized beds).

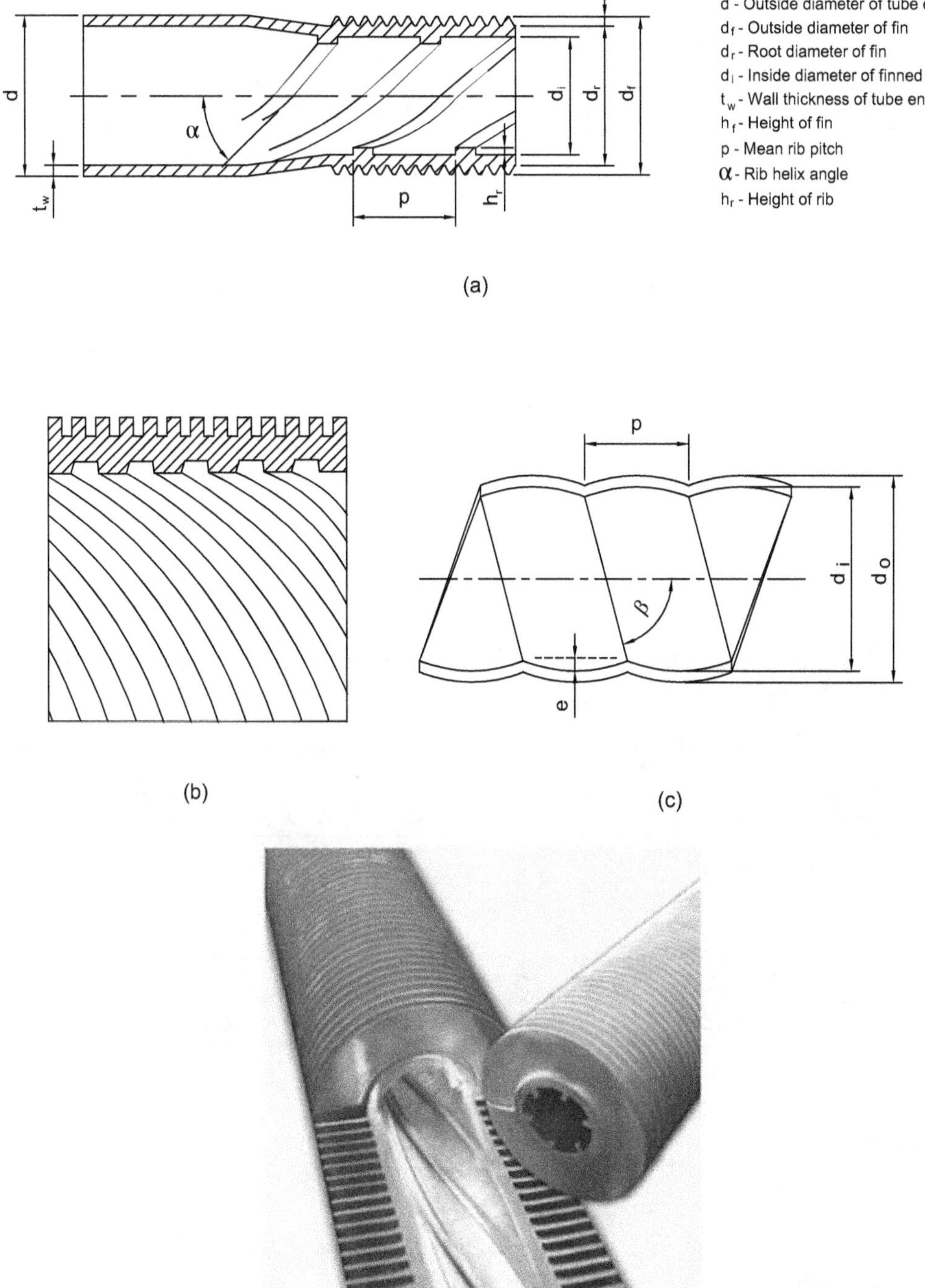

FIGURE 8.23 (a, b and c) Doubly enhanced tubes: and (d) thermofin integral turbulator. Note. – 'e' is depth of corrugation. (From Thermofin; www.thermofin.net)

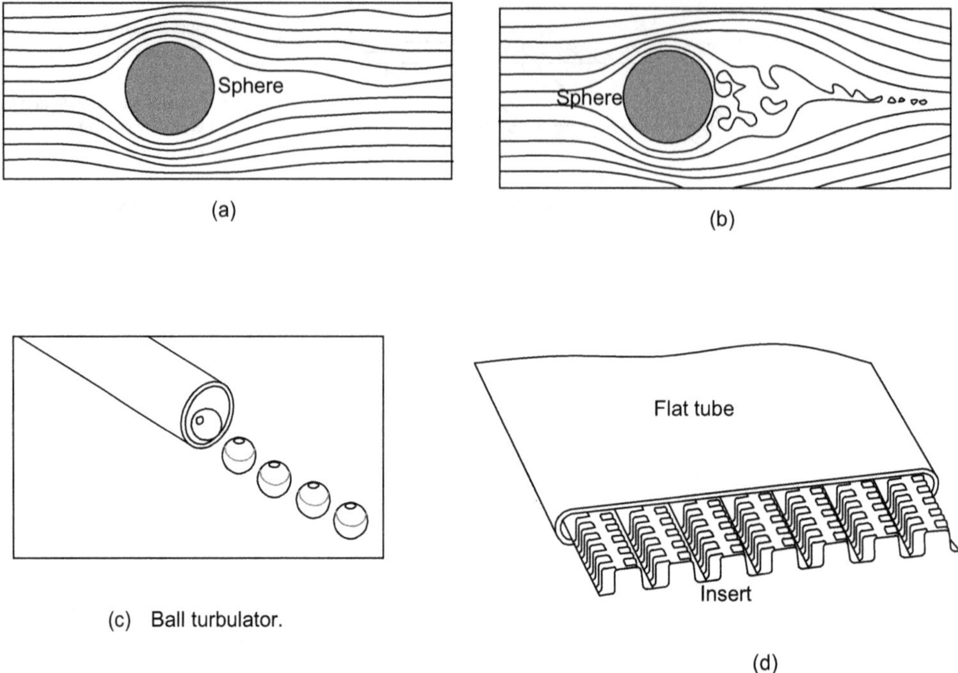

FIGURE 8.24 Turbulators. (Adapted and modified from www.superradiatorcoils.com/blog/heat-exchanger-turbulators-types-and-purposes.)

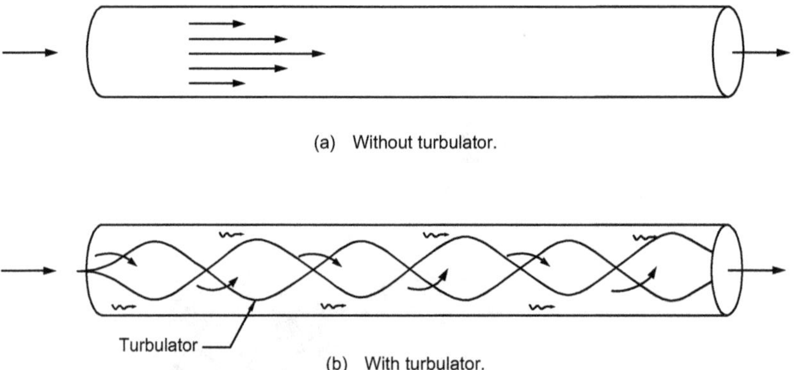

FIGURE 8.25 Flow over helically shaped strips. (Adapted and modified from www.jdturbulators.com/technology/.)

8.8 ACTIVE TECHNIQUES

The active techniques (after Bergles [1,25]) are described next.

1. *Mechanical aids* stir the fluid by mechanical means or by rotating the surface or by surface scraping. Surface scraping is widely used for viscous liquids in the chemical process industry.
2. *Surface vibration* (which thins or restarts the boundary layer and/or induces secondary flows) at either low or high frequency has been used primarily to improve single-phase heat transfer. Possible modes of heat transfer applications are single-phase flow, boiling, and condensing.
3. *Fluid vibration* is the most practical type of vibration enhancement, given the mass of most heat exchangers. The vibrations range from pulsations of about 1 Hz to ultrasound. Single-phase

fluids are of primary concern. Possible modes of heat transfer applications are single-phase flow, boiling, and condensing.

4. *Electrostatic fields* (dc or ac) are applied in many different ways to dielectrics to cause greater bulk mixing of fluid in the vicinity of the heat transfer surface. An electrical field and a magnetic field may be combined to provide forced convection via electromagnetic pumping.
5. *Boundary layer injection* involves supplying gas to a flowing liquid through a porous heat transfer surface or injecting similar fluid upstream of the heat transfer section [25]. (Enhancement is primarily for multiphase flows.)
6. *Boundary layer suction* (this involves removal of the boundary layer and restarts the boundary layer downstream) involves either vapor removal through a porous heated surface in nucleate or film boiling, or fluid withdrawal through a porous heated surface in single-phase flow [25].
7. Electrohydrodynamic (EHD): High voltage (>1 kV) applied to an electrode near a plate which induces secondary flows in the boundary layer (applicable for liquid flows only) [25].

The active techniques have not found commercial interest because of the capital and operating costs of the enhancement devices, complex nature, and problems associated with vibration or noise [12] or high voltage.

8.9 ENHANCEMENT IN PHASE CHANGE APPLICATIONS

References [1,6,7] detail heat transfer enhancement techniques for phase change. One major area of application of EHT surfaces for phase change applications is in refrigeration and air-conditioning applications. Bergles [1], Ref. [5], and Pate et al. [13] detail the application of heat transfer enhancement techniques in the refrigeration and air-conditioning industries.

8.10 CONDENSATION ENHANCEMENT

The heat transfer resistance in condensation of pure components is primarily due to conduction across the condensate film thickness. Condensation enhancement is generally obtained by taking advantage of surface tension forces to obtain a thin condensate film or to strip condensate from the heat transfer surface. Special surface geometries or treated surfaces are effective in attaining this goal. Heat transfer enhancement is possible on either the shellside or tubeside of a condenser. Since the tube orientation will affect its condensate drainage characteristics, one must distinguish between horizontal and vertical orientations.

8.10.1 Horizontal Orientation

Shellside enhancement

1. Short radial fins, fluted tubes, wires attached to tubes, modified annular fins with sharp tips on tubes [7].
2. Surfaces of doubly augmented tube geometries such as helically corrugated tubes, helically deformed tubes having spiral ridges on the outer surface and grooves on the inner surface, or corrugated tubes formed by rolling in helical shape followed by seam welding are used [5].
3. Extended surfaces for film condensation on horizontal tubes include Hitachi Thermoexcel-C, and the spine fin surface.

Tubeside enhancement: Closely spaced helical internal fin, twisted tape insert, repeated rib roughness, and sandgrain type roughness may be used.

8.10.2 Shellside Condensation on Vertical Tubes

Shellside condensation enhancement on a vertical surface can be achieved by finned tubes, fluted surfaces, and loosely attached spaced vertical wires [5].

8.10.3 Modes of Condensation

Condensation is the heat transfer process by which a saturated vapor is converted into a liquid by means of removing the latent heat of condensation. Condensation occurs when a vapor contacts a solid surface or a fluid interface whose temperature is below the saturation temperature of the vapor. Condensation can occur on the outside of surfaces, such as on plates, horizontal or vertical tubes, and tube bundles. The following section provides a treatment of intube condensation, in which the process takes place inside enclosed channels typically with forced flow conditions. Four basic mechanisms of condensation are generally recognized: drop-wise, film-wise, direct contact, and homogeneous. Refer to Chapter 5 for mechanisms of condensation.

8.10.4 Condensation on Low-Finned Tubes and Tube Bundles

Integral low-finned tubes have been utilized for enhancing condensation for more than half a century. The geometry of a low-finned tube is illustrated in Figure 8.9.

8.11 EVAPORATION ENHANCEMENT

The mechanisms important for evaporation enhancement are thin film evaporation, convective boiling, and nucleate boiling [7]. Nucleate and transition pool boiling are usually dependent on the surface condition as characterized by the material, the surface finish, and the surface chemistry. Certain types of roughness have been shown to reduce wall superheats, increase peak critical heat flux, and destabilize film boiling [1]. Heat transfer augmentation techniques for thin film evaporation and nucleation are as follows:

1. Thin film vaporization: Fluted vertical tubes.
2. Nucleation: For nucleation enhancement, structured surfaces are used. Structured surface refers to fine-scale alteration of the surface finish. A coating may be applied to the plain tube, or the surface may be deformed to produce subsurface channels or pores [1]. Typical structured surfaces are *enhanced surface cavities*—a pore or reentrant cavity within a critical size range, interconnected cavities, and nucleation sites of a re-entrant shape; and *integral roughness*—the three commercially used boiling surface of this type are Trane bent fins, Hitachi bent sawtooth fins, and Weiland flattened fins (T-shaped fin forming) [5].
3. Convective boiling: On the shellside, integral fins and enhanced boiling surfaces are used. On the tubeside, there are internal ridges, and axial and helical fins.
4. Flow boiling inside tubes: Porous coating and special internal roughness geometries.

8.11.1 Enhanced Boiling Surfaces

Numerous enhanced boiling surfaces have been proposed and patented over the years. Figure 8.26 depicts a diagram showing a few of their geometries. The first type of enhanced boiling tube to become successful was the low-finned tube.

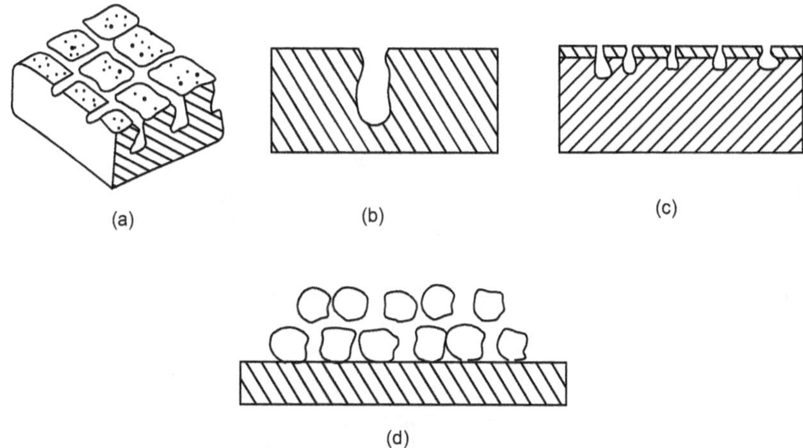

FIGURE 8.26 Some of boiling heat transfer enhanced surface geometries.

8.11.2 Boiling on Enhanced Tubes

Boiling heat transfer coefficients on smooth surfaces can be increased by roughening the surface, but this is not normally practical (and perhaps temporal if the surface fouls). To achieve significant enhancement, numerous types of geometries have been proposed and patented. Tubes with external porous coatings were apparently the next important enhancement to be proposed, yielding augmentations of up to 10–15 times the boiling performances on plain tubes at optimum conditions.

8.11.3 Bundle Boiling

An important heat transfer process is evaporation on the outside of horizontal tube bundles. This process is generic to refrigerant flooded evaporators, waste heat boilers, fire-tube steam generators, kettle and thermosyphon reboilers, feed effluent heat exchangers, etc.

8.11.4 Boiling Heat Transfer inside Enhanced Tubes

Evaporation in both vertical tubes and horizontal tubes is addressed. The types of enhancements discussed include microfin tubes, twisted tape inserts, corrugated tubes, and internally porous coated tubes. Other enhancements exist but are either not widely used anymore (aluminum star-inserts and internally high-finned tubes, for example), are not appropriate for enhancing boiling heat transfer, or do not have very much published about their thermal performance. Several prediction methods are presented for microfin tubes and tubes with twisted tape inserts [5].

In vertical tubes, the most important applications and potential benefits for use of enhancements for evaporation are in the petrochemical industry and hence for vertical thermosyphon reboilers. In horizontal tubes, the most important applications of enhancements are to direct-expansion evaporators in refrigeration, air-conditioning, and heat pump units.

8.11.5 Evaporation

The enhancement techniques for evaporation include microfin tubes, twisted tape inserts, corrugated tubes, and tubes with porous coatings.

8.11.6 Microfin Tubes

Numerous applications for microscale flow boiling are emerging: high heat flux cooling of computer microprocessor chips and power electronics, precise cooling of micro-reactors, rapid and uniform cooling of LED displays, development of automotive evaporators with multi-port aluminum tubes, etc. [5]. Application of microfin tubes for flow boiling is discussed in Chapter 20 of Ref. [5]. Microfin tubes are available primarily in copper. For instance, Wolverine Tube Inc. is a major manufacturer of copper microfin tubes for the air-conditioning and refrigeration industries. Figure 8.8 shows the characteristic geometry, which is defined by the maximum internal diameter df, number of fins, their helix angle α_f (or axial pitch), their height e_f, their thickness, their cross-sectional shape, and the internal area ratio.

Most microfins have approximately a trapezoidal cross-sectional shape with a rounded top and rounded corners at the root. Other shapes are triangular, rectangular, and screw (no root area between fins). The thickness of the fins is not of much importance since the fin efficiencies are close to 1.0 even in alloy tubes since the fins are typically only from 0.2 to 0.3 mm (0.008–0.012 in.) high.

8.11.7 Microfin Grooved Copper Tubes

Internally grooved copper tubes, also known as "microfin tubes," are a small-diameter coil technology for modern air conditioning and refrigeration systems. Grooved coils facilitate more efficient heat transfer than smooth coils. This increases the surface area to volume ratio, mixes the refrigerant, and homogenizes refrigerant temperatures across the tube. Tubes with MicroGroove technology can be made with copper or aluminum. Copper fins are an attractive alternative to aluminum due to the better corrosion resistance of copper and its antimicrobial benefits. The MicroGroove copper tube heat exchanger coil is discussed in Chapters 1 and 3.

8.11.8 Twisted Tape Inserts

Heat transfer augmentation ratios are typically in the range from 1.2 to 1.5, while two-phase pressure drop ratios are often as high as 2.0 since the tape divides the flow channel into two smaller cross-sectional areas with smaller hydraulic diameters. Twisted tapes have seen some applications in both horizontal and vertical units. Twisted tape inserts and corrugated tubes have been commercially available for many years, although both have been largely supplanted by microfin tubes. Figure 8.17 depicts a twisted tape insert and a diagram of its geometry. The twist ratio is defined as the axial length of tape necessary to make a 180° turn divided by the tube internal diameter. Twist ratios of 3–5 are typically used, while a twist ratio of infinity represents a straight tape without any twist. As twisted tape inserts fit loosely inside the tubes, the surface area of the tape is not considered to be a heat transfer surface area since little heat is conducted into the tape from the tube wall.

8.11.9 Corrugated Tubes

Heat transfer ratios are usually between 1.2–1.8, with performances matching microfin tubes at high mass velocities but with much larger two-phase pressure drop ratios, which are on the order of two times those of a plain tube. Corrugated tubes are manufactured in many metals: copper, copper alloys, carbon steels, stainless steels, and titanium. The corrugations are defined by the corrugation pitch, corrugation depth, and number of corrugations, i.e. the number of starts. Multi-start corrugated tubes with large corrugation depths are typically referred to as fluted tubes. Twisted tape inserts are available in most metals, diameters, and lengths. They are made by twisting a metal strip and fit loosely in the tubes to allow for standard tube wall dimensional tolerances. Figures 8.19 and 8.20 illustrate corrugated tubes.

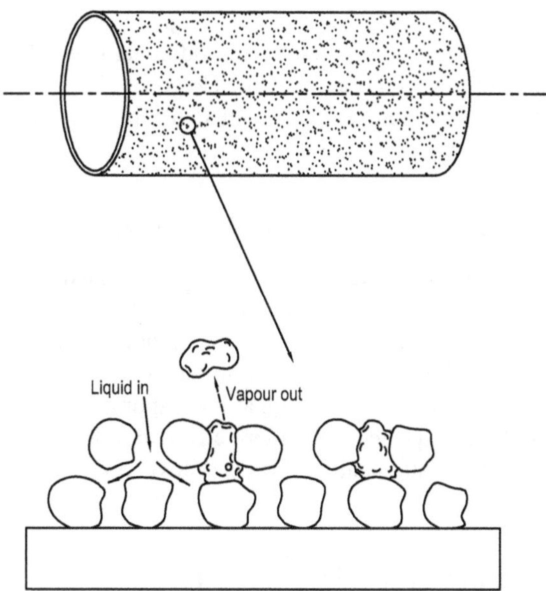

FIGURE 8.27 Treated heat transfer surface.

Most corrugated tubes have a single start of corrugation that is defined by its depth e and its axial pitch p (or helix angle). The maximum internal diameter is typically used to define the internal heat transfer coefficient. The internal area ratio is slightly larger than 1.0 but is not often cited in publications.

8.11.10 Treated Surfaces

Treated surfaces involve fine-scale alternation of the surface finish or provision of either continuous or discontinuous coating. Treated surfaces such as hydrophobic coatings and porous coatings (Figure 8.27) are most effective for phase changes but are not applicable to single-phase convection.

8.11.11 Porous Coated Tubes

These have heat transfer performances similar to those for nucleate pool boiling and are on the order of 5–10 times plain tube performance in vertical tubes. For evaporation in horizontal tubes, the porous coating is only effective for annular flows but not for stratified flows where part of the tube perimeter is dry. Not much information is available about two-phase pressure drop penalties, which can be expected to be only marginal in vertical units where the static head dominates, while it is significant in horizontal tubes where the frictional pressure drop is typically dominant.

8.11.12 Flow Boiling in Vertical Microfin Tubes

The Gregorig effect enhances film evaporation by drawing the liquid film from the microfin tips toward their roots, similar to film condensation on low-finned tubes and corrugated surfaces. Enhancement of microfin flow boiling heat transfer coefficients relative to plain tube values can be attributed to [5]:

- An increase in the convective contribution by the single-phase effect of the microfins
- A rise in the nucleate boiling contribution by the additional internal surface area
- Enhancement of annular film evaporation by the Gregorig effect
- Conversion of stratified-wavy flow (partially wetted tube perimeter) to annular flow (complete wetting) at low mass velocities

8.11.13 Flow Boiling in Vertical Tubes with Twisted Tape Inserts

Twisted tape inserts have the distinct advantage that they can be used to increase the thermal capacity of existing evaporators without replacing the tube bundle. Twisted tapes fit loosely within the tubes and hence they enhance heat transfer by the swirl that they impart on the flow but do not provide any additional heat transfer surface area. Most of the augmentation is observed at the higher vapor qualities, where it is as much as 40%, while at low vapor qualities it is only on the order of 10%. The swirl effect would be expected to be more effective at lower mass velocities and lower heat flux where the potential for augmenting the convective contribution to the two-phase flow boiling coefficient is most evident [5].

8.11.14 Flow Boiling in Vertical Tubes with an Internal Porous Coating

Applying a thin porous coating to the inside of vertical evaporator heat transfer tubes is very successful in improving performance. Heat transfer augmentation is on the order of 10-fold. It has been noted that single-tube nucleate pool boiling heat transfer data are suitable for predicting the flow boiling performance of the High Flux tube since the nucleate boiling contribution dominates the convective boiling contribution.

8.11.15 Flow Boiling of Pure Fluids in Enhanced Horizontal Tubes

The pressure drops are the total pressure drops for both the evaporating and superheating zones. The enhancement factors are the respective heat transfer and pressure drops relative to the plain tube. A parametric study shows that the twisted tape is much less effective than the microfin tubes while having a larger pressure drop. The microfin tubes augment heat transfer substantially in both the boiling and superheated zones.

8.12 PERTINENT ISSUES

8.12.1 Friction Factor

Due to the form drag and increased turbulence caused by the disruption, the pressure drop with flow inside an enhanced tube always exceeds that obtained with a plain tube for the same length, flow rate, and diameter (after Ref. [20]). In most cases, the increase over that of a plain tube is about 70–500%, and usually exceeds the increase in heat transfer, which is about 50–200%. However, for Reynolds number in the range of 5,000–10,000 and for fluids with Prandtl numbers greater than about 2, the heat transfer enhancement can exceed the pressure drop increase.

8.12.2 Application-Related Issues

The issues related to enhanced surfaces in respect of their applications are

(1) Performance evaluation and testing methods
(2) Fouling

(3) Performance evaluation criteria (PEC)
(4) Market factors
(5) Mechanical design and construction considerations

8.12.3 Testing Methods

The basis for selecting an "optimum" heat transfer surface is easy, provided generalized heat transfer and friction correlations and quantitative relations for performance evaluation are available. The multitude of augmentation devices with various surface geometries made it impossible to generalize specific correlations for *j* and *f* factors. However, this can be overcome, by devising a reliable experimental technique to determine the performance. A common method for determining turbulent-flow, single-phase coefficients from heat-exchanger tests is the Wilson method [7]. In this method, the coefficient on one side is held constant, while the coefficient to be determined is varied by making a series of tests at different flow rates. The equation for Nusselt number describing the coefficient to be measured is of the form

$$\mathrm{Nu} = k_1 \mathrm{Re}^{k_2} \mathrm{Pr}^{k_3} \tag{8.12}$$

where the constants k_2 and k_3 are known in advance. The objective of the test is to determine k_1. As shown next, the total resistance R_T is composed of a constant resistance R_c and a variable resistance R_v, which is to be determined:

$$R_T = R_c + R_v = R_c + \frac{1}{k_1}\left(\frac{1}{k\,\mathrm{Re}^{k_2}\mathrm{Pr}^{k_3}/D_h}\right) \tag{8.13}$$

where D_h is the tube hydraulic diameter and k is the thermal conductivity of the fluid.

8.12.4 Fouling

An important consideration for whether enhanced surfaces should be used is the effect of fouling on the heat exchanger performance. Fouling is a very difficult phenomenon to model and predict in general terms since its exact mechanism is specific to the process conditions. Many research programs are ongoing to study the fouling of enhanced surfaces. There are some techniques that are not affected by fouling; in some applications, fouling can be reduced by using enhanced surfaces. In general, the same factors that influence fouling on unenhanced surfaces will influence fouling of high-performance surfaces [7].

8.12.5 Performance Evaluation Criteria

The basic thermohydraulic characteristic of an enhanced surface is defined by curves of the *j* and *f* versus Reynolds number. Numerous factors decide the merit of enhanced surfaces developed for augmentation of heat transfer. A PEC establishes a quantitative statement to account for the heat transfer enhancement, taking into account the pressure drop characteristics of the enhanced surface. Possible performance objectives of interest include [5,8,9] the following:

1. Reduced heat transfer size and hence low material requirement for fixed heat duty and pressure drop.

2. Increased UA; this may be exploited in two ways: to obtain increased heat duty and to secure reduced LMTD for fixed heat duty.
3. Reduced pumping power for fixed heat duty.

Bergles et al. [26] reviewed these criteria, which were classified into the following three objectives: (1) increase in heat transfer; (2) reduced pumping power for fixed heat duty; and (3) reduce exchanger size for fixed heat duty and pressure drop. Shah [27] also critically reviewed more than 30 dimensional or nondimensional comparison methods.

8.12.5.1 Webb's PECs: Performance Comparison with a Reference

Webb [28] developed a number of PECs of interest with flow inside enhanced and smooth tubes of the same envelope diameter. Generally, these PECs are defined by the following three different geometry constraints:

1. *FG criteria*: The cross-sectional envelope area and tube length are held constant.
2. *FN criteria*: These criteria maintain a fixed-flow frontal area and allow the length of the heat exchange to be a variable. These criteria seek reduced surface area or reduced pumping power for constant heat duty.
3. *VG criteria*: These criteria maintain a constant exchanger flow rate.

The concepts of PEC analysis are explained for the case of a prescribed wall temperature or a prescribed heat flux in Ref. [9]. The heat transfer enhancement, $hA/h_pA_p = K/K_p$, in terms of Stanton number St and mass velocity G, is given by [8]

$$\frac{K}{K_p} = \frac{\mathrm{St}}{\mathrm{St}_p}\frac{A}{A_p}\frac{G}{G_p} \tag{8.14a}$$

or

$$\frac{K}{K_p} = \frac{j}{j_p}\frac{A}{A_p}\frac{G}{G_p} \tag{8.14b}$$

where the subscript p refers to the reference or plain smooth surface, St is Stanton number, G is mass velocity, and A is flow area. The relative friction power (P) is

$$\frac{P}{P_p} = \frac{f}{f_p}\frac{A}{A_p}\left(\frac{G}{G_p}\right)^3 \tag{8.15}$$

It is assumed that the smooth tube operating conditions (G_p, f_p, j_p, and St_p) are known.

The advantages of this comparison method for compact heat exchangers include the following [27]: (1) the designer can select his or her own criteria for comparison; (2) then the performance of a surface can be compared with the reference surface; and (3) there is no need to evaluate the fluid properties since they drop out in computing the comparison ratios.

8.12.5.2 Shah's Recommendation for Surface Selection of a Compact Heat Exchanger with Gas on One Side

Four of the criteria recommended by Shah [27] are discussed here. These criteria require the *j* and *f* factors, and Reynolds number, together with the geometry specification. These criteria are employed for comparison of flat-tube radiators with plain fin and ruffled fin performances by Maltson et al. [29].

1. *Flow area goodness factor comparison*: The ratio of *j* factor to the friction factor *f*, against Reynolds number, generally known as the "flow area goodness factor," was suggested by London [30]

$$\frac{j}{f}=\frac{\mathrm{Nu}\,\mathrm{Pr}^{-1/3}}{f\,\mathrm{Re}}=\frac{1}{A_c^2}\left[\frac{\mathrm{Pr}^{1/3}}{2g_c}\frac{\mathrm{NTU}\,M^2}{\Delta p}\right] \tag{8.16}$$

   A surface having a higher *j/f* factor is "good" because it will require a lower free flow area and hence a lower frontal area for the exchanger.

2. *Volume goodness factor (E_{std}) comparison*: This is the standardized heat transfer coefficient against the pumping power per unit of heat transfer surface area, suggested by London and Ferguson [31]. Two types of comparison are suggested: (1) h_{std} versus E_{std}, and (2) $\eta_o h_{std}\beta$ versus $E_{std}\beta$ as given in the following:

$$h_{std}=\frac{j\,\mathrm{Re}\,\mu_{std}C_{p.std}}{D_h\mathrm{Pr}_{std}^{2/3}} \tag{8.17}$$

   and

$$E_{std}=\frac{P_p}{A}=\frac{f\,\mathrm{Re}^3\mu_{std}^3}{2g_c\rho_{std}^2D_h^3} \tag{8.18}$$

3. The performance of the heat exchanger per unit volume, the criterion suggested by Shah [27]. This method includes the effect of the fin effectiveness. A good performance using this criterion gives the best heat exchanger to use where the size of the unit is an important consideration. Two types of comparison are suggested as follows: (1) h_{std} versus E_{std}, and (2) $\eta_o h_{std}\beta$ versus $E_{std}\beta$ as given in the following:

$$\eta_o h_{std}\beta=\frac{j\,\mathrm{Re}(4\sigma)\,\eta_o C_{p.std}\mu_{std}}{\mathrm{Pr}_{std}^{2/3}D_h^2} \tag{8.19}$$

   and

$$E_{std}\beta=\frac{f\,\mathrm{Re}^3(4\sigma)\mu_{std}^3}{2g_c\rho_{std}^2D_h^4} \tag{8.20}$$

4. The actual heat transfer rate performance versus the gas side fan power, suggested by Bergles et al. [32].

8.13 MARKET FACTORS

8.13.1 Alternate Means of Energy Savings

The extent to which the EHT devices and techniques that affect design variables such as heat transfer efficiency and pressure drop and maintenance issues such as corrosion and fouling as compared with other performance enhancement options ultimately affects its ability to penetrate the market and appear in products. For example, heat transfer enhancement options may compete with the use of larger conventional heat exchangers and higher efficiency compressors to increase unit efficiency [25].

8.13.2 Adoptability to Existing Heat Exchangers

Manufacturers may be cautious about introducing novel heat transfer enhancement approaches/techniques into products. If approaches require appreciable modifications and hence incur heavy costs to existing heat exchangers in service, the technique may not be acceptable to industries [25].

8.13.3 Proven Field/Performance Trials

To avoid deterioration of performance in service, manufacturers must carry out extensive laboratory and field testing to verify that the enhancements perform reliably over the range of expected operating conditions.

8.14 MECHANICAL DESIGN AND CONSTRUCTION CONSIDERATIONS

Mechanical stress calculations: Burst tests with internal pressure on integral low-finned tubes have shown that the plain ends of the tube are the weakest point along the tube because the external helical fins act as reinforcement rings [25]. Choice of the minimum wall thickness under the augmentation

FIGURE 8.28 Enhanced tubes (TRU-TWIST™ tubes) with pain length at end. (Courtesy of Turbotec Products, Inc., Windsor, CT.)

has a direct impact on the amount of metal per meter of tubing. Thus, this choice is economically important, especially for expensive alloy materials. For most applications, the base wall thickness under the augmentation must be used for the mechanical stress calculations defined by various pressure vessel codes, such as that of ASME Code Section VIII Div. 1 [33].

Heat exchanger fabrication: Enhanced tubes normally have plain length at each end, as shown in Figure 8.28. This allows these tubes to be rolled and/or welded into the tubesheet. The outside diameter over the augmentation (such as a low-finned tube) is equal to or slightly less than that of the plain ends. Thus, these tubes can be drawn into the tube bundle during assembly without any problems. Tube inserts normally have pull rings or attachments to install them, fix them in place, and remove them for cleaning.

U-tubes: Nearly all integral heat transfer augmentations can be bent into U-tube heat exchangers. For the minimum bending radius of a particular type of tube, one should refer to the manufacturer's recommendations or heat exchanger standards like TEMA [34].

8.15 THERMAL AND ECONOMIC ADVANTAGES OF HEAT TRANSFER AUGMENTATIONS

The thermal and economic advantages of heat transfer augmentations are discussed in Chapter 2 of Ref. [5]. The principal advantage of introducing an augmentation is the possibility of substantially increasing thermal duty to meet the needs of new process conditions or production goals. This can be achieved either by [5]:

1. Installing removable inserts inside the tubes
2. Replacing a removable tube bundle with a new enhanced tube bundle
3. Replacing the heat exchanger with a new enhanced tube heat exchanger of the same size or smaller

8.15.1 Cost Savings

The cost savings for appropriate applications of enhanced tubes to shell and tube heat exchangers is to be assessed. When utilizing high-alloy tubes in heat exchangers (stainless steel, titanium, nickel alloys, duplex stainless steels, etc.), applying the appropriate augmentation can very significantly reduce their initial cost. The augmentation may not only reduce the cost of the tubing, but also those of the heads and tubesheets (smaller diameters, smaller wall thicknesses, fewer tube holes to drill, less alloy cladding material, etc.).

8.16 MECHANICAL DESIGN AND CONSTRUCTION CONSIDERATIONS

8.16.1 Mechanical Stress Calculations

Burst tests with internal pressure on integral low-finned tubes have shown that the plain ends of the tube are the weakest point along the tube because the external helical fins act as reinforcement rings. The choice of the minimum wall thickness under the augmentation has a direct impact on the amount of metal per meter of tubing. Thus, this choice is economically important, especially for expensive alloy materials. For most applications, the base wall thickness under the augmentation must be used for the mechanical stress calculations defined by various pressure vessel codes, such as that of ASME [33]. Consult the applicable pressure vessel code for guidance.

8.16.2 Heat Exchanger Fabrication

Enhanced tubes normally have plain lengths at each end that are a little longer than the tubesheet thickness. This allows these tubes to be rolled and/or welded into the tubesheet. The outside diameter over the augmentation (such as a low-finned tube) is equal to or slightly less than that of the plain ends. Thus, these tubes can be drawn into the tube bundle during assembly without any problems. Tube inserts normally have pull rings or attachments to install them, fix them in place, and remove them for cleaning.

U-Tubes. Nearly all integral heat transfer augmentations (i.e., those augmentations that are an integral part of the tube wall) can be bent into U's for U-tube heat exchangers. For the minimum-bending radius of a particular type of tube, one should refer to the manufacturer's recommendations.

8.16.3 Mean Metal Temperature Differences

Some fixed tubesheet plain tube designs result in very large mean metal temperature differences between the tubes and the heat exchanger shell. This large temperature difference causes unequal thermal expansion (or contraction) and large stresses. Since augmenting one fluid stream of an exchanger almost always shifts the controlling thermal resistance to the other fluid stream, an enhancement can be used to reduce the mean metal temperature difference and avoid using an expansion joint where they are undesirable [5]. When installing inserts in existing fixed tubesheet units, it is good practice to compare the new mean metal temperature difference value against that used for the mechanical design of the exchanger.

8.17 MAJOR AREAS OF APPLICATIONS

8.17.1 Refrigeration and Air-Conditioning System Applications

For in-tube boiling (or condensation), the aluminum star inserts that were once widely used have been nearly completely abandoned in favor of internally microfinned tubes [5].

For evaporation in flooded evaporators, many external augmentations are available commercially, most of which are normally produced in doubly enhanced versions.

For chilled water flowing inside tubes, highly efficient internal rib and fin designs provide a good tradeoff between heat transfer and pressure drop.

For chilled water on the shell-side, the best tube selection may be a low-finned tube with internal microfins where one pays for more tube metal per meter but will use only about one-half the meters of tubing.

8.17.2 Shellside Evaporation of Refrigerants

Enhanced tubes for shellside evaporation of refrigerants and their commercial names include (1) modified structured surfaces with porous coatings, such as High Flux (Union Carbide) and (2) modified low-fin tubes such as are used for shellside evaporation of refrigerants, such as GEWA-T (WeilaandWerke), GEWA-TX, GEWATXY, Thermoexcel-E (Hitachi), Turbo-B (Wolverine) [13], EverFin-40 (Furukawa Electric), and Thermoexcel-HE.

8.17.3 Shellside Condensation of Refrigerants

Recognition of surface tension as the dominant force for determining condensate thickness has led to the development of surfaces designed to produce thin films, and to remove condensate from the heat transfer surface. Profiles such as Thermoexcel-C (Hitachi) and Turbo-V-shaped fin with a steep angle help in maintaining a thin refrigerant film.

8.17.4 In-Tube Evaporation of Refrigerants

In-tube enhancement devices for evaporation of refrigerants include (1) rough surfaces like helical wire inserts, internal thread, and corrugated tubes; (2) extended surfaces like microfin tubes and high-profile fins; and (3) swirl flow devices such as twisted tape inserts.

8.17.5 Refinery and Petrochemical Plant Applications

Refinery and process industries—single-phase exchangers [5]:

1. Use of integral low-finned tubes in heat exchangers when the limiting thermal resistance is on the shellside
2. Use of tube inserts (wire mesh or twisted tape types) is highly effective in laminar flows inside tubes
3. Installation of inserts on the tubeside of heat recovery units

8.17.6 Applications in Lubricating Oil Coolers

Normally, an increase in tubeside heat transfer performance is obtained with an insert while meeting the same pressure drop limitations as for the plain tube unit. This is achieved by using fewer tube passes.

8.17.7 Power Plant Operations and Other Areas

Integral low-finned tubes are becoming widely used in power plant main condensers. Use of external fins shifts the controlling thermal resistance to the cooling waterside. Thus, low-finned tubes with internal ribs are particularly suitable for this application, especially when utilizing alloys. Corrugated tubes are also beneficial in these applications. Installation of inserts on the tubeside (i.e., waterside) of an existing steam condenser (plain or low-finned tube) increases the overall heat transfer coefficient and the vacuum in the steam chest.

Fossil fuel power boilers are designed with internally ribbed tubes by some steam generator manufacturers. The swirl flow created by the ribs keeps the tube wall better wetted in this asymmetrical heat flux environment, which increases the critical heat flux before passing into the film boiling regime. Twisted tape inserts can be installed in tube sections prone to this problem in existing plain I.D. units.

Heating, ventilating, and refrigeration and air-conditioning: In the common evaporator or condenser coil, airside heat transfer is improved with louvered, corrugated, or serrated fins. Microfin configurations are usually found in evaporator or condenser applications.

Heat exchangers of automobiles: Radiators, charge air coolers, and intercoolers of automobiles have been provided with various forms of extended surfaces: louvered fin, offset strip fin, and low-fin tubes. On the tubeside, twisted tape inserts have received considerable attention for oil cooler applications [1].

Air coolers. For cooling viscous fluids in air-cooled heat exchangers, tube inserts inside high-finned tubes can significantly reduce the number of parallel units and the plot size required (some air cooler manufacturers have already been building units with wire mesh type inserts, for instance).

Process industries: The chemical process industry has sparingly adopted enhancement technology because of concerns about fouling. Wire loop inserts and vapor sphere matrix fluted spheres (tubeside) or solid spheres (shellside) not only improve heat transfer, but reduce fouling with typical process fluids [1].

Industrial heat recovery: Ceramic tubes that are enhanced externally and/or internally for high-temperature waste heat recovery are used.

Aerospace: Improved gas turbine blade cooling is achieved by transverse repeated ribs and pin fins cast into the blade, thereby reducing the wall temperature.

Nomenclature

A	total heat transfer area, m^2 (ft^2)
A_c	minimum free flow area, m^2 (ft^2)
C_p	specific heat, J/kg °C (Btu/lbm · °F)
D_h	hydraulic diameter, ($4r_h$), m (ft)
E_{std}	P_p/A, volume goodness factor or the standardized heat transfer coefficient against the pumping power per unit of heat transfer surface area as defined by Eq. 8.18
f	Fanning friction factor
G	mass velocity based on minimum flow area, $kg/m^2 \cdot s$ ($lbm/h \cdot ft^2$)
g_c	acceleration due to gravity or proportionality constant 9.81 m/s^2 (32.17 ft/s^2) = 1 and dimensionless in SI units
h	convective heat transfer coefficient, $W/m^2 \cdot °C$ ($Btu/ft^2\ h \cdot °F$)
hA	thermal conductance W/°C (Btu/h. °F)
j	Colburn heat transfer factor
K	the heat transfer enhancement factor or thermal conductance, hA as defined by Eq. (8.14a)
k	thermal conductivity of the fluid, W/m °C (Btu/h · ft · °F)
k_1, k_2, k_3	constants in Equation 8.13
M	mass flow rate of the fluid, kg/s (lbm/h)
NTU	number of transfer units, hA/C_p
Nu	Nusselt number
P	fluid friction power
p	pressure, Pa (lbf/ft^2)
Pr	Prandtl number
Q	total heat duty of the exchanger, W · s (Btu)
q	heat transfer rate, W (Btu/h)
Re	Reynolds number
r_h	hydraulic radius, m (ft) = $D_h/4$
St	Stanton number = $h/Gc_p = h/\rho u c_p$
U	overall heat transfer coefficient, $W/m^2 \cdot °C$ ($Btu/h \cdot fr^2 \cdot °F$)
Δp	pressure drop, Pa (lbf/ft^2)
M	viscosity of the fluid at bulk mean temperature, kg/m. s or Pa. s (lbm/h · ft)
ρ	fluid density, kg/m^3 (lbm/ft^3)
β	area density, m^2/m^3 (ft^2/ft^3)
η_o	fin efficiency or the overall surface efficiency
σ	the ratio of minimum free flow area to frontal area

Subscripts

p	the reference or plain smooth heat transfer surface
std	standardized parameter

REFERENCES

1a. Bergles, A.E. (1988) Some perspective on enhanced heat transfer—Second generation heat transfer technology. *Transactions of ASME, Journal of Heat Transfer* 110: 1082–1096.

1b. Bergles, A.E. (2003) High-flux processes through enhanced heat transfer. Keynote at the *Fifth International Conference on Boiling Heat Transfer*, Montego Bay, Jamaica, May 4–8, 2003, pp. 1–13.

2. Webb, R.L., Bergles, A.E. (1983) Heat transfer enhancement: Second generation technology. *Mechanical Engineering* 115(6): 60–67.

3. Marner, W.J., Bergles, A.E., Chenoweth, J.M. (1983) On the presentation of performance data for enhanced tubes used in shell and tube heat exchangers. *Transactions of ASME, Journal of Heat Transfer* 105: 358–365.

4. Bergles, A.E. (1985) Techniques to augment heat transfer. In: *Handbook of Heat Transfer Applications* (W.M. Rohsenow, J.P. Haitnett, E.N. Ganic, eds.), pp. 3-1–3-79. McGraw-Hill, New York.

5. Thome, R.T. (2004) Enhanced single phase laminar tube side flows and heat transfer, Chapter 4, pp. 4.1–4.27. *Wolverine Heat Transfer Engineering Data Book III*, Wolverine Division of UOP Inc., Decatur, AL.

6. Webb, R.L. (1980) Special surface geometries for heat transfer augmentation. In: *Developments in Heat Exchanger Technology— 1* (D. Chisholm, ed.), pp. 179–215. Applied Science Publishers, London, UK.

7. Kohler, J.A., Staner, K.E. (1984) High performance heat transfer surfaces. In: *Handbook of Applied Thermal Design* (E.C. Guyer, ed.), pp. 7-37–7-49. McGraw-Hill, New York.

8. Yang, W.-J. (1983) High performance heat transfer surfaces: Single phase flows. In: *Heat Transfer in Energy Problems* (Mizushina, T., Yang, W.J., eds.), pp. 109–116. Hemisphere Publications, Washington, DC.

9. Webb, R.L. (1987) Enhancement of single phase heat transfer. In: *Handbook of Single Phase Convective Heat Transfer* (S. Kakac, R. K. Shah, W. Aung, eds.), pp. 17.1–17.62. John Wiley & Sons, New York.

10. www.nuclear-power.com/nuclear-engineering/fluid-dynamics/flow-regime/laminar-turbulent-flow/

11. www.grc.nasa.gov/www/k-12/BGP/boundlay.html

12. Ravigururajan, T.S., Rabas, T.J. (1996) Turbulent flow in integrally enhanced tubes, Part 1: Comprehensive review and database development. *Heat Transfer Engineering* 17: 19–29.

13. Pate, M.B., Ayub, Z.H., Kohler, J. (1990) Heat exchangers for the airconditioning and refrigeration industry. In: *Compact Heat Exchangers—A Festschrift for A. L. London* (R.K. Shah, A.D. Kraus, D. Metzger, eds.), pp. 567–590. Hemisphere, Washington, DC.

14. Khanpara, J.C., Bergles, A.E., Pate, M.B. (1986) Augmentation of R-113 intube evaporation with microfin tubes. *ASHRAE Transactions,* 92(Part 2B): 506–524.

15. Thomas, C. (1997) Recent developments in the manufacturing and application of integral low fin tubing in titanium and zirconium. *Proceedings of Zirconium/Organics Conference*, pp. 169–177. Gleneden Beach, OR.

16. Minton, P.E. (1990) Process heat transfer. *Proceedings of the 9th International Heat Transfer Conference on Heat Transfer 1990—Jerusalem*, Paper No. KN-22, Vol. 1, pp. 355–362. Jerusalem, Israel.

17. Ravigururajan, T.S., Bergles, A.E. (1985) General correlations for pressure drop heat transfer for single phase turbulent flow in internally ribbed tubes. *Conference Proceedings of the Augmentation of Heat Transfer in Energy Systems*. ASME, Miami Beach, FL.

18. Chiou, J.P. (1987) Experimental investigation of the augmentation of forced convection heat transfer in a circular tube using spiral spring inserts. *Transactions of ASME, Journal of Heat Transfer* 109: 300–307.

19. Bergles, A.E., Joshi, S.D. (1982) Augmentation techniques for low Reynolds number in-tube flow. In: *Low Reynolds Number Flow Heat Exchangers* (S. Kakac, R.K. Shah, A.E. Bergles, eds.), pp. 695–720. Hemisphere, Washington, DC.

20. Ravigururajan, T.S., Rabas, T.J. (1996) Turbulent flow in integrally enhanced tubes, Part 2: Analysis and performance comparison. *Heat Transfer Engineering* 17: 30–40.

21. Sethumadhavan, R., Rao, R. (1986) Turbulent flow friction and heat transfer characteristics of single and multistart spirally enhanced tubes. *Transactions of ASME, Journal of Heat Transfer* 108: 55–61.
22. www.superradiatorcoils.com/blog/heat-exchanger-turbulators-types-and-purposes
23. www.jdturbulators.com/technology/
24. Junkhan, G.H., Bergles, A.E., Nirmalan, V., Ravigururajan, T. (1985) Investigation of turbulators for fire tube boilers. *Transactions of ASME, Journal of Heat Transfer* 107: 354–360.
25. Westphalen, D., Roth, K., Brodrick, J. (2006) Heat transfer enhancement—Emerging technologies. *ASHRAE Journal* 48: 68–71.
26. Bergles, A.E., Webb, R.L., Junkhan, G.H., Jensen, M.K. (1979) *Bibliography on Augmentation of Convective Heat and Mass Transfer*. Engineering Research Institute, Iowa State University, Ames, IA.
27. (a) Shah, R.K. (1978) Compact heat exchanger surface selection methods. In: *Heat Transfer*, Vol. IV, pp. 193–199. Hemisphere, Washington, DC. (b) Shah, R.K. (1978) Compact heat exchanger surface selection methods. In: *Heat Transfer*, Vol. 4, pp. 279–284. Hemisphere, Washington, DC.
28. Webb, R.L. (1981) Performance evaluation criteria for use of enhanced heat transfer surfaces in heat exchanger design. *International. Journal of Heat & Mass Transfer* 24: 715–726.
29. Maltson, J.D., Wilcock, D., Davenport, C.J. (1989) Comparative performance of rippled fin plate fin and tube heat exchangers. *Transactions of ASME, Journal of Heat Transfer* 111: 21–28.
30. London, A.L. (1964) Compact heat exchangers, Part 2: Surface geometry. *Mechanical Engineering* 86: 31–34.
31. London, A.L., Ferguson, C.K. (1949) Test results of high performance heat exchanger surfaces used in aircraft intercoolers and their significance for gas turbine regenerator design. *Transactions of ASME* 71: 17–26.
32. Bergles, A.E., Blumenkrantz, A.R., Taborek, J. (1974) Performance evaluation criteria for enhanced heat transfer surfaces. In: *Heat Transfer*, Vol. II, pp. 239–243. JSME.
33. ASME Boiler and Pressure Vessel Code, Section VIII, Division 1—Pressure Vessels, American Society of Mechanical Engineers, New York, 2021.
34. Standards of the Tubular Exchanger Manufacturers Association (TEMA), 10th edn., 2019, Tubular Exchanger Manufacturers Association, Inc., Tarrytown, NY.

BIBLIOGRAPHY

Arshad, J., Thome, J.R. (1983) Enhanced boiling surfaces—Heat transfer mechanism and mixture boiling. *ASME/JSME Joint Thermal Engineering Conference Proceedings*, Vol. 1, pp. 191–197. Honolulu, HI.

Bergles, A.E. (1978) Enhancement of heat transfer. In: *Heat Transfer—1978*, Vol. VI, pp. 89–108. Hemisphere, Washington, DC.

Bergles, A.E., Junkhan, G.H., Webb, R.L. (1978) Energy conservation via heat transfer enhancement, Paper No. EGY205. *1978 Midwest Energy Conference*, pp. 10–21. Chicago, IL, November 19–21, 1978,.

Carnavos, T.C. (1974) Some recent developments in augmented heat exchange elements. In: *Heat Exchangers: Design and Theory Source Book*, pp. 441–489. Scripta, Washington, DC.

Deng, S.-J. (1990) *Heat Transfer Enhancement and Energy Conservation*. CRC Press.

Dewan, A., Mahanta, P., Sumithra Raju, K., Suresh Kumar, P. (2004) Review of passive heat transfer augmentation techniques. *Proceedings of the Institution of Mechanical Engineers Part A: Journal of Power and Energy* 218: 509–527.

Fujie, K., Nakayama, W., Kuwahara, H., Kakizaki, K. (1977) *Heat transfer wall for boiling liquids*. U.S. Patent 4060125, November 29, 1977.

Gupta, J.P. (1986) *Fundamentals of Heat Exchanger and Pressure Vessel Technology*. Hemisphere, Washington, DC.

Takahashi, K., Nakayama, W., Senshu, T., Yoshida, H. (1984) Heat transfer analysis of shell-and-tube condensers with shell-side enhancement. *ASHRAE Transactions* 90(Part IB): 60.

Thome, J.R. (1990) *Enhanced Boiling Heat Transfer*. Hemisphere, Washington, DC.

Webb, R.L. (1972) *Heat transfer surface having a high boiling heat transfer coefficient*. U.S. Patent 3696861, October 10, 1972.

Webb, R.L. (1979) Toward a common understanding of the performance and selection of roughness for forced convection. In: *A Festschrift for E. R. G. Eckert* (J.P. Hartnett, T.F. Irvine, Jr., E. Pfender, E. M. Sparrow, eds.). Hemisphere, Washington, DC.

Webb, R.L. (1983) Enhancement for extended surface geometries used in air cooled heat exchangers. In: *Low Reynolds Number Flow Heat Exchangers* (S. Kakac, R.K. Shah, A.E. Bergles, eds.), pp. 721–734. Hemisphere, Washington, DC.

Webb, R.L., Bergles, A.E. (1982) Performance evaluation criteria for selection of heat transfer surface geometries used in low Reynolds number heat exchangers. In: *Low Reynolds Number Flow Heat Exchangers* (S. Kakac, R. K. Shah, A.E. Bergles, eds.), pp. 735–752. Hemisphere, Washington, DC.

Index

A

B

C

D

E

F

G

H

R

S

T

U

W

For Product Safety Concerns and Information please contact our EU representative GPSR@taylorandfrancis.com
Taylor & Francis Verlag GmbH, Kaufingerstraße 24, 80331 München, Germany

www.ingramcontent.com/pod-product-compliance
Lightning Source LLC
Chambersburg PA
CBHW080342220326
41598CB00030B/4583